Insects as Sustainable
Food Ingredients

Insects as Sustainable Food Ingredients
Production, Processing and Food Applications

Edited by

Aaron T. Dossey
All Things Bugs LLC
Griopro cricket powder
Athens, GA, United States

Juan A. Morales-Ramos
USDA-ARS National Biological Control Laboratory
Stoneville, MS, United States

M. Guadalupe Rojas
USDA-ARS National Biological Control Laboratory
Stoneville, MS, United States

AMSTERDAM • BOSTON • HEIDELBERG • LONDON • NEW YORK • OXFORD • PARIS
SAN DIEGO • SAN FRANCISCO • SINGAPORE • SYDNEY • TOKYO

Academic Press is an imprint of Elsevier

Academic Press is an imprint of Elsevier
125 London Wall, London EC2Y 5AS, United Kingdom
525 B Street, Suite 1800, San Diego, CA 92101-4495, United States
50 Hampshire Street, 5th Floor, Cambridge, MA 02139, United States
The Boulevard, Langford Lane, Kidlington, Oxford OX5 1GB, United Kingdom

Copyright © 2016 Elsevier Inc. All rights reserved.

Copyright protection is not available in the United States for J.A. Morales-Ramos and M.G. Rojas's contribution to the Work as they are US government employees.

No part of this publication may be reproduced or transmitted in any form or by any means, electronic or mechanical, including photocopying, recording, or any information storage and retrieval system, without permission in writing from the publisher. Details on how to seek permission, further information about the Publisher's permissions policies and our arrangements with organizations such as the Copyright Clearance Center and the Copyright Licensing Agency, can be found at our website: www.elsevier.com/permissions.

This book and the individual contributions contained in it are protected under copyright by the Publisher (other than as may be noted herein).

Notices
Knowledge and best practice in this field are constantly changing. As new research and experience broaden our understanding, changes in research methods, professional practices, or medical treatment may become necessary.

Practitioners and researchers must always rely on their own experience and knowledge in evaluating and using any information, methods, compounds, or experiments described herein. In using such information or methods they should be mindful of their own safety and the safety of others, including parties for whom they have a professional responsibility.

To the fullest extent of the law, neither the Publisher nor the authors, contributors, or editors, assume any liability for any injury and/or damage to persons or property as a matter of products liability, negligence or otherwise, or from any use or operation of any methods, products, instructions, or ideas contained in the material herein.

Library of Congress Cataloging-in-Publication Data
A catalog record for this book is available from the Library of Congress

British Library Cataloguing-in-Publication Data
A catalogue record for this book is available from the British Library

ISBN: 978-0-12-802856-8

For information on all Academic Press publications
visit our website at https://www.elsevier.com/

Publisher: Nikki Levy
Acquisition Editor: Megan Ball
Editorial Project Manager: Billie Jean Fernandez
Production Project Manager: Caroline Johnson
Designer: Victoria Pearson

Typeset by Thomson Digital

Dedication

We would like to, in part, dedicate this book
to the late Professor Gene R. DeFoliart
(University of Wisconsin, Madison, WI, USA)
whose pioneering work and outreach activities
for insects as a food source set the stage
for the industry that is emerging today.

Disclaimer

As two of the editors of this book and some of the chapter authors are also employees of the US federal government, we are obligated to provide the following disclaimer:

The mention of any products or companies in this book does not constitute an endorsement by the US Department of Agriculture.

Contents

List of Contributors xiii
Acknowledgments xv

1. Introduction to Edible Insects

F.V. Dunkel, C. Payne

Introduction 1
Aims 2
Historic Relevance of Edible Insects 2
Reevaluation of Our Sources of Protein Worldwide 2
Population Growth and a Rising Demand for Animal-Derived Protein 3
Land Use 4
Urban and Vertical Agriculture 5
Climate Change and Agricultural Productivity 6
Aquaculture and the Environment 7
Limits to Nonrenewable Energy 7
Water Use 10
Insects as a Living Source of Protein in Space 10
Insects Are an Important and Feasible Solution 11
Worldwide Acceptance of Insects as Food 12
Funding and Legislation 14
Increasing Recognition in the Academic Sector 15
Current Trends in Using Insects as Food 17
Psychological Barriers and Disgust 18
Definition of Terms 20
Summary of Book 20
A Call to Action 23
References 24

2. Insects as Food: History, Culture, and Modern Use around the World

E.M. Costa-Neto, F.V. Dunkel

Introduction 29
Edible Insects of the World 31
History of Insects as Human Food 33
   Archeological Data 33
   Early Historic Times (~3600BC–500AD) 34
   Age of Reason and Emergence of the Scientific Era (1700s/18th Century) 35
Modern Cultural Uses 36
   Edible Insects Around the World: Selected Classic Examples 37
   Indirect or Unintentional Presence in Foods 42
Cultural Restrictions 43
   Food Taboos and Religious/Dietary Restrictions 43
   Disgust Factor 44
   Educational Campaigns 47
Edible Insects as Nutraceuticals 48
Ethnoentomology 50
Harvesting and Cultivation 51
   Farming of Edible Insects 52
Final Comments and Recommendations 53
References 54

3. Nutrient Content and Health Benefits of Insects

J.P. Williams, J.R. Williams, A. Kirabo, D. Chester, M. Peterson

Nutrient Content 61
Insect Physiology and Functionality 78
Insects as a Food Ingredient 81
Insect Protein Functionality 81
Conclusions 83
References 84

4. Edible Insects Farming: Efficiency and Impact on Family Livelihood, Food Security, and Environment Compared With Livestock and Crops

R.T. Gahukar

Introduction 85
Food Security/Family Livelihood 87
Biodiversity and Availability of Insects 89
Consumption of Insects versus Other Livestock 90
Cost of Cultivation 91
Possibility of Replacing Livestock With Insects as Human Food 93
Environmental Impact 94
Industrial Perspective 96

ix

Commercial Insect Farming for Mass Production	97	Supply Chain Needs: Feed, Farms, Insects, Transportation, Processing, and Manufacturing	146
Indoor Farming	97	Feed	146
Outdoor Farming	101	Farms and Farmed Species	147
Farming in Space	102	Transportation, Storage, and Distribution	147
Market Potential	103	Processing and Manufacturing Infrastructure	148
Retail/Local Marketing	103	Intriguing the Larger Food Industry: Uses of Insects as Industrial Food Ingredients	148
Export	104	Conclusions	149
New Products From Farmed Insects	104	References	150
Safety Regulations	105		
Current Challenges and Conclusions	105		
References	106		

5. Modern Insect-Based Food Industry: Current Status, Insect Processing Technology, and Recommendations Moving Forward

A.T. Dossey, J.T. Tatum, W.L. McGill

Introduction	113
Efficiency	114
Biodiversity	115
Current Insect Farming Industry	117
Modern Industrial Mass Production of Insects	118
Recommendations and Considerations for Selection for Aspiring Insect-Based Food Producers and Insect Farmers	122
Feed Formulations and Biomass Sources for Farming Insects: Considerations for Insect Feed Formulations	122
Underutilized Biomass Amenable as Feed Ingredients for Mass-Farmed Edible Insects	122
Diseases Affecting Mass Produced/Farmed Insects	123
The NASCAR Jacket of Food: Food Labeling Requirements and Aspects of Consumer Demand. How do I label it? Is it Organic? GMO? Gluten Free? Paleo? Other?	123
Words Matter: Terminology Recommendations	124
Animal Welfare	125
Getting Past the "Fear Factor": Important Considerations for Normalizing Insects as a Mainstream Food Ingredient Beyond the Novelty Niche	125
Insect Processing Considerations: Considerations for Insect-Based Food Production, Processing, and Safety	126
The Real Pioneers: Entrepreneurs in the Insect-Based Food Space	135
Summary of Current Companies, Farms, and Other Organizations	140
Subindustry Niches: Importance of Finding One's Focused Role in the Industry and Product Selection	141

6. Insect Mass Production Technologies

J.A. Cortes Ortiz, A.T. Ruiz, J.A. Morales-Ramos, M. Thomas, M.G. Rojas, J.K. Tomberlin, L. Yi, R. Han, L. Giroud, R.L. Jullien

Introduction	153
Mass-Produced Insect Species and Their Respective Applications	154
Insects for Food and Feed	154
Insects for Medicinal Use	155
Insects for Other Applications	155
Potential of Using Conventional Feedstock for Rearing Insects	156
Principals of Feed Production for Insect Mass Production	156
Manufacture of Insect Feed	157
Nutritional Requirements for Farmed Insects	158
Macronutrients	158
Micronutrients	160
Considerations for Insect Mass Rearing Equipment and Mechanization	162
Production and Operation Management	162
Rearing Area	163
Feeding and Watering	164
Separation and Sorting Room and Product Traceability	165
Cleaning Room	165
Compost Area	166
Production Techniques by Species	166
Mealworm Production Technologies	166
Black Soldier Fly, *Hermetia illucens*, Production	173
Housefly, *Musca domestica*, Production	176
Cricket Production in the United States	179
Waxworm Production	182
Environmental Control for Efficient Production of Insects in General	184
Estimating Optimal Conditions Required for Design for Insects in General	184
Air Flow Design	186
Equipment for Climate Control	191
Automatic Control and Artificial Intelligence	194
Concluding Remarks	196
References	196

7. Food Safety and Regulatory Concerns

P.A. Marone

Introduction	203
Regulatory Considerations for Insects-as-Food Ingredient	204
Present History of Use and Regulations	204
Labelling Regulation and Health Claims Applicable to Insects	207
Safety Considerations for Insects as Food	207
Species Identity and Characterization of the Insect-Based Product for Whole Body or Processed Feed	209
Toxicological Assessment	210
Clinical Evaluation of Safety	211
Toxicological Hazards of Insect-Based Foods and Food Ingredients	211
Chemical	211
Physical	213
Farming and Novel Considerations Driving Insect Food/Feed Safety	214
Processing, Preparation, Packaging, and Transport of Insect-Based Foods and Food Ingredients	216
Conclusions on the Use of the Insects as Food and Feed	217
References	218

8. Ensuring Food Safety in Insect Based Foods: Mitigating Microbiological and Other Foodborne Hazards

D.L. Marshall, J.S. Dickson, N.H. Nguyen

Introduction	223
Microbes Associated with Insects	224
Insects as a Vector of Foodborne Disease Hazards	225
Bacterial Infections	229
Salmonellosis	230
Shigellosis	230
Vibriosis	230
E. coli	231
Yersiniosis	231
Campylobacteriosis	231
Listeriosis	232
C. perfringens	232
Other Bacterial Foodborne Infections	232
Nonbacterial Foodborne Infections	232
Infectious Hepatitis	233
Enteroviruses	233
Multicellular Parasites	233
Prions	234
Foodborne Bacterial Intoxications	234
Staphylococcus aureus Enterotoxin	234
B. cereus Enterotoxin	234
Botulism	235
Chemical Intoxications	235
Physical Hazards	236
Administrative Regulation	236
United States Department of Agriculture (USDA)	237
United States Food and Drug Administration (FDA)	237
Milk Sanitation	237
International Administration	237
Prerequisite Programs	238
Good Manufacturing Practices	238
Training and Personal Hygiene	238
Pest Control	238
Sanitation	239
Sanitary Facility Design	239
Sanitary Equipment Design	239
Cleaning and Sanitizing Procedures	240
Hazard Analysis Critical Control Point System	241
HACCP Plan Development	242
Food Safety Modernization Act	243
Validation	244
Food Preservation	244
Principles of Food Preservation	245
Asepsis/Removal	245
Modified Atmosphere Conditions	246
High-Temperature Preservation	246
Low-Temperature Preservation	247
Drying	248
Preservatives	248
Irradiation	249
Fermentation	249
Conclusions	250
References	250

9. Insects and Their Connection to Food Allergy

M. Downs, P. Johnson, M. Zeece

Introduction	255
Food Allergy	255
Insects and Food Allergy	257
Insect Allergens	259
Tropomyosin	259
Arginine Kinase	260
Sarcoplasmic Calcium Binding Protein	261
Myosin Light Chain	262
Troponin C	263
Sarcoplasmic Endoreticulum Calcium ATPase	263
Hemocyanin	263
Phospholipase	264
Other Allergens	264

Known Aero-Allergens	264
Novel Allergens	264
Effects of Processing	264
Methods of Allergen Detection	266
Conclusions	267
References	268

10. Brief Summary of Insect Usage as an Industrial Animal Feed/Feed Ingredient

M.J. Sánchez-Muros, F.G. Barroso, C. de Haro

Overview	273
Justification of Using Insects in Animal Feed	273
Current Overview of the Use of Insects in Animal Feeding	274
Examples of Livestock Fed With Insects as Feed Ingredients	277
Poultry	277
Pigs (*Sus* sp.)	282
Fish	283
Hybrid Fish	290
Polyculture	290
Crustaceans (Shrimp, Crabs, Lobsters and Their Relatives)	291
Mollusks (Clams, Oysters, Snails and Their Relatives)	292
Overview	292
Other Animals	292
Benefits and Constraints Associated with Using Insects as Livestock Feed Ingredients	293
Nutritional	293
Feed Security and Safety	295
Animal Welfare	297
Promising Opportunities for Research and Technological Advancement	298
Ecological Aspects and Sustainability	298
Environmental Enrichment for Livestock Animals	299
Chitin	299
Insect Nutritive Value Improvement Using Different Rearing Systems	299
Physical and/or Chemical Treatments of Insect Meals to Improve Their Assimilation	300
Conclusions	300
References	301

Appendix 311

Documented Information for 1555 Species of Insects and Spiders	312
References	372

Subject Index 377

List of Contributors

F.G. Barroso, Department of Biology and Geology, University of Almeria, Carretera de Sacramento, Almería, Spain

D. Chester, USDA/NIFA, Washington, DC, United States

J.A. Cortes Ortiz, Entomotech, S.L. PITA, Agroingroindustrial Technological Park of Almería, Almería, Spain

E.M. Costa-Neto, Feira de Santana State University, Department of Biology, Bahia State, Brazil

C. de Haro, Department of Biology and Geology, University of Almeria, Carretera de Sacramento, Almería, Spain

J.S. Dickson, Department of Animal Science, 215F Meat Laboratory, Ames, IA, United States

A.T. Dossey, All Things Bugs LLC, Griopro cricket powder, Athens, GA, United States

M. Downs, Department of Food Science and Technology, Food Allergy Research and Resource Program, University of Nebraska-Lincoln, Food Innovation Center, Lincoln, NE, United States

F. Dunkel, Department of Plant Sciences and Plant Pathology, Montana State University, Bozeman, MT, United States

F.V. Dunkel, Montana State University, Department of Plant Sciences and Plant Pathology, Bozeman, Montana, United States

R.T. Gahukar, Arag Biotech Private Limited, Nagpur, Maharashtra, India

L. Giroud, Insagri, S.L. Coin, Málaga, Spain

R. Han, Guangdong Entomological Institute, Guangzhou, China

P. Johnson, Department of Food Science and Technology, Food Allergy Research and Resource Program, University of Nebraska-Lincoln, Food Innovation Center, Lincoln, NE, United States

R.L. Jullien, Khepri, and FFPIDI, French Federation of Producers Importers and Distributors of insect, France

A. Kirabo, Uganda Industrial Research Institute, Kampala, Uganda

P.A. Marone, Department of Pharmacology and Toxicology, Virginia Commonwealth University, Toxicology and Pathology Associates, LLC, Richmond, VA, United States

D.L. Marshall, Eurofins Microbiology Laboratories, Inc., Fort Collins, CO, United States

W.L. McGill, Rocky Mountain Micro Ranch and PhD Candidate, National University of Ireland at Galway, Ireland

J.A. Morales-Ramos, USDA-ARS National Biological Control Laboratory, Stoneville, MS, United States

N.H. Nguyen, Department of Environmental Science, Policy & Management, University of California, Berkeley, CA, United States

C. Payne, Nuffield Department of Population Health, University of Oxford, British Heart Foundation Centre on Population Approaches for Non-Communicable Disease Prevention, Oxford, United Kingdom

M. Peterson, USDA/NIFA, Washington, DC, United States

M.G. Rojas, USDA-ARS National Biological Control Laboratory, Stoneville, MS, United States

A.T. Ruiz, Entomotech, S.L. PITA, Agroingroindustrial Technological Park of Almería, Almería, Spain

M.J. Sánchez-Muros, Department of Biology and Geology, University of Almeria, Carretera de Sacramento, Almería, Spain

J.T. Tatum, Ripple Technology LLC, Atlanta, GA, United States

M. Thomas, Zetadec B.V., Wageningen, The Netherlands

J.K. Tomberlin, Department of Entomology, Texas A&M University, College Station, TX, United States

J.P. Williams, USDA/NIFA, Washington, DC, United States

J.R. Williams, University of Maryland, College Park, MD, United States

L. Yi, Zetadec B.V., Wageningen, The Netherlands

M. Zeece, Department of Food Science and Technology, Food Allergy Research and Resource Program, University of Nebraska-Lincoln, Food Innovation Center, Lincoln, NE, United States

Acknowledgments

For their respective contributions and support in life, his career, and this book, Dr. Aaron T. Dossey would like to thank the following: God and his savior Jesus Christ; his family (grandpa Jerry C. Dossey, grandma Emma A. Dossey, mother Teresa M. Scott, aunt Sandra Dossey, and cousin Candace N. Kane) for their love, support, teaching, and inspiration; Laurie Keeler (University of Nebraska's Food Processing Center, Lincoln, NE, USA) for believing in him and his research in insect-based foods early on and going beyond the call of duty providing support during various research projects; Thomas Dietrich (consultant, Dietrich's Milk Products, Reading, PA, USA) for many supportive conversations helpful to this book and Dr. Dossey's company and career; John Tyler Tatum (business partner, Ripple Technology, Atlanta, GA, USA) for going out of his way for help with Dr. Dossey's business and last-minute help with figures in this book; Jack Armstrong (Armstrong Cricket Farm, West Monroe, LA, USA) and Steve Hederman (Millbrook Cricket Farm) for their helpful advice and information about cricket farming ; Jeff Armstrong (Armstrong Cricket Farm, Glenville, GA, USA) for allowing Dr. Dossey to tour their farm and utilize photographs of cricket, mealworm, and superworm farming; Professor Frank Franklin (University of Alabama, Birmingham, AL, USA) for help getting started in the field of insects as a food source via assistance with Dr. Dossey's Bill & Melinda Gates Foundation grant application in 2011; and Dave Gracer for inspiring conversations and initial introduction to the field of insects as food. Dr. Dossey also is greatly appreciative of the Bill and Melinda Gates Foundation for their initial grant support of his work in this field, the USDA NIFA for their SBIR research grant support for his company and to USDA NIFA director Dr. Sonny Ramaswamy for his overall support of our field and emerging insect-based commodity industry. Additionally, Dr. Juan Morales-Ramos and Dr. Guadalupe Rojas would like to thank Dr. Edgar King for his input on the production of waxworms and for the review of Chapter 6 of this book.

Chapter 1

Introduction to Edible Insects

F.V. Dunkel*, C. Payne**

*Department of Plant Sciences and Plant Pathology, Montana State University, Bozeman, MT, United States; **Nuffield Department of Population Health, University of Oxford, British Heart Foundation Centre on Population Approaches for Non-Communicable Disease Prevention, Oxford, United Kingdom*

Chapter Outline

Introduction	1	Insects as a Living Source of Protein in Space	10
Aims	2	Insects Are an Important and Feasible Solution	11
Historic Relevance of Edible Insects	2	Worldwide Acceptance of Insects as Food	12
Reevaluation of Our Sources of Protein Worldwide	2	Funding and Legislation	14
Population Growth and a Rising Demand for Animal-Derived Protein	3	Increasing Recognition in the Academic Sector	15
		Current Trends in Using Insects as Food	17
Land Use	4	Psychological Barriers and Disgust	18
Urban and Vertical Agriculture	5	Definition of Terms	20
Climate Change and Agricultural Productivity	6	Summary of Book	20
Aquaculture and the Environment	7	A Call to Action	23
Limits to Nonrenewable Energy	7	References	24
Water Use	10		

INTRODUCTION

This book is about optimism. This optimism is emerging from forward-thinking scientists and entrepreneurs who have a strong social conscience. Carefully and deliberately, the book will demonstrate to the reader the feasibility and value of insects as a sustainable commodity for food, feed, and other applications. It will weave a tapestry describing the richness of this source of protein, the numbers from nutritional tables, the numbers from mass rearing processes, and the safety and concerns of microbial associations in comparison to typical Western culture protein sources. As a prelude to the sustainability, large-scale farming, and production aspects of food insects, this book will walk the reader through the psychological aspects that separate many humans from this important and delicious protein source.

The time is ripe for a complete, detailed treatment of edible insects. Academia, policy makers, leaders in food security, food safety officers, animal scientists, and the business community need to have a single, foundational, go-to book for their professional teaching and learning. This book presents the basics of edible insects.

The background to this topic has its roots in a small group of researchers worldwide who have been advocating the tremendous potential of edible insects for the past few decades. It was in 1978 that Professor Gene DeFoliart, a forward-looking, environmentally concerned medical entomology professor, began to investigate the feed potential of edible insects for chickens. Seeing the potential being far wider than chickens and beetles, he launched *The Food Insects Newsletter* in 1988 (DeFoliart et al., 2009). In 1995, DeFoliart appointed Professor Florence Dunkel, a natural product entomologist, the new editor of the *Newsletter*, and DeFoliart went on to publish an online book on the subject. Dunkel and a handful of other Western culture entomology professors meanwhile had begun to serve insects at local and national insect fairs, events, and even a dinner party filmed on the Discovery Channel in 1996. Both the research and the outreach was considered a "fringy" activity. For 20 years, during the "dark ages of insects for food and feed," this publication was physically sent 3 times a year to a readership of 1000 in 87 countries. As such, *The Food Insects Newsletter* kept researchers in the academic world connected and the public informed. Meanwhile, similarly forward-thinking entrepreneurs worldwide began to produce and market insect-based food products and develop insect farming techniques that would prove even more resource-efficient than initial research had suggested.

The movement toward using insects as a food ingredient then accelerated considerably with the publication of the Forestry Paper 171 of the Food and Agricultural Organization of the United Nations (FAO) (Van Huis et al., 2013), which served as the catalyst for many additional companies entering this industry, for new researchers entering this field, and for academics putting edible insects on their curriculum change agenda. It was the most downloaded FAO document to date, currently numbering 6 million downloads, with 1 million in the first 24 h after release on May 13, 2013. There is no doubt that edible insects have taken their place at the table of serious solutions to food security for the future. Given this evidence from the response to the FAO publication, this textbook serves as the next level of basic information for the world of food, nutrition, health, and environmental sustainability.

Thus, we can say that the industrial sector has entered a new phase, supported by a surge in academic research. It is time now to build on the knowledge of scientists, local environmentalists, Indigenous peoples' traditional ecological knowledge, and the general public to help design plans for land use going forward that recognize the efficiency of insect protein production on a very small footprint. Insects need to be part of the collective envisioning of a new way to live sustainably.

AIMS

In this introductory chapter, we aim to contextualize "edible insects" within a contemporary setting. We briefly describe the historic relevance of insects used as food for humans. Then, we turn to the potential of insects at the present time, reviewing the place of edible insects within our contemporary global food system. The current epoch has been termed the "Anthropocene"—an era in which we must recognize the irreversible impacts that our existence has on the environment in which we live, and the imperative need for us to acknowledge this and take responsibility (Purdy, 2015). We describe some of the major challenges that our planet currently faces, and the part that the food system has played in contributing to these challenges. For each challenge we show that a food system that incorporates edible insects and insects as ingredients may have the potential to contribute to the solutions. We conclude the chapter with a description of the book, and a brief synopsis of what each chapter can teach us about this underutilized, but exciting and diverse, family of food ingredients.

HISTORIC RELEVANCE OF EDIBLE INSECTS

Insects have been eaten throughout the history of humanity, and are one of the many food sources that humans have relied upon in our journey toward developing our unique technological capability and complex social systems. Insects, despite their absence in a conventional Western diet, remain popular today on a global scale.

Perhaps the greatest irony of the cultural avoidance of edible insects as human food lies in the fact that insects may well have played a major role in our collective past, shaping what it means to be human. Every primate is, to some degree, insectivorous (McGrew, 2014). Insects are a crucial source of fat, protein, and micronutrients in diets that are otherwise heavily plant-based (Rothman et al., 2014). The desire to obtain edible insects in larger quantities appears to have been a key driver of the discovery of tool use. Our closest living primate relatives, the chimpanzees, are famous for their ability to use tools, and many of these tools are used to forage efficiently for edible insects (Sanz et al., 2010; Koops et al., 2015; Whiten et al., 2005). There is even evidence from early human sites that some of the first bone tools were developed for termite extraction (Lesnik, 2011). Therefore, developing innovative technology to increase our access to edible insects will be a logical extension of our past and critical aspect of our future success.

Edible insects, however, are not merely a relic of our prehistoric past. Humans have continued to forage for, to semicultivate and even to fully domesticate insects for food throughout history. Insects are currently estimated to play a role in the diet of at least two billion people worldwide, the majority of whom are in the global south (FAO, 2013).

REEVALUATION OF OUR SOURCES OF PROTEIN WORLDWIDE

While the ability of the earth overall to withstand human impact is incontrovertible, the ability of many species, including humans, to withstand our own actions is questionable. The environment that we have come to rely upon is composed of interconnected biomes with animals and plants that facilitate high intensity food production. Yet we allow cultural prejudices to dictate which of these are prioritized, thus ignoring the connections between them and jeopardizing the system as a whole.

Interconnectedness forms the basis of many philosophical systems worldwide, and the interconnectedness of all living and nonliving matter on Earth and its atmospheric layers is seeing increasing recognition in Western cultural thought. Throughout the past century, cultural prejudice and resulting economic pressures have taken precedence over these concepts, and this is reflected in the organization of our global food system (Berry, 2000; Norberg-Hodge, 2000).

Prior to industrialization, the productivity of agricultural systems depended on the strengths of the connections within them, exploiting the benefits that accrue from mutualistic relationships between species. Rice-fish farming systems in Asia facilitate nitrogen fixation in soils, diminishing the need for additional fertilizer (Lu and Li, 2006) and optimizing water use (Frei and Becker, 2005). Rice-duck farming systems in China eliminate the need for pesticide use (Zhang et al., 2009). Agroforestry systems—in which trees and crops are cultivated alongside one another—are used worldwide to maintain biodiversity while also providing important additional sources of income (McNeely and Schroth, 2006). The chitemene agricultural systems of Tanzania are based around the distribution of *Macrotermes* mounds, which improve soil fertility and also provide valuable by-products in the form of food, fertilizer, and building materials (Mielke and Mielke, 1982). Finally, across the world, insects that feed on crops are also collected as food, exemplifying the multiple benefits of interconnected systems (Yen, 2009).

Commoditization of several staple crops during the 20th century meant that it became economically preferable for many farmers to reject traditional, interconnected agriculture in favor of monoculture. Monoculture requires prioritizing the yield of a single species, and as a result, foodstuffs that were once "by-products" or perhaps even "coproducts," are increasingly treated as hindrances to one-crop production.

The click beetles (Elateridae) of Montana are one such example of this. They feed on wheat crops, grow extremely well in subterranean environments, their larvae have up to 14 instars, and they reach lengths of up to 25 mm (Morales-Rodriguez et al., 2014). Their ecological success is considered a blight by researchers, who work to develop new pesticides that will eliminate them. Yet why is the question not raised: which is the higher quality protein? Which is the crop that is more efficient, water-wise, land-wise, energy-wise in production of that protein? Is it the wheat or the insect? The answer, of course, is the insect. Although data specific to this insect are not available, another coleopteran species, *Tenebrio molitor* (mealworms) is known to contain 20.9 g of crude protein per 100 g ($n = 10$) (Payne et al., 2015), while in contrast, 100 g of wheat grain contains just 13.53 g ($n = 3$) (Xiang et al., 2008). This insect also has a favorable amino acid composition compared to wheat, meaning that 100 g of the insect contains up to twice the quantity of essential amino acids such as lysine and leucine, which cannot be synthesized by the human body, compared to 100 g of wheat grain. Why, then, do we invest our energy, land, even pristine land, and water in growing a lower quality protein that uses more inputs than the insect? The answer in this case appears to be by and large: "We eat wheat here, and we have grown it in Montana for many generations." What, then, is the answer for farmers in less wealthy parts of the world, who used intercropping systems within living memory yet now cultivate imported staple crops for a fluctuating market?

There is increasing evidence that such answers to these questions are not sustainable. The following sections discuss the evidence that our current agricultural systems are inadequate, particularly in light of the changing dynamics of demography and demand, and show that the consideration of agriculture that harnesses insect biology may be one of our best options for a brighter future.

POPULATION GROWTH AND A RISING DEMAND FOR ANIMAL-DERIVED PROTEIN

By 2050, the global population is predicted to grow to nine billion people, and the demand for animal-derived protein is expected to increase at an even higher rate (Godfray et al., 2010). While this forecast is not without controversy (Tomlinson, 2013), it highlights issues of capacity, demand and resource depletion that cannot be ignored.

Increasing population and consumption are placing unprecedented demands on agriculture and natural resources. In the past 40 years, technological advances and land clearance have enabled food production to rise in tandem with population increase (FAOSTAT, 2015). Yet even if current trends in yield increase continue in the coming decades, this will not be sufficient to meet the demands of a population that is growing in wealth at the same time it is growing in numbers (Ray et al., 2013). We can also expect a continued rise in the demand for high-quality, animal-derived protein, with most people demanding higher quality protein sources than their parents and grandparents had available on a routine basis. This will require a projected 72% rise in meat production over the next 35 years (Wu et al., 2014). Yet food from animal origins is increasingly expensive, both in economic and in environmental terms. "Livestock systems"—that is, including the production of feed for livestock—currently occupy a staggering 45% of the world's surface area (Thornton et al., 2011) and contribute 18% of global greenhouse gas emissions (GHGEs) (Steinfeld et al., 2006). Feed production represents 45% of these emissions (Gerber et al., 2013), and principal protein sources for animal feed are fishmeal and soybean meal, both fraught with multiple environmental challenges: soy production is a major driver of rainforest clearance in the Amazonian basin (Salazar et al., 2015), and fishmeal production is depleting marine stocks, with the result that fishmeal prices in real terms are expected to rise by 90% in the period 2010–30 (FAO, 2013).

Overall, to meet the world's future food security and sustainability needs, it is clear that food production must grow substantially while, at the same time, agriculture's environmental footprint must shrink dramatically. Thus, significant progress could be made by halting agricultural expansion for livestock production, closing "yield gaps" on underperforming lands, and increasing cropping efficiency. This will require shifting diets and reducing waste (Gustavsson et al., 2011; Lundqvist et al., 2008; Parfitt et al., 2010). By forming an alliance with insects instead of an adversary relationship, the dilemma of increasing population and consumption can be solved. Together, these strategies could double food production. Meanwhile, farmed insects can be used to process preconsumer food waste on the road to creating a more nutritious feed for cattle, chickens, fish, pigs, and other vertebrate livestock. This will greatly reduce the environmental impacts of agriculture. In India, Premalatha et al. (2011) summed up well the supreme irony of populations and the growing demand for complete proteins, "All over the world, money worth billions of rupees is spent every year to save crops that contain no more than 14% plant protein by killing another food source (insects) that may contain up to 75% high-quality animal protein."

Individuals and groups of people all over the world, including new parents, are seeking a healthier and more sustainable world. They are increasingly interested in the best sources of essential amino acids, favorable ratios of essential fatty acids (high omega-3 fatty acids in comparison to omega-6 fatty acids), and other specific nutrients that lead to strong minds and bodies. At the same time, many people increasingly demand that the sources of these nutrients are natural, pesticide free, and produced in locally environmentally friendly ways.

LAND USE

Agricultural production is responsible for ongoing land clearance, with devastating implications for the environment including contributing to climate change (Kurukulasuriya and Rosenthal, 2003) and mass species extinction (Dirzo et al., 2014), both of which are in turn likely to prove detrimental to long term agricultural productivity. Some say this deliberate environmental degradation is an immoral process (Pope Francis, 2015). It is imperative that we find new ways to use land efficiently for food production (Godfray and Garnett, 2014). Edible insects provide one such opportunity.

Food production takes up almost half of the planet's land surface and threatens to consume the fertile land that still remains. In 2005, 40% of the Earth's land was under cultivation for agriculture (Owen, 2005). The global impact of farming on the environment that this use of land represents should be a serious concern for all humans. Navin Ramankutty, a land-use researcher with University of Wisconsin-Madison's Center for Sustainability and the Global Environment (SAGE) and his coworkers compiled maps using satellite images and crop and livestock production data from countries around the world. This satellite data indicates where cultivation is occurring with good spatial accuracy, while the census data indicates what is being grown there. Roughly an area the size of South America is used for crop production, while even more land—7.9–8.9 billion acres (3.2–3.6 billion hectares)—is being used to raise livestock. With the world's population growing so rapidly, the pressure is on farmers to find new land to cultivate. Production of food on land cannot be separated from negative environmental consequences such as deforestation, water pollution, and soil erosion (Ramankutty et al., 2008).

According to the Food and Agriculture Organization of the United Nations, total farmland increased by 12.4 million acres (5 million hectares) *annually* between 1992 and 2002. A major part of this expansion has been from soybeans, and soybean production has been fueled by demand for soy from China. To meet this demand, in Brazil, for example, huge areas of rainforest have been replaced by soybean plantations. Soy is not a traditional crop in South America. University of Maryland researchers found that 72% of land cleared for crops in that region between 2001 and 2003 was previously pasture for livestock (Morton et al., 2006). It is likely that these pastures were formerly rainforest. So the transition may have been from forest to pasture to soybeans (Ramankutty et al., 2002, 2008).

Amato Evan, a Center for Sustainability and the Global Environment (SAGE) researcher, has warned us, "If current trends continue, we should expect to see increased agricultural production at the cost of increased tropical deforestation" (Owen, 2005). Crops used as feed for cattle are driving tropical cropland expansion (DeFries et al., 2010, 2004). In 2015, a decade later, the need to stop further land claimed for agriculture is even more urgent.

Forests, particularly tropical rainforests, are a reservoir of biodiversity of plants, animals (including insects), and many other types of unique life forms. Within this diversity are important food insects, plants of commercial value, and a storehouse of pharmaceuticals we may need in the future (DeFries and Rosenzweig, 2010). Even beyond what critical utility biodiversity may hold for human kind, the current mass extinction of these species and their habitats is simply immoral. Indeed, even the Pope of the Holy See recently published an encyclical warning of the dangers and moral imperative of protecting the earth and its natural environment (Pope Francis, 2015).

Countries with the least suitable agricultural land are likely to be the ones hardest hit by increased food demand. However, these are also countries where edible insects form part of their recent history. At-risk regions were identified by superimposing remaining potential arable land with locations with high projected population growth. Based on this analysis, 16 regions were identified to be trouble spots over the next 45 years (Owen, 2005). The regions include several parts of Asia, North Africa, and the Middle East. Remaining unexploited areas are not the best suited for agriculture, but are presently supporting valuable natural ecosystems. Certainly, new food production technologies, such as GPS-based precision farming, can improve productivity while reducing the use of water and the application of fertilizer and other potentially harmful chemicals while using the land more effectively. Since farming of insects indoors (and potentially vertically in multistory facilities) would not be tied to arable land or to land suited for agriculture, the utility of food products from insects seem able to solve the main land use issues without high technology based on managing fertilizers, potentially harmful herbicides and insecticides, and water use.

The time has come to look at which crops, plants, livestock, and minilivestock have the best natural efficiency of protein production in tandem with high nutritional quality and low environmental impact. It is time to take an integrated, honest look at land use.

Producers in the industrialized world are concerned with producing more food on less land. This is imperative because, across the world, land clearance for farming is having a major impact on species extinction rates (Dirzo et al., 2014), and "land sparing"—that is, increasing yields in order to minimize demand for farmland—is recommended as the best strategy to address this (Green et al., 2005).

The feed conversion efficiency for many species of insects is well above that of beef (Van Huis et al., 2013). Recent research shows that some species are also more efficient than soy production: Insect rearing trials conducted by the EU initiative PROteINSECT, using the housefly (*Musca domestica*) and black soldier fly (*Hermetia illucens*), found that 1 hectare of land could produce at least 150 tons of insect protein per year in comparison to less than a ton of soybeans for the same area (Zanolli, 2014). This is just one example of the land-sparing, indoor, intensive insect protein production that is described in this book, in the hope that it will inspire farmers, researchers, and investors alike to give serious consideration to the potential of insects as a food ingredient in order to alleviate pressure on land for agriculture. This rich source of information could provide a way forward to avert more forests and prairies from being claimed for agricultural land, even in the face of growing population numbers.

URBAN AND VERTICAL AGRICULTURE

A recent surge in urban agriculture, in both the developed and developing world, is a trend that is forecast to continue due to a growing interest in household-level food sovereignty, environmental advantages in terms of transport and energy costs, and continued urbanization. Conventional livestock are not suited to such conditions. Insects, however, thrive in indoor environments and are ideal candidates for urban agriculture, even at a household level.

Urban agriculture has been proposed as a key strategy for policymakers looking to improve global food security (Bakker et al., 2000). Indoor farming within cities; vertical farming; and farming that makes use of underutilized spaces within urban environments such as idle land, rooftops, and unused water bodies is seeing increasing interest (Smit and Nasr, 1992). Growing crops and livestock in multistory facilities is the *coup-de-gras* of land use efficiency, and insects are uniquely capable of being mass produced in multistory buildings. Indoor vertical farming can be conducted on multiple levels under one roof, and although energy and water costs are currently high, once these systems are optimized this may be offset by the drop in food transportation costs from rural to urban areas. Furthermore, although still in its infancy, involvement in urban agriculture correlates with dietary adequacy in developing countries (Zezza and Tasciotti, 2010). Urban agriculture in many places is primarily practiced by women and research suggests this can contribute to redressing gender-based inequalities (Hovorka et al., 2009), and is likely to form part of a globally relevant strategy for a more sustainable and food-secure future.

Meanwhile, urban agriculture in the form of backyard food production for the individual household is also becoming popular among wealthier consumers. Yet in this context, because backyards now are not attainable or desired by many, container food production is on the rise. Container gardens change the food system in two basic ways. First, containers do not need land beyond one's living quarters. The corner of a room, patio, a table top, a roof, the side of a fence, or house wall will suffice, making this mode of food production accessible even to those living in the most densely populated areas of the world. Second, this means that consumers who can afford to do so are rediscovering the joys of producing their own food, the flavor bursts from eating their own fresh-picked food, and spiritual wholeness of farm-to-table eating within their own home.

This trend is relevant to the promotion of insects as an alternative protein source for humans. Insects are the only livestock that it is possible to cultivate not only in your own backyard, but even in your home kitchen (McGrath, 2015). Already, grow-your-own indoor farming kits are available for mealworm and black soldier fly production. Marketed as a healthy and very local sustainable protein source, these enclosed systems could be the future of home cultivation, particularly in the context of a rapidly increasing urban population.

Similarly, insect-rearing facilities may also play a key role in the future of commercial scale urban agriculture. Farmed insects such as crickets, mealworms, and black soldier fly larvae thrive in closed, controlled environments typical of indoor agriculture. Unlike avian and mammalian livestock, the welfare and productivity of farmed insects do not seem to be compromised by crowded conditions and a lack of sunlight. In theory, then, they can be grown anywhere on earth. Consequently, an urban setting is ideal for minimizing overall environmental impact because it ensures that the food is produced as close as possible to the consumers themselves. Indeed, it may be the case that insects are the only livestock that really work in an urban setting.

Overall, then, while a grand diversity of food plants are appropriate to container gardens and modular food production systems in cities, insects are the only animal livestock that are amenable to this environment. However, while fruits and vegetables generally pose no real challenge for the householder harvesting from their containers, this may not be the case for insects. We foresee a time in the near future when householders will be growing insects on their countertop.

CLIMATE CHANGE AND AGRICULTURAL PRODUCTIVITY

Climate change is exacerbated by excessive land clearance, and by the greenhouse gas emissions generated by food production. Changes in the earth's temperature and in sea levels are in turn expected to have a severe impact on global agricultural production (Kurukulasuriya and Rosenthal, 2003), particularly in some of the world's poorest areas (Winsemius et al., 2015). Insects are a nutrient source that is robust in the face of climatic variation, and that have a low global warming impact when farmed.

The need to change our food system to combat climate change is now unequivocal, particularly since climate change itself is predicted to have devastating effects on global agriculture (Nelson et al., 2014). Even seemingly slight temperature changes cause changes in weather patterns, contributing to the melting of the polar ice caps, loss of glaciers, and a resultant rise in sea level that threatens to engulf large areas of inhabited land. Meanwhile, producers in both coastal and inland regions are already experiencing the effects of changing climates. One such example can be found in the Great Plains of the USA, a region that has already experienced an extension of the growing season from 90 to 120 days in the past 20 years. These longer growing periods, however, mean that the temperatures of the earth and of ambient air will rise, meaning more evaporation of water following melting of the glaciers, frozen water that remains stored in the mountains over 1 year. So, this region has been experiencing not only less water but also a longer growing season. The Rocky Mountain regions (as well as the Midwestern United States) had a drought during 2002, which was accompanied by dry conditions, wildfires, and hot temperatures in the western and midwestern areas (NOAA, 2003). The US drought of 2002 turned a "normal" fire season into a very dangerous, treacherous, and violent season. Adjacent Canadian provinces were also affected. The 2001–02 rain season in Southern California was the driest since records began in 1877. Records were broken in an even worse drought just 5 years later, during the 2006–07 rain season in Los Angeles. The California drought continued through 2010 and did not end until Mar. 2011. The drought shifted east during the summer of 2011 to affect a large portion of the southwestern United States and Texas (NOAA, 2015).

As a result, farmers will have to rethink their strategy regarding the type of crops to grow. Irrigated crops, such as alfalfa, corn, sugar beets, some wheat, and some barley will become increasingly costly in terms of water use. On the other hand, native grasses and other vegetation for browsing wild animals will thrive. How can we access the nutrients we need without using irrigated crops and using minimal water? Is it more efficient to eat wheat, or to eat the animals, such as grasshoppers and Mormon crickets that eat the wheat? Alternatively, is it more efficient to farm the insects directly, particularly given their low water usage (Oonincx and DeBoer, 2012; Van Huis et al., 2013)?

The most significant component of agriculture that contributes to climate change is livestock. Globally, beef cattle and milk cattle have the most significant impact in terms of greenhouse gas emissions (GHGEs), and are responsible for 41% of the world's CO_2 emissions and 20% of the total global GHGEs (Gerber et al., 2013). The atmospheric increases in GHGEs caused by the transport, land clearance, methane emissions, and grain cultivation associated with the livestock industry are the main drivers behind increases in global temperatures.

In contrast to conventional livestock, insects as "minilivestock" are low-GHGE emitters (Oonincx et al., 2010), use minimal land, can be fed on food waste rather than cultivated grain, and can be farmed anywhere thus potentially also avoiding GHGEs caused by long distance transportation. If there was a dietary shift toward increased insect consumption and decreased meat consumption worldwide, the global warming potential of the food system would be significantly reduced.

AQUACULTURE AND THE ENVIRONMENT

Environmental degradation is not limited to land; a growing demand for fish is causing the rapid depletion of marine fisheries, seriously threatening marine environments and biodiversity. The main cause for this is the use of wild fish stocks for fishmeal—a feed ingredient that could be replaced by insects.

In 2007 the US National Academy of Sciences published a report summarizing the change in fish production globally (Brander, 2007). The report concluded that current global fisheries production of ≈160 million tons was rising as a result of increases in aquaculture production. Aquaculture, however, depends on of the production of fishmeal—the feed used for farmed fish. Fishmeal is composed of wild fish. These are sourced from wild, ocean fisheries and are small, bony, pelagic fish such as anchoveta, sardines, pilchard, blue whiting, sandeel, sprat, and capelin. These fish constitute roughly one-third of the global annual fisheries catch, and are processed to produce fishmeal and fish oil used in fish, poultry, and livestock feeds (Jacobson et al., 2001). So, as aquaculture, poultry, and livestock rise in production, ocean fisheries are depleted.

Brander (2007) identified a number of climate-related threats to both capture fisheries and aquaculture, but placed low confidence in his predictions of future fisheries production because of uncertainty over future global aquatic net primary production and the transfer of this production through the food chain to human consumption. Commercial fishing exacerbates the effects of climate because fishing reduces the age, size, and geographic diversity of populations and the biodiversity of marine ecosystems, making both more sensitive to additional stresses such as climate change. Inland fisheries are threatened by changes in precipitation and water management. The frequency and intensity of extreme climate events is likely to have a major impact on future fisheries production in both inland and marine systems. Brander concluded that reducing fishing mortality in the majority of fisheries, which are currently fully exploited or overexploited, is the principal feasible means of reducing the impacts of climate change. For this to be a viable solution, it will be necessary to seek alternative sources of feed for fish and livestock, and for animal-based protein to feed a growing human population.

Many commercially farmed fish are natural predators of insects, and therefore fishmeal can be replaced by insect meal in fish feed. This replacement would have little adverse effects on fish growth, but by alleviating pressure on marine fish stocks, would have significant benefits for the future of marine biodiversity.

LIMITS TO NONRENEWABLE ENERGY

Current global energy production relies on nonrenewable energy sources. The global food system, and livestock production in particular, contributes disproportionately to energy use. Insects are a low-energy-intensive source of protein that could contribute to the sustainability of world food production.

Fossil fuels and fuels like uranium are "spent" once they are used to obtain energy. These are called nonrenewable sources of energy. Although new plants can be planted that eventually turn into coal, the process takes millions of years and that is why coal and other fossil fuels are considered nonrenewable. Fossil fuels (coal, oil, gas) result from a transformation of plant and animal material over millions of years. The solar energy originally stored in the plant or animal is eventually converted into energy stored in carbon and hydrogen bonds of the fossil fuel. The fuels that took millions of years to make are being used at an enormously rapid rate.

What does insect farming and industrial processing have to do with limiting nonrenewable energy? To begin with, using the diverse 2000 or more species of edible insects in the locations where they are naturally found will reduce nonrenewable energy currently used for transportation of high-quality protein. The greatest food needs at this time, and predicted greatest growth needs are in the same areas where food insects are already appreciated.

Energy use in each human activity has grown exponentially since the early days of human civilization. For example, technological capabilities enable us to travel more and process more food. Fig. 1.1 shows the amount of energy (in calories) we spend for each calorie of food we obtain. It shows that technologies have mechanized and made large production systems of cultivation and fishing. These systems involve large expenditures of energy (Fig. 1.1). Wet rice production in Asian countries takes just 0.02–0.1 calories of human generated energy to produce 1 calorie worth of rice as food (not including natural energy utilized by rice plants for photosynthesis). In contrast, large-scale production of animal-derived food consumes enormous amounts of energy. For example, it takes over 2 calories of energy input to produce 1 calorie worth of eggs in large-scale farms, and it takes 10–15 calories of input for every calorie worth of beef produced in the United States. The intensity of energy consumption for US food production has grown almost 10-fold in the 20th century, and yet these are the very systems that we will need to expand to meet the demands of our growing population. This demand is exacerbated by the fact that with industrialization, average daily calorie intakes also rise, resulting in an enormous increase in calorie consumption throughout human history (Table 1.1).

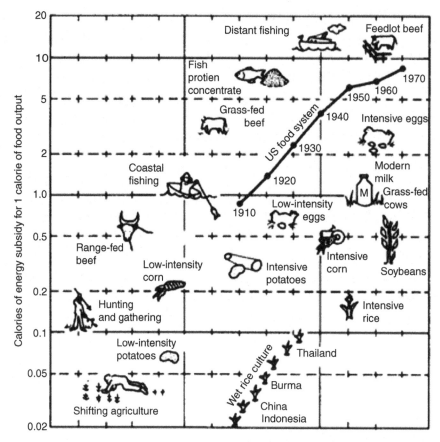

FIGURE 1.1 Summary of the energy required for various types of food production. *(Clark, M.E., 1989. Ariadnes Thread. St. Martin's Press, New York, NY)*

TABLE 1.1 Per Capita and Global Energy Consumption for Different Types of Human Economies

Economic Systems	Years Ago	Maximum Global Population[a] (Approx.)	Daily Available Calories/Person[b]	Global Daily Calories Consumed by Human Population
Hunter-gathering (before cooking)	1,000,000–500,000	1 million	3,000	3×10^9
Hunter-gathering (after cooking)	500,000–10,000	10 million	8,000	8×10^{10}
Early agriculture	10,000–2,000	300 million	15,000	4.5×10^{12}
Middle ages	1,000	500 million		$\sim 8 \times 10^{12}$
		Europe 10%	23,000	
		Rest 90%	15,000	
Today	0	5,000 million		2.8×10^{14}
Today by region: North America Europe, USSR, Japan Third World		5% 18% 77%	314,000 157,000 15,000+	

[a]Leakey, R., Lewin, R., 1977. Origins. E.P. Dutton, New York, NY, p. 143; Weeks, J., 1981. Population: An Introduction to Concepts and Issues, second ed. Wadsworth, Belmont, CA, p. 46.
[b]Brown, H., 1978. The Human Future Revisited. W.W. Norton, New York, NY, pp. 30–33, with per capita figures for industrialized nations upgraded from 1970 to 1980 levels.
Source: Clark, M.E., 1989. Ariadne's Thread. St. Martin's Press, New York, NY, p. 102.

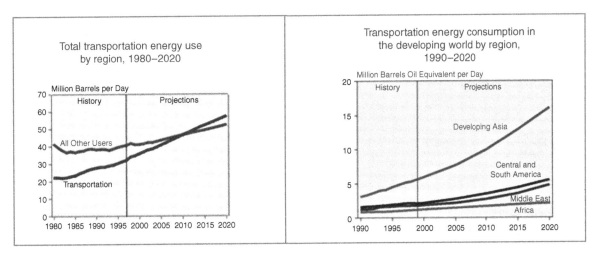

FIGURE 1.2 Use of crude oil for transportation needs. *(DOE/EIA, 2001)*

Liquid fuels such as gasoline and diesel have made our lives and our food supply more mobile. Intercontinental travel has become increasingly commonplace, and we now enjoy a global trade in agricultural produce that enables us to eat food from around the world. Yet the crude oil driving our mobility is a nonrenewable energy source, and therefore if we wish to maintain this lifestyle, it is imperative that we develop new technologies before our current travel habits deplete the remaining crude oil stores on Earth. The global production and movement of food plays a key role in this dilemma. Energy use for transportation is the least efficient use of fossil fuels (DOE/EIA, 2001). Use of crude oil is escalating as developing countries emulate industrialized mobility (Fig. 1.2). This increase in oil consumption specifically for transportation not only impacts the environment, it depletes the limited oil reserves.

Most of the energy planning is done by looking at the supply side, rather than by asking how the demand side (all our uses of energy) can be managed. Energy availability and use are good indicators of the standard of living in our technological world. In the United States, the average annual consumption per capita is 55 barrels of oil. In economically poorer countries, consumption is 6 barrels per year. Fig. 1.3 shows the projections of world energy supplies from 1970 to 2020.

Finally, nonrenewable fuels are currently essential for production of grains and oilseeds. The demand for meat is rising steeply. As the demand for meat rises, demand for grain and protein feeds grows proportionately more quickly (Trostle, 2008). At this point, with no decrease in meat demand in sight, we need to pay careful attention to feed-to-product conversion ratios (Table 1.2).

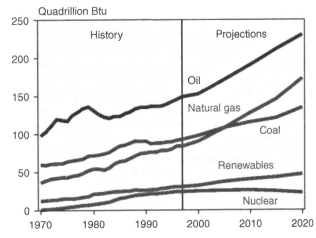

FIGURE 1.3 World energy consumption by fuel type, 1970–2020. *[History: Energy Information Administration (EIA), Office of Energy Markets and End Use, International Statistics Database and International Energy Annual 1997, DOE/EIA-0219 (97) (Washington, DC, April 1999). Projections: EIA, World Energy Projection System, 2000].*

TABLE 1.2 Comparison of Feed Needed to Produce a Pound of Meat From Microlivestock (Such as Insects) or Macrolivestock (Such as Beef)

Food Commodity	Pounds of Feed Needed to Produce Pound of Meat
Insects	0.0 (If side stream, eg, preconsumer food waste, is used)
	1.0 (If separately produced feed is used)
Chicken (Trostle, 2008)	2.6
Pork (Trostle, 2008)	6.5
Beef (Trostle, 2008)	7.0

Source: Trostle (2008).

BOX 1.1 Hawaii

Hawaii's geographic isolation makes its energy infrastructure unique among the US states. Recently, more than one-tenth of the state's gross domestic product has been spent on energy, most of that for imported crude oil and petroleum products (US Energy Information Administration, 2013a,b). More than four-fifths of Hawaii's energy comes from petroleum, making it the most petroleum-dependent state in the nation (Glick, 2015). Hawaii's largest industry is tourism. Hawaii's second major sector is agriculture. Transportation accounts for about half of all energy consumed (US Energy Information Administration, 2015). One large user of agricultural oceanic transportation is for animal feed, particularly for early weaned pigs. When one puts feed insects, such as black soldier flies, into the equation (Newton et al., 2005), this energy cost for transportation in, for example, Hawaii, can be greatly reduced. This can be and should be replicated worldwide.

The purpose of this discussion of nonrenewable energy has been to set the stage for a rethinking of food production. Box 1.1 illustrates how putting insects into the equation of nonrenewable energy use changes the bottom line of use in a positive way. In 1999 the price for food commodities and the prices for oil had been stable for a decade, but during that year, this began to change. In the next 10 years, food commodity prices rose 98% and the price of crude oil rose over fivefold (547%) (Trostle, 2008). It seems logical that it is time to search for a food and/or feed supply that is similarly or more nutritious than the one we currently utilize but that is overall more efficient and requires much less energy for production or transportation. If by-products from that food and/or feed supply could also be used to produce fuel, this would of course be an added bonus.

Insects, as it happens, do indeed have the potential to address both issues. First, because food insects can thrive on sidestreams or preconsumer food waste, such as "ugly" fruits and vegetables, and produce high quality protein without the use of crude oil, the use of insects as food and/or feed would increase overall food production while reducing nonrenewable energy use. Second, oil extracted from edible insects is one potential by-product of insectmeal production for feed, and research suggests that this can be used as an energy source in the form of biodiesel (Manzano-Agugliaro et al., 2012).

WATER USE

In addition to the environmental impacts of livestock in terms of land use, greenhouse gas emissions, and use of nonrenewable energy, the water footprint of mammalian and avian food production is extortionate. Depletion of water resources worldwide calls for a concerted effort to develop future food products that are less water-dependent, and insects are an ideal candidate.

To put 1 kg of corn-fed beef steak or hamburger on the table requires 22,000 L water. Much of this is due to the water footprint of feed crops, which have had devastating effects on natural river systems worldwide. For example, the alfalfa and sugar beets grown on the Great Plains of the United States have resulted in the dewatering of the Colorado River, parts of the Wind River, and other rivers in the west. Equally serious is the diminishing of the Ogalalla Aquifer due to intensive water use for agriculture (Siebert and Döll, 2010). It is time to think creatively. Box 1.2 gives an example of this.

We are at a crossroads (Gleick et al., 2009; Postel et al., 1996; Gordon et al., 2005). The problems are wicked (not easily solved) and food security is the mother of all wicked problems (Ramaswamy, 2015). This calls for extraordinary action and creative solutions.

INSECTS AS A LIVING SOURCE OF PROTEIN IN SPACE

One strength of insects as "minilivestock" that has already been discussed is that they can be farmed in almost any setting. With this in mind, ongoing research is investigating the suitability of insect farming systems for space travel (Katayama et al., 2008).

> **BOX 1.2 Eurasian Milfoil as Insect Feed**
>
> In the Great Lakes area of the United States, water scarcity is not as much a daily concern. The pollutants and eutrophying compounds entering water systems in the Midwestern United States are a daily concern. Some of these eutrophying compounds encourage growth of invasive weeds, such as Eurasian milfoil, *Myriophyllum spicatum* (Creed, 1998) and now its even more invasive hybrid form, along with native milfoil, are a concern (LaRue et al., 2013). This is an opportunity for weed scientists, plant pathologists, entomologists, botanists, animal scientists, and sportspeople to think out of the box and look at the opportunity this might provide for growing more nutritious food without nonrenewable energy use or use of water. Consider feed insects. Midwestern ponds and lakes are doing a great job of producing biomass. Along with investing in more research to develop new herbicides or even insect or fungal biocontrol (Shearer, 2013) for the milfoil, are there ways to just use the milfoil? Instead of using increasingly more and more water and fertilizer on cropland to produce alfalfa and corn for livestock feedlots, what are the possibilities of using the milfoil as a substrate to raise high-quality feed for livestock produced from black soldier fly larvae fed on milfoil? Although we do not yet know the omega-6 and omega-3 fatty acid ratio of this aquatic grass, the fact that it is used as fish food (Catarino et al., 1997) and preferred by insects indicates that it is likely to have nutritional content favorable for growth of black soldier fly larvae, making the larvae nutritious feed for livestock, and so forth to be appreciated by humans.

Insects have long been known to be able to reproduce inside shuttles and enclosed stations in the zero gravity environment of outer space and can serve both as a source of protein and other nutrients and as a tool for recycling materials and producing soil fertilizer. Mars is the second target of our manned space flight next to the Moon, and possibly the most distant extraterrestrial body to which we could travel, land, and explore within the next half century. Requirements and design of life support for a Mars mission are quite different from those being operated on near Earth orbit or for a lunar mission, because of the long mission duration (2.5 years for round trip travel) (Yamashita et al., 2009). With insects as a protein source that can be cultivated on board a spaceship as well as a station on Mars or another planet, they are the ideal food for astronauts. A recent trial held in a sealed laboratory in Beijing, China, found that three astronauts could subsist healthily and happily with mealworms as their staple protein source for 105 days, in a simulation of a space mission conducted for research purposes (BBC, 2015).

INSECTS ARE AN IMPORTANT AND FEASIBLE SOLUTION

Insects are nutritious, delicious, and viable food choice. Their potential is growing due to current trends toward a heightened appreciation of cultural diversity, and a global recognition of the imperative need to address environmental impacts of contemporary agricultural systems.

As we have discussed in this chapter, farmed insects are an ideal solution to address many of our health and environmental concerns going forward. Insects can be efficiently farmed in urban settings. It is estimated that over 2000 insect species are already a part of human diets and the nutrition offered by several of the species matches or surpasses that contained in traditional diets (Premalatha et al., 2011).

Additionally, there is now a synergistic and frenzied interest in cultures other than one's own. Other cultures may have solutions to feeding the future populations and solutions to being able to afford the energy costs of producing the food. This is not a new concept. We need to remember the story of corn or maize, *Zea mais*, and how it made its way around the world starting from the Mayan and Aztec civilizations of Central America as a way to quickly harvest large seeds on a cob without a pod or husk covering each individual seed. Maize was orders of magnitude more efficient to harvest than millet or sorghum or even rice for these early agriculturally based people. As large scale agriculture developed to include corn as a main grain, people were not paying attention to the drain corn was on local water supplies, soil fertility, and on human and domestic animal nutrition that supported strong minds and bodies. Maize was high in omega-6 fatty acids and low in omega-3 fatty acids, thereby setting the stage for a population where chronic disease, of metabolic and cardiovascular origins, would thrive (Simopoulos, 2002). We chose large scale efficiency and profit as the focus for agriculture rather than health, either environmental health, *or* human health.

We are now asking our biosphere for the very best choices in nutritious foods that also support environmental health. We are asking for foods that supply essential nutrients everywhere, to avoid transcontinental and transoceanic transport, but that do not require much land or water to produce. Western cultures are noticing that over one-third of the world's population *are* consuming insects. Western cultures are further noticing that this interest is not just related to poverty. In Thailand, Korea, Japan, Botswana, and Mexico, edible insects are sought-after delicacies in both the local marketplaces and pace-setting restaurants. There is a serious upscale interest—light industrial interest in Europe and North America, particularly in the United States, with the Netherlands, England, Denmark, Switzerland, Germany, France, and Italy following suit.

Western cultures are also noticing that edible insects represent a rich source of biodiversity worldwide, with over 2000 species in use, compared to our 15–20 typical food crops. In addition, new taste experiences are being discovered within those 2000 species. This species diversity means most all human communities can find edible insects on their own land or neighboring waterways.

To prepare for the predicted increase in urban insect farms that will supply high-quality protein where the need is the greatest, and to prepare for the products made with insects as an ingredient, specialized information in sectors such as industrial farming, food processing and production, bioengineering, microbiology, medicine, biochemistry, food law, and regulatory processes are needed. And, yes, it is even time to begin to think seriously about the important role that insects for food and feed will no doubt play in preparing for space travel.

WORLDWIDE ACCEPTANCE OF INSECTS AS FOOD

There is a growing interest in insects as a food source, and this trend can be found in the number of companies now selling insect-based food products. Despite concerns about cultural acceptance, examples from around the world attest that "disgust" is relative and can be subject to rapid change. It is likely that the diversity and versatility of insects will facilitate wider acceptance, as will their biochemical properties that enable a broad range of potential uses. Historical precedents in entrepreneurial activity related to previous food movements are worth considering at this stage, and raise some important questions.

There is a growing interest in edible insects worldwide. This is seen not only in the academic sphere, but also in the availability of food products containing insects as an ingredient. Müller et al. (2015) conducted a systematic review of companies selling insects online. This yielded approximately 50 results, showing that insect foods are rapidly becoming available to online consumers, although prices remain high, at an average of $18.33 per 30 g portion (Fig. 1.4).

It is also significant to note that many insect-based food products—certainly the majority of those available online—are sourced from countries in the global north. Fig. 1.5 shows the geographical distribution of companies selling edible insects in terms of registration, nationality of owner, and geographical origin of the insects themselves. This is an exemplary clear shift toward promotion of edible insects in cultures that do not have a long term history of insect consumption.

Why are edible insects gaining a newfound popularity among the wealthier sections of society, who do not have a history of insect consumption? One reason may be the evidence suggesting that insects can combat environmental challenges and contribute to human health, particularly against a cultural backdrop that emphasizes the importance of sustainable consumption. Nevertheless, the "disgust factor" remains an important barrier. In the past few years, researchers have investigated the factors that encourage or dissipate levels of "disgust" when confronted by insect foods (Looy et al., 2013).

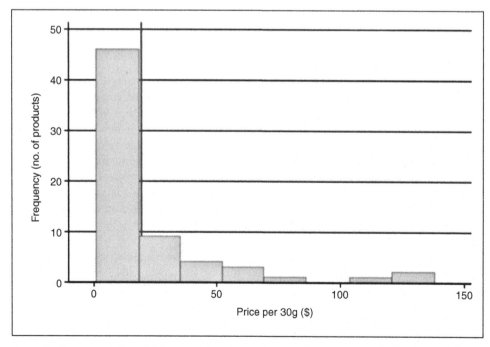

FIGURE 1.4 **Price distribution of products containing edible insects, for 27 companies selling online in Oct. 2015.** *(Müller et al., 2015)*

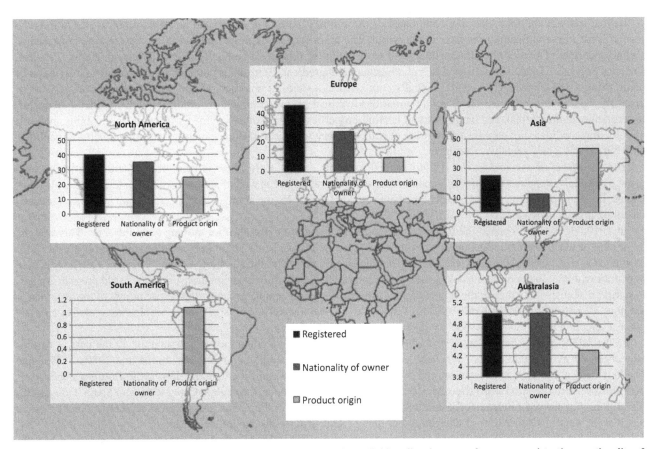

FIGURE 1.5 **Geographical distribution of edible insect products currently available online, in terms of company registrations, nationality of company owner, and original source of insects for sale.** *(Müller et al., 2015)*

However, acceptance of edible insects is growing among forward-thinking people of all generations. It is now necessary to carefully introduce public relations and branding that addresses this looming "disgust factor." It cannot be ignored, it must be dealt with directly, compassionately, with care, empathy, and understanding. There are many aspects to "disgust sensitivity training." The best is to eat with a group and discover how delicious insects are. Think about lobster and how it moved from being a prison food to an elite, special, high-priced delicacy with wide acceptance in Western cultures. Lobster is a disgusting bottom feeder. Insects, cousins of the lobster, but with more refined eating habits, will also make this transition, which is already underway now in Western cultures. In cultures that have a long history of edible insects such as those in Uganda, lobsters and shrimp, their sister crustaceans, the opposite transition is not as likely. Many Ugandans who grew up eating insects learned at an early age that lobster and shrimp were "dirty, disgusting animals" and they were not food.

When humans begin to think historically, probably subconsciously, they will realize that insects were the bread and butter of their early existence, not the big mammals that were a chance stroke of good luck from a successful hunt. To keep good quality protein in the digestive systems of the mothers and young children of the early humans, insects were a food of choice. This should tell us something.

Switching to insects is not as difficult as it seems. They are versatile. They can be introduced as a powdered ingredient in many otherwise widely accepted food and beverage products such as bars, snacks, baked goods, smoothies, shakes, tortillas, pastas, and meat alternatives. They can serve as an appetizer, a salad topping, a pizza ingredient, a main course, an ingredient in the sauce, a dessert, and a great snack. They can be used in many parts of the cuisine of one's culture.

Large-scale processing of insects is available now at a near industrial scale. Working with whole insects is not amenable to many types of food products or food processing equipment. Perhaps the best for now is to work with insects raised in large indoor farms by one company, which are then processed (whole) into a fine powder by another (such as the currently commercially available cricket powder). New entrepreneurs will want to: (1) study the current industrial insect farming practices to learn what works; (2) innovate far beyond the current farming practices, which are largely manual; (3) keep very careful microbial control in mass culture practices; and (4) innovate and think outside the box as far as the various potential food and other products that can be made from insect material.

Beyond protein and food, the fiber component of insects may have a very wide and lucrative set of uses. Chitin is the second most important natural polymer in the world. The main sources exploited for chitin are two marine crustaceans, shrimp and crabs. The uses of chitin are many and more are in the process of being discovered (Rinaudo, 2006). Chitosan, a natural polysaccharide, is widely used as a pharmaceutical ingredient. Chitosan is obtained by the partial deacetylation of chitin (Singla and Chawla, 2001) and is actually composed of a series of polymers varying in their degree of acetylation, viscosity, and other properties. There are a myriad of biological actions exhibited by chitosan, including being hypocholesterolemic, antimicrobial, and having wound healing properties. Low toxicity coupled with wide applicability makes it a promising candidate not only for the purpose of drug delivery for a host of drug moieties (antiinflammatories, peptides, etc.) but also as a biologically active agent.

The high level of entrepreneurial activity related to edible insects at this time is incredible. There is an unexpected growth of small companies and small companies that are growing larger, as well as an emergence of a few large-scale companies. For example, in the United States alone between 2012 and 2015, the number of registered small businesses focused on edible insects increased 200% from 10 to 20.

In 1910, Henry Ford opened the 60-acre Highland Park automotive plant with a moving assembly line. This was the beginning of what eventually became an enormous use for fossil fuels. Fossil fuels were used not only to propel the automobiles that were made at the plant, but also to generate electric power for the automotive plant. This met a need for people to be more mobile. In the form of trucks which soon replaced the horse-drawn wagon, transportation could provide oranges to northern Minnesota in the middle of winter and strawberries to any remote grocery store every month of the year. There are now 1.2 billion cars on the world's roads (Tencer, 2011). We now know limits to nonrenewable energy will eventually place a selection pressure on this industry to change its source of energy.

In the 1950s in the United States, pizza first became something one could buy in a restaurant. In the past, it had always been a way to use leftovers with cheese, as learned from one's Sicilian grandmother. In the mid-1980s, I (Florence Dunkel) was eating pizza in a restaurant run by French Canadians in Kigali, Rwanda. By Jan. 1989, Domino's opened its 5000th store, moving into Puerto Rico, Mexico, Guam, Honduras, Panama, Colombia, Costa Rica, and Spain. US sales hit $2 billion (http://www.referenceforbusiness.com/history2/60/Domino-s-Pizza-Inc.html#ixzz3r7rCRc60). Quickly, pizza touched the palate of Americans of all ages and of many ethnic groups throughout the world.

A similar story emerged in the United States with sushi in the 1980s. Sushi is an Indo-Chinese meal reinvented in 19th-century Tokyo as a cheap fast food. Pioneers who opened the Kawafuku Restaurant in Los Angeles' Little Tokyo brought sushi to the United States in the late 1960s (Bhabha, 2013) and now this unlikely meal is exploding into the American heartland. Sushi is particularly interesting because, unlike pizza, sushi is made up of foods your "mother and grandmother told you never to eat—raw fish." Since for Europeans and Euro-Americans, food insects are in this same category of "not food," there are many lessons that can be learned from the rise of sushi in the United States.

Yet, sushi is not entirely like the edible insect entrepreneurial explosion happening at this time. Insects are microlivestock which can be produced anywhere and are incredibly versatile. They are even now used in sushi (for example in the Sushi Masi restaurant in Portland, Oregon). Food insects can be both a commodity and an ingredient, a powder, a coarse meal, or a whole animal eaten like a shrimp or prawn. In addition, insects are also feed for farmed fish, chickens, piglets, and cattle. As such, insects for food and feed, in contrast to sushi, can have a wide-reaching influence in land use, energy use, and water and air quality, in addition to their well appreciated taste and nutrient contents. Sushi uses about 10 different species of fish (Corson, 2008), whereas food insects include more than 2000 species.

We should be aware of some caveats. We do not want to unintentionally promote a solution that is naive to its potential unintended negative environmental or other consequences, particularly if not executed with care to these issues. Of what do we need to be particularly careful? What is in the future that could be harmful with respect to these insect-based products for food and feed? What have we learned from our mistakes with large-scale agriculture with corn, wheat, cows, and chickens? Answers to these and many other related questions are waiting for you in the next 10 chapters of this book as well as the upcoming years of exciting research, innovation, and entrepreneurial activity exploring and implementing insects as sustainable food and feed ingredients.

FUNDING AND LEGISLATION

There is increased funding available for research and development in the field of insects as food and feed. Furthermore, there is a concerted effort to combat current legislative barriers to developments in the commercial insect industry.

National and international research and funding support is now available in North America and Europe.

Major funding sources are now selecting proposals for projects focusing on edible insects and insects for feed. Among these are the Bill and Melinda Gates Foundation, the USDA Small Business and Innovative Research (SBIR) program

(REF), and the European Union, who have already funded edible insect projects. The European Union funded PROteIN-SECT, which has portions of its project in Mali, Kenya, and in other countries. In 2011 the US company All Things Bugs, LLC was founded, and shortly thereafter, received a grant from the Bill and Melinda Gates Foundation to utilize insects as a protein source in a food product to alleviate malnutrition in children (Crabbe, 2012). In 2013, 2014, and 2015, the USDA Small Business Innovation Research program awarded grants to All Things Bugs, LLC in order to explore the properties and functionality of insects as a food ingredient, improve processing methodologies, and mechanize cricket farming to increase efficiency and drive down cost (Mermelstein, 2015; Day, 2015; Tarkan, 2015). Since 2008, the United Nations Food and Agriculture Organization (FAO), Forestry Section in Rome, Italy has funded and managed an exploratory project in insects for food and feed ending in 2014. Primarily this project was funded by the government of the Netherlands. It is now urgent that this global approach to edible insects integrating efforts in academia, world governments, and the private sector be continued. It is time for other nations, such as the United States, to step up to the plate and take their place in supporting the FAO's work in edible insects and insects for feed.

The legislation process is underway in Western cultures to regulate and provide safety parameters for edible insects and insect protein

The European Commission has addressed the edible insect issue by requiring vendors selling insects in a commercial setting to prove these insects have a history of being eaten by humans or undergo rigorous testing for a classification as a "novel food" item. Indeed, in Groningen and Haren, The Netherlands, insects have been on sale in the supermarket since Nov. 1, 2014 with plans to expand to 500 other locations within a year. In Belgium, supermarkets Carrefour and Delhaize are selling insect meals and snacks. In the United Kingdom, the Food Standards Agency (FSA) is reaching out to edible insect vendors to ask for information about the history of consumption of their insects. The commission requires evidence that all food products have been widely eaten since before 1997, or else they must be authorized as novel foods. Currently 13 UK companies sell insects in an unflavored, seasoned, or candied form, including giant toasted ants and the house cricket.

In the United States, insects for human consumption must pass the same standards and criteria that any other food for human consumption does that is put into interstate commercial trade. It is currently legal in the United States to sell whole insects for human consumption if they are raised for that purpose and if the scientific name is printed on the packaging. Powder of the house cricket, *Acheta domestica* has been commercially available in the United States since late 2013. Insects as feed for macrolivestock such as chickens and pigs is more problematic in the United States. Insects can be added to feed and used within the farm gate for consumption by any domestic animal, but this feed cannot yet be put into commercial arenas. It is not yet legal in the United States to sell insects as animal feed for livestock that will be consumed by humans. It is also not yet legal to grind up insects and extract the protein for sale as a stand-alone product.

INCREASING RECOGNITION IN THE ACADEMIC SECTOR

Despite the developments described previously, food policy makers are in serious need of good quality information right now. To date, much of the information available is coming from the entrepreneurial sector. Going forward, policy makers, industry, and the consumer public will require this need to be further met by greater coverage of this topic by academic researchers. Indeed, academic research on insects as food and feed has seen a sharp increase in recent years. Academic institutions are starting to develop education courses in the area. This book is a part of this movement toward knowledge accumulation, in what remains a relatively understudied field.

Edible insect research being published in the peer-refereed literature has been accumulating at an exponential rate since the year 2000, in parallel with an increase in patents related to insects as food and feed. This can be seen in Fig. 1.6, which shows the rise in peer-reviewed journal articles and patents relating to edible insects over the past 500 years.

Patterns of published academic articles after 2000 is particularly revealing. Müller et al. (2016) conducted a systematic review on peer-reviewed original research looking explicitly at insects as food. They searched three major databases (Web of Science, Embase, and Scopus) and found that since 2000, the publication of peer-reviewed original research on this topic has also seen a sharp increase, the majority of the sharp increase occurring in the 2 years following the publication of the landmark FAO report on insects as food and feed (Van Huis et al., 2013) (Fig. 1.7).

Furthermore, the nature of these articles shows a gradual shift in interest toward farmed insects (Fig. 1.8), suggesting that research priorities are supporting an increase in the scope and scale of insect production.

Edible insect research is also showing an increasingly global representation. Fig. 1.9 (reproduced with permission from Müller et al., 2015) shows the number of research articles published on entomophagy by the location of the author's institution, and the location of the research itself. All areas of the world, including those without a recent history of insect consumption such as Europe and North America, are showing increasing interest in this topic.

16 Insects as Sustainable Food Ingredients

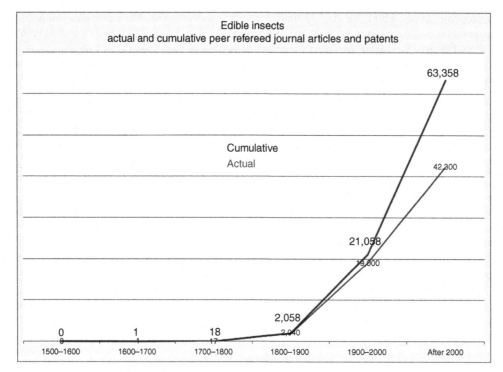

FIGURE 1.6 History of peer-refereed articles and patents related to edible insects.

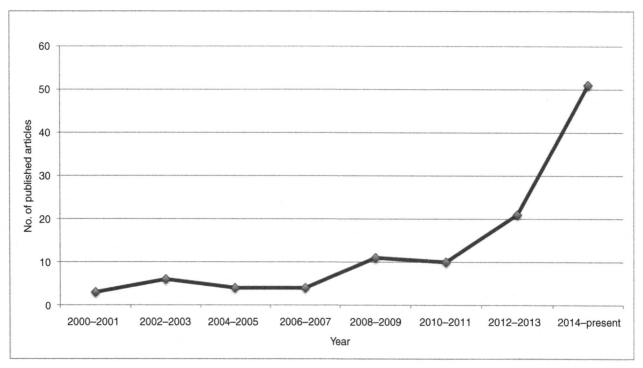

FIGURE 1.7 The number of peer-reviewed research articles published on insects as food, by year, since 2000. *(Reproduced with permission from Müller et al., 2016)*

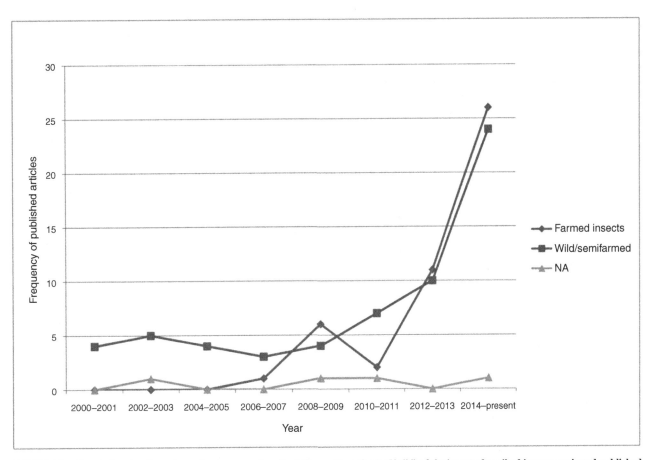

FIGURE 1.8 Graph showing the nature of procurement method (farmed/semifarmed/wild) of the insects described in peer-reviewed published research articles, by year, since 2000. *(Reproduced with permission from Müller et al., 2016)*

At this time in the United States, there are portions of courses, 1 lecture, 1 week of lecture, and a lab devoted to edible insects in a handful of the 50 1862-Land Grant University Entomology curricula. At this time, one full semester, senior level course devoted entirely to edible insects and insects for feed is under construction at Montana State University. It will be a joint offering for students in the College of Agriculture, College of Health and Human Development, and in the College of Business. It will be a lecture-less, texts-and-critics format with a research and laboratory component that includes participating in planning and holding the annual Bug Buffets. It has also been suggested that this course be offered in an online format as well. At Wageningen University in the Netherlands, one course devoted to just edible insects and insects for feed is currently being offered. Overall, then, the curricula in universities across the United States and globally is slowly being adapted to meet the needs of this growing entrepreneurial sector. This book will serve as a text for the development of new courses, and its publication will no doubt spur development of more such courses, worldwide, in academic departments of various disciplines including entomology, food science, agricultural related fields, and potentially more.

CURRENT TRENDS IN USING INSECTS AS FOOD

Pace-setting culinary explorations are underway throughout Western cultures as well as in other cultures in Africa, Asia, and South and Central America. Many of these draw upon indigenous knowledge and recipes to develop palatable, healthy products using traditional seasonings and preservation methods. Meanwhile, others are attempting to develop a new cuisine entirely, by introducing insects as ingredients in recipes that would traditionally have used very different ingredients.

An increasing appreciation of global cuisine and the quest for "authenticity" of food products means that the market is ripe for exploring the potential of insects as a food ingredient. Many cuisines that are popular worldwide, such as Mexican, Japanese, Thai, and Chinese food cultures do include insect dishes within their culinary repertoires. Yet these dishes are rarely available outside of their countries of origin. However, with a growing interest in edible insects worldwide, some forward-thinking restaurants that represent these cuisines on a global scale are introducing insect foods to their menus.

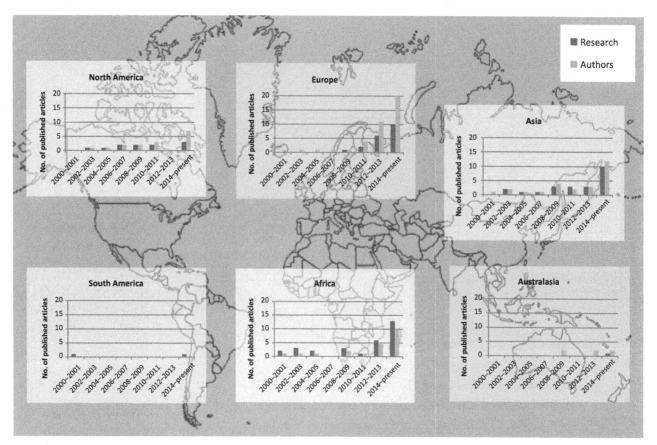

FIGURE 1.9 World map showing the number of published articles during 2000–2015, per continent, by the location of the research and the location of the author's institution. *(Müller et al., 2015)*

Meanwhile, entrepreneurs from cultures without a history of insect consumption are developing twists on their own recipes that use insects in the place of other food ingredients. Examples of this range from cookies and sweets to pasta and cocktail bitters.

The availability of cricket powder for human consumption has unleashed amazing culinary creativity, such as, at Montana State University (MSU) where the MSU Catering Service's most frequent edible insect request is for chirpy smoothies. Pace-setting restaurant, Noma, in Copenhagen, has two Michelin stars and was voted the "best restaurant in the world" by *Restaurant* magazine in 2010, 2011, 2012, and 2014. One of their exploratory cuisines is edible insects. To develop high caliber insect cuisine, they have traveled to indigenous communities harvesting insects worldwide to explore the "deliciousness" of insects as part of a project that will result in a documentary film and a book on insect gastronomy. We invite the readers to run wild with all the possibilities of using insects as a sustainable, nutritious, delicious commodity!

Insects "fit the bill" in all these cases: nutritious, delicious, good for the environment, versatile culinary ingredients, grown anywhere without land or water, useful for nongluten diets, acceptable for some "vegetarian philosophies," and time-tested through millennia of human civilizations. Manufacturing practices, the wholesale marketplace, and end-use consumer products are emerging at an unprecedented rate. Distribution, though, has not been as universal as the most recent Harry Potter book. Why?

The biggest limitations at this time are about scale, availability, and price. Price is related to scale and innovation. Many mainstream food companies are very interested, but only if the insect component were less expensive. In addition, a few large food companies would require a larger amount of the insect material than can currently be produced right now, at least in the United States. If scale-up is successful and current production costs could be cut by 50%, many companies would be using insect products and edible insects and insect products would "take off."

PSYCHOLOGICAL BARRIERS AND DISGUST

Psychological barriers to insect consumption are found worldwide, but only to an extent that is comparable to some other nonplant food products. "Disgust" can often be rooted in irrational logic or misinformation, and here we show examples of how this irrationality can be exposed in the context of insects as food. Food taboos are less robust than they may seem, as evidenced by the rapid acceptance of raw fish in Western cultures with the promotion of sushi as a health food, for example. Importantly, psychological barriers are also found in cultures with a history of insect consumption, and this is likely to be a result of Western cultural influence. This reflects the existence of power inequalities which could be challenged by a growing acceptance of insects as food.

The "eeeuwh factor," (or "yuck factor") known professionally as the disgust factor, is not as big in the industry and with consumers as most outsiders in academia think. In national policy-making circles, however, the disgust factor is still a major push-back force, as well as in the food and agricultural sciences of academia, and among particularly older Western culture members. The disgust factor is formed early in one's life, generally before one is 6 or 7 years old (Looy et al., 2013). To pry lose this concept of insects being only a negative factor to squash, repel, or avoid once the psychological "open period" is passed, requires patience.

For example, ask anyone who has tried honey if they like the taste, and almost 100% of the responses will be "sure." Then explain the disgusting way in which honey is produced. First the bees collect various sweet materials such as flower nectar and the excrement of aphids (also called honeydew). Then they swallow it and spit it up. Another bee ingests this emesis and then the second bee spits it up and so forth, until it is just right. So, honey is essentially insect excrement mixed with the digestive enzymes from the stomach of the bees. Now ask the person whom you are interviewing if they will ever eat honey again. The typical answer is "of course!" I (Florence Dunkel) use this example to help people who are caught in the "disgust factor" stage to understand that the condition of "disgust" they are in is not something that can be consciously willed away even though a person may want to enjoy edible insects.

Another approach to helping folks out of the disgust factor stage is to remind them how delicious food is that automatically has insects in it. A good example of this approach is to ask a person in a Western culture if they like figs, especially Fig Newtons (commercial fig bars popular with young children in the United States). Again the answer is usually "oh, yes." I usually respond with "Then you must like edible insects!" I explain that the fig has an internal flower and in order to be an edible fig fruit, the fig must attract and trap the minute fig wasps that do the pollination. These fig wasps also have parasitoids (minute, microscopic wasps) who are trapped inside the fig. So fig wasps were probably the most common food insects in Western cultures until crickets, mealworms, and wax moth larvae became the popular insects for feasts, energy bars, chips, and gluten free-powder or coarse meal. Again, it is important to ask your interviewee if this information on fig behavior and how figs develop will now cause them to avoid eating figs. Dispelling the disgust factor is an important signpost on the roadmap to getting insects accepted as a large-scale commodity. Indeed, alleviating the disgust factor existing with food lawyers and food policy makers is on the critical path to large-scale use of insects as food and feed.

In the material resource-poor world, however, there is another equally important limiting factor in setting a place at the table of the future with food insects is the "embarrassment factor." I (Florence Dunkel) first observed the "embarrassment factor" and coined the term when living in a small Bambara farming village in southern Mali. I had already known these women and men farmers for 12 years. My Malian colleague, Keriba Coulibaly, also an agricultural scientist, and I were visiting with the elders of the village and mentioned how concerned we were that the children could not make use of the grasshoppers in the fields by the village anymore because the fields were now mainly planted in cotton. For example, cotton with its high load of pesticides made it dangerous for the children to hunt hoppers in those fields so the mothers and grandmothers of the village said no more hopper hunting. The conversation continued that afternoon and Keriba and I began to reminisce about his hopper feasting as a child in his village in Mali and I about my grasshopper feasts with my students for the past couple decades. Then we began to talk about our new recipes that Keriba and I designed with crickets, mealworms (beetle larvae), and wax moth larvae now that hoppers were not available to us because of safety rules of the USDA ARS for insects reared in their laboratories. Keriba and I enjoyed reminiscing and he translated all into Bambara so the entire group gathered could join in our conversation. Not much was said by the villagers, but the next morning as we were sitting in the same chairs in the courtyard beside the family's mud huts having breakfast, one by one, women farmers joined us, each with a reminiscence about a different food insect they enjoyed as a child. Not until it was clearly a safe environment, even after knowing me for 12 years, did the women feel comfortable sharing their food insect stories with us. From the stories, though, it was clear that the "embarrassment factor" had, in effect, squelched any further thought of insects as food in this village.

This story has several important threads. First, there was the fact that pesticides had the capability of plunging this village into a far deeper percentage of malnutrition, particularly the lack of complete proteins. Mothers were concerned about pesticides, but equally important was finding another source of complete protein to replace that of the grasshoppers. The women of this village had no formal education, and, until we supplied them with some amino acid information, they were not aware of how important the insects were to their children's health. This we easily remedied with some carefully crafted stories and models about the building blocks and amino acids, but thanks to the "embarrassment factor," the knowledge that the mothers now needed for the health of their children, was all but lost.

The challenges posed by insects are not only technical; they are cultural and psychological. Overcoming psychological barriers and cultural practices is a complex process, but not impossible, and it is quite possible that soon insects will be welcomed into Western culture homes from which they were once ousted. Equally important will be the time when Western culture visitors to other countries will approach differences in food preferences in an ethnorelative manner (Bennett, 2004) and be accepting of cultural practices that include edible insects.

DEFINITION OF TERMS

Finally, we would like to address the term that has dominated the topic of insects as food and feed, *entomophagy*. Naming plays an important psychological role in the acceptance of edible insects, and loaded terms should not be used lightly. What does "entomophagy," then, allude to?

Entomo is a Greek stem that has to do with cutting or dividing into sections or segments. Sure, insects have segments but so do other invertebrate animals. The stem "entomo-" has stuck, so we folks who study six-legged insects are called entomologists and not hexapodists who study hexapodology. Hexapod-phagy would be more correct for a term referring to the eating of insects since insects are the only animals that have six legs in their adult form.

Phagy is the process of eating. Entomophagy is technically the term for the process of eating insects. An *entomophagist* is an insect eater. But, do we talk about bovine-phagy for those who eat beef or porcine-phagy for those who eat pork?

This book is for those working in the insects as food and feed sector and is specifically about industrialization of insects as a sustainable commodity for food and feed, and other applications. We suggest moving on and abandoning the term entomophagy, an issue that is dealt with in greater detail by Evans et al. (2015) and in Chapter 5 of this book.

SUMMARY OF BOOK

Having put some of the broad spectrum picture-framing comments regarding edible insects on the table, we now summarize the technical content and detail that the rest of the book will walk you, the reader, through in each of the following chapters.

History, geography, and cultural aspects of edible insects beginning with paleontologic evidence. In Chapter 2, the reader is taken first to a historical review, beginning with the amazingly rich paleontological records of the harvesting and use of edible insects. From there both the geographical and cultural aspects are covered in great breadth and depth. Even the stages of development of intercultural competency with respect to accepting edible insects into the food industry, the marketplace, and the home are described.

Following this detailed biodiversity, biogeographical, and psychologically focused opening, the book then moves very logically through other technical topics from the nutrient content and health benefits to the sustainability an environmental benefits, the current status and considerations for mass production technologies and methods, the current insect-based food industry, and its future. Food safety considerations (regulatory, microbiologic, allergens) and the current knowledge on that complete the book, followed by a brief chapter to keep in mind the possible feed applications and how those might tie into the food applications, and then a conclusion/summary chapter.

More specifically, what follows are the main ideas from each of the subsequent chapters.

Nutritional advantages of including insects in one's diet. Chapter 3 will emphasize the basic details such as the warning that it is incorrect to make all-encompassing statements about the nutrient content of insect. Of the thousands of species of insects, each has its own nutritional profile. Even within species, the metamorphic stage and diet of the insects affects their nutritional content. Most insects do have significant levels of both macronutrients (protein, etc.) and micronutrients, although in many cases the micronutrient content is likely to be due to dietary factors, depending on their substrate. Iron deficiency anemia is worldwide one of the most common and widespread nutritional disorder. In material-resource-poor countries, one in two pregnant women and about 40 preschool children are believed to be anemic (FAO/WHO, 2001). Several insect species are known to have high iron content. Zinc is a similar core problem, globally. Deficiencies lead to stunting, delayed bone and sexual maturation, diarrhea, violent behavior, increased susceptibility to infections mediated by the immune system such as malaria (FAO/WHO, 2001). To whet your appetite for Chapter 3, consider the palm weevil larvae. Beef averages 12.5 mg zinc per 100 g dry weight, while the palm weevil larvae, *Rhynchophorus phoenicis*, for example, contains 26.5 mg zinc per 100 g (Bukkens, 2005). It is time to consider edible insects in emergency food relief programs.

Efficiency and sustainability of insects versus traditional vertebrate livestock: predicted human population, level of food insecurity, feed conversion rations of current livestock and insects as livestock farming. Chapter 4 brings the reader around the world to look at farmed insects and the benefits. One of the most important takeaways from this chapter is that beef, pigs, and chickens contribute 15–24% of the current greenhouse gas emissions (GHGE), most of which are methane and nitrous oxide, whereas mealworm larvae, crickets, and locusts emit 100 times less GHG and no methane. Therefore insects, just based on this statistic, are a far better choice for human food than vertebrate livestock.

Considerations for introducing insect protein into the industrial manufacturing sector and into the marketplace and the path to becoming a mainstream food ingredient or family of commodities. The story told in Chapter 5 is told from the perspective of trail blazing entrepreneurs seeking to disrupt food and agriculture by adding insects to their portfolio of products and ingredients. At this time, there are approximately 27 insect-based food and edible insect companies registered in the United States and about 26 in Europe. Just 2 years ago, this number for the United States was 10 and 5 years ago, the number was zero. Clearly this is a logarithmic growth curve. Labor is still quite manual for the farming of insects. Breeding efforts to create improved food insects are in their infancy. Humane euthanasia by freezing is easy and natural. This simple picture gains dimensions when we look further back in history. Insect farms and factories for biocontrol agents like *Trichogramma* spp. and for screwworm (a fly) eradication have been around for almost a half a century. Bee colonies producing bee brood for human consumption in addition to honey have been in operation for centuries. The silk industry has been harvesting pupae for food and feed as a by-product for millennia. Details needed by companies that are part of the exponential growth of edible insects for food and feed will be found in Chapter 5. Fuel for continued growth is also part of the story told in this chapter. Here are several examples of future growth opportunities. Insects that are not fed on wheat are gluten free, and when they are fed grass or algae or other green vegetation, they are high in omega-3 fatty acids. The reptile, amphibian, and bird pet food industry in the United States harvests $150 million in revenues each year from farming edible insects. Fish and mammals raised for human food are using insects such as the black soldier fly larvae. What will create the greatest push for continued growth of edible insect companies are the simple conversion equations. To produce 1 kg of animal, crickets need 1.7 kg of feed, less than 1 L of water, and 15 m^2. To produce the same amount of beef requires 10 kg of feed, 22,000 L of water (including irrigation of feed grains), and 200 m^2 (Jongema, 2015; Van Huis et al., 2013).

Current methodologies of insect mass production and areas ripe for improvement. Chapter 6 creates the stage to play out all scenarios the interior farm or insect factory needs to become a large-scale operation. From the laboratory to industrial mechanized and automated insect mass production, this chapter has it all. To use insects as an ingredient requires synchrony in oviposition and egg collection. Separation of frass from insects and various insect stages from each other is the main tedious, incredibly time-consuming part of mass production. Knowledge of dietary needs for insects when developing new and improved (and more sustainable/efficient) feeds is critical. The authors of this chapter come from decades of experience in insect mass rearing as well as mechanization of separation and will have suggestions for you that will amaze you (ie, the shape of the aperture of the separation sieves). I (Florence Dunkel) can attest to the importance of this seemingly small fact in mass rearing insects. Choosing the just right feed in order to balance cost and nutrition is another piece of elder-advice the reader will gain from this chapter. While most treatments of this topic deal with mealworms and crickets, this chapter will also provide readers with specific information on wax moth larvae, *Galleria melonella*, that is not available elsewhere. Climate control is the other pillar of a superior large-scale insect rearing facility. Airflow design and climate control are dealt with in the detail necessary to set up such a facility.

Food Safety and regulatory issues related to insects as a food ingredient. Chapter 7 thoroughly details the regulatory picture for food insects and feed insects in the major production areas of the world. Uncertainty in the regulatory framework is a deterrent to acquiring the monetary start-up capital necessary to grow these industries that produce and that rely on the insect ingredients for their products. The goal of this chapter is to detail the regulatory considerations for potential safety concerns of insects as food and feed. The urgency expressed in this chapter is that this development of regulations needs to take place immediately and needs to operate in cooperation with international regulatory food safety authorities to establish a scientifically sound, globally harmonized set of rules, standards, and criteria. A great step forward in signaling to particularly the European Union and the United States and Canada's regulatory agencies occurred in May 2014 when the first international conference of "Insects to Feed the Future" was held in The Netherlands. The conference organized by FAO, Wageningen University, and a global organizing committee, drew 450 participants which included regulatory officers from Western cultures. The worldwide enthusiasm was a clear signal that insects are a commodity and an ingredient that must be dealt with in the regulatory arena quickly and efficiently. Thailand is the country with the largest thriving insect farming sector. China, one of the most populous and economically prosperous countries in the world, has many food insect species as well as spiders and scorpions, but there is no coordinated industrial or government effort to develop this sector. Neither Thailand nor China have governmental oversight. FAO does have insect food and feed guidelines according to its Codex Alimentarus, which apply as voluntary standards in Thailand and China. The European Union (EU) is a consortium of 28 member countries and has the European Food Safety Authority (EFSA). This group considers food insects a "novel food" regardless of its paleo-origins as food. Novel food is the category used by regulatory bodies in Canada, Australia, and New Zealand. Finland, presently, has a different approach in forbidding importation, selling, marketing or growing of whole or processed insects for use as food. The United States has two agencies poised to regulate food and feed insects: the Food and Drug Administration (FDA) and the United States Department of Agriculture (USDA), through its Animal and Plant Health Inspection Service (APHIS). Research consortia such as PROteinINSECT and GREEiNSECT in Europe are models for exchanging technical information between Europe, Asia, and Africa (Halloran, 2014). Interesting parts of the chapter deal with safety concerns from bioaccumulation, neurodepressants, inhibitory enzymes, metabolic steroids, and contaminants from dietary exposure, and determining if a food insect is pesticide-free. Laboratories already exist to determine species through taxonomic DNA identification. At times it seems that the details involved in elevating food insects from the paleo-diet to a government sanctioned commodity or ingredient is a redundant exercise. For Western culture governments, it is, at this point, a very complicated, but an essential rite of passage for edible insects.

Microbial considerations of insects. Chapter 8 opens with a summary of microorganisms found in insects that have human pathogenic histories. Some of these include food insects. Recognizing the biodiversity of nutrient contents of even a sample of the 2000 food insects, we must consider the variety of microbial communities found in insects. Almost a half a century ago, we learned that some insects are microbially quite sterile (Dunkel and Boush, 1968; Boush et al., 1968). Food insects are known to have many interesting commensal and mutualistic microorganisms (Klunder et al., 2012). The termite, a widely appreciated, delicious food insect and important candidate for a source of food ingredients has its hindgut populated by a dense and diverse community of microbial symbionts working in concert to transform lignocellulosic plant material and derived residues into acetate, to recycle and fix nitrogen, and to remove oxygen. Although much has been learned about the breadth of microbial diversity in the hindgut, the ecophysiological roles of its members is less understood. We do know that this microbiome does not hold human pathogens (Isanapong et al., 2013). Neither the gregarines of Coleoptera nor the flagellates and other microorganisms of termites are able to survive in humans that ingest the insect hosts. Grasshoppers, a well-appreciated food insect worldwide, have their own parasitic nematodes. These are mermithid nematodes that do not infect mammals, birds, fish, or any vertebrate. Another example of the lack of microbial cross-reactivity between mammals and insects is in the bacterial genus *Bacillus*. *Bacillus thuriengensis* is a natural and also commercial entomo-pathogenic bacterium that has no effect on humans or other vertebrates. Similarly, *Bacillus cereus* and *Bacillus subtilis* are well-known human pathogens. Some insects are known to contain antibacterial peptides. Larvae of the common housefly, *Musca domestica*, for example, produced a novel peptide (Hf-1) that inhibits strains of food pathogens such as *Escherichia coli*, *Pseudomonas aeruginosa*, *Salmonella typhimurium*, *Shigella dysenteriae*, *Staphylococcus aureus*, and *B. subtilis* (Hou et al., 2007). Most of Chapter 8 is a detailed summary of the incubation periods, human symptoms, and course of the disease of the most common foodborne diseases, not specifically associated with food insects. This section also includes information on other contaminants of food. The concluding one-third of the chapter provides detailed information on how to disinfest food in food processing and food preparation facilities as well as how to store food safely when canning, drying, refrigerating, and freezing. For the industrialization of food insects, the detailed section on the seven principles of Hazard Analysis Critical Control Point (HACCP) will be quite useful. The FDA has moved regulations for seafood to a HACCP-based system for production. Since one route for the upscaling of edible insect farming and processing will be to emulate that used for seafood, insect entrepreneurs should take note.

Insects and their connection to allergies. Chapter 9 lays out for the reader a full discussion of allergies related to food insects: cross-reactivity with house dust mites, crustaceans, and nonfood insects; food allergens and inhalant allergens; worker's sensitivity; and ways to diminish crustacean allergens that have potential for working on insect allergens. This chapter describes the food allergy risks associated with insect consumption and the proteins responsible for these risks. For example, tropomyosin in insects is similar to tropomyosins found in shellfish and house dust mites. It was not a surprise when scientists found most shrimp-allergic patients reacted to *Tenebrio molitor* as a food challenge (Broekman et al., 2015) and so labeling edible insects and food with insects as a deliberate ingredient for seafood allergies will be necessary. This confirms the practice of those holding events with insects or dishes with insect ingredients to ask guests if they have seafood allergies, just as we would inquire about vegan choices, nut allergies, or gluten-sensitivity. There may be some utility in using the recently coined, common term for edible insects, "land shrimp," to signal to the public that edible insects are related to "seafood," and hence to the allergic response. One compound causing allergic reactions is arginine kinase. In this chapter, we learn that arginine kinase isolated from the cockroach retains 50% of its activity after heating to 50°C for 10 min (Brown and Grossman, 2004). Since arginine kinase is a major allergen in crustaceans such as the snow crab, mud crab, crayfish, and white shrimp that has caused asthma and fatal anaphylaxis in fishermen and plant processing workers, we know that this relationship is an important aspect to pay attention to in food and feed insects as the industry scales up. It is noteworthy to also consider the "allergy–hygiene hypothesis." Allergies are increasing in Western populations in contrast to material-resource poor countries. The hygiene hypothesis states that the difference is due to lack of exposure to pathogens, including intestinal parasites. Most parasites contain chitin. It is hypothesized this variation in exposure to chitin may be a key to explaining the asymmetric prevalence of allergies in populations (Roos, 2013). However, it is likely that proteins from chitin containing organisms are responsible for the allergic reaction rather than the chitin (a polysaccharide) itself. Also, the ability of chitin to induce asthma and allergies appears to depend on the particle size of the chitin. Medium sized-particles induce allergic inflammation, while small-sized chitin particles have the reverse effect of reducing the inflammatory response (Brinchmann et al., 2011). Chitin is a polysaccharide, the second-most abundant polysaccharide in nature, whereas tropomyosins and arginine kinase are proteins found in insect muscle. By inducing nonspecific host resistance against infections of bacteria and viruses, chitin has shown potential for boosting immune system functioning, making it a promising alternative to antibiotics (Van Huis et al., 2013).

Insect farming for use as sustainable vertebrate livestock feed protein/ingredient is a well-developed area, but still fraught with many government regulations. Chapter 10 presents an exhaustive summary of the poultry, pigs, fish (including polyculture and hybrid fish), crustaceans, mollusks, and other animals for which feeding trials with insects have been published. Sanchez-Muros et al., conclude with insightful observations and recommendations for the way forward with substituting insect meal or whole insects for soybean and/or fishmeal. The authors are methodical in presenting this rich set of data and as such, it is a very useful document. Some of the large scale challenges have been met and success has resulted. Companies are producing feed in large scale from insect colonies and have already met and overcome many of these challenges. For example, the Los Angeles waste management situation recognizes insects as a solution to their issue of water scarcity as well as saving land and energy used to put the municipal waste into landfills (Los Angeles City Food Council, personal communication. May, 2015). At least one company is utilizing insects as food for piglets in the Hawaiian pork industry with defatted black soldier fly larvae, using the oil by-product as a biofuel. Factories for producing larger black soldier fly larvae already exist in the United States and large-scale factories are being built this year in South Africa. Scientific interest in insects as fish food is high. Whereas livestock feeding experiments doubled in the past 15 years, the number of published feeding experiments in aquaculture has quadrupled. Twenty-six fish species have been the focus of feeding trials with insects. In many studies detailed in this chapter, protein levels and protein quality in insects were found to be higher than in fishmeal or soy meal. The key to greater industrialization of insects as feed is to remember that insects and fish are living systems and between species there are fine differences in acceptability. Details of actual experiments showed some fish species thrived on insects and others did not. With the great diversity of both fish species and local insects that will work in feed formulations, the future is bright for mainstream feed industries.

Value proposition of insects versus other ingredients and other roadblocks in use of insect protein and how they are being overcome. This book, *Insects as Sustainable Food Ingredients: Production, Processing, and Applications*, represents an entirely new addition to the disciplines of Entomology, Human Nutrition, Food Science, and Animal Science. Insects for food and feed are now an academic subspecialty within Entomology itself, as well as a serious, recognizable source of human nutrients, and a commodity within food sciences, which has its own processing and storage technology and engineering considerations. Food insects are, at the same time, an animal production system which is taking its place as a subspecialty within the study of animal science along with poultry science. At the same time, feed insects are a new cropping system and an animal feed production system. They are, along with corn, alfalfa, oats, barley, and ocean fish, an important source of nutrients for chickens, piglets, and farmed fish.

A CALL TO ACTION

This book provides readers with the breadth and depth of edible insects from the wild-collected moments to the food industry level with all of the appropriate biochemical, engineering, and microbial details in between. This book will also provide the reader with a source of information for those concerned about food safety, regulatory, allergen, and microbial contamination issues. This is a collection of information and guiding tables and figures for entomologists, human nutritionists, food scientists, and animal scientists. As such, it is a rich source for the exponentially increasing number of young entrepreneurs who are choosing insects for food and feed as a sector to focus their business skills, either in manufacturing, marketing, or distribution. The importance of this book for both the academicians and for the entrepreneurs is that it is an encyclopedic and well-integrated volume. Going forward, if we can maintain the insects for food and feed sector as a strongly integrated sector that has not lost touch with its origins in traditional ecological knowledge (TEK), we will be well on the way to a more sustainable world.

The significance that is represented by the publishing of this book is that it is a far-reaching example of how TEK can overcome intercultural prejudice and permeate the hallowed halls of academia, not only in agricultural disciplines, but in health and human development as well. The legacy of insect foods reaches far back into the history of humanity, yet also has a message for our future. This book will guide the reader in great detail through all contemporary issues surrounding the historical and future potential of insects as a food ingredient, on a scale that ranges from the personal to the cultural to the industrial. In choosing insects as a way forward, we hope to take an important cultural step toward world peace. We recognize insects as a valid, valuable, and important nutritional component in the diet of over two billion of the world's people, many of whom are in the rural, economically impoverished parts of the world. We recognize that their nutritional security and cropping systems can support them without switching to unfamiliar foods or production systems. With this recognition on the part of Western cultures, much of the food security needs and undernutrition of the world may yet be solved.

The publishing of this book is also a landmark in application of TEK to sustainable food and feed production. Insects for food and feed represent a clear path to accomplishing the sustainable challenge goals set by the United Nations for the next 15 years (UN, 2015). In choosing from over 2000 species of insects currently supplying high quality protein for the world's people, we are moving toward a significant decrease in greenhouse gas emissions.

The importance of this moment in history can be fully experienced in the following 10 chapters, but we also encourage you to enrich your reading experience by tasting. It is all about tasting. It is about beginning to see insects as part of a tasting option served with other foods. Don't just read about edible insects, put them on the table, on your plate, and in your palate. Enjoy!

REFERENCES

Bakker, N., Dubbeling, M., Gündel, S., Sabel Koschella, U., Zeeuw, H.D., 2000. Growing Cities, Growing Food: Urban Agriculture on the Policy Agenda. A Reader on Urban Agriculture. DSE, Feldafing, Germany.

BBC, 2015. China: volunteers test worm diet for astronauts. Available from: http://www.bbc.com/news/blogs-news-from-elsewhere-27515900

Bennett, M., 2004. Becoming interculturally competent. In: Wuzel, J. (Ed.), Toward Multiculturalism: A Reader in Multicultural Education. second ed. Intercultural Resource Corp, Newton, MA, pp. 62–77.

Berry, W., 2000. The whole horse: the preservation of the agrarian mind. In: Kimbrell, A. (Ed.), Fatal Harvest: The Tragedy of Industrial Agriculture. Island Press, Washington, DC, pp. 7–12, 384 p.

Bhabha, L., 2013. The history of sushi in the U.S. Available from: http://food52.com/blog/9183-the-history-of-sushi-in-the-u-s

Boush, G.M., Dunkel, F.V., Burkholder, W.E., 1968. Progeny suppression of Attagenus megatoma and Trogoderma parabile by dietary factors. J. Econ. Entomol. 61, 644–646.

Brander, K.M., 2007. Global fish production and climate change. Proc. Natl. Acad. Sci. USA 104, 19709–19714.

Brinchmann, B.C., Bayat, M., Brogger, T., Muttuvelu, D.V., Tjonneland, A., Sigsgaard, T., 2011. A possible role of chitin in the pathogenesis of asthma and allergy. Ann. Agric. Environ. Med. 18, 7–12.

Broekman, H., Knulst, A., Den Hartog Jager, S., Gaspari, M., De Jong, G., Houben, G., Verhoeckx, K., 2015. Shrimp allergic patients are at risk when eating mealworm proteins. Clin. Transl. Allergy 5 (Suppl. 3), P77. Poster presentation from Food Allergy and Anaphylaxis Meeting 2014. Dublin, Ireland, 9–11 October, 2014.

Brown, A.E., Grossman, S.H., 2004. The mechanism and modes of inhibition of arginine kinase from the cockroach (*Periplaneta americana*). Arch. Insect Biochem. Phys. 57 (4), 166–177.

Bukkens, S.G.F., 2005. Insects in the human diet: nutritional aspectsIn: Paoletti, M.G. (Ed.), Ecological Implications of Minilivestock: Role of Rodents, Frogs, Snails, and Insects for Sustainable Development. Science Publishers, Inc., Enfield, NH, pp. 545–577, (Chapter 28).

Catarino, L.F., Ferreira, M.T., Moreira, I.S., 1997. Preferences of grass carp for macrophytes in Iberian drainage channels. J. Aquat. Plant Manage. 36, 79–83.

Corson, T., 2008. The Story of Sushi: An Unlikely Saga of Raw Fish and Rice. Harper Perennial, New York, NY, 416 p.

Crabbe, N., 2012. Local expert gets funding to develop insect-based food for starving children. Gainesville Sun, 1B–6A, May 10, 2012.

Creed, Jr., R.P., 1998. A biogeographic perspective on Eurasian water milfoil declines: additional evidence for the role of herbivorous weevils in promoting declines. Amer. J. Aquat. Plant Manage. 36, 16–22.

Day, A.C., 2015. Insectpreneur Series: Interview with Dr Aaron T. Dossey of All Things Bugs, LLC. August 5, 2015. Available from: http://4ento.com/2015/08/05/insectpreneur-series-interview-aaron-dossey-all-things-bugs/

DeFoliart, G., Dunkel, F.V., Gracer, D., 2009. Chronicle of a changing culture. The Food Insects Newsletter. Aardvark Global Publishing, Salt Lake City, UT, 414 p.

DeFries, R., Rosenzweig, C., 2010. Toward a whole-landscape approach for sustainable land use in the tropics. Proc. Natl. Acad. Sci. USA 107, 19627–19632.

DeFries, R.S., Foley, J.A., Asner, G.P., 2004. Land-use choices: balancing human needs and ecosystem function. Front. Ecol. Environ. 2, 249–257.

DeFries, R.S., Rudel, T., Uriarte, M., Hansen, M., 2010. Deforestation driven by urban population growth and agricultural trade in the twenty-first century. Nat. Geosci. 3, 178–181.

Dirzo, R., Young, H.S., Galetti, M., Ceballos, G., Isaac, N.J., Collen, B., 2014. Defaunation in the Anthropocene. Science 345 (6195), 401–406.

DOE/EIA, 2001. World energy projection system. DOE/EIA 0219(99), Washington, DC. January 2001.

Dunkel, F.V., Boush, G.M., 1968. Biology of the gregarine *Pyxinia frenzeli* in the black carpet beetle, Attagenus megatoma. J. Invert. Path. 11, 281–288.

Evans, J., Alemu, M.H., Flore, R., Frøst, M.B., Halloran, A., Jensen, A.B., Maciel-Vergara, G., Meyer-Rochow, V.B., Münke-Svendsen, C., Olsen, S.B., Payne, C., Roos, N., Rozin, P., Tan, H.S.G., van Huis, A., Vantomme, P., Eilenberg, J., 2015. Entomophagy: an evolving terminology in need of review. J. Insect. Food Feed 1 (4), 293–305.

FAO, 2013. Fish to 2030 Prospects for fisheries and aquaculture. Aquaculture and environmental services discussion paper 03. World Bank Report No. 83177-GLB.

FAO/WHO, 2001. Human vitamin and mineral requirements. Rome.

FAOSTAT, 2015. Available from: http://faostat3.fao.org/

Frei, M., Becker, K., 2005. Integrated rice fish culture: coupled production saves resources. In: Natural Resources Forum, vol. 29, No. 2. Blackwell Publishing, Ltd., Malden, MA, pp. 135–143.

Gerber, P.J., Steinfeld, H., Henderson, B., Mottet, A., Opio, C., Dijkman, J., Falcucci, A., Tempio, G. (Eds.), 2013. Tackling Climate Change through Livestock: A Global Assessment of Emissions and Mitigation Opportunities. Food and Agricultural Organization of the United Nations, Rome, xxi+115.

Gleick, P.H., Cooley, H., Morikawa, M., 2009. The World's Water 2008–2009: The Biennial Report on Freshwater Resources. Gleick, P.H. et al. (Ed.), Island Press, Washington, DC, pp. 202–210.

Glick, M., 2015. State Energy Administrator, Dept. of Business, Economic Development, and Tourism, State of Hawaii, Testimony before U.S. Senate Committee on Energy and Natural Resources, July 14, 2015, p. 1.

Godfray, H.C.J., Garnett, T., 2014. Food security and sustainable intensification. Philos. Trans. R. Soc. Lond. B Biol. Sci. 369 (1639), 20120273.

Godfray, H.C.J., Beddington, J.R., Crute, I.R., Haddad, L., 2010. Food security: the challenge of feeding 9 billion people. Science 327, 812–818.

Gordon, L., Steffen, J.W., Jönsson, B.F., 2005. Human modification of global water vapor flows from the land surface. Proc. Natl. Acad. Sci. USA 102, 7612–7617.

Green, R.E., Cornell, S.J., Scharlemann, J.P., Balmford, A., 2005. Farming and the fate of wild nature. Science 307 (5709), 550–555.

Gustavsson, J., Cederberg, C., Sonesson, U., van Otterdijk, R., Meybeck, A., 2011. Global Food Losses and Food Waste Section 3.2 (Study conducted for the International Congress "Save Food!" at Interpack 2011, Düsseldorf, Germany). FAO, Rural Infrastructure and Agro-Industries Division.

Halloran, A., 2014. Discussion paper: regulatory frameworks influencing insects as food and feed. Preliminary Draft. Version December 5, 2014.

Hou, L., Shi, Y., Zhai, P., Le, G., 2007. Inhibition of foodborne pathogens by Hf-1, a novel antibacterial peptide from the larvae of the housefly (*Musca domestica*) in medium and orange juice. Food Control 18 (11), 1350–1357.

Hovorka, A., de Zeeuw, H., Njenga, M. (Eds.), 2009. Women Feeding Cities: Mainstreaming Gender in Urban Agriculture and Food Security. Practical Action Publishing, Rugby.

Isanapong, J., Hambright, W.S., Willis, A.G., Boonmee, A., Callister, S.J., Burnum, K.E., kPasa-Tolic, L., Nicora, C.D., Wertz, J.T., Schmidt, T.M., Rodrigues, J.L.M., 2013. Development of an ecophysiological model for Diplosphaera colotermitum TAV2, a termite hindgut Verrucomicrobium. ISME J. 7, 1803–1813.

Jacobson, L.D., De Oliveira, J.A.A., Barange, M., Cisneros-Mata, M.A., Félix-Uraga, R., Hunter, J.R., Kim, J.Y., Matsuura, Y., Niquen, M., Porteiro, C., et al., 2001. Surplus production, variability, and climate change in the great sardine and anchovy fisheries. Can. J. Fish Aquat. Sci. 58, 1891–1903.

Jongema, Y., 2015. List of edible insect species of the world. Laboratory of Entomology, Wageningen University, Wageningen. Available from: http://www.wageningenur.nl/en/Expertise-Services/Chair-groups/Plant-Sciences/Laboratory-of-Entomology/Edible-insects/Worldwide-species-list.htm

Katayama, N., Ishikawa, Y., Takaoki, M., Yamashita, M., Nakayama, S., Kiguchi, K., Force, S.A.T., 2008. Entomophagy: a key to space agriculture. Adv. Space Res. 41 (5), 701–705.

Klunder, H.C., Wolkers-Rooijackers, J., Korpela, J.M., Nout, M.J.R., 2012. Microbiological aspects of processing and storage of edible insects. Food Control 26, 628–631.

Koops, K., Schöning, C., McGrew, W.C., Matsuzawa, T., 2015. Chimpanzees prey on army ants at Seringbara, Nimba Mountains, Guinea: predation patterns and tool use characteristics. Am. J. Primatol. 77 (3), 319–329.

Kurukulasuriya, P., Rosenthal, S., 2003. Climate change and agriculture: a review of impacts and adaptations. Climate Change Series Paper No. 91, World Bank, Washington, DC.

LaRue, E.A., Zuellig, M.P., Netherland, M.D., Heilman, M.A., Thum, R.A., 2013. Hybrid watermilfoil lineages are more invasive and less sensitive to a commonly used herbicide than their exotic parent (Eurasian watermilfoil). Evol. Appl. 6 (3), 462–471.

Lesnik, J.J., 2011. Bone tool texture analysis and the role of termites in the diet of South African hominids. PaleoAnthropology 268, 281.

Looy, H., Dunkel, F.V., Wood, J.R., 2013. How then shall we eat? Insect-eating attitudes and sustainable foodways. Agric. Hum. Values 31, 131–141.

Lu, J., Li, X., 2006. Review of rice–fish-farming systems in China—one of the globally important ingenious agricultural heritage systems (GIAHS). Aquaculture 260 (1), 106–113.

Lundqvist, J., De Fraiture, C., Molden, D., 2008. Saving water: from field to fork: curbing losses and wastage in the food chain. Stockholm International Water Institute. Policy paper, Working brief, pp. 20–23.

Manzano-Agugliaro, F., Sanchez-Muros, M.J., Barroso, F.G., Martínez-Sánchez, A., Rojo, S., Pérez-Bañón, C., 2012. Insects for biodiesel production. Renew. Sustain. Energy Rev. 16 (6), 3744–3753.

McGrath, J., 2015. Lunching on larvae: this desktop farm lets you grow and harvest edible mealworms. Available from: http://www.digitaltrends.com/home/livin-farms-desktop-hive-makes-edible-mealworms/#ixzz3sZyIl

McGrew, W.C., 2014. The "other faunivory" revisited: insectivory in human and non-human primates and the evolution of human diet. J. Hum. Evol. 71, 4–11.

McNeely, J.A., Schroth, G., 2006. Agroforestry and biodiversity conservation: traditional practices, present dynamics, and lessons for the future. Biodivers. Conserv. 15 (2), 549–554.

Mermelstein, N., 2015. Crickets, mealworms, and locusts, oh my!. Food Technology Magazine, p. 69–73. Available from: http://www.ift.org/food-technology/past-issues/2015/october.aspx.

Mielke, H.W., Mielke, Jr., P.W., 1982. Termite mounds and chitemene agriculture: a statistical analysis of their association in southwestern Tanzania. J. Biogeogr. 9, 499–504.

Morales-Rodriguez, A., O'Neill, R.P., Wanner, K.W., 2014. A survey of wireworm (Coleoptera: Elateridae) species infesting cereal crops in Montana. Pan-Pac. Entomol. 90 (3), 116–125.

Morton, D.C., DeFries, R.S., Shimabukuro, Y.E., Anderson, L.O., Arai, E., del Bon Espirito-Santo, F., Freitas, R., Morisette, J., 2006. Cropland expansion changes deforestation dynamics in the southern Brazilian Amazon. Proc. Natl. Acad. Sci. USA 103 (39), 14637–14641.

Müller, A., Evans, J., Payne, C., Roberts, R., 2015. Entomophagy and Power, poster presentation, Insects as Food and Feed Workshop, Oxford Martin School, University of Oxford, December 4, 2015.

Müller, A., Evans, J., Payne, C., Roberts, R. 2016. Entomophagy and power. J. Insects Food Feed. Submitted for publication.

Nelson, G.C., Valin, H., Sands, R.D., Havlík, P., Ahammad, H., Deryng, D., et al., 2014. Climate change effects on agriculture: economic responses to biophysical shocks. Proc. Natl. Acad. Sci. USA 111 (9), 3274–3279.

Newton, G.L., Booram, C.V., Barker, R.W., Hale, O.M., 2005. Dried *Hermetia illucens* larvae meal as a supplement for swine. J. Anim. Sci. 44, 229–243.

NOAA, 2003. Climate of 2002—Annual Review. Drought: National Paleoclimatic (Pre-Instrumental) Perspective. National Oceanic and Atmospheric Administration. National Climatic Data Center. January 23, 2003. Available from: http://www.ncdc.noaa.gov/oa/climate/research/2002/ann/paleo-drought.html

NOAA, 2015. Drought reports from 1999 to 2015. Available from: https://www.ncdc.noaa.gov/sotc/drought/200710

Norberg-Hodge, H., 2000. Global monoculture: the worldwide destruction of diversity. In: Kimbrell, A. (Ed.), Fatal Harvest: The Tragedy of Industrial Agriculture. Island Press, Washington, DC, pp. 13–16, 384 p.

Oonincx, D.G., DeBoer, I.J., 2012. Environmental impact of the production of mealworms as a protein source for humans: a life cycle assessment. PLoS One 7 (12), e51145.

Oonincx, D.G.A.B., van Itterbeeck, J., Heetkamp, M.J.W., van den Brand, H., van Loon, J., van Huis, A., 2010. An exploration on greenhouse gas and ammonia production by insect species suitable for animal or human consumption. PLoS One 5 (12), e51145.

Owen, J., 2005. Farming claims almost half earth's land, new maps show. National Geographic News. December 9, 2005. Available from: http://news.nationalgeographic.com/news/2005/12/1209_051209_crops_map.html

Parfitt, J., Barthel, M., Macnaughton, S., 2010. Food waste within food supply chains: quantification and potential for change to 2050. Phil. Trans. R. Soc. Lond. B Biol. Sci. 365, 3065–3081.

Payne, C.L.R., Scarborough, P., Rayner, M., Nonaka, K., 2015. Are edible insects more or less 'healthy' than commonly consumed meats? A comparison using two nutrient profiling models developed to combat over-and under nutrition. Eur. J. Clin. Nutr. 70, 285–291.

Pope Francis, 2015. Laudato Si': On care for our common home. Encyclical Letter.

Postel, S.L., Daily, G.C., Ehrlich, P.R., 1996. Human appropriation of renewable fresh water. Science 271, 785–788.

Premalatha, M., Abbasi, T., Abbasi, T., Abbasi, S.A., 2011. Energy-efficient food production to reduce global warming and ecodegradation: the use of edible insects. Renew. Sustain. Energy Rev. 15 (9), 4357–4360.

Purdy, J., 2015. Coming into the Anthropocene. Harvard Law Review. Available from: http://scholarship.law.duke.edu/faculty_scholarship/3504

Ramankutty, N., Foley, J.A., Norman, J., McSweeney, K., 2002. The global distribution of cultivable lands: current patterns and sensitivity to possible climate change. Global Ecol. Biogeogr. 11, 377–392.

Ramankutty, N., Evan, A.T., Monfreda, C., Foley, J.A., 2008. Farming the planet: 1. Geographic distribution of global agricultural lands in the year 2000. Global Biogeochem. Cycles 22, GB1003.

Ramaswamy, S.B., 2015. Setting the table for a hotter, flatter, more crowded earth: insects on the menu? J. Insects Food Feed. 1, 171–178.

Ray, D.K., Mueller, N.D., West, P.C., Foley, J.A., 2013. Yield trends are insufficient to double global crop production by 2050. PLoS One 8 (6), e66428.

Rinaudo, M., 2006. Chitin and chitosan: properties and applications. Prog. Polym. Sci. 31 (7), 603–632.

Roos, N., 2013. The allergy-hygiene hypothesis. Box 10.4. In: Van Huis, A., Van Itterbeeck, J., Klunder, H., Mertens, E., Halloran, A., Muir, G., Vantomme, P., (Eds.), Edible Insects: Future Prospects for Food and Feed Security. FAO Forestry Paper 171. Food and Agriculture Organization of the United Nations, Rome.

Rothman, J.M., Raubenheimer, D., Bryer, M.A., Takahashi, M., Gilbert, C.C., 2014. Nutritional contributions of insects to primate diets: implications for primate evolution. J. Hum. Evol. 71, 59–69.

Salazar, A., Baldi, G., Hirota, M., Syktus, J., McAlpine, C., 2015. Land use and land cover change impacts on the regional climate of non-Amazonian South America: a review. Global Planet. Change 128, 103–119.

Sanz, C.M., Schöning, C., Morgan, D.B., 2010. Chimpanzees prey on army ants with specialized tool set. Am. J. Primatol. 72 (1), 17–24.

Shearer, J.F., 2013. Evaluation of a New Biological Control Pathogen for Management of Eurasian Watermilfoil. Technical Note. Engineer Research and Development Center, Vicksburg, Mississippi. Aquatic Plant Control Research Program. PDF Url: ADA583422. Report Date: June, 2013.

Siebert, S., Döll, P., 2010. Quantifying blue and green virtual water contents in global crop production as well as potential production losses without irrigation. J. Hydrol. 384, 198–217.

Simopoulos, A.P., 2002. The importance of the ratio of omega-6/omega-3 essential fatty acids. Biomed. Pharmacother. 56 (8), 365–379.

Singla, A.K., Chawla, M., 2001. Chitosan: some pharmaceutical and biological aspects—an update. J. Pharm. Pharmacol. 53 (8), 1047–1067.

Smit, J., Nasr, J., 1992. Urban agriculture for sustainable cities: using wastes and idle land and water bodies as resources. Environ. Urban. 4 (2), 141–152.

Steinfeld, H., Wassenaar, T., Castel, V., Rosales, M., Haan, C., 2006. Livestock's Long Shadow: Environmental Issues and Options. FAO, Rome.

Tarkan, L., 2015. Why these startups want you to eat bugs. Fortune Magazine, August 25, 2015. Available from: http://fortune.com/2015/08/25/edible-insects-bug-startups/

Tencer, T., 2011. Number of cars worldwide surpasses 1 billion: can the world handle this many wheels? Huffington Post Canada. Business Section. Posted August 23, 2011, updated February 19, 2013.

Thornton, P., Herrero, M., Ericksen, P., 2011. Livestock and climate change. ILRI. Available from: http://go.nature.com/wYaVA6.

Tomlinson, I., 2013. Doubling food production to feed the 9 billion: a critical perspective on a key discourse of food security in the UK. J. Rural Stud. 29, 81–90.

Trostle, R., 2008. Global agricultural supply and demand: factors contributing to the recent increase in food commodity prices. Report from Economic Research Service. Available from: www.ers.usda.gov. WRS-0801.

UN, 2015. 2030 Agenda: transforming our world: the 2030 sustainable development goals. Available from: https://sustainabledevelopment.un.org/post2015/summit

US Energy Information Administration, 2013a. State of Hawaii Energy Data System, Table F15, Petroleum Consumption Estimates.

US Energy Information Administration, 2013b. State of Hawaii Energy Data System, Table F30, Total Energy Consumption, Price, and Expenditure Estimates.

US Energy Information Administration, 2015. Hawaii. Available from: http://www.eia.gov/state/analysis.cfm?sid=HI - 9

Van Huis, A., Van Itterbeeck, J., Klunder, H., Mertens, E., Halloran, A., Muir, G., Vantomme, P., 2013. Edible Insects: Future Prospects for Food and Feed Security. FAO Forestry Paper 171. Food and Agriculture Organization of the United Nations, Rome.

Whiten, A., Horner, V., De Waal, F.B., 2005. Conformity to cultural norms of tool use in chimpanzees. Nature 437 (7059), 737–740.

Winsemius, H.C., Jongman, B., Veldkamp, T., Hallegatte, S., Bangalore, M., Ward, P., 2015. Disaster risk, climate change, and poverty: assessing the global exposure of poor people to floods and droughts. World Bank Policy Research Working Paper, p. 7480.

Wu, G., Bazer, F.W., Cross, H.R., 2014. Land based production of animal protein: impacts, efficiency, and sustainability. Ann. NY Acad. Sci. 1328 (1), 18–28.

Xiang, X.L., Tian, J.C., Zhi, H.A.O., Zhang, W.D., 2008. Protein content and amino acid composition in grains of wheat-related species. Agric. Sci. China 7 (3), 272–279.

Yamashita, M., Hashimoto, H., Wada, H., 2009. On-site resources availability for space agriculture on Mars. In: Badescu, V. (Ed.), Mars: Prospective Energy and Material Resources. Springer, Berlin, Heidelberg, pp. 517–542.

Yen, A.L., 2009. Edible insects: traditional knowledge or western phobia. Entomol. Res. 39 (5), 289–298.

Zanolli, L., 2014. Insect farming is taking shape as demand for animal feed rises. Available from: http://www.technologyreview.com/news/529756/insect-farming-is-taking-shapte-as-demand-for-animinal-feed-rises

Zezza, A., Tasciotti, L., 2010. Urban agriculture, poverty, and food security: empirical evidence from a sample of developing countries. Food Policy 35 (4), 265–273.

Zhang, J., Zhao, B., Chen, X., Luo, S., 2009. Insect damage reduction while maintaining rice yield in duck-rice farming compared with mono rice farming. J. Sustain. Agric. 33 (8), 801–809.

Chapter 2

Insects as Food: History, Culture, and Modern Use around the World

E.M. Costa-Neto*, F.V. Dunkel**
*Feira de Santana State University, Department of Biology, Bahia State, Brazil; **Montana State University, Department of Plant Sciences and Plant Pathology, Bozeman, Montana, United States*

Chapter Outline

Introduction	29
Edible Insects of the World	31
History of Insects as Human Food	33
Archeological Data	33
Early Historic Times (~3600BC–500AD)	34
Age of Reason and Emergence of the Scientific Era (1700s/18th Century)	35
Modern Cultural Uses	36
Edible Insects Around the World: Selected Classic Examples	37
Indirect or Unintentional Presence in Foods	42
Cultural Restrictions	43
Food Taboos and Religious/Dietary Restrictions	43
Disgust Factor	44
Educational Campaigns	47
Edible Insects as Nutraceuticals	48
Ethnoentomology	50
Harvesting and Cultivation	51
Farming of Edible Insects	52
Final Comments and Recommendations	53
References	54

INTRODUCTION

Insects have played a far greater role in the cultural histories of the people on Earth than is generally recognized (Posey, 1976; Meyer-Rochow, 2004). The influence of insects in humans' lives can be felt in so many culturally-related connections, ranging from basic living needs, such as food, medicine, and self-preservation to more humanistic uses, such as aesthetics, arts, cosmetics, ethics, mythology, religion, economics, and science (Clausen, 1971; Berenbaum, 1995; Iroko, 1996; Costa Neto, 2002, 2014; Lauck, 2002; Hogue, 1987; Hoyt and Schultz, 1999).

One of the oldest connections with insects is their use as food resource (known broadly as "entomophagy"). The definition of entomophagy is the dietary consumption of insects by any organism; but it is commonly used to refer specifically to the human consumption of insects. The word entomophagy is derived from the Greek words "entomon" meaning insect and "phagy" meaning to eat. The term insectivory is sometimes used for the consumption of insects by animals besides humans as well as by some carnivorous plants. Recently, a new term was coined to refer strictly to the use of insects and insect-derived products for human consumption: *anthropoentomophagy* (Costa Neto and Ramos-Elorduy, 2006). However, as the early term is more widespread in the literature, throughout this chapter we will use it interchangeably with and in referring only to the dietary consumption of insects (and spiders and scorpions, as well) by humans.

Insect consumption by humans is historically and geographically an old, widespread phenomenon. From the earliest Chinese annals to Mexican codices, through the chronicles of naturalists and travelers and the old papyrus of ancient Egypt, we have records of insect-eating peoples. Nowadays, as many as 3071 ethnic groups in 130 countries utilize insects as essential elements of their diet (FAO, 2008; Ramos-Elorduy, 2009; Srivastava et al., 2009; Vantomme, 2010; Yen, 2009a,b). It is estimated that insect eating is practiced regularly by at least 2 billion people worldwide (Pal and Roy, 2014). The use of edible insects varies by local preference, sociocultural significance, and region (Fig. 2.1). Edible insects are often regarded as cultural resources and this place-based knowledge reflects a rich biodiversity. Likewise, people who eat insects have

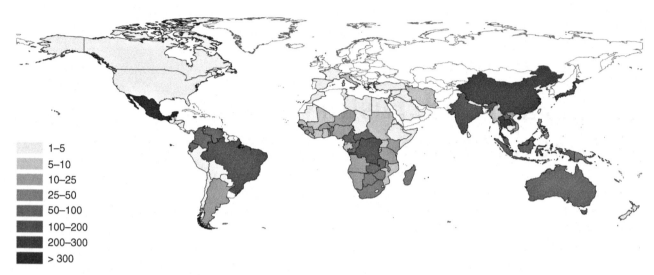

FIGURE 2.1 World map showing numbers of insect species eaten in various countries around the world. *(Reproduced with permission from Centre of Geo information by Ron van Lammeren, Wageningen University, The Netherlands. Based on data compiled by Yde Jongema, 2015.) Downloaded on Nov. 29, 2015: http://www.wageningenur.nl/upload_mm/8/6/5/81d32d6d-4bc7-43bd-8149-59d65904cc35_geodata_edible_grH300.jpg. See also: http://www.wageningenur.nl/upload_mm/8/6/5/81d32d6d-4bc7-43bd-8149-59d65904cc35_geodata_edible_grH300.jpg.*

established a broad variety of methods for their collection and preparation (Nonaka, 2009). Almost all insect orders (taxonomic orders within the class *Insecta*) have represented species that are consumed, but some groups are only consumed within a few localities. For example, while saturniid moth (silk moth) caterpillars, grubs (larvae of beetles), grasshoppers, and crickets are popular almost everywhere, adult dragonflies are consumed only in southeast Asia and certain salt flies only by certain Native American communities in the United States (Bristowe, 1932; Essig, 1934; Bergier, 1941; Bodenheimer, 1951; Malaisse, 1997; Banjo et al., 2003, 2004; Ramos-Elorduy, 2004; Yen, 2009b). In some cases, only selected developmental stages of an insect species are consumed, that is, termite adults or wasp larvae and pupae. In other cases, almost all stages are exploited, for example, juvenile and adult grasshoppers and crickets or ant eggs, larvae, pupae, and adults (DeFoliart, 1989).

As living and vital elements around the kitchen, insects have always been present for their great environmental adaptability and excessive power of reproduction. The utilization of insects as a sustainable and secure source of animal-based food for the human diet has continued to increase in popularity in recent years (Ash et al., 2010; Crabbe, 2012; DeFoliart et al., 2009; Dossey, 2013; Dzamba, 2010; FAO, 2008; Gahukar, 2011; Katayama et al., 2008; Nonaka, 2009; Premalatha et al., 2011; Ramos-Elorduy, 2009; Smith, 2012; Srivastava et al., 2009; van Huis, 2013; van Huis et al., 2013; Vantomme et al., 2012; Vogel, 2010; Yen, 2009a,b). Indeed, insects have numerous attributes that make them highly attractive, yet underexplored sources of highly nutritious and sustainable food. When compared to conventional production animals, insects are suggested to be an interesting protein source because they have a high reproductive capacity, high nutritional quality, very low water and land utilization, and high feed conversion efficiency, they can use waste as feed, and they are suggested to be produced more sustainably (Dossey, 2013; Shockley and Dossey, 2014; Oonincx, 2015).

The Food and Agriculture Organization of the United Nations (FAO) believes that the specific role of edible insects and their potential in food security, quality diet and poverty alleviation is severely underestimated (van Huis, 2012). Recently, this organization held a meeting on entomophagy in Asia (Durst et al., 2010), showing both diversity and potential of insect consumption across Asia. However, most data from that conference and its resulting report focus on insects captured from the wild. Still, this conference created more awareness in seriously considering entomophagy as a part to solving the food supply problems to come. In 2012, the UN FAO held an expert consultation in Rome, Italy to explore this topic and published a resulting report titled "Assessing the Potential of Insects as Food and Feed in Assuring Food Security." Based on discussions at the 2012 meeting, a conference in May 2014 organized by the UN FAO and Wageningen University was launched titled "International Conference Insects to Feed the World." It was a global epideictic moment, a milestone. The gathering opened with a keynote address by Sonny Ramaswamy, Director of the US Department of Agriculture National Institute of Food and Agriculture (Ramaswamy, 2015). During the week, we witnessed an amazing coming together of food insect academicians, feed industry leaders, NGO representatives, insect breeders, large and small-scale insect farmers, and government officers to share professional knowledge and policy needed for this new, developing food sector. By the end of the week, it was clear that there was an entrepreneurial explosion taking place worldwide with the aim of producing and commercializing insects for food and there were epicenters of research to support these start-up companies. The main epicenter was Wageningen University, but Asia, Africa, Australia, and North and South America each had strong research programs. Voices from the conference worldwide were captured in a video now on the FAO website and Food Insects Newsletter site (Chaikin, 2014). Gathered for the first time, these 450 experts from 45 countries agreed on a clear message: insects for feed and food are viable solutions for the protein deficit problem.

EDIBLE INSECTS OF THE WORLD

The taxonomic knowledge for most invertebrates is still preliminary and potentially thousands of species may be part of the human diet. The actual number of species of insects used as food and the level of their consumption are still unknown. Given the diversity of species consumed, it is difficult to assign a local name for the species, genus, family, or specific order, especially when dealing with immature specimens (DeFoliart, 1997). In this regard, just how many species of insects are suitable for human food or animal feed is not known (Yen, 2015). There are over 1 million species described and 4–30 million species estimated to exist on Earth, living in every niche inhabited by humans and beyond. With this diversity and their collective adaptability, they are a much safer source for future food security than are vertebrate animals such as cattle, fowl, or even fish. The number of known edible species is only a fraction of the estimated number of species, but we have no idea of just how many are suitable as food or feed.

Taking as starting point the inventory of edible insects (and spiders) of the world, prepared by entomologist Yde Jongema of the Wageningen University (Jongema, 2015), we propose here a similar analysis and present an updated alternative list (see Table A1 in the Appendix).

The orders of insects follow the phylogenetic classification proposed by Wheeler et al. (2001). Using the biogeographical realms proposed by Udvardy (1975), high insect consumption of ~290 to ~530 edible insect species can be observed in the African, Neotropical, Oriental, and Palearctic Realms (Figs. 2.2 and 2.3). Table A1 in the Appendix provides detailed information on 1555 species of insects and spiders used as food. These bioresources are given by order, family, genus, and species. Data about their common names, faunal realm, distribution, and references are also provided. In this new inventory, only those insects with complete taxonomic identification were considered. There are several major orders of insects utilized in entomophagy worldwide, including Lepidoptera, Hemiptera, Coleoptera, Orthoptera, Isoptera, and Hymenoptera. However, it is very difficult to quantify the information any further because of the lack of reliable information from each country (Yen, 2015).

Worldwide, the most common orders of insects eaten by humans are Coleoptera (19 families and 467 species), Lepidoptera (29 families and 296 species), Hymenoptera (6 families and 268 species), and Orthoptera (9 families and 219 species). At the species level, the insect families that are most commonly consumed by humans worldwide belong to the Coleoptera and Orthoptera, with the Scarabaeidae (201 species) and Acrididae (150 species) being the most commonly eaten families of insects. As for Lepidoptera, the most commonly consumed families are Saturniidae (86 species), Hepialidae (41 species), and Sphingidae (25 species). Among Hemiptera, the most commonly edible families are Cicadidae (52 species), Pentatomidae (27 species), Nepidae (9 species), and Belostomatidae (9 species).

The following list outlines edible insect and spider species consumed by humans by biogeographical realm, from those with the highest to the lowest frequency: Neotropical (523 species), Oriental (388 species), African (324 species), Palearctic (290 species), Nearctic (72 species), and Australian (56 species) (Fig. 2.2). More specifically, in the Neotropical realm, 14 orders, 76 families, and 523 species of insects and five species of spiders are edible. As for the Oriental realm, 15 orders, 60 families, and 388 species of insects and three species of spiders are incorporated into human diets. In the African realm, 11 orders, 45 families, and 324 species (two of these are spiders) are eaten. In the Palearctic realm, 14 orders, 68 families, and 290 species of insects are edible. In the Nearctic realm, 8 orders, 24 families, and 72 species of edible insects have been recorded. Last, in the Australian realm, 9 orders, 22 families, and 56 species of insects and one species of spider are consumed by humans (Fig. 2.2). This information is consistent with Western bias and Western influence against entomophagy, with the Nearctic and Australian realms consuming the lowest number of species of insects. It is estimated that as much as 80% of the world's population eats insects intentionally, and 100% do so unintentionally (Srivastava et al., 2009).

Diet often reflects the socioeconomic conditions of people. Dietary habits and taste perception are closely bound to a people's history and their geographic origin and evolve in relation to life style, tradition, and education. This may explain that in some cultures insects are looked upon as a primitive food, whereas some other cultures consider them to be a valuable and integral part of the diet (Ramos-Elorduy, 1997). European populations and European-derived populations in North

32 Insects as Sustainable Food Ingredients

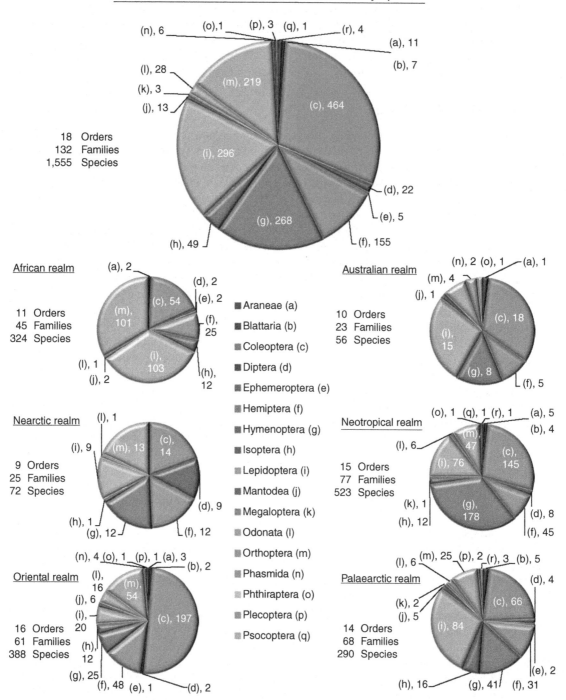

FIGURE 2.2 Edible insects and spiders of the world and by region. *(Figure courtesy of book editor Dr. Aaron T. Dossey, president, founder, and owner of All Things Bugs LLC / Griopro cricket powder; www.cricketpowder.com)*

America are accustomed to eating and willing to eat food items that they perceive as safe and of which they are inherently unafraid. Even before the mid-20th century, it was recognized in the United States that for insect eating to be popularly accepted, this perception of "primitive" must be replaced by a more ethno-relative world view (Lewin, 1943). In the past 5 years, this "low status" view of edible insects has rapidly changed in the United States, Canada, and Europe. However, in recent years, primitive or "paleo" diets have been advised by nutritional biochemists and seen as particularly desirable by a growing segment of Western cultures.

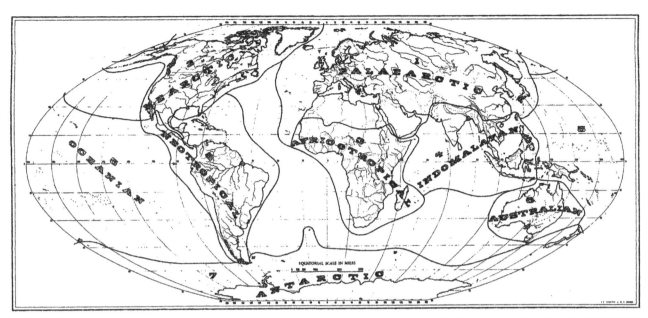

FIGURE 2.3 Map showing geographic realms of the world referred to in Fig. 2.2. *(Reproduced with permission from Udvardy, M.D.F., 1975. A Classification of the Biogeographical Provinces of the World. IUCN, Morges, Switzerland.)*

HISTORY OF INSECTS AS HUMAN FOOD

Archeological Data

Insects have been used as foodstuffs since the dawn of humanity (Sutton, 1995). In fact, these arthropods also played a significant role in the diet of early hominids, especially on the livelihoods of females and their offspring (Sutton, 1990). *Paranthropus* (=*Australopithecus*) *robustus* Broom used bone tools to dig into termite mounds 500,000 years ago (van der Merwe et al., 2003). It is interesting to point out that all the tools found were very similar to each other, which implies a technological advance culturally transmitted (Lizama, 2004). It has been proposed that a diet with a high energy density, containing large amounts of polyunsaturated fatty acids (PUFAs), was needed to facilitate the expansion of the hominid brain (Tommaseo-Ponzetta, 2005).

Based on the records of insect fossils found in coprolites from different archeological sites, insects were usually processed and widely used as food for humans (Fenenga and Fisher, 1978; Elias, 2010). Very often these insects have been misidentified. For example, the term "locust" was applied indiscriminately to refer to grasshoppers, crickets, or cicadas, and the term "worm" was not only used to describe some beetle larvae but also other holometabolous insects and caterpillars. Thus, the insects were poorly recognized and included in publications and much less specified by ethnographers. Some archaeological records on insects are disputed and certainly very unrepresentative (Sutton, 1988; Ramos-Elorduy, 2007).

Another example of early insect consumption comes in the form of an orthopteran (*Troglophilus* sp.) sculptured on a bison bone by Cro–Magnon man more than 10,000 years ago, found in the Grotto "Les Trois Frères" in Ariège, France (Bellés, 1997; Meyer-Rochow, 2004). Other Paleolithic evidence of the importance of edible insects is the enigmatic scutiform (shield shape) and escaleriform (ladder shape) signs painted in the cave of Altamira de Santillana de Mar, in Cantabria, which Pager (1976) interpreted as nests of bees, honeycombs, and stairs, which would relate with honey collection scenes (Bellés, 1997).

From Mesolithic cave art, there are realistic scenes of gathering honey, where the collectors and the methods they used are represented. Certainly, one of the places where the Mesolithic cave art has been particularly well studied is the Iberian Levante (Bellés, 1997). In this area, the pioneering work of Eduardo Hernández-Pacheco and Franciso Hernández-Pacheco led to the discovery of the beautiful scene of gathering honey in the Araña Caves, in Bicorp, Valencia, which is the most famous prehistoric illustration (Hernández-Pacheco, 1921). A similarly complex scene composition has been described by Dams (1978) in the Cingle de l'Ermita, in the Barranc Fondo, adjacent to the Vallorta in Castelló de la Plana, Spain. One can see five anthropomorphic figures on a ladder leading to honeycomb, and a group of men and women at the foot of it, perhaps expecting to share the honey collected. This likely demonstrates that not only honey, but bee larvae and pupae were also collected and consumed (Ramos-Elorduy, 1982).

In Africa, many paintings are culturally associated with the Levantine art, yet are difficult to prove dates of creation. The most interesting are concentrated in southern Africa, highlighting various representations of gathering honey, particularly in South Africa, Zimbabwe, and Namibia (Pager, 1973; Bellés, 1997). There are several paintings of swarms of bees, others of honeycomb, bees' nests, and human honey gatherers with variously constructed ladders. The extent to which prehistoric peoples used honey is strongly suggested by the ethnographic evidence for modern traditional groups (Brothwell and Brothwell, 1969). Why the honeybee has been the insect most used in the folklore, customs, symbolism, art, and religious rites is because honey was human's only sweetening material for thousands of years.

Early Historic Times (~3600BC–500AD)

Both the Old and the New Testament of the Bible give us some records of the use of insects as food (Kritsky, 1997). For example, in the Old Testament, the "manna" described in the Bible's book of Exodous may have been provided to the Jews crossing the desert in the form of the excretion of the mealybug *Trabutina mannipara* Hemprich and Ehrenberg (Homoptera: Pseudococcidae). As with many Homopterans, females of this species eliminate a sugary liquid that in arid climates dries over leaves and builds up in layers (Buzzi and Miyazaki, 1993). In the Middle East, people today still collect the sweet excretion of scale insects that feed on tamarisk trees (*Tamarix mannifera* Ehrenberg and *T. gallica* L.). They call it "man" suggesting that it may have been the manna described in the Old Testament (Hölldobler and Wilson, 1994).

Probably the most prominent descriptions of edible insects in the Bible is given in the Old Testament by Moses, who described four kinds of locusts which the Hebrews were permitted to eat: "Yet then may ye eat of all winged creeping things that go upon all four, which have legs above their feet, to leap withal upon the earth; even these of them ye may eat; the locust after its kind, and the bald locust after its kind, and the cricket after its kind, and the grasshopper after its kind" (Leviticus, XI: 21–22). Additionally, in Mark 1:6 (the New Testament), we read: "John's clothing was of camels' hair, with a leathern girdle, and his food consisted of locusts and wild honey" (in reference to John the Baptist).

In the Talmud (the body of Jewish civil and religious law developed on the scriptural law after the closing of the Pentateuchal text at about 400 BC), there are several references to eating locusts (referred to as *chagavim*) and gathering them for food (Zivotofsky, 1996). The view that people of the time did not display a feeling of disgust towards insects, at least not towards grasshoppers, is supported by the Talmudic reference (Shabbath 90b) to certain grasshopper species commonly kept as pets for children to play with (Harpaz, 1973). Isman and Cohen (1995), Taylor (1975), and others questioned why the Bible lawmakers have released the consumption of locusts (eg, *Schistocerca gregaria* Forskål) and related orthopterans, while all other insects were excluded. The criterion of the Talmud seems to be that insects which spend little or no time on the soil are acceptable to eat (ie, are clean). It is unclear why adult Lepidoptera were not also considered part of this criterion.

Isman and Cohen (1995) bring up an interesting misconception about biblical dietary laws. That the rationale for the dietary restrictions is rooted in ancient hygiene and health regulations is a common misconception that has no basis in the biblical text. Dietary laws were more spiritually based, part of the Israelites' obsession with discovering the line separating the divine from the profane aspects of things. Acceptable land animals were graminivorous species, so it follows suit that the graminivorous insects that are not predators would also be preferred among insect species.

The Koran, the Muslims' holy book, mentions insects for their virtue and harmful effects on humans, as well as lessons on the relationship which the Muslim must maintain with these creatures. Honey is regarded not only as a valuable food, but also as a means of healing (Benhalima et al., 2003). Considering Orthoptera, we have: "Whoever does not eat my locusts, my camel and my turtle, is not worthy of me, says the Prophet" (Bergier, 1941). There are some restrictions, however. It is held, for example, that locusts ought not to be eaten if they have died of cold, and one segment of the religion holds that they are lawful only if their heads are cut off (Taylor, 1975). It is said that Mohammed's wives used to send him trays of locusts as presents.

There are widespread and well-attested references of locust/grasshopper-eating in the ancient Near East (Kelhoffer, 2004). A bas-relief (c. 700 BC) shows two servants of King Sennacherib of Assyria carrying strings of granates and grasshoppers probably to offer them to the king. Near Ninevah, in the palace of Assurbanipal, there is a contemporary bas-relief depicting a similar scene (Berenbaum, 1995).

In ancient Greece, cicadas constituted a very popular dish and, according to Aristotle, females full of eggs were juicier than males. However, according to the writings of Plutarch, in Hellenistic Greece, cicadas were regarded as sacred and not generally eaten (Brothwell and Brothwell, 1969).

A character in a play by Aristophanes remarks: "Are locusts superior in flavor to thrushes? Why! Do you want to fool me? Everybody knows that locusts taste much better!" His compatriot, Alexis, mentions the locust (a grasshopper, whereas cicadas are often erroneously called "locusts" or "locuses") among the provisions of a poor Athenian family:

For our best and daintiest cheer,
Through the bright half of the year,
Is but acorns, onions, peas,
Ochros, lupines, radishes,
Vetches, wild pears nine and ten, With a locust now and then.

Herodotus (c. 484 BC–c. 425 BC) and Diodorus of Sicily in the first century BC described a people in Ethiopia, which Diodorus called Acridophagi (Diodorus Siculus, 1814), whose diet consisted of locusts: "… which fed on locusts that came in swarms from the southern and unknown districts" (Herodotus, 2008).

Pliny the Elder (23 AD–79 AD) often referred to the larva of an insect known as "cossus" (probably the larvae of the stag beetle species *Lucanus cervus* L. and similar species as *Cerambyx cerdo* L.), considered a delicacy by the Romans. He tells that the epicures of his time considered insects on a par with the daintiest meats and even fed them on flour and wine in order to fatten them and heighten their flavor (Bequaert, 1921).

Considering the Asia Pacific region, there has been a long history in the use of insects as human food or as animal feed. According to Yi et al. (2010) cited by Yen (2015), insects were consumed in China 3200 years ago. The silkworm breeding in the Celestial Empire for obtaining textile is very emblematic. Although legend attributes the invention of the method of rearing silkworms and silk production to Shih Hsi Ling, wife of the Yellow Emperor Huang Ti, around the year 3000 BC, recent archaeological findings place the origin of this industry earlier, between 6000 and 7000 years old. To obtain the silk, cocoons are boiled to facilitate the silk extraction and in the process, killing the caterpillar or pupa inside. As a byproduct of this activity, huge amounts of silk pupae become available, which are edible and widely consumed by those individuals who live close to places where the silk industries are settled. The pupae of the silkworm are considered a great delicacy dish in Japan and Korea and are served in restaurants. They are also canned for consumption by the local population and for export to the United States and other countries (DeFoliart, 1989).

Although not covered by the early historic times, we do have examples from America before the settlement by Europeans. In pre-Hispanic Mesoamerica, the food was very heterogeneous, as all available resources were used. Before the arrival of the conquistadors, different Mexican native peoples had a very balanced and diversified diet (Wicke, 1982). Edible insects were one of the sources of animal protein. The story records that when the Aztecs arrived at the Anahuac lake in search for the promised land, they lived for some time in the Chapultepec Hill (from the Nahatl *chapul* = grasshopper and *tepēc* = mountain), named due to the abundance of grasshoppers. The plague did not discourage the travelers who started to eat them, removing their wings, legs and antennae. However, a series of changes took place after the conquest. As food also has an emotional and/or affective meaning and has a certain prestige, some of the resources used by the Aztec people were rated negatively and therefore depreciated and/ or forgotten (Ramos-Elorduy and Pino Moreno, 1989).

Age of Reason and Emergence of the Scientific Era (1700s/18th Century)

Aldrovande, in his book *De Animalibus Insectis Libri Septem* published in 1644, comments on the use of several insects as food in different parts of the world, such as ants in India and bee pupae in China. Erasmus Darwin in his *Phytologie* (1800) says that the larvae of *Melolontha* sp. (Coleoptera, Scarabaeidae) (also known as May Beetles or June Beetles, or sometimes erroneously as "June Bugs"), and even the adult beetle, are worthy of being served at our table for the reason that they feed on roots and leaves. *An Introduction to Entomology* published in 1863 by Kirby and Spence gives several examples of the direct benefits derived from insects, including their use as food items in a worldwide context. An 1877 book by Alpheus Spring Packard Junior dedicated a chapter to edible insects, describing insects such as grasshoppers of the East, eaten by Arabs and in other parts of Africa. He also described bees and ants eaten in Mexico and used in Sweden to flavor brandy, and discussed the Chinese eating silkworms and the larvae of hawkmoths. Last, he discussed the palm weevils eaten in the West Indies (Packard, 1877). Another book, titled *Why Not Eat Insects?* author Vincent M. Holt shares his angst about "a long-existing and deep-rooted public prejudice" to insects as food (Holt, 1885).

Honeybees are insects that have played a major role in human culture throughout history. In one example, cameos of the honeybee and ancient human-built hive designs play a prominent role in the decoration of the Roman Catholic Pope's throne in St. Peter's Basilica, Rome, Italy. They appear here because they were a symbol in the coat of arms of Pope Urban VIII (papacy from 1623–1644) and his family (the Barberini family, a prominent Italian family of nobility at the time). The adult honeybees are anatomically correct and decorate the carved marble base of each of the four columns supporting the canopy over the throne. The fact that the honeybee was chosen to hold such a prominent place in the Basilica is clear testament to the importance of edible insects in Europe 500 years ago (Hogue, 1987).

In a 1916 publication, Leland Ossian Howard and members of the United States Bureau of Entomology and the United States Bureau of Biological Survey discussed practical suggestions concerning any new cheap foods such as insects (Howard, 1916). During World War I, food prices worldwide were increasing, and many nations were facing very serious shortages. He and his colleagues prepared and personally ate various edible insect dishes. Howard concluded that colleges of agriculture, with their departments of home economics and entomology, were in an excellent position to conduct entomophagy research (Howard, 1916). Why this did not happen during this time of great need when there was such a long history of edible insects is extremely interesting in entomological psychology! It seems to us that the dominant culture (European and Euro-Americans) sometime in the 1800s switched from rearing their children with an understanding of the beneficial as well as the destructive sides of insects and spiders and just focused on the destructive and dangerous aspects. Along with that change in rearing practices, the useful aspects of entomophagy were lost. This is well documented in journals of the early pioneers that settled around the Great Salt Lake after their lives were saved by the Ute Native Americans who prepared "prairie cakes," pemmican with local service berries and a local edible insect species, the Mormon cricket, *Anabrus simplex* Haldeman. The "disgust factor" was already dominant at that time among European-Americans. In parallel, in the late 19th century the condition of delusory parasitosis, a tactile hallucination about small insects or other arthropods, was first identified in patients by Western culture psychologists (Berrios, 1982).

Bergier (1941) wrote a book called *Peuples Entomophages et Insectes Comestibles*, where he tells about the insects that are eaten in the various continents of the earth, exposing the species consumed by continent, gathering techniques, methods of preparation and consumption, and illustrates with numerous photographs and anecdotes. Ten years later, Friedrich Simon Bodenheimer's book *Insects as Human Food* cites more than 500 human entomophagy references, from prehistory to the 1950s (Bodenheimer, 1951; Dufour, 1990). For an overview of the history of human entomophagy since biblical times to the year 2007, readers should consult Ramos-Elorduy and collaborators' comprehensive review (Ramos-Elorduy et al., 2007).

In spite of this wealth of existing and published knowledge about human use of insects as food over the millennia, insects and other arthropods are still insufficiently investigated by anthropologists and entomologists, being generally regarded as marginal resources in studies on the uses of various available resources (Jara, 1996; Nonaka, 2009). As Sutton, 1988 points out: "The incomplete treatment of the use of insect resources results in only a partial understanding of the subject [as an] economic system. From an ecological standpoint, an understanding of, or at least a delineation of, all parts of an economic system are necessary for an understanding of the system as a whole and of its interactions with other systems. In this regard, an understanding of the role of insects in an economic system is necessary before the role of other resources, and so the entire economic and sociopolitical system, can be fully understood."

MODERN CULTURAL USES

Utilization of insects as a food resource is currently accepted and practiced by many cultures around the world and constitutes a major source of nutrition for many people (DeFoliart, 1995; Nonaka, 2009; Ramos-Elorduy, 2009) (Fig. 2.1). For a respectable number of species, at least some nutritional data are available, showing great variation according to species and instar (life stage); while protein prevails in many adults (imagos), many juvenile instars [maggots (larvae of flies), grubs (larvae of beetles), caterpillars (larvae of butterflies and moths), and pupae] also tend to contain elevated levels of fat. Vitamins and mineral content also varies. Interesting to note that larval and pupal stages of holometabolous insects tend to be higher in fat because they need the energy while they develop into adult, during pupal stages, and are not eating. At this point a short clarification is needed to define some entomological terms used throughout this book. Holometabolous insects develop through the egg, larva, pupa, and adult stages (eg, butterflies wasps, bees, ants, and beetles), while hemimetabolous insects develop through the egg, larva, and adult stages (eg, grasshoppers, katydids, crickets, and cicadas).

A Mexican study revealed that in dry matter, grasshoppers may yield up to approximately 80% of protein, longhorn beetle grubs 60% of fat and 35% of fiber, and pierid caterpillars 10% of ashes (Ramos-Elorduy and Pino Moreno, 1989; rounded values). The chitinous exoskeleton is not digestible by humans (functioning as a source of fiber similar to the skin of the apple), but the exoskeleton is only a small part of the total biomass (about 4% in caterpillars) and does not affect the nutritional value of insects as food (Berenbaum, 1995). See also Chapter 3 of this book for more discussion on the nutritional content and value of various insects.

And what about sensory acceptability? According to Ramos-Elorduy (1997), the organoleptic characteristics of insects, such as the crunchy texture, absence of odor, and white color of most of the larvae, as well their flavor, could favor their acceptance among Westerners. Those individuals who already consume insects know well how they need to prepare them to make a good and tasty meal. Some insect species are said to have a flavor resembling apple, almonds, or peanuts. Others do not have a special flavor and take on the flavor of the oil used to fry them or the relish used to dress them (eg, garlic, onion, lemon, or chilli). The aquatic insects often have a flavor similar to that of fish when they are eaten fresh, and a flavor similar to that of dry shrimp when eaten dried" (Ramos-Elorduy, 1997, p. 259).

Bristowe (1932), who wrote about the consumption of insects by the Laos natives of Siam, tells us that the meat of the giant water-bug *Lethocerus* (Belostomatidae) has a strong flavor, reminiscent of Gorgonzola cheese. The bugs are also cooked, dried, and pulverized to add zest for curries or sauces. Combined with shrimp, lime juice, garlic, and pepper they form a popular native sauce known as namphla. Ancona (1933) records a similar use in Mexico of stink-bugs of the genus *Euschistus* (Pentatomidae). They are toasted, ground up with chile or pimiento pepper, dashed with tomato sauce, and used as a substitute for maple syrup on the native pancakes or tortillas. A strong flavor may be avoided if the scent glands are extracted from the more innocuous contents of the body by pinching off the tip of the abdomen to which they are attached.

Most traditionally-consumed insects may be picked, handled, and converted into nutritious foodstuff easily (Johnson, 2010). Traditionally, insects are consumed either in a raw state or heated in some fashion. Depending on the culture, heating occurs in hot ashes, toasting devices, or by deep-frying them in oil. In dry areas, preservation is done by heating or sun-drying; eventually, evisceration was performed before, as in the case of grubs and caterpillars (Ramos-Elorduy et al., 2008). Some works suggested using wild pest species as food, especially grasshoppers (Cerritos and Cano Santana, 2008), while others express their concern regarding insect consumption from the wild (Ramos-Elorduy, 2006; Yen, 2009a). A series of works contemplates the health risks associated with managing insects (Adamolekun et al., 1997; Akinnawo et al., 2002; Zagrobelny et al., 2009). See also Chapters 7–9 of this book, which discuss food safety issues relating to consuming insects (such as regulatory, microbiological, and allergenicity).

Currently, it is widely recognized that damage caused to the environment by the indiscriminate use of biocides in agriculture and livestock directly impacts the food chain, and affects animal life, including insects. This factor may pose some additional risks and limits on supply of wild harvested edible insects, since they also suffer the impacts of the use of insecticides both in the countryside and urban domestic households. While pesticides are generally designed to kill insects and thus insects found live (in the wild or in farms) are unlikely to have been exposed, pesticides affect different species differently and wild harvested insects may sequester certain pesticides in their bodies or harbor traces of pesticides on their outer surface, if exposed to them, for a period of time until the insect dies. It is well known the current recommendation of care and limits on the consumption of certain fruits and vegetables that can accumulate significant amounts of systemic pesticides. FAO (1981 cited by Santos, 2014) estimates that the amount of harvested product that is lost all over the world due to the insects, pests, diseases, and weeds ranges from a minimum of 5–10% to a maximum of 10–40%. Thus, pesticides have been used on high value crops such as cotton, rice, fruits, and vegetable crops. Half of the pesticides produced in the world are used in four main crops: small grain cereals (wheat and barley), corn, rice, and cotton. Therefore, the excessive use of pesticides during the growing season agricultural activity and harvesting without respecting the withdrawal period (vesting period) are sources of contamination for foods of plant-origin. This is undoubtedly one of the limiting factors to the widespread and indiscriminate use of edible insects, because together with plant food, they are the most affected by the massive use of biocides in industrial agriculture worldwide (Santos, 2014).

Edible Insects Around the World: Selected Classic Examples

Although insects can cause harm to our food supply, the beneficial uses of insects far outweigh the negative examples. There are over 2 billion people of the world for whom insects are an important source of protein and/or a desired tasty addition to their diet as well as an important cultural component of family gatherings. We have selected the following examples of widely appreciated food insects, some from each of the regions of the world.

Widely appreciated by visitors and the native aboriginal peoples of Australia are witchitty grubs, the bogong moths, and honeypot ants. One of the world's most energy dense edible insect is found in Australia. This is the green weaver ant, *Oecophylla smaragdina* Fabr., which contains 1272 kcal/100 g fresh weight (FAO, 2012). The best known Australian edible insect is the witchitty grub (locally known as witjuti) which is the most important insect food in the desert of Australia and is a staple in the diets of Aboriginal women and children as a high-fat, high-protein food. Traditional preparation is simply eaten raw or lightly cooked in hot ashes. The raw witchetty grub or larva tastes like almonds. When cooked the skin of the larva becomes crisp like roast chicken and the inside becomes light yellow in color (Isaacs, 2002). Witchetty grubs are a relatively broad category of large, white, wood-eating larvae of several moth species. This term particularly applies to larvae of the cossid moth *Endoxyla leucomochla* (Turner), which feed on roots of the Witchetty bush (named after the grubs) that is found in central Australia. The term may also apply to larvae of other cossid moths, ghost moths (Hepialidae), and longhorn beetles (Cerambycidae). The term is used mainly when the larvae are being considered as food. The different larvae are said to taste similar, probably because they have similar wood-eating habits. Witchetty grubs live about 60 centimeters (24 in.) below ground and feed upon the roots of river red gum (*Eucalyptus camaldulensis* Dehn). They can also be found under black wattle trees, and are attributed as the reason why wattles die within 10–15 years. Roots of the *Acacia kempeana* F. Muell. shrub are another source of the grubs.

In the Australian outback, honeypot ants or simply honey ants are eaten by Aboriginal people as sweets and are considered a delicacy. These ants have specialized workers (repletes) that are gorged with food by workers to the point that their abdomens swell enormously, a condition called plerergate. Other ants then extract nourishment from them. They function as living larders. Honeypot ants belong to any of five genera, including *Myrmecocystus* (Morgan, 1991). They were first documented in 1881 by Henry C. McCook. These ants can live anywhere in the nest, but in the wild, they are found deep underground, literally imprisoned by their huge abdomens, swollen to the size of grapes. They are so valued in times of little food and water that occasionally raiders from other colonies, knowing of these living storehouses, will attempt to steal these ants because of their high nutritional value and water content. These ants are also known to change colors. Some common colors are green, red, orange, yellow, and blue. Honeypot ants such as *Camponotus inflatus* Lubb. are edible by humans and form a well appreciated part of the diet of aboriginal people in Australia. This important food insect has even become part of the traditional stories of the aboriginal peoples of Australia. Papunya, in Australia's Northern Territory, is named after a honey ant creation story, or dreaming, which belongs to the people there, such as the Warlpiri. The name of Western Desert art movement, Papunya Tula, means "honey ant dreaming." *Myrmecocystus* nests are found in a variety of arid or semiarid environments. Some species live in extremely hot deserts, others reside in transitional habitats, and still other species can be found in woodlands where it is somewhat cool but still very dry for a large part of the year. For instance, the well-studied *Myrmecocystus mexicanus* Wesmael resides in the arid and semiarid habitats of the southwestern United States and Mexico and is also an important edible insect in Mexico.

Mopane, termites, and food katydids are several of the most popular edible insects that are currently in use in sub-Saharan Africa. The edible katydid, *Ruspolia differens* (Serville) (or *R. baileyi* or other species of the genus sometimes called "Nsenene"), is particularly appreciated in eastern and southern Africa. In the Lake Victoria region, *R. differens* forms a major part of food culture (Kinyuru et al., 2012). In Tanzania (Bahaya ethnic group in the Bukoba district) these katydids are a delicacy (van Huis et al., 2013). In Uganda, *R. differens* are collected in the early morning by women and children. Recently, at Makerere University, this species is the focus of an insect farming and food processing program to provide nutritional fortification to products in the urban areas of Uganda. A market study of another food katydid, *Ruspolia nitidula* (Scopoli, 1786) in Kampala and Makaka districts of Uganda with 70 traders and 70 consumers found that this katydid was well appreciated in the marketplace and a strong source of income (Agea et al., 2008). The retail price was five times higher ($2.80 USD/kg) than the wholesale price ($0.56 USD/kg), and the retail price of 1 kg katydids was 40% higher than 1 kg of beef.

Mopane, *Cirina forda* (Westwood), a traditional delicacy in Botswana and other areas in southern Africa, bring similarly high prices in comparison to beef. Mopane are a regular part of the diet in Zimbabwe, Namibia, Botswana, and northern South Africa where they flourish in their specific habitat, the mopane tree woodlands. Great numbers of people partake in the mopane harvest, since this insects often generates incomes higher than that generated by conventional agricultural crops and certainly greater nutritional content (Munthali and Mughogho, 1992; Chidumayo and Mbata, 2002). People travel hundreds of kilometers across the woodlands of southern Africa in search of these insects (Kozanayi and Frost, 2002). Unfortunately, now with commercialization in full swing since the 1990s, mopane populations are dwindling (Sunderland et al., 2011).

Termites are appreciated as food throughout Africa. In West Africa, villagers use various collecting devices to collect the small, worker termites and feed them to chickens. Alates, which are larger, are collected for human food and generally fried. In Kenya, a study conducted by scientists in cooperation with the Kenya Industrial Development Organization revealed that simple mechanization of collection devises and lights could facilitate the mass collection of *Macrotermes subhyalinus* (Rambur), locally known as "agoro." The importance of this combination of modern technology with an indigenous practice for collection of termites (Ayieko et al., 2011) is that the area where Ayieko and colleagues were working, western Kenya near Lake Victoria, has that "double whammy" of compromised nutrition because their main grain is missing two of the essential amino acids, tryptophan and lysine, and their growing season, and hence the "hungry season," exactly coincides with the malaria season. This location in western Kenya, Kakamega County, is where the highest transmission of malaria takes place in the country. A good quality protein source is sorely needed in this area.

In the United States there are 54 species of edible insects documented as being used by the 564 federally recognized Native American Tribes. We have chosen four of these species to highlight here. The Western Shoshone in the central Great Basin area of the United States, where large mammals were relatively scarce and not moving in herds, focused on smaller meat sources, such as lizards, rodents, and insects (Steward, 1938 cited in Dyson-Hudson and Smith, 1978). Grasshoppers are abundant and there are many species. For example, just in Montana there are over 100 species of grasshoppers, probably all edible. Although these wild grasshoppers are well-known to be delicious, because of their univoltine reproduction (one life cycle per year), they are not easily adapted to insect farming. Mormon crickets, *Anabrus simplex* Haldeman, which are actually katydids, are also a popular Native American, edible insect in the Great Plains and Rocky Mountains (Sutton, 1988). This is the insect that became part of the pemmican that the Ute Nation living along the shores of the Great Salt Lake offered to the pioneers who came to them for help to keep from starving the first summer after the settlers arrived and their crops failed. No statue was built to the Mormon cricket even though it saved the lives of those pioneers. Algonquin nations in the eastern United States have festivals and prepare special food dishes when the cicadas emerge. Women of the Paiute Nation in California, such as at Mono Lake California, often harvest brine flies, *Ephydra hians* Say, from the shores of lakes for food. These insects are called "kutsavi" (Steward, 1933).

Just south of the US border in Mexico there are many species of delicious food insects. Of these, the most well-known are the chapulines (Fig. 2.4). The other two species nearly as well-known are escamoles (ant eggs) and agave larvae. A study conducted by Katia Figueroa, entitled "Insectos, alimento del futuro?" was conducted (Figueroa Rodriguez, 2001). Figueroa Rodriguez and her team conducted a market study of processed grasshoppers, *Sphenarium* spp., in Oaxaca and the community of Santa Maria Zacatepec, Puebla, Mexico. The *Sphenarium* spp. are the Mexican grasshoppers in highest demand and they also have the best nutritional characteristics. Their reproductive rate is high and at times they can become agricultural pests. The market in Oaxaca moves the greatest volume and the community of Santa Maria Zacatepec, Puebla, is the most important supplier of this market. This study found that in 1996 only 22% of household budgets in Mexico were spent for food. Of this 85% was spent for food prepared at home. Of the food prepared at home, grasshoppers, agave larvae, and other insects represented $1.4 million USD per year. Food budgets spent on edible insects was nearly the same for those of the lower and middle income classes as for those of the highest economic class! Some insects should be processed before they are sold. Some insects are sold live, such as the red agave larvae, chinicuil, which is collected from beneath agave plants (Fig. 2.4). Larval trails guide the collector to an agave which must then be turned over. Obtaining a liter of larvae is usual from one agave plant. A liter of larvae was worth $45 USD in 2000 at a roadside stand (Rehder and Dunkel, 2009). We found price escalation of 300% between harvesters and middle people and then again between these groups and the restauranteurs. The chapulin, *Sphenarium purpurascens* Charpentier (see Section 18.5), is controlled by harvesting them as a food item when they reach economic injury levels in field crops. This activity provides an annual profit of about $3000 USD per family and a total of 100 metric tons of edible insect mass (Premalatha et al., 2011).

Thailand has a unique history with respect to edible insects (Fig. 2.4). Its cultures have transitioned, from mainly agrarian lifestyles to a thriving urban population. Edible insects never stopped being widely accepted in Thailand, despite the arrival of "planners and developers" from Western cultures. The disgust factor they often carry with them did not result in an "embarrassment" factor in response that caused the edible insect tradition to disappear. There are now 20,000 cricket farmers in Thailand who supply both urban and rural requests for crickets, but many other species of edible insects are appreciated in Thailand. Palm weevils and bamboo larvae are also farmed and wild collected. Interestingly, edible insect species preferred by urban-dwellers in Thailand are different than the species generally preferred for food by rural populations in Thailand (Hanboonsong et al., 2013).

In addition to the wide diversity of edible insects in Thailand, throughout Asia, the most typical food insects are of course the silk moth pupae as well as bee brood and wasps in Japan; metdugi or rice-field grasshoppers in Korea. It is likely that the first edible insect on the global market was the silk moth pupae. The nutritious and delicious silk moth pupa is the industrial by-product of the ancient practice of making silk which evolved in China over 5000 years ago. This edible insect is already in commercial scale production with the majority, 99% of commercial silk is produced from *Bombyx mori* L., but there are several other popular silk moth species, such as the Chinese (oak) tussah moth, *Antheraea pernyi* (Gérin-M.), the camphor silkwmoth, *Eriogyna pyretorum* (Westw.), the Thai or eri silk moth, *Samia cynthis* (Drury), and the Japanese oak silk moth, *Antheraea yamamai* (Gérin-M.). Pupae are traditionally eaten and sold in many markets and by vegetable grocers in northeastern China (Zhang et al., 2008). Silk moth pupae are commonly eaten in other Asian countries, including Japan, Thailand, and on the Korean Peninsula. The Thai silk moth is particularly prized for its taste and considered a delicacy in China, Japan, Thailand, and Vietnam. About 137,000 households raise silk moths in Thailand, thereby contributing to their own household nutrients as well as income. This accounts for 80% of the silk produced in Thailand. In 2004 this silk production created an income for these Thai households of $50.8 million USD. Thus, the silk moth is one of the most serious nutritional sources for people of Asia as well as an income producer for small-holder farmers, particularly for women.

At this time, Europeans and European-Americans are choosing a small set species of edible insects from which new traditions are being formed. European edible insects available commercially are primarily the mealworm, *Tenebrio molitor* L., and the migratory locust, *Locusta migratoria* L. For European-Americans, the main food insects are the larval stage of the wax moth, *Galleria mellonella* L., the house cricket, *Acheta domesticus* L., as well as larvae of *T. molitor*. The traditional

FIGURE 2.4 Photos of edible insects from various parts of the world. (A) "Mopane worms" (caterpillars of a large African silk moth) shown live and freshly harvested in the city of Lubumbashi, in the Provence of Haut-Katanga, Democratic Republic of Congo, Africa (Apr., 2015), (B) "Mopane worms" shown traditionally prepared in the city of Lubumbashi, in the Provence of Haut-Katanga, Democratic Republic of Congo, Africa (Apr., 2015), (C) Chapulines (grasshopper in Spanish) traditionally prepared in Mexico; boiled with lemon or garlic, dried crickets are sold in the traditional market of "Cholula," a pre-Hispanic town in Puebla, Mexico, (D) Red agave larvae, chinicuil (also known as "red worms of Magey") sold in the roads of Mexico City (and in areas like Puebla) after the middle of Sep., when the caterpillars leave the roots, emerge from the soil, and walk in search of a place to transform into butterflies. People who know their lifecycle search for them in that time of the year and catch them for family consumption or to sell in the roads at high prices, (E) "Escamoles" (the larvae and pupae of ants), traditionally prepared on a tortilla in Mexico City, 2010. These were consumed as exquisite meals in precolonial times by Aztec emperors and are served today by the chefs in exclusive restaurants in Mexico. Ciudad de Mexico, 2010, and (F) Photo of a local food market in Bangkok, Thailand, featuring several types of prepared edible insects (taken in 2009). (Part A: Photograph courtesy of Tshikwata Sam from Entomofood Africa; Part B: Photograph courtesy of Tshikwata Sam from Entomofood Africa; Part C: Photograph courtesy of Carolina Pardo Torres; Part D: Photograph courtesy of Carolina Pardo Torres; Part E: Photograph courtesy of Carolina Pardo Torres; Part F: Photograph courtesy of Patrick Durst)

insects eaten in Europe are associated with fermentation processes in making cheese, primarily in Italy, Corsica, Sardinia, and France. Derived from Pecorino (family of hard, Italian/Sicilian/Sardinian cheeses made from ewe's milk), Casu marzu (also called casu modde, casu cundídu, casu fràzigu in Sardinian, or, in Italian, formaggio marcio, "rotten cheese"), is a traditional Sardinian sheep milk cheese, notable for containing live insect larvae (maggots). Although found mostly in the island of Sardinia, the cheese (casgiu merzu) is also found in the nearby island of Corsica (Dessa, 2008). In 2005, cooperation between sheep farmers and University of Sassari researchers resulted in a hygienic production method aiming to allow the legal selling of the cheese (Dessa, 2008). However, as of November 2013, *casu marzu* or *casu frazigu* is not listed as a recognized traditional food in the Database of Origin and Registration (European Commission, 2013) or in the list of the Italian Ministero delle politiche agricole (Ministero delle politiche agricole alimentary e forestali, 2015). Therefore, the legal status of casu marzu in the European Union remains questionable. All of the four main varieties of Pecorino, though, have Protected Designation of Origin (PDO) status under EU law.

Similar milk cheeses notable for containing living insect larvae are produced in several Italian regions and in Corsica, France:

- Cacie' Punt (formaggio punto) in Molise
- Casgiu merzu in Corsica
- Casu du quagghiu in Calabria
- Gorgonzola co-i grilli in Liguria
- Frmag punt in Apulia
- Furmai nis (formaggio Nisso) in Emilia Romagna
- Marcetto or cace fraceche in Abruzzo
- Salterello in Friuli Venezia Giulia

In Europe, there are several other regional varieties of cheese with fly larvae. For example, goat-milk cheese is left to the open air until *Piophila casei* L. larvae are naturally laid in the cheese. Then it is aged in white wine, with grapes and honey, preventing the larvae from emerging, giving the cheese a strong flavor. Digestive actions of the larvae cause an advanced level of fermentation, breaking down the cheese's fats. Other regions in Europe have traditional cheeses that rely on live arthropods (cheese mites) for ageing and flavoring, such as the German *Milbenkäse* and French *Mimolette*.

The palm weevil is a delicacy in Asia, Africa, and Latin America. Even the Dutch scientist, Linneaus, noted this in his 1758 work, *Systema Naturae,* remarked that palm weevil larvae were delicious *"Larvae assate in deliciis habentur"* (in translation: fried larvae are delicious) (van Huis et al., 2013). The delicious flavor of these large larvae is from their high fat content. Palm weevils are found wherever their hosts are found in the world. Generally, the weevils invade damaged or otherwise stressed palm trees. High populations of weevils can be collected from these trees creating a semicultivation process. This traditional ecological knowledge is well-honed indigenous peoples harvesting these excellent sources of fat and protein throughout the tropics (van Itterbeeck and van Huis, 2012). In Africa, each adult female palm weevil lays a few hundred eggs and in 7 weeks the fully extended larva is 10.5 cm long and 5.5 cm wide and weighs 6.7 g (Fasoranti and Ajiboye, 1993). The reader is encouraged to do their own exploratory library research on the nutrient content of this delicious food insect. Unfortunately, this tasty treat was not understood by the pest management patrols in Sicily and Italy when the red palm weevil invaded their 13,000 date palm trees (Mormino, 2012) and inhabitants chose the insecticide-, not food-route for management. Had they read this chapter and done their own exploratory nutritional research, they would have remembered that harvesting and consuming is often the most efficient, nutritious, safest for the environment, and the most delicious.

Of course, not all insects are edible. As with many types of plants and animals, bright colors, especially red, or showy behavior such as slow, deliberate flight may suggest that an insect contains toxins, or is unpalatable, and should be avoided. The use of bright colors and flamboyant displays in nature to advertise real (or feigned) danger as toxic or venomous is known as Aposematism (and organisms utilizing aposematism are known as aposematic). The toxicity of various insects (and other arthropods or animals) may come intrinsically from their own physiology (defense or venom glands, internal de novo production of toxins, etc.) or indirectly sequestered via their diet if they feed on toxic plants, algae or other toxic organisms (Dossey, 2010). In this regard, Blum (1994) discusses the toxicity of insects eaten by humans, providing several examples of species that should be avoided as food, such as cyanogenic glucosides species (eg, butterflies of the families Nymphalidae and Heliconidae), vesicants species (eg, *Lonomia* moths), those ones that produce steroids (eg, *Ilybius fenestratus* [Fab.], Dysticidae) and corticosteroids hormones (eg, *Dytiscus marginalis* L., Dysticidae), necrotoxic alkaloids (eg, fire ants *Solenopsis* spp.) and toluene and O-cresol (eg, longhorn beetles of the genera *Stenoceutras* and *Syllitus*). According to this author, the scientific knowledge about the toxic effects of most natural insects is still very scarce. He classifies toxic insects into two groups: cryptotoxic and phanerotoxic species.

Phanerotoxic insects include those that are venomous, presenting a poison apparatus that includes a venom gland, a reservoir, a duct, and an apparatus for injecting the venom. Representatives of this group are insects of the orders Lepidoptera (urticating caterpillars), Hymenoptera (ants, wasps, and bees), and Hemiptera (assassin bugs), whose secretions are distributed both by retractable stingers, mouthparts for piercing, or stinging arrows. The toxins produced by phanerotoxic species only become active by injection, becoming inactive in the gastrointestinal tract. Nevertheless, some caution is advised.

Cryptotoxic insects are those that produce nontoxic exocrine secretions, whose toxicity is manifested only when they are ingested. These species require more care in their selection as items to be consumed. Staphylinid beetles of the genus *Paederus*, for example, produce vesicants which are only detected when they are crushed. One of these is the pederin, a nonprotein compound that is a potent inhibitor of protein synthesis and mitosis.

For each traditionally-consumed species, there is a history of safe use that intends to minimize the consumption risks. The history of safe use is also important in order to tell inoffensive species from toxic ones (van Huis, 2012). As edible insects have become more popular day by day around the world, further investigations on their potential toxic effects should be conducted (ncekara and Türkez, 2009). For example, the consumption of the African silkworm *Anaphe venata* Butler by traditional societies in Nigeria is associated with a seasonal ataxic syndrome, which is characterized by sudden onset of severe muscular tremors and gait ataxia. *Cirina forda* Westwood is another caterpillar widely eaten in Nigeria as an essential ingredient of vegetable soup. Morphometrical and histopathological studies carried out with albino rats have shown that the raw larva was toxic and produced a vascular circulatory disturbance resulting in organ damage in the animals. Processing the larvae by boiling and sun-drying reduced the toxicity on the liver and heart but not in the kidney (Akinnawo et al., 2005).

Some societies have managed to determine how to make apparently inedible species edible. Yen (2009a) cites some examples: the stink bug *Encosternum delegorguei* Spinola is made palatable by washing with warm water to make it release its pheromones, then boiled in water and sun-dried; the armored ground cricket *Acanthoplus spiseri* Brancsik is made edible by removing its gut after carefully pulling off its head, boiling for a minimum of five hours, frying in oil and serving with a relish of tomato and onion. Boiling is important as it removes the toxic substances that can cause severe bladder irritation. Yen (2009a) stresses that the methods developed to make inedible species edible are a significant intellectual property of the traditional societies that discovered them.

Scorpions are also a delicious, but dangerous food item. In China, they are raised in homes, deep fried, and served on skewers (Menzel and D'Aluisio, 1998). The Japanese way is a tempura batter and deep fat fried resulting in a light, earthy version of crab (Martin, 2014). In the 2000, Chef Nicole-Anne Gagnon of the Food Insect Festival in Montreal, Canada, at the Insectarium, gave scorpions a place on their menu among seven other entrees. This event typically draws over 20,000 people (Dunkel, 1997). In Zambia, for scorpion consumption safety, one must remove the poison sack at the tip of the abdomen before cooking and eating.

Indirect or Unintentional Presence in Foods

Insects are also consumed indirectly through ingestion of processed foods from bread to tomato ketchup. This is due to the impossibility of complete removal of insect parts from these products (Posey, 1986). According to Myers (1982), the Food and Drug Administration of the United States (FDA) puts out a list of permissible levels of insect infestation or damage, that is, a maximum number of insect adults, eggs, immatures, droppings, or fragments, which can be present in a food to be sold. For example, it is acceptable to encounter no more than five insects or insect parts in every 100 grams of apple butter and 30 insect fragments per 100 grams of peanut butter (Berenbaum, 1995) (see Chapter 7 of this book for more on the regulatory status of insects in food). Indian vegetarians get their dose of vitamin B_{12} from insects and bacteria that usually contaminate their food (Allport, 2000).

They are also eaten through the substances produced by them and used by us such as honey, certain food colorants, and so forth. Today, most culinary dyes have synthetic origin, but in the past the red and pink colors were obtained from insects. The pigment phenoxazine (orange) extracted from *Pachilis gigas* L. (Hemiptera: Coreidae) showed high dyeing power; it can be used as a colorant to ice creams, cheese, mayonnaise, and cream (Pérez et al., 1989). The red carmine pigment, commercially known as E120, is extracted from *Dactylopius coccus* Costa (Homoptera: Coccidae) and used in many different products such as cosmetics and food (Table 2.1).

Consumption of figs is one of the most common ways that insects are ingested by humans. The fig is an inside-out flower and it is not self-pollinated. The fig tree has evolved an obligatory mutualistic relationship with a hymenopteran, the fig wasp (several species of the superfamily Chalcidoidea). The female wasp carries the pollen from fig to fig and then lays her clutch of eggs in the one fig she pollinates. The females that emerge from those eggs mate with the wingless males that have hatched from other eggs and fly out of the fig. The males stay behind. These fig communities in the inside flower

TABLE 2.1 Industrial Use of Cochineal Derived Coloring

Cosmetics	Lipsticks, blush, soaps, and fingernail polish
Pharmaceutical industry	Various medications including cough syrups, throat lozenges, antibiotics, combination syrup-antibiotics, nasal solutions, mouthwashes, and the sugary covering on pills and ointments
Food industry	Various pastries and candies, sausages, hamburgers, soups, various canned vegetables, jellies and marmalades, chewing gums, ice cream, dehydrated cocktails, milk products (especially cheeses), soft drinks, and pills
Textiles	Wool and silk covers, carpets, and other cloths
Wine	Vermouth Campari
Soft drinks	Various soft drinks

Source: Pino Moreno and Ramos-Elorduy (2002).

of the fig are attacked by other wasps, parasitoids that use the bodies of the fig wasps to incubate their own eggs. So the unsuspecting human insect consumer eats these along with the fig.

CULTURAL RESTRICTIONS

The choice and consumption of a food item depends on a set of factors of ecological; historical; cultural; nutritional; organoleptic (flavor, texture, aroma, etc.); indulgence; general enjoyment; and social and economic orders, some of which are linked to a network of representations, symbols, and rituals (Álvarez, 2002). The ideas that a given population has about food resources are reflected in the meanings attributed to them or peoples' eating habits, as well as by the interaction human beings maintain with natural resources, technology, likes and dislikes, experiences, feelings, the economy, and many other elements that develop from childhood. All of these received multiple influences throughout one's life, being conditioned by ethnic, family, social, and cultural aspects, and of course, for both regional and seasonal availability of food (Long, 2002; León-Portilla, 2002; Looy et al., 2013).

In general, the determinants of the use of insects as a food resource for humans result from four variables: (1) physical environment; (2) availability and accessibility of insects that, in turn, depend on their life cycle, host plants, behavioral adaptations and general ecology; (3) mode of production and subsistence methods of a given culture; (4) dietary restrictions, both nutritionally and temporally (Miller, 1997). Miller (1997) states that an insect is suitable for use as a food source when it is available in large quantities and is easily collected. The importance of a particular animal as a food source for humans is also determined by the efficiency by which this animal converts food that it consumes into its own body mass. The highest amount of weight that is gained per gram of food ingested corresponds to the more efficient animal in terms of feed conversion (Conconi, 1984). Thus, edible insects are highly efficient in this process, only competing with chicken.

Food Taboos and Religious/Dietary Restrictions

Very often, insects are considered inedible items due to taboos, cultural or religious doctrine, and/or dietary restrictions. As mentioned previously, Judaism (faith of the Jewish people) forbids eating any but a few particular species (ie, those which are considered "Kosher", a term used by Jews for food allowed to be eaten). Similar doctrines exist in Islam (the Muslim faith). Hindus (mostly from India) tend to maintain a vegetarian diet. Additionally, many people follow dietary restrictions based on nutritional or animal welfare reasons. For example vegetarians typically do not eat flesh from animals (meat) or anything requiring the death of an animal. However, many vegetarians will eat dairy, eggs, and other materials derived from animals (eg, ovo-lacto vegetarians). Vegans, however, have a more strict diet which is devoid of any animal-derived material. Pescetarians have a similar diet to vegetarians, but their diet also includes fish and other seafood. The rationale for these various dietary restrictions can vary and may include religious, nutritional, ethical, or other reasons. Some people following these diets have little, no, or a seemingly arbitrary rationale. Thus it is certainly possible that many people identifying with these forms of dietary restriction may also be open to eating insects and/or insect based foods. For example, vegetarians concerned with the environment, animal welfare, or food being more natural may be attracted to eating insects because of their relatively low environmental impact resulting for their production and the ability to farm and harvest them in relatively humane ways, compared with other animals.

In Central Australia, where totemism exists in its most pure and complete form, the Arunta tribe has no less than six groups associated with insect totems (cicada, beetle larvae, honey ants, and caterpillars). These insects are prohibited as food, but the exception occurs during the ceremonies that induce fertility, known as *intichiuma* (Berenbaum, 1995). Fasoranti and Ajiboye (1993) say that the members of the Ire clan of the Yoruba tribe do not eat the crickets *Brachytrupes membranaceus* Drury. These people are predominantly worshipers of the iron god Ogun. According to local belief, Ogun does not accept animals or other creatures that have no blood. Thus, the children and other relatives of blacksmiths (who constitute the majority of Ogun worshipers) are forbidden to touch or eat crickets. On the other hand, the eating of crickets in Omido region is believed to make individuals smarter. The white fluid inside these insects, known as *moyiomoyio*, presumably enables the consumer to calculate more accurately and solve arithmetic problems. Children of this clan that taste the larvae of *Rhynchophorus phoenicis* Fab. (Curculionidae) can become intoxicated because adults teach them that these larvae feed heavily on palm wine during their development.

The Yoruba people also believe that eating termites [*Macrotermes natalensis* (Haviland)] at night can cause constipation and there is even the fear that their consumption is carcinogenic. In Uganda, certain species are held in high esteem and therefore reserved for the most important members of the community. If someone is found eating insects that traditionally are regarded as taboo he may experience unpleasant consequences (Owen, 1973). Among the Tukano Indians (Colombia), a diet based on ants and termites is related to the following situations: sickness, adolescents' initiation rites, and young menstruating (Dufour, 1987).

In Australia, there are beliefs associated with the collection of honey ants (*Camponotus inflatus* Lubbock and *Melophorus bagoti* Lubbock). Collectors should speak softly and avoid whistling when they are searching or digging nests because they believe that the ants are shy and run away from the noise. Some Aborigines believe that the nest extends itself to the heart of the earth; others burst an ant on their foreheads during excavation to ensure they will find many workers full of honey (Conway, 1994).

Other times, insect consumption is also released by traditional customs. The Chuh Indians of Guatemala, for example, collect wasp nests of *Polistes* in search of pupae, since they believe that the wasps' pigmented black eyes will give them procreative powers, enabling them to have children with big eyes (Spradbery, 1973). In the Quintana Roo Mayan culture, Mexico, pregnant women eat the wasp larvae known as *ek* because they believe that their children will have an aggressive and wild character, as the wasps do (Ruiz and Castro, 2000). In Thailand, pregnant women buy pupae and larvae of the ant *Oecophylla* sp. because they believe that consumption is healthy for babies (Chen et al., 1998). Indeed, the weaver ant, *Oecophylla smaragdina* Fab., was found to have polyunsaturated fatty acids (PUFA) as the most predominant fatty acid ranging from 213 to 1514 mg/100 g (Raksakantong et al., 2010). The five PUFAs detected were; $18:3n-3$, $20:3n-6$, $20:4n-6$, $20:5n-3$ and $22:6n-3$, $20:3n-6$. In addition $20:4n-6$ were found in all analyzed insects. PUFA have many health benefits for pregnant women. Docosahexaenoic acid ($22:6n-3$), which is a vital component of the phospholipids of cellular membranes, especially in the brain and retina, is necessary for their proper functioning (Connor, 2000). The omega-3 ($n-3$) fatty acids favorably affect atherosclerosis, coronary heart disease, inflammatory disease, and perhaps even behavioral disorders (Connor, 2000).

Disgust Factor

Until recently, despite the growing number of articles, reviews, and books published on the potential of insects as human food, the use of edible insects remained generally unknown among Western cultures. Some people in Western cultures associated the consumption of insects as a practice of "primitive people" (Costa Neto, 2003, 2004, 2011). In 2004, Professor Julieta Ramos-Elorduy observed that it is a disservice to state that insects are eaten only in times of scarcity and famine, as these organisms provide a significant amount of calories and nutrients that are available to people (Ramos-Elorduy, 2004). Since 2011, the environment with respect to food insects in Western cultures has changed remarkably in academic spheres, as well as in the US government, and in the US marketplace. A similar change has taken place in Europe. Now many products are available on the market as energy bars, chips, insect powder, coarse insect meal, and other products. Food insects are now accepted by a growing number of young people. Many people, however, in academia, government, and among the general European and European-American public still have a specific aversion to edible insects. This is the "disgust factor," essentially, a cultural approach to identifying insects as a nonfood item, learned as a young child.

According to Maheu (2011), the greatest obstacles to consumption of insects are cultural and without a deeper understanding of the limitations and possibilities crystallized by culture it will not be possible to promote entomophagy. The idea that we are what we eat is present in all cultures. These beliefs complicate the incorporation of insects in the diet because all associations and assignments related to these arthropods influence our perception of them as food, much more than their nutritional value.

What are the origins of the disgust factor? Many of our most valuable wild animals that were domesticated were large terrestrial mammalian herbivores and omnivores in Eurasia, each over 45 kg. These areas where the large mammals were mainly domesticated also overlapped with one of the areas of the world where agriculture arose. This area is known as the Fertile Crescent, comprising areas in Western Asia, the Nile Delta of Northeast Africa, and Nile Valley. These animals yielded tools, fabric, traction, warmth, milk, transportation, and a meat protein source. Together, agricultural practices and these relatively large animals spread quickly through Europe (Diamond, 2005). It is hypothesized that because of the utility of these large animals, that insects, with the exception of honeybees, silk moths, and scale insects, failed to gain much traction (van Huis et al., 2013). There also is a parallel hypothesis that from the influence of Aristotle, there came a hierarchical view of living things with humans at the top of the hierarchy and using the other (lower) living things for their human survival (Dunkel, 2000). We see at this point in the history of biological philosophy, a digression from the Eastern cultures view and that of most indigenous peoples that humans are only one part of the world of living and nonliving matter and that each is interconnected and dependent on each other. The beneficial roles of insects began to be ignored culturally even though honey, silk, and the cocconial dye of insects remained highly valued insect products.

By the 19th century in Europe and the United States, people of Western cultures generally considered this food habit as disgusting, primitive, or a sign of material poverty. This example is currently followed by those sectors of traditionally entomophagous societies that seek to copy a Western lifestyle; being able to afford meat of domestic animal species is one of the ways to show social and economical progress and to mark a difference to economically weaker parts of their societies (Ramos-Elorduy, 2011; Verspoor et al., 2013). Globalization is resulting in the use of more fast foods and preprepared foods and the loss of traditional ways of life (Yen, 2009a). In Thailand, a country that has a strong tradition of entomophagy, individuals who have migrated to urban centers and received a formal education "learned" to despise local entomophagous resources when they expose themselves to a cosmopolitan culture that has prejudices against insects (Chen et al., 1998). More recent research indicated that edible insects were appreciated by both urban and rural communities in Thailand. Different species, however, were preferred by urban dwellers than by rural inhabitants (Hanboonsong, 2010; Hanboonsong et al., 2013). The Aztec people used to feed on 91 species of insects, preparing them in different ways: baked, fried, boiled, in sauces, or as a condiment for any dish. Some species were even stored dried. With the arrival of the Spanish conquerors, however, many of the indigenous foods were negatively qualified and then forgotten and/or depreciated (Ramos-Elorduy and Pino Moreno, 1996). Moreover, use of edible insects may have served as a form of national identity as they turned into elements of identification of individuals belonging to the same culture (Ramos-Elorduy, 1996).

Many people dislike or hesitate to consume insects because they consider these animals to be dirty. However, some of the major edible species, such as grasshoppers, crickets, cicadas, and lepidopteran and coleopteran larvae, mostly eat fresh plant leaves or wood and are therefore cleaner and more hygienic than crabs, lobsters, or some shrimp species, which eat carrion (Mitsuashi, 2010). Distaste or hesitation for the consumption of insects causes that a considerable amount of animal protein becomes unavailable to those people who suffer from hunger and malnutrition. In Descola's (1998) opinion, the habit of eating or not eating insects depends on the variability of individual choice within an accepted norm, as well as the accessibility of the animal. The author complements saying that even within tribal societies there is great variability of individual food preferences and attitudes toward animals. Marvin Harris, a cultural materialistic, explained the repulse on the consumption of insects through a cost/benefit ratio. He said there are three reasons why a food is to be banned from the menu: when food becomes too expensive to be obtained or prepared, when there is a more nutritious and inexpensive replacement, or when there is a negative impact on the environment. Over time the food becomes culturally repelled as a food "bad to eat" (Harris, 1999). Considering the use of insects, he states "The reason we do not eat insects is not because they are dirty and disgusting. Instead, they are dirty and disgusting because we do not eat them."

Human–insect relationship in Western cultures is generally centered on negative attitudes, that is, once the word "bug" is present, it is often associated with the image of venomous animals with unsightly, dirty bodies; filthy, disease vectors; and pests of food, causing reactions of disgust, fear, and loathing (Costa Neto, 2002; Arana, 2007). Different explanations for the consistent human aversion addressed to insects and other invertebrates are available in the literature. One may say that humans are biologically predisposed or prepared to fear certain animals, like snakes, scorpions, and spiders (Seligman, 1971). Another explanation says that common and nonclinical entomophobias may be closely related to the human reaction of "disgust," whose benefits are considered as an adaptive prevention of disease transmission (Davey, 1993, 1994; Matchett and Davey, 1991; Ware et al., 1994). The psychologist Paul Rozin masterfully captured the disgust psychology. He shows how foods are primarily classified as pleasant or unpleasant, categories that depend largely on individual variations, sometimes genetic, and secondly as appropriate, inappropriate, disgusting, dangerous, or beneficial, categories that are basically determined by culture. Smell and touch are also repulsive. Disgust prevents people from eating certain things or, if it is too late, causes them to spit or vomit (Rozin, 1995). The author also shows how some things can be considered edible as well as taboo, and how certain foods are more prestigious than others. Our way of classifying insects and the incorporative beliefs form the basis of a reaction of disgust toward them has resulted on blocking the idea of consuming insects in the majority of the Western cultures' populations. The aversion involves deep emotions and our own sense of identity in a way that defies any kind of rationalization (Looy et al., 2013; Miller, 1997; Rozin and Fallon, 1987). In Asian and Native American cultures, insects are respected and understood to be part of the ecosystem along with humans, thus, they do not engender fear or negative emotions.

A third interpretation was suggested by the notion of human alienation to creatures different and distinct from our own species (Kellert, 1993). According to this author, more positive attitudes directed at invertebrates are observed when these animals have aesthetic, utilitarian, ecological, or recreational values.

Although edible insects have a high nutritional value for humans, they have not been considered a significant source of food by Western cultures, and studies on their use as animal feed and waste recycling are still scarce. It is clear that we cannot consume what we do not know about. To introduce insects in the diet, we need to know which species are edible, and how they need to be prepared and cooked for consumption. The fact is that knowing about the nutritional and ecological advantages of entomophagy, as well as the consumption of insects by indigenous peoples, can contribute to arise some sympathy among enlightened people; however, this gain of knowledge will never be enough to eliminate blockages and reinvent the insect as a privileged food where it is initially viewed with despise. To the statements of economic, ecological, and nutritional importance of entomophagy to human beings, we also need to take into consideration the very nature of our psycho-cultural limitations (Maheu, 2011).

In a society that rejects the consumption of insects there are some individuals who overcome this rejection, but most will continue with this attitude. It may be very difficult to convince an entire society that insects are totally suitable for consumption. However, there are examples in which this reversal of attitudes about certain foods has happened to an entire society. Several examples in the past 120 years from European-American society are: considering lobster a luxury food instead of a food for servants and prisoners; considering sushi a safe and delicious food (Corson, 2008); and considering pizza not just a food for the rural poor of Sicily. In Latin American countries, where insects are already consumed, a portion of the population despises their consumption and associate it with poverty and "indianness" due to ethnocentric reasons (Katz, 2011). There are also examples of people who have had the habit of consuming them and abandoned that habit due to shame, and because they do not want to be categorized as indians or poor (Costa Neto, 2011). According to Katz (2011), if the consumption of insects as a food luxury is to be promoted, there would be more chances that some individuals who do not present this habit overcome ideas under which they were educated. And this could also help to revalue the consumption of insects by those people who already eat them.

Multiple attempts have been made by entomologists to make insects more broadly appealing. A popular example is Ronald Taylor's (1975) book *Butterflies in My Stomach*, and the accompanying recipe guide, *Entertaining with Insects* (1976) (Taylor, 1975; Taylor and Carter, 1976). Several subsequent entomophagy cookbooks have been published, including *Des Insectes à Croquer: Guide de Découvertes*, produced by the Montreal Insectarium (Thémis, 1997); and *La Cuisine des Insectes: À la Découverte de l'Entomophagie*, by Gabriel Martinez (2000), a French culinary guide offering professional cooking advice. Others, such as the humorous *The Eat-a-Bug Cookbook*, offer readers familiar American recipes such as pancakes, pizza, and alphabet soup altered with the addition of edible invertebrates (Gordon, 1998). Due to recent popularity and interest in entomophagy in the United States, a second edition of *The Eat-a-Bug Cookbook* was released in 2013 (Gordon, 2013).

In the mid-1970s, Gene DeFoliart, a medical entomologist who as an administrator was filling a request for a speaker on edible insects, stumbled on a neglected but globally important area of food security. By 1988, DeFoliart considered it his ethical duty to share his discovery of the importance of entomophagy internationally and to connect those with knowledge in this area. In July 1988, he launched *The Food Insects Newsletter*. By 1995, there were 1000 subscribers in 63 countries who received a 12-page newsletter four times a year. In 1995, DeFoliart selected Florence Dunkel to take over the editorship of the *Newsletter*. In 2000, the *Newsletter* became an electronic communication, a website, continuously updated. Simultaneously, DeFoliart published an e-book collecting all the literature on edible insects by geographic area. In 2009, DeFoliart and Dunkel joined with David Gracer to create a bound volume of 22 years of these printed *Newsletters*, entitled *Chronicle of a Changing Culture* (DeFoliart et al., 2009). The *Newsletter* represented an awakening in Western culture of the global interest in edible insects and the growing curiosity of younger generations in this food source.

Edible insects are also featured in Peter Menzel and Faith D'Aluisio's photo-essay volume, *Man Eating Bugs: The Art and Science of Eating Insects*, including diverse entomophagy scenes and dishes from around the globe (Menzel and D'Aluisio, 1998). Julieta Ramos-Elorduy's *Creepy Crawly Cuisine* is an introduction to the world of edible insects, complete with recipes and photographs (Ramos-Elorduy and Menzel, 1998). It includes a historical look at the use of edible insects in indigenous cultures and provides information on where to obtain insects and how to store and prepare them. *Het insectenkookboek* (*The Insects Cookbook*) was recently published in The Netherlands and features various insect recipes (van Huis et al., 2012). Recently, these authors released an English version, titled *The Insect Cookbook: Food for a Sustainable Planet* (van Huis et al., 2014). *Edible Bugs: Insects on Our Plate* by Chad Peterson was published in 2012 and is available as an electronic book (Peterson, 2012). And *Das Insektenkochbuch: Der etwas andere Geschmack* (*The Insect Cookbook: The Slightly Different Taste*) by Fritzsche and Gitsaga (2002) provides recipes and photographs of Thai-style insect dishes. *Edible* by Daniella Martin (2014) not only presents a compelling argument for adding insects to our diets and predicts this will happen, but she also provides recipes and a basic bio-review of edible insects.

Educational Campaigns

There is a general informal consensus among entomophagy educators and activists that incorporating edible insects into entomology outreach programs, as a form of informal education to introduce the general public to entomophagy, may be vital for the acceptance of entomophagy (Looy and Wood, 2006). We have observed that the world of edible insects is beyond the realm of entomology and has been taken over by the food entrepreneurs.

In general, nonentomophagous societies are increasingly being exposed to entomophagy phenomenon through documentaries, films, media interviews, lectures, food festivals, and the like. (Dunkel, 1998). For example, in December 2000, more than 2,680 entomologists were able to taste different delicacies insect-based at the Insectarium in Montreal, Canada, and later that year over 20,000 visitors sampled these delicacies. The theme has also been incorporated into many courses in several university campuses and research institutes. The Canadian Broadcasting Company will air a full-length documentary on edible insects in 2016. The British Broadcasting Company (BBC) aired an edible insect documentary in 2014. Since 2011, a number of TEDx and TED talks have been given on edible insects. Examples are http://www.ted.com/talks/marcel_dicke_why_not_eat_insects and https://www.youtube.com/watch?v=W5GGKoYuXHs. Websites on edible insects abound, as for example www.foodinsectsnewsletter.org; http://www.girlmeetsbug.com/ and http://www.proteinsect.eu/index.php?id=37.

Many "bug fests" and "bug feasts" have been taking place each year in some countries (Table 2.2). In the United States, a few of these include: events for the public held on university campuses such as the annual Bug Bowl sponsored by the Department of Entomology, Purdue University and held on campus in West Lafayette, Indiana since 1990, the Fear Film Festival at University of Illinois, and the annual Bug Buffet at Montana State University since 1989 (Dunkel, 1998); events for the public sponsored by insectaria or nature centers such as the BugFest at Garfield Park Nature Center in the Garfield Park Reservation, Cleveland, Ohio; and portions of national meetings or special conferences such as the Food Insect Workshop held as part of the Insects in Captivity Conference, Tucson, Arizona. Additionally, the largest BugFest in the United States, in Raleigh, North Carolina, typically features a buffet of various insects cooked by several local and invited chefs and restaurants. They usually also have an insect-based food tasting contest judged by local media and prominent figures. Also, the Los Angeles County Natural History Museum (Los Angeles, California, USA) has an annual insect-themed festival at which there is an insect cooking competition. Additionally, the USA Science and Engineering Festival, in 2012 and 2014, had an invertebrate section (hosted by All Things Bugs, LLC in 2012 and Invertebrate Studies Institute in 2014) at which there were tables with various cooked and flavored insects for people to try.

TABLE 2.2 Organizations that Host Insect-Eating Events

Country	Organization
Australia	New South Wales Entomological Society
Canada	Alberta State Museum; Ontario Joint Entomological Meeting; L'Insectarium de Montreal
Japan	Tokyo Tama Zoo, Insectarium, Tokyo
Republic of South Africa	South Africa Entomological Society
United States	Audubon Zoo Insectarium, LA; Buffalo Museum of Natural History, NY; Cincinati Zoo, OH; Crowley's Ridge State Park, AK; Garfield Park Nature Center, OH; Invertebrate in Captivity Conference, AZ; Los Angeles Museum of Natural History, CA; Iowa State University, Department of Entomology, IA; New York Entomology Society, NY; North Carolina Museum of Natural History, NC; Northwest College, WY; Oregonridge Nature Center, OR; Pennsylvania State University, Department of Entomology, PA; Provincial Museum of Alberta, AB; Purdue University, Department of Entomology, IN; Montana State University, Bozeman MT; San Francisco Zoo, CA; Smithsonian Institute, Museum of Natural History, Washington, DC; State Botanical Garden, GA; The Jonathan Ferrara Gallery, LA; University of Illinois, Department of Entomology, IL; Invertebrate Studies Institute, GA; International Society for Food and Feed Insects, GA; USA Science and Engineering Festival, Washington, DC.

Source: Dunkel (1997), Mitsuashi (2010) and discussions with colleagues.

During the First National Symposium on Anthropoentomophagy held at Feira de Santana State University, Brazil, in March 2009, this scenario could be observed. In the workshop of insect cuisine, participants tasted salty and sweet culinary preparations containing some insects (crickets, mealworms, and leaf-cutting ants). During this workshop, "in nature" presented insects brought greater resistance to consumption, while preparations with crickets (Cantonese stew, pizza, brochette, and cheese rolls), strawberries with mealworms, and dulce de leche ant canapés were readily accepted and tasted by all participants. It might be noted also that most people, after consumption, reacted positively to the flavor and texture of the insects, while the look and sociocultural issues were factors of lower receptivity (Linassi and Borghetti, 2011).

We must change the Western culture idea that insects cannot be included as food items in the human diet (educational campaigns), and we should also encourage the marketing strategies coming from the Millennial generation worldwide to continue to launch products based on edible species. Edible insects are now part of the Western culture marketplace: protein bars; granola; chips; cricket powder; cricket meal; hoppers in the meat section of European grocery stores. Edible insects are also a growing feature in restaurant entrees: Sushi Mazi in Portland, Oregon; VIJ's in Vancouver, B.C., and Noma in Copenhagen, Denmark are several examples. In San Francisco, for example, artist and designer Monica Martinez operates the "Don Bugito Prehispanic Snackeria" food cart with delicacies like chocolate covered crickets and spicy superworms. Methods for mass production of insects in adequate sanitary conditions for human consumption have been created in the United States and Europe. This new food industry will not depend on the collection of specimens directly from nature. Edible insects are "in" now and are becoming hipster-faddish in some maxi-trending venues. Clearly there has been an "entrepreneurial explosion."

EDIBLE INSECTS AS NUTRACEUTICALS

Insects are not only eaten for their abundance, taste, nutritive value, or other organoleptic characteristics, but they are also used as treatment in medicine. This topic has been well reviewed in the literature and is somewhat tangential to the scope of the current chapter, so we will provide only a brief summary of the topic here (Dossey, 2010, 2011; Costa Neto, 2005). Based on the traditional use of insects as a food source and medicine at the same time, Costa Neto and Ramos-Elorduy, 2006 have postulated the hypothesis of nutraceutical entomofauna. According to this hypothesis, many species of insects are sources of functional foods thus providing health improvements against some illnesses and diseases. Indeed, at the time when the insects were prescribed for therapeutic purposes by healers and practitioners of traditional medicine, people were familiar with the idea of ingestion (Holt, 1885). It is interesting to mention that the word medicine owes its origin to honey, because the first syllable has the same root as mead, an alcoholic drink made from honeycomb, which was often consumed as an elixir (Hogue, 1987).

The practice of using insects as nutraceuticals occurs in many parts of the world (Ramos-Elorduy, 2004; Meyer-Rochow and Chakravorty, 2013; Zimian et al. 2005; Pemberton, 2005; Yhoung-Aree and Viwatpanich, 2005). Folk logic appears to be the basis for some arthropod drug uses. For example, in Ijebuland, Nigeria, crickets (*B. membranaceus*) are consumed as food items in order to aid mental development and for pre- and postnatal care purposes (Banjo et al., 2003).

The peanut beetle *Ulomoides dermestoides* (Fairmaire) is commonly used in Asia for the treatment of asthma, arthritis, tuberculosis, and sexual impotence. In Brazil, people use this species as a fortifier and to treat impotence, eye irritation, and rheumatism (Costa Neto, 2005). To be used as treatment, adult beetles should be reared with peanuts and eaten alive. Ingestion of this beetle has become popular as a way to combat various diseases, especially asthma and inflammatory processes. Its antiinflammatory properties have been already proven in the laboratory (Santos, 2004).

Periplaneta americana L. is commonly used as a nutraceutical resource. In the Brazilian folk medicine, people use this cockroach for the treatment of several maladies. For example, the whole toasted insect is turned into a tea, which is drunk three times a day to treat asthma and colic. Patients should not know what they are drinking. The Pankararé Indians from northeastern Brazil toast a whole roach, then grind it and make a tea to treat asthma and constipation. The water in which a cockroach has been cooked is drunk in order to treat heartburn in the city of Feira de Santana (Costa Neto, 2005).

In Feira de Santana, some *Atta* ants are toasted and ground to make a tea, which is drunk to treat asthma and throat ache. In the county of Matinha dos Pretos, people eat its abdomen (buttock) for healing throat ache. This ant is also eaten raw or fried for the treatment of tuberculosis (Costa Neto, 2005).

A liquid preparation from blister beetles, popularly known as Spanish fly, has been used for millennia as a female aphrodisiac (Karras et al., 1996). It is mixed with liquid and taken as a drink. The active ingredient, Cantharidin, when taken in negligible amounts has debatable effect on those consuming it (Dossey, 2010), but noncommercial preparations can cause severe poisoning which include mucosal erosion, renal dysfunction, hematuria, and GI hemorrhage associated with diffuse injury of the upper GI tract (Karras et al., 1996).

Indigenous knowledge of medicinal compounds from natural resources is still one of the most important means for discovery of unknown biotic drug sources (Oldfield, 1989). The investigation of folk medicine has also proven a valuable tool in the developing art of bioprospecting for pharmaceutical compounds (Costa Neto and Ramos-Elordy, 2006; Kunin and Lawton, 1996). It is known that insects are very prolific in the synthesis of chemical compounds, such as alarm and mating pheromones, defensive sprays, venoms and toxins sequestered from plants or their prey and later concentrated or transformed for their own use (Pemberton, 1999). This huge diversity of chemicals includes compounds that are emetic, vesicant, irritating, cardioactive, or neurotoxic (Berenbaum, 1995). Chemical screening applied to 14 insect species has confirmed the presence of proteins, terpenoids (triterpenoids and steroids, carotenoids, iridoids, tropolones), sugars, polyols and mucilages, saponins, polyphenolic glicosides, quinones, anthraquinones glycosides, cyanogenic glycosides, and alkaloids (Andary et al., 1996; see Table 2.3). Chitosan, a compound derived from chitin, has been used as an anticoagulant and to lower serum cholesterol level, as well as to repair tissues, and even in the fabrication of contact lenses (Goodman, 1989). Kunin and Lawton (1996) have recorded that promising anticancer drugs have been isolated from the wings of Asian sulphur butterflies [*Catopsilia pomona* (=*crocale*) Fab., Pieridae] and from the legs of Taiwanese stag beetles (*Allomyrina dichotomus* L., Scarabaeidae). These compounds were isoxanthopterin and dichostatin, respectively. Oldfield (1989) records that about 4% of the extracts evaluated in the 1970s from 800 species of terrestrial arthropods (insects included) showed some anticancer activity. Although the antiviral and antitumor activities of chitin, the second most abundant polysaccharide in nature, have been known for some time, the immunological effects of chitin have only recently been recognized (Lee et al., 2008). Small-sized chitin particles have been implicated in reducing the inflammatory response (Brinchmann et al., 2011). This is one of the examples of how traditional knowledge obtained from a process of

TABLE 2.3 Examples of Chemical Substances and Products Derived from Some Insect Species Pharmacologically Tested

Species	Substance	Pharmacological Action
Acheta domestica L.	Iridoids	Antimicrobial, tonic, antiinflammatory
	Cumarins	Anticoagulant
Allomyrina dichotomus L.	Dicostatin	Anticancer
Anoplius samariensis Pal.	Pompilidotoxin	Neurotoxic
Anterhynchium flavomarginatum micado Kirsch	Eumenine mastoparan-AF	Peptide that acts on degradation of mastocytes
Apis mellifera L.	Propolis	Anticancer, anti-HIV
Bombyx mori L.	Attacin, moricin, drosocin	Antibacterial
Catopsilia crocale (Cramer)	Isoxantopterin	Anticancer
Drosophyla melanogaster Meigen	Defensin, diptericin	Antibacterial
Edessa cordifera Walker	Cumarin	Anticoagulant
	Alkaloids	It increases muscle tone and contractility
Euschistus crenator S.	Tannins	Antitoxic, antitumoral, antiviral
Hyalophora cecropia L.	Cecropin A y B	Antibacterial
Lonomia obliqua Walter	"lopap" protein	Antithrombotic
Lytta vesicatoria L.	Cantharidin	Vesicant
Phoenicia sericata (Meigen)	Allantoin	Antibacterial
Polybia occidentalis nigratella Oliv.	Saponins	Antiinflammatory, it helps to resist "stress," antihepatotoxical
Prioneris thestylis Doubleday	Isoguanine	Anticancer
Pseudagenia (*Batozonellus*) *maculifrons* Sm.	Pompilidotoxin	Neurotoxic
Sarcophaga peregrine (Robineau-Desvoidy)	Sarcotoxin IA, IB, IC, sapecin	Antibacterial
Sphenarium purpurescens Ch.	Iridoids, carotenoids	Antiinflammatory, tonic, antimicrobial, activity of provitamin A
Tetragonisca angustula angustula Latreille	Honey	Antibacterial

Source: Costa Neto (2005).

testing through millennia by many generations conducting empirical science, is often verified by modern Western science procedures.

Most nutraceuticals from insects as well as most of the 1571 known species of edible insects have come to us as traditional ecological knowledge from the process of native empirical science. Native empirical science is a process of making seasonal and bio-cycle linked observations over long periods of time (Cajete, 1999). Storytelling is the reservoir. Modern Western science uses the scientific method, which begins with short-term observations, question asking, and hypothesis generation followed with tests, experiments designed and conducted to negate hypothesis statement. Truth is determined by the results of the experimental procedure. Generally the experiments are repeated at least two more times. The process of knowing often starts with empirical science and modern science verifies. In European-American culture, there often is a distain for traditional knowledge, and all forms of empirical science, among the Western medicine and the Western scientific communities. However, once Western scientists look at side-by-side comparisons of the two processes and the results from each, there is often a growing appreciation for the results of native empirical science.

ETHNOENTOMOLOGY

Ethnoentomology is the subdiscipline that establishes the functional relationships between human societies and the world of insects (Costa Neto, 2002). Like all science, ethnoentomology is involved in a complex web of social and political relations, which are linked to global processes (Alexiades, 1997). Posey (1976) points out that insects are one of the most important ecological factors affecting any human population: site selection, horticultural systems, house types, and numerous aspects of daily life have been affected by adaptations to the coexistence with insects. Why, then, cultural anthropologists and ecologists almost completely ignore the insects in human societies? Maybe because, as he claims, "we do not conceive of insects as being anything other than incidental pests, for we have subdued the really harmful ones. We find it difficult, therefore, to imagine any reason for other peoples to be aware of or concerned with such creatures" (p. 147).

It is essential to record the rich amount of traditional knowledge, customs, and practices of indigenous and nonindigenous peoples since, with no doubt, people are living representatives of their own culture and they have significant knowledge on the biological resources they depend upon (Diegues and Arruda, 2001). Peoples' knowledge and perception toward insects vary in quality and quantity depending on their interest in the subject, the environment and the importance of insects to their lives. Obviously, the interest, motivation, ability, opportunity and prior knowledge as well as experience play a tremendous impact on the acquisition of peoples' knowledge and perception (Gurung, 2003).

The corpus of ethnoentomological knowledge is generally transmitted from generation to generation through oral tradition, which is an important vehicle for the dissemination of biological information (Posey, 1987). Much of this knowledge is encoded, for example, in myths and legends, with the result that not all researchers are academically prepared to decode them, and the complete relearning sometimes can only be obtained through multidisciplinary research. In this regard, Western scientists, whether deprived of their cultural ethnocentrism, could learn from the natives a rich set of information about tinctures, oils, dyes, pesticides, natural essences, medicines, foods, repellents, and so forth, all of which are natural and many times organic (Posey, 1982).

The literature contains several examples of how the ethnoentomological knowledge can be decoded and used by academic scientists. Blake and Wagner (1987) drew the attention of entomologists to the importance of this knowledge as a significant source of information on the species, their life cycle, and behavior. According to these authors, Paiute Indians' traditional entomological knowledge was used by decision-makers in the Inyo National Forest in 1981. At that time, a severe defoliation of the Jeffrey pine forest by the Pandora moth (*Coloradia pandora lindseyi*) Barnes and Benjamin (Saturniidae) caused a public outcry and appeared to threaten the trees with growth loss and mortality. As mature larvae are a traditional food for the Owen Valley-Mono Lake Paiutes, the elders told the forest service personnel that caterpillars would not harm the healthy trees, but only the very "sick" trees might die. This "sickness" is caused by the dwarf mistletoe. Taking into consideration the recommendation of the Paiute Indians, the forest service decided not to undertake to control the insects, especially to avoid the possibility of contaminating a traditional Paiute food source.

Traditional knowledge of native foods and techniques to get them are crucial to the survival of many communities around the world (Somnasong et al., 1998). In general, people who consume insects know when, how, and where to collect the species they use, and they have a number of ways to prepare them and keep them to count as a food source in times when this is scant (Ramos-Elorduy, 1982). Consumption of edible insects is related to the life cycle and the geographic location of species and depends on both the abiotic conditions (temperature, humidity, soil type, latitude, altitude, light, and climate) and biotic conditions (vegetation, host, type of feeding, and physiology of reproduction), as well as the fact that a species is aquatic or terrestrial. All these factors, according to Ramos-Elorduy and Pino Moreno (1996), influence the collection times and consumption of insects. For example, the Desana Indians who inhabit the upper Rio Negro (northern Brazil) reconcile the appearance of insects with the constellation cycle (Ribeiro and Kenhíri, 1987).

As Posey (1976) points out, using insects and their products in any quantity presupposes a sophisticated folk knowledge about these organisms and their behavior, which is revealed through the analysis of oral traditions that somehow encompass insects. Indigenous peoples have an intimate knowledge of their biological environment, since they are expert manipulators of processes and ecological relationships, and tend to adopt sustainable management practices. Researchers studying the traditional entomological knowledge of indigenous and tribal societies almost always get impressed by the consistency of knowledge that these communities have on the species of insects with which they live and interact.

Ekagi people (from Indonesia) have a general knowledge of the biology of cicadas. They know that the nymphs emerge from the soil during the night and that ecdysis occurs by opening of the nymphal skin on the dorsal surface. […]. Cicadas are collected early in the morning or during the night at the light of flares, on forested mountain slopes and on trees along the river banks; newly emerged cicadas, with a soft white body, are the most highly appreciated (Duffels and van Mastrigt, 1991).

Villagers (from the Kipushi Territory of Zaire) are knowledgeable about the hosts of the edible species and the season when each is ready for harvest. This knowledge among indigenous peoples of the host plants and seasonal history of local food insects has been noted by writers in other tropical countries (DeFoliart, 1989).

Bees explode from a nest, fleeing the smoke of Bahadur's oka with a buzz that rises to a roar. Though made less aggressive by smoke, these bees, known for their ferocity, in mass attack can kill a man. To avoid their stings, other honey hunters cut three limbs so that combs come smashing down. Raji leave limbs intact: Within an hour the bees return, as do the Raji, season after season (Valli, 1998).

Traditional rural people know insects more intimately than anyone except entomologists, but few entomologists know how to harvest wasp honey or are aware that leafcutter ants host lizard lodgers (Bentley and Rodríguez, 2001).

The Lao knowledge of life-histories surprised me, and in some cases at least this knowledge has led them to devise ingenious methods of capturing insects (Bristowe, 1932).

The potential utility of insects represents an important contribution to the debate of biodiversity, as it opens a perspective for economic and cultural valuation of animal usually considered useless. It is important to record and maintain traditional knowledge about edible insects while respecting traditional ways of life (Costa Neto, 2002; Yen, 2009a).

Thus, the researcher interested in performing ethnoentomological studies need to obtain theoretical and methodological training from both the social sciences and the natural sciences, investigating the phenomena of his area of interest in order to formulate more sophisticated hypotheses and generate new paradigms (Posey, 1986). In doing this, the researcher will be able to work with tools and intrinsic issues of both disciplines to, according to Toledo (1991), record the cognitive means (corpus) that allow people to take ownership of natural resources (praxis) without fear of being "ritually polluted" to address issues unrelated to traditional paradigms (Posey, 1986). To paraphrase the Brazilian anthropologist Ordep Serra (2001), an ethnoentomologist should know and become expert in ethnology, but also attain technical knowledge as an entomologist.

The role that insects play in any economic system is important, although they represent only a small part of total resources. However, due to ethnocentric reasons, since insects are not considered appropriately "mainstream" in dominant Western cultures, there has been a tendency to ignore or omit the relevance that edible insects have to other cultures. As previously mentioned, we have only their indigenous or generic names, making it difficult to know which species they belong to. Therefore, it is necessary for researchers to conduct studies using an interdisciplinary approach paying attention to the values and knowledge of traditional peoples.

HARVESTING AND CULTIVATION

In the fieldwork there is a gender gap, and generally, nondangerous species are collected by women accompanied by their children. Difficult locations to search, for example, where there are rocks and ravines, and dangerous species to collect (bees, wasps, and stinging ants), are searched and collected by men. Likewise, men check their crops and finding some immature stages called "pests," they may collect and consume them.

In fact, according to Barth (2002), local human populations exhibit a surprising diversity of ethnographic knowledge sharing. The traditional division of labor contributes to understanding the differences between men and women. For example, in northeastern Thailand women normally collect adults, pupae, and eggs of red ants; men participate only when insects are collected for sale (Somnasong et al., 1998). Aboriginal women are also those who know where and how to collect honey ants, this teaching being transmitted by elders in the community (Conway, 1994). Among the Araweté, men are responsible for getting the honey while women collect larvae (Viveiros de Castro, 1992).

In addition to the nutritional importance, economic importance related to the sale and/or purchase of edible insects and derived products is also highlighted in all the local, national, and international level. Through the sale of edible insects in semiurban and urban areas, many individuals from rural communities manage to increase their income and therefore can purchase several items in addition to the most essential (Chen et al., 1998). According to Vantomme (2010), in urban Bangui, Central African Republic, edible insects contributed up to one-third of the protein intake during the rainy season when supplies of bushmeat and fish declined, and dried *Imbrasia* caterpillars were sold for up to $14 USD per kg, making them a major source of cash for rural women. Mbata et al. (2002) demonstrated that the dried caterpillar trade is economically profitable, reporting that the Bisa people of Zambia (Africa) process and market the caterpillars of the species *Gonimbrasia zambesina* (Walker) and *Gynanisa maja* (Klug) (both Saturniidae). Ramos-Elorduy (1997) says that in Mexican small towns, insects are sold door to door, by measures (small logs of different size). At the village markets, they are even sold by handfuls or by heaps. In Rwanda, when brown locusts, *Locustana pardalina* Walker, are in their gregarious phase, it is a joyous opportunity for young children to collect them and sell them. In Bamako, it is especially expedient for the children to collect them in plastic bags. One liter sold for 500 CFA (Central Africa Franc) (the equivalent of 85 cents of a US dollar) (Dunkel, 2012).

Farming of Edible Insects

The traditional use of insects as food exists in both protocultures and formal cultures (Ramos-Elorduy, 2009). The mass production of insects has great potential to provide animal proteins for human consumption, either directly, or indirectly as livestock feed (Yen, 2009a) (see also Chapters 5 and 6 of this book). The species currently used for food are mainly collected from natural populations or are grown in mini-farm systems. A few species have been highly domesticated. This is what happens, for example, with the silkworm (*Bombyx mori* L.), which is no longer able to survive in nature without human interference (DeFoliart, 1995). According to the author, the pupae of this species have been widely used as food and feed in almost all Asian countries. However, instead of relying on the collection of native populations, artificial breeding methods should be sought to reduce the pressure collection of individuals in nature. Insect mass production would ensure a continuous and abundant supply of insects for food or feed. Many insect species can be grown more efficiently than the commonly consumed mammals. The farming of the mealworm beetle larvae (*Tenebrio*) is ideal for housework, as it is clean, does not require special equipment, and takes up very little space.

Encouraging the consumption of insects in many rural and indigenous societies, particularly in those where there is a history of insect use, can help, diversify, and complement the diet of people and provide revenue as well. Furthermore, the mass collection of insect pests could be incorporated into integrated pest management (IPM) campaigns, so the use of pesticides could be reduced (Idowu and Modder, 1996; Ramos-Elorduy, 2005). In addition, insects should be less environmentally harmful than livestock, which devastated forests and native grasses, to grow protein (Gullan and Cranston, 2005). However, DeFoliart (1995) comments that although the insects collected in nature are widely sold in tropical countries, few have the characteristic of being domesticated and kept under cultivation for sale and use. There are few species bred through systems of "miniranches" as the silkworm (*B. mori*) and honeybees (*Apis* spp.) which are bred on a large scale worldwide. Weevils of the genus *Rhynchophorus*, are nondomesticated insects, but widely grown in Asia, Africa, and Latin America (DeFoliart, 1995). We have found that there are countless species of stingless bees (Meliponinae) that have been cultivated for thousands of years in different parts of the world and that a "proto-culture" exists in the field on the use of many species like some ants, wasps, beetles, cockroaches, and the like.

Although the exploitation of edible insect resources is considered negatively by many governments (Mbata et al., 2002), successful examples are recorded supported by the United States' and the Netherlands' governments. An additional protein supplement derived from insects is incorporated in feed for our vertebrate meet animals such as fish, swine, and chickens. Insect larvae fed on organic household waste can provide an inexpensive source of protein for animal feed. An experimental study carried out with the domestic fly (*Musca domestica* L.) recorded the capacity and efficiency that this insect has to recycle different organic sources of waste. Its pupal stage was used as feed for rainbow trout fingerlings (Ramos-Elorduy et al., 1982). Hanping and Changzhen (1993) corroborate the importance of mass production of fly larvae (*M. domestica vicina* Macquart), since high rates of protein and fat can be considered as a good source of animal protein.

Commercial and small-holder farm operations have both experienced the effectiveness of using black soldier fly larvae for improving nutrients in their feed economically (Makkar et al., 2014). In several parts of the world, black soldier fly larvae have served as an important processor of preconsumer food waste and, at the same, time an important supplier of fish feed (Barroso et al., 2014), chicken feed, and piglet feed (Rumpold and Schluter, 2013; Sanchez-Muros et al., 2014).

Now, insects in places like the United States and Europe are still not widely available and come at a high cost as compared to other animal-based food ingredients. However, as the production of insects as human food and animal feed increases worldwide, particularly in Western countries, many aspects related to the feasibility of insect-based foods, such as cost, safety, the efficiency of insect mass production, and availability, will improve. The resulting increase in demand will synergistically drive the status of insects as a standard mainstream food ingredient.

As was the case in the emergence of other new food products into the western world, there are many barriers that need to be overcome for insects to be part of our daily lives. The particularities of taste barriers are very interesting, as it is mainly visual cues that trigger disagreeable responses. As food ingredients however, they pose significant nutritional advantages that, somewhat disappointingly, hold less interest for the mainstream consumer than for the nutritional enthusiast. Indeed, the fact remains that the drab taste of most edible and commercially easier-to-market insects can only pose an opportunity window for the insect-producing industry when they are combined with another source of flavor.

While the insect-producing community is united in key milestones such as the legal acceptance of insects in Europe, in particular applications like creating new products, there are no united efforts. The variety of food pairings and further applications beyond food for which insects can serve disperses current efforts and does little to fill the lack of definition and basic knowledge consumers need to form opinions and buying preferences. A basic level of primary demand needs to exist in each of the Western countries for markets to function, and this will require a great deal of educating consumers about the uses and flavors of insects.

FINAL COMMENTS AND RECOMMENDATIONS

In cultures that eat insects on a regular basis, they are considered valuable and are viewed as nutritious, medicinal, environmental, and a sustainable and secure food item. Once insects become more widely recognized as a respectable food item in industrialized countries, the monetary implications will have a profound positive impact on businesses, industry, governments, and research. Replacing vertebrate livestock-derived foods and food ingredients with those derived from insects will also considerably improve the health of the earth's natural environment. Incorporating insects into Western food habits will enlarge and expand food production, increase food supply and availability, as well as meet the nutritional needs of resource-limited households (Yen, 2009a,b). Insects will form a whole new class of foods for low-input, small-business, and small-farm production, with tremendous potential to be mass produced for human consumption. The future of worldwide acceptance of entomophagy relies on the commercialization of new food products but must be coupled with patents and revised regulations. International trade in edible insects would almost certainly increase as well (DeFoliart, 1992).

In doing this, we should try to understand the cultural restrictions addressed to the consumption of insects, and all the biased views people have about these creatures, at least in western world. If the public's positive perception about insects as a mainstream edible food source is our intended ending, knowledge about why people choose the foods they do is essential while simultaneously addressing the specific processes involved. A multidisciplinary approach is needed to apprehend the substantial potential benefits of human consumption of insects on a global level. Engineers are needed to develop suitable rearing systems for different environments and insects. Food scientists are needed to study the nature of insect foods and nutritional content, the causes of deterioration of insect food products, the principles underlying insect food processing, and the improvement of insect foods. Family and consumer scientists are needed to address the relationship between individuals, families, and communities in relation to food habits and food choices. Nutritionists are needed to advise consumers on insect foods and the nutritional impacts of insect foods on human health. Marketing, promotion, and advertising specialists are needed to promote the nutritional benefits of edible insects while marketing and selling insects for human consumption.

In order to realize their potential as a major source of human food, there are several constraints that must be overcome. These include production cost and efficiency, commercialization, technology, regulation, and social change regarding food habits and food choices. The need for development of multiple-product food–insect systems is imperative (Gahukar, 2011). Limited research and government funding are focused on mass-rearing or mass-processing methodologies of insects for human consumption.

However, several small companies (in the United States, Europe, and other areas of the world) are beginning to produce food products containing insects as well as develop methods of processing insects for use as safe, efficient food ingredients (see also Chapter 5 of this book for more on the emerging insect based food industry).

Westerners will likely need to be exposed to entomophagy in informal environments such as festivals, fairs, museums, nature centers, parks, and restaurants as well as formal teaching and research environments in order to be more accustomed to the concept of accepting insects as food. DeFoliart (1999) emphasized that Westerners should become more aware of their negative impact on the global natural environment and their bias against consuming insects, and should increase their acceptance of insects as an alternative food source. In order to achieve greater recognition of insects as a viable alternative food source, a more positive social attitude about insects as human food must occur. With an increasing human population and environmental degradation, many people face a major problem in obtaining adequate protein levels in their diet.

Westernized societies are reluctant to use insects, despite being major consumers of other animal proteins. We now need to consider insects as a source of food for humans in a manner that acknowledges both the role of entomophagy in indigenous societies and the need for Westernized societies to diminish the size of their environmental footprint with regard to food production, in part by replacing vertebrate livestock with insects wherever and whenever possible.

There is a need to eradicate or greatly reduce the Western-driven stigma over the use of insects as food. This will help to provide increased opportunities for research on large- and small-scale mass production as well as optimization of ecological benefits and the nutritional benefits of insects. In our global society, entomophagy must play a role in decision making and policies related to agriculture, nutrition, and food security. In order to realize the potential benefits that insects can offer to our food security as a human food and as animal feed, Shockley and Dossey (2014) call for the following: (1) greater support and attention from government funding, agricultural, and regulatory agencies for research on insect production and use as a human food ingredient; (2) support from industry to provide the means to move insect-based foods and other products from the laboratory to the market; and (3) the establishment of a formal international society, an industry association, and a journal for researchers, academics, industry partners, and other practitioners in the field of food and feed insect production. These will help tremendously in moving this emerging field forward. A formal international society would promote and present how entomophagy research could be beneficial to science, society, and industry and, thus, set the stage for what might be one of the most substantial revolutions in modern agriculture and food production: the human utilization of insects for food.

REFERENCES

Adamolekun, B., McCandless, D.W., Butterworth, R.F., 1997. Epidemic of seasonal ataxia in Nigeria following ingestion of the African silkworm *Anaphe venata*: role of thiamine deficiency? Metab. Brain Dis. 12 (4), 251–258.

Agea, J.G., Biryomumaisho, D., Buyinza, M., Nabanoga, G.N., 2008. Commercialization of *Ruspolia nitidula* (Nsenene grasshoppers) in Central Uganda. Afr. J. Food Agric. Devel. 8 (3), 319–332.

Akinnawo, O.O., Abatan, M.O., Ketiku, A.O., 2002. Toxicological study on the edible larva of *Cirina forda* (Westwood). Afr. J. Biomed. Res. 5, 43–46.

Akinnawo, O.O., Taiwo, V.O., Ketiku, A.O., Ogunbiyi, J.O., 2005. Weight changes and organ pathology in rats given edible larvae of *Cirina forda* (Westwood). Afr. J. Biomed. Res. 8, 35–39.

Alexiades, M.N., 1997. Ethnobotanical research and the fieldwork enterprise: some thoughts from the field. Am. J. Bot. 84 (Suppl. 6), 111–112.

Allport, S., 2000. The Primal Feast: Food, Sex, Foraging, and Love. Harmony Books, New York, NY.

Álvarez, M., 2002. La cocina como patrimonio (in)tangible. In: La Cocina como Patrimonio (in)tangible. Vol. 6. Comisión para la Preservación del Patrimonio Histórico Cultural de la Ciudad de Buenos Aires, Buenos Aires, pp. 11–25.

Ancona, L.H., 1933. Los jumiles de Cuautla *Euschistus sopilotensis*. Ann. Ins. Biol. Mexico 4, 103–108.

Andary, C., Motte-Florac, E., Ramos-Elorduy, J., Privat, A., 1996. Chemical screening: updated methodology applied to medicinal insects. In: Abstracts of the Third European Colloquium on Ethnopharmacology, and First International Conference of Anthropology and History of Health and Disease. Erga Edizione, Genes.

Arana, F., 2007. Insectos Comestibles: Entre el Gusto y la Aversión. UNAM, Mexico, DF.

Ash, C., Jasny, B.R., Malakoff, D.A., Sugden, A.M., 2010. Feeding the future. Science 327, 797.

Ayieko, M.A., Obonyo, G.O., Odhiambo, J.A., Ogweno, P.L., Achacha, J., Anyango, J., 2011. Constructing and using a light trap harvester: rural technology for mass collection of agoro termites (*Macrotermes subhylanus*). Res. J. Appl. Sci. Engin. Technol. 3 (2), 105–109.

Banjo, A.D., Lawal, O.A., Owolana, O.A., Olubanjo, O.A., Ashidi, J.S., Dedeke, G.A., Soewu, D.A., Owa, S.O., Sobowale, O.A., 2003. An ethno-zoological survey of insects and their allies among the Remos (Ogun State) south western Nigeria. Indilinga Afr. J. Indigenous Know. Sys. 2 (1), 61–68.

Banjo, A.D., Lawal, O.A., Olubanjo, O.A., Owolana, A.O., 2004. Ethno-zoological knowledge and perception of the value of insects among the Ijebus (south western Nigeria). Glob. J. Pure Appl. Sci. 10 (1), 1–6.

Barroso, F.G., de Haro, C., Sanchez-Muros, M., Venegas, E., Martinez-Sanchez, A., Perez-Banon, C., 2014. The potential of various insect species for use as food for fish. Aquaculture 422–423, 193–201.

Barth, F., 2002. An anthropology of knowledge. Curr. Anthropol. 43 (1), 1–18.

Bellés, X., 1997. Los insectos y el hombre prehistoric. Bull. Entomol. Soc. Aragonesa 20, 319–325.

Benhalima, S., Dakki, M., Mouna, M., 2003. Les insectes dans le Coran et dans la société islamique (Maroc). In: Motte-Florac, É., Thomas, J. (Eds.), Les "Insectes" dans la Tradition orale. Peeters, Paris, France, pp. 533–540.

Bentley, J.W., Rodríguez, G., 2001. Honduran folk entomology. Curr. Anthropol. 42 (2), 285–301.

Bequaert, J., 1921. Insects as food: how they have augmented the food supply of mankind in early and recent times. J. Am. Museum Nat. Hist. 21, 191–200.

Berenbaum, M.R., 1995. Bugs in the System: Insects and their Impact on Human Affairs. Perseus Books, Cambridge, MA.

Bergier, E., 1941. Peuples Entomophages et Insectes Comestibles: Étude sur les Moeurs de l'Homme et de l'Insecte. Imprimerie Rullière Frères, Avignon, France.

Berrios, G.E., 1982. Tactile hallucinations: conceptual and historical aspects. J. Neurol. Neurosur. Ps. 45, 285–293.

Blake, E.A., Wagner, M.R., 1987. Collection and consumption of Pandora moth, *Coloradia pandora lindseyi* (Lepidoptera: Saturniidae), larvae by Owens Valley and Mono Lake Paiutes. Bull. Entomol. Soc. Am. 33, 23–27.

Blum, M.S., 1994. The limits of entomophagy: a discretionary gourmand in a world of toxic insects. Food Insects Newslett. 7 (1), 6–11.

Bodenheimer, F.S., 1951. Insects as Human Food. W. Junk Publishers, The Hague, Netherlands.

Brinchmann, B.C., Bayat, M., Brogger, T., Muttuvelu, D.V., Tjonneland, A., Sigsgaard, T., 2011. A possible role of chitin in the pathogenesis of asthma and allergy. Ann. Agr. Env. Med. 18, 7–12.

Bristowe, W.S., 1932. Insects and other invertebrates for human consumption in Siam. Trans. Entomol. Soc. London 80, 387–404.

Brothwell, D., Brothwell, P., 1969. Food in Antiquity: A Survey of the Diet of Early Peoples. Frederick A. Praeger, New York, NY.

Buzzi, Z.J., Miyazaki, R.D., 1993. Entomologia Didática. Editora da UFPR, Curitiba.

Cajete, G.C., 1999. Igniting the Sparkle: An Indigenous Science Education Model. Kivaki Press, Skyand, NC.

Cerritos, R., Cano Santana, Z., 2008. Harvesting grasshoppers *Sphenarium purpurascens* in Mexico for human consumption: a comparison with insecticidal control for managing pest outbreaks. Crop Prod. 27, 473–480.

Chaikin, E., 2014. Insects to Feed the World. Available from: http://www.fao.org/forestry/edibleinsects/84962/en/.

Chen, P.P., Wongsiri, S., Jamyanya, T., Rinderer, T.E., Vongsamanode, S., Matsuka, M., Sylverter, H.H., Oldroyd, B.P., 1998. Honey bees and other edible insects used as human food in Thailand. Am. Entomol. 41 (1), 24–29.

Chidumayo, E.N., Mbata, K.J., 2002. Shifting cultivation, edible caterpillars and livelihoods in the Kopa area of northern Zambia. Forests Trees Livelihoods 12, 175–193.

Clausen, L.W., 1971. Insect Fact and Folklore, ninth ed. The Macmillan Company, New York, NY.

Conconi, J.R.E., 1984. Los insectos como un recurso actual y potencial. In: Memórias del Seminário sobre Alimentación en México. Instituto de Geografía de la UNAM, México, D.F., pp. 126–139.

Connor, W.E., 2000. Importance of n-3 fatty acids in health and disease. Am. J. Clin. Nutr. (1 Suppl.), 171S–175S.

Conway, J.R., 1994. Honey ants. Am. Entomol. 40 (4), 229–234.

Corson, T., 2008. The Story of Sushi: An Unlikely Saga of Raw Fish and Rice. Harper Perennial, New York, NY.

Costa Neto, E.M., 2002. Manual de Etnoentomología. Manuales & Tesis SEA, 4. Sociedad Entomológica Aragonesa, Aragón.

Costa Neto, E.M., 2003. Insetos como fontes de proteínas para o homem: valora T de recursos considerados repugnantes. Interciência 28, 136–140.

Costa Neto, E.M., 2004. Insetos como recursos alimentares nativos no semi-árido do estado da Bahia, nordeste do Brasil. Zonas Áridas 8, 33–40.

Costa Neto, E.M., 2005. Entomotherapy, or the medicinal use of insects. J. Ethnobiol. 25 (1), 93–114.

Costa Neto, E.M., 2011. Antropoentomofagia: sobre o consumo de insetos. In: Costa Neto, E.M. (Ed.), Antropoentomofagia: Insetos na Alimentação Humana. UEFS Editora, Feira de Santana, pp. 17–37.

Costa Neto, E.M. (Org.), 2014. Entomologia Cultural: Ecos do I Simpósio Brasileiro de Entomologia Cultural 2013. UEFS Editora, Feira de Santana.

Costa Neto, E.M., Ramos-Elorduy, J., 2006. Los insectos comestibles de Brasil: etnicidad, diversidad e importancia en la alimentación. Bull. Entomol. Soc. Aragonesa 38, 423–442.

Crabbe, N., May 10, 2012. Local expert gets funding to develop insect-based food for starving children. Gainesville Sun, 1B–6A.

Dams, L.R., 1978. Bees and honey-hunting scenes in the Mesolithic rock art of eastern Spain. Bee World 59 (2), 45–53.

Davey, G.C.L., 1993. Factors influencing self-rated fear to a novel animal. Cognition Emotion 7 (5), 461–471.

Davey, G.C.L., 1994. The "disgusting" spider: the role of disease and illness in the perpetuation of fear of spiders. Soc. Anim. 2 (1), 17–25.

DeFoliart, G.R., 1989. The human use of insects as food and as animal feed. Bull. Entomol. Soc. Am. 35 (1), 22–35.

DeFoliart, G.R., 1992. Insects as human food. Crop Prot. 11, 395–399.

DeFoliart, G.R., 1995. Edible insects as minilivestock. Biodivers. Conscrv. 4, 306–321.

DeFoliart, G.R., 1997. An overview of the role of edible insects in preserving biodiversity. Ecol. Food Nutri. 36, 109–132.

DeFoliart, G.R., 1999. Insects as food: why the western attitude is important. Ann. Rev. Entomology 44, 21–50.

DeFoliart, G.R., Dunkel, F.V., Gracer, D., 2009. The Food Insects Newsletter: Chronicle of a Changing Culture. Aardvark Global Publishing Company, Salt Lake City, UT.

Descola, P., 1998. Estrutura ou sentimento: a relação com o animal na Amazônia. Mana 4 (1), 23–45.

Dessa, M.A., 2008. Casu Marzu: la rivincita di un grande formaggio Premiata Salumeria Italiana. Available from: http://www.pubblicitaitalia.com/cocoon/pubit/riviste/articolo.html?Testata=2&idArticolo=8001.

Diamond, J., 2005. Guns, Germs, and Steel: A Short History of Everybody for the Last 13,000 years. UK, Vintage.

Diegues, A.C., Arruda, R.S.V., 2001. Saberes Tradicionais e Biodiversidade no Brasil. Ministério do Meio Ambiente, Brasília; USP, São Paulo (Biodiversidade, 4).

Diodorus Siculus, 1814. Historical Library in Fifteen Books. (G. Booth, Trans.), 2 vols. London, UK.

Dossey, A.T., 2010. Insects and their chemical weaponry: new potential for drug discovery. Nat. Product Rep. 27 (12), 1737–1757.

Dossey, A.T., 2011. Chemical defenses of insects: a rich resource for chemical biology in the Tropics. In: Vivanco, J., Weir, T. (Eds.), Chemical Biology of the Tropics: An Interdisciplinary Approach. Springer, New York, NY, pp. 27–57.

Dossey, A.T., 2013. Why insects should be in your diet. Scientist 27, 22–23.

Duffels, J.P., van Mastrigt, J.G., 1991. Recognition of cicadas (Homoptera, Cicadidae) by the Ekagi people of Irian Jaya (Insonesia), with a description of a new species of *Cosmopsaltria*. J. Nat. Hist. 25, 173–182.

Dufour, D.L., 1987. Insects as food: a case study from the Northwest Amazon. Am. Anthropol. 89, 383–397.

Dufour, D.L., 1990. Insects as food: aboriginal entomophagy in the Great-Basin. Am. Anthropol. 92, 214–215.

Dunkel, F.V., 1997. Food insect festivals of North America. Food Insects Newslett. 10 (3), 1–7.

Dunkel, F.V., 1998. Chronicle of a changing culture: the food insect newsletter in its second decade. Food Insects Newslett. 11 (3), 1–3.

Dunkel, F.V., 2000. The Issues of Insects and Human Societies: Course Notes and Laboratory Manual. Montana State University, Bozeman, MT.

Dunkel, F.V., 2012. Eat less meat, more insects. TEDx Bozeman MT. Available from: http://www.proteinsect.eu/index.php?id=37.
Durst, P.B., Johnson, D.V., Leslie, R.N., Shono, K., 2010. Edible Forest Insects: Humans Bite Back. Proceedings of a Workshop on Asia-Pacific Resources and their Potential for Development. FAO Regional Office Bangkok, Chiang Mai, Thailand.
Dyson-Hudson, R., Smith, E.A., 1978. Human territoriality: an ecological reassessment. Am. Anthropol. 80 (1), 21–41.
Dzamba, J., 2010. Third Millennium Farming. Is it Time for Another Farming Revolution? Architecture, Landscape and Design, Toronto, CA. Available from: http://www.thirdmillenniumfarming.com/.
Elias, S., 2010. The use of insect fossils in archaeology. Adv. Quaternary Entomol. 12, 89–121.
Essig, E.O., 1934. The value of insects to the California Indians. Sci. Mon. 38, 181–186.
European Commission, 2013. Agriculture and Rural Development. Agriculture and food. (DOOR). Denomination Information.Application Type: PDO; Type of product: Class 1.3. Cheeses. Dossier Number IT/PDO/0217/01127 Ec.europa.eu.
FAO UN, 2008. In: Durst, P.B., Johnson, D.V., Leslie, R.N., Shono, K., (Eds.), Forest Insects as Food: Humans Bite Back. Regional Office for Asia and the Pacific, Chiang Mai, Thailand.
FAO. 2012. Composition database for biodiversity version 2, BioFoodComp2. (Latest update: 10 January 2013). Available from: www.fao.org/infoods/infoods/tables-and-databases/en/.
Fasoranti, J.O., Ajiboye, D.O., 1993. Some edible insects of Kwara State, Nigeria. Am. Entomol. 39 (2), 113–116.
Fenenga, G.L., Fisher, E.M., 1978. The Cahuilla use of piyatem, larvae of the white-lined sphinx moth (*Hyles lineata*) as food. J. CA Anthropol. 5 (1), 84–90.
Figueroa Rodriguez, K.A., 2001. Estudio de comercializacion en la ciudad de Oaxaca, Oaxaca y la comunidad de Santa Maria Zacatepec, Pueblo; de chapulin (*Sphenarium* spp.) industrializado. Thesis presented in partial fulfillment of the M.S. degree at Instituto de Socioeconomia Estadistica e Informatica, Colegio de Postgraduados, Montecillo, Texcoco, Mexico.
Fritzsche, I., Gitsaga, B., 2002. Das Insektenkochbuch: der etwas andere Geschmack. Natur und Tier, Verlag GmbH, Munich, Germany.
Gahukar, R.T., 2011. Entomophagy and human food security. Int. J. Trop. Insect Sci. 31, 129–144.
Goodman, W.C., 1989. Chitin: a magic bullet? Food Insects Newslett. 2 (3), 6–7.
Gordon, D.G., 1998. The Eat-a-Bug Cookbook. Ten Speed Press, Berkeley, CA.
Gordon, D.G., 2013. The Eat-a-Bug Cookbook, second ed. Ten Speed Press, Berkeley, CA.
Gullan, P.J., Cranston, P.S., 2005. The Insects: An Outline of Entomology, fifth ed. Chapman and Hall, London, UK.
Gurung, A.B., 2003. Insectas–a mistake in God's creation? Tharu farmer's perception and knowledge on insects: a case study of Gobardiha Village Development Committee, Dang-Deukhuri, Nepal. Agr. Hum. Val. 20, 337–370.
Hanboonsong, Y. 2010. Edible insects and associated food habits in Thailand. In: Durst, P.B., Johnson, D.V., Leslie, R.N., Shono, K., (Ed.), Edible Forest Insects: Humans Bite Back. Proceedings of a Workshop on Asia-Pacific Resources and their Potential for Development. FAO Regional Office Bangkok, Chiang Mai, Thailand. pp. 173–182.
Hanboonsong, Y., Jamjanya, T., Durst, P.B., 2013. Six-legged livestock: edible insect farming, collection and marketing in Thailand. RAP Publication 2013/03. FAO regional office for Asia and the Pacific.
Hanping, L., Changzhen, Z., 1993. Preliminary studies on housefly (*Musca domestica vicina* Macq.) larval protein production. I. Modeling for factors influencing fly oviposition. J. Huazhong Agri. Univ. 12, 231–236.
Harpaz, I., 1973. Early entomology in the Middle East. In: Smith, R.F., Mittler, T.E., Smith, C.N. (Eds.), History of Entomology. Annual Reviews, Inc., CA, pp. 243–278.
Harris, M., 1999. Bueno para Comer: Enigmas de Alimentación y Cultura. Alianza, Madrid, Spain.
Hernández-Pacheco, E., 1921. Escena pictórica con representaciones de insectos de época paleolítica. Memoria de la Real Sociedad de Historia Natural, tomo del cincuentenario, Madrid, Spain, pp. 66–67.
Herodotus, 2008. The Histories. Oxford University Press, Oxford.
Hogue, C.L., 1987. Cultural entomology. Ann. Rev. Entomol. 32, 181–199.
Hölldobler, B., Wilson, E.O., 1994. Journey to the Ants: A Story of Scientific Exploration. Harvard University Press, Cambridge, MA.
Holt, V.M., 1885. Why Not Eat Insects? Field & Tuer, London, UK.
Howard, L.O., 1916. *Lachnosterna* larvae as a possible food supply. J. Econ. Entomol. 9, 390–392.
Hoyt, E., Schultz, T., 1999. Insect Lives: Stories of Mystery and Romance from a Hidden World. John Wiley & Sons, New York, NY.
Idowu, A.B., Modder, W.W.D., 1996. Possible control of the stinking grasshopper, *Zonocerus variegatus* (L) (Orthoptera: Pyrgomorphidae) in Ondo State, through human consumption. The Nigerian Field 61, 7–14.
ncekara, Ü., Türkez, H., 2009. The genotoxic effects of some edible insects on human whole blood cultures. Munis Entomol. Zool. 4 (2), 531–535.
Iroko, A.F., 1996. L'Homme et les Termitières en Afrique. Karthala, Paris, France.
Isaacs, J., 2002. Bush Food: Aboriginal Food and Herbal Medicine. New Holland Publishers (Australia), Frenchs Forest, New South Wales.
Isman, M.B., Cohen, M.S., 1995. Koshner insects. Am. Entomol. 41 (2), 100–102.
Jara, F., 1996. La miel y el aguijón. Taxonomía zoológica y etnobiología como elementos en la definición de las nociones de género entre los Andoke (Amazonia colombiana). J. Soc. Am. 82, 209–258.
Johnson, D.V., 2010. The contribution of edible forest insects to human nutrition and to forest management. In: Durst, P.B., Johnson, D.V., Leslie, R.N., Shono, K. (Eds.), Forest Insects as Food: Humans Bite Back. FAO Regional Office Bangkok, Chiang Mai/Thailand, pp. 5–22.
Jongema, Y., 2015. List of Edible Insects of the World. Wageningen University, Wageningen, The Netherlands. Available from: http://www.ent.wur.nl/UK/Edible+insects/Worldwide+species+list/.
Karras, D.J., Farrell, S.E., Harrigan, R.A., Henretig, F.M., Gealt, L., 1996. Poisoning from "Spanish fly" (cantharidin). Am. J. Emerg. Med. 14 (5), 478–483.

Katayama, N., Ishikawa, Y., Takaoki, M., Yamashita, M., Nakayama, S., Kiguchi, K., Kok, R., Wada, H., Mitsuhashi, J., Force, S.A.T., 2008. Entomophagy: a key to space agriculture. Adv. Space Res. 41, 701–705.

Katz, E., 2011. Prefácio. In: Costa Neto, E.M. (Ed.), Antropoentomofagia: Insetos na Alimentação Humana. UEFS Editora, Feira de Santana, Brazil, pp. 9–16.

Kelhoffer, J., 2004. Did John the Baptist eat like a former Essene? locust-eating in the ancient Near East and at Qumran. Dead Sea Discoveries 11, 293–314.

Kellert, S.R., 1993. Values and perceptions of invertebrates. Conserv. Bio. 7 (4), 845–853.

Kinyuru, J.N., et al., 2012. Identification of traditional foods with public health potentional for complementary feeding in Western Kenya. J. Food Res. 1 (2), 148–158.

Kozanayi, W., Frost, P., 2002. Marketing of the Mopane Worm in Southern Zimbabwe. Institute of Environmental Studies, Harare, Zimbabwe.

Kritsky, G., 1997. The insects and other arthropods of the Bible, the new revised version. Am. Entomol. 43 (3), 183–188.

Kunin, W.E., Lawton, J.H., 1996. Does biodiversity matters? Evaluating the case for conserving species. In: Gaston, K.J. (Ed.), Biodiversity: A Biology of Numbers and Differences. Blackwell Science, Oxford, UK, pp. 283–308.

Lauck, J.E., 2002. The Voice of the Infinite in the Small: Re-visioning the Insect-Human Connection. Shambhala Publications, Boston, MA.

Lee, K.P., Simpson, S.J., Wilson, K., 2008. Dietary protein-quality influences melanization and immune function in an insect. Funct. Ecol. 22 (6), 1052–1061.

León-Portilla, M., 2002. Alimentación de los antiguos mexicanos. In: Alarcón, S.D., Bourges, R.H. (Eds.), La Alimentación de los Mexicanos. Colegio Nacional, México, D.F, pp. 13–24.

Lewin, K., 1943. Forces behind Food Habits and Methods of Change: The Problem of Changing Food Habits, Report of the Committee on Food Habits. National Research Council, National Academy of Sciences, Washington, DC. Bulletin No. 108.

Linassi, R., Borghetti, B., 2011. Antropoentomofagia: um estudo sobre as potencialidades dos insetos como alimento no Brasil. In: Costa Neto, E.M. (Ed.), Antropoentomofagia: Insetos na Alimentação Humana. UEFS Editora, Feira de Santana, Brazil, pp. 55–75.

Lizama, J.C., 2004. Entomofagía: Alimentación con Insectos. El Nibelungo, Madrid, Spain.

Long, T.J., 2002. La riqueza culinaria del altiplano. In: Alarcón, S.D., Bourges, R.H. (Eds.), La Alimentación de los Mexicanos. Colegio Nacional, México, D.F, pp. 79–96.

Looy, H., Wood, J.R., 2006. Attitudes toward invertebrates: are educational "bug banquets" effective? J. Environ. Educ. 37 (2), 37–48.

Looy, H., Dunkel, F.V., Wood, J.R., 2013. How then shall we eat? Insect-eating attitudes and sustainable foodways. Agriculture and Human Values. Available from: http://link.springer.com/article/10.1007%2Fs10460-013-9450-x.

Maheu, E., 2011. Onívoros? Limitações e possibilidades do comestível e do palatável diante das fronteiras culturais: o caso dos insetos. In: Costa Neto, E.M. (Ed.), Antropoentomofagia: Insetos na Alimentação Humana. UEFS Editora, Feira de Santana, Brazil, pp. 39–54.

Makkar, H.P.S., Tran, G., Heuze, V., Ankers, P., 2014. State-of-the-art on use of insects as animal feed. Anim. Feed Sci. Tech. 197, 1–33.

Malaisse, F., 1997. Se Nourir en Foret Claire Africaine: Approche Ecologique et Nutritionnelle. Les Presses Agronomiques de Gembloux, Gembloux, Belgium.

Martin, D., 2014. Edible: An Adventure into the World of Eating Insects and the Last Great Hope to Save the Planet. Houghton Mifflin Harcourt, New York, NY.

Martinez, G., 2000. La Cuisine des Insectes: À la Découverte de l'Entomophagie. Jean-Paul Rocher, Paris, France.

Matchett, G., Davey, G.C.L., 1991. A test of a disease-avoidance model of animal phobias. Behav. Res. Ther. 29 (1), 91–94.

Mbata, K.J., Chidumayo, E.N., Lwatula, C.M., 2002. Traditional regulation of edible caterpillar exploitation in the Kopa area of Mpika district in northern Zambia. J. Insect Conserv. 6, 115–130.

Menzel, P., D'Aluisio, F., 1998. Man Eating Bugs: The Art and Science of Eating Insects. Ten Speed Press, Berkeley, CA.

Meyer-Rochow, V.B., 2004. Traditional food insects and spiders in several ethnic groups of northeast India, Papua New Guinea, Australia, and New Zealand. In: Paoletti, M.G. (Ed.), Ecological Implications of Minilivestock: Rodents, Frogs, Snails, and Insects for Sustainable Development. Science Publ., Inc., Boca Raton, FL, pp. 385–409.

Meyer-Rochow, V.B., Chakravorty, J., 2013. Notes on entomophagy and entomotherapy generally and information on the situation in India in particular. Appl. Entomol. Zool. 48, 105–112.

Miller, C.A., 1997. Determinants of the use of insects as human food within the Great Basin. Food Insects Newslett. 10 (1), 1–4.

Mitsuashi, J., 2010. The future use of insects as human food. In: Durst, P.B., Johnson, D.V., Leslie, R.N., Shono, K., (Eds.), Edible Forest Insects: Humans Bite Back. Proceedings of a Workshop on Asia-Pacific Resources and their Potential for Development. FAO Regional Office Bangkok, Chiang Mai, Thailand, pp. 115–122.

Morgan, R.C., 1991. Natural history, field collection and captive management of the Honey ant *Myrmecocystus mexicanus*. Int. Zoo Yearbook 30, 108–117.

Mormino, V., 2012. Evil weevils attack Sicily! Best of Sicily. Available from: www.bestofsicily.com/mag/art325.htm.

Munthali, S.M., Mughogho, D.E.C., 1992. Economic incentives for conservation: bee-keeping and Saturniidae caterpillar utilization by rural communities. Biod. Cons. 1, 143–154.

Myers, N., 1982. Homo insectivorus. Sci. Dig. 90 (5), 14–15.

Nonaka, K., 2009. Feasting on insects. Entomol. Res. 39, 304–312.

Oldfield, M.L., 1989. The Value of Conserving Genetic Resources. National Park Service, Washington, DC.

Oonincx, D.G.A.B., 2015. Insects as Food and Feed: Nutrient Composition and Environmental Impact. Wageningen University, Wageningen, The Netherlands.

Owen, D.F., 1973. Man's Environmental Predicament. An Introduction to Human Ecology in Tropical Africa. Oxford University Press, London, UK.

Packard, Jr., A.S., 1877. Half Hours with Insects. Estes and Lauriat, Boston, MA.

Pager, H., 1973. Rock paintings in southern Africa showing bees and honey hunting. Bee World 54, 61–68.
Pager, H., 1976. Cave paintings suggest honey hunting activities in ice age times. Bee World 57 (1), 9–14.
Pal, P., Roy, S., 2014. Edible insects: future of human food: a review. Int. Lett. Nat. Sci. 21, 1–11.
Pemberton, R.W., 1999. Insects and other arthropods used as drugs in Korean traditional medicine. J. Ethnopharmacol. 65, 207–216.
Pemberton, R.W., 2005. Contemporary use of insects and other arthropods in traditional Korean medicine (*hanbang*) in South Korea and elsewhere. In: Paoletti, M.G. (Ed.), Ecological Implications of Minilivestock: Potential of Insects, Rodents, Frogs and Snails. Science Publishers, Enfield, NH, pp. 459–471.
Pérez, R.M., Conconi, J.R.E., Yesgas, G., Muñoz, J.L., 1989. Aislamiento de fenoxazina apartir del insecto *Pachilis gigas* L. (Insecta-Hemiptera-Coreidae). Mexican Sci. Technol. Act 6 (21-24), 71–73.
Peterson, C., 2012. Edible Bugs: Insects on Our Plates. Amazon Digital Services, Inc; New York, 23 pp.
Pino Moreno, J.M., Ramos-Elorduy, J., 2002. Pragmatic uses of cochineal (Homoptera Dactylopiidae). In: Stepp, J.R., Wyndham, F.S., Zarger, R.K. (Eds.), Ethnobiology and Biocultural Diversity: Proceedings of the Seventh International Congress of Ethnobiology, Athens, Georgia, USA, October 2000, pp. 353–364.
Posey, D.A., 1976. Entomological considerations in southeastern aboriginal demography. Ethnohistory 23 (2), 147–160.
Posey, D.A., 1982. Indigenous knowledge and development: an ideological bridge to the future. Sci. Cult. 35 (7), 877–894.
Posey, D.A., 1986. Etnobiologia de tribos indígenas da Amazônia. In: Ribeiro, D. (Ed.), Suma Etnológica Brasileira. Etnobiologia. Vozes/Finep, Petrópolis, pp. 251–272.
Posey, D.A., 1987. Temas e inquirições em etnoentomologia: algumas sugestões quanto à geração de hipóteses. Newslett. Paraense Emilio Goeldi Museum 3 (2), 99–134.
Premalatha, M., Abbasi, T., Abbasi, T., Abbasi, S.A., 2011. Energy-efficient food production to reduce global warming and ecodegradation: the use of edible insects. Renew. Sustain. Energ. Rev. 15, 4357–4360.
Raksakantong, P., Meeso, N., Kubala, J., Siriamornpun, S., 2010. Fatty acids and proximate composition of eight Thai edible terricolous insects. Food Res. Int. 43, 350–355.
Ramaswamy, S.B., 2015. Setting the table for a hotter, flatter, more crowded earth: insects on the menu? J. Insects Food Feed 1 (3), 171–178.
Ramos-Elorduy, J., 1982. Los Insectos como Fuente de Proteína en el Futuro, first ed. Limusa, México, D.F.
Ramos-Elorduy, J., 1996. Insect consumption as a mean of national identity. In: Jain, S.K. (Ed.), Ethnobiology in Human Welfare. Deep, New Delhi, India, pp. 9–12.
Ramos-Elorduy, J., 1997. Insects: a sustainable source of food? Ecol. Food Nutri. 36, 247–276.
Ramos-Elorduy, J., 2004. La etnoentomología en la alimentación, la medicina y el reciclaje. In: Llorente, J.B., Morrone, J., Yañez, O.O., Vargas, I.F. (Eds.), Biodiversidad, Taxonomía y Biogeografía de Artrópodos de México: Hacia una Síntesis de su Conocimiento. Vol. 4. UNAM, México, D.F., pp. 329–413.
Ramos-Elorduy, J., 2005. Insects: a hopeful food source. In: Paoletti, M.G. (Ed.), Ecological Implications of Minilivestock: Potential of Insects, Rodents, Frogs and Snails. Science Publishers, Inc., New Hampshire, pp. 263–292.
Ramos-Elorduy, J., 2006. Threatened edible insects in Hidalgo, Mexico and some measures to preserve them. J. Ethnobiol. Ethnomed. 2, 51.
Ramos-Elorduy, J., 2007. Evolución de la antropoentomofagia. In: Navarrete-Heredia, J.L., Quiroz-Rocha, G.A., Fierros-López, H.E. (Coords.), Entomología Cultural: Una Visión, Iberoamericana, Universidad de Guadalajara, Guadalajara, pp. 285–307.
Ramos-Elorduy, J., 2009. Anthropo-entomophagy: cultures, evolution and sustainability. Entomol. Res. 39, 271–288.
Ramos-Elorduy, J., 2011. Evolución de la comercialización de insectos comestibles. In: Costa Neto, E.M. (Ed.), Antropoentomofagia: Insetos na Alimentação Humana. UEFS Editora, Feira de Santana, Brazil, pp. 103–122.
Ramos-Elorduy, J., Bourges, R.H., Pino, J.M., 1982. Cambios del valor nutritivo de Liometopum apiculatum y Liometopum occidentale var. luctuosum W. en los estados inmaduros de la casta obrera y de la reproductora. Fol. Ent. Mex. 54, 118–120.
Ramos-Elorduy, J., Menzel, P., 1998. Creepy Crawly Cuisine: The Gourmet Guide to Edible Insects. Park Street Press, Rochester, VT.
Ramos-Elorduy, J., Pino Moreno, J.M., 1989. Los Insectos Comestibles del México Antiguo. AGT Editor, México, D.F.
Ramos-Elorduy, J., Pino Moreno, J.M., 1996. El consumo de insectos entre los Aztecas. In: Long, J. (Coord.), Conquista y Comida: Consecuencias del Encuentro de dos Mundos. UNAM, México, D.F., pp. 89–101.
Ramos-Elorduy, J., Pino Moreno, J.M., Martínez, V.H., 2007. Historia de la antropoentomofagia. In: Navarrete-Heredia, J.L., Quiroz-Rocha, G.A., Fierros-López, H.E. (Coords.), Entomología Cultural: Una Visión Iberoamericana. Universidad de Guadalajara, Guadalajara, pp. 239–284.
Ramos-Elorduy, J., Landero-Torres, I., Murguía-González, J., Pino Moreno, J.M., 2008. Biodiversidad antropoentomofágica de la región de Zongolica, Veracruz, México. J. Trop. Biol. 56 (1), 303–316.
Rehder, M., Dunkel, F.V., 2009. For Mexico: Insects, food of the future? In: DeFoliart, G., Dunkel, F.V., Gracer, D. (Eds.), The Food Insects Newsletter. Chronicle of a Changing Culture. Aardvark Global Publishing Company, Salt Lake, pp. 369–370.
Ribeiro, B.G., Kenhíri, T., 1987. Calendário econômico dos índios Desâna. Sci. Today 6 (36), 26–35.
Rozin, P., 1995. Des goûts et des dégoûts. In: Bessis, S. (Ed.), Mille et une Bouches: Cuisines et Identités Culturelles. Autrement, Paris, France, pp. 96–105.
Rozin, P., Fallon, A.E., 1987. A perspective on disgust. Psychol. Rev. 94, 23–41.
Ruiz, D.C.A., Castro, A.E.R., 2000. Maya ethnoentomology of X-Hazil Sur y anexos, Quintana Roo, Mexico. In: Abstracts of the Seventh International Congress of Ethnobiology. University of Georgia, Athens, GA.

Rumpold, B.A., Schluter, O.K., 2013. Potential and challenges of insects as an innovative source for food and feed production. Innov. Food Sci. Emerg. 17, 1–11.

Sanchez-Muros, M., Barroso, F.G., Manzano-Agugliaro, F., 2014. Insect meal as a renewal source of food for animal feeding: a review. J. Cleaner Prod. 65, 16–27.

Santos, R.C.V., 2004. Estudos in vivo e in vitro da atividade antiinflamatória de *Ulomoides dermestoides* (Fairmaire, 1893). Dissertação (mestrado em Zoologia), Pontifícia Universidade Católica do Rio Grande do Sul, Porto Alegre, Brazil.

Santos, N.M.M., 2014. In: Costa Neto, E.M. (Ed.), Antropoentomofagia: Insetos na Alimentação Humana, second ed.. UEFS Editora, Feira de Santana Brazil, pp. 123–141.

Seligman, M.E.P., 1971. Phobias and preparedness. Behav. Ther. 2 (3), 307–320.

Serra, O., 2001. As atribuições teórico-práticas do etnocientista. In: Costa Neto, E.M., Souto, F.J.B. (Orgs.), Anais do I Encontro Baiano de Etnobiologia e Etnoecologia 1999. UEFS, Feira de Santana Brazil, pp. 55–59.

Shockley, M., Dossey, A.T., 2014. Insects for human consumption. In: Morales-Ramos, J., Rojas, G., Shapiro-Ilan, D.I. (Eds.), Mass Production of Beneficial Organisms. Invertebrates and Entomopathogens. Academic Press, New York, NY, pp. 617–652.

Smith, A., 2012. Get Ready to Pay More for Your Steak. CNN Money, New York, NY.

Somnasong, P., Moreno-Black, G., Chusil, K., 1998. Indigenous knowledge of wild food hunting and gathering in North-East Thailand. Food Nut. Bull. 19 (4), 359–365.

Spradbery, J.P., 1973. Wasps: An Account of the Biology and Natural History of Solitary and Social Wasps. University of Washington Press, Seattle, WA.

Srivastava, S.K., Babu, N., Pandey, H., 2009. Traditional insect bioprospecting as human food and medicine. Ind. J. Trad. Knowledge 8, 485–494.

Steward, J.H., 1933. Ethnography of the Owens Valley Paiute. University of California Publications in American Archeology and Ethnology 33.

Sunderland, T.C.H., Ndoye, O., Harrison-Sanchez, S., 2011. Non-timber forest products and conservation: what prospects? In: Shackleton, S., Shackleton, C., Shanley, P. (Eds.), Non-timber Forest Products in the Global Context. Springer, Heidelberg, Germany, pp. 209–224.

Sutton, M.Q., 1988. Insect as Food: Aboriginal Entomophagy in the Great Basin. Ballena Press, CA.

Sutton, M.Q, 1990. Insect resources and Plio-Pleistocene hominid evolution. In: Posey, D.A., Overal, W.L., Clement, C.R., Plotkin, M.J., Elisabetsky, E., Mota, C.N., Barros, F. (Eds.). Proceedings of the First International Congress of Ethnobiology, vol. 2. Museu Paraense Emílio Göeldi, Belém, pp. 195–207.

Sutton, M.Q., 1995. Archaeological aspects of insect use. J. Archaeol. Method Th. 2, 253–298.

Taylor, R.L., 1975. Butterflies in My Stomach. Woodbridge Press, Santa Barbara, CA.

Taylor, R.L., Carter, B.J., 1976. Entertaining with Insects. Woodbridge Press, Santa Barbara, CA.

Thémis, J.L., 1997. Des Insectes à Croquer: Guide de Découvertes. Les Éditions de l'Homme, Montreal, Canada.

Toledo, V.M., 1991. El Juego de la Supervivencia: Un Manual para la Investigación Etnoecológica en Latino-América. CLADES, Berkeley, CA.

Tommaseo-Ponzetta, M., 2005. Insects: food for human evolution. In: Paoletti, M.G. (Ed.), Ecological Implications of Minilivestock: Potential of Insects, Rodents, Frogs and Snails. Science Publishers, Enfield, NH, pp. 141–161.

Udvardy, M.D.F., 1975. A Classification of the Biogeographical Provinces of the World. IUCN, Morges, Switzerland.

Valli, E., 1998. Golden harvest of the Raji. Natl. Geogr. 193, 86–103.

van der Merwe, N.J., Thackcray, J.C., Lee-Thorp, J.A., Luyt, J., 2003. The carbon isotope ecology and diet of *Australopithecus africanus* at Sterkfontein, South Africa. J. Hum. Evol. 44, 581–597.

van Huis, A., 2012. Potential of insects as food and feed in assuring food security. Rev. Adv. 58, 563–583.

van Huis, A., 2013. Potential of insects as food and feed in assuring food security. Ann. Rev. Entomol. 58, 563–583.

van Huis, A., Van Gurp, H., Dicke, M., 2012. Het Insectenkookboek. Uitgeverij Atlas, Amsterdam, The Netherlands.

van Huis, A., Itterbeeck, J.V., Klunder, H., Mertens, E., Halloran, A., Muir, G., Vantomme, P., 2013. Food and Agriculture Organization of the United Nations. Edible Insects: Future Prospects for Food and Feed Security, Food and Agriculture Organization of the United Nations, Rome, Italy.

van Huis, A., Van Gurp, H., Dicke, M., 2014. The Insect Cookbook: Food for a Sustainable Planet. Columbia University Press, New York, NY.

Vantomme, P., 2010. Edible forest insects, an overlooked protein supply. Unasylva 61 (236), 19–21.

Vantomme, P., Mertens, E., van Huis, A., Klunder, H., 2012. Assessing the Potential of Insects as Food and Feed in Assuring Food Security. United Nations Food and Agricultural Organization, Rome, Italy.

Verspoor, R., Riggi, L., Veronesi, M., MacFarlane, C., 2013. Bugs for Life 2013: exploring the practices, perceptions and possibilities of edible insects in Northern Benin. Antenna 37 (2), 63–66.

Viveiros de Castro, E., 1992. Araweté: O Povo do Ipixuna. CEDI, São Paulo, Brazil.

Vogel, G., 2010. For more protein, filet of cricket. Science 327, 811–1811.

Ware, J., Kumud, J., Burgess, I., Davey, G.C.L., 1994. Disease-avoidance model: factor analysis of common animal fears. Behav. Res. Ther. 32 (1), 57–63.

Wheeler, W.C., Whiting, M., Wheeler, Q.D., Carpenter, J.M., 2001. The phylogeny of the extant hexapod orders. Cladistics 17, 113–169.

Wicke, C., 1982. Así comían los aztecas. Esplendor del México antiguo 2, 983–994.

Yen, A.L., 2009a. Edible insects: traditional knowledge or western phobia? Entomol. Res. 39, 289–298.

Yen, A.L., 2009b. Entomophagy and insect conservation: some thoughts for digestion. Insect Conserv. 13, 667–670.

Yen, A.L., 2015. Insects as food and feed in the Asia Pacific region: current perspectives and future directions. J. Insects Food Feed 1 (1), 33–55.

Yhoung-Aree, J., Viwatpanich, K., 2005. Edible insects in the Laos PDR, Myanmar, Thailand, and Vietnam. In: Paoletti, M.G. (Ed.), Ecological Implications of Minilivestock: Potential of Insects, Rodents, Frogs and Snails. Science Publishers, Enfield, NH, pp. 415–440.

Yi, C., He, Q., Wang, L., Kuang, R., 2010. The utilization of insectresources in Chinese rural area. J. Agric. Sci. 2 (3), 146–154.

Zagrobelny, M., Dreon, A.L., Gomiero, T., Marcazzan, G.L., Glaring, M.A., Møller, B.L., Paoletti, M.G., 2009. Toxic moths: source of a truly safe delicacy. J. Ethnobiol. Ethnomed. 29 (1), 64–76.

Zhang, C.X., Tang, X.D., Cheng, J.A., 2008. The utilization and industrialization of insect resources in China. Entomol. Res. 38 (1), 38–47.

Zimian, D., Yonghua, Z., Xiwu, G., 2005. Medicinal terrestrial arthropods in China. In: Paoletti, M.G. (Ed.), Ecological Implications of Minilivestock: Potential of Insects, Rodents, Frogs and Snails. Science Publishers, Enfield, NH, pp. 481–489.

Zivotofsky, A.Z., 1996. Further clarifications on kosher insects. Am. Entomol. 42 (4), 195–196.

Chapter 3

Nutrient Content and Health Benefits of Insects

J.P. Williams*, J.R. Williams*,**, A. Kirabo[†], D. Chester*, M. Peterson*
*USDA/NIFA Washington, DC, United States; **University of Maryland, College Park, MD, United States;
[†]Uganda Industrial Research Institute, Kampala, Uganda

Chapter Outline

Nutrient Content	61	Insect Protein Functionality	81
Insect Physiology and Functionality	77	Conclusions	82
Insects as a Food Ingredient	79	References	82

NUTRIENT CONTENT

Much of our current food and nutritional supply, particularly animal protein, is produced unsustainably. Insects offer a substantial and largely unexplored opportunity to provide for the world's nutritional needs while requiring much less water, feed, and land to produce compared to other forms of livestock (Chapters 1 and 4; Dossey, 2013; Shockley and Dossey, 2014; van Huis et al., 2013). Insects have numerous attributes that make them highly attractive, yet underexplored sources of nutrients. Anthropological literature as well as historic references point to use of insects as food, medicines, and for other uses throughout history (Chapter 2). Indeed, insects have nearly always been a critical nutrient source for humans as well as a large fraction of animal species on earth. Even today it is accepted and practiced by many people around the world (Chapters 2 and 5; www.cricketpowder.com; Dossey, 2013; Shockley and Dossey, 2014; Defoliart, 1995; Nonaka, 2009; Ramos-Elorduy, 2009). About 3071 ethnic groups in 130 countries (Ramos-Elorduy, 2009) utilize insects as food (Chapter 2; Durst et al., 2010; Srivastava et al., 2009; Yen, 2009). As much as 80% of the population eats insects intentionally and 100% unintentionally (Srivastava et al., 2009). In recent years, particularly in North America and Europe, there has been an increasing interest in insect utilization as food ingredients as well as an emerging insect-based food industry worldwide (Chapter 5) (Gahukar, 2011; Durst et al., 2010; van Huis et al., 2013).

Insects in the United States will most likely be consumed as an ingredient in a processed product, not whole or raw. Thus, it is critical to review the nutritional value and functional properties of insects as ingredients. In this chapter, we will summarize the importance of exploring insects as a source of sustainable nutrients for the human population. Subsequently, we will provide a compilation of data on nutrient content from various species of insects specifically, proximate analysis, mineral content, amino acid, vitamins, and fatty acid composition. This data not only provides nutritional information, but also adds to the body of knowledge for insect protein functionality of which there is currently minimal published work.

This chapter also identifies functional properties associated with insect protein and successes as well as challenges in utilizing insect protein as a functional food ingredient in baked goods, snack products, and meat analogs. It also highlights the importance of continued research in this area to further expand insect ingredient utility in the United States and European countries.

With rapidly increasing human population, it is critical to decrease utilization of natural resources. Humans consume 40% of biomass that land and coastal seas produce (Safina, 2011). The world adds about 70 million people each year. The United Nations expects the population to grow to more than 9 billion people by 2050, adding approximately twice the current population of China (Dzamba, 2010; Safina, 2011; Vogel, 2010). Additionally, the UN FAO reports that approximately 18% (7100 teragrams of CO_2 eq. per year) of anthropogenic greenhouse gases (GHGs) are directly and indirectly related to the world's livestock production (Steinfeld et al., 2006). Expanding the amount of land used for livestock production is neither a feasible nor a sustainable solution to cover the food/protein needs of the projected increases in population. Thus, it is important to use sources of high-quality animal protein, which reduce the amount of pollution, habitat destruction, and abuse of natural resources.

Food reserves are the lowest they've been in 40 years (Tatum, 2014), yet demand for food (particularly protein) is increasing rapidly (expected increase of 50% by 2030). About 70% of agricultural land and 30% of the total land on earth is used to raise livestock (Steinfeld et al., 2006). This not only stresses our already overtaxed natural resources, but also means ever-increasing amounts of pollution (pesticides, fuel, etc.). In North America about 70% of the protein comes from animal-derived sources (Wardlaw and Smith, 2006). However, these animal-based products are becoming increasingly scarce and not sustainable to produce, particularly for the burgeoning human population. Estimates are that the global market for protein products was $15.2 billion in 2012. By 2017, these markets are projected to increase 30% to almost $20 billion dollars. Due to the unsustainable nature of current sources of protein, the costs for everything from meat products to whey powder are on the rise. It is unlikely these conditions will change toward cheaper prices given that the resources (eg, food, water, and land) required to grow these sources of protein will only be increasing. Now appears to be a great time to introduce alternative sources of protein. Indeed, in the recent 3–5 years over 30 companies in North America and over 50 in Europe are already beginning to offer a wide variety of food products with insects, primarily crickets, as a key ingredient (Chapter 5; www.cricketpowder.com). Currently, over 2 billion people worldwide eat insects (Cather, 2016). By 2021, the market for edible insects is projected to grow to $1.53 billion from $105 million today. A 30 times increase is expected for processed, powdered, and packaged insect products (Digitaljournal.com, 2016). Other reports indicate the current market at $423 million, growing to $722 million by 2024 (Marketwatch, 2016).

The potential for insects to contribute to sustainable food security has taken the notice of multiple organizations, particularly the UN Food and Agricultural Organization (UN FAO) (Gahukar, 2011). Three major meetings on the feasibility and benefits of insects as a food source have resulted from this initiative: (1) a February 2008 workshop in Thailand which produced an important book in the field called "*Forest Insects as Food: Humans Bite Back*" (Durst et al., 2010), (2) a technical consultation (see also Chapters 1 and 5) held January 2012 at UN FAO headquarters in Rome, Italy to plan a global summit on insects and food security for 2014 (Vantomme et al., 2012), and more recently (3) an international conference on insects as food and feed titled "Conference Insects to Feed the World" which hosted approximately 450 attendants from at least 45 different countries (May 14–17, 2014; Wageningen University, the Netherlands). Thus, it behooves the agricultural and food industry enterprise, to begin now developing methods, intellectual property and strategies to create some of the first insect-based food products for these emerging markets.

Increased utilization of insects in food products rather than ingredients from vertebrate livestock will significantly reduce the human impact on the natural environment, including our contribution to climate change and clean water depletion. Toward reducing climate change, insects possess attractive features including: (1) high feed conversion ratio and (2) producing less GHGs (Oonincx et al., 2010). However, new technologies for improving food security, such as production and processing insects as human food, take some time for application on large scale, so it is important to make investments in these innovations sooner rather than later (Gahukar, 2011). Insects have numerous attributes that make them highly attractive, yet underexplored sources of highly nutritious and sustainable food. The general categories where insects provide the most substantial benefits for a sustainable and secure food supply are: (1) *efficiency* (Chapters 4 and 5) and (2) *diversity (eg, biodiversity)* (Chapters 2 and 5, and Table 1A in the Appendix).

Meanwhile, food reserves are at a 40-year low (Tatum, 2014). Currently, there are approximately 870 million undernourished people in the world (FAO, 2012). Approximately 178 million children under the age of 5 years are stunted from malnutrition—most of those live in sub-Saharan Africa (Bhutta et al., 2008; Black et al., 2008). About 55 million children are "wasted," including 13–19 million having severe wasting or severe acute malnutrition (Bhutta et al., 2008). It contributes to over 50% of the 10–11 million children under the age of 5 years who die from preventable causes annually (Black et al., 2003; Caulfield et al., 2004; Pelletier and Frongillo, 2003; Rice et al., 2000). The World Health Organization (WHO) estimates that malnutrition accounts for 54% of child mortality worldwide (Duggan et al., 2008). About 60 million children suffer from moderate acute and 13 million from severe acute malnutrition (Collins et al., 2006). Fatality rates from severe acute malnutrition have averaged 20–30% since the 1950s (Schofield and Ashworth, 1996). Doctors Without Borders/Medecins Sans Frontieres (MSF) estimates that only 3% of the 20 million children suffering from severe acute malnutrition receive the lifesaving treatment they need. Malnutrition results in wasting reduced ability to fight infection, impaired cognitive and other developmental disorders which can be permanent (Victora et al., 2008). Such severe and lasting consequences impact a community's long-term prospects to improve their health, economy, and social wellbeing (Michaelsen et al., 2009; Victora et al., 2008). However, with the proper access to appropriate nutrition, including access to animal-derived dietary protein, the situations in these areas can be greatly improved.

The good news is insects can be farmed using less land, feed, water, and other resources than other livestock at any scale and there is substantial data available demonstrating their nutritional content. Generally, insects are a readily available source of protein, lipids, carbohydrates, certain vitamins, and minerals, such as calcium, iron, or zinc (Tables 3.1 and 3.2). Edible insects as a human food source could help developing countries to support their needs for elementary components of diets (Defoliart, 1992; Ramos-Elorduy et al., 1997), but could also provide complementary food for developed countries' populations. It is well known that insects are very high in protein. Protein is a critical source of nutrients for humans. On average, people require about 50 g of high-quality protein per day (Dolson, 2014; Cdc.gov., 2014). Animal-based nutrients are critical for health. Animals, including insects, are important or even the sole source of numerous necessary nutrients such as the eight essential amino acids (Table 3.3), vitamin B12, riboflavin, the biologically active form of vitamin A (retinol, retinoic acid, and retinaldehyde) (Table 3.4), essential and healthy fatty acids (Table 3.5), and several minerals

TABLE 3.1 Proximate Analysis Data (Moisture, Protein, Fat, Ash, and Fiber) for Selected Insect Species and Common High Protein Commodities

Food Insects	Common Name	Preparation	Moisture (%)	Protein (Crude Measured as N × 6.25)	Fat (% Crude Fat)	Fiber (a, acid detergent fiber; b, crude fiber)	Ash
Lepidoptera							
Agrotis infusa (larva)	Bogong moth	Roasted	49.2	52.7	39.0	5.3	–
Anaphe panda (larva)	African moth	Intestinal contents and hair removed	73.9	45.6	35.0	6.5b	3.7
Anaphe Venata (larva)	African moth	Dried without hairs	6.6	60.0	23.2	3.2	–
Ascalapha odorata (larva)	Black witch moth	Whole raw, not fasted	56.0	15.0	–	12.0b	6.0
Bombxy mori (larva fed artificial diet)	Silkworm	Whole, raw, not fasted	82.7	53.8	8.1	6.4a	6.4
Bombyx mori (larva fed mulberry leaves	Silkworm	Whole raw, not fasted	76.3	64.7	20.8	–	–
Bombyx mori (larva fed mulberry leaves)	Silkworm	Whole raw, intestinal contents removed	69.9	62.7	14.2	–	–
Bombyx mori (pupa)	Silkworm	Whole raw (dried)	18.9	60.0	37.1	–	10.6
Callasomia promethea (larva)	Silk moth	Whole raw, freeze-dried	4.5	51.7	10.5	11.3b	7.2
Catasticta teutila (larva)	Pure banded Dartwhite moth	Whole raw, not fasted	60.0	19.0	–	7.0b	7.0
Chilecomadia Moorei (larva)	Tebo worms	Whole raw, fasted	60.2	15.5	29.4	1.4	1.2
Conimbrasia belina (larva)	Mopani worm	Intestinal contents removed, dried	62.0	16.0	–	11.4b	7.6
Galleria mellonella (larva)	Waxworm	Whole raw, fasted	58.5	34.0	60.0	8.1a	1.4
Heliothis zea (larva fed Broad Beans)	Corn earworm	Whole raw, fasted	77.4	18.2	–	–	–
Heliothis zea (larva fed artificial diet)	Corn earworm	Whole raw, fasted	77.5	30.2	–	–	–
Hyalophora cecropia (larva)	Cecropia moth	Whole raw, freeze-dried	2.6	56.2	10.5	15.1b	6.1
Imbrasia epimethea (larva)	African moth larva	Smoked and dried	7.0	62.5	13.3	–	4.0

(Continued)

TABLE 3.1 Proximate Analysis Data (Moisture, Protein, Fat, Ash, and Fiber) for Selected Insect Species and Common High Protein Commodities (cont.)

Food Insects	Common Name	Preparation	Moisture (%)	Protein (Crude Measured as N × 6.25)	Fat (% Crude Fat)	Fiber (a, acid detergent fiber; b, crude fiber)	Ash
Imbrasia ertli (larva)	African moth larva	Viscera removed then boiled or roasted. Dried and salted	9.0	52.9	12.2	–	15.8
Imbrasia truncate (larva)	African moth larva	Smoked and dried	7.3	64.7	16.4	–	4.0
Manduca sexta (larva fed artificial diet)	Carolina sphynx moth	Whole raw, freeze-dried	4.7	61.0	21.7	9.9b	7.8
Manduca sexta (larva fed fresh plant material)	Carolina sphynx moth	Whole raw, freeze-dried	4.7	60.7	17.3	8.8b	8.5
Nudaurelia oyemensis (larva)	–	Smoked and dried	7.0	61.1	12.2	–	3.8
Porthetria dispar (adult with eggs)	Gypsy moth	Whole raw, not fasted	68.6	80.0	44.6	8.0	–
Pseudaletia unipuncta (larva)	Army worm	Whole raw, freeze-dried	2.0	55.5	15.2	5.1b	7.0
Spodoptera eridania (larva)	Fall army worm	Whole raw, freeze-dried	4.5	57.3	14.6	7.4b	10.3
Spodoptera frugisperda (larva fed artificial diet)	Fall army worm	Whole raw, freeze-dried	2.1	59.0	20.6	6.8b	5.7
Spodoptera frugiperda (lkarva fed fresh plant material)	Fall army worm	Whole raw, freeze-dried	3.6	59.3	11.7	12.4b	11.6
Usta terpischore (larva)	African moth	Viscera removed then boiled or roasted, dried and salted	9.2	48.6	9.5	–	13.0
Xyleutes redtenbacheri (larva)	Carpenter moths	Whole raw, not fasted	43.0	48.0	–	6.0b	2.0
Hylesia frigida	–	–	–	42	10	–	–
Arsenura armida	Giant silk moth	–	–	52	8	–	–
Phasus triangularis	moth	–	–	15	77	–	–
Coleptera							
Aplagiognathus spinosus (larva)	–						
Callipogon barbatus (larva)	–		41.0	34.0	–	23.0b	2.0
Oileus rimator (larva)	Beetle	Whole raw, not fasted	26.0	36.0	–	15.0b	3.0
Passalus punctiger (larva)	Beetle	Whole raw, not fasted	26.0	44.0	–	15.0b	3.0
Rhyncophorus ferrugineus (larva)	Red palm weevil	–	70.5	20.7	44.4	–	–
Rhyncophorus palmarum (larva)	Red palm weevil	Whole raw, not fasted	71.7	25.8	38.5	–	2.1
Rhyncophorun phoenicis (larva)	Red Palm weevil	Incised, fried in oil	10.8	22.8	46.8	–	2.7

TABLE 3.1 Proximate Analysis Data (Moisture, Protein, Fat, Ash, and Fiber) for Selected Insect Species and Common High Protein Commodities (cont.)

Food Insects	Common Name	Preparation	Moisture (%)	Protein (Crude Measured as N × 6.25)	Fat (% Crude Fat)	Fiber (a, acid detergent fiber; b, crude fiber)	Ash
Scyphophorus acupunctatus (larva)	Agave weevil	Whole raw, not fasted	36.0	52.0	–	6.0b	1.0
Tenebrio molitor (adult)	Mealworm beetle	Whole raw, fasted	63.7	65.3	14.9	20.4a	3.3
Tenebrio molitor (larva)	Mealworm beetle worm	Whole raw, fasted	61.9	49.1	35.0	6.6a	2.4
Zophobas morio (larva)	Darkling beetle	Whole raw, fasted	57.9	46.8	42.0	6.3z	2.4
Orthoptera							
Acheta domesticus (adult)	House cricket	Whole raw, fasted	69.2	66.6	22.1	10.2a	3.6
Acheta domesticus (nymph)	House cricket	Whole raw, fasted	77.1	67.2	14.4	9.6a	4.8
Blatella germanica (not specified)	German cockroach	Whole raw, not fasted	71.2	78.8	20.0	–	4.3
Blatta Lateralis (nymphs)	Turkestan cockroach	Whole raw, fasted	69.1	19	10	2.2	1.2
Brachytrupes sp.	Cricket	Fresh: blanched, inedible parts removed	73.3	47.9	21.3	13.5b	9.4
Crytacanthacris tatarica	–	Fresh: blanched, inedible parts removed	76.7	61.4	14.2	17.2b	4.7
Gryllotalpa Africana	Mole cricket	Fresh: blanced, inedible parts removed	71.2	53.5	21.9	9.7b	9.4
Oxya verox	–	Whole raw, dried	29.8	64.2	2.4	–	3.4
Oxya yezoensis	–	Whole raw, not fasted	65.9	74.7	5.7	–	6.5
Sphenarium histro (nymphs and adults)	Grasshoppers	Whole raw, not fasted	77.0	4.0	–	12.0b	2.0
Zonocerus sp.	Grasshoppers	Whole raw, not fasted	62.7	71.8	10.2	6.4b	3.2
Sphenarium purpurascens Ch	Grasshoppers	–	–	75.9	6.02	7.1	4.8
Taeniopodaques B	Horse lubber grasshopper	–	–	71.1	5.9	10.6	9.6
Melanoplus femurrubrum D	Red-legged grasshopper	–	–	74.7	5.2	10.0	6.7
Schistocerca	Bird grasshoppers		–	62.5	16.0	10.1	7.0
Isoptera							
Cortaritermes silvestri (worker)	South American termites	Whole raw, not fasted	77.8	48.6	6.9	–	8.5
Macrotermes bellicosus (alate)	African termites	Dewinged, raw	6.0	34.8	46.1	–	10.2
Macrotermes subhyalinus (alate)	African termites	Dewinged, fried in oil	0.93	8.8	46.5	–	6.6
Nasutitermes corniger (soldier)	Central American tree termite	Whole raw, not fasted	69.6	58.0	11.2	34.8a	3.7

(Continued)

TABLE 3.1 Proximate Analysis Data (Moisture, Protein, Fat, Ash, and Fiber) for Selected Insect Species and Common High Protein Commodities (cont.)

Food Insects	Common Name	Preparation	Moisture (%)	Protein (Crude Measured as N × 6.25)	Fat (% Crude Fat)	Fiber (a, acid detergent fiber; b, crude fiber)	Ash
Nasutitermes corniger (worker)	Central American tree termite	Whole raw, not fasted	75.3	66.7	2.2	27.1a	4.6
Procornitermes araujoi (worker)	–	Whole raw, not fasted	78.1	33.9	16.1	–	3.5
Syntermes ditus (worker)	–	Whole raw, not fasted	79.7	43.2	3.4	–	17.1
Hymenoptera							
Apis mellifera (adult female)	European honeybee	Whole raw, not fasted	65.7	60.0	10.6	–	17.4
Apis mellifera (adult male)	European honeybee	Whole raw, not fasted	72.1	64.4	10.5	–	17.8
Apis mellifera (larva)	European honeybee	Whole raw, not fasted	76.8	40.5	20.3	1.3a	3.4
Atta mexicana (reproductive adult)	Leaf-cutter ant	Whole raw, not fasted	46.0	39.0	–	11.0b	4.0
Olecophylla smaragdina	Weaver ant	Fresh: blanched, inedible parts removed	74.0	53.5	13.5	6.9b	6.5
Oecophylla virescens	Green tree ant/ weaver ant	Inedible parts removed	78.32	41.0	26.7	–	6.0
Polybia sp. (adult)	Wasp	Whole raw, not fasted	63.0	13.0	–	15.0b	6.0
Trigon sp.	–	–	–	28	41	–	–
Parachartegus apicallis	–	–	–	55	–	–	–
Brachygastra azteca	Paper wasp	–	–	63	24	–	–
Brachygastra melifica	Mexican honey wasp	–	–	53	30	–	–
Vespula squamosa	Southern yellow jacket	–	–	63	22	–	–
Polistes instabilis	Paper wasp	–	–	31	62	–	–
Diptera							
Copestylum anna & *C. haggi* (larva)	–	Whole raw, not fasted	37.0	31.0	–	15.0b	8.0
Drosophila melanogaster (adult)	Common fruit fly	Whole raw, not fasted	67.1	56.3	17.9	–	5.2
Hermetia illucens (larva)	Black soldier fly	Dried, ground, not fasted	3.8	47.0	32.6	6.7b	8.6
Musca autumnalis (pupa)	Face fly/ autumn house fly	Dried, ground, not fasted	51.7	11.4	28.9	–	–
Musca domestica (pupa)	Common house fly	Dried, ground, not fasted	61.4	9.3	11.9	–	–

TABLE 3.1 Proximate Analysis Data (Moisture, Protein, Fat, Ash, and Fiber) for Selected Insect Species and Common High Protein Commodities (cont.)

Food Insects	Common Name	Preparation	Moisture (%)	Protein (Crude Measured as N × 6.25)	Fat (% Crude Fat)	Fiber (a, acid detergent fiber; b, crude fiber)	Ash
Hemiptera							
Edessa petersii (nymphs and adults)	–	Whole raw, not fasted	37.0	42.0	–	18.0b	2.0
Euchistus egglestoni (nymphs and adults)	–	Whole raw, not fasted	35.0	45.0	–	19.0b	1.0
Pachilis gigas (nyumphs and adults)	–	Whole raw, not fasted	64.0	22.5	–	7.5b	3.5
Hoplophorion monograma (nymphs and adults)	Treehopper	Whole raw, not fasted	64.0	14.0	–	18.0b	3.0
Umbonia reclinata (nymphs and adults)	–	Whole raw, not fasted	29.0	33.0	–	13.0b	11.0
Callipogon barbutus	–	–		41	34	–	–
Commonly Consumed Protein Source							
Beef	–	Ground/raw	65.81	17.37	17.07	0	0.86
Pork	–	Ground/raw	64.46	15.41	17.18	0	0.79
Chicken	–	Ground/raw	73.24	17.44	8.10	0	1.17
Egg	–	Whole/raw	76.15	12.56	9.51	0	1.06
Salmon	–	Wild/raw	68.50	19.84	6.34	0	2.54
Milk	–	Dry/whole	2.47	26.32	26.71	0	6.08
Milk	–	Fluid/whole	87.69	3.28	3.66	0	0.72

Source: USDA National Nutrient Database, 2015; Bukkens, 1997; Finke, 2007, 2013; Ramos-Elorduy et al., 1997.

TABLE 3.2 Mineral Content of Selected Insect Species and Common High-Protein Commodities (mg/100 g Dry Matter)

Insect	Na	K	Ca	P	Fe	Mg	Zn	Cu	Mn	Cl	Se	I
Lepidoptera												
Galleria mellonella	16.5	221.0	24.3	195.0	2.09	31.6	2.54	0.38	0.13	–	–	–
Imbrasia ertli	2418	1204	55	600	2.1	254	–	1.5	3.4	–	–	–
Usta Terpsichore	3340	3259	391	766	39.1	59	25.3	2.6	6.7	–	–	–
Nadaurelia oyemensis	140	1107	149	871	9.7	266	102	1.2	5.5	–	–	–
Imbrasia truncate	183	1349	132	842	8.7	193	11.1	1.4	3.2	–	–	–
Imbrasia epimethea	75	1258	225	666	13.0	402	11.1	1.2	5.8	–	–	–
Conimbrasia belina	1032	1024	174	543	31	160	14	0.91	3.95	–	–	–
Anaphe venata B.	30	1150	40	730	10	50	10	1	40	–	–	–
Agrotis infusa (whole)	43	554	431	–	23.6	260	14	1.6	–	–	–	–
Agrotis infusa (Abdomen)	40	488	174	–	11	123	16.9	1.1	–	–	–	–
Bombyx mori	47.5	316.0	15	641	3.1	49.8	3.07	0.36	0.43	–	–	–

(Continued)

TABLE 3.2 Mineral Content of Selected Insect Species and Common High-Protein Commodities (mg/100 g Dry Matter) (cont.)

Insect	Na	K	Ca	P	Fe	Mg	Zn	Cu	Mn	Cl	Se	I
Witchetty gubs	5.7	293	29	–	15.0	37	0.4	3.4	–	–	–	–
Sago grub	13	304	18	96	–	29	4.9	3.1	0.55	–	–	–
Rhynchophorus phoenicis	44.8	2209	208	352	14.7	33.6	26.5	1.6	0.8	–	–	–
Grub (aruk)	16.1	550	21	402	13.3	49.7	3.61	0.55	0.21	–	–	–
Chilecomadia moorei (larvae)	<198	2590	125	2250	14	278	35.7	2.95	0.71	1160	<0.03	<0.1
Orthoptera												
Brachytrypes membranaceus (raw crickets)	–	–	75	–	54	–	–	–	–	–	–	–
Locusta migratoria manillensis (roasted)	55	545	90	424		62	8.4	3.0	1.46	–	–	–
Acheta domesicus (nymphs)	135.0	352.0	27.5	225.0	2.12	22.6	6.80	0.51	0.89	–	–	–
Blatta lateralis (nymphs)	744	2240	385	1760	14.8	2.64	32.7	7.93	2.64	1600	0.30	0.30
Isoptera												
Macrotermes subhyalinus	1988	480	40	442	7.6	421		13.7	64.4	–	–	–
Macrotermis bellicosus	–	117	44.6	–	–	28.0	–	–	–	–	–	–
Hymenoptera												
Oecophylla sp.	180	541	48	517	21.8	70	10.1	0.87	9.06	–	–	–
Oecphylla virescens	270	957	79.7	936	109.0	122.1	16.9	2.17	6.30	–	–	–
Diptera												
Chaborus povilla – stone ground	433	1106	296	1220	1442	187	14.5	5.8	16.4	–	–	–
Hermetia illucens (larva)	887	4530	9340	3560	66.6	1740	56.2	4.03	61.8	1160	–	–
Musca domestica (pupa)	1350	3030	765	3720	125	26.6	85.8	12.9	26.6	1760	<0.1	0.15
Coloptera												
Zophobas morio larvae	47.5	316.0	17.7	237.0	1.65	49.8	3.07	0.36	0.43	–	–	–
Tenebrio molitor larvae (Giant)	48.9	297.0	18.4	272.0	2.15	86.4	4.45	0.64	0.36	–	–	–
Tenebrio molitor (larvae)	53.7	341.0	16.9	285.0	2.06	80.1	5.20	0.61	0.52	–	–	–
Tenebrio molitor (Adult)	63.2	340.0	23.1	277.0	2.18	60.6	4.62	0.75	0.40	–	–	–
Commonly Consumed Protein Sources												
Beef	57	246	7	122	1.69	17	3.59	0.062	0.015	–	12.1	–
Pork	58	297	14	181	0.91	20	2.28	0.047	0.010	–	24.6	–
Chicken	60	522	6	178	0.82	21	1.47	0.065	0.016	–	10.2	0.016
Egg	142	138	56	198	1.75	12	1.29	0.072	0.028	–	30.7	–
Salmon	44	490	12	200	0.8	29	0.64	–	–	–	–	–
Whole milk –dry	371	1330	912	776	0.47	85	3.3	0.250	0.016	–	36.5	–
Whole milk –fluid	49	151	119	93	0.05	13	0.38	0.010	0.004	–	2.0 µg	–

Source: USDA National Nutrient Database, 2015; Bukkens, 1997; Finke, 2007, 2013.

TABLE 3.3 Essential Amino Acid Content for Selected Insect Species and Common High Protein Commodities (mg/100 g Dry Matter)

Food Insect	ILe	Leu	Lys	Met	Cys	SAA	Phe	Tyr	AAA	Thr	Trp	Val	Arg	His	Limiting AAA	Amino Acid Score
Diptera																
Copestyla anna and C. haggi	40	74	55	19	18	37	54	66	120	49	7	61	63	29	Trp	64
Lepidoptera																
Caterpillar, Nadurelia oyemensis	25.6	82.7	79.8	23.5	19.7	43.2	58.6	75.7	134	44.5	16.0	96	63.5	18.1	ILe	91
Caterpillar, Imbrasis truncata	24.2	73.1	78.9	22.2	16.5	38.7	62.2	76.5	139	46.9	16.5	102	55.5	17.4	ILe	86
Caterpillar, Imbrasia epimethea	28.6	81.0	74.2	22.4	18.7	41.1	65.0	75.0	140	48.0	16.0	102	66.2	19.7	ILe	102
Caterpillar, Imbrasia ertli	36.0	36.7	39.3	15.8	13.4	29.2	17.4	13.2	30.6	40.5	8.1	41.9	–	–	–	49
Caterpillar, Usta terpischore	108.7	91.3	91.0	11.3	12.9	24.2	55.9	33.0	88.9	50.8	6.6	75.8	–	–	Trp	60
Maguey Worm, Aegiate sp.	49	52	36	10	–	–	37	42	79	33	9	47	30	16	Lys	49
Gusano rojo de maguey, Cossus retenbachi	51	79	49	8	13	21	40	53	93	47	6	61	60	16	Trp	55
Caterpillar meal Bombycomorpha sp.	46.1	62.1	64.5	17.9	29.9	47.8	59.8	81.4	141.3	42.9	13.0	60.5	41.8	8.31	Leu	94
Caterpillars cooked Bombycomorpha sp.	44.3	77.4	47.8	16.2	21.8	37.9	93.4	63.4	157	49.1	9.4	55.0	37.0	16.2	Lys	82
African silkworm larvae Trp (Anaphe venata)	21.4	13.12	8.8	–	–	–	21.4	24.9	46.4	3.8	–	17.6	3.20	7.8	Trp	–
Spent silkworm pupae	57	83	75	46	14	60	51	54	105	54	9	56	68	25	Trp	82
Xyleutes redtembacheri	51	79	49	21	13	34	93	53	146	47	6	61	60	16	Trp	55
Ascalapha odorata	41	69	63	23	21	44	95	44	139	40	4	48	28	67	Trp	36
Arsenura armida	43	69	54	24	19	43	93	52	145	42	4	48	63	29	Trp	36
Hylesia frigida	44	71	57	26	54	80	64	52	116	41	5	49	20	68	Trp	45
Phasus Triangularis	46	80	57	22	13	35	72	95	167	38	4	57	57	25	Trp	36
Chilecomadia Moorei (larvae)	6.5	10	8.7	2.5	0.87		5.47	7.95		5.74	1.56	9.71	11.7	4.08		

(Continued)

TABLE 3.3 Essential Amino Acid Content for Selected Insect Species and Common High Protein Commodities (mg/100 g Dry Matter) (cont.)

Food Insect	ILe	Leu	Lys	Met	Cys	SAA	Phe	Tyr	AAA	Thr	Trp	Val	Arg	His	Limiting AAA	Amino Acid Score
Coleoptera: Curculionidae																
Palm weevil larvae, Rhynchophorus phoenicis	77.5	58.9	63.9	12.0	10.6	22.6	32.8	13.6	46.4	28.6	5.1	54.9	–	–	Trp	46
Larvae of Sciphophonts acupunctatus 1	48.2	78.2	53.5	20.2	26.7	46.9	46.1	63.5	109.6	40.4	8.1	62.0	44.0	14.7	Trp	74
Scyphophorus acupunctatus	48	78	55	20	22	42	46	64	110	40	8	62	44	15	Trp	72
Callipogon barbutus	58	100	57	20	20	40	47	42	98	40	9	90	59	22	Trp	64
Orthoptera: Acrididae																
Chapulin, Sphenarium histrio'	53	87	57	7	13	20	44	73	117	40	6	51	66	11	Trp	55
"Chapulin," Sphenarium purpurascens6	42	89	57	25	18	43	103	63	166	38	6.5	57	60	22	Trp	59
Mexican "Chapulines"	46	64	52	8	–	–	36	32	68	49	10	54	42	21	Lys	90
Schistocerca sp.	53	79	51	–	–	–	–	–	–	41	4	51	–	–	–	–
Taeniopoda eques B	41	72	31	–	–	–	–	–	–	45	6	54	–	–	–	–
Melanoplus femurrubrum D	47	88	27	–	–	–	–	–	–	41	5	52	–	–	–	–
Blatta Lateralis (nymphs)	7.73	12	12.8	3.35	1.44	–	7.67	14.3	–	7.89	1.66	12.3	14	5.49	–	–
Brachytrupes sp.	3.1	5.5	4.8	1.9	1	2.9	2.9	3.9	–	2.75	–	4.42	3.7	1.94	–	–
House crickets, Acheta domesticus	36.4	66.7	51.1	14.6	8.3	–	30.2	44	–	31.1	6.3	48.4	57.3	23.4	–	–
Hymenoptera: Formicidae																
Ants, Atta mexicana'	53	80	49	19	15	34	41	47	88	43	6	64	47	25	Trp	55
Escamol," Liometopum apiculatum'	49	76	58	18	14	32	39	68	10	42	8	60	50	29	Trp	73
Atta Mexicana	53	80	49	34	15	49	88	47	135	43	6	64	47	25	Trp	55
Trigana sp.	48	73	79	13	23	36	75	64	139	48	6	53	60	22	Trp	55
Apis mellifera	41	66	60	25	9	34	70	41	111	44	7	59	64	33	Trp	64
Parachartegus apicalis	42	77	58	20	24	44	43	71	114	47	5	57	43	29	Trp	45
Brachygastra Azteca	51	85	61	14	16	30	41	65	106	44	7	64	44	28	Trp	64
Brachygastra melifica	44	78	36	18	20	38	40	75	115	44	7	54	57	36	–	–

Vespula squamosal	49	63	51	17	28	45	49	63	112	44	7	57	42	30	Trp	64
Polybia sp	45	78	74	21	29	50	33	56	99	40	7	59	57	30	Trp	64
Polistes instabilis	64	115	43	21	17	38	42	66	108	49	3	67	34	22	Trp	72
Isoptera: Termitidae																
Termites, Macrotermes bellicosus	51.1	78.3	54.2	7.5	18.7	26.2	43.8	30.2	74.0	27.5	14.3	73.3	69.4	51.4	Thr	81
Termites, mature alates, Macrotermes subhyalinu	37.1	79.7	35.4	12.9	9.0	21.9	43.1	36.8	79.9	41.9	7.7	51.4	—	—	Lys	61
Hemiptera																
"Ahuahutle" or Mexican caviar, eggs of water bugs (Corixidae)	50	80	35	15	—	—	34	111	145	40	11	60	77	33	Lys	60
"Axayácatl," adults and nymphs of water bugs (Corixidae and notonectidae)	59	80	43	16	—	—	32	45	77	44	16	55	55	24	Lys	74
"Jumiles de Taxco," nymphs of Atizies taxcoensis (Pentatomidae)	41	77	31	17	10	27	36	66	102	42	1	73	51	18	Trp	9
"Jumiles," nymphs of several species of Pentatomidae	45	62	38	15	—	—	25	40	65	28	15	48	29	30	Lys	66
Edessa petersii	40	71	40	28	10	37	67	115	182	45	5	64	45	23	Trp	45
Umbonia reclinata	38	68	57	19	14	33	59	68	127	47	60	40	46	37	Trp	55
Diptera																
Aquatic insect flour	52	79	78	24	8.5	32.5	47	58	105	46	—	47	69	32	Trp	0
Hermetia illucens	7.6	12	12	3.4	1.02	—	7.56	12.1	—	6.82	3	12.9	12.3	5.9	—	—
Musca domestica (pupa)	8.14	12.4	12.6	5.84	1.4	—	7.91	9.26	—	7.54	2.40	11	12.1	5.71	—	—
Commonly Consumed Protein Sources																
Beef	0.902	1.71	1.85	0.601	0.219	—	0.803	0.731	—	0.933	0.236	0.953	1.39	0.680	—	—
Pork	1.02	1.76	1.92	0.569	0.238	—	0.868	0.786	—	0.927	0.217	1.08	1.37	0.892	—	—
Chicken	0.739	1.36	1.51	0.446	0.188	—	0.683	0.604	—	0.727	0.147	0.826	1.128	0.529	—	—
Egg	0.671	1.09	0.912	0.380	0.272	—	0.680	0.499	—	0.556	0.167	0.858	0.820	0.309	—	—
Salmon	0.914	1.613	1.82	0.587	0.213	—	0.775	0.670	—	0.870	0.222	1.02	1.19	0.584	—	—
Whole milk—dry	1.59	2.58	2.09	0.660	0.243	—	1.27	1.27	—	1.19	0.371	1.76	0.953	0.714	—	—
Whole milk—fluid	0.198	0.321	0.260	0.082	0.030	—	0.158	0.18	—	0.148	0.046	0.220	0.119	0.089	—	—

Source: USDA National Nutrient Database, 2015; Melo-Ruiz et al., 2015; Bukkens, 1997; Finke, 2007, 2013; Ramos-Elorduy et al., 1997.

TABLE 3.4 Vitamin Content of Selected Insect Species and Common High-Protein Commodities (mg/100 g Dry Weight Unless Otherwise Noted)

Food Insect	Retinol (Vitamin A)	β-carotene	Thiamin (B1)	Riboflavin (B2)	Niacin	Pyridoxine (B6)	Folic Acid	Pantothenic Acid	Biotin	Cyanocobalamin (B12)
Lepidoptera										
Bombyx mori	1580 IU/kg	<0.02	0.33	0.94	2.63	0.16	0.071	2.16	0.025	<1.2 µg/kg
Galleria mellonella	<1000 IU/kg	<0.02	0.23	0.73	3.75	0.13	0.044	2.02	0.029	<1.2 µg/kg
Comimbrasia belina	21.6 IU	1.71 IU	0.58	4.98	11.9	—	—	—	—	—
Comimbrasia belina (dried)	0.053	0.634	0.55	1.99	11.6	—	—	—	—	—
Nudaurelia oyemensis	32 µg	6.8 µg	0.21	3.4	10.1	54 µg	21.5 µg	9.5	32.0	0.015 µg
Imbrasia truncata	33 µg	7.1 µg	0.32	5.5	11.8	151 µg	40.0 µg	11.0	48.5 µg	0.027 µg
Imbrasia epimethea	47.3 µg	8.2 µg	0.21	4.3	11.8	86 µg	6.8 µg	7.8	24.7 µg	0.016 µg
Chilecomadia moorei (larva)	<300 µg	0.57	<0.01	64.5	33.6	3.29	0.83	26.5	0.46	—
Colodoptera										
Zophobas morio larvae	<1000 IU/kg	<0.02	0.06	0.75	3.23	0.32	0.0066	1.94	0.035	0.42
Tenebrio molitor (giant larvae)	<1000 IU/kg	<0.02	0.12	1.61	4.13	0.58	0.117	1.45	0.0037	0.13
Tenebrio molitor (adult)	<1000 IU/kg	<0.02	0.10	0.85	5.64	0.81	0.139	2.40	0.028	0.56
Tenebrio molitor (larvae)	<1000 IU/kg	<0.02	0.24	0.81	4.07	0.81	0.157	2.62	0.030	0.47
Orthoptera										
Oxya verox	356 µg	78 µg	0.34	7.84	10.0	—	—	—	—	—
Acheta domesticus (Adult)	<1000 IU/kg	<0.02	0.04	3.41	3.84	0.23	1.50	2.30	0.017	5.37
Acheta domesticus (nymphs)	<1000 IU/kg	<0.02	0.02	0.95	3.28	0.17	0.145	2.63	0.005	8.72
Blatta lateralis (nymphs)	<300 µg/kg	<0.2	0.9	15.6	43.8	3.10	1.11	37	0.37	—
Isoptera (dried, smoked, fried)										
Termes sp. Dried	—	—	0.03	6.07	5.9	—	—	—	—	—
Termes sp. Smoked	—	—	0.11	0.07	1.95	—	—	—	—	—
Termes sp. Fried	—	—	0.14	3.79	9.81	—	—	—	—	—

Hymenoptera										
Oecophylla sp.	—	—	0.44	0.98	—	—	—	—	—	—
Vespa singulata (canned)	—	—	0.70	1.08	11.3	—	—	—	—	—
Diptera										
Chaoborus sp.	—	—	1.5	4.1	21.7	—	—	—	—	—
Chaoborus sp., and Povilla sp. stone ground mixture	—	—	1.8	8.9	28.8	—	—	—	—	—
Hermetica illucens	<300 µg/kg	—	7.77	16.2	71	6.01	2.7	38.5	0.35	—
Musca domestica	<300 µg/kg	—	11.3	77.2	90.5	1.72	1.82	45.3	0.68	—
Commonly Consumed Protein Sources										
Beef	0	—	0.056	0.232	4.21	0.241	7	—	—	1.91
Pork	2 µg	—	0.558	0.214	3.96	0.350	5	0.610	—	0.64
Chicken	0	0	0.109	0.241	5.58	0.512	0	1.09	—	0.56
Egg	160 µg	0	0.040	0.457	0.075	0.170	0	1.53	—	0.89
Salmon	12 µg	—	0.226	0.380	7.86	0.818	0	1.664	—	3.18 µg
Whole milk—dry	253 µg	55 µg	0.283	1.21	0.646	0.302	0	2.27	—	3.25 µg
Whole milk—fluid	31 µg	—	0.038	0.161	0.084	0.042	0	0.313	—	0.36 µg

Source: USDA National Nutrient Database, 2015; Bukkens, 1997, 2005; Finke, 2013.

TABLE 3.5 Fatty Acid Composition of Selected Insect Species and Common High-Protein Commodities

Food Insect	C10:0	C12:0	C14:0	C14:1	C15:0	C16:0	C17:0	C17:1	C18:0	Other SFA	Total SFA
Lepidoptera											
Smoked caterpillar *Nudaurelia oyemensis*	–	–	0.2	–	–	21.8	–	–	23.1	0.2	45.3
Smoked caterpillar *Imbrasia trúncala*	–	–	0.2	–	–	24.6	–	–	21.7	trace	46.5
Smoked caterpillar *Imbrasia epimethea*	–	–	0.6	–	–	23.2	–	–	22.1	0.2	46.1
Caterpillar *Imbrasia ertli*	–	–	1.0	–	–	22.0	–	–	0.4	38.0	–
Caterpillar, *Usía terpsichore*	–	–	2.3	–	–	27.4	–	–	0.1	29.7	7.0
Waxworms, *Galleria mellonella*	–	<0.2	0.4	–	<0.2	79.6	<0.2	0.3	3.4	–	–
Witchetty grub, *Xyleutes* sp.	–	–	–	–	–	29.4	–	–	3.1	–	32.5
Silkworms, *Bombyx Mori*	–	<0.2	<0.3	–	<0.4	1.7	<0.2	<0.2	1.2	–	–
Spent silkworm pupae, *Bombyx mori*	–	–	–	–	–	26.2	–	–	7.0	–	33.2
Tebo worms, *Chilecomadia moorei*	<0.10	0.93	0.95	0.20	<0.10	69.3	<0.10	<0.10	2.19	–	–
Coleoptera											
Mealworm (larvae), *Tenebrio molitor*	–	<0.2	2.9	–	<0.2	22.9	<0.2	0.3	3.9	–	–
Mealworm (adult), *Tenebrio molitor*	–	<0.1	0.8	–	<0.1	8.5	0.2	0.2	2.6	–	–
Palm weevil larvae *Rhynchophorus phoenicis* 1	–	–	2.5	–	–	36.0	–	–	0.3	2.1	–
Palm worm (entire larvae) *Rhynchophorus phoenicis*	–	–	1.8	–	–	38.0	–	–	4.5	–	44.3
Superworms (larvae), *Zophobus morio*	–	<0.2	1.7	–	0.4	52.8	0.7	0.6	12.6	–	–
Orthoptera											
Crickets (adult), *Acheta domesticus*	–	<0.2	0.4	–	<0.2	15.6	0.1	0.1	2.9	–	–
Crickets (nymph), *Acheta domesticus*	–	<0.2	0.2	–	<0.2	6.1	0.1	0.1	2.9	–	–
Turkestan cockroaches, *Blatta lateralis*	<0.20	<0.20	0.48	<0.20	<0.20	17.4	<0.20	<0.20	4.22	–	–
Isoptera											
Termites, *Macrotermes bellicosus*	–	–	0.18	–	–	46.54	–	–	–	–	46.7
Termites, mature alates, *Macrotermes subhyalinus*	–	–	0.9	–	–	33.0	–	–	1.4	3.8	–
Termites, boiled	–	–	1.3	–	–	28.0	–	–	8.5	–	37.8
Diptera											
Soldier Fly, *Hermetica illucens*	0.69	51.2	12	0.50	0.12	16.1	0.20	<0.08	2.45	–	–
House Fly, *Musca domestica*	<0.01	0.02	0.32	0.02	0.17	3.72	0.10	<0.01	0.40	–	–
Commonly Consumed Protein Sources											
Beef	0.020	0.020	0.470	–	–	–	–	–	2.04	–	6.81
Pork	0.008	0.016	0.267	–	–	3.87	–	–	2.07	–	6.38
Chicken	0.000	0.000	0.041	0.016	0.000	1.79	0.007	0.000	0.456		2.301
Egg	0.006	0.000	0.033	0.007	0.008	2.23	0.022	0.012	0.811	–	3.13
Salmon	–	–	0.137	–	–	0.632	–	–	0.212	–	0.981
Whole milk—dry	0.596	0.614	2.82	–	–	–	–	–	2.85	–	16.74
Whole milk—fluid	0.092	0.103	0.368	–	–	0.963	–	–	0.444	–	2.28

Source: USDA National Nutrient Database, 2015; Bukkens, 1997, 2005; Finke, 2002, 2013.

C16:1	C18:1	Other MUFA	Total MUFA	C18:2	C18:3	C20:0	C20:1	C20:2	C20:4	C22:0	Other PUFA	Total PUFA	All Others
0.6	5.6	–	6.2	5.7	35.6	–	–	–	–	–	2.1	43.4	–
0.2	7.4	–	7.6	7.6	36.8	–	–	–	–	–	–	44.4	–
0.6	8.4	–	9.0	7.0	35.1	–	–	–	–	–	0.4	42.5	–
22.0	2.0	0.8	–	20.0	11.0	–	–	–	–	–	0.2	–	–
27.4	1.7	0.2	–	27.2	2.8	–	–	–	–	–	0.1	–	–
5.1	124.0	–	–	15.2	1.1	0.3	–	–	–	–	–	–	<0.2
–	67.1	–	67.1	0.4	–	–	–	–	–	–	–	0.4	–
0.1	3.2	–	–	3.5	1.4	0.1	–	–	–	–	–	–	0.1
–	36.9	–	36.9	4.2	25.7	–	–	–	–	–	–	29.9	–
14.7	149	–	–	6.99	0.45	0.24	0.19	<0.10	<0.10	<0.10	–	–	–
3.5	53.9	–	–	34.8	1.4	0.3	–	–	–	–	–	–	0.2
0.6	17.9	–	–	13.7	0.4	0.2	–	–	–	–	–	–	<0.1
36.0	30.0	0.6	–	26.0	2.0	–	–	–	–	–	trace	–	–
2.5	46.2	–	48.7	5.0	1.5	–	–	–	–	–	–	6.5	–
0.7	66.0	–	–	32.9	1.1	0.4	–	–	–	–	–	–	0.2
0.9	15.4	–	–	22.9	0.6	0.4	–	–	–	–	–	–	0.5
0.3	6.4	–	–	11.0	0.4	0.3	–	–	–	–	–	–	0.1
1.21	40.9	–	–	21.6	0.71	<0.20	0.25	0.66	0.35	<0.20	–	–	–
2.09	12.84	–	14.9	34.42	3.85	–	–	–	–	–	–	38.3	–
33.0	9.5	1.2	–	43.1	3.0	–	–	–	–	–	3.7	–	–
3.4	48.0	–	51.4	9.5	1.4	–	–	–	–	–	–	10.9	–
4.96	15.6	–	–	16.9	0.65	0.16	<0.08	<0.10	<0.08	0.09	–	–	–
1.96	2.89	–	–	4.15	0.45	0.04	0.01	0.02	0.04	0.03	–	–	–
0.680	6.40	–	7.413	0.053	0.070	–	0.020	–	0.080	–	–	0.710	–
0.527	6.89	–	7.65	1.35	0.057	–	0.130	–	0.065	–	–	1.55	–
0.536	2.99	–	3.61	1.32	0.014	0.005	0.025	0.011	0.074	0.000	–	1.51	–
0.201	3.41	–	3.66	1.56	0.048	0.003	0.027	0.018	0.188	0.004	–	1.91	–
0.251	1.35	–	2.103	0.172	0.295	–	0.223	–	0.267	–	–	2.54	–
1.20	6.19	–	7.92	0.460	0.204	–	0.00	–	0.00	–	–	0.665	–
0.082	0.921	–	1.06	0.083	0.053	–	0.00	–	0.00	–	–	0.136	–

(Table 3.2) (Bukkens, 1997, 2005; Hoppe et al., 2008; Michaelsen et al., 2009; Singh and Singh, 1991). In particular, it is broadly accepted that animal-sourced dietary protein superior to that derived from plants (Babji et al., 2010; Hoppe et al., 2008; Michaelsen et al., 2009; Singh and Singh, 1991). Insects can be grown highly efficiently and in many areas not amenable to dairy cattle and, thus, help provide a robust alternative to milk as well as a potential alternative income source for farmers. Insects present a substantial, yet extremely underexplored, alternative opportunity to provide much needed animal-sourced nutrients (Bukkens, 1997, 2005; Defoliart, 1992; Gahukar, 2011; Michaelsen et al., 2009). For example, insects are generally high in protein and fat at levels comparable to meat, such as beef and milk. As with beef and chicken, insects are a source of "complete" animal protein containing all of the eight essential amino acids (Table 3.3), which is generally nutritionally superior to protein from plant sources (Hoppe et al., 2008; Michaelsen et al., 2009; Singh and Singh, 1991). A review by Bukkens concludes that the amino acid composition of insects compares favorably with the reference standard recommended by UN FAO, WHO, and UNU (United Nations University) (Bukkens, 2005). Many insect species contain a higher portion of protein per 100 g of dry weight (ie, 68.7 g for house crickets) than ground beef (27.4 g) or broiled cod fish (28.5 g) (Gahukar, 2011). Some estimate that the digestibility of flour fortified with insect powder or meal is as high as 91% (Bukkens, 1997). Actually, many commercial food products are enriched with protein derived from legumes but insect protein is better in terms of nutritional properties, since insect protein contains all the essential amino acids. Moreover, insects are richer in protein than beans (23.5% of protein), lentils (26.7%), or soybean (41.1%) (Ramos-Elorduy et al., 2012).

Insects are also particularly rich in fat and often with higher levels of healthier (unsaturated) fats and lower levels of unhealthy saturated fats than other animal products (Table 3.5) (Bukkens, 1997, 2005; Defoliart, 1992; Finke, 2002, 2013; Gahukar, 2011), and, thus, can supply a high caloric contribution for such energy dense foods. The energy content of insects is on average comparable to that of meat (on a fresh weight basis) except for pork because of its particularly high fat content (Sirimungkararat et al., 2010). What is important, insects are a good source of essential amino acids and polyunsaturated fatty acids (Rumpold and Schlüter, 2013). Expectedly, holometabolous insects, particularly their larval stages, appear to be higher in fat and correspondingly lower in protein than hemimetabolous insects which lack larval and pupal stages. This is expected since during the pupation transition from larva to adult, holometabolous insects do not eat (pupae do not eat) and yet need tremendous stores of energy for execution of their metabolic changes transforming their bodies from larva into adult form. Larval stages also contain less-fibrous chitin exoskeleton than nonlarval stages and are softer bodied, thus possibly being easier to process and digest. For example, mealworms are higher fat, have softer chitin exoskeleton, and their proteins may have very different functional properties as a food ingredient. In the reviews by Bukkens, all insect species were found to be a "significant source of the essential fatty acids linoleic and linolenic acid" (Bukkens, 1997, 2005). Some insects can also provide a higher caloric contribution to the diet than soy, maize, and beef (Gahukar, 2011). Insects are also particularly high in omega 3 fatty acids. Additionally, many insect species are significantly higher in B vitamins, such as thiamin and riboflavin than whole meal bread and hen's eggs (Bukkens, 1997, 2005). Retinol (a biologically active form of vitamin A) and β-carotene content of many insect species is also high, with levels in some species as high as 356 and 1800 μg/kg, respectively (Bukkens, 2005). A recent study has even shown that insects may be a better and more bioavailable source of iron compared with other protein sources including beef (Gladys et al., 2016).

There are other good reasons why it is critical to begin innovation and evaluation around insects as a sustainable ingredient/commodity. Insects are a promising source of high-quality animal protein, oils, chitin, nutrients, and other bioproducts with a substantially lower environmental footprint than vertebrate livestock (Chapters 4 and 5; Dossey, 2013; Shockley and Dossey, 2014; van Huis et al., 2013). Insect-based food ingredients (powders, pastes, or liquids) can have a very wide range of commercial applications in the human food, animal/pet feed, and neutraceutical industries. Products utilizing these ingredients might include: fortified dry goods, fortified protein supplement powders, high-protein fortified porridges and cereals, "meat" substitutes, chitosan (neutraceutical derived from chitin), high-protein beverages, protein-fortified bars and powders for athletes as well as numerous types of snack foods (Chapter 5). In general, insects are a very promising sustainable source of food as well as numerous other biomaterials (plastics, neutraceuticals, antimicrobials/antibiotics, pharmaceuticals, cosmetics, etc.). Byproducts from edible insect farming may eventually be useful for such applications. One of the most promising byproducts is the fibrous chitin exoskeleton of insects. Recent research discussed the use of shrimp shells and silk to produce a material very similar to plastic called "Shrilk" (Fernandez and Ingber, 2012). Chitin has the potential to be used in biomedical application, including surgical sutures, gauzes, and a short-term scaffold for tissue (see later for a more detailed discussion on insect chitin). Additionally, insects are known to contain many antimicrobial peptides that could eventually be cheap antibiotics once insect mass production for food applications achieves sufficient scale (Dossey, 2010). Many other possibilities for insect-derived products/byproducts are currently being explored. Also, insect farming operations may eventually utilize a wide variety of untapped and underutilized sources of biomass and agricultural byproducts as feed.

INSECT PHYSIOLOGY AND FUNCTIONALITY

Physiology has a significant impact on the fiber and nutrient content of insects. Insects are comparable to conventional livestock meat in terms of nutritional content. In general, the crude protein content of insects ranges from 40% to 75% on dry weight basis, largely depending on species and stage in the life cycle. Protein quality in relation to human requirement is measured by amino acid profiles and digestibility. Physiology has a significant impact on the fiber and nutrient content of insects. The fiber of insects contain both chitin and significant amounts of amino acid cuticular proteins (Finke, 2007). Published studies show that whole insects contain significant amounts of fiber, but the fiber concentrations vary depending upon analysis due to the high levels of chitin. Chitin is a long-chain polymer of N-acetyl glucosamine, a derivative of glucose, and is found in many places throughout the natural world; it is one of the most important biopolymers in nature and is in fact the second most abundant after cellulose (Shadidi et al., 2011). It is a characteristic component of the cell walls of insects, but can also be found in crustaceans and shellfish. In terms of functionality, it may be compared to the protein keratin. In insects, it serves as a scaffold material that supports the cuticles of the epidermis, trachea, and the peritrophic matrices lining the gut epithelium. Insect growth is highly dependent on the ability to remodel chitin containing structures; because of this, insects continuously produce chitin synthases and chinolytic enzymes in different tissues of the insect body. To enable growth and development, insects periodically replace old cuticle with a new looser one in a highly controlled process known as molting or ecdysis (Merzendorfer and Zimoch, 2003). Chitin compares highly to cellulose, the most important plant source of fiber. Unlike cellulose, chitin possesses a nitrogen atom which allows increased hydrogen bonding between adjacent polymers giving the chitin–polymer matrix increased strength (Fig. 3.1).

Obesity in humans is a significant problem for the young and old alike. Overweight and obesity both contribute to cardiovascular disease, diabetes mellitus, and several cancers and generally diminish the quality of an overall healthy life. Estimations of the global burden of disease attributable to excess body weight indicate that high body mass index (BMI) is a leading cause of loss of healthy life, causing two and a half million deaths and over 30 million lost years of healthy life

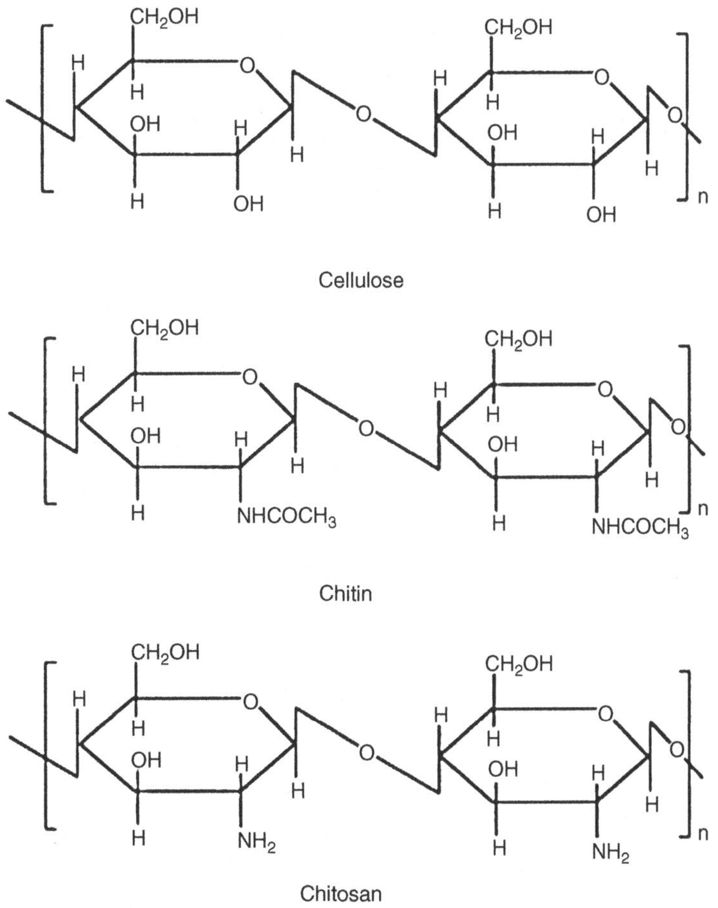

FIGURE 3.1 Cellulose, chitin, and chitosan structures (Kumar 2000).

TABLE 3.6 Insect Fiber Compared to Other High-Fiber Plant Sources

Food Item	Fiber Content (g/100 g)
Insects	4.9–12.1
Lettuce	1.3
Oats	10.1
Peas (cooked)	5.5
Beans (cooked)	2.9
Pineapples (raw)	1.4

in 2000 (Mhurchu et al., 2004). Increased intake of dietary fiber, coupled with other lifestyle changes, forms a big fraction of the solution to this problem. Studies have indicated that chitin and its derivative chitosan are high in fiber and have a lipid binding function in the human gastrointestinal tract. As in cellulose, chitin is largely indigestible and as such may act as dietary fiber. Cellulose is the most important source of dietary fiber by far. USDA dietary guidelines 2010 state that most Americans greatly underconsume dietary fiber, and yet diets high in fiber lead to increased nutrient density, healthier lipid profiles, increased glucose tolerance, better cardiovascular health, and normal gastrointestinal function. Finke (2007) found that by determining the amino acid composition of the ADF fraction, the percentage of ADF that is composed of amino acids can be estimated. This estimation can then be used to calculate chitin concentration by correcting for the amino acid content of the ADF fraction and assuming the remainder of the ADF fraction is chitin. Finke (2007) estimated the chitin content of insects species bred as food for insectivores to range from 2.7 mg to 49.8 mg/kg (fresh) and from 11.6 mg to 137.2 mg/kg (dry matter). Only about 40% of insect chitin is composed of fiber. Data on insect fiber content is scarce but a study conducted by Bukkens (1997, 2005) showed that free range insects with hard exoskeleton had fiber contents of 4.9–12.1 g/100 g dry weight while insects with soft exoskeleton have fiber content ranging from 6.5–11.4 g/100 g dry weight. Both studies indicate high fiber contents in insects bred for food and those that grow wildly (Table 3.6).

Insects clearly exhibit higher amounts of fiber as compared to plant sources and since they are high-protein foods, their high fiber content makes them superior to other high-protein foods such meat, fish, and eggs (Table 3.7).

Chitin-chitosan, when used as a food supplement, may lower plasma cholesterol and triglycerides and improves the HDL-cholesterol/total cholesterol ratio. A study carried out at the New York Center for Biomedical Research stated that the beneficial effect of chitin-chitosan as a food component is the reduction of plasma cholesterol and triglycerides due to its ability to bind dietary lipids, thereby reducing intestinal lipid absorption (Koide, 1998). The study also stated that the hypolipidemic influence of chitosan may also be due to interruption of the enterohepatic bile acid circulation. Bile acid composition and short-chained fatty acid content in the cecum are altered by chitosan, which impedes lipid emulsification and absorption. However, the plasma cholesterol in animals on a cholesterol-free diet is not affected, indicating that endogenous biosynthesis of cholesterol remains intact. In effect, therefore, chitin should be able to reduce intestinal absorption of fat and flush it through the body before it is absorbed, a process that leads to weight loss. Studies on this subject are inconclusive, but it is being used widely in the United States as an over-the-counter weight loss dietary supplement. Chitosan acts by forming gels in the intestinal tract which not only entrap lipids but also other nutrients, including

TABLE 3.7 Insect Fiber Content as Compared to Other High-Protein Food Sources

Food Item	Fiber Content (g/100 g)
Insects	4.9–12.1
Soy milk	0.6
Fish (salmon)	0.0
Beef (lean)	0.0
Chicken	0.0
Eggs	0.0

Source: USDA National Nutrient Database, 2015.

TABLE 3.8 Nutrient Content at Varying Developmental Phases for Select Insects

	Juvenile Crickets	Adult Crickets	Mealworm Larvae	Adult Mealworms
Crude protein (g/100 g)	67.2	66.6	49.1	65.3
Fat (g/100 g)	14.4	22.1	35.0	14.9
Chitin (g/100 g)	9.6	10.2	6.6	20.4

Source: Finke, 2007.

fat-soluble vitamins and minerals, thus interfering with their absorption, a side effect that is still being examined. Certain medical precautions should also be observed with long-term ingestion of high doses of chitosan to avoid potential adverse metabolic consequences.

Ascorbic acid enhances gel formation of chitosan, thereby potentiating the plasma cholesterol lowering activity. In Japan, dietary cookies, potato chips, and noodles enriched with chitosan are manufactured because of its hypocholesterolemic effect. Vinegar products containing chitosan are also manufactured and sold, again because of their cholesterol lowering ability. In the United States, the Food and Drug Administration (FDA) approved the use of chitin and its derivative chitosan in 1983 in food and food supplements as well as processing aids.

However, some researchers argue that chitin is a feeding deterrent as it adds bulk to the diet but is not digestible by primates unless they possess chitanases. Studies have shown that insect-eating populations in Asia, Africa, and Central America possess small amounts of chitanases in their gut compared to non–insect-consuming populations that did not possess any traces of chitanases. Studies show that prolonged ingestion may alter the normal flora of the intestinal tract which may result in the growth of resistant pathogens. Arguments have also been made that presence of chitanases in the gastrointestinal tract (GIT) and chitin itself may bring about an imbalance to the natural micro-flora in the GIT.

Despite its widespread occurrence in insects, the main commercial sources of chitin are crab and shrimp shells. In industry, chitin can be extracted from the insect exoskeleton by acid treatment to dissolve calcium carbonate followed by alkaline extraction to solubilize its protein contents. This is then usually proceeded by a decolorization step to remove leftover pigments and obtain a clearer product. When the degree of deacetylation of chitin reaches about 50% (depending on the origin of the polymer), it becomes soluble in aqueous acidic media and is called chitosan. These treatments must be adapted to each chitin source, owing to differences in the ultrastructure of the initial materials. The resulting chitin needs to be graded in terms of purity and color since residual protein and pigment can cause problems for further processing procedures. Wang et al. (2004) compared cricket, shrimp, and crab chitin and found that the qualities of chitin and chitosans from the field cricket were better than those from the shells of shrimp and crab, which is critical for incorporation of insect meal or powder as an ingredient in food products. Because humans do not possess chitinase, the enzyme necessary for digesting chitin, it has been suggested that insect protein has poor digestibility. Finke (2007) also identified significant differences in chitin and fat as well as protein in cricket nymphs and adult crickets, mealworm adults, and larvae. Table 3.8 shows Finke's research, which identified that protein and fat content was increased in adults, and chitin content was higher in nymphs. Understanding these differences at varying developmental phases will be critical as the industry continues to grow and identify uses for the various ingredients.

Chitin has been found to be very useful in the food processing industry, such as in the formation of biodegradable films, food preservation, water purification, and clarification and deacidification of fruit juices. However, due to its low solubility, its derivative chitosan is preferred for use, or a combination of the two. Chitosan is more soluble and less crystalline than chitin itself, making it more suitable for use in industrial processes.

INSECTS AS A FOOD INGREDIENT

There are three ways insects are consumed: whole insects, whole insects processed into a granular powder or paste, or an extract such as a protein isolate. Internationally, whole, recognizable insects are prepared and consumed in a variety of ways, most popularly as a fried or dried snack. As nonrecognizable form, farmed insects can be processed into a dried and milled product, typically a powder which can be incorporated into food products for increased nutritional value or functionality. Finally, proteins can be extracted from the insects and applied in food and/or feed as an alternative to soy or meat to increase protein content and/or to add functionality and remove other ingredients that may reduce the nutritional profile of foods. Supplementing food products in this way requires increased knowledge of the functional properties of these extracted proteins to include amino acid profile, thermal stability, solubility, gelling, foaming, and emulsifying capacity, and how these properties vary across species. There are several methods for extracting protein groups, solubility in solvents, enzymatic processes, ultrafiltration, and fluidized bed chromatography that are available, but many are cost prohibitive to the industry at this point (van Huis et al., 2013).

Traditionally, insects have been minimally processed to maintain quality and to improve shelf life by canning, refrigeration, roasting, toasting, drying, and baking. All these processing methods, like in many other foods, modify the nutritional value, sensory quality, and shelf life of the product. The storage and processing practices are as varying as the quality parameters that are affected by them. Even minimal processing can impact quality as research conducted by Kinyuru et al. (2010) has shown. Their research showed that processing methods affected their nutrient potential of the insects as identified by changes in protein digestibility and vitamin content. There were significant differences in percent protein concentrations found in green grasshoppers that were toasted and fresh dried. As we continue to research and identify these issues, we need to remain mindful of the fluctuations in nutritional value due to processing as these insect powders are incorporated into nontraditional food products to encourage entomophagy. Therefore, various areas of continued Research and Development (R&D) are needed to investigate functional properties, optimal preparation and processing methods even as we promote commercialization of insects as sustainable food ingredients (Fig. 3.2).

In addition to changes in nutritional value and functionality, organoleptic and rheological properties may also be modified by further processing. Consumer acceptance will be closely tied to the development and implementation of appropriate processing strategies including appropriate extraction and purification methods to utilize insect protein as a food additive. In order for this to be achieved, there are certain protein characteristics that need to be determined outside of the nutritional research that has already been conducted to evaluate protein solubility, thermal stability, gelling capacity, emulsification, and foaming properties.

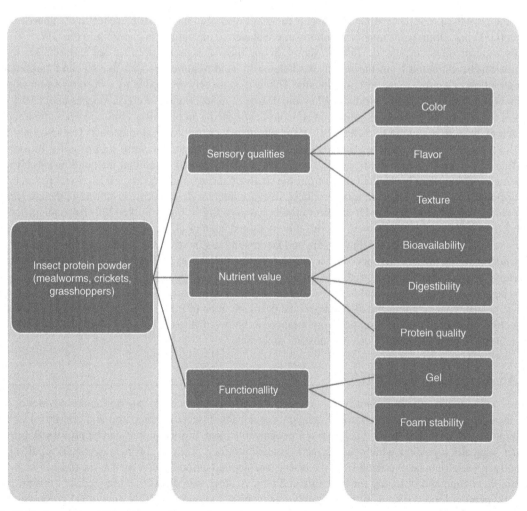

FIGURE 3.2　Critical areas for additional research.

Functional properties are physical and chemical properties that affect the behavior of proteins in food systems during processing, storage, preparation, and consumption (Kinsella, 1982). Protein solubility is measured as the concentration (%) of protein in aqueous solution. Protein solubility is used to predict many functional properties of proteins including its ability to form and stabilize emulsions, foams, and gels. Foam formation and stability are functions of the type of protein, pH, processing methods, viscosity, and surface tension, most of which are or can be modified when developing prepared foods.

All of these properties will significantly impact the sensory properties of these insect powders in terms of mouthfeel, palatability, and flavor. Very few researchers have published peer-reviewed articles that look at the sensory aspects of the processed food products that have incorporated dried or milled insect powder. Aguilar-Miranda et al. (2002) found that by incorporating dried, roasted mealworm powder into a tortilla, the protein concentration increases by fivefold over the traditional maize flour. Results from the Triangle Sensory test with trained judges determined the taste, color, and flavor to be acceptable. Ayieko et al. (2010) incorporated mayflies (Diptera and Ephimeroptera) and termites (Isoptera: termitidae) into crackers, meatloaf, muffins, and sausages. Based on consumer acceptance panels, all the products were highly rated in taste, flavor, smell, texture, and appearance. Although these particular insects are unlikely to be consumed in the United States, it does provide encouragement that the incorporation of insect powders into commonly consumed foods can be palatable and consumable.

INSECT PROTEIN FUNCTIONALITY

Very little information or research currently exists on the functionality of insect protein, but as entomophagy becomes increasingly popular, more research in this area is being conducted. To that end, Yi et al. (2013) worked with the WHO/FAO to propose research that studied the functional properties of extracted insect proteins from the yellow mealworm, the superworm, the lesser mealworm, the house cricket, and the Dubia cockroach. Their research determined that overall, the foaming capacity is very poor, likely due to the presence of oil in supernatant fractions. Gels were formed for all five insect species, using solutions from an aqueous extraction procedure. Additional research needs to be conducted to better understand the mechanism of gelation, and what happens to these gels when exposed to shear as well as the sensory and rheological properties. In addition, additional work needs to be completed to better understand the foaming instability and if it is simply the oil present in the supernatant. This can be remedied by utilizing alternative extraction and purifying techniques to ensure the quality of the functional protein source.

Table 3.9 summarizes the currently available research on functional properties of a variety of insect species. A variety of insect species under certain processing conditions will solubilize in water, have water and fat absorption capacity, and have the capacity to form and stabilize gels and foams. The lipid concentration in many of these powders also allows for significant flavor binding to improve the flavor of the final product.

In addition to protein functionality, insect chitin and chitosan may also have significant functional properties as a food-grade material. Chitin and chitosans have proven to be very important as a processing and preservation aid in the edible film industry. Edible films can provide supplementary and sometimes essential means of packaging certain food stuffs. These films can control physiological changes in food products, thereby extending their shelf life and maintaining quality. Chitosan can be used to control enzymatic browning in produce, to clarify fruit juices during processing, and to reduce food spoilage, due to its inherent antimicrobial nature.

TABLE 3.9 Functional Properties of Varying Insects

Functional Properties	Insect Species	Properties
Solubility	C. forda larvae, T. molitor, A. diapeinus, Z. morio, A. domesticus, B. dubia	X
Water absorption and binding	C. forda larvae	X
Viscosity	–	–
Gelation	C. forda larvae, T. molitor, A. diapeinus, Z. morio, A. domesticus, B. dubia	X
Elasticity		
Emulsification	C. forda larvae	X
Fat absorption	C. forda larvae	X
Flavor binding	–	–
Foaming	C. forda larvae, T. molitor, A. diapeinus, Z. morio, A. domesticus, B. dubia	X

Source: Omotoso, 2006; Yi et al., 2013.

High-density polyethylene (PET) films are extensively used in the packaging industry. However, with increasing concern for biodegradable packaging and natural solutions, there is room for novel product incorporation that may satisfy these consumer demands. In addition, PET films encourage fermentation due to the depletion of oxygen and condensation of water due to fluctuation of storage temperatures, which promotes fungal growth and reduces shelf life. Chitosan products are tough, long lasting, flexible, and very difficult to tear. Most of these mechanical properties are comparable to many medium-strength commercial polymers (Shadidi et al., 2011). Chitosans also have moderate water permeability values and, in combination with other processing techniques such as heating and modified atmosphere packaging, can be used to increase the storage life of fresh fruits and vegetables. Various vegetables and fruits such as cucumbers, bell peppers, strawberries, and tomatoes can maintain good postharvest quality for longer time periods after coating with chitosan. It has been documented that chitosans decrease respiration rates, inhibit fungal development, and delay ripening due to the reduction of ethylene and carbon dioxide evolution thereby prolonging shelf life (Muzzarelli and Muzzarelli, 2006).

Mechanical injury during postharvest handling and processing causes browning of fruits and vegetables with increased loss of quality and value. Phenolic compounds, together with the activity of polyphenoloxidase (PPO), are responsible for this phenomenon and will affect the color, taste, and nutritional value of fruits and vegetables. In recent years, concern over the adverse health effects of sulfite, a widely used browning inhibitor, has spurred research in alternate antibrowning compounds. Application of chitosan film coating leads to delayed changes in contents of anthocyanins, flavonoids, and total phenolics in litchi (*Litch chinensis* Sonn). It also delays the increase in polyphenol oxidase activity and partially inhibits the increase in peroxidase activity. Application of chitosan may form a layer of film on the outer pericarp surface, thus resulting in less browning (Seddiki et al., 1997).

Clarification of processed juices usually involves the use of clarifying agents, such as gelatin, bentonite, silica sol, tannins, and potassium caseinate, and polyvinyl pyrrolidone. Chitosan salts have been found to be effective as dehazing agents due to their strong positive charge, and they may also be used to control acidity in fruit juices. Chitosan is a good clarifying agent for grapefruit juice either with or without the use of pectinase treatment, as well as apple juice and white wines.

There has been a shift in recent years for consumer demand towards foods without chemical preservatives, with more research dedicated to the discovery of novel natural antimicrobials. In this context, the unusual antimicrobial nature of chitin, chitosan, and their derivatives against different groups of microorganisms, such as bacteria, yeast, and fungi has received considerable attention. Chitosan is more soluble and has better antimicrobial activity than chitin. The exact mechanism of the antimicrobial action of chitin, chitosan, and their derivatives is still not clearly known, but varying mechanisms have been proposed. Interaction between positively charged chitosan molecules and negatively charged microbial cell membranes leads to the leakage of proteinaceous and other intracellular constituents. Chitosan also acts as a chelating agent that selectively binds trace metals and thereby inhibits the production of toxins and microbial growth. It also activates several defense processes in the host tissue, acts as a water-binding agent, and inhibits various enzymes. Binding of chitosan with DNA and inhibition of mRNA synthesis reoccurs via chitosan penetrating the nuclei of the microorganisms and interfering with the synthesis of mRNA (Shadidi et al., 2011).

CONCLUSIONS

Insects may not be a very common source of food in many parts of the world, but research has proven their potential to make an impact in both the nutrition and processing aspects of the food industry. The dietary and functional uses of the protein and fibrous chitin-chitosan from insect sources is a field that needs to be further explored and researched.

REFERENCES

Aguilar-Miranda, E.D., Lopez, M.G., Escamilla-Santana, C., Barba de la Rosa, A.P., 2002. Characteristics of maize flour tortilla supplemented with ground *Tenebrio molitor* larvae. J. Agric. Food Chem. 50 (1), 192–195.

Ayieko, M.À., Oriamo, V., Nyambuga, I.A., 2010. Processed products of termites and lake flies: improving entomophagy for food security within the Lake Victoria region. Afr. J. Food Agric. Nutr. Dev. 10, 2085–2098.

Babji, A.S., Fatimah, S., Ghassem, M., Abolhassani, Y., 2010. Protein quality of selected edible animal and plant protein sources using rat bio-assay. Int. Food Res. J. 17, 303–308.

Bhutta, Z.A., Ahmed, T., Black, R.E., Cousens, S., Dewey, K., Giugliani, E., Haider, B.A., Kirkwood, B., Morris, S.S., Sachdev, H.P., Shekar, M., 2008. What works? Interventions for maternal and child undernutrition and survival. Lancet 371, 417–440.

Black, R.E., Allen, L.H., Bhutta, Z.A., Caulfield, L.E., De Onis, M., Ezzati, M., Mathers, C., Rivera, J., 2008. Maternal and child undernutrition: global and regional exposures and health consequences. Lancet 371, 243–260.

Black, R.E., Morris, S.S., Bryce, J., 2003. Where and why are 10 million children dying every year? Lancet 361, 2226–2234.
Bukkens, S.G.F., 1997. The nutritional value of edible insects. Ecol. Food Nutr. 36, 287–319.
Bukkens, S.G.F., 2005. Insects in the human diet: nutritional aspects. In: Paoletti, M.G. (Ed.), Ecological Implications of Minilivestock: Potential of Insects, Rodents, Frogs, and Snails. Science Publishers, Enfield, NH, pp. 545–577.
Cather, A. 2016. Two billion people eat insects and you can too. Food Tank. N.p.
Caulfield, L.E., De Onis, M., Blossner, M., Black, R.E., 2004. Undernutrition as an underlying cause of child deaths associated with diarrhea, pneumonia, malaria, and measles. Am. J. Clin. Nutr. 80, 193–198.
Cdc.gov. 2014. Nutrition for Everyone: Basics: Protein DNPAO CDC. Available from: http://www.cdc.gov/nutrition/everyone/basics/protein.html
Collins, S., Dent, N., Binns, P., Bahwere, P., Sadler, K., Hallam, A., 2006. Management of severe acute malnutrition in children. Lancet 368, 1992–2000.
Defoliart, G., 1992. Insects as human food. Crop Protect. 11, 395–399.
Defoliart, G.R., 1995. Edible insects as minilivestock. Biodivers. Conserv. 4, 306–321.
Digitaljournal.com. 2016. Global edible insects market projected to reach $1.53 billion in 2021—Press Release–Digital Journal. N.p.
Dolson, L. 2014. Protein Info—How Much Protein Do You Need. Available from: http://lowcarbdiets.about.com/od/nutrition/a/protein.htm
Dossey, A.T., 2010. Insects and their chemical weaponry: new potential for drug discovery. Nat. Prod. Rep. 27, 1737–1757.
Dossey, A.T., 2013. Why insects should be in your diet. Scientist 27, 22–23.
Duggan, C., Watkins, J.B., Walker, W.A., 2008. Nutrition in Pediatrics: Basic Science, Clinical Application. BC Decker, Hamilton, xvii, 923 p.
Durst, P.B., et al., 2010. Forest Insects As Food: Humans Bite Back. RAP Publication, Regional Office for Asia and the Pacific, Chiang Mai, Thailand.
Dzamba, J., 2010. Third Millennium Farming: Is it time for Another Farming Revolution? Architecture. Landscape and Design, PL Toronto, CA, Available from: http://www.thirdmillenniumfarming.com/
FAO, 2012. Assessing the potential of insects as food and feed in assuring food security. In: Vantomme, P., Mertens, E., van Huis, A., Klunder, H. (Eds.), Summary Report of Technical Consultation Meeting, Rome, Italy.
Fernandez, J.G., Ingber, D.E., 2012. Unexpected strength and toughness in chitosan-fibroin laminates inspired by insect cuticle. Adv. Mater. 24, 480–484.
Finke, M.D., 2002. Complete nutrient composition of commercially raised invertebrates as food for insectivores. Zoo Biol. 21, 269–285.
Finke, M.D., 2007. Estimate of chitin in raw whole insects. Zoo Biol. 26, 105–115.
Finke, M.D., 2013. Complete nutrient content of four species of feeder insects. Zoo Biol. 32, 27–36.
Gahukar, R.T., 2011. Entomophagy and human food security. Int. J. Trop. Insect Sci. 31, 129–144.
Latunde-Dada, Gladys O., Yang, Wenge, Aviles, Mayra Vera, 2016. In Vitro Iron Availability from Insects Sirloin Beef. J. Agric. Food Chem. 64 (44), 8420–8424.
Hoppe, C., Andersen, G.S., Jacobsen, S., Molgaard, C., Friis, H., Sangild, P.T., Michaelsen, K.F., 2008. The use of whey or skimmed milk powder in fortified blended foods for vulnerable groups. J. Nutr. 138, 145S–161S.
Kinsella, J.E., 1982. Structure and functional properties of food proteins. In: Fox, P.F., Condon, J.J. (Eds.), Food Proteins. Applied Science Publishers, New York, NY.
Kinyuru, J.N., Kenji, G.M., Njoroge, S.M., Ayieko, M., 2010. Effect of processing methods on the in-vitro protein digestibility and vitamin content of edible winged termite (*Macrotermes subhylanus*) and Grasshopper (*Ruspolia differens*). Food Bioprocess Tech. 3 (5), 778–782.
Koide, S.S., 1998. Chitin-chitosan: properties, benefits and risks. Nutr. Res. 18 (6), 1091–1101.
Kumar, M., 2000. A review of chitin and chitosan applications. React. Funct. Polym. 46, 1–27.
Marketwatch. 2016. Global edible insects market value set to reach US$ 423.8 Mn in 2016 end—Persistence Market Research. N.p.
Melo-Ruiz, V., Sandoval-Trujillo, H., Quirino-Barreda, T., Sanchez-Herera, K., Diza-Garcia, R., Calvo-Carrillo, C., 2015. Chemical composition and amino acids content of five species of edible Grasshoppers from Mexico. Emirates J. Food Agr. 27 (8), 654–658.
Merzendorfer, H., Zimoch, L., 2003. Chitin metabolism in insects: structure, function and regulation of chitin synthases and chitinases. J. Exp. Biol. 206, 4393–4412.
Mhurchu, C., Dunshea-Mooij, C., Bennett, D., Rodgers, A., 2004. The effect of the dietary supplement, chitosan, on body weight: a systematic review of randomized controlled trials. Obes. Rev. 6, 35–42.
Michaelsen, K.F., Hoppe, C., Roos, N., Kaestel, P., Stougaard, M., Lauritzen, L., Molgaard, C., Girma, T., Friis, H., 2009. Choice of foods and ingredients for moderately malnourished children 6 months to 5 years of age. Food Nutr. Bull. 30, S343–S404.
Muzzarelli, R.A., Muzzarelli, C., 2006. Chitosan as a dietary supplement and a food technology agent. In: Billaderis, C.G., Izydorczyk, M.S. (Eds.), Functional Food Carbohydrates. CRC Press, Florida, pp. 215–238.
Nonaka, K., 2009. Feasting on insects. Entomol. Res. 39, 304–312.
Omotoso, O.T., 2006. Nutritional quality, functional properties and anti-nutrient compositions of the larva of *Cirina forda* (Westwood) (Lepidoptera: Saturniidae). J. Zhejiang Univ. Sci. B 7 (1), 51–55.
Oonincx, D.G., Van Itterbeeck, J., Heetkamp, M.J., Van Den Brand, H., Van Loon, J.J., Van Huis, A., 2010. An exploration on greenhouse gas and ammonia production by insect species suitable for animal or human consumption. PLoS One 5, e14445.
Pelletier, D.L., Frongillo, E.A., 2003. Changes in child survival are strongly associated with changes in malnutrition in developing countries. J. Nutr. 133, 107–119.
Ramos-Elorduy, J., 1997. Insects: A sustainable source of food? Ecol. Food Nutr. 36, 247–276.
Ramos-Elorduy, J., 2009. Anthropo-entomophagy: cultures, evolution and sustainability. Entomol. Res. 39, 271–288.
Ramos-Elorduy, J., Moreno, J.M.P., Prado, E.E.P., Manuel, O.A., Otero, J.L., de Guevara, L.O., 1997. Nutritional value of edible insects from the state of Oaxaca, Mexico. J. Food Compos. Anal. 10, 142–157.

Ramos-Elorduy, J., Pino Moreno, J.M., Martinez Camacho, V.H., 2012. Could grasshoppers be a nutritive meal? Food Nutr. Sci. 3, 164–175.

Rice, A.L., Sacco, L., Hyder, A., Black, R.E., 2000. Malnutrition as an underlying cause of childhood deaths associated with infectious diseases in developing countries. Bull. World Health Organ. 78, 1207–1221.

Rumpold, B.A., Schlüter, O.K., 2013. Nutritional composition and safety aspects of edible insects. Mol. Nutr. Food Res. 57 (5), 802–823.

Safina, C., 2011. Why Are We Using Up the Earth? CNN Opinion: Carbon Dioxide. CNN, New York.

Schofield, C., Ashworth, A., 1996. Why have mortality rates for severe malnutrition remained so high? Bull. World Health Organ. 74, 223–229.

Seddiki, N., Mbemba, E., Letourneur, D., Ylisastigui, L., Benjouad, A., Saffar, L., Gluckman, J.C., Jozefonvicz, J., Gattegno, L., Zhang, D., Quantick, P.C., 1997. Effects of chitosan coating on enzymatic browning and decay during postharvest storage of litchi (*Litchi chinesnsis* Sonn.) fruit. Postharvest Biol. Tech. 12 (2), 195–202.

Shadidi, F., Arachchi, J., Jeon, Y., 2011. Food applications of chitin and chitosans. Trends Food Sci. Tech. 10 (2), 37–51.

Shockley, M., Dossey, A.T., 2014. Insects for human consumption. In: Morales-Ramos, J.A., Rojas, M.G., Shapiro-Ilan, D.I. (Eds.), Mass Production of Beneficial Organisms, Invertebrates and Entomopathogens. Academic Press, Waltham, MA, pp. 617–652.

Singh, B., Singh, U., 1991. Peanut as a source of protein for human foods. Plant Foods Human Nutr. 41, 165–177.

Sirimungkararat, S., Saksirirat, W., Nopparat, T., Natongkham, A., 2010. Forest Insects as Food: Humans Bite Back. In: Durst, P.B., Johnson, D.V., Leslie, R.N., Shono, K. (Eds.). FAO, Bangkok, Thailand, pp. 189–200.

Srivastava, S.K., Babu, N., Pandey, H., 2009. Traditional insect bioprospecting—as human food and medicine. Ind. J. Trad. Know. 8, 485–494.

Steinfeld, H., Gerber, P., Wassenaar, T.D., Castel, V., Rosales, M.M., Haan, C.D., Food and Agriculture Organization of the United Nations, and Livestock Environment and Development (FIRM), 2006. Livestock's Long Shadow: Environmental Issues and Options. Food and Agriculture Organization of the United Nations, Rome, xxiv, 390 p.

Tatum, T. 2014. Global Food Reserves Lowest in 40 Years. Available from http://planetsave.com/2012/10/19/global-food-reserves-lowest-in-40-years/

van Huis, A., Itterbeeck, J.V., Klunder, H., Mertens, E., Halloran, A., Muir, G., Vantomme, P., Food, and Agriculture Organization of the United Nations, 2013. Edible Insects: Future Prospects For Food And Feed Security. Food and Agriculture Organization of the United Nations, Rome, 187 p.

Vantomme, P., Mertens, E., Van Huis, A., Klunder, H., 2012. Assessing the Potential of Insects as Food and Feed in Assuring Food Security. United Nations Food and Agricultural Organization, Rome, Italy.

Victora, C.G., Adair, L., Fall, C., Hallal, P.C., Martorell, R., Richter, L., Sachdev, H.S., 2008. Maternal and child undernutrition: consequences for adult health and human capital. Lancet 371, 340–357.

Vogel, G., 2010. For more protein, filet of cricket. Science 327, 811.

Wang, D., Bai, Y., Li, J., Zhang, C., 2004. Nutritional value of the field cricket (*Gryllus testaceus* Walker). Insect Sci. 11 (4), 275–283.

Wardlaw, G.M., Smith, A.M., 2006. Contemporary Nutrition, sixth ed. New York: McGraw Hill Higher Education.

Yen, A.L., 2009. Entomophagy and insect conservation: some thoughts for digestion. J. Insect Conserv. 13, 667–670.

Yi, L., Lakemond, C.M., Sagis, L.M., Eisner-Schadler, V., Huis, A.V., van Boekel, M.A., 2013. Extraction and characterization of protein fractions from five insect species. Food Chem. 141 (4), 3341–3348.

Chapter 4

Edible Insects Farming: Efficiency and Impact on Family Livelihood, Food Security, and Environment Compared With Livestock and Crops

R.T. Gahukar

Arag Biotech Private Limited, Nagpur, Maharashtra, India

Chapter Outline

Introduction	85	Indoor Farming	97
Food Security/Family Livelihood	87	Outdoor Farming	101
Biodiversity and Availability of Insects	89	Farming in Space	102
Consumption of Insects versus Other Livestock	90	**Market Potential**	**103**
Cost of Cultivation	91	Retail/Local Marketing	103
Possibility of Replacing Livestock With Insects as Human Food	93	Export	104
		New Products From Farmed Insects	104
Environmental Impact	94	**Safety Regulations**	**105**
Industrial Perspective	96	**Current Challenges and Conclusions**	**105**
Commercial Insect Farming for Mass Production	97	**References**	**106**

INTRODUCTION

With increasing human population (expected to be more than 9.6 billion by 2050) and per capita demand for protein, there is need to double the protein production and increase global food production by an estimated 60% from current levels (Steinfeld et al., 2006; Pelletier and Tyedmers, 2010). The world human population increases by about 70 million people each year. The United Nations expects the human population to grow to more than 9 billion by 2050, adding twice the current population of China (Dzamba, 2010; Safina, 2011; Vogel, 2010). Humans consume roughly 40% of the biomass that the land and the coastal seas produce (Safina, 2011). Approximately 70% of agricultural land and 30% of total land on Earth is used to raise livestock (Steinfeld et al., 2006). Food reserves are at a 50-year low, yet demand for food is expected to increase by 50% by 2030. Relying on food strategies including livestock production to feed our ever-growing human population seems to be impossible. To cope with this situation, food sources other than traditional food crops and animals are being explored, and insects could be an affordable and nutritious alternative and can address the current gloomy picture of food security (Gahukar, 2011a).

Fortunately, insects provide a very promising alternative to other livestock. They can be farmed using less resources such as land, water, energy, and feed than livestock while providing more protein and contributing less to climate change due to generating lower greenhouse gas emissions and lower levels of pollutants such as ammonia (Fig. 4.1).

Insects form part of the diets of at least 2 billion people in nearly 100 countries of the world particularly in Southeast Asia and the Pacific, sub-Saharan Africa, and Latin America (Hanboonsong et al., 2013). Human entomophagy (the practice of humans utilizing insects as food) is nearly absent in the Western world (van Huis, 2013). However, in the future, incorporat-

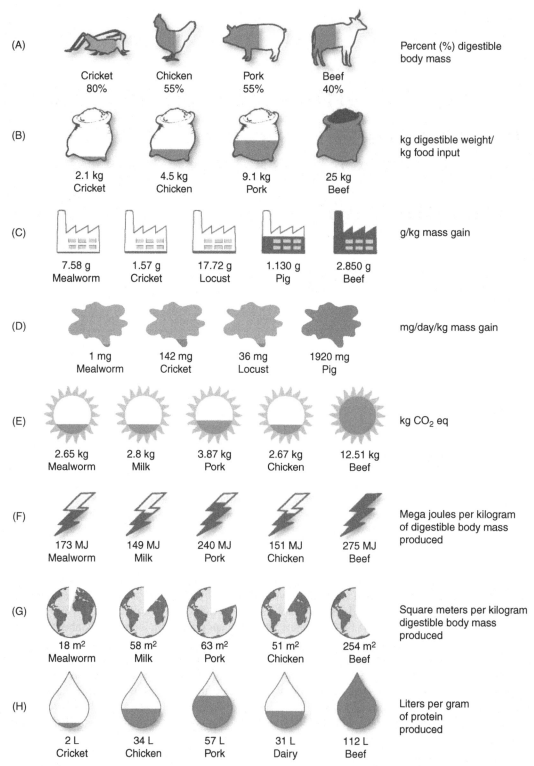

FIGURE 4.1 Resource use and environmental impact parameters of insect farming versus other livestock. (A) Percentage of digestible biomass (van Huis et al., 2013); (B) feed conversion ratio (van Huis, 2013); (C) production of greenhouse gas equivalents per kg of body mass gain (Oonincx et al., 2010); (D) production of ammonia pollution per kg of body mass gain (Oonincx et al., 2010); (E) global warming potential (Oonincx and de Boer, 2012; the GWP of 1 kg of fresh mealworms was 2.7 kg of CO_2- eq); (F) energy use (van Huis et al., 2013); (G) land use (Oonincx and de Boer, 2012); and (H) water use (adjusted according to food input per protein output for dairy and crickets; Earth-policy.org, 2014; Pimentel and Pimentel, 2003; A Strategic Look at Protein, 2014; Ansc.purdue.edu, 2014; Waterfootprint.org, 2014; News.cornell.edu, 2014; van Huis et al., 2013). *(Figure courtesy of book editor Dr. Aaron T. Dossey, president, founder, and owner of All Things Bugs LLC / Griopro cricket powder; www.cricketpowder.com)*

ing insects into the mainstream Western diet will not only be a healthy idea for inhabitants, but also a feasible way to produce greater amounts of food in a more sustainable way (Schosler et al., 2012). Indeed this is already being explored (Nonaka, 2005; Gahukar, 2011a; FAO, 2013; van Huis, 2013; Hanboonsong et al., 2013; see also chapters 1 and 5 of this book).

More than 1900 insect species have reportedly been used as human food (FAO/WUR, 2012; van Huis, 2013). Ramos-Elorduy (2009) reported 2086 insect species consumed by 3071 ethnic groups in 130 countries due to high content of protein and minerals, and also because of their taste and palatability. For example, female house crickets, *Acheta domesticus* L. are described as having pleasant taste and delightfully crunchy texture owing to the large number of eggs inside their abdomen (Hanboonsong et al., 2013). In northeastern India, the nutty flavored "polu leta fry" (made from silkworm pupae) and fried silkworms are regular dishes of the Assamese tribes who are also fond of the ants and ant egg dish due to distinctly minty citrus taste that is imparted to ants by foraging on lemon trees (Doley and Kalita, 2012). A dish prepared from ground larvae mixed with chilli pepper and salt popularly known as "Gran Mitla" is very much appreciated in Mexico (Gray, 2014). The stir-fried coconut rhinoceros beetle is a delicacy in the tourist areas of Thailand (Ratcliffe, 2006). Crickets, silkworm larvae/pupae, ant pupae, giant water bugs, and cicadas cooked with rice and vegetables are served in local eateries in the USA (Gracer, 2010). The traditional dishes can be encouraged by disseminating information to locals on the benefits, particularly as a rich source of protein. Thus, there is great precedent on which to build for extending insects as food ingredients to regions and countries where food insecurity is a persistent problem of local communities for whom farming/rearing prolific and adaptable species forms a part of daily work (FAO, 2013). Mass production of edible insects has also been envisaged by a rapidly growing number of enterprises in Western nations (particularly Europe and North America). Means to make insect farming more effective and remunerative, rearing techniques, feed conversion ratios, and cost of production compared with crops and livestock are reviewed in this chapter with regard to future perspective and concerned legislation. Constraints of insect farming and certain possible measures to improve insect farming as a large-scale commodity for the Western world will also be discussed.

FOOD SECURITY/FAMILY LIVELIHOOD

In rural areas, insect farming facilitates food provision, particularly against seasonal shortage of food; it also provides additional cash for basic expenditures including food, farm implements, and education. The system also offers an opportunity for improvement of local diets, employment, and earning income to landless dwellers and women who are involved in the gathering, farming/cultivation, processing, and sale of insects (Agea et al., 2008; Hope et al., 2009).

Mopane caterpillars (*Gonimbrasia belina* Westwood) are available whole year and consumed by about 70% of population in Congo PDR and about 95% population in Central African Republic (FAO, 2004). In Papua New Guinea, sago pancakes are fortified with sago larvae (*Rhynchophorus ferrugineus* Olivier) to increase their nutritional value (Mercer, 1997). In Australia, food containing witchetty grubs [*Endoxyla leucomochla* (Turner)], scale insects, beetle larvae, ants and Bogong moths (*Agrotis infusa* Boisduval) forms an integral part of the daily diet of aborigines who enjoy these foods during religious rituals and social functions (Yen, 2005). For family livelihood, some insects considered crop pests are also collected from the fields for home consumption. For example, stink bugs (*Tessaratoma papillosa* Drury, *Tessaratoma* spp.) provide not only nutrition to locals but also generate income for family (Cerritosa, 2009). In Uganda, 1 kg of tettigonid grasshoppers, *Ruspolia nitidula* (Scopoli) fetches in local market about 40% higher price than 1 kg of beef (Agea et al., 2008). In Thailand, farmers grow corn to feed Bombay locust, *Patanga succincta* L., collect and sell them rather than cultivating the corn for sale (Hanboonsong, 2010). Mechanical harvesting and eating of red grasshopper, *Sphenarium purpurascens* Charpentier (known locally as Chapulines) in Mexico (Ramos-Elorduy, 2009), and weaver ants (*Ocophylla smaragdina* Fab.) in Thailand (Offenberg and Wiwatwitaya, 2009) is a common practice.

Nowadays, large scale farming has been adopted by farmer groups. For example, about 137,000 households in Thailand farm silkworms contributing to 80% of the country's silkworm production (Sirimungkararat et al., 2010). Besides larvae, silkworm pupae are processed, packaged, labeled, and exported. The exports earned were worth US $50.8 m for the Thai government in 2004 (Sirimungkararat et al., 2010). Similarly, larvae and pupae of weaver ants are routinely collected by women, who can get about 30% of yearly income from this sale. Fresh larvae and pupae provide 7 g of protein and 79.2 kcal of energy per 100 g of produce that is routinely consumed by locals in Thailand (Yhoung-Aree et al., 1997). Additionally, Offenberg (2011) identified Asian weaver ant as a very promising insect to farm on a commercial basis. In India, for 15 years farmers have been able to earn net profit of US $1,843–1,905 (Rs. 110,600–114,340) from farming domesticated mulberry silkworm by planting mulberry crop only on 0.4 ha. This village-level activity is gaining popularity in central and southern states (A.D. Jadhav, personal communication). In Southeast Asia, the price of one bag (100–120 grubs of palm weevil) fetches US $2.11 in local markets, which is comparable in value to 20 chicken eggs and 3 kg of rice (Ramandey and van Mastrigt, 2010).

Utilization of insects as a food ingredient along with insect farming can support rural livelihood (Gahukar, 2011b). To achieve this, it is essential that farming should be a sustainable, low-cost practice to produce insects in a large number. It should offer delicious and affordable dishes to consumers and create a demand for sustainably produced food. This is possible because in nature, even after collecting/harvesting edible insects from neighboring areas and forests, some insect populations survive and multiply even in harsh conditions due to their abundance and prolific nature. Insects are thus always available for certain communities dependent on wild harvested insects, and the family need for food is fulfilled year-round. However, destruction of the forest and natural lands has a severe and negative effect on the food security of such communities.

In tribal communities, traditional knowledge of farming and collecting insects is generally passed on to next generation and whole family is assured of food availability, at least during the period of distress or hardship (Gahukar, 2011a). Therefore, economic value of insects can be ascertained by integrating traditional and ecological knowledge (Losey and Vaughan, 2006). Since farming insects is often done by women and children for whom it can be a low-input sustainable practice, mass collection of the favorite/preferred species can eventually contribute to rural food stocks if insects are dried and preserved. Later, processed insects can be utilized for family food and for rituals and religious functions when live insects are not easily available (Munyuli Bin Mushambanyi, 2000).

Insects are farmed and collected in traditional manner by several tribal communities. There is need to improve and intensify these practices so that locals can subsist on insect farming not only for family livelihood but also for earning some income so that they can afford other essentials for their livelihood. Those who rely on wild harvested insects as the only reliable source of food typically have to travel long distance away from their homes to collect enough insects. Otherwise, farmers have to earn money by marketing of farm produce and other agricultural commodities. For example, selling peal millet, a major food crop in sub-Saharan Africa, may fetch enough money for a family (van Huis, 2003); consequently, family members are not interested in insect farming. On the other hand, farmers in northeast Thailand leave paddy fields barren whenever rains are insufficient for cultivation and face food penury. In these times, they rear insects on chicken feed, pumpkins, and other vegetables. This type of insect farming is practiced by nearly 2,000 registered farms, and 217,529 rearing pens are producing nearly 7,500 tons of crickets, grasshoppers, and red ant eggs every year. From this activity, the per capita annual income for families is the equivalent of US $5000 year^{-1} versus US $2200 from crop cultivation (Gray, 2014).

For many, wild harvested insects are mainly seasonal and the abundance of insects depends upon several environmental factors. Habitat management including prevention from bush fire, attack of predators, and biodiversity conservation is necessary for continuous sustainable harvesting of insects. In periods of nonavailability of primary hosts, alternative food plants play an important role in the survival and development of edible insects. In northeastern India, three wild silkworm species (eri silkworm, *Philosamia ricini* Donovan, muga silkworm, *Antheraea assamensis* Helfer, and tasar silkworm, *Antheraea mylitta* Drury) are being successfully farmed on wild host plants with technical guidance and financial assistance from the government (Gahukar, 2014). In Zambia, Chidumayo and Mbata (2002) recommended certain measures for maintaining populations of mopane caterpillars including studying the ecology and monitoring the population dynamics of tree species, protecting host plants against bush fire, and restricting insect harvesting. Moreover, community forestry with collaborative management by locals can be a viable solution to stop current overexploitation of edible insects. This would facilitate forest insect industry as livelihood program with active support for the conservation of insect biodiversity (Bhatia and Yousuf, 2013). Nowadays, there is an increasing initiative for organic cultivation of host plants of edible insects. Likewise, inventory of new host plants in agroecosystems and forest areas, and maintaining natural ecological balance would be ideal measures. Apart from traditional collection, insects are farmed under controlled conditions by farmers and eventually, they can create employment in both rural and urban areas (Gray, 2014). Furthermore, Hanboonsong et al. (2013) recommended small-scale cricket farms for students in primary school as a cheap and authenticated source of protein for the children at midday snacks or lunch. Insects can be farmed also by resource-poor producers for generating reasonably sufficient income for rural folk. For this purpose, locals should be trained in improved rearing techniques and quality control for those insect species that have high nutritional values. Small units and cottage industries can be converted into industrial units with adoption of economic insect production on a large scale as it has been done in Korea (Kim et al., 2008). At the regional level, opening of insect banks would facilitate exchange of insects and insect products, as well as sharing of information. At government level, inclusion of edible insects in projects and policies would be a welcome step for small families who subsist on edible insects.

Indoor or semiindoor farming is the next stage in sophistication and toward larger scale insect production in a more controlled and reliable environment. Insect farming in structures prepared from easily available indigenous material has been successful in India (Sathe et al., 2008), China (Feng et al., 2009), and Korea (Kim et al., 2008). Sathe et al. (2008) constructed rearing sheds in fields with mud wall, and used nylon nets and bamboo poles for erecting rearing beds for mulberry silkworm. This technique proved helpful in maintaining humidity and temperature and is reasonably economical. Other low-cost techniques include wooden hive boxes for giant hornets/wasps (*Vespula* spp., *Vespa* spp.) in Japan (Nonaka, 2010), cement tanks or wooden containers covered with plastic sheets for crickets in Thailand (Hanboonsong et al., 2013), and small clay jars kept near the walls of houses for stingless bees (Meliponines) in Mexico (Ramos-Elorduy, 2009). Successful farming in nature has been reported from India by Ronghang and Ahmed (2010) for the giant honeybee, *Apis dorsata* Fb., Indian honeybee, *Apis cerana indica* Fb., spider wasp, *Pompilus atrox* Dahlbo, pollen beetle, *Meligethes aeneus* Fb., and from Latin America by Cerda et al. (2001) for the South American palm weevil, *Rhyncophorus palmarum* L.

There are clear possibilities to support a family livelihood with insects as food or ingredient in food recipes because they are efficient, sustainable, prolific, and can be grown in small places even in urban areas (farming in skyscrapers or low-rent industrial areas in cities). Biodiversity is usually high and robust enough for ideal species selection in each location. Therefore, insects are viewed as an industrial product, farmed on a large scale, indoors, in clean facilities, utilizing mechanization and automation for lowering cost.

BIODIVERSITY AND AVAILABILITY OF INSECTS

The UN FAO estimates that there are well over 1000 edible insects currently used (Vogel, 2010), and others estimate that number to be over 2000 (Ramos-Elorduy, 1997, 2009; Jongema, 2012). There are over 1 million species of insects described and 4–30 million species estimated to exist on Earth, living in every niche inhabited by humans and beyond (Dossey, 2010). With this diversity and their collective adaptability, they are a much safer source for future food security than are vertebrate animals such as cattle, fowl, or even fish. Development of more diversity in animal livestock/protein sources is critical to human food security going forward. Insects, being the largest and most diverse group of organisms on Earth, certainly have a substantial role to play. For example, since there are insects of some sort on nearly every patch of land on earth, chances are that some local species in every area can be farmed as human food without the need to import nonnative species for the same purpose. Since insects are prolific in the wild, they can be readily tapped to replenish gene pools of farmed stock, unlike most terrestrial vertebrate livestock. That capability, along with a reliance on multiple farmed insect species, can greatly reduce the impact of livestock shortages on food security.

Previously, Shockley and Dossey (2014) have described in detail the worldwide biodiversity of edible insects. A more complete and up-to-date summary of the diversity of insect use as food worldwide can be found in Chapter 2. In continental distribution, Americas have the highest number of 276 of edible insects followed by Asia with 264 species, Australia with 182 species, and Africa with 83 species, the predominant regions being the Latin America and Southeast Asia (DeFoliart, 1997). Since the late 1990s, several surveys have shown a significant increase in edible insect populations due to identification of new species. For example, Ramos-Elorduy (1997) reported 348 insects in Mexico and later, this number increased to 549 (Ramos-Elorduy, 2009). On a global scale, a majority of edible insect species belong to Coleoptera (31%) followed by Lepidoptera (18%), Hymenoptera (14%), and Orthoptera (13%) (see Chapter 2 and Tables 4.1 and 4.2).

On a regional level, there are differences in biodiversity, for example, 428 edible insects in the Amazon basin in Latin America (Paoletti and Dufour, 2005), 164 insects in Southeast Asia (Yhoung-Aree and Viwatpanich, 2005), and 250 in the sub-Saharan region of Africa (van Huis, 2003). Similar reports are available for country figures. Table 4.1 shows a wide variation from 549 insects in Mexico to 40 insects in Nigeria. This variation is natural due to geographic situation, topography, climate, survival of native species, and so forth. In India, Chakravorty (2014) reported for the northeast region, a total

TABLE 4.1 Which Insects Are We Eating?

Coleoptera (beetles) = 31%
Lepidoptera (caterpillars) = 18%
Hymenoptera (bees, wasps, ants) = 14%
Orthoptera (grasshoppers, locusts, crickets) = 13%
Hemiptera (leafhoppers, planthoppers, scale insects, true bugs) = 10%
Odonata (dragon flies) = 3%
Isoptera (termites) = 3%
Diptera (flies) = 2%
Others = 6%
Source: Jongema (2012).

TABLE 4.2 Number of Insects Used as Food in Certain Countries

Countries	No. of Insects	References
Mexico	549	Ramos-Elorduy (2008)
China	187	Chen et al. (2009)
Thailand	150	Hanboonsong et al. (2013)
Brazil	135	Costa-Neto (2012)
Central African Republic	96	Roulon-Doko (1998)
Ecuador	83	Onore (1997)
Japan	55	Nonaka (2005)
Nigeria	40	Banjo et al. (2006b)

of 255 species: Coleoptera = 86, Orthoptera = 62, Hemiptera = 43, Hymenoptera = 26, Odonata = 20, Lepidoptera = 11, Isoptera = 5, and Ephimeroptera = 2. In a survey in Nigeria, Adeoye et al. (2014) caught only 14 species as follows: Coleoptera = 4, Orthoptera = 4, Lepidoptera = 3, Hemiptera = 1, Hymenoptera = 1, and Isoptera = 1. From the same region, Alamu et al. (2013) reported 22 species: Coleoptera = 6, Lepidoptera = 6, Orthoptera = 5, and Isoptera + Hemiptera + Hymenoptera = 5. In the Central African Republic, although insect numbers differed in each community, overall percentages are: Orthoptera = 40%, Lepidoptera = 36%, Isoptera = 10%, Coleoptera = 3%, and others = 8% (Roulon-Doko, 1998). These data showed that diversity of edible insects differs in each country and regions within countries.

The host plants, cropping systems, natural enemies, and weather factors can affect insect diversity and, indirectly, can influence the availability and choice of insects farmed and eaten in the region (Yen, 2009b; Tikader and Kamble, 2010). When plentiful, some insects are eaten more than others, for example, beetle grubs (31%); caterpillars (18%); wasps, honeybees, and ants (15%); crickets, grasshoppers, and locusts (13%); cicadas, leafhoppers, plant hoppers, scale insects, and true bugs (11%); termites (3%); dragonflies (3%); flies (2%); and others (5%) (Cerritosa, 2009; FAO, 2014). This information is important because the choice/preference of local communities can be positively influenced by easy availability of insect species such as, *Omphisa fuscidentalis* Hmps., *Anaphe venata* Butler, *Imbrasia* spp., *Gynanisa* sp., *Oryctes rhinoceros* F., and *Macrotermes* spp. in sub-Saharan Africa; *Lethocerus indicus* L., *Camponotus* spp., *Bombyx mori*, *Philosamia ricini* in the Indian subcontinent; and *Apis* spp., *Vespa* spp., *Acheta domesticus*, *Gryllotalpa* spp. and grasshoppers in Southeast Asia (Gahukar, 2011a).

Insect farming in cultivated and uncultivated land under established crops or horticultural systems has been successfully explored in Nigeria (Banjo et al., 2006b). Ronghang and Ahmed (2010) reported reduction in biodiversity due to overexploitation and practice of *zhum* cultivation in the state of Assam in India. On the contrary, shifting crop cultivation has improved caterpillar farming and thereby livelihood in the Kopa area of northern Zambia (Chidumayo and Mbata, 2002). Unfortunately, biodiversity of plants that serve as hosts for many insects is declining due to conversion and degradation of habitats (Hanboonsong et al., 2013). Consequently, some species of edible insects have become extinct. For example, 14 species in Mexico have been considered in this category (Ramos-Elorduy, 2006). This fact goes often neglected and many species may have vanished from various ecosystems. A detailed survey would be useful in knowing such species and in planning for their rejuvenation.

Biodiversity is important to maintain a continuous availability of farmed insects in sufficient quantity. For example, if one species dies of disease or cannot be reared in one location, another species or subspecies or close relative can be used, as this is already being done for the cricket industry in the USA (Weissman et al., 2012). Another advantage of biodiversity is that native or local species can be used because those species are well adapted to the local environment. Different insects can work differently for urban and rural populations concerning different recipes, high protein versus high oil, different flavors, and so forth. Even different species can be utilized for human food and animal feed, which may help lower the cost of insect feed.

CONSUMPTION OF INSECTS VERSUS OTHER LIVESTOCK

Insects can be produced more sustainably and with a much smaller ecological footprint than most vertebrate livestock such as cattle and swine. They are very efficient at biotransformation of a wide variety of organic matter into edible insect biomass (a high feed conversion ratio) (Nakagaki and DeFoliart, 1991; Oonincx et al., 2010). For example, cows consume

8 g of food mass per gram of weight gained, whereas insects can require less than 2 g (Vogel, 2010). This is partly due to insects being poikilothermic (ectothermic or "cold blooded"), thus using less energy for body warmth since they utilize their environment for body temperature regulation (Premalatha et al., 2011). House crickets (*A. domesticus*) have an "efficiency of conversion of ingested food (ECI)" twice as high as that of pigs and chickens, 4 times that of sheep, and 6 times that of steers (Gahukar, 2011a). This efficiency can also lead to less usage of pesticides on animal feed, thus providing additional environmental, health, and economic incentives.

In addition to their highly efficient feed conversion ratios and diet variability, rapid growth rates and short life cycles add to the greater efficiency with which they can be produced as a human food ingredient compared with vertebrate livestock. Insects tend to reproduce quickly, are highly adaptable, have large numbers of progeny per individual (high fecundity) and have a large biomass. For example, a female house cricket can produce over 1500 eggs in its lifetime. In general, insects tend to produce progeny numbers in the hundreds, so insect farms can scale up and recover from losses much more rapidly than other types of livestock farms.

As food source, insects can be eaten raw or processed, whole or ground into a powder or meal or paste, with or without condiments, or as a supplement to a major dish; they can also be incorporated into the diet of livestock used for meat products (Gahukar, 2013). Some of the main advantages of edible insects include their reasonably high protein content (Lundy and Parrella, 2015) and greater feed conversion efficiency (FCE) (Collavo et al., 2005) compared to meat animals. The FCE is a measure of efficiency in converting feed mass into body mass (protein), expressed as ratio or kilogram body mass/kilogram feed intake. Protein production depends upon what insects are fed and with which systems of livestock production the comparison is made. For example, 12.7, 5.9, and 1.7–2.3 kg of dry feed is needed to produce 1 kg of live weight in beef cattle, pigs, and chickens, respectively, compared to only 1.3–1.8 kg for house crickets (Lundy and Parrella, 2015). Similarly, protein conversion efficiency is 5, 13, and 25% in beef cattle, pigs, and chickens, respectively, compared to house crickets with 23–35% (Lundy and Parrella, 2015). These values ranged from 2.5–10% for chicken and beef, respectively (van Huis, 2013). Additionally, low CO_2 production per kg of mass gain in insects means higher FCE for insects than for livestock (Oonincx et al., 2010). Another advantage is the high growth efficiency or ECI, that is, the amount of food ingested by an insect and converted into biomass. It is measured by the weight gain and represents efficiency of both digestion and metabolic activity. In insects, the edible and digestible weight is 100% for larvae (such as caterpillars) and about 80% for adult house crickets (without legs and exoskeleton), whereas it is 55% for chickens and only 40% for meat animals. The efficiency is particularly high in the last instars of immature life stages when a high proportion of growth takes place (Nakagaki and DeFoliart, 1991). Overall, for the whole body of the insect, the daily weight gain is 7.2–19.6% (Lundy and Parrella, 2015), compared to only 3.2% for pigs (Cabaraux et al., 2009), and 0.3% for beef cattle (Harper et al., 2009).

The ECI is an important index for establishing economic feasibility of farming insects and other animals. This measure can also help to identify unnecessary financial investments in insect farming. For example, ECI for beef cattle is 10%, compared to 19.3% for silkworms and 44% for cockroaches (Capinera, 2004). When losses in carcass trim and dressing percentage are accounted for, the house cricket has an ECI twice as efficient as pigs and broiler chickens, 4 times greater than that of sheep, and 6 times higher than steers (oxen), and 20 times higher than beef (Capinera, 2004).

On a dry-weight basis, the ECI for crickets is 21.1%, compared to 9.9% for broiler chickens, 8.5% for pigs, 4.8% for lamb, and only 3.5% for steers (Capinera, 2004). Likewise, insects are efficient at transforming a wide variety of organic matter into edible body mass. For example, a cow consumes 8 g of feed to gain 1 g in weight, whereas insects require less than 2 g of feed for the same weight gain (Vogel, 2010). When crickets were fed with solid filtrate from food waste processed with an enzymatic digestion-achieved feed, the protein efficiencies were similar to that of chickens (Lundy and Parrella, 2015). Similarly, the production cost can be lowered further if insects are farmed on waste materials and rejected organic matter from other food production processes (Smithers, 2015).

Among insects, crickets have been preferred for farming because they are prolific breeders (a female can produce 1200–1700 eggs), have a short life cycle (30–45 days), and can be grown in small places (on small farms or in small modules in larger urban farms, etc.). These are important benefits of insect farming compared to livestock (eg, one cannot rear dairy cows in a skyscraper, but one can fill a skyscraper with crickets).

COST OF CULTIVATION

Insect production also uses much less water than production of vertebrate livestock (van Huis, 2013) because insects obtain their water directly from food. Additionally, the higher feed conversion ratios for insects, described earlier in this chapter, also contribute to this water use efficiency since a lower feed requirement means less water is used to grow that feed. Lower water usage also reduces the energy needed to pump or recycle more clean water for crops and vertebrate livestock, adding to the benefits of farming insects rather than those larger animals. In fact, many insects can be produced with almost no additional feed crop production. For example, many insects, such as the black soldier fly (*Hermetia illucens* L.) (Bondari and Sheppard, 1981; Popa and Green, 2012) can eat organic biomass such as agricultural and food by-products (see section "Production techniques by species" of Chapter 6 for more details on the black soldier fly). Such organic biomass could include corn stalks; pulp from fruit juicing or wine making operations; expired produce from grocery stores; yeast from wine or beer production; and portions of crops which cannot be converted into human food and other types of clean, safe, low-, or no-value biomass. For simplicity, this type of biomass will be referred to as "agricultural or food by-products," "nonfood crops," or simply "by-products." Insects are able to convert these by-products and other biomass otherwise not useful for human food production, such as switchgrass and algae, into edible insect mass. Hence, the production of some insects as human food in many cases may neither require feed to be grown especially for that purpose nor will compete with the existing human food supply, unlike vertebrate livestock, such as cows and chickens, which are often fed a diet primarily consisting of grain, such as corn. In addition to production efficiency, insects may provide nutritional value directly to the consumer more efficiently, per kilogram of foodstuff consumed, than other food resources (Ramos-Elorduy, 2008; Ramos-Elorduy et al., 2008). This adds another level of environmental impact reduction from the use of insects as human food over other more currently popular alternatives such as vertebrate livestock. For example, edible insects provide 217–777 kcal/100 g (insects raised on organic by-products provide 288–575 kcal/100 g), whereas energetic values for livestock are 165–705 kcal/100 g and vegetables are 308–352 kcal/100 g. Fats provide the majority of the energy necessary for sustaining life. Immature stages of holometabolous insects have high quantities of polyunsaturated fat, which is stored in preparation for the pupal stage during which they do not eat and are developing into adults. While the energy contents of edible insects vary according to the species and region found, coleopteran and lepidopteran species tend to provide more energy. Additionally, the energetic cost of collecting edible insects can be lower than that for vertebrates. Hence, insects may efficiently provide the necessary energy for the vital functions of the human body (Ramos-Elorduy, 2008; Ramos-Elorduy et al., 2008).

Cost of production is often calculated on the basis of raw materials, and therefore direct cost figures are not available. For 25 kg (50 lb) of formulated cricket feed, the cost is US $18, whereas mealworm feeds cost less and often vegetable refuse serve as cheap feed (such as potatoes, cabbage, and carrots) to further reduce production cost. Agriculture consumes 70% of fresh water worldwide and livestock consumes about 30% of crops and 8% of freshwater (Steinfeld et al., 2006). Chapagain and Hoekstra (2003) have revealed the fact that producing 1 kg of animal protein requires 5–20 times more water than generating 1 kg of grain protein; it means about 100 times more water for forage and grain production (Pimentel and Pimentel, 2003). Production of a kilogram of chicken, pork, or beef requires 2,300; 3,500; or 22,000–43,000 L of water, respectively (Pimentel et al., 2004). On the other hand, insects obtain hydration directly from food and use significantly less water than vertebrate livestock (Vogel, 2010). The per-head requirement of cattle, sheep, and goat is 40 mL (40,000 m^3), 5 mL, 500 mL, and 5 mL, respectively (Swaminathan, 2015). This is also true when compared with feed crops. For example, when producing 1 kg of wheat, corn, rice, soybean, potatoes, oilseeds, and fruits, the water requirement is 900; 1,400; 1,900; 2,000; 500; 2,000; and 1,000 L, respectively, and 1 kg of bread, eggs, cheese, milk, broiler chicken, and beef would require 40; 3,300; 5,000; 738; 3,500; and 15,500 L of water, respectively (Swaminathan, 2015). In another estimate, Gray (2014) reported that 454 g (1 lb) of cricket requires 908 g (2 lb) of feed + 3.78 L (1 gallon) of water, whereas 454 g of beef would require 12.5 times (11.35 kg) more feed and 2,900 times (10,962 L) more water. Similarly, 3,682 L water is needed in the USA to produce 1 kg of boneless meat by the way of irrigation practices (Beckett and Oltjen, 1993), whereas van Huis (2013) evaluated these values to 22,000–43,000 L/kg of meat. (These figures are from different farm locations and countries and therefore may not be comparable.) More data are expected in coming years. Jongema (2012) reported water requirement of only 0.8–1.6 g/g protein for insects compared to 16.8 g for cows (Table 4.3; Fig. 4.1).

Regarding feed, for producing 1 kg of live weight, feed requirement for insects, chickens, pigs, and beef cattle are 1.7, 2.5, 5, and 10 kg, respectively (Collavo et al., 2005). In the USA, livestock production consumed more than 7 times as much grain as was consumed directly by the entire American population in 2003 and for every kilogram of high-quality animal protein produced, 6 kg of plant protein was consumed (Pimentel and Pimentel, 2003). Land used for feed crops and animal production occupies nearly 70% of the world's cultivated land and 30% of the earth's whole land surface for protein-producing systems (Premalatha et al., 2011). A 220 g beef steak costs, in terms of CO_2 production, the same as driving 16 km with an 11 km/L gasoline car (Fiala, 2008). These examples demonstrate that use of land, water, and other natural resources for the production of insects is much lower than for producing meat animals.

It is true that insect food is not cheap, but when all aspects of insect production are taken into account, the overall cost is less than livestock production, mainly due to high feed conversion efficiency, low water use, low land use, low energy use, and high productivity. Therefore, in the coming years, insect production will be the cheapest of all sources of protein.

TABLE 4.3 How Much Water is Needed per Gram of Protein?

Cow = 16.8 g
Pig = 5.8 g
Broiler chicken = 5.2 g
Mealworm = 1.6 g
House cricket = 0.8 g
Source: Jongema (2012).

POSSIBILITY OF REPLACING LIVESTOCK WITH INSECTS AS HUMAN FOOD

In the near future, costly livestock protein can be replaced by insect protein. Indeed, changes are happening as people are becoming more convinced of the value of insects as a resource for family livelihood and also for food security. For example, consumption of insects in Africa has estimated to fluctuate between 2–30% of total meat consumption in a year, depending on the availability of insects (FAO/WUR, 2012), and local diets containing insects can be improved to fulfil family needs. Despite this fact, a great amount of money is spent every year to save crops that contain no more than 14% of plant protein by killing insects that contain up to 75% of high-quality animal protein (Premalatha et al., 2011). Further investigations showed that for every 1 ha of land required to produce mealworm protein, 2.5 ha would be required for producing an equal quantity of milk protein, 2–3.5 ha for pork or chicken protein and 10 ha for beef protein. Thus, insects are a more environmentally friendly source of animal protein than milk, chicken, pork, and beef (Oonincx and de Boer, 2012).

Animal production has to be increased due to increasing demand for meat products. As such, the census data revealed that worldwide meat consumption (per year/person) during 1997–99 through 2030 could be from 88.2 to 100.1 kg in developed countries, 25.5 kg–36.7 kg in developing countries, and from 9.4 kg to 13.4 kg in sub-Saharan Africa (excluding South Africa) (FAO, 2014). If animal meat as a protein source can be avoided, there would be a surplus of 2700 mha (mega hectares) of pasture and 100 mha of crop land, resulting in a large carbon uptake from vegetative growth (Stehfest et al., 2009). Therefore, current meat consumption can be brought down by replacing it with edible insects (Schosler et al., 2012). Indeed, it is of urgent importance to feed an ever-increasing human population, particularly those who are underfed or malnourished. For example, there are 825 m people still undernourished with 32.9% in India and 12.8% in China, the 2 top countries in terms of human population (UN, 2014).

Overall, replacing livestock with insects is certainly possible for the following reasons:

1. Insects are extremely prolific with rapid growth rates and very short life cycles. For example, house crickets have been extensively used as human food as they are prolific breeders, whereas four breeding animals are required for each animal marketed (Capinera, 2004).
2. Because insects are ectothermic, metabolic warming is not needed like it is for endothermic animals. Thus, insects need to consume fewer calories from feed.
3. Applications of chemical pesticides can be reduced drastically or may not be needed in field operations, and risk of soil and water contamination is minimized if crop damaging insects (such as grasshopper) can be hand-picked and consumed (Cerritosa and Cano-Santana, 2008).
4. Natural ecological balance is maintained, as natural enemies survive on native insect populations in agroecosystems.
5. Insects possess a greater harvestable proteins portion than chicken, pigs, and cattle, and they recover crude protein more efficiently (Collavo et al., 2005).
6. The amount of animal feed required to produce meat is enormous, thus water used for irrigation of land to generate feed crops for livestock and pesticides used in animal husbandry can be drastically reduced (Vogel, 2010).
7. Grasshoppers, crickets, flies, and beetles can consume agricultural waste or plants such as grasses and silage that humans and traditional livestock cannot consume. Beetles feeding on fallen trees in the forest help decomposition and the balance of the forest is restored (Gray, 2014).
8. By converting biomass that is not consumable by humans into edible animal biomass, insects do not compete with the human food supply or the feed required for pigs and chickens, which are primarily fed with grains (Vogel, 2010).
9. Insects can be reared on commercial diets. For example, a diet containing sugar and protein or kitchen waste has been used for rearing weaver ants, *Oncophylla smargdina* Fb., in Thailand (Offenberg, 2011).
10. In Western countries, insects are not eaten due to cultural taboo and beliefs associated with dirt, fear of contamination and disease, and psychological and biased thinking regarding taste, odor, and color (Deroy, 2015). However, several facts clearly indicate that insects are clean, tasty, and nutritious (Gahukar, 2013).

Thus, there are opportunities to explore farming of human-grade insects for consumption and to increase the possibility of replacing animal products with insects.

ENVIRONMENTAL IMPACT

Global average temperature has increased by 1.4°F over the last century and it is predicted to continue rising, a trend that is expected to lead to significant negative consequences. Global warming is primarily caused by greenhouse gas (GHG) emissions related to human activity (Smith et al., 2001). Several studies have indicated the negative effect of agriculture on environment, even at farm level (van der Warf and Petit, 2002; Rodriguez et al., 2004; Killbrew and Wolff, 2010). Table 4.4 shows examples of the deleterious effects of agriculture technology on soil, water, air, and biodiversity, all of which are important components of environment.

Global warming will have significant impacts on ambient temperature, rainfall, and crop production, which is, in fact, decreasing. Current land use for cattle ranching, heavy logging, bioenergy crops, slash-and-burn farming practices, and the

TABLE 4.4 Impact of Agricultural Technology on Environment (Soil, Water, Air/Climate, and Biodiversity)

Technology	Impacts
Monoculture	Reduces habitat for insects and wildlife, leading to increased need for pesticides
	Example: reduced bird population in monocropped coffee fields in Colombia and Mexico
Continuous cropping	Soil fertility declines due to nutrient mining; reduces farmers' ability to use natural pest cycles, leading to increased need for pesticides
	Example: nutrient offtake in reduced fallow cassava farms in Kenya and Uganda
Conventional tillage	Reduces soil organic matter, leading to increased erosion; contributes to CO_2 emissions due to decomposition of soil organic matter
	Example: Soil compaction due to tillage in maize fields in Nigeria
Intensive hillside cultivation	Increases erosion, leading to soil degradation
	Example: significant soil loss rates due to erosion in Ethiopian highlands
Intensive livestock systems	Increases erosion and soil compaction due to overgrazing and hoof action; untreated livestock waste degrades water quality; water usage competes with other needs; degrades grassland habitat due to overgrazing; contributes to CH_4 and N_2O emissions due to enteric fermentation and manure management
	Example: soil degradation and erosion caused by overgrazing in the Irangi hills in Tanzania
Inorganic fertilizers	Increases soil acidification due to nitrate leaching; reduces oxygen levels due to run-off, harming aquatic ecosystems, impairs water for human uses; contributes to smog, ozone, acid rain, and N_2O emissions
	Example: eutrophic dead zones in the Baltic Sea, Black Sea, and west coast of India
Pesticides	Harms animal and human health by accumulating in soils and leaching into water bodies
	Example: Use of unauthorized pesticide recipes in maize fields in Ethiopia
Irrigation systems	Inadequate drainage and over-irrigation causes waterlogging and salinization; degrades downstream ecosystems due to polluted run-off and overextraction of water
	Example: shrinking of Aral Sea due to overextraction for irrigation, particularly for cotton cultivation
New seed varieties	May increase need for inputs that negatively impact soils; may increase need for inputs that negatively impact water quality and quantity; reduces maintenance of genetic diversity in landrace varieties; may increase need for fertilizer, leading to increased greenhouse gas emissions
Intensive rice production	Inadequate drainage and continuous flooding causes waterlogging, salinization, and nutrient problems; degrades downstream ecosystems due to polluted run-off and overextraction of water; contributes to CH_4 emissions due to anaerobic conditions in paddy fields
	Example: overextraction for rice irrigation in Tamil Nadu, India
Industrial crop processing	Degrades downstream ecosystems due to water requirements and discharge of untreated wastewater; contributes to CO_2 emissions due to energy requirements of machinery
	Example: water pollution near coffee processing plants in Mexico

Source: Killbrew and Wolff (2010).

like force deforestation and desertification. The fertility levels of soils are decreasing and intensive farming with synthetic fertilizers and pesticides is spoiling the environment at a great speed. Chemical pesticides are used against pests, weeds, and plant diseases to safeguard food and feed crops from damage (Oerke and Dehne, 2004). Major methods of chemical application are soil incorporation, seed treatment, foliar sprays, and use of aerosols. When chemicals are wrongly used (overdose of active ingredient, spurious formulations, unwarranted applications, faulty equipment), there are several side effects that negatively impact the environment (Gahukar, 2011a). Also, foliage from host plants treated with chemicals, when fed to tropical silkworms, induces significant changes in mortality and alters the levels of different bioconstituents and metabolic processes in the silkworms (Gahukar, 2014). Due to chemical pesticides that adversely affect health of both the silkworms and the field workers in China, annual mulberry silk production decreased by up to 30% (Li et al., 2010). Therefore, mulberry foliage should be tested for chemical residue level before feeding to silkworms. Also, minimum use of chemical pesticides should be advocated while planning silkworm farming. Otherwise, plant-derived products including pure azadirachtin are safe to silkworms due to low inherent contact toxic action and minimum residual toxicity and, therefore, can be used in mulberry plantation (Gahukar, 2014).

In animal husbandry, chemicals used for disinfection and animal treatment have a great influence on environment. For example, livestock are responsible in the USA for an estimated 55% of erosion and sediment, 37% of pesticide use, 50% of antibiotic use, and one-third of the loads of nitrogen and phosphorus into freshwater resources (Steinfeld et al., 2006). Agriculture accounted in 2005 for 10–12% of the total global emissions of GHGs, according to the Intergovernmental Panel on Climate Change reports. Ammonia emitted from manure and urine of livestock leads to nitrification and acidification of soil (Oonincx et al., 2010). Animal production emits as much as 18% of GHGs, a higher share than the transport sector, and is responsible for 65% of nitrous oxide (N_2O), 35–40% methane (CH_4), and 9% carbon dioxide (CO_2) emission (Steinfeld et al., 2006). Indeed, methane and nitrous oxide have greater global warming potential than CO_2. In terms of emission, 1 kg of beef cattle causes the emission equivalent of 14.8 kg of CO_2, while emissions are lower for pigs and chickens, 38 and 1.1 kg, respectively (Fiala, 2008). Among insects, only cockroaches, termites, and scarab beetles emit methane whereas mealworm larvae, crickets, and locusts emit 100 times fewer GHGs and 10 times less ammonia than pigs and beef cattle (Oonincx et al., 2010). Therefore, to reduce environmental problems, insects are a far better choice as human food and animal feed. There are data showing incompatibility of animals in agricultural production, too. According to FAO (2014), meat production generates 15–24% of current GHG emissions. Considering the demand for meat by 2050, the emissions are predicted to increase by 39% and biomass appropriation will increase by 21% (Pelletier and Tyedmers, 2010). Ultimately, these trends will result in environmental pollution expectedly attributable to livestock production (Steinfeld et al., 2006). Therefore, a significant reduction in the methane and nitrous oxide emissions and about a 50% reduction in the mitigation cost has been recommended for the CO_2-equivalent (=$CO_2 + CH_4 + N_2O$) by 2050 (Stehfest et al., 2009). By raising insects rather than cows, current GHG emissions can certainly be reduced by up to 10% (Oonincx et al., 2010). Amounts for emission of methane, nitrous oxide, carbon dioxide, and ammonia are given in Fig. 4.1 and Tables 4.5 and 4.6. The rate of emission of these gases depends upon the conditions in which insects or animals are farmed (Cabaraux et al., 2009; Harper et al., 2009; Oonincx et al., 2010; Lundy and Parrella, 2015).

Comparatively, insects produce much smaller quantities of GHG per kilogram of meat than conventional livestock (pigs and beef cattle) (Oonincx et al., 2010). This fact is evident in the ratio between body growth and CO_2 production and, therefore, it serves as an indicator of environmental impact. The University of Wageningen (The Netherlands) found that the CO_2 production per kilogram of mass gain for the five edible insect species was 39–129% compared to pigs and

TABLE 4.5 Average GHG Equivalent Production per Kilograms of Body Mass per Day for Three Edible Insects, Pigs, and Beef Cattle

Species	Methane (g/kg BM per day)	Nitrous Oxide (mg/kg BM per day)	Carbon Dioxide (g/kg BM per day)	Ammonia (mg/kg BM per day)
Tenebrio molitor	0.00	1.5	0.45	0.0
Acheta domesticus	0.00	0.1	0.05	5.4
Locusta migratorita	0.00	8.0	2.37	5.4
Pigs	0.049[a]	2.7[a]	2.03[a]	4.8[a]
Beef cattle	0.239[a]	NA	5.98[a]	14.0[a]

[a]Minimum value.
Source: Oonincx et al. (2010).

TABLE 4.6 Average GHG Equivalent Production per Kilograms of Body Mass Gain for Three Edible Insects, Pigs, and Beef Cattle

Species	Methane (g/kg BM per day)	Nitrous oxide (mg/kg BM per day)	Carbon dioxide (g/kg BM per day)	Ammonia (mg/kg BM per day)
Tenebrio molitor	0.1	25.5	7.58	1.0
Acheta domesticus	0.0	5.3	1.57	1.42
Locusta migratorita	0.0	59.5	17.72	36.0
Pigs	1.92[a]	106.0[a]	79.59[a]	1140.0[a]
Beef cattle	114.0[a]	NA	2.850[a]	NA

[a]*Minimum value.*
Source: Oonincx et al. (2010).

12–54% compared to cattle (Oonincx et al., 2010). For 1 kg of poultry, about 2.7 kg CO_2 is associated (Oonincx and de Boer, 2012). On the contrary, the global warming potential of insects is less resource-intensive than conventional livestock, which depends upon protein conversion efficiency when a grain-based diet is used (Oonincx and de Boer, 2012). These results clearly showed that insects can be a more environmentally friendly alternative to other protein sources. Recently, Deroy (2015) contested this point of view and mentioned that insect farming and their utilization as food may not help environment safety. This is rather debatable and additional research and experimentation are needed on this matter.

Currently, energy used in agriculture to cultivate feed crops for livestock is also responsible for pollution (Oonincx et al., 2010). For example, pumping irrigation water makes the highest contribution (2.52–15.72 t/ha) to CO_2 emissions in India (Reddy et al., 2015). Likewise, Table 4.2 shows comparative emission of methane, carbon dioxide, and ammonia. Beef cattle produces 100 times the methane and a pig produces 20 times the nitrous oxide, 50 times the CO_2, and 8 times the ammonia that house crickets produce. In yellow mealworm (*T. molitor*), direct GHG emissions only totaled 0.29%, whereas the mixed grain and carrot diet they consumed accounted for 56% of the total quantity (Oonincx and de Boer, 2012). It can therefore be concluded that mealworms are more sustainable in comparison with farm animals for land use and global warming potential (CO_2 equivalent). Based on a life cycle analysis (LCA) that examines the entire production process, the global contribution by the food-producing animal sector is 9% for CO_2, 35–40% for CH_4, and 65% for N_2O, whereas the most polluting component, nitrogen dioxide (NO_2), is emitted from farm manure and urine (Steinfeld et al., 2006).

In Mexico, the egg density of a red grasshopper (*S. purpurascens*) was less in fields where mechanical picking of insects was practiced than in those fields with insecticide applications (Ramos-Elorduy, 2009). In Thailand, hand collection of the Bombay locust (*P. succinta*) from corn fields significantly reduced pest infestation levels (Boongrid, 2010). By these practices, insecticidal applications can be substantially reduced or altogether avoided and, consequently, risk of water and soil contamination is minimized.

From the perspective of environmentally clean farming, insects are more suitable when compared with livestock because (1) their farming on organic side streams is possible; (2) they emit less GHGs and little ammonia; (3) high feed conversion efficiency is possible; (4) compared to mammals and birds, insects pose less risk of transmitting zoonotic infections to humans, livestock, and wildlife; and (5) an increase in animal production will require additional cropland and feed and may trigger deforestation of land used for grazing. For example, forest land in the Amazon basin would be reduced in the future nearly by 70% when used for pasture or feed crops (Steinfeld et al., 2006).

INDUSTRIAL PERSPECTIVE

Insect farming in a closed or indoor environment is an important means to making food available continuously year-round, since many insects are available in nature only during certain seasons or months. It is particularly important to increase scale (as well as lower cost) and efficiency of insect and insect-based food production and use for maximum impact to improve food supply and reduce environmental impact. This is especially critical for urban populations. This aspect will become more important in the coming years as the food supply will decrease, government aid schemes will become more restricted, and the human population will continue to increase, posing a serious threat of survival. The number of malnourished and unfed low-income people will increase in both rural and urban areas. In fact, the amount required to treat these anomalies is much more than for prevention. See Chapter 6 for an additional detailed discussion on various industrial insect mass rearing techniques.

For indoor rearing, ambient environmental control (temperature, relative humidity, photoperiod), quality feed, and prevention of parasitoids and diseases are needed for proper insect growth and development. Feed biofortification with products containing protein, vitamins, and essential nutrients has been recommended for rearing of silkworms for better mass production (Gahukar, 2014). In such cases, sustained intensive farming of edible species can be undertaken under technical supervision. In Thailand, about 20,000 medium- and large-scale enterprises are successful in the farming of crickets, grasshoppers, and other insects (Hanboonsong et al., 2013). Protein production and sustainability depends upon the efficiency of any insect production system, quality and quantity of its protein contribution, and ecological impact, which is determined by the quality of insect diets (Lundy and Parrella, 2015).

Worldwide, insect farming is still manual. Large-scale cricket, mealworm, and waxworm [*Galleria mellonella* L.] production, as well as for some biocontrol species (particularly flies), have been around in the USA and Europe. Some of these industries are forming the basis for the emerging insect-based food industry. The concept and applications of new strategies are now gearing up and moving toward industrialization of insect products. Currently, a few industrial enterprises are in various stages of development for insect farming. For large-scale production, critical elements including research on insect biology, suitable rearing conditions, and diet formulas are required. To achieve commercial mass production, current farming systems need automation of some key processes to make them economically competitive with the production of meat from livestock.

COMMERCIAL INSECT FARMING FOR MASS PRODUCTION

The use of insects as a major human food source presents two important technological challenges: (1) how to turn insects into safe, healthy, and tasty food products and (2) how to cheaply, efficiently, and sustainably produce enough insects to meet market demand. Many also advocate the wild harvesting of insects as human food as a solution to hunger and global food security. While this may be a reasonable short-term solution in a number of isolated situations in certain localities, we feel strongly that for full realization of the potential benefits derived from utilization of insects as a safe, reliable, and sustainable alternative to vertebrate livestock, the ultimate goal must be the efficient farming and/or mass production of insects. Dependence on wild harvested insects for feeding large populations involves serious risks such as: overharvesting, ecological damage, consuming insects contaminated by pesticides or environmental contaminants and/or exposure to pathogens, parasites, and other disease-causing agents which may exist in the environment but can be eliminated or controlled in farmed or captive-reared stocks. Additionally, data on nutrient content varies widely by species, so using homogenously farmed stock of individual species provides much greater product quality control (Bukkens, 1997, 2005). As a consequence, this chapter focuses primarily on the latter.

Insects have numerous attributes which make them highly attractive, yet underexplored, sources of highly nutritious and sustainable food (FAO, 2008; Katayama et al., 2008; Nonaka, 2009; Ramos-Elorduy, 2009; Srivastava et al., 2009; Yen, 2009a,b; Vogel, 2010; Dzamba, 2010; Gahukar, 2011a,b; Premalatha et al., 2011; Smith, 2012; Crabbe, 2012; FAO/WUR, 2012; van Huis, 2013). The general categories where insects provide the most substantial benefits for sustainable and secure food supplies are: (1) efficiency and (2) biodiversity. With these advantages, insect farming on an industrial scale is picking up in India, China, and Korea with community initiatives and government support. The current farming techniques are quite simple, easy, and remunerative because: (1) insects can be reared and multiplied easily in small spaces in containers or insect beds; (2) they have short life cycles with high fecundity, intrinsic growth, and survival rates; (3) they are not fed on food grains, but on plant foliage, artificial diet, or organic waste; (4) manipulation is easy due to the insects' small size; (5) most of the edible insects are culturally accepted for taste, odor, and nutrition; (6) insects live in high population densities; (7) insect farming can be carried out in available space in urban, periurban, and rural areas (Oonincx and de Boer, 2012); (8) no special infrastructure besides some nets, plastic sheets, and containers is required; and (9) high financial return is possible because insects are easy to manage and transport, and no specialized training for technical personnel is required.

Next, major insect species have been described to promote their farming and marketing in countries where edible insects have been oriented for large-scale production.

Indoor Farming

Regarding the insects currently farmed for protein, house crickets and yellow mealworms are probably the major ones, primarily because they are already a commercial success. They are mostly used as pet food in Europe, North America, and parts of Asia. An increasing share of insect protein content in commercial food products has been successfully undertaken by a few enterprises in the developing world. Other industrial uses such as extract of protein, fat, or chitin for fortifying food should be initiated and marketed.

FIGURE 4.2 (A) Cabinet for village level farming of mulberry silkworm; (B) ant farming on indigenous substrate; and (C) grasshopper farming cages. *(Courtesy: Part A—Jadhav, A.D., 2015. Personal Communication; Part B—Plester, L., 2015. Antenna 39 (1); and Part C— Smithers, P., 2015. Antenna 39 (1))*

Mulberry Silkworm

Among domesticated silkworms, the mulberry silkworm is a monophagous insect and is most extensively farmed under indoor conditions due to its rising demand as edible larvae and pupae. In India, certified eggs [disease-free layings (Dfl)] are procured by government agencies. Rearing beds with compartments of different sizes are erected by using wooden material or bamboo poles in permanent cement buildings or temporary mud houses (Fig. 4.2).

Low-cost rearing houses with nylon mesh or permanent houses in cement are currently in use with satisfactory yield of larvae and cocoons. Beds, trays, and appliances are disinfected. Fresh, tender, chopped leaves are provided to hatching larvae in small plastic or metal trays until the 2nd–3rd instars. A diet containing protein Nutrid has been recommended due to the overall positive results showing improvements in survival rates of early instars, development of later instars, weight of larvae/cocoon, and so forth on laboratory diets, compared with natural host plants (Sundarraj et al., 2000; Gahukar, 2014). Larval feces and used leaves are separated and replaced by fresh feeding material daily. However, the success of mass production depends upon abiotic and biotic factors (Gahukar, 2014).

Salient Points

1. Modern silkworm rearing on chopped leaves has been successful for the last four decades. The recent development is providing shoots instead of chopped leaves. This system of feeding insects saves labor costs up to 50%, requires nearly 15% less foliage, reduces leaf drying in rearing beds, reduces contamination of disease pathogens, and provides better aeration in rearing beds (Tamil Nadu Agricultural University, 2014). Furthermore, two feedings/day at equal intervals instead of several irregular feedings have given satisfactory results. It is now proved that tender succulent and nutritious

leaves result in better growth and development of young larvae, whereas progressively mature leaves with less moisture are required for later larval instars (Tamil Nadu Agricultural University, 2014).
2. Among abiotic factors, temperature plays a major role on growth and productivity of silkworms. The optimal conditions for larval rearing are 25–26°C + 65–70% RH and 12 h light. Late instar larvae do better at relatively lower temperatures than young larvae and fluctuation of temperature during larval development is more favorable. Good quality cocoons are produced within a temperature range of 22–27°C. Recent studies on tolerance of silkworms to high temperatures showed that compatible multivoltine and bivoltine breeds can be reared at relatively higher temperatures, as they can withstand the stress.
3. Several silkworm hybrids have been developed and are available for laboratory rearing (Gahukar, 2014). Bivoltine silkworms are preferred due to high larval weight, less development period, high fecundity, and so on. Currently, double hybrids such as, (CSR50 × CSR52) × (CSR51 × CSR53) are preferred for mass production.
4. Leaf quality of mulberry varieties influences the uniformity and rate in production (Bhede et al., 2013). Therefore, selection of variety should be based on high yield of nutritious foliage. Plant nutrition, cultivation practices, soil type, and so forth can affect leaf moisture as well as nutritional and biochemical constituents. Also, infestation of insect, nematode, and mite pests affects the nutritive value of leaves and feeding of such leaves adversely influence the growth and development of silkworms (Malik and Reddy, 2007). Foliage biofortification with commercial supplements such as, protein powder, herbal tonic (Kanafi et al., 2007; Rani et al., 2011), jaggery solution, soybean powder (Sundarraj et al., 2000), and royal jelly (Nguku et al., 2007) has shown beneficial effects on biological characteristics and has increased the efficiency of silkworms for consumption. Efficiency of conversion of ingested and digested food are important criteria in larval development; both are significantly reduced when mulberry leaves sprayed with 0.01% fenvalerate 20EC and fed in rearing rooms (Wijayanthi and Subramanyam, 2002). Organophosphorous insecticides such as monocrotophos 35EC, fenitrothion 50EC, and dichlorvos 76EC also adversely affected the biology of silkworms (Gahukar, 2014).

Constraints and Measures

If chemically treated feed is used in insect farming, toxic metals are accumulated in the insect body and residual toxicity of a chronic nature is experienced on the human body after consumption. For example, mulberry cultivated in soil with lead, cadmium, chromium, or other toxins can be transferred to humans through consumption of silkworm (Zhao et al., 2013; Zhou et al., 2015). Toxic metals can come from soil treated with fertilizers and municipal solid waste compost (Zhao et al., 2013). Lead coming from polluted soils can be detoxified in different trophic levels (Zhou et al., 2015). However, drying of insects may contribute to increased metal content. Precautions are therefore needed to select mulberry clones and also silkworm strains that are lead resistant so that this metal is detoxified and eliminated before it reaches the human consumer. Dead insects resulting after a disinfection program in Thailand (DeFoliart, 1999) and locusts sprayed with organophosphorus pesticides in Kuwait (Saeed et al., 1993) were sold in open markets. Foliage treated with pesticides can indirectly affect human health through the food chain (mulberry leaves–silkworm–human food).

Among natural enemies, the uzi fly, *Exorista sorbillans* (Lewis) is a major larval ectoparasitoid which kills up to 80% of silkworm larvae during farming, followed by *Exorista japonica* Towsend and *Blepharipa zebina* Wlk. Two hymenopterous parasitoids (*Xanthopimpla predator* Fb., *Apanteles* sp.) are occasional but can bring a loss of 9–15% in cocoon production (Gahukar, 2014). Precautionary measures including sanitation and hygiene in the rearing room; covering windows and doors with wire mesh; collection and proper destruction/disposal of parasitized larvae, fly maggots, and pupae; and the use of water traps with chemical attractants are quite efficient practices. Spraying of 1% benzoic acid against parasitoid fly eggs laid on silkworms and dusting of 2.5% diflubenzuron (diluted with China clay) against uzi fly maggots and pupae can reduce fly attack. Dusting of uzi powder on silkworms on the second day during third instar, second and fourth day during fourth instar, and on the third and fourth day during fifth instar has proven effective against parasitoids (Tamil Nadu Agricultural University, 2014).

Plant-derived products, particularly 5% water extract of leaves or seeds, 0.05% seed oil, 10% alcoholic extract, or commercial neem products gave up to 100% mortality in parasitoid fly populations (Gahukar, 2014). If possible, release of gregarious hyperparasitoids should be adopted in order to maintain silkworm population for large-scale production. An ecto-pupal parasitoid, *Nesolynx thymus* Girault, is efficient in reducing fly populations if released in three doses corresponding to fourth and fifth instar larvae (Tamil Nadu Agricultural University, 2014).

Silkworms are infected with bacteria (*Bacillus, Pseudomonas*) that cause flacherie disease. Apart from maintenance of sanitary and hygienic conditions in the rearing rooms, disinfection of appliances and the rearing room should be done. After collecting and burning of infected larvae, fecal matter, and bed refuses, spraying beds with streptocycline (1 g/10 L) has been recommended (Tamil Nadu Agricultural University, 2014). White and green muscardine fungi (*Beauveria bassiana* (Bals.-Criv) Vuill. *Spicaria prasina* (Maubl.); Sawada, dark green fungus, *Aspergillus flavus* Link; rusty brown fungus,

Aspergillus tamarii Kita all attack silkworms. The management strategy is to disinfect the rearing room and trays/utensils with 2% bleaching powder; early diagnosis of the disease and rejection of the infected lots can prevent further spread of disease (Tamil Nadu Agricultural University, 2014). Also, aqueous or alcohol extract of plants are effective against fungal diseases (Isaiarasu et al., 2011). Infection of the nuclear polyhedrosis virus and cytoplasmic polyhedrosis virus causes jaundice or grasserie disease. For effective control, sun-drying of appliances for 1–2 days and disinfection with 2% bleaching powder of the rearing room, trays, and appliances has been recommended (Tamil Nadu Agricultural University, 2014). Proper ventilation and air circulation should be maintained. Overcrowding and accumulation of feces in the rearing beds can be avoided by keeping proper bed spacing. Early diagnosis and rejection of the infected lots are effective preventive measures (Tamil Nadu Agricultural University, 2014). Along with plant-derived products, bed disinfection with 2% lime powder can prevent the spread of both flacherie and grasserie diseases (Reddy and Rao, 2009). The pebrine, or microsporidiosis, disease is caused by a microsporidian, *Nosema bombycis* Nageli. This chronic and highly virulent disease is present in all life stages of the silkworm. As preventive measures, early diagnosis, production of healthy eggs, disinfection of rearing rooms and appliances, maintenance of hygienic conditions during rearing, and surface disinfection of eggs in 2% formalin for 2 min before incubation are to be followed (Tamil Nadu Agricultural University, 2014).

House Cricket

The house cricket is most extensively reared in indoor conditions in Thailand, Vietnam, and Lao People's Democratic Republic (Hanboonsong et al., 2013). Farming is done for human consumption, pet food, or for protein extraction by industries. Crickets can be reared continuously and all year in small containers in a short period due to their small size and life cycle of 30–45 days. Growth is rapid and development occurs without diapause at 30°C. Higher production has been reported by exposing crickets to 24 h light/day (Collavo et al., 2005). They can be multiplied in different sized glass aquaria, plastic boxes, metal troughs, or reusable cardboard gallon cartons with screen lids (Clifford and Woodring, 2009).

Currently, crickets are farmed in 4 types of containers in Thailand. For small farms, concrete cylinder pens (80 cm diameter × 50 cm height) are inexpensive, easy to maintain, and sufficient for 20–150 pens/unit with 2–4 kg of crickets/pen. Plastic bottles are used to provide water or a plate with water and stones is used to prevent the crickets from drowning. Sticky tape or plastic tablecloths are glued on the inner side of the walls just below the edge to keep the crickets from crawling out of the arena. Rectangular concrete black pens (1.2 m long × 2.4 m wide × 0.6 m high) with a capacity of 25–30 kg/pen are quite convenient for a medium farm of 100 blocks. Risk of disease outbreak or overheating is possible due to overcrowding. In such cases, pens can be replaced by plywood boxes of the same size elevated off the ground by 15–20 cm by fixing wooden legs. Boxes are movable and easy to clean and overheating does not occur, but wooden boxes are sensitive to hot, cold, or dump weather. Plastic drawers or trays are a recent introduction (0.8 m long × 1.8 m wide × 0.3 m high). They are stacked on metal or wooden shelves. Whenever quality deteriorates, particularly due to overheating, these trays are replaced with new ones. Currently, cement tanks or wooden containers covered with a plastic sheet are popular. A layer of sandy loam soil is added and covered with dry grasses, bamboo shoots, or egg cartons to provide shade for the crickets.

Crickets are introduced in the breeding enclosures (2.2 m long × 4.8 m wide × 0.6 m high) or (2.5 m long × 8 m wide × 0.5 m high) and are provided with bedding of burned rice husk and sand. Enclosures are covered with nylon/mosquito net to prevent escape and entering of predators. Bowls are covered with layer of rice hull to maintain a suitable incubating temperature. The rearing arena is surrounded by a narrow strip of water containing very small fish to prevent the entrance of ants (Yhoung-Aree and Viwatpanich, 2005). One egg bowl has the potential to produce about 3 kg of adult crickets and the reproductive cycle can be repeated 1–3 times for each generation. A diet consisting of chicken feed (14–21% protein) along with grasses or weeds and water is provided. After 4–6 weeks, adult crickets are ready to be harvested (Jamjanya et al., 2001). Chicken feed is replaced with vegetables after 21 days up to 45 days and a few days before harvesting. The vegetables used are cassava leaves, pumpkin, morning glory leaves, water melon, or similar foods to improve taste and to reduce use of expensive feed. Indeed, even up to 50% profit is possible by reducing costly feed (Hanboonsong et al., 2013). A yearly income of US $468–1,093 [150,000–350,000 Thai Baht (ThB)] per year can be generated considering a harvesting cycle of 45 days, a harvest of 750 kg insects, and a market price of 15–300 ThB for 1 kg of produce (Hanboonsong et al., 2013). With this experience in the recent past, a total of 19,961 cricket farmers with 217,529 cricket pens produced 1,087,645 kg of crickets in Thailand (Hanboonsong et al., 2013). In the USA, the largest 10 producers of crickets collectively produce approximately 2 billion crickets annually.

Various researchers, governments, and entrepreneurs are seeking to identify and develop a suitable regional waste substrate of sufficient quantity as a feed ingredient to produce insects. Such waste by-products have no direct competition from existing food production systems and, at the same time, minimize ecological impact (Lundy and Parrella, 2015). Patton (1967) developed oligidic diets and obtained satisfactory insect growth and development in the laboratory. The cost of laboratory diets was only 0.21 versus US $2.55 kg^{-1} for commercial diets, and performance on laboratory diets was better in terms of production of the house cricket (Nakagaki and DeFoliart, 1991). However, Clifford and Woodring (2009) suggested commercial feeds such as Purina Cricket Chow because they provide all essential nutrients for continuous rearing to have several generations. Thus, with further research and development (R&D) programs, insects can be produced far more efficiently and with lower environmental impact than today, even from feed improvement research alone.

Constraints: Nearly half of the production cost goes toward high-protein feed; it is particularly the chicken feed that undermines profitability. Disease/infection may become a problem due to overcrowding or contaminated feed. Inbreeding is common wherever breeding stocks are produced in farms close to each other. There is a need for standard farm management practices from nursery to cricket harvest. Government support to the cricket farming sector is lacking. Marketing is not well organized for the farmers and awareness of cricket food items is not viewed as a business opportunity. Improvement in farming techniques would solve these problems to some extent and would result in higher insect production.

Yellow Mealworm

The yellow mealworm is next to the house cricket in popularity in the farming sector. About 500 beetles can be farmed in 5 g wooden or plastic containers with holes in the lids for aeration. Cornmeal, bonemeal, crushed bran flake meal, and "cricket chow" are generally considered in the industry to be the best bulk feeds currently available. On top of the feed, a piece of fruit or vegetable is placed for moisture, and changed every couple of days to keep the substrate dry and to prevent any mold development (Hanboonsong et al., 2013). A diet high in yeast-derived protein appears favorable compared to other diets used by commercial breeders in order to shorten larval development time, reduce insect mortality, and to increase weight gain (van Broekhoven et al., 2015). Mealworms can also be reared on diets composed of organic by-products without any effect on larval protein content (van Broekhoven et al., 2015).

Constraints: Except moisture and dampness, which attracts fungal pathogens, there are no major constraints in the mealworm farming process.

Outdoor Farming

Sometimes, the term "semicultivation" is used for outdoor farming because insects are farmed in the wild and not kept in captivity. Because insects are naturally available for the locals, habitat conservation is facilitated and food security is enhanced. For example, palm weevil grubs in the Amazon basin, Indonesia, Malaysia, Papua New Guinea, Thailand, and sub-Saharan Africa; mopane caterpillars and termites in tropical Africa; bamboo caterpillars in Thailand; and eggs of aquatic bugs in Mexico. Cerda et al. (2001) demonstrated a few simple cottage methods adopted by locals in Venezuela to farm the South American palm weevil on several crops including banana, vegetable refuse, and fruits. Likewise, semicultivation of termites on dry sorghum stems or other cereals is common in West Africa (Farina et al., 1991).

Palm Weevil or Sago Larva

The red or sago palm weevil (*R. ferrugineus*) is farmed on sago palms (*Metroxylon sagu*, *M. rumphii*) which are their natural food plants and, therefore, the insects are often known as sago larva. Insects are left in the palm trunk to grow, which later is cut into 50 cm long pieces and 10 holes are drilled 5 cm deep into the trunk. Adult weevil pairs are released on the trunk pieces and the top is covered with tree bark. Fully grown larvae are harvested 40–45 days after adult release. The average yield is about 2 kg of larvae/trunk. The farming is common in Indonesia, Malaysia, Papua New Guinea, and Thailand (Mercer, 1994; Hanboonsong et al., 2013). The sago larvae can also be farmed in plastic containers indoors (45 cm diameter × 15 cm height), which are filled with ground palm stalk mixed with pig feed. Adult weevil pairs are released into the container and after 25–30 days, a yield of 1–2 kg larvae/container can be obtained. The production cost of 1 kg of larvae (nearly 200 individuals) is about US $0.4 (15 ThB/kg). The breeding cycle lasts 35–45 days and produces 400–600 kg larvae. A net profit of US $2,625–3,937 (84,000–126,000 ThB) is earned from each harvesting cycle (Hanboonsong et al., 2013).

Constraints: Since the insect is farmed on its natural source, its farming is limited to areas where palm trees are abundant. However, because the palm weevil is a major pest of the coconut tree, collecting them is a natural means of controlling the pest populations. Natural farming is practiced in traditional ways and indoor farming is still not available for farmers' use.

Bamboo Caterpillar

Bamboo (*Dendrocalamus* spp., *Thyrsostachys* spp.) is a natural host plant of the caterpillar *Omphisa fuscidentalis* Hampson. For farming the caterpillars in earlier days, farmers used to cut bamboo clumps to take out caterpillars. The current improved technique consists of slicing the infested internode to detect the presence of caterpillars because caterpillars live and eat inside the bamboo shoot. Then caterpillars are taken out by cutting a rectangular hole (9 cm × 13 cm) at specific internodes, preferably in the upper part of the young shoot. After 45–60 days, caterpillars congregate in the internode and remain there for about 8 months. To produce infestations, adult moths are allowed to mate in nylon net cages covering the bamboo shoots or are released into the bamboo plantation for mating to occur naturally.

Generally, caterpillars are packed in plastic boxes and sold in local markets for about US $6–8 (200–250 ThB) kg^{-1}. Recently, private companies have developed a new product of cooked caterpillars that are sealed in containers (Hanboonsong et al., 2013). Caterpillars are eaten after the containers are heated in a microwave oven.

Constraints: Caterpillars are farmed on bamboo in a seminatural habitat and a sizeable bamboo plantation is needed for insect farming. Proper management and harvesting techniques for sustainable use of this insect would facilitate and encourage locals to become businessmen with a reasonable profit margin. Efficient insect breeding techniques and product processing are still traditional and further improvement will certainly encourage locals to start commercial enterprises.

Weaver Ants

Weaver ants are used as food by some local communities, and they are particularly preferred by traditional healers due to their medicinal uses. Also, their eggs are a popular delicacy (condiment). The ant workers construct nests by weaving leaves together using larval silk. The colonies can be maintained in host plants (particularly mango trees) if protected from predators and substrate and water are provided, as the ants need it to produce acetic acid (Fig. 4.2). In order to accelerate multiplication, highways are made from jute or cotton woven rope or rattan cane. Ants are harvested once a year by using a long bamboo pole with a bag or basket attached with strings to the tip. A hole is poked into the nest with the tip of the pole and it is shaken. In this way, larvae and pupae fall down into the bag. The content of the bag is poured into a big plastic container in which some rice and tapioca from cassava (*Manihot esculenta*) flour is added to prevent the ants from climbing and escaping. A branch is inserted into the container so that adult ants can climb up. Then, the branch is whipped against a tree to release them and the larvae and pupae left in the bag are collected for human consumption. Ants can be farmed in home gardens also by feeding them with food scraps and water (Hanboonsong et al., 2013).

About 300–400 g of larvae and pupae and about 2–8 kg of adults can be collected every day. On average, a collector can earn US $8–15 (250–500 ThB) per day. With current market demand for fresh produce in Thailand, sellers can earn more from farmed insects than farm conventional crops of rice or cassava (Hanboonsong et al., 2013).

Constraints: Because of the market demand, ant populations are decreasing in the wild. This trend has a negative impact on the ecology, as weaver ants are predators of crop pests. Collecting queens to start new colonies is a hard task because queens are found in small nests at the highest point of the tree, where they are difficult to reach.

Grasshoppers

Farming various species of grasshoppers (*Patanga succincta* Johannson, *Locusta migratoria* L., *Acrida* spp., *Cyrtacanthacris tatarica* L., etc.) can be undertaken in caged potted plants or those rooted out in the open (Fig. 4.2). Of course, insect farming depends upon the nature and growing habits and life span of the major or alternative host plants. Only small plants can be maintained indoors. Enclosures with wire mesh and timely preventive measures can reduce the mortality of edible insects (Banjo et al., 2006a).

Constraints: Grasshopper farming on a commercial scale is difficult, as it is often not economically viable unless there is market demand. Farming of grasshoppers generally takes a year for completing a generation and insects require natural food for growing. Residue of lead through "chapulines" (dried grasshoppers) has been reported from Mexico (Handley et al., 2007). Metal toxicity is hazardous to human health if a contaminated product is sold and consumed.

Other Insects

Eri silkworm and muga silkworm in India (Sarmah, 2011; Gahukar, 2014) and Thailand (Hanboonsong, 2010), the giant hornets/wasps in Japan (Nonaka, 2010), and termites in East Africa (Ayieko et al., 2010) are farmed in the wild. Although the local techniques of insect farming are economical and profitable, commercial production has not yet been practiced in rural areas. If improved techniques are available, entrepreneurs may be interested in the large-scale farming and marketing of these insects.

Farming in Space

Silkworms, termites, and flies have been successfully farmed in space (Mitsuhashi, 2007). Silkworm rearing was successful on mulberry (*Morus alba*) or lettuce (*Lactuca sativa*) grown in space, or on larval powder (ground- or freeze-dried) containing 71.4% protein and amino acids, 12 essential vitamins, 9 minerals, and 12 fatty acids, and producing energy of 359 kcal/100 g powder (Tong et al., 2011). Approximately 105 silkworms could satisfy the food requirements of one person/day by acting as "bioregenetative life support" (Yu et al., 2008) and "controlled ecological life support" (Yang et al., 2009). Other insects, such as *Macrotermes subhyalinus* (Rambur), *Stegobium paniceum* L., and *Agrius convolvuli* L., may also serve as an alternative food source for astronauts and as a key to space agriculture in the future (Katayama et al., 2005, 2008).

MARKET POTENTIAL

Insects as a food product and the insect-based food industry are only now in their infancy. For many decades, insects have been sold in many areas of the world as locally sourced and eaten whole. Only in the past few years have astute entrepreneurs began to take notice that insects can be mass produced, processed, and marketed similarly to any other animal biomass commodity with all of the aforementioned benefits to a sustainable human population while creating an entirely new industry. In this section, a discussion of current diverse examples of edible insect commerce and considerations is provided. For a detailed discussion on the more modern and future insect-based food industry now developing, and the considerations entrepreneurs, food companies, and others should keep in mind as they enter this industry, see Chapter 5.

Retail/Local Marketing

If the sale of insects can yield more revenue than food crops as reported by van Huis (2003) for termites in the sub-Saharan region and by Hanboonsong et al. (2013) for weaver ants in Thailand, then at least a few small commercial units can be started by locals. There is potential to make insect farming a more remunerative business if farming is moved from small units to an industrial phase, with adoption of economical farming on a large scale (Hardouin, 1995), as happened in Thailand where 15 insect species are marketed locally, both fresh and processed (Hanboonsong et al., 2013).

Until recently in North America and Europe, foods containing insect ingredients have existed as novelty items such as lollipops with ants, insects as a condiment in modern recipes and fish, and dead insects as decorative items in homes and museums (Fairman, 2010). Generally, insects can be treated like any other high-protein agricultural product or raw ingredient. Many insects are sold live, however, insect-processed products and by-products probably account for the majority of insect commercialization (Kampmeier and Irwin, 2003). For processing, insects can be boiled or sun-dried, and then packed and frozen for storing in cool and dry places to increase shelf-life. They can also be mass processed into powders, meals, pastes, or liquids as well as canned, pickled, and otherwise transformed into alternative forms of more traditional foodstuffs. More information in this domain can be found in Chapter 5.

Local restaurants are trying innovative cuisine with insects. As large and diverse markets as well as new enterprises are appearing, customers are seeking more natural and environmentally friendly food/ingredient choices such as gluten-free, protein-rich foods, as an alternative to meat. The most logical recipes may not only contain insect powder or meal but may utilize it like any other dry ingredient by mixing in a reasonable proportion (at least 5%) without adversely impacting nutrition, taste, or flavor. Smithers (2015) mentioned some restaurants in the United Kingdom where insect menus have been recently introduced, for example, crickets as a snack, orthopteran burger in a fresh bread roll with polenta chips, grubs for dinner, bread spread with tomatoes, carrots, and mealworms, and so forth. Some of the industrially mass-produced insect-based products are finding their place in supermarkets and grocery stores, such as chips made with cricket meal, cricket-powder baked goods, and cricket energy bars, which are becoming as popular as roadside potato chips and trail-nut mixes in a rapidly growing number of markets within the USA. In the United Kingdom, dung beetles are being reared at bug farms and delivered to restaurants, where they are served fresh to customers. Additionally, right now in many parts of the world (and potentially in the coming years in the United States and Europe) selling fresh farmed insects is convenient and remunerative.

Fried grasshoppers and chocolate-covered ants are sold in local markets in Mexico; chocolate chirpy chips or popcorn with roasted crickets and grasshoppers (Chapulines), stir-fried mealworms, and caterpillar crunch are popular in Central America and ants with popcorn is a local delicacy in Colombia. Recipes containing garlic-smashed potatoes with cicadas, chocolate-covered cicadas, deep-fried cicadas used as croutons on salad, and maggot cheese in Italy are popular snacks for the urban population (Kittler and Sucher, 2008). In the USA, crickets, silkworm larvae/pupae, ant pupae, giant water bugs, and cicadas cooked with rice and vegetables are served in eateries (Gracer, 2010).

Local markets in Australia are flooded with insects packed in plastic bags and sold as food items during the period when food is not easily available (Yen, 2009a). Likewise, whenever there are social, political, or natural disasters, food packets containing insects can be supplied. For this purpose, protein bars containing peanuts, fruit, soy, and bugs have been recently introduced in local markets in the USA (Williams, 2015), which can be an encouraging step toward the general distribution of insect products.

Export

Currently, the international trade of farmed insects is insignificant compared to that of food grains. In developed countries, trade depends upon the demand from immigrants who like insect foods often surpassing the demand for exotic foods (Tabuna, 2000). Of course, a great deal of trading takes place on border countries, particularly those in Central Africa and Southeast Asia. For example, mopane caterpillars are exported from Zimbabwe to Botswana, Congo PDR, South Africa, and Zambia (Kozanayi and Frost, 2002).

The extent of trade of edible insects by commercial companies varies considerably from one region to another. Once the produce is procured and brought to enterprise, the lots are packed, frozen, and stored in cool storage (temperature up to $-20°C$). The giant water bugs are preserved with salt to extend shelf life. The edible insect industry thus has much potential for providing protein sources and for income generation through sale. Establishment of farming centers on both a small and large scale could reduce the pressure of wild populations of insects collected in the wild and forests. But in the case of grasshoppers and locusts, wild harvesting is still feasible, as field crops are always available as sources of host plants. Appropriate methodology for the farm-based industry needs to be developed for all edible insects collected in the wild and discussed in this chapter because from the viewpoint of food safety, farmed insect sources are reliable and preferable.

Opportunities can be explored regarding advanced technologies for packaging, processing, and marketing (Ramos-Elorduy, 1998). Products such as beetle juice, canned silkworm pupae, caterpillars of hesperid butterflies, and immature stages of ants are exported by food industries in the developed countries (Ramos-Elorduy, 1998). In South African countries, mopane caterpillars are degutted, cooked in brine, sun-dried, and preserved or canned and exported annually, with 1600 m kg from South Africa alone, and Botswana earns about US $8 m (Mpuchane et al., 2001; Mulhane et al., 2001). In China, 200 t/year of the dried mealworms are exported to Australia, Europe, North America and Southeast Asia. Since insect farming is considered an export-oriented business, new enterprises adopting large-scale insect production are emerging in several countries. In South Africa, farming houseflies on low-cost waste material has been successful. Before being exported, harvested maggots are dried, milled into flakes, and packed and exported to the extent of around 100 t/day. Maintaining the integrity of the food industry would be necessary to have high market penetration, as evidenced in Thailand, where more than 20,000 insect-farming enterprises are registered, most of which are small-scale operations (FAO, 2014).

The export-oriented insects include indoor farmed crickets, palm weevil/sago larvae, wild harvested bamboo caterpillar, weaver ant, grasshoppers, and the giant water bug. For most of the exported insect-based products, the import/export figures are not readily available. Japan regularly imports wasp-foods from Korea, China, and New Zealand (Nonaka, 2010). France and Belgium import dried caterpillars from the Democratic Republic of Congo (Johnson, 2010), wasp foods by Japan from Korea, China, and New Zealand (Nonaka, 2010), tortillas by USA from Mexico (Munoz, 2008). Beer is served to tourists with bugs as a snack in the Lao People's Democratic Republic (Boulidam, 2010).

Edible insects can be marketed for industrial purposes. For example, mulberry silkworms, when fed with dye-added artificial diet, produce colored cocoons, which are in demand by the silk fabric industry (Kang et al., 2011). Chitin/chitosan, which can be extracted from the cuticle, comprises 10% of whole dried insects (Duan, 1998). Chitin is an economical source of fiber and calcium (Paulino et al., 2006) and protein concentrate from dechitinized insects can be produced on a large scale and fed to animals. There are possibilities for local enterprises to sell insect products to tourists. The stir-fried coconut rhinoceros beetle grubs are sold as local delicacy in the tourist spots in Krabi Province in Thailand (Ratcliffe, 2006). With such opportunities, gastronomic enterprises such as Nordic Food Lab in Copenhagen is enthusiastically engaged in processes to make insects and their products more interesting and appealing to consumers and would probably be the first company to market insect-processed products on an industrial scale.

New Products From Farmed Insects

The market share and consumers' demand for new insect products are unknown, but these products can be brought into the mainstream of the food supply in the future (van Huis, 2014). Processed mealworms are currently the most extensively used in such food products as bread, noodles, pastries, biscuits, candy, and condiments. Other promising products that could have reasonable market demand include SOR-mite (sorghum porridge enriched with termites) and termite crackers and muffins (termite-based crackers, muffins, meatloaf, sausages) in Africa; Bugadilla (snack of spicy chick pea with mealworm powder) and Crikizz (Spicy/popped snacks based on mealworms and cassava), and chocolate topped with crickets dipped in gold paint in Europe; fried insects embedded in chocolate or hard candy and fried, and seasoned larvae in the USA. By 2020, the European market for insect products is estimated to be US $73 m; this opportunity can be seized by several private enterprises because one-fifth of meat eaters are now ready to adopt insects as food, with men being 2.17 times more likely than women to do so. Thus, there is worldwide potential for insect products, particularly as a powder (compared to meal, paste, frozen, canned, or liquid) due to extended shelf life. Initiatives in this respect have been launched in North America and Europe (for more information, see Chapter 5).

SAFETY REGULATIONS

With globalization, consumer concern over food quality has become of prime importance and accordingly, insect production methods have been improved. However, hygiene and health hazards arising from insect eating are still a problem (Giaccone, 2005). In developing and less developed countries, legislation is nearly absent because a formal census system on edible insects is lacking. In developed countries, on the contrary, food complaint actions are formulated and executed (Gahukar, 2013; Belluco et al., 2013). Overall, the Codex Alimentarius is applied for trading of farmed insects and to reinforce international standards of food. However, in many countries, establishing guidelines of regulatory reforms is a major difficulty in industrial development of farming insects. Only in recent times has this novel food concept emerged, and today rules for insect foods are being established (Belluco et al., 2013). This codification should attract investment, which will lead to the development of international trade of farmed insects.

There is need for cheap production methods that are under regulatory supervision in order to produce healthy food and maintain the environment. This can be achieved with collaborative effort of multiple disciplines whether working on local, national, or global levels. Precautions during farming and collection are necessary to avoid contamination by pathogens (bacteria, fungi, and nematodes) and any residue of synthetic pesticides. Processing and storage of farmed insects and insect products should follow the same health and sanitation regulations and norms as for any other traditional food. Generally, regulatory laws are governed in every country by the Food and Drug Administration under the Ministry of Health. In Europe and USA, safety measures have been formulated and implemented with strict legislation (Belluco et al., 2013).

CURRENT CHALLENGES AND CONCLUSIONS

1. Consumers should be told about the nutritive value of edible insects. Marketing, large-scale production units, and networks can promote this idea in at least some urban areas. Harvesting, processing, and packaging with modern technology need to be adopted. In Indonesia, insect farming has proved profitable since public interest is increasing due to consumer acceptance of nutritional value of insects (Offenberg, 2011). Thus, public awareness is necessary to change prevailing attitudes of rejecting insect-based diets in the Western world. Education in schools, organizing shows, street plays, exhibitions, and the like are effective means of promoting insect farming with emphasis on environmental benefits, nutritional value, marketing prospects, and food security during periods of distress or natural calamity (Fairman, 2010).
2. In the wild, insects are often destroyed due to commercial exploitation, bush fire, unfavorable climatic conditions, and attack by natural enemies (Yen, 2009b). Conservation of forest-farmed insects in the wild can be facilitated by habitat management and biodiversity preservation with local participation (Mbata et al., 2002). Research on management potential of wild edible insects will encourage locals to undertake insect farming if sustainability in insect production is assured. Similarly, postharvest technology is important for both insects and consumers (Rumpold and Schluter, 2013).
3. Validating or revising feed conversion efficiency for insect species intended for rearing at an economically relevant production scale and density is an essential step toward accurately assessing their potential as alternative source of dietary protein. An international network can facilitate R&D activity and motivate innovative entrepreneurs. Once technology of insect farming is proven to be economically and ecologically sustainable, such examples can be replicated in other countries, since the production system depends upon the consumers and local and international industries (van der Meer, 2004).
4. To develop farming of new native edible insect species, identification, ecology, and nutritive value of those edible species that have a short life cycle and are suitable for farming need to be studied. This is a very promising new area of research to expand insect utilization as food by providing a wider choice of insects for farming. It will also be important to enhance the possibility of farming local insect species to prevent transporting nonnative insects into new areas. A better knowledge of the nutritional requirements of farmed insects may help to increase the production efficiency by enabling the development of efficient and cheap insect feeds. Cost is the most important criterion for consumers who would prefer cheap and high-quality food produced in a controlled environment. In the future, efficient, cost-effective, and low-carbon farming methods would be required to deliver high yields.
5. Human food containing insects is gaining momentum in the general public and the addition of insects as an ingredient could be a general practice within just a few years worldwide (including North America and Europe) (Fairman, 2010). Since business models for insect-based food and feed products do not exist at present, creating products for human

consumption can help entrepreneurs to build business models because insects can also be used as food additives, antimicrobials, and biomaterials (Dossey, 2013). However, accurate selection of insect species should be envisaged before farming, particularly where automation is needed for mass production (Ohura, 2003). For small-scale farming by family members, rearing kits and cheap local sources of feed (including organic waste) would encourage villagers to adopt insect farming. For large-scale production and marketing, certain measures would expand the current insect farming business and insect products, for example, creation of an international society of producers, development of a code of standard price fixation, development of marketing strategies, recommendations on insect species, information networks, liaison with policy makers and researchers, and the like (FAO, 2014).

6. Legislation to control and maintain standards of food products containing insects to safeguard consumers is lacking or not adopted equally throughout the world. As products are introduced in the market, testing and confirmation of contents, nutritive value, raw material storage, distribution chain, and so forth should be established at major marketing points. Hygienic conditions must be compulsory during farming and processing because insects can come into contact with infected soil or plant parts and become contaminated with pathogenic microorganisms including bacteria, fungi, protozoa, and others (Vega and Kaya, 2012). Infection may cause spoilage, or contaminated insects can serve as vectors or intermediate hosts for vertebrate pathogens, and certain diseases can therefore be transmitted to humans. There are also chances of potential recontamination or cross-contamination (Banjo et al., 2006a). Along with these measures, the Hazard Analysis Critical Control Points (HACCP) should be used as this list is a scientific and systematic tool which identifies specific hazards and establishes control systems to ensure safety through preventive measures. Insects alone and also those used as ingredients for protein enrichment in fermented or processed food, particularly the "ready-to-eat" food sold in markets, should be inspected and approved for sale as safe foods (Zhou and Han, 2006; Klunder et al., 2012).

7. Recommendations on diet preparation and doses, specific to various insect species, have to be managed to reduce exposure to contamination and other health risks associated with feed. Traditional legislation exists in some countries where local communities take care of overexploitation of insects and management of natural resources (Mbata et al., 2002). Other appropriate legislative measures would also facilitate banning species that are recognized as dangerous to human health whenever insect-based diets are consumed. Protein extracted from farmed edible insects can be of interest to food technologists and also to the pharmaceutical industry, apart from human consumption.

8. The census data on species of edible insects and their numbers in the wild are not available. A simple survey in each agroecosystem would reveal the richness of nature that can be used for years to come by local communities not only for their livelihood but for small cottage industries in the localities. Government financial assistance and technical guidance would be of immense help for commercialization of farmed insects. This step is urgently needed because most banks are not aware of business patterns and potential income earned from edible insects (wild collection and farmed lots) and, therefore, they are not interested in sanctioning loans to those farmers who are active in commercial activities. Along the same lines, international organizations, particularly FAO, IFAD, and UNDP, may promote farming of edible insects and insects as a potential new or alternative food protein. Strong support for potential farmers and entrepreneurs, as well as extensive awareness campaigns for consumers, would certainly increase consumption demand. Some attention to processing and product development, particularly in the form of instant products, new food recipes, and microwavable items, will also appeal to new consumers.

9. Finally, insects can be considered as a potential substitute for animal protein, especially when produced worldwide, on industrial scale. This will help improve food security, human health, and environmental safety. In the coming years, efficient and cost-effective management practices, food safety issues, and international trade would be important issues for businesses involved in farming and marketing of edible insects. Of course, further research on production of farmed insects, particularly postharvest processing, is urgently needed for widening the scope of existing enterprises. Currently, the insect food industry is largely overlooked by policy makers and the concepts and applications of insects and insect products are not included in food products by government agencies.

REFERENCES

Adeoye, O.T., Oyelowo, O.J., Adebisi-Fagbohungbe, T.A., Akinyemi, O.D., 2014. Eco-diversity of edible insects of Nigeria and its impact on food security. J. Biol. Life Sci. 5, 175–187.

Agea, J.G., Biryomumaisho, D., Buyinza, M., Nabanoga, G.N., 2008. Commercialization of *Ruspolia nitidula* (Nsenene grasshopper) in Central Uganda. Afr. J. Agric. Dev. 8 (3), 319–332.

Alamu, O.T., Amao, A.O., Nwokedi, C.I., Oke, O.A., Lawa, I., 2013. Diversity and nutritional status of edible insects in Nigeria: a review. Int. J. Biodivers. Conserv. 5, 215–222.

Ayieko, M.A., Oriamo, V., Nyambuga, I.A., 2010. Processed products of termites and lake flies: improving entomophagy for food security within the Lake Victoria region. Afr. J. Food Agric. Nutr. Dev. 10 (2), 2085–2098.

Banjo, A.D., Lawal, O.A., Songonuga, E.A., 2006b. The nutritional value of fourteen species of edible insects in southwestern Nigeria. Afr. J. Biotechnol. 5, 298–301.

Banjo, A.D., Lawal, O.A., Adeyemi, A.J., 2006a. The microbial fauna associated with the larvae of *Oryctes monoceros*. J. Appl. Sci. Res. 2, 837–843.

Beckett, J.L., Oltjen, J.W., 1993. Estimation of the water requirement for beef production in the United States. J. Anim. Sci. 77, 818–826.

Belluco, S., Losasso, C., Maggioletti, M., Alonzi, C.C., Paoletti, M.G., Ricci, A., 2013. Edible insects in a food safety and nutritional perspective: a critical review. Compr. Rev. Food Sci. Food Saf. 12 (3), 296–313.

Bhatia, N.K., Yousuf, M., 2013. Forest insect industry in collaborative forest management: a review. Int. J. Ind. Entomol. 27, 166–179.

Bhede, B.V., Lande, U.L., Pathrikar, D.T., 2013. Evaluation of different mulberry varieties for commercial rearing of silkworm hybrid in Maharashtra. J. Entomol. Res. 37, 17–24.

Bondari, K., Sheppard, D.C., 1981. Soldier fly larvae as feed in commercial fish production. Aquaculture 24, 103–109.

Boongrid, S., 2010. Honey and non-honey foods from bees in Thailand. In: Forest Insects as Food: Humans Bite Back. Proceedings of a Workshop on Asia-Pacific Resources and Their Potential for Development, February 19–21, 2008. FAO, Chiang Mai, FAO Regional Office for Asia and the Pacific, Bangkok (Publication No. 2010/02), pp. 165–172.

Boulidam, S., 2010. Edible insects in a Lao market economy. In: Forest Insects as Food: Humans Bite Back. Proceedings of a Workshop on Asia-Pacific Resources and Their Potential for Development, February 19–21, 2008. FAO, Chiang Mai, FAO Regional Office for Asia and the Pacific, Bangkok (Publication No. 2010/02), pp. 131–140.

Bukkens, S.G.F., 1997. The nutritional value of edible insects. Ecol. Food Nutr. 36, 287–319.

Bukkens, S.G.F., 2005. Insects in the human diet: nutritional aspects. In: Paoletti, M.G. (Ed.), Ecological Implications of Minilivestock: Potential of Insects, Rodents, Frogs and Snails. Science Publishers, Enfield, NH, pp. 545–577.

Cabaraux, J.F., Philippe, F.X., Laitat, M., Canart, B., van den Heede, M., 2009. Gaseous emission from weaned pigs raised on different floor systems. Agric. Ecosyst. Environ. 130, 86–92.

Capinera, J.L., 2004. Encyclopedia of Entomology. Kluwer Academic Publishers, Dordrecht, The Netherlands, 258 p.

Cerda, H., Martinez, R., Briceno, N., Pizzoferrato, L., Manzi, P., Tommaseo Ponzetta, M., Martin, O., Paoletti, M.G., 2001. Palm worm (*Rhynchophorus palmarum*), traditional food in Amazonas, Venezuela: nutritional composition, small scale production and tourist palatability. Ecol. Food Nutr. 40, 13–32.

Cerritosa, R., 2009. Insects as food: an ecological, social and economical approach. CAB Rev. 4 (27), 1–10.

Cerritosa, R., Cano-Santana, C., 2008. Harvesting grasshoppers, *Sphenarium purpurascens* in Mexico for human consumption: a comparison with insecticidal control for managing pest outbreaks. Crop Protect. 27, 473–480.

Chakravorty, J., 2014. Diversity of edible insects and practices of entomophagy in India: an overview. J. Biodivers. Bioprospect. Dev. 1 (3), 124.

Chapagain, A.K., Hoekstra, A.Y., 2003. Virtual flows between nations in relation to trade in livestock and livestock products. Value of Water Research Report series no. 13, United Nations Educational, Scientific and Cultural Organization, Paris.

Chen, X., Feng, Y., Chen, Z., 2009. Common edible insects and their utilization in China. Entomol. Res. 39, 299–303.

Chidumayo, E.N., Mbata, K.J., 2002. Shifting cultivation, edible caterpillars and livelihoods in the Kopa area of northern Zambia. Insects Trees Livelihoods 12, 175–193.

Clifford, C.W., Woodring, J.P., 2009. Methods for rearing the house cricket, *Acheta domesticus* (L.), along with baseline values for feeding rates, growth rates, development times and blood composition. J. Appl. Entomol. 109 (1–5), 1–14.

Collavo, A., Glew, R.H., Huang, Y.S., Chuang, L.T., Bosse, R., 2005. House cricket small-scale farming. In: Paoletti, M.G. (Ed.), Ecological Implications of Minilivestock: Role of Rodents, Frogs, Snails and Insects for Sustainable Development. Science Publishers, Enfield, NH, pp. 519–544.

Costa-Neto, E.M., 2012. Estudos etnoentomologicos no estado da Bahia: uma homenagen aons do compo de pesquisa. Biotemas 17 (1), 117–149.

Crabbe, N., 2012. Local expert gets funding to develop insect-based food for starving children. Gainsville Sun, 1B–6A, May 10, 2012.

DeFoliart, G.R., 1997. An overview of the role of edible insects in preserving biodiversity. Ecol. Food Nutr. 36, 109–132.

DeFoliart, G.R., 1999. Insects as food: why the western attitude is important. Ann. Rev. Entomol. 44, 21–50.

Deroy, O., 2015. Eat insects for fun, not to help the environment. Nature 521 (7553), 395.

Doley, A., Kalita, J., 2012. Traditional uses of insects and insect products in medicine and food by the Mishing tribe of Dhemaji district, Assam, North-East India. Soc. Sci. Res. 1 (2), 11–21.

Dossey, A.T., 2010. Insects and their chemical weaponry: new potential for drug discovery. Nat. Prod. Rep. 27, 1737–1757.

Dossey, A.T., 2013. Why insects should be in your diet. Scientist, February 1, 2013.

Duan, X., 1998. Introduction of research situation about chitin and chitosan and their application in agriculture and forestry. World Forest Res. 11, 9–14.

Dzamba, J., 2010. Third Millennium Farming: Is It Time for Another Farming Revolution?, Architecture, Landscape and Design, Toronto, Canada.

Fairman, R.J., 2010. Instigating an education in insects: the eating creepy crawlies' exhibition. Antenna 34, 169–170.

FAO, 2004. Contribution des insectes de la forêt à la sécurité alimentaire: L'exemple des chenilles d'Afrique Centrale. NTFP Working document No. 1; FAO, Rome. Available from: www.fao.org/docrep/007/j3463f/j3463f3400.htm

FAO, 2008. Forest Insects as Food: Humans Bite Back. In: Durst, P.B., Robin, D.V.J., Leslie, N., Shono, K. (Eds.), FAO Regional Office for Asia and the pacific, Chiang Mai, Thailand.

FAO, 2013. Edible Insects: Future Prospects for Food and Feed Security. In: van Huis, A., van Itterbeeck, J., Klunder, H., Mertens, E., Halloroan, A., Muir, G., Vantomme, P. (Eds.), FAO Forestry Paper No. 171, FAO, Rome, 187 p.

FAO, 2014. Corporate Document Repository, Food and Agriculture Organization of United Nations, Rome.

FAO/WUR, 2012. Expert committee meeting: assessing the potential of insects as food and feed in assuring food security. In: Vantomme, P., Mertens, E., van Huis, A., Klunder, H. (Eds.), Summary Report, 23–25 January 2012, FAO, Rome.

Farina, L., Demey, F., Hardouin, J., 1991. Production de termites pour l'aviculture villageoise au Togo. Tropicultura 9 (4), 181–187.
Feng, Y., Zhao, M., He, Z., Chen, Z., Sun, L., 2009. Research and utilization of medicinal insects in China. Entomol. Res. 39, 313–316.
Fiala, N., 2008. Meeting the demand: an estimation of potential future greenhouse gas emissions from meat production. Ecol. Econ. 67, 412–419.
Gahukar, R.T., 2011a. Entomophagy and human food security. Int. J. Trop. Insect Sci. 31, 129–144.
Gahukar, R.T., 2011b. Entomophagy can support rural livelihood in India. Curr. Sci. 103, 10.
Gahukar, R.T., 2013. Insects as human food: are they really tasty and nutritious? J. Agric. Food Info. 14, 264–271.
Gahukar, R.T., 2014. Impact of major biotic factors on tropical silkworm rearing in India and monitoring of unfavourable elements: a review. Sericologia 54, 150–170.
Giaccone, V., 2005. Hygiene and health features of "minilivestock". In: Paoletti, M.G. (Ed.), Ecological Implications of Minilivestock: Role of Rodents, Frogs, Snails and Insects for Sustainable Development. Science Publishers Inc., Enfield, NH, pp. 579–598.
Gracer, D., 2010. Filling the plates: serving insects to the public in the United States. In: Durst, P.B., Johnson, D.V., Leslie, R.N., Shono, K. (Eds.), Forest Insects as Food: Humans Bite Back. Proceedings of a Workshop on Asia-Pacific Resources and Their Potential for Development, February 19–21, 2008. FAO, Chiang Mai, Thailand, FAO Regional Office for Asia and the Pacific, Bangkok (Publication No. 2010/02), pp. 217–220.
Gray, D.D., 2014. Edible insects a boon to Thailand farmers. Asian Diversity News, August 26, 2014.
Hanboonsong, Y., 2010. Edible insects and associated food habits in Thailand. In: Durst, P.B., Johnson, D.V., Leslie, R.N., Shono, K. (Eds.), Forest Insects as Food: Humans Bite Back. Proceedings of a workshop on Asia-Pacific Resources and Their Potential for Development, February 19–21, 2008. FAO, Chiang Mai, Thailand, FAO Regional Office for Asia and the Pacific, Bangkok (Publication No. 2010/02), pp. 171–182.
Hanboonsong, Y., Jamjanya, T., Durst, P.B., 2013. Six-Legged Livestock: Edible Insect Farming, Collection and Marketing in Thailand. Food and Agriculture Organization, Regional Office for Asia and the Pacific, RAP publication 2013/03, Bangkok, Thailand.
Handley, M.A., Hall, C., Sanford, E., Diaz, E., Gonzalez-Mendez, E., Drace, K., Wilson, R., Villalobas, M., Croughan, M., 2007. Globalization, binational communities, and imported food risks: results of an outbreak investigation of lead poisoning of Monterey County, California. Am. J. Public Health 97 (5), 900–906.
Hardouin, J., 1995. Minilivestock: from gathering to controlled production. Biodivers. Conserv. 4, 220–232.
Harper, L.A., Flesch, T.K., Powell, J.M., Coblentz, W.K., Jokela, W.E., Martin, N.P., 2009. Ammonia emissions from daily production in Wisconsin. J. Dairy Sci. 92, 2326–2337.
Hope, R.A., Frost, P.G.H., Gardiner, A., Ghazoul, J., 2009. Experimental analysis of adoption of domestic mopane worm farming technology in Zimbabwe. Dev. S. Afr. 26, 29–46.
Isaiarasu, L., Sakthivel, N., Ravikumar, R., Samuthiravelu, P., 2011. Effect of herbal extracts on the microbial pathogens causing flacherie and muscardine disease in the mulberry silkworm, *Bombyx mori* L. J. Biopesticides 4 (2), 150–155.
Jamjanya, T., Thavornaukulkit, C., Klibsuwan, V., Totuyo, P., 2001. Mass Rearing of Crickets for Commercial Purpose. Faculty of Agriculture, Khon Kaen University, Khon Kaen, Thailand.
Johnson, D.V., 2010. The contribution of edible forest insects to human nutrition and to forest management: Current status and future potential. In: Forest Insects as Food: Humans Bite Back. Proceedings of a Workshop on Asia-Pacific Resources and Their Potential for Development, February 19–21, 2008. FAO, Chiang Mai. FAO Regional Office for Asia and the Pacific, Bangkok (Publication No. 2010/02), pp. 5–22.
Jongema, Y., 2012. List of edible insects of the world. Wageningen University, The Netherlands. Available from: http://www.ent.wur.nl/UK/Edible+insects/Worldwide+species+list/
Kampmeier, G.E., Irwin, M.E., 2003. Commercialization of insects and their products. In: Resh, V.H., Carde, R.T. (Eds.), Encyclopedia of Insects. Academic Press, Burlington, MA, pp. 252–260.
Kanafi, R.R., Edabi, R., Mirhosseini, S.Z., Seidevi, A.R., Zolfaghari, M., Etebari, K., 2007. A review on nutritive effect of mulberry leaves enrichment with vitamins on economic traits and biological parameters of silkworm, *Bombyx mori* L. Indian J. Seric. 46, 86–91.
Kang, P.D., Kim, M.J., Jung, I.Y., Kim, K.Y., Kim, Y.S., Sung, G.B., Sohn, B.H., 2011. Production of coloured cocoons by feeding dye-added artificial diet. Int. J. Ind. Entomol. 22, 21–23.
Katayama, N., Ishikowa, Y., Takaoki, M., Chi, K., Yamashita, M., Nakayama, S., Kiguchi, K., Kak, R., Wada, H., Mitsuhashi, J., 2008. Entomophagy: a key to space agriculture. Adv. Space Res. 41 (5), 701–705.
Katayama, N., Yamashita, M., Wada, H., Mitsuhashi, J., 2005. Entomophagy as part of a space diet for habitation on Mars. J. Space Technol. Sci. 21 (2), 27–38.
Killbrew, K., Wolff, H., 2010. Environmental impacts of agricultural technologies. EPAR Brief No. 65, Evans School of Public Affairs, University of Washington, Washington, DC, 18 p.
Kim, S.A., Kim, K.M., Oh, B.J., 2008. Current status and perspectives of the insect industry in Korea. Entomol. Res. 38, 79–85.
Kittler, P.G., Sucher, K., 2008. Food and Culture, fifth ed. Wadsworth, Belmont, CA.
Klunder, H.C., Wolkers-Rooijackers, J., Korpela, J.M., Nout, M.R., 2012. Microbial aspects of processing and storage of edible insects. Food Control 26, 628–631.
Kozanayi, W., Frost, P., 2002. Marketing of Mopane Worm in Southern Zimbabwe. Institute of Environmental Studies, Harare, Zimbabwe.
Li, B., Wang, Y., Liu, H., Xu, Y., Wei, Z., Chen, Y., Shen, W., 2010. Resistance comparison of domesticated silkworm (*Bombyx mori* L.) and wild silkworm (*Bombyx mandarina* L.) to phoxim insecticide. Afr. J. Biotechnol. 9, 1771–1775.
Losey, J., Vaughan, M., 2006. The economic value of ecological services provided by insects. Bioscience 56, 311–323.
Lundy, M.E., Parrella, M.P., 2015. Crickets are not a free lunch: Protein capture from scalable organic side streams *via* high-density populations of *Acheta domesticus*. PLoS One 10, e0118785.
Malik, F.A., Reddy, Y.S., 2007. Role of mulberry nutrition on manifestation of post cocoon characters of selected races of the silkworm, *Bombyx mori* L. Sericologia 47, 63–67.

Mbata, K.J., Chidumayo, E.N., Lwatula, C.M., 2002. Traditional regulation of edible caterpillar exploitation in the Kopa area of Mpika district in northern Zambia. J. Insect Conserv. 6, 115–130.

Mercer, C.W.L., 1994. Sago grub production in Luber swamp near Lae-Papua New Guinea. Klinkii 5 (2), 30–34.

Mercer, C.W.L., 1997. Sustainable products of insect for food and income by New Guinea villagers. Ecol. Food Nutr. 36 (2–4), 151–157.

Mitsuhashi, J., 2007. Use of insects in closed space environment. Biol. Sci. Space 21, 124–128.

Mpuchane, G., Gashe, B.A., Allotey, J., Ditihogo, M.K., Siame, B.A., Teferra Impanya, M.F., 2001. Phane: Its Exploitation and Conservation in Botswana. Technical Bulletin No. 6, University of Botswana, Gaborone, Botswana, 46 p.

Mulhane, S., Gashe, B.A., Allotey, J., Siame, A.B., Teferra, G., Ditlhogo, M., 2001. Quality determination of Phane, the edible caterpillar of an Emperor moth, *Imbrasia belina*. Food Control 11, 453–458.

Munoz, C.B., 2008. Transnational Tortillas (Race, Gender and Shop-floor Politics in Mexico and the United States). Cornell University Press, Ithaca, NY, 202 p.

Munyuli Bin Mushambanyi, T., 2000. Etude preliminaire orientee vers la production des chenilles consommables par l'elevage des papillons *Anaphe infracta* (Thaumetopoeidae) a Lwiro, Sud-Kivu, Republique Democratique du Congo. Tropiculture 18, 208–211.

Nakagaki, B.J., DeFoliart, G.R., 1991. Comparison of diets for mass rearing *Acheta domesticus* (Orthoptera: Gryllidae) as a novelty food and comparison of food conversion efficiency with values reported for livestock. J. Econ. Entomol. 84, 891–896.

Nguku, E.K., Muli, E.M., Raina, S.K., 2007. Larval, cocoon and post-cocoon characteristics of *Bombyx mori* L. (Lepidoptera: Bombylidae) fed on mulberry leaves fortified with Kenyan royal jelly. J. Appl. Sci. Environ. Manage. 11 (4), 85–89.

Nonaka, K., 2005. Ethnoentomology: Insect Eating and Human–Insect Relationship. University of Tokyo Press, Tokyo.

Nonaka, K., 2009. Fasting on insects. Entomol. Res. 39, 304–312.

Nonaka, K., 2010. Cultural and commercial roles of edible wasps in Japan. In: Durst, D.B., Johnson, D.V., Leslie, R.N., Shono, K. (Eds.), Forest Insects as Food: Humans Bite Back. Proceedings of a Workshop on Asia-Pacific Resources and Their Potential for Development, February 19–21, 2008. FAO, Chiang Mai, Thailand, FAO Regional Office for Asia and the Pacific, Bangkok (Publication No. 2010/02), pp. 123–130.

Oerke, E.C., Dehne, H.W., 2004. Safeguarding production: losses in major crops and the role of crop protection. Crop Protect. 23, 275–285.

Offenberg, J., 2011. *Oecophylla smaragdina* food conversion efficiency: prospects for ant farming. J. Appl. Entomol. 135 (8), 575–581.

Offenberg, J., Wiwatwitaya, D., 2009. Weaver ants convert pest insects into food: prospects for the rural poor. Paper presented at the International Conference on Research for Food Security, Natural Resources Management and Rural Development, University of Hamburg, Germany, October 6–8, 2009.

Ohura, M., 2003. Development of an automated warehouse type silkworm rearing systems for the production of useful materials. J. Insect Biotechnol. Sericol. 72 (3), 163–169.

Onore, G., 1997. A brief note on edible insects in Ecuador. Ecol. Food Nutr. 36, 277–285.

Oonincx, D.G., de Boer, I.J., 2012. Environmental impact of the production of mealworms as a protein source for humans: a life cycle assessment. PLoS One 7 (12), e51145.

Oonincx, D.G., van Itterbeeck, J., Heetkamp, M.J., van den Brand, H., van Loon, J.J., van Huis, A., 2010. An exploration on greenhouse gas and ammonia production by insect species suitable for animal or human consumption. PLoS One 5 (12), e14445.

Paoletti, M.G., Dufour, D.L., 2005. Edible invertebrates among Amazonian Indians: a critical review. In: Paoletti, M.G. (Ed.), Ecological Implications of Minilivestock: Potential of Insects, Rodents, Frogs and Snails. Science Publishers, Enfield, NH, pp. 293–342.

Patton, R.L., 1967. Oligidic diets for *Acheta domesticus* (Orthoptera: Gryllidae). Ann. Entomol. Soc. Am. 60 (6), 1238–1242.

Paulino, A.T., Simionato, J.I., Garcia, J.C., Nozaki, J., 2006. Characterization of chitosan and chitin produced from silkworm chrysalides. Carbohydr. Polym. 64, 98–103.

Pelletier, N., Tyedmers, P., 2010. Forecasting potential of global environmental costs of livestock production 2000–2050. Proc. Natl. Acad. Sci. USA 107, 18371–18374.

Pimentel, D., Berger, B., Filiberto, D., Newton, M., Wolfe, B., Karabinakis, E., Clark, S., Poon, E., Abbett, E., Nandgopal, S., 2004. Water resources: agricultural and environmental issues. Bioscience 54 (10), 909–918.

Pimentel, D., Pimentel, M., 2003. Sustainability of meat-based and plant-based diets and the environment. Am. J. Clin. Nutr. 78 (Suppl.), 660–663.

Popa, R., Green, T.R., 2012. Using black soldier fly larvae for processing organic leachates. J. Econ. Entomol. 105, 374–378.

Premalatha, M., Abbasi, T., Abbasi, T., Abbasi, S.A., 2011. Energy-efficient food production to reduce global warming and ecodegradation: the use of edible insects. Renew. Sustain. Energy Rev. 15, 4357–4360.

Ramandey, E., van Mastrigt, H., 2010. Edible insects in Papua, Indonesia: from delicacies snack to basic need. In: Durst, P.B., Johnson, D.V., Leslie, R.N., Shono, K. (Eds.), Forest Insects as Food: Humans Bite Back. Proceedings of a Workshop on Asia-Pacific Resources and Their Potential for Development, February 19–21, 2008. FAO, Chiang Mai, Thailand, FAO Regional Office for Asia and the Pacific, Bangkok (Publication No. 2010/02), pp. 105–114.

Ramos-Elorduy, J., 1997. Importance of edible insects in the nutrition and economy of people of the rural areas in Mexico. Ecol. Food Nutr. 36, 347–366.

Ramos-Elorduy, J., 1998. Creepy Crawly Cuisine: The Gourmet Guide to Edible Insects. Park Street Press, Rochester, VT, 150 p.

Ramos-Elorduy, J., 2006. Threatened edible insects in Hidalgo, Mexico and some measures to preserve them. J. Ethnobiol. Ethnomed. 2, 51.

Ramos-Elorduy, J., 2008. Energy supplied by edible insects from Mexico and their nutritional and ecological importance. Ecol. Food Nutr. 47 (3), 280–297.

Ramos-Elorduy, J., 2009. Anthropo-entomophagy: cultures, evolution and sustainability. Entomol. Res. 39, 271–288.

Ramos-Elorduy, J., Landero-Torres, I., Murgula-Gonzalez, J., Pino, M.J.M., 2008. Anthropoentomophagic biodiversity of the Zongolica region, Veracruz, Mexico. Revista de Biologia Tropical 56, 303–316.

Rani, A.G., Padmalatha, C., Raj, R.S., Ranjit Singh, A.J., 2011. Impact of supplementation of Amway protein on the economic characters and energy budget of silkworm, *Bombyx mori* L. Asian J. Anim. Sci. 10, 1–10.

Ratcliffe, B.C., 2006. Scarab beetles in human culture. Coleopts. Soc. Mono. 5, 85–101.

Reddy, K.S., Kumar, M., Maruthi, V., Umesha, B., Vijayalaxmi, Nageswar Rao, C.V.K., 2015. Dynamics of well irrigation systems and CO_2 emission in different agroecosystems of South Central India. Curr. Sci. 108, 2063–2070.

Reddy, B.K., Rao, J.V.K., 2009. Seasonal occurrence and control of silkworm diseases: grasserie, flacherie and muscardine, and insect parasitoid uzi fly in Andhra Pradesh, India. Int. J. Ind. Entomol. 18, 57–61.

Rodriguez, E., Sultan, R., Hilliker, A., 2004. Negative effects of agriculture on an environment. Traprock 3, 28–32.

Ronghang, R., Ahmed, R., 2010. Edible insects and their conservation strategy in Karbi Anglong district of Assam, Northeast India. Bioscan 2 (Special issue), 515–521.

Roulon-Doko, P., 1998. Les Activites de Cueillete: Chasse, Cueillette et Culture chez les Gbaya De Centrafrique. Edition Harmattan, Paris.

Rumpold, B.A., Schluter, O.K., 2013. Potential and challenges of insects as an innovative source of food and feed production. Innov. Food Sci. Emerg. Technol. 17, 1–11.

Saeed, T., Dagga, F.A., Saraf, M., 1993. Analysis of residual pesticides present in edible insects captured in Kuwait. Arab Gulf J. Sci. Res. 11 (1), 1–5.

Safina, C., 2011. Why are we using up the Earth? CNN Opinion: Carbon Dioxide. CNN, New York.

Sarmah, K., 2011. Eri pupa: a delectable dish of North East India. Curr. Sci. 100, 279.

Sathe, T.V., Jadhav, A.D., Kamadi, N.G., Undale, J.P., 2008. Low cost rearing technique for mulberry silkworm (PM $\times$ NB_4D_2) by using nylon and indigenous shelves. Biotechnol. App. Entomol. 5, 205–211.

Schosler, H., de Boer, J., Boersema, J.J., 2012. Can we cut out the meat of the dish? Constructing consumer-oriented pathways towards meat substitution. Appetite 58, 39–47.

Shockley, M., Dossey, A., 2014. Insects for human consumption. In: Morales-Ramos, J.A., Rojas, M.G., Shapiro-Ilan, D.I. (Eds.), Mass Production of Beneficial Organisms: Invertebrates and Entomopathogens. Academic Press, Salt Lake City, UT, pp. 617–652.

Sirimungkararat, S., Saksirirat, W., Notongkham, A., 2010. Edible products from eri and mulberry silkworm in Thailand. In: Durst, P.B., Johnson, D.V., Leslie, R.N., Shono, K. (Eds.), Forest Insects as Food: Humans Bite Back. Proceedings of a Workshop on Asia-Pacific Resources and Their Potential for Development, February 19–21, 2008. FAO, Chiang Mai, Thailand, FAO Regional Office for Asia and the Pacific, Bangkok (Publication No. 2010/02), pp. 189–200.

Smith, A., 2012. Get ready to pay more for your steak. CNN Money, New York City, New York, USA.

Smith, A., Schellnhuber, H.J., Mirza, M.M.Q., 2001. Vulnerability to climate change and reasons for concern: a synthesis. In: McCarthy, J.J., White, K.S., Canziani, O., Leary, N., Dokken, D.J. (Eds.), Climate Change 2001 Impacts, Adaptation and Vulnerability. Contribution of Working Group II to the Third Assessment Report of the Intergovernmental Panel on Climate Change.

Smithers, P., 2015. Insects are the business: the challenges to insect farming for human consumption in the UK. Antenna 39 (1), 29–30.

Srivastava, S.K., Babu, N., Pandey, H., 2009. Traditional insect bioprospecting; as human food and medicine. Indian J. Trad. Know. 8, 485–494.

Stehfest, E., Bouwman, L., van Vuuren, D.P., den Elzen, M.G.J., Eickhout, B., Kabat, P., 2009. Climate benefits of changing diet. Clim. Change 95, 83–102.

Steinfeld, H., Gerber, P., Wassenaar, T., Castel, V., Rosales, M., Haan, C.D., 2006. Livestock's Long Shadow: Environmental Issues and Options. FAO, Rome.

Sundarraj, S., Nangia, N., Chinnaswamy, K.P., Sannappa, B., 2000. Influence of protein supplement on performance of PN $\times$ NB_4D_2 silkworm breed. Mysore J. Agric. Sci. 34, 302–307.

Swaminathan, C., 2015. Hidden waters. Kisan World 42 (2), 27–28.

Tabuna, H., 2000. Evaluation des Echanges des Produits Forestiers Non-ligneux entre l'Afrique et l'Europe. FAO regional Office for Africa, Accra, Ghana.

Tamil Nadu Agricultural University, 2014. AgriTech Portal. Tamil Nadu Agricultural University, Coimbatore, India.

Tikader, A., Kamble, C.K., 2010. Seri-biodiversity with reference to host plants in India. Asian Austral. J. Plant Sci. Biotechnol. 4 (1), 1–11.

Tong, L., Yu, X.H., Liu, H., 2011. Insect food for astronauts: gas exchange in silkworms fed on mulberry and lettuce and the nutritional value of these insects for human consumption during deep space flights. Bull. Entomol. Res. 101 (5), 613–622.

UN, 2014. Millennium Development Goals Report. United Nations, New York, NY.

van Broekhoven, S., Oonincx, D.G.A.B., van Huis, A., van Loon, J.J.A., 2015. Growth performance and feed conversion efficiency of three edible mealworm species (Coleoptera: Tenebrionidae) on diets composed of organic by-products. J. Insect Physiol. 73, 1–10.

van der Meer, K., 2004. Exclusion of small scale farmers from coordinated supply chains: market failure, policy failure or just economies of scale? Paper presented at the workshop "Is There a Place for Small Holder Producers in Coordinated Supply Chains," organized by World Bank, December 8, 2004, Washington, DC.

van der Warf, H., Petit, J., 2002. Evaluation of the environmental impact of agriculture at the farm level: a comparison and analysis of 12 indicator-based methods. Agric. Ecosyst. Environ. 93, 131–145.

van Huis, A., 2003. Insects as food in sub-Saharan Africa. Int. J. Tropical Insect Sci. 23, 163–185.

van Huis, A., 2013. Potential of insects as food and feed in assuring food security. Ann. Rev. Entomol. 58, 563–583.

van Huis, A., Itterbeek, J.V., Klunder, H., Mertens, E., Halloran, A., Muir, G., Vantomme, P., 2013. Edible insects: future prospects for food and feed security. FAO Forestry paper no.171. Food & Agriculture Organization, Rome Italy.

Vega, F., Kaya, H., 2012. Insect Pathology. Academic Press, London.

Vogel, G., 2010. For more protein, filet cricket. Science 327 (5967), 811.

Weissman, D.B., Gray, D.A., Pham, H.T., Tijssen, P., 2012. Billions and billions sold: pet-feeder crickets (Orthoptera: Gryllidae), commercial cricket farms, an epizootic densovirus, and government regulations make for a potential disaster. Zootaxa 3504, 67–88.

Wijayanthi, N., Subramanyam, M.V., 2002. Effect of fenvalerate 20EC on sericigenous insects. 1. Food utilization in the late-age larva of the silkworm, *Bombyx mori* L. Ecotoxicol. Environ. Saf. 53, 206–211.
Williams, J., 2015. Local business makes protein powder from bugs. The Red & Black, September 23, 2015.
Yang, Y., Tang, L., Tong, L., Liu, H., 2009. Silkworms culture as a source of protein for humans in space. Adv. Space Res. 43 (8), 1236–1242.
Yen, A.L., 2005. Insects and other invertebrates food of the Australian aborigines. In: Paoletti, M.G. (Ed.), Ecological Implications of Minilivestock: Potential of Insects, Rodents, Frogs and Snails. Science Publishers, Enfield, NH, pp. 367–388.
Yen, A.L., 2009a. Edible insects: traditional knowledge or western phobia? Entomol. Res. 39, 289–298.
Yen, A.L., 2009b. Entomophagy and insect conservation: some thoughts for digestion. J. Insect Conserv. 13, 667–670.
Yhoung-Aree, J., Puwastien, P.P., Attig, G.A., 1997. Edible insects in Thailand, an unconventional protein source? Ecol. Food Nutr. 36, 133–149.
Yhoung-Aree, J., Viwatpanich, K., 2005. Edible insects in the Laos PDR, Myanmar, Thailand and Vietnam. In: Paoletti, M.G. (Ed.), Ecological Implications of Minilivestock: Potential of Insects, Rodents, Frogs and Snails. Science Publishers, Enfield, NH, pp. 415–440.
Yu, X.H., Liu, H., Tong, L., 2008. Feeding scenario of the silkworm *Bombyx mori* L. in the BLISS. Acta Astronaut. 63 (7–10), 1086–1092.
Zhao, S., Shang, X., Duo, L., 2013. Accumulation and spatial distribution of Cd, Cr and Pb in mulberry from municipal solid waste compost following application of EDTA and $(NH4)_2SO_4$. Environ. Sci. Poll. Res. 20 (2), 967–975.
Zhou, J., Han, D., 2006. Safety evaluation of protein of silkworm (*Antheraea pernyi*) pupae. Food Chem. Toxicol. 44, 1127–1130.
Zhou, L., Zhao, Y., Wang, S., Han, S., Liu, J., 2015. Lead in the soil—mulberry (*Morus alba* L.)—silkworm (*Bombyx mori*) food chain: translocation and detoxification. Chemosphere 128, 171–177.

Chapter 5

Modern Insect-Based Food Industry: Current Status, Insect Processing Technology, and Recommendations Moving Forward

A.T. Dossey*, J.T. Tatum**, W.L. McGill[†]

*All Things Bugs LLC, Griopro cricket powder, Athens, GA, United States; **Ripple Technology LLC, Atlanta, GA, United States;
[†]Rocky Mountain Micro Ranch and PhD Candidate, National University of Ireland at Galway, Ireland

Chapter Outline

Introduction	113
Efficiency	114
Biodiversity	115
Current Insect Farming Industry	117
Modern Industrial Mass Production of Insects	118
Recommendations and Considerations for Selection for Aspiring Insect-Based Food Producers and Insect Farmers	122
Feed Formulations and Biomass Sources for Farming Insects: Considerations for Insect Feed Formulations	122
Underutilized Biomass Amenable as Feed Ingredients for Mass-Farmed Edible Insects	122
Diseases Affecting Mass Produced/Farmed Insects	123
The NASCAR Jacket of Food: Food Labeling Requirements and Aspects of Consumer Demand. How do I label it? Is it Organic? GMO? Gluten Free? Paleo? Other?	123
Words Matter: Terminology Recommendations	124
Animal Welfare	125
Getting Past the "Fear Factor": Important Considerations for Normalizing Insects as a Mainstream Food Ingredient Beyond the Novelty Niche	125
Insect Processing Considerations: Considerations for Insect-Based Food Production, Processing, and Safety	126
The Real Pioneers: Entrepreneurs in the Insect-Based Food Space	135
Summary of Current Companies, Farms, and Other Organizations	140
Subindustry Niches: Importance of Finding One's Focused Role in the Industry and Product Selection	141
Supply Chain Needs: Feed, Farms, Insects, Transportation, Processing, and Manufacturing	146
Feed	146
Farms and Farmed Species	147
Transportation, Storage, and Distribution	147
Processing and Manufacturing Infrastructure	148
Intriguing the Larger Food Industry: Uses of Insects as Industrial Food Ingredients	148
Conclusions	149
References	150

INTRODUCTION

As the human population grows, it is ever more important to decrease the levels of materials we consume and harvest from the earth and its ecosphere. Food reserves are the lowest they've been in 40 years. Given that the population is increasing at 70 million people per year (Dzamba, 2010; Safina, 2011; Vogel, 2010) and that humans are consuming roughly 40% of the biomass that the land and the coastal seas produce (Safina, 2011), the increasing need for protein is not sustainable (Steinfeld et al., 2006). The historic US drought of 2012 also throws into sharp relief our need for more sustainable agricultural practices (Smith, 2012). Number 7 on the UN MDGs (United Nations Millennium Development Goals) is "ensuring environmental sustainability..." The destruction of the natural environment, the reduction in biodiversity, the mass extinctions that have occurred, and the dangerous impact the hand of humans has had on the earth's climate has even been denounced by many (including Dr Aaron T. Dossey, one of this chapter's authors and book editor) as immoral, most notably by the Pope of the Holy See in his 2015 encyclical titled "Laudato Si': On care for our common home. Encyclical Letter" (Pope Francis, 2015). Climate change, reduced productivity of agricultural lands, overfishing, dwindling freshwater resources, pollution from fertilizers and pesticides, and a host of other factors mean that current and future population increases will place a disproportionate burden on Earth's ecosphere. Something has to change.

The good news amid these gloomy facts: insects are a very promising component to a more sustainable future. Insects (Kingdom Animalia, Phylum Arthopoda, Class Insecta) make up the largest and most diverse group of organisms on the planet, with 4–30 million species estimated to exist and 1 million species described by biologists to date (Dossey, 2010). While Insecta is the largest and most diverse group, it is also the most underutilized by humans. Human interaction with insects today is nearly all negative. Humans are solely focused on trying to eliminate them instead of looking for ways to utilize them. However, as you see in this book, it now behooves us to begin looking at insects as a valuable resource that is important for our planet rather than as a nuisance or pest. Indeed, insects hold many beneficial attributes that make them attractive as a large-scale sustainable food resource (and suite/family of commodities for many applications). They can be farmed using less land, feed, water, energy, and other resources than other livestock and at any scale (small or large) (more on the efficiency and environmental benefits of insects over other livestock in Chapters 1 and 4). Insects have numerous attributes which make them highly attractive, yet underexplored sources of highly nutritious and sustainable food. Two general categories where insects provide substantial benefits for a sustainable and secure food supply are: (1) efficiency and (2) biodiversity.

Efficiency

Insects can be produced more sustainably and with much smaller ecological footprint than most vertebrate livestock such as cattle and swine (Dossey, 2013; Shockley and Dossey, 2014; van Huis et al., 2013). They are very efficient at biotransformation of a wide variety of organic matter into edible insect body mass (eg, a high feed conversion ratio) (Nakagaki and Defoliart, 1991; Oonincx et al., 2010; van Huis et al., 2013). For example, cows consume 8 g of mass to gain 1 g in weight, whereas insects can require less than 2 g (Vogel, 2010). This is partly due to insects being poikilothermic (cold blooded), thus using less energy for body warmth (Premalatha et al., 2011). House crickets (*Acheta domesticus* L.) have an "efficiency of conversion of ingested food" (ECI) that is twice that of pigs and chickens, 4 times that of sheep and 6 times that of steer (Capinera, 2004; Gahukar, 2011). This efficiency leads to less usage of pesticides on animal feed, thus providing additional environmental, health, and economic incentives. Compared to all other animals on Earth, insects are substantially more prolific (higher fecundity) and have shorter life spans, so they can be grown rapidly. For example, house crickets can lay 1200–1500 eggs in a 3–4-week period, whereas beef cattle require about 4 breeding animals for each animal marketed (Capinera, 2004; Gahukar, 2011). Insect production also uses much less water than vertebrate livestock (Table 5.1) (Capinera, 2004; Shockley and Dossey, 2014) (see also Chapters 1 and 4). Insects also give off lower levels of greenhouse gases than do cows (Oonincx et al., 2010). Additionally, many insects can eat plant materials no suitable for human consumption or agricultural byproducts, thus they don't compete with the human food supply like vertebrate livestock such as cows, chickens, and pigs.

Insects can also be grown in small containers, small spaces, modularly, and even vertically in sky scrapers. In fact, insects may be the only animal feasible to farm or produce in mass scale in tall vertical farms in cities, other densely

TABLE 5.1 Comparison of Protein Sources

	Land Use	Food (Grams of Food Input per 1 g Body Mass)	Water (Liters per Gram of Protein)
Insect	1	1.25	2
Chicken	3	2	34
Pork	3	4	57
Dairy	2.5	5.5[a]	31
Beef	10	7	112

[a]Adjusted according to food input per protein output for dairy and crickets.
Sources: Earth-policy.org (2014), Pimentel and Pimentel (2003), A Strategic Look at Protein (2014), Ansc.purdue.edu (2014), Waterfootprint.org (2014), News.cornell.edu (2014), van Huis et al. (2013)

populated areas, or really anywhere vertical farming is desired. Thus, the emerging insect-based food industry could be a great crop for the emerging vertical and urban farming efforts taking hold in cities around the world. In general, insect farming has the potential to become a great productive opportunity for current and future farmers and food entrepreneurs all over the world. This novel food source has the potential to provide new alternative crops or additional side crops that allow farmers to maximize overall production in both the on-season and the off-season and to maximize overall farm productivity. Additionally, insects can provide exciting new food products never before available.

Biodiversity

The UN FAO estimates that there are well over 1000 edible insects currently used (Shockley and Dossey, 2014; Vogel, 2010), and others estimate that number to be over 2000 (Ramos-Elorduy, 1997, 2009) (see also Chapter 2). There are over 1 million species described and 4–30 million species estimated to exist on Earth, living in every niche inhabited by humans and beyond (Dossey, 2010). With this diversity and their collective reproductive capacity, they are a safer prospect for future food security than are vertebrate animals. Increasing diversity in animal livestock and protein sources is critical to human food security going forward. For example, since there are insects of some sort on nearly every patch of land on Earth, chances are that some local species in every area can be farmed as human food without transporting nonnative species into the area for the same purpose. Additionally, the large numbers of edible species mean that an insect disease affecting a farm's initial species can likely switch to another species which is resistant. This has already been done at some US cricket farms (Weissman et al., 2012).

In addition to being very efficient to produce, insects are very nutritious. They are well known to be very high in protein. They are also high in healthy fats and oils and low in unhealthy types: low in saturated fat, high in omega-3 fatty acids. Insects are also high in other valuable nutrients, depending on species and in some cases their diet (Table 5.2) (see also Chapter 3 for more detail on the nutrient content and nutritional value of insects). It is worth noting that the larval stages of holometabolous insects (those with larval, pupal, and adult stages such as mealworms, butterflies, moths, flies, etc.) tend to be higher in fats and oils and lower in protein than their respective adult stages or compared with most hemimetabolous insects (such as crickets, grasshoppers, etc.). This is because the larval stages of holometabolous insects must store their fats for use as metabolic energy and for supporting other metabolic functions during their often immobile pupal stages. As insects in their pupal stage do not eat yet they must transform into their respective adult reproductive stages.

Protein is a critical source of nutrients for humans. On average, we require about 50 g of high-quality protein per day (Dolson, 2014; Cdc.gov, 2014). In North America about 70% of our protein comes from animal-derived sources (Wardlaw, 2006). Meat, fish, and poultry contribute about 40% while dairy contributes about 20%. Beef accounts for 53% and chickens account for 21% of meat purchased (Montanabeefcouncil.org, 2014). However, these animal-based products are becoming increasingly scarce and not sustainable to produce, particularly for the burgeoning human population. Food input to weight increase for cattle is 7:1, for pork is 4:1, for poultry is 2:1, and for fish is less than 2:1 (Earth-policy.org, 2014; Pimentel and Pimentel, 2003). By contrast, crickets create approximately 453.6 g (1 lb) of body mass for every 680.4 g (1.25 lb) of feed (Table 5.1). The feed conversion ratio for milk is 1:1, however, milk is 87% water. Additionally, dry milk powder is only 30% protein. Compared to milk, crickets create 4.4 times more protein output per food input (A Strategic Look at Protein, 2014; Ansc.purdue.edu, 2014). Water and land requirements for animal-derived protein versus insect protein output are equally disproportionate (Pimentel and Pimentel, 2003). Other studies have found that to produce 1 kg weight of animal, crickets need 1.7 kg of feed, less than 1 L of water and 15 m^2. Raising the same amount of beef requires

TABLE 5.2 Nutritional Content of Insects Compared With Other High-Protein Foods (Per 100-Gram Serving)

Food Item	Protein (g)	Fat (g)	Calories (kcal)	Omega-3 Fatty Acids (g)	Iron (mg)
Griopro™ Cricket Powder (All Things Bugs LLC)[a]	63	19	447	0.25	5.9
Beef	25.6	18.7	278	0.009	2.4
Milk Powder	26.3	26.7	496	0	0.47
Chicken	39	7.4	190	0.05	1.2

[a]Cricket powder data from: All Things Bugs LLC (www.cricketpowder.com and www.allthingsbugs.com).
Current USDA National Nutrient Database for Standard Reference (beef, ground, 75% lean meat, 25% fat, patty, cooked, broiled; milk, dry, whole, without added vitamin D (USDA); chicken, chicken, broilers or fryers, meat only, roasted)

10 kg of feed, 22,000 L of water (including irrigation for feed grains) and 200 m^2 (Jongema, 2015; van Huis et al., 2013). The tremendous benefits of insects in ensuring environmental sustainability are also discussed in detail in other chapters of this book (particularly Chapters 1 and 4).

Contrary to popular belief, particularly in Western cultures, insects can be a very clean and potentially low pathogen risk source of food. At least two primary research articles describing studies on this topic have been published so far (Giaccone, 2005; Klunder et al., 2012) (see also Chapter 8 for more on pathogen risk from edible insects). These have shown that the human pathogen loads in farmed insects (crickets and mealworms) are very low and standard foodborne pathogens appear to be absent. For example, one study failed to isolate *Salmonella* spp. and Listeria monocytogenes in samples of the following commercially farmed cricket and mealworm species: [*Zoophobas morio* Fab. (superworm), *Tenebrio molitor* L. (yellow mealworm), *Galleria mellonella* L. (greater waxworm), and *A. domesticus* (house cricket)] (Giaccone, 2005). Additionally, insects are biologically more separated from humans than vertebrate livestock, so the risk of an insect viral pathogen or parasite jumping to humans is exceedingly low (van Huis et al., 2013). Thus, pathogen risk transmission or infection appears to be very low for farmed insects.

The market demand for protein is rapidly increasing. Estimates are that the global market for protein products was $15.2 billion in 2012. By 2017, these markets are projected to increase 30% to almost $20 billion. The market for protein has expanded well beyond the traditional weight gain market into every area of life including weight loss, wellness, and sports nutrition, among others (A Strategic Look at Protein, 2014). Markets for protein supplements ($1.8 billion) and protein meal replacement and weight-loss products ($7 billion) are equally as attractive. By 2017, these markets are projected to increase 40% for sports nutrition and protein supplements and 19% for meal replacement products. This could bring the total for these markets to almost $20 billion within the next 5 years. Market research firms, such as Mintel, state that 19% of all food and beverage products are claiming high-protein contents. The market for protein has expanded well beyond the traditional weight gain market into every area of life including weight loss, wellness, and sports nutrition among others.

Due to the unsustainable nature of current sources of protein, the costs for everything from meat products to whey powder are on the rise. Whey protein concentrate, for example, has tripled in price to almost $4.50 per pound since 2008. Beef prices are up 26% primarily due to adverse environmental conditions such as drought. Other meat products are seeing prices remain above historical averages. The increase in prices for meat is twice that of other food products. It is unlikely these conditions will change towards cheaper prices given that the cost of resources (eg, food, water, and land) required to grow these sources of protein will only be increasing. Now appears to be a great time to introduce alternative sources of protein.

The consumption of insects as food, is accepted and practiced by many cultures around the world (Shockley and Dossey, 2014; Dossey, 2013; DeFoliart, 1995; Nonaka, 2009; Ramos-Elorduy, 2009) (see also Chapter 2). As many as 3071 ethnic groups in 130 countries (Ramos-Elorduy, 2009) utilize insects as essential elements of their diet (Durst et al., 2008; Srivastava et al., 2009; Yen, 2009). In fact, it is estimated that as much as 80% of the world's population eats insects intentionally, and 100% do unintentionally (Srivastava et al., 2009). Even in the United States there has been a rapidly increasing interest in insect-based foods (Gahukar, 2011; Durst et al., 2008) (see also Chapter 1). Insect-based food ingredients, such as powders (fine), meals (course or roasted), pastes, slurries or liquids have a very wide range of commercial applications in the human food, animal/pet feed, and nutraceutical industries. Products utilizing these ingredients might include: fortified dry goods, fortified protein supplement powders, high-protein fortified porridges and cereals, "meat" substitutes, chitosan (nutraceutical derived from chitin), high-protein beverages, protein fortified bars and powders for athletes, as well as numerous types of snack foods.

Indeed, the world is beginning to take notice of insects as a valuable, nutritious, sustainable yet vastly underutilized food source. Consumers, large and midsized companies, startups, the food and agricultural industries, academia, government, and other entities have begun to get involved in the emerging revolution of insect utilization as a sustainable food source. In the past few years, there has been a steady and rapid increase in popular press and major media outlets covering the rise of this budding industry (Table 5.3) (Crabbe, 2012; Mermelstein, 2015; Day, 2015; Tarkan, 2015) (see also Chapter 1 for more on the increase in publications on insects as a food and feed source).

The United Nations (UN) has placed heavy emphasis on alleviating hunger and malnutrition in children. Two of their eight "Millennium Development Goals" (MDGs) are directly related to this area: The first MDG is to "eradicate extreme poverty and hunger" and the fourth is to "reduce child mortality rates." Number 7 on the MDG list is "ensuring environmental sustainability." Insects have the ability to address all of these MDG's and, in fact, the United Nations Food and Agricultural Organization (FAO) in recent years has conducted multiple conferences, meetings, and even published a book encouraging increased utilization of insects as both human food and animal feed ingredients (see also Chapter 1 for more on the UN FAO's efforts in this area). The topic of the potential for insects to contribute to sustainable human food security has taken the notice of multiple organizations. For example, the UN Food and Agricultural Organization (UN FAO) has taken

TABLE 5.3 Popular Press Articles on the Insect-Based Food Industry

Huffington Post, 2016. "Reinventing Entomophagy for the 21st Century"
http://www.huffingtonpost.com/olena-kagui/post_10945_b_9076028.html?utm_hp_ref=green&ir=Green

Fortune, 2015. "Why These Startups Want You to Eat Bugs"
http://fortune.com/2015/08/25/edible-insects-bug-startups/

Newsweek, 2015. "Weak Oversight is Holding Back Edible Insects"
http://www.newsweek.com/2015/04/10/weak-oversight-holding-back-edible-insects-317518.html

PRI (Public Radio International), 2015. "The Next Big Thing in Protein Will Likely Make You Squirm"
http://www.pri.org/stories/2015-08-26/next-big-thing-protein-will-likely-make-you-squirm

Huffington Post, 2015. "Add Insects to Your Diet to Get Your Protein, Scientists Say"
http://www.huffingtonpost.co.uk/2015/07/17/insects-protein-source-diet_n_7816950.html

Forbes, 2014. "The Next New Miracle Superfood: Insects, Scientists Say"
http://www.forbes.com/sites/melaniehaiken/2014/07/11/the-next-new-miracle-superfood-insects-scientists-say/#2715e4857a0b5710faf11eda

Fast Company, 2014. "Inside the Edible Insect Industrial Complex"
http://www.fastcompany.com/3037716/inside-the-edible-insect-industrial-complex

Seattle Weekly, 2014. "The Rise of the Edible Insect"
http://www.seattleweekly.com/home/955089-129/the-rise-of-the-edible-insect

The Scientist, 2013. "Why Insects Should Be in Your Diet"
http://www.the-scientist.com/?articles.view/articleNo/34172/title/Why-Insects-Should-Be-in-Your-Diet/

The New Yorker, 2011. "Grub: Eating Bugs to Save the Planet"
http://www.newyorker.com/magazine/2011/08/15/grub

initiative and proposed a program of feeding people with alternative food sources including insects (Gahukar, 2011). Two major meetings on the feasibility and benefits of insects as a food source have resulted from this initiative: (1) a Feb. 2008 workshop in Thailand which produced an important book in the field called *Forest Insects as Food: Humans Bite Back* (Durst et al., 2008) and (2) a technical consultation (to which one of this chapter's authors and book editor Dr Aaron T. Dossey was invited and attended) held Jan. 2012 at UN FAO headquarters in Rome, Italy to plan a global summit on insects and food security for 2014 (Vantomme et al., 2012). Additionally, UN FAO published a book in 2013 touting the vast potential benefits of utilizing insects as an alternative food and feed source (van Huis et al., 2013). Even celebrities and the media are catching on! Angelina Jolie, Salma Hayek, and Shailene Woodley, to name a few, have all recently touted eating insects in the media (Joseph, 2015; Calderone, 2015; Shira and Oh, 2010).

Now the time is ripe for the insect industry to rise and take its logical place among the other animal-based commodities utilized for food and other applications. In the following sections, this chapter: (1) briefly summarizes the current and historic US cricket industry to date, (2) gives some recommendations and guidance for aspiring entrepreneurs creating insect-based food and farming enterprises as well as established food companies and insect farms wishing to enter this exciting new and transformative industry, (3) provides valuable tips and important considerations for the processing of insects and the production of insect-based foods and food ingredients, (4) summarizes the current nascent insect-based food companies and provides a list and description of the many companies already marketing edible insects and insect-based foods, (5) describes recommendations on industry structure, areas of specialization, and potential applications and products to consider, (6) discusses ways which insect-based foods might help alleviate malnutrition in the world, and (7) closes with general recommendations and paths forward seeking to inspire new and future practitioners in the insect-based food and farming industry, which we believe will pave the road to success most effectively and efficiently. Simply put, industrialization of insects will revolutionize food and agriculture for the better. This is the most exciting time to be involved in this emerging transformative industry, as we work toward establishment of a family and suite of new commodities derived from the largest and most diverse, yet most underutilized, group of organisms on the planet: Class Insecta.

CURRENT INSECT FARMING INDUSTRY

From fishing boats to pet shops, and now to the dinner table, the US and European insect farming industry has gone through quite an evolution over the past several decades (Weissman et al., 2012). In the United States, some of the first insect farms (mostly raising crickets) started appearing just after World War II (WWII) in the 1940s. As the United States began expanding its highway and lake infrastructure, and as soldiers and others began returning to more normal livelihoods with more disposable income and access to items previously reserved "for the war effort" such as metal, fuel, and boats, one favorite pastime which emerged was fishing. Some of the first cricket farms in the United States started in the southeastern United States in places like Georgia. Farms such as Armstrong Cricket Farm (founded around 1945 in Glennville, GA, USA; now with a second farm in West Monroe, LA, USA) and Ghann's Cricket Farm (founded about 1952 in Aubrey, GA, USA) are two of the oldest cricket farms in the United States that are still operating. Following over the years, various other farms started farming mostly crickets, waxworms, and superworms. Some of the larger US cricket and mealworm farms currently in business include Armstrong Cricket Farm (Farm 1: Glennville, GA, USA; Farm 2: West Monroe, LA, USA), Ghann's Cricket Farm (Augusta, GA, USA), Timberline (in Marion, IL, USA), Top Hat Cricket Farm (Portage, MI, USA), American Cricket Ranch (Lakeside, CA, USA), Rainbow Mealworms and Crickets (Compton, CA, USA), The Bug Company (Ham Lake, MN, USA), Fluker's Cricket Farm (Port Allen, LA, USA), Millbrook Cricket Farm (Richland, MS, USA), Five Points Cricket Farm (Kempton, PA, USA), Henderson Cricket Farm (Lancaster, KY, USA), Lazy H Bait (LaBelle, FL, USA), and Bassetts Cricket Ranch (Visalia, CA, USA). While the fish bait market has always been and is still a major market for these farms, in the 1970s and 1980s a new, larger, and much less seasonal market emerged. As people (particularly in North America and Europe) began to have more access to exotic pets from around the world such as reptiles and other insectivores, the need and demand for live feeder insects (and other live feeder animals such as mice, rats, and rabbits) exploded. Not only did the pet owners themselves need regular supplies of clean, safe, and easy to access live insects such as crickets and mealworms, but so did the pet shops (large chains, small mom-and-pop shops, and everything in between) as well as the breeders/producers of these pets. Thus, in those years the market for cricket farms went from being mostly fish bait to mostly (some farms say around 80%) live pet feeder insects (see Chapter 6 for additional information on the history of, applications for and techniques used in insect farming in North America, Europe, and beyond). Today, we seek to push the insect farming industry into a vastly (orders of magnitude) larger and more impactful market: supplying insects to the human food industry to both supplement human population growth and important alternatives to augment the need for protein from less sustainable sources such as certain vertebrate livestock (meat, eggs, and dairy) and wild-harvested fish and seafood (meat, shelfish, fishmeal for animal feed, etc.).

While the insect farming industry to date provides a wealth of knowledge and an excellent base from which to innovate and grow, the food industry is a much more vast and diverse market and application opportunity. Additionally, current insect farms are very small and lack technology and mechanization that food farms, ranches, dairies, and food processing plants have. At present, industrial insect farming at any scale and for all species remains dependent on manual labor, generally lacking in technological processes and innovation. In many ways, industrial insect rearing differs little from household processes beyond scale and size of operations. This reliance on human labor is credited as being a primary driver in keeping insect prices high despite natural advantages of high feed to meet conversion ratio and overall low resource needs for most popular insect species. Similarly, little research has yet focused on modifications of the insects themselves to increase efficiencies, nutrition, or other positive attributes. When compared to a more mature industry such as large-scale poultry farming, which has reduced the life span of a fryer chicken to a similar amount of time as the natural life span of a house cricket, there is much room for improvements that will enhance the ability to raise large quantities of insects sustainably and without compromising nutrition and safety.

Thus, moving insects into the realm of food ingredients and commodities is ripe for innovation and in need of more research and development (including research funding from governments and private organizations) to make a truly transformative and positive impact in the world for human food security, environmental health of the earth, and just adding some delicious new ingredients to our food supply! Insects are truly "low crawling fruit" for the food and agricultural industries!

Modern Industrial Mass Production of Insects

Summary of Current Industrial Insect Farms and Predominant Farmed Species

The majority of insect farms worldwide currently consist of pet feed (primarily for reptiles and amphibians), fish food, zoos, pest control companies (particularly for biocontrol), research labs, and a handful of aquaculture companies. There are examples of medium- to large-scale farms in Thailand and China that produce insects for human food and medicinal applications, respectively, but these remain specific to, and isolated within, their regions. In the United States and Europe, an increasing number of companies are growing edible insect species for human food and animal feed (see also Chapters 2 and 6 for more information on insects eaten around the world and insect mass production methods, respectively).

As mentioned earlier, in the United States and indeed worldwide, the majority of industrial scale insect farms are currently supplying the pet feed and fishing bait industries with live, frozen, and dried insects as pet food, primarily for lizards, snakes, and other exotic species (Weissman et al., 2012). In the United States alone, an estimated 71 million households, or over 60% of all households, own pets and of these, 4.7 million own lizards. The FAO estimates that more than 1 billion people own pets around the world. Industry experts estimate that companies raising insects for pet food and fish bait have revenues exceeding $ 150 million. Estimates for the global revenues for the industry vary but there are companies of significant size operating, for example, in China, where an estimated 100 or more farms raise cockroaches at an industrial scale for traditional medicine and cosmetic use (Demick, 2013). Additionally, European farms are raising similar amounts of insects for pet food, research, and aquaculture, with a vibrantly growing subsector of insect farms for human food.

There are an estimated 30 large farms (a few of the largest listed previously) in the United States rearing pet feeder insects with certain species making up the vast majority of insect production including crickets [usually House Cricket, *A. domesticus*, or Banded Cricktet, *Gryllodes sigillatus* (Walker)], mealworms (larvae of the darkling beetle species *T. molitor*), superworms (larvae of the darkling beetle species *Z. morio*), and waxworms (larvae of *G. mellonella*) (Weissman et al., 2012). For example, the 10 largest producers of crickets in the United States are estimated to collectively produce well over 2 billion crickets annually. This amounts to about 1.36 million kg (3 million pounds), or 1360 tonnes (metric tons, or 1500 tons) of total cricket mass produced each year, assuming a weight of 680 g (1.5 lb) per 1000 crickets. Similar quantities of mealworms are also produced in the United States, with superworms being slightly lower in production and slightly less waxworms (Shockley and Dossey, 2014).

There are many more small companies producing from only a few thousand of various feeder species to hundreds of millions of insects per year. Many insect farms in the United States are located in rural communities and contribute to local economies. There is also substantial interest in, and new companies being formed for the purpose of, mass producing other feeder insects, such as black soldier flies (Shockley and Dossey, 2014). Here we provide a few prominent examples of insects that are mass reared in the United States, which merit further examination as potential targets for use as insect-based food ingredients. The next few sections provide a summary of some of the more commonly mass produced/farmed insects which currently have the greatest potential for the food industry and commoditization. See Chapter 2 for a more detailed list of edible insects worldwide and Chapter 6 for a more detailed discussion of insect mass rearing, production, and farming.

Crickets

There are over 900 described species of crickets to date. Around the world, 12 families and 278 species of crickets, grasshoppers, and katydids are recorded as being consumed by humans. Orthoptera is the fifth most consumed insect order worldwide (Fig. 5.2). Acrididae represent the highest frequency of human consumption (171 species) followed by Gryllidae (34 species) and Tettigoniidae (30 species) (Jongema, 2015) (see also Chapter 2).

As mentioned previously, the 10 largest producers of crickets in the United States already collectively produce approximately 2 billion crickets annually. This amounts to about 1.36 million kg (3 million lb), or 1360 tonnes (metric tons, or 1500 tons) of total cricket mass produced each year, assuming a weight of 680 g (1.5 lb) per 1000 crickets. The house cricket (*A. domesticus*) probably has the longest industry history of any insect mass produced in the United States, with the first large cricket farms having started in the 1940s–50s. Originally from Southeast Asia, crickets are also one of the most commonly mass-reared insects on the planet (Shockley and Dossey, 2014). The house cricket is also considered the most commonly raised edible insect, and an examination of applications for its use as a whole insect and from powder follow in this chapter. The popularity of crickets at the beginning of the current emerging insect-based food industry has been likely due to a combination of multiple factors including but not limited to: (1) their current market price per pound compared with mealworms, (2) their higher protein content and lower fat content compared with mealworms, (3) their endearing (so-called "cute") attributes in cultural venues such as children's cartoons ("Jiminy Cricket," etc.), that they hop and jump and sing, (4) their relatively short lifecycle allowing many to be produced on demand rather quickly (eg, within 5 weeks from egg to adult) and (5) their overall wide availability on the market (including availability as frozen from certain farms as early as 2011 or before).

House crickets are likely one of the least expensive insects to farm since their mass rearing methods have been refined for several decades. Their nutrient content is also well established (Finke, 2013). Crickets and other insects are proving more efficient and sustainable to produce than several types of vertebrate livestock including chickens, pigs, lamb, and steers (Nakagaki and Defoliart, 1991). Currently, a typical wholesale cost per 1000 feeder house crickets in the United States, which weighs about 453–680 g (1–1.5 lb), is approximately $4–$10 (personal conversations with cricket farm owners). This is the price point even without the considerably larger potential demand for these insects, which will arise as insect-based foods become more popular and widespread. Minimal shifts in industry practices, scale, and innovation, which are already beginning to happen with the emerging insect-based food industry, will almost certainly cause prices of crickets marketed for human consumption to become more competitive in the near future.

Mealworms, Superworms, and Buffalo Worms (Lesser Mealworms)

At the species level, the insects most commonly consumed by humans worldwide are the Coleoptera with 661 documented species being consumed amongst 26 families (see also Chapter 2). This makes sense, since Coleoptera is by far the largest group of any organism on Earth. Scarabaeidae (247 species) demonstrate the greatest diversity followed by Dytiscidae (55 species), and Cerambycidae (129 species), as being the most commonly eaten by humans (Jongema, 2015). In the United States, mealworms (*T. molitor*) and superworms (*Z. morio*) are currently mass-produced for the pet industry as live feeder animals. In Europe, a number of farms rear these and also the lesser mealworm (*A. diaperinus*) (sometimes also referred to as "buffalo worms"). Their nutrient content is also well established (Finke, 2013) (see also Chapter 3). Interest in using mealworms and superworms as human food has increased recently in the United States. They are featured as edible insects in many entomology outreach programs and are being purchased by various businesses and incorporated into hard candies as well as baked goods such as cakes, cookies, and cupcakes, including companies like Don Bugito and Hotlix.

In addition, in the pet food and research industries in North American and Europe, mealworms (*T. molitor*), lesser mealworms (*A. diaperinus*), and superworms (*Z. morio*) are being explored by a small but growing number of companies for production and use as food ingredients, in particular to create mealworm powder or to use the larvae in ground meat substitute products such as tofu alternatives. Mealworms, buffalo worms, and superworms are the larvae of beetles and each has a life span of approximately 90 days when raised indoors or in warm climates. In the wild, life spans can be longer when the larval stage occurs during winter. Mealworms are also commonly raised as supplements to chicken feed, especially for chickens kept at the household level, that is, not industrial egg producing facilities (Shockley and Dossey, 2014). One main difference between rearing mealworms and superworms is that while mealworms will pupae and turn to adult beetles in a large group, superworms must be isolated as individuals when the larvae mature and placed in an undisturbed container/place before they will pupate and complete their metamorphosis as holometabolous insects.

Grasshoppers

As with the species previously mentioned, grasshoppers are primarily grown in industrial settings for the pet food industries, but in lower quantities than crickets, and also far less commonly than crickets, mealworms, buffalo worms, and super worms. The vast majority of consumed and commercially available grasshoppers worldwide are currently wild harvested. Many grasshopper species have a longer life cycle than crickets, making them more difficult to raise at an industrial scale. Containment and risk of crop infestation are more problematic for grasshoppers than crickets since grasshoppers are usually either wild collected or must feed on live plants (that are still in the soil they were grown in). Thus, it may be good to consider preconsumer fresh leafy plant materials (cut and rinsed away from the soil and roots), plant seed sprouts, or hydroponically grown plants as farmed edible grasshopper feed. Also, the grasshopper's physical ability to jump and fly (compared with crickets) adds to production challenges that require newer innovations to overcome. Nonetheless, research is being conducted on captive rearing of grasshoppers and some advancements in artificial dry feed use have been made and several indoor grasshopper farms are appearing in Europe, Israel, and elsewhere.

While there is a sizable, though primarily informal, industry collecting, processing, and selling the Mexican grasshopper called "chapulines," there are few companies producing foods from grasshoppers globally. Some of the larger insect producers in the pet feeder sector offer grasshoppers among other species.

Other Species (Wax Worms; Silk Worms; Horn Worms; Fruit, Black Soldier, and Other Flies; Cockroaches; Termites; etc.)

Worldwide, more families of Lepidoptera (36) are consumed by humans than any other insect (see also Chapter 2 for more detail on diversity of insect species consumed worldwide). The most commonly consumed families are Saturniidae (109 species), Hepialidae (47 species), and Sphingidae (36 species). Their frequency of human consumption is highest in the neotropics and the palaearctic biogeographical realms, where they are the dominant order of insect consumed (Jongema, 2015) (see also Chapter 2). In the United States, a commonly consumed lepidopteran insect by humans is the waxworm (*G. mellonella*). They are mass reared for the animal feed industry as well as for fish bait. Their nutrient content is well established (Finke, 2013). Due to their holometabolous life cycle and lack of noticeable appendages during the larval stage, they are also a great candidate for roasting, grinding into powder or meal, and incorporating into various food products. They are very high in fat, which makes them an appealing food supplement in resource-limited areas where people are malnourished and underfed.

Humans are thought to have raised silk worms perhaps longer than any other species, with a history that goes back to 2700 BC. In countries with significant silk worm farms, including South Korea, China, and India, sericulture companies sell the chrysalis of the silk worm as food after harvesting silk and eggs. The larvae of the silk moth *Bombyx mori* L. is considered a completely domesticated insect species and none are thought to live in the wild any longer. Silk industry practices allow the silk worm to complete development into the adult moth only for renewal of the productive larval stage. Adults are born blind and are unable to fly after living in domestication for thousands of years and untold generations. Fertilized eggs are extracted manually (by pressing their abdomens) from the adult females, which do not oviposit by themselves.

In China, cockroaches are raised for medicinal and cosmetic applications, with a smaller emphasis on raising them for food, although the two main varieties of cockroaches are considered edible. *Periplaneta americana* L., or the so-called American cockroach, is the most commonly raised cockroach in China and *Blattella germanica* L., the German cockroach, is the other commonly raised type of edible cockroach.

Flies (Order Diptera) are the fourth most consumed insect order by humans, with 16 families and 39 total edible examples documented in the literature (Jongema, 2015). They are probably the insects with the largest reproductive capacity, shortest life cycles, rapid growth rates, and which are able to eat the widest variety of organic material as feed input for mass production of insect biomass. These and other features make them some of the most attractive insects for applications to increase world food security from a production perspective. A number of applied laboratories and companies are already beginning to focus on scaling up production of black solider flies for use as animal feed, composting, waste mitigation, and other nonfood applications. The common house fly (*Musca domesticus* L.) has been studied a great deal on its mass production and has been produced industrially for several applications such as for research, live pet feeder insects, and has recently been explored by some firms as a larger scale animal feed protein source. Another very popular fly in the emerging insect mass production industry for food and feed is the black soldier fly [*Hermetia illucens* L.]. Black soldier fly nutrient content is well established. For example, their larvae are very high in fat and calories (Finke, 2013) (see Chapter 2), since, like other holometatolous insects, they store it for their immobile noneating pupal development stage. Black soldier fly larvae have very little hard chitin and are easy to process. This means that we can efficiently remove the chitin without losing protein. Black soldier fly is also high in lauric acid (Finke, 2013; Huang et al., 2011), a fatty acid with significant antimicrobial activity (Huang et al., 2011) and typically found in milk (Beare-Rogers et al., 2001). Fly larvae are known to contain other antimicrobial compounds, some of which might improve shelf-life of food ingredients made from them (Dossey, 2010). Additionally, several aspects of this fly give it potential for highly efficient and sustainable production. Black soldier fly larvae can develop on almost any kind of nontoxic organic matter including a wide array of agricultural by-products (Bondari and Sheppard, 1981; Sheppard et al., 2002; Popa and Green, 2012). Thus, their production costs are likely to decrease over time as methodologies for mass rearing and low-cost feed are identified. Additionally, black soldier fly rearing involves several processes that are highly amenable to automation. Black soldier flies are beginning to be raised on industrial scales to exploit their natural ability to convert food waste and manure to nutritious feed for poultry, aquaculture, and pigs. According to some researchers, black soldier flies also have the ability to reduce pathogens and diseases common in animal manure, as well as speed the compost process for other organic waste. Private companies and universities are testing how to best raise this species, including a company in the United States, one company in South Africa, and the University of Stirling in the United Kingdom, which has an ongoing project researching commercial and international development applications for converting waste streams into livestock feed.

Currently, some of the most substantial examples of mass reared Diptera in the United States and around the world are flies. These include various tephritid fruit flies (Family Tephritidae), mass reared for sterile male releases. Mass production methods and facility/equipment designs for these insects have been heavily researched and are already very efficient and highly refined. Thus, these production methods could likely be very easily adapted for human food or animal feed applications. These flies are typically produced as pupae which are irradiated and then the resulting reproductively nonviable adults are released in orchards in farms in order to reduce wild populations of these pests in a technique known as sterile insect technique, or SIT (Dowell et al., 2005). Some of the most prominent examples include: the Caribbean fruit fly [or "carib fly," *Anastrpha suspensa* (Loew)], the Mediterranean fruit fly [or "med fly," *Ceratitis capitata* (Widemann)], the Mexican fruit fly [or "mex fly," *Anastrepha ludens* (Loew)], and the oriental fruit fly [*Bactrocera dorsalis* (Hendel)]. However, in the United States and many other places, only government facilities produce these insects and only for SIT. As a consequence, the potential of these insects is tremendously underrealized. While regulatory constraints may make actual pest species impractical for commercial production in areas where they are not already established, the methods developed to grow pest species are easily adaptable to many nonpest species of flies. Thus, commercial enterprises producing these insects as human food, animal feed, or other biomass applications (drugs, biomaterials, nutraceuticals, etc.), even if only in places where these "pests" are already established, can be a tremendous benefit to the local economies of those areas as well as contributing to global sustainability, food security, human health, and technological advances.

RECOMMENDATIONS AND CONSIDERATIONS FOR SELECTION FOR ASPIRING INSECT-BASED FOOD PRODUCERS AND INSECT FARMERS

Here we provide a discussion on various aspects of the insect-based food industry, from farm to table, which merit close consideration by the insect-based food industry, both current and future. These are based on our experiences as entrepreneurs, scientists, and innovators in this industry over the past several foundational years of the industry. They include a comprehensive discussion of the industry including aspects such as farming, processing, product considerations, safety, marketing, and communication.

Feed Formulations and Biomass Sources for Farming Insects: Considerations for Insect Feed Formulations

One of the most important needs for any animal producing farm is biomass that the farmer will use for feed. As insect farming is a relatively new and currently small (but fast growing) enterprise, there is not much accumulated knowledge about the optimal nutritional content each farmed species needs for maximal productivity. There have been no exhaustive studies on what sort of feed ingredients will most effectively allow various insect species (crickets, mealworms, others) to thrive on and grow efficiently. There is some basic knowledge about insect feed available from academic and research labs as well as developed by the industry over the last 60 years of farming these species for pet food (see Chapter 6 for more information on formulation and other considerations for mass produced insect feed development). Additionally, the edible insect industry is rapidly learning both the needs of the insects they are farming as well as aspects of consumer demand which drive what feeds or feed ingredients are appropriate and which are not. For example, consumers are already demanding everything from organic to gluten free. While some insect farmers formulate their own feeds, there are already several animal feed companies offering commercially available cricket and mealworm feeds which are specially formulated for those species. One of the best known in the industry is the Cricket Chow produced by the major animal/pet feed company Purina. It has been around for a number of years (possibly decades) and was formulated by Purina's subsidiary Mazuri. It has been reported that Purina has sold the product since at least 1985, a year when over 445 metric tons of the product was sold to the cricket farming industry (Walker and Masaki, 1989). There is even a certified organic version available, currently offered through another subsidiary or partner called Nature's Grown Organics in Wisconsin, USA (produced by and sold through Premier Cooperative, Westby, WI, USA). There are other companies (Coyote Creek Farm—an organic farm and feed mill—Elgin, TX, USA) and Homestead Organics (Berwick, Ontario, Canada, among others) also offering cricket feeds, and even some offering certified organic, gluten-free, and GMO-free blends. It is very interesting to see the feed companies already well prepared to meet the coming demand for the potentially massive, specialized human consumption insect farms, even before the industry has reached mainstream scale.

Probably the most important aspect of what feed or feed ingredients to use for farmed insects is the nutrient content of the feed and its components. Again, there is not nearly as much information from detailed scientific studies on farmed insect nutritional needs as there is for other livestock, but some is known from both the academic literature and the industry at large. Additionally, vitamin A seems to be an important component of the feeds to reduce mortality and possibly aid in resistance to cricket viruses. Additional details on various aspects and parameters of insect mass rearing can also be found in Chapter 6. In addition to feed nutrient content, moisture level and shelf life are important factors as well, which is similar to what is expected for many products including perishable items such as human food and animal feed. These can be optimized via a number of parameters such as ingredient choice, processing methods, storage conditions, and packaging, among others.

Underutilized Biomass Amenable as Feed Ingredients for Mass-Farmed Edible Insects

One of the great aspects of insects as a human food ingredient is that insects can typically consume a wider variety of biomass and plant types than most vertebrate livestock, such as cattle and chickens. This fact gives insect tremendous potential for providing a sustainable source of protein. Insects' ability to convert various forms of feed is actually a very cutting edge area of scientific research as well as an important focus of the industry. Various sources of biomass as insect feed ingredients are being discussed in both industry and academia. Some of the most promising will likely be underutilized leftover material from the existing food and agriculture industry. Some popular examples among these include: yeast and spent grains from alcohol brewery or ethanol production operations, fines or other by-products from food manufacturing that do not end up as part of finished products (dust from cereal production, etc.), preconsumer waste produce (sometimes called "ugly fruit" or "ugly vegetables" not suitable for the shelves of grocery stores or the right shape for processing machines as a whole fruit or vegetable product), the unused parts of various crop plants (plant tops of peanut or potato, various vegetable leaves and stalks, etc.) and silage (the unused parts of grain plants such as corn, wheat, millet, milo, sorghum, corn cobs, etc). By-products from the food and beverage industry and other areas of agribusiness (ethanol production, etc.) may also be sources of valuable insect feed ingredients such as spent grains and yeast from brewing operations. Additionally, many nonfood crop plants and algae may be ideal ingredients for various mass produced insects such as crickets, mealworms, and grasshoppers. These might include various grasses (switchgrass, etc.) and algae as well as plants grown currently for low cost, high nutrient animal feed (alfalfa, Bermuda grass, Johnson grass, etc.). Many of these also make biological sense since various mealworm (Family Tenebrionidae) and cricket species will sometimes consume dried as well as fresh leaves of various plants and grasses in the wild. Of course, many of these materials could be utilized in live and/or fresh form to feed fresh grass/plant material to more selective insects such as grasshoppers (imagine vast screen cages or greenhouses with layers of grasses and other plants feeding billions of grasshoppers). Though some insects such as caterpillar (larvae of Order Lepidoptera, butterflies and moths) and grasshoppers prefer live/fresh plant material, artificial/dry diets have been developed for some such obligate herbivore and species such as grasshoppers and caterpillars such as the silkworm moth, *B. mori*, hornworm caterpillars, *Manduca sexta* L., and at least 35 species of other lepidopterans of agricultural importance (Singh, 1977; Singh and Moore, 1985) (see additional information on this topic in Chapter 6).

Diseases Affecting Mass Produced/Farmed Insects

As with any farm (raising plants, animals, or any other living thing), pathogens and diseases of the organism being produced can be a concern. Given the nascent status of the insect farming industry compared to other farms, there is only a relatively shallow knowledge of what sorts of disease risks to insect livestock exist in the context of indoor farms. There is a substantial amount of literature on various pathogens which affect insects in general, but little of it is relevant to the species which are farmed or mass produced. As such, much of that literature is beyond the scope of the current chapter, so we will focus more on pathogens likely to be a concern to edible insect farms. Among some of these are various fungi and microbial pathogens (see also Chapter 6 for considerations for farming insects, including farm hygiene).

One major example of a disease affecting insect farms in recent years has been certain viruses affecting cricket farms (Weissman et al., 2012). Additionally, there are several rather potent viruses which have been known to affect insects, particularly crickets and caterpillars (larvae of butterflies and moths). Moth larvae tend to be particularly sensitive to viruses well known to cause problems in the cricket industry, which include *A. domesticus* Densovirus (AdDNV) and Cricket Paralysis Virus (CrPV). These viruses seem to be somewhat species selective; with the house cricket *A. domesticus* (historically the most commonly farmed species) most affected. Indeed many farms were forced to close permanently after having been affected by these viruses. It is apparently difficult, given current knowledge and technology, to resanitize a cricket farm once these viruses have infected the majority of the crop. Nonetheless, the closing of certain farms could also have been a combination of economic and other factors precipitated by the virus. With enough resources (particularly financial) it could have theoretically been possible for the farms to recover, even if they decided to grow *A. domesticus* again. Some farms were able to recover and continue to grow *A. domesticus*, while others were able to switch to another species (usually the banded cricket, or tropical cricket, *Gryllodes sigillatus*) and get back up to previous production capacity within 1–3 years. Though the cricket industry does not readily disclose information about their production scale, methods, financial, and other business information, our interviews with various cricket farms indicate that farms in Canada and Europe seem to be some of the hardest hit by the virus, and Europe may have been the original source of the epidemic. Many of these farms, especially in Canada, have switched to the banded cricket.

The NASCAR Jacket of Food: Food Labeling Requirements and Aspects of Consumer Demand. How do I label it? Is it Organic? GMO? Gluten Free? Paleo? Other?

First, it is critical to alert aspiring insect-based food entrepreneurs, food companies, consumers, and anyone seeking to incorporate insects into food products that insects are a potential food allergen for some people and are related to shellfish crustaceans. Thus, it is critical that all food products containing insects in any quantity as a known or intentional ingredient be labeled as such so that consumers with food allergens can be aware. Proper labeling will allow these consumers to avoid these products, if necessary. The finer details of insects as a food allergen risk are areas in need of more research, and are beyond the scope of this current chapter. However, a detailed discussion and review of the current knowledge on insect derived allergenicity can be found in Chapter 9.

There is a growing demand for food products made with insects (especially crickets, mealworms, and grasshoppers) in the United States, Europe, and around the world. This demand is largely driven by the knowledge that insects are sustainable and environmentally friendly to produce. Further, current farming practices do not utilize chemical agents like antibiotics, steroids, hormones, pesticides, or other nonnatural chemical components often utilized in vertebrate livestock operations. Thus, it seems that the demand for insect-based foods exist independently of any special certifications, much as the large majority of consumers require no special certification on other foods. Nonetheless, there is also a growing demand for foods with various certifications such as organic, gluten-free, and non-GMO. Just as more consumers are demanding these specialty labels in other foods, a significant portion of insect-based food consumers are demanding these labels and the standards they represent for insect-based products. However, organic certifications can be lengthy, expensive, laborious, and inefficient for any farm to achieve. Given the small scale, low budget, and nascent state of the edible insect industry, full organic certification is certainly an insurmountable task for most farms. The good news is that, as described previously, (1) there is substantial demand for nonorganic insect-based foods due to their many other benefits and (2) the insect-based food consumers are flexible and willing to accept a wide area of middle ground given the startup nature of the industry. One alternative which various consumers or company buyers of farmed insects are willing to accept is simply utilizing a certified organic feed for the insects. As described previously, several feed companies are already offering certified organic feed designed for crickets, and are developing such feeds for other insects. Some of these include Coyote Creek Farm (an organic farm and feed mill) (Elgin, TX, USA), Nature's Grown Organics (Premier Cooperative, Westby, WI, USA), and Homestead Organics (Berwick, Ontario, Canada).

When fed the right feed, another very attractive aspect of insect-based foods and ingredients is that they can be a very nutritious source of gluten-free protein. In addition to gluten-free being a food/health trend, a rapidly increasing segment of the population in recent years is discovering that they have sensitivity to gluten (a protein typically found in wheat and some other grains). Gluten sensitivity causes primarily symptoms of the digestive tract such as abdominal pain and cramping, nausea, diarrhea, excessive gas, bloating, and in severe cases, malnutrition. People with celiac disease are particularly sensitive to gluten. In addition to those who are already sensitive to gluten, many more are concerned about becoming sensitized to gluten. Thus, a large and growing portion of the population is making an effort to reduce or eliminate gluten from their diet. The great news is that insects do not naturally produce their own gluten. Thus, insects fed on a gluten-free feed (ie, free from wheat) will indeed be gluten free. Thus, insects fed a gluten-free diet are acceptable and attractive to a very wide and growing market segment. For more information on food safety and allergenicity (including information on foodborne pathogens and food regulatory concerns), see Chapters 7–9. While food safety and allergenicity are large important topics in their own right, they are beyond the scope of this current chapter.

An additional food trend that insects are very well positioned to capitalize on is the growing interest in foods seen as "Paleo." The Paleo diet is one made up of foods which are thought to have been eaten by early people (eg, living in the Paleolithic age). Many believe that such foods, mostly including whole fruits, vegetables, and meats, are what our bodies are naturally adapted to digest. It is believed that these foods provide the optimal nutrients our bodies have adapted to need most. The definition of what Paleo food or ingredients are is quite nebulous. There is no regulatory agency or certification for what specifically constitutes "Paleo food." Additionally, there is little if any hard evidence for what foods were truly eaten by Paleolithic peoples. This could pose certain challenges in marketing foods as Paleo. However, the lack of certainty could make marketing foods as Paleo more flexible and thus easier. To date there does seem to be a general consensus that insects indeed meet the demands of a Paleo diet. Given this and the fact that the Paleo foods have enjoyed a massive and rapidly growing following, it behooves the insect-based food industry to pursue this market as aggressively as possible.

Words Matter: Terminology Recommendations

As with any industry or human endeavor, the language, terminology, and other linguistic tools employed have a tremendous impact on success and efficacy. The entire field of marketing is largely dedicated to the cleaver use of language and visual symbols to effectively and positively convey the real or perceived need for products and services. This is also quite true of the insect-based food industry, particularly in Western and industrialized areas of the world. In introducing this valuable, nutritious, and sustainable resource to the industrialized world where its benefits can have maximal impact, we must tread lightly on how it is introduced. The first exposure many will have to insect-based food, as with other "new" things, will be through verbal communication. Thus, here we will discuss some recommendations of terminology which we believe will smooth the transition to acceptance of insects as an industrial food ingredient. See also Chapter 1 for more discussion on language use in this field.

First, we would like to address terminology used to refer to the basic concepts for which this book is written to address. One overarching term used to date has been *entomophagy*. The actual/technical definition of this word is simply "the consumption of insects." It can refer to any organism which consumes an insect—animals (including mammals; birds; reptiles; and insects, such as mantises and other predators/parasites), plants (pitcher plants, Venus fly catchers, etc.), nematodes, fungi, and so forth. It is also used to describe the practice of human consumption of insects (more accurately termed *human entomophagy*). However, this is a very technical, scientific, and almost clinical term which is not appetizing to the ear. It is also rather unusual to utilize this type of term to describe an eating or a culinary experience. For example, we do not say we will "go to engage in porcinophagy." We simply say we are going to eat some bacon or other pork product derived from pigs (porcine animals). Thus, we would like to recommend abandoning the use of the word "entomophagy" in the industry for purposes of improving acceptance of insects as a sustainable, quality, tasty, nutritious, and very useful food ingredient. We feel that this term is better left to deeply technical scientific articles and other academic works.

Additionally, we would like to make a recommendation regarding terms utilized for dry ingredients made from insects. In the early stages of the recently emerging insect-based food industry in North America and Europe, the main ingredient utilized in most products was referred to as "cricket flour." This clearly was meant to harken to or remind one of the various dried and ground up food ingredients commonly utilized traditionally such as wheat/grain flours, corn flours, and the like. However, as this concept has grown, more and more people are exploring the utilization of insect-based ingredients (largely powders or courser meals, but sometimes whole insects or pastes) in a rapidly increasing repertoire of foods and beverages. Again, language comes into play. When a product is called a "flour," many people, including home cooks and industrial food producers, bakers and restaurant chefs, will want to utilize the product in ways identical to how they utilize products like grain flours. However, insects are almost entirely made of protein (sometimes as high as 70% by dry weight or higher!) and fat/oil with only small amounts of fiber (mostly chitin), carbohydrate, sugar, and micronutrients. On the other hand, true flours (made primarily from plant seeds such as wheat and other grains, nuts, and plant fibers) are made mostly of starches and fibers with only smaller quantities of protein, with some being somewhat high in fat/oil. Thus, true plant flours perform very differently compared to dried products made from insects. For example, a chocolate chip cookie made by replacing wheat flour completely with cricket powder (67% protein) might taste more like a burger patty with chocolate chips in it. Indeed, since cricket powder is made from the tissue of an animal (ie, muscle, etc., or one might say "cricket meat") it reconstitutes well in water and can be cooked into a patty having an aroma, texture, and flavor very similar to ground meat (beef, pork, or chicken). Thus, we recommend utilizing the term "powder" (ie, cricket powder) when referring to dry, insect-based ingredient material consisting of a fine particle size (ie, 100 μm max particle size) and "meal" (ie, cricket meal) when referring to courser dried or roasted ground insect material. This would be consistent to terminology utilized with other products and commodities such as fish meal, corn meal, milk powder, and the like.

Animal Welfare

Animal welfare is considered to be high for industrial insect farming as compared to traditional livestock operations, in particular when compared to confined animal feeding operations (CAFO). Many insect species naturally live in large groups in small amounts of space, meaning that industrial farming schemes that raise large amounts of insects in small spaces are close to natural conditions and thus the microlivestock are not stressed from overcrowding. Further, when living conditions are sanitary, industrial insect farming does not have additional risk of disease beyond what is normally present for the respective species. In this way, by mimicking natural conditions, insect species do not require feed additives or medicines to prevent health consequences from their living conditions as do most CAFO's, which create significant risk of contamination and animal stress. See later sections of this chapter on insect harvesting and processing for an additional discussion on insect animal welfare considerations, and Chapters 7 and 8 for a full discussion of food safety concerns for the edible insect industry.

Further, the most widespread industrial method of harvesting insect species is by chilling them to freezing temperatures. This process causes the insects to enter a state of sleep much like a coma as their body temperature lowers. After an extended period of being frozen, which varies by species but is generally 2–3 days, the insects die without regaining consciousness. As compared with modern methods of slaughtering traditional livestock, the pain levels are believed to be drastically lower than those of cows, pigs, and chickens; however, we lack the full understanding of the way insects experience pain or if they indeed do at all.

Getting Past the "Fear Factor": Important Considerations for Normalizing Insects as a Mainstream Food Ingredient Beyond the Novelty Niche

Insect-based food companies and advocates are actively attempting to shift Western consumer perceptions to favorably consider insects as a legitimate and healthy food that is a smart choice to support long-term sustainability. In addition to some religious influences, for example, insects are not considered a clean food in Hinduism, but are kosher according to Judaism, cultural considerations in Western countries tend to identify insects as dirty and "pests" to be avoided or killed, rather than eaten. It is in fact those in North America and Europe who do not regularly consume insect who are the outliers in this respect, as an estimate 80% of the world's population regularly consume insects as food, and the only continent without the practice is the one with no real insect population to speak of, Antarctica. Studies examining demographic and psychological traits related to eating or trying insect-based foods showed that the top concerns include:

- Disgust sensitivity
- Beliefs about the risk of consuming insects
- Beliefs about the benefits of consuming insects
- Desire to have new and stimulating experiences
- Risk tolerance, food neophobia (resistance to try new foods)

The majority of educational efforts to change Western consumer perception of insects as a regular and good food choice can broadly be classified under two umbrella themes:

1. Environmental concerns
2. Nutrition and health

Under environmental concerns, there are many questions around climate change effects on, and by, agriculture, global population growth, and the environmental degradation caused by a worldwide growing demand for meat. Generally speaking, an overall scientific observation is that the human population is simultaneously facing the need for increasing amounts of food but a decreasing ability to produce more food with climate change and current cultivation methods, crops, and livestock. Nutrition and health issues as they relate to edible insects contribute to the "value proposition" for consumers to add insects to their diets. These concerns stand alone on the virtues of many insect species' nutritional qualities, as well as linking eating insects to other food trends. Some of the most closely linked popular eating trends include:

- Low carbohydrate or so-called Paleo eating style
- Gluten-free foods
- Natural and/or organic foods

The call to more implicitly add insects to diets was emphasized in the 2013 United Nations' Food and Agricultural Organization report, "Edible Insects: Future Prospects for Food and Feed Security." This 200+-page report focused on both the environmentally positive aspects of raising insects for food and feed, as well as the potential edible insects have to contribute to food and nutrition security in the present, particularly in places that experience high levels of food insecurity and malnutrition. Additionally, the report emphasized the future concerns for the global human population to adopt eating patterns that are sustainable given the assumed future of higher populations and lower ability to produce sufficient and healthful food for all people (see also Chapters 1 and 2 for more detailed discussions on the public perception of edible insects). In addition to the psychological, nutritional, and environmental considerations impacting public perception, it is also clear that consumers largely value taste and overall sensory appeal (visual, flavor, texture, aroma, and other organoleptic qualities) in their food choices. Put another way: "Does it taste good, look good and do I generally enjoy consuming it?" Later sections of this chapter discuss processing and culinary considerations for edible insects and insect-based food and ingredient product recommendations which can greatly improve public consumer acceptance of insects as part of their diet.

Insect Processing Considerations: Considerations for Insect-Based Food Production, Processing, and Safety

Full realization of aforementioned benefits of insects as a suite of commodities requires development and implementation of appropriate food processing and preparation methods. Processing is a critical aspect of any food or food ingredient for incorporation of insects into more standard foodstuffs at all scales including large-scale industrial, cottage industry, restaurant and professional culinary, as well as at the household level. Good processing maintains or enhances the nutritional, organoleptic (taste, aroma, texture, etc.), color, and functionality of raw materials converted into food ingredients while also destroying or removing potential safety hazards such as pathogens and maintaining the longest possible shelf life. Functionality and format are two particularly critical considerations of processing. Functionality means how the material performs in the food and interacts with the other ingredients through the process of converting raw materials into finished products such as foods and beverages. Oil binding, moisture binding, gel formation, and ability to remain solid without falling apart are all examples of functionality. Little is known or published about the functionality of insects or insect-based food ingredients, so this is a potentially important area for research and technological innovation. Formats of ingredients include powders, pastes, liquids, and so forth. Often in modern culture, the term *processing* has a bad connotation as it is often (sometimes erroneously) connected to utilization of unsafe chemicals, preservatives, or other methods which are believed to make the food less healthy or less safe (see Chapter 8 for some discussion on this). However, the term itself simply means "to perform a series of mechanical or chemical operations on (something) in order to change or preserve it." Thus, processing is a neutral term which is neither good nor bad—each process must be evaluated independently by the consumer, business, or regulatory group to determine if it meets their own standards for nutrition and safety. Additionally, for most of our food, processing it is critical to improve quality, make it palatable, and reduce foodborne safety risks such as pathogens (see Chapter 8 for more on this topic). Cooking food, for example, is an act of processing. Consider the difference between what a chef in a very good 5-star Michelin rated restaurant can do versus the worst restaurant at which you have ever eaten. Often they begin with the same or similar raw materials, but the difference in end product quality is quite different. Any steak can be burned or overcooked or undercooked.

Currently there is very little literature on studies demonstrating the capabilities of food processing methodologies utilizing insects as starting raw materials. Nonetheless, we can begin here to discuss the important features of food processing methods and considerations for other high-protein food ingredients in general, which are most likely to be important for handling and processing of insects from farm to table.

There are numerous procedures and methods already utilized in the food industry dealing with various ingredient formats. The most common formats for ingredients are powders, pastes, and liquids so it is likely that little additional innovation, if any, will be developed to feasibly use many of these methods with insects. Various solid forms also exist including semisolids such as meats and tofu; larger component materials such as whole or chopped nuts, grains, fruits, and vegetables; and materials made into various solid shapes such as grains, protein crisps, and the like, which may be more amenable to working with whole insects. Insects are largely protein and oil, so most processes will seek to exploit the usefulness of these for making foods. The other major component of insects is the chitin fiber material. This, if ground finely enough (ie, less than 100 µm in particle size) they likely will not affect the quality or texture of foods or beverages to which they are added (provided they are not added in abundance, of course). However, the chitin in insects may be undesirable for some food products due to its indigestibility, possible function as fiber which limits nutrient absorption, or simply adding a disagreeable texture to foods which are ideally low in fiber. Thus, methods which efficiently extract the valuable nutritional content from insects while leaving behind chitin are highly desirable.

Shelf life is another important consideration for any food product. There is currently very little information available on the shelf-life properties of various insects, insect components (such as fats, oils, etc.), and insect-based food ingredients (powders, etc.). Nonetheless, there are some known features of insects which suggest that insect-based food ingredients may have an advantage of longer shelf life over similar noninsect alternatives, particularly for dry products or pastes with high levels of insect content. For example, a number of insects have been shown to contain antimicrobial substances such as peptides, fatty acids, and other secondary metabolites (Dossey, 2010; Finke, 2013; Huang et al., 2011), which may be helpful protecting the shelf-life of insect-containing foods. Additionally, the chitin from insects is known to have antimicrobial properties (Tharanathan and Kittur, 2003). Thus, even though, in many cases, the texture of insect-based foods is improved by removing chitin as mentioned in this section, in other food products, there may be advantages in leaving the chitin. In those cases, the texture issues with chitin might be alleviated by grinding the insects into a fine powder, while maintaining the fiber and antimicrobial benefits of chitin.

Food safety is one of the most important considerations for food processing toward developing methods for incorporating insects into the food industry. While many people around the world currently consume whole prepared insects in significant numbers without any apparent negative health effects, it is important to point out some potential hazards. Proper processing and preparation of insects as a food ingredient can alleviate many hazards which insects may present. Nearly all insect-based foods and food ingredients made from insects will contain the whole insect in some form or another, including the digestive tract and its contents. Most insect-based food ingredient production methods will involve, at some stage, simply grinding the whole animal (insect) into a powder, course meal, paste, or liquid. This is different from more traditional vertebrate livestock-derived food products whereby the intestines and other bacteria-rich portions are removed in early stages of processing. It has even been suggested by some that removing the digestive tract (gutting) of insects is important to prepare them for human consumption, but not if fed to livestock because the livestock benefit from fiber (Shackleton, 2004). However, robust processing methods and microbial "kill-steps," such as pasteurization and cooking, will likely mitigate the need for this sort of tedious and inefficient process (see Chapter 8 for more on food born pathogen risk related to insect-based foods). Such kill-steps, which eliminate bacteria and other microbes (and their spores) from processed insect material, are quite necessary and it is highly desirable to incorporate these early in any insect-derived food product operations/protocols. Additionally, as insects become a more popular and widely used as food ingredient, regulations on how they are handled will need to evolve, particularly in the United States (see Chapters 7 and 8 for a detailed discussion on the worldwide food regulatory landscape). The Federal Food, Drug, and Cosmetic Act (FD&C Act) prohibits the "adulteration of any food in interstate commerce…if it consists in whole or in part of any filthy, putrid, or decomposed substance, or if it is otherwise unfit for food" (Gorham, 1979). Therefore, Title 21, Code of Federal Regulations, Part 110.110 allows the FDA to establish maximum levels of natural or unavoidable defects (such as insect parts) in foods for human use that present no health hazard and are known as *The Food Defect Action Levels* (FD&C Act, 1976). Finally, educational programs, publications (such as this book), and materials containing protocols, recipes, and methods for safe and effective processing and cooking can assure the safe and efficient use of insects as food ingredients in general, particularly in the developing world where more effective food processing equipment and methods are not readily available (Amadi et al., 2005).

The following subsections will discuss various food processing considerations and methods to help guide the beginning insect-based food entrepreneur as well as demonstrate to the larger food industry how insects might be processed for incorporation into their existing or new lines of food or beverage products.

Product Decision: Whole Insect, Powder, Liquid, Paste, or Oil?

When deciding on an insect-based food product (or any insect-based product), one of the first things to consider is the format of the product, and thus the insect-based ingredient. Do you want to see the whole insects, or is the product a whole insect itself? Is your end product dry? Is it wet?

For most products, a fine particle size dry powder will be the ideal insect-based ingredient format. This is true for many reasons: (1) powders have the longest shelf life (typically more than a year, depending on production/processing method used, packaging, and storage environment), (2) powders can be blended with many other ingredients effectively without compromising the texture or structural integrity of the product, (3) depending on the processing method, a powder can have the most mild flavor and aroma as well as the lightest color, (4) in general, a powder is the best way to introduce insect into a product without changing it or "noticing the insect," which is ideal for market acceptability and (5) powders tend to be ideal for most food equipment as they can be poured into, flowed through, and overall, utilized in the widest variety of food equipment (extruders, etc.). The general usefulness and long shelf life of powders make them ideal for any product—even when the final product might not be a dry one—as the ingredient can be shipped, stored for long periods, and then used as needed. This can help a company purchase larger quantities of the insect-based ingredient at a time without worrying about having to use it quickly.

The next most useful insect-based ingredient format would be liquids as well as possibly slurries or pastes. Liquids (especially if the insect material is ground finely enough) can blend well with other ingredients and into other products and are amenable to use in a wide variety of food production equipment. They can be pasteurized and even shipped as an ingredient, though the shelf life of a liquid will always be much shorter than that of a dry product. Liquid may be desirable over powder in a limited number of instances, primarily if all aspects of product manufacturing are done at a single location from raw insect to finished product (ie, no shipping or long-term storage of the liquid). Using a liquid might be best in those situations since the insects are already mostly water (like most organisms), and thus the need to dry them just to add water later is avoided. Aside from the shorter shelf life and more care needed in storage and handling to avoid microbial growth, care must also be taken with processing insect-based liquids. Since insects contain a large amount of protein and nonliquid tissue, insect material in liquid form can denature and turn into a paste or solid when heated or with other treatments (with protein denaturing salts, reagents, enzymes, etc.). If the final product needs to be very smooth, is a beverage, or the liquid needs to flow into small spaces or through pipes very efficiently, this may cause some limitations. Though these limitations can be easily overcome (Dossey, unpublished). In fact, in some applications, such as an insect-based meat, tofu, or other solid or semisolid wet product, the solidification due to heating or treatment may be a beneficial or even a necessary part of production. Nonetheless, powders are still quite advantageous because most often, insect-based powders can be reconstituted at the time they are needed and maintain the functional properties they had when they were in liquid form.

In addition to ingredients made from the whole processed insects (the powders, course roasted meals, pastes, and liquids discussed in this section), like any commodity, mass-produced insects will eventually also be fractionated into their various components. As mentioned earlier in this chapter, insects predominantly consist of protein, fat/oil, and fiber (chitin). Thus, the vast majority of extracts and isolates derived from insects will be protein concentrates, protein extracts, protein isolates, oils, and the chitin fiber.

Oils from insects may be particularly desired as many species tend to be lower in saturated fat, low or devoid of trans fat, and high in omega-3 fatty acids (see Chapter 3 for more on the nutrient value and content of various insect species). Oils are essential to the food industry for many uses: cooking, frying, blending with other ingredients, texture, flavor, nutrient value, and the like. As discussed throughout this book, insects are the most diverse group of organisms and, likewise, represent a diverse source of varying types of fats and oils. Oils from different species, and a species fed on different diets, can have drastically different physical properties, flavors, and nutritional value. For example, as mentioned early in the chapter, the larvae of holometabolus insects (such as mealworms, butterflies, moths, flies, etc.) tend to be higher in fats and oils than their adult stages or hemimetabolous insects (crickets, grasshoppers, etc.). Thus, one wishing to produce an insect oil may consider the larval stage of insects, although it also behooves one to consider other attributes such as fatty acid content (for nutritional and other purposes, omega-3 content, saturated versus unsaturated fat, cholesterol, etc). One important consideration for fats and oils is melting point and density. Some fats have low melting points and are thus liquids at ambient temperatures. Others may be solids at ambient temperatures and thus impart a different texture and firmness on products in which they are incorporated. Thus, it is important to carefully select the fat/oil of choice for products to get the desired consistency and "flowability" [for a paste product like Ready-to-Use Therapeutic Food (RUTF), etc.], general firmness and oiliness (appearance, oil on hands/packaging when used by consumer, ability to flow easily through food manufacturing machines, etc.). One may even wish to utilize insect oils as blends of different insect species, or insect oil plus different plant based oils to achieve the desired properties (physical, flavor, nutritional, etc.) of one's product of choice (bar, cookie, etc.). Additionally, different oils, and oils blended in different types of formulations (bound to other ingredients well, or not, etc.) will have varying degrees of susceptibility to oxidation when stored alone or in food products. Thus, different insect oils will have different shelf lives, or impart different shelf lives in the products they are utilized in, so it will be critical to obtain shelf stability studies in different insect oils and the products in which they are used. There is a great need for published studies in this area of shelf life of insect-based food ingredients in general, particularly for whole insect powders and insect oils. An additional consideration for insect oils can be their use as frying or cooking oils, or even as biofuels. In fact, some researchers and companies have begun exploring the potential utility of insect-derived oils as biofuels from species such as black soldier flies (Li et al., 2011; Zheng et al., 2012a,b).

When considering fractionation of insects into various components, also consider the potential value of the chitin fiber. The chitin eventually may be either a by-product or a high-value product leaving the protein and fats/oils and other nutrients as the lower value bulk commodities. See the later section on chitin as a potential commercial product or commodity derived from insects.

For all of these, the food industry possesses a plethora of equipment, methodologies, and technologies for extracting and separating biological materials into their various components. For example, protein concentrates, extracts, and isolates often involve extraction of the fats and oils away from the material as a major component of the process. Given that insects are nearly completely fat and protein, very high percentages of protein can be obtained for products which are simply made from insects from which the oil has been extracted (also referred to as "defatted" product). There are various standard industrial methods for this, including utilization of organic solvents (hexane, ethanol, etc.). However, since consumers are demanding less utilization in food processing now than ever before, newer nonsolvent methods for oil and protein extraction are being developed with varying levels of cost, efficacy, efficiency, and end-product quality. While a review of food industry extraction methodologies is beyond the scope of this chapter, we would recommend that companies wishing to utilize various protein or other concentrates, extracts, or isolates from insects fully explore their options and seek to utilize the most "clean" and least chemical-intensive methodology available. Nonetheless, there are numerous solvents approved for extraction of foodstuffs which allow for lower cost and higher efficiency oil extraction.

While the vast majority of food, feed, and other insect-based applications will certainly call for bulk and finely ground commodity-like ingredients such as powders, meals, pastes, slurries or liquids, protein isolates, extracts, and/or oils, there is certainly a market and need for products utilizing whole insects. Whole insects come in a wide variety of shapes, sizes, and textures (some with hard/crunchy exoskeletons such as crickets and grasshoppers, others with very soft bodies covered by a thin plastic-like skin such as caterpillars and some other larvae). Additionally, whole insects tend to be fragile, as legs and antennae tend to fall off easily for many species and larvae (caterpillars, etc.) or are soft-bodied with skins that are easily damaged causing the liquid contents to leak out. Thus, each type of whole insect will require a rather specialized protocol and set of methods, equipment, processing, handling, packaging, and storage procedures. Many types of existing food processing equipment, packaging, or storage situations may not be amenable to working with whole insects, which can be seen in one of two ways: as a limitation or an opportunity. The limitation is as stated previously: existing facilities and methods may not be amenable to the fragile, varying, and oddly shaped whole insect bodies. Also, many people may not want to snack on whole insects for aesthetic reasons or simply because having to chew on chitin or juicy larvae is not appealing. However, this also represents an opportunity: there exists a wide open window and need for more innovation in designing machines, packaging, processing, and handling procedures for dealing with various marketable whole edible insects. The markets for these will largely be culinary (high end, ethnic, or exotic restaurants—possibly mainstream restaurants in the future), novelties, home use, sacks (packaged dried or preserved and flavored insects, etc.), canned, and possibly picked insects. Given the limited utility of whole insect bodies as food or food ingredients in the current food industry, the proportion or percentage of the insect-based food and edible insect industry in which insects are utilized whole will likely always remain relatively small. However, as the insect-based food industry grows and consumers become more comfortable with purchasing insects as food in stores and restaurants, many companies selling whole-insect products among their other offerings or even specializing in whole edible insect products will likely find lucrative market niches (regional, area of specialization such as high-end restaurant, etc.) that can be exploited and leveraged.

Harvesting, Packaging, and Storage
Freezing

Harvesting the live insects, packaging them raw, and possibly some initial processing or temporary preservation step (ie, freezing, etc.) is primarily the task of the insect farming company. The processing/ingredient companies and even end product companies should be aware of, and may have preferences for, the types of these activities done with the insects they buy from farmers. There are various considerations one may take into account when determining the optimal methods for farmed insect harvesting and preparation for market they may prefer for their farm, or the farm from which they will purchase. In the vast majority of cases, the best technique for harvesting and initial temporary preservation will be to freeze the insects, to be then packaged, purchased, and shipped to processing/ingredient companies (who make insect powders, etc.), restaurants, or individual households. There are several reasons why freezing is ideal where this technique is available. First, since insects are ectothermic (sometimes referred to as "cold blooded") and poikilothermic, and individuals of many species die naturally with the coming of winter (at least in temperate and subtropical zones), a consensus among the current insect-based food industry, and its most supportive advocates, seems to be that the most humane and efficient way to kill them for use in food is by freezing. This is because their metabolisms decrease with cold, they "go to sleep," then die frozen in a very similar manner as to how many die in nature. Second, most insect preparations will likely involve using whole insects or grinding them into a paste or powder. Thus, freezing is the ideal euthanasia method because it is also a very good and immediate preservation and storage method. When choosing a freezing method or apparatus, it is important, as with any other food or raw biological material, to lower the temperature and freeze the insects as quickly as possible for shelf-life reasons. Insects naturally contain internal bacteria and, as with any food or biological material, have bacteria and other microbes and fungi on the outside. The "danger zone" for microbial growth in general (particularly for food) is from 4.5°C to 60°C (40–140°F). Thus, products in the process of being frozen from warm or ambient temperatures are in this range until they are near frozen. Thus, methods such as flash freezing or using small food grade containers (such as plastic bags, etc.) to maximize surface area to volume ratio will be ideal for freezing quickly and to avoid microbial growth or degradation of your product before frozen. For example, some farmers prefer to pack no more than 2.26 kg (5 lb) of crickets into a bag to be frozen at a time, and/or use large, flat bags instead of conventional bags that tend to accumulate into a thick pile. This maximizes the speed at which the cold can penetrate into (or really, where they heat can escape from) the core of the bagged insects. Insects, again like other biological materials, also contain enzymes which can degrade their bodies when they die, which are also temporarily inactivated when they are frozen (another reason to freeze quickly). Additionally, it is best not to stack the bags into a freezer too densely before they are frozen, since this largely defeats the purpose. One can also freeze the insects in open containers or separate containers (again, in thin layers) separate from those used ultimately for storage (ie, freeze in a big box before bagging in a walk-in freezer), then bag them after they are frozen. Flash freezing is ideal because the insects can be frozen rapidly then packaged in large bags, boxes, or containers in bulk (limitless container sizes: 50-pound bags, 1 ton crates, etc.). Innovations are currently being made in this area through the United States Department of Agriculture Small Business Innovation Research (USDA SBIR) funding (Dossey, All Things Bugs LLC, unpublished).

Oxygen Deprivation

As with just about any animal, insects must consume oxygen regularly for survival. Thus, some harvesting methods which reduce their access to oxygen (ie, suffocate them) may be useful in areas where freezing is either expensive or not easily accessible. However, this is not ideal, as it is not as humane as freezing and one must still employ an additional step after suffocation to preserve the insects, as they will begin to decompose (microbial growth, enzymes, etc.) immediately once they are deceased. Little has been done with this method and we are not aware of any farms currently utilizing this method, but it could be a rapid and efficient euthanasia method in some situations. There is little, if any, published information on how rapid or humane this method may be or what methods/gases might be utilized, though there is some consensus on use of CO_2 due to its low cost and wide availability. Additionally, in some research laboratories, CO_2 is often utilized to anesthetize insects for experiments which require them to be immobile or resting (while attaching probes, making injections, taking hemolymph samples, etc.), and presumably CO_2 may also be used to euthanize insects in farm settings as well. There are no standards on how long the CO_2 exposure or oxygen deprivation must be for total euthanasia of insects and this will vary per species.

Other gasses may be eventually utilized to harvest mass volumes of insects for industrial use, for example, those which paralyze the insects. However, these will need to be vetted very carefully and will likely require many expensive and long-term safety trials to be approved for use on insects destined for the human food industry. It is not known at all how effective or efficient this would be, or the time exposure parameters required for very large-scale insect harvests (many tons at a time), which are beginning to be done at farms for the new insect-based food market and will inevitably grow even larger in the future.

Heating and Other Methods

Heating of insects, assuming they are able to perceive pain as many other animals do, is likely to be, or at least is believed to be by many, the least humane method of euthanasia during harvest. Heating/cooking/burning is not a normal part of the natural life cycle or experience of insects in nature (unlike freezing). Nonetheless, some believe that rapid heating (eg, immediate immersion into boiling water) may be reasonably humane because of the very short time of exposure until the insects die. Historically, in areas of the world without access to freezing or other technologies mentioned previously, this has been utilized to euthanize insects. Heating can be effective both as a killing and as a preservation technique, as the heat (if high enough and for a long enough time) will kill the microbes in and on the insects as well as their enzymes. However, once insects are heated to this extent (roasted, steamed, boiled, etc.), much of their chemical and internal physical properties change, as do their functional properties. This may make them less useful to many processors or end-product companies who have their own heating processes that achieve the same microbe/enzyme killing in more gentle ways for higher quality products. Thus, heating as a killing/harvest method is not ideal for farmers to use for the vast majority of the edible insect market unless absolutely necessary.

The "Kill Step": Mitigating Microbiological Risk

All raw food materials must be processed in some form between the field and the consumer, even if it is as simple as the washing of a piece of fruit. At some point during insect processing prior to consumption, there must be a step or procedure which kills the microbial flora in and on the insects, as with any other food. Depending on the product, this can take many forms. Before raw insects are further cooked or processed by that firm's clients (insect-based food or ingredient company, restaurant, end consumer, etc.), there must be some preservation step involved, such as freezing, by the farm or other supplier. Additionally, the microbial and pathogen levels of all food products, including edible insects and products made from them, must be monitored to guarantee that they are safe for human consumption (see Chapter 8 for more on microbial risks in food and their mitigation). A standard panel of micro tests for foods to be considered, which many companies do on each batch of food or ingredient produced, include: total plate count (or aerobic plate count), coliforms, *Escherichia coli* (Migula) (general and pathogenic forms such as O157:H7), *Salmonella*, *Staphylococcus*, *Listeria*, yeast, and mold. For processing to mitigate microbial content, consider first what your starting material is and the quality standards of your desired end product. Microbiological risk at insect farms can be mitigated by keeping facilities and equipment clean via good farm hygiene, and not utilizing contaminated feed ingredients (ie, not utilizing postconsumer waste, manure, etc.). However, even the cleanest farms will have both safe and pathogenic bacteria, yeast, and mold. Even nonpathogenic microbes can cause product degradation that affect product quality and/or safety. Microbes which are generally harmless but affect product quality or shelf life are known as "spoilage" microbes. Many foodborne pathogens and cases of microbial contamination come from within food processing and handling facilities, often from the environment, employees, or other products being processed at the same location. Thus, it is prudent to always take care to have a heat or another type of microbial kill-step or additive (acidification, etc.) in the processing of insect-based foods or food ingredients with the assumption that some microorganism is present, which may be mitigated. Eventually, as the industry matures, there may be reliable supplies of certified fully sterile insect-based ingredients and whole insects. In the food industry, such microorganism-free food products exist (ie, those used in infant formulas, ready-to-eat and shelf-stable foods and beverages not requiring refrigeration, etc.). Thus, the technology exists to achieve this but to date there are no verified sterile insect-based food ingredients on the market. For example, of course cooking (steaming, boiling, roasting, etc.) is generally the most effective method for microbe mitigation. For other food commodities, the US Food and Drug Administration (FDA) even has time and temperature regulations for various commodities such as milk, dairy products, meats, and other fools (http://www.fda.gov/food/guidanceregulation/guidancedocumentsregulatoryinformation/milk/ucm063876.htm; http://www.fda.gov/food/guidanceregulation/guidancedocumentsregulatoryinformation/milk/ucm063876.htm). For milk. the standards for proper pasteurization are often referred to in the industry as a "legal pasteurization." While such standards (or even data on which they could be based) do not yet exist for the insect-based food industry, this is a ripe area in much need of research and these standards need to be, and eventually will be, established. For those wanting to mitigate heat exposure to their insects or insect-based products, there are numerous alternatives for microbial mitigation depending on the desired end product.

For example, nonheat or less-heat alternatives to reduce or eliminate microbial loads include: acidification (addition of citric or other acids), preservatives, High Pressure Processing (HPP), irradiation, and others. Even dry products/powders can be pasteurized using longer time at moderately high temperatures (like egg powder), for products in which liquid pasteurization would hurt the functionality or otherwise lower product quality. Technology even exists to pasteurize a raw egg without cooking it. Acidification also kills many types of microbes. Full microbial mitigation (ie, complete sterilization) is not necessary for most food products, but elimination of pathogenic species is critical. Nonpathogenic species may not be harmful to human health, but they may be "spoilage" microbes which may cause degradation and reduction in quality of the product on the shelf. Insects sold raw or in need of cooking, steamed insects sold refrigerated or frozen, and so forth may be sold with higher acceptable microbial levels (with pathogens fully mitigated) than insect-based foods sold as ready-to-eat. While the microbiology of food as a topic is an entire field of study, and thus beyond the scope of this chapter, it is a critical matter to consider for anyone interested in food and agriculture. Likewise, more information on food microbiology and food-borne pathogen mitigation and regulation can be found in Chapter 8, and information on regulations regarding food safety can be found in Chapter 7.

Drying Methods
Whole Insects

Drying is primarily a preservation step to extend shelf life, but it also makes a product more efficient and cost effective to store (usually dry food products do not need to be refrigerated or kept cool) and ship (because dry things weigh less—shipping water in a product adds a lot to its weight and thus, adds to transport cost). Insects can be dried in their whole form or as pastes, slurries, or liquids. Generally the more water is in a product, the less efficient and more costly and time-consuming the drying process is. Thus, minimizing the addition of water to a product beyond what occurs in it naturally is optimal, but sometimes adding water is unavoidable. Also, the longer it takes to dry a product with some methods, the higher the risk of microbial growth and other forms of product degradation. When drying, it is important to measure both the total moisture level as well as the "water activity" in the product. The water activity is a measurement of how much water is "free" versus bound in a product. Typically, the free water is what contributes to microbial growth. The FDA and other agencies and organizations provide guidance and have recommendations on ideal water activity and moisture levels in dry products to mitigate microbial growth during storage (http://www.fda.gov/ICECI/Inspections/InspectionGuides/InspectionTechnicalGuides/ucm072916.htm).

Currently, roasting insects whole is one of the most popular and common methods for drying them, both in the industry as well as for home use. This is often done for insects which will ultimately be eaten whole or for production of dry ground insect meals or powders. While this is a very popular, widely available, and low-tech method, it is far from ideal. Roasting is typically more expensive at the industrial scale than other drying methods (on a perunit or weight basis), utilizes more heat and energy, and causes a greater level of product degradation (due to the extended time exposed to higher temperatures) than other drying methods. Insects are designed to resist desiccation (not dry out) in nature. Their exoskeleton and cuticle are designed to keep moisture in. This makes them very efficient animals to farm because they are able to conserve water efficiently. However, this also means that removing water from insects via heating and evaporation is very inefficient. Also, the heat required to achieve this is in a reasonable amount of time tends to be high (and long roasting times are required as well). High heat exposure over time degrades the quality of all foods, food ingredients, and biological materials. Fats and oils oxidize (leading to lower shelf life), proteins denature or agglomerate, nutrients such as vitamins degrade, undesirable flavors and aromas can become stronger, and so on. Roasted insects become darker in color, and have stronger aromas and flavors than insects dried in other ways. Additionally, proteins form "shellack" like solids, which stick to the insect chitin exoskeletons and become impossible to remove. This causes the overall functionality of products made from roasted insects to be much less desirable than insects dried in other ways. The proteins also separate from the fats and oils, which melt at high temperatures, rather than blend with them homogeneously, and thus become more exposed to oxidation and make the product generally more "oily." If roasted insects are ground into meals or powders, the product is less free flowing, clumps more, adds an oil film to the containers in which it is packed, and does not blend well with other ingredients. Also, reconstituting such roasted insect meals/powders in water is less effective.

Freeze-drying is another option for drying insects in their whole, unground form as well as in paste, slurry, or liquid form. Freeze-drying is probably the best possible method for preserving the chemical and other properties of any material, biological material or food. However, freeze-drying is typically very expensive on a perunit basis (ie, perpound), takes a long time, is not very efficient, and requires access to highly specialized and energy-intensive (electricity-intensive) equipment that is not available in many areas of the world. This is largely why in the food industry, freeze-drying is not the first choice of food companies for product drying.

Drying Pastes, Slurries, and Liquids to Produce Powders and Meals

While drying insects whole can be done, it is not ideal for the production of powders (fine) or meals (course) made from high moisture (wet) raw materials. Insects are indeed a high moisture material, with most species naturally containing around 50–70% moisture (water). In the food industry, high moisture materials are typically (when possible) ground into some sort of slurry and subsequently dried. Thus, particle reduction (at least to a certain size) is taken care of in the wet phase. This has a number of advantages. First, liquids, slurries, and pastes are typically easy to manipulate and move through pumps, hoses, and pipes. Thus, it makes them easier to process compared with the original whole materials (whole insects, beans, seeds, corn, meats, etc.). Additionally, slurries, liquids, and pastes are a more homogeneous mixture, yielding a much more consistent dried product at the end. In the case of high protein/fat, or high fat/protein/starch food products (such as insect powders, milk powders, etc.), this allows the fats, oils, proteins, and other components to remain together throughout the drying process. When they dry together, the nonprotein/starch components can become microencapsulated into the proteins and starches and somewhat protected from oxidization and other elements that affect shelf life. Homogeneous powders also tend to be more free-flowing, clump less, blend better with other ingredients, and be generally more functional than powders produced in ways that do not keep the components well homogenized.

There are various methods for drying liquids, pastes, and slurries into powders or meals. These include drum drying, spray drying, fluid bed drying, tray drying, freeze-drying, cooking, roasting, vacuum drying, and dehydrating. Freeze-drying, cooking (heating) and roasting were discussed earlier. The same limitations of these methods exist for slurries, liquids, and pastes as do with whole insects. A full discussion of food processing/drying methods used in the industry is extensive and beyond the scope of this chapter. Full books/volumes have been published on this topic. However, here we will consider two drying methods for slurries, liquids, and pastes utilized in the food industry to make high-quality powders and meals: spray drying and drum drying.

The best available drying method for manufacturing of powders, considering cost as well as efficiency and product quality, is typically spray drying. Spray drying is also one of the techniques with the largest scale capacity based on size of equipment used, throughput, efficiency, and magnitude of facilities available. Spray drying equipment exists from laboratory scale (capable of drying only a few grams per hour) all the way up to some of the largest powder producing equipment in the world (producing a few thousand pounds of powder per hour). The large investments that the food and other industries have made into spray drying and spray-dry facilities and capacity alone attest to its value as a drying method. Typically, if something can be spray dried, then that is the technique of choice for drying. To date only one company has demonstrated the feasibility of and technology designed to spray dry insects at industrial scale (All Things Bugs, LLC) (Patent Pending) (Dossey, 2015) with technologies developed during its research projects funded by the USDA (Crabbe, 2012; Mermelstein, 2015; Day, 2015; Tarkan, 2015).

How spray drying works is basically simple: a product (liquid, paste, or slurry) is prepared and propelled into the air as a spray or mist using either spray atomization (with a high pressure spray nozzle) or rotary atomization (where material is spun or "flung" into the air at a high rate of speed by spinning to create fine droplets of spray). This spray or mist is suspended and circulated in the air, usually warm dry air, and allowed to drop to the bottom of the spray drying apparatus as the water is rapidly removed. When the droplets/particles reach the end of the process, the water has been removed from droplets, leaving only the solid nonvolatile (and nonwater) components. The result is a fine solid particle.

The efficiency of spray drying is largely due to the very large surface area to volume ratio of the product during drying—nearly all of the product's surface is instantly exposed to dry air on a microscopic level (rather than most of the product being within a solid or liquid as with roasting, drum drying, freeze-drying, and other techniques). Since the components were initially in a homogeneous well-mixed slurry or liquid initially, all of the spray-dried solid powder particles are consistent with one another both chemically and physically. They also tend to be homogeneous within the particle, with fats oils and other components being microencapsulated into the larger quantity protein and/or starch components of the particle evenly mixed rather than separated (with oils on the surface of the particles), like other drying methods.

Spray drying can also be more energy efficient and generally efficient and may result in the product being exposed to less heat than other methods. This is in part because the particles, while being dried, benefit from rapid evaporative cooling and thus do not experience the full temperature of the warm dry air being utilized to dry them, unlike with roasting or other drying methods. Thus, fats, oils, vitamins, and other components do not heat, cook, oxidize, or degrade as much in spray drying as with some other methods, such as roasting. This often results in lighter color and more neutral aroma and flavor of products, which are desirable traits of bulk food ingredients such as protein powders. In spray drying, the lower heat exposure as well as the microencapsulation of components such as fats and oils can also impart a better shelf life and overall better functionality and usefulness in a wider range of food and beverage products than products dried using other techniques. The major downsides to spray drying include: need for large expensive equipment, need for large-scale production for economic feasibility, access to sufficient electrical power and other industrial resources (though not significantly more than other similarly scaled technologies), and the fact that some products (such as those with very high oil or sugar content) are very difficult to spray dry.

In areas of the world where spray drying is not available, drum drying is typically a viable alternative for producing powders and meals with acceptable quality and is still better than roasting. Drum drying involves the tumbling of a material (liquid, slurry, paste, or even wet solid) in a large "drum" while blowing warm air over it until it is dry. It differs from spray drying in certain fundamental ways, including that the material is exposed to hot or warm surfaces (typically metal) during the drying process and remains together as a unit and is not propelled into fine particles while being dried. Thus, drum drying does not benefit from the large surface area to volume ratio that spray drying, or even roasting (for whole insects) do. The exposure to hot surfaces means that some of the material eventually dries/cooks onto these surfaces and often requires being scraped off, resulting in large flake particles or cakes that need to be further milled into powder. Since the material agglomerates, forms flakes or clumps versus fine particles, and is all in one pile while being dried, fats oils and other components can and often do separate during drying so that the resulting dried material particles are not as homogeneous and are less controlled in size and content as compared with spray drying or even, to some extent, freeze-drying. Thus, drum drying very often begets a lower product quality for powders than does spray drying. Nonetheless, drum drying remains a popular drying technique for companies or areas where spray drying is not an option, and can be an alternative at smaller scales than are often feasible for spray-drying operations.

Grinding/Milling

The vast majority of food ingredients, and thus the vast majority of applications for insects in food, call for the materials to be in a fine powder, paste, slurry, or other homogeneous format with small particle size. Indeed, while there are many things to consider for milling or grinding, the primary goal is to reduce the maximum particle size of the product to the desired, and ideally, a consistent size (all particles close to the same size or a tight distribution). With milling, one must consider both the characteristics of the starting material and those desired for the end product. Insects, as living organisms, have a high content of water; therefore the timing for grinding, milling, or particle reduction should be selected to take place before, during, or after drying, heating, microbial mitigation (kill-step) or other steps in the process toward the desired product. Other ingredients may also be blended in along with the insects during the grinding or milling. Here we will briefly discuss some considerations for both wet and dry milling of the insects themselves.

Wet Milling

For the milling of whole raw insects, which have not yet been dried, or insects previously dried then reconstituted into water, the process of choice is wet grinding or wet milling. For example, when conducting the heating kill-step (cooking, pasteurization, etc.) after grinding (ie, pasteurizing a liquid or slurry) or before (ie, steaming or boiling insects before grinding), wet grinding and milling take a product containing a significant content of water or other liquid and reduce it into a liquid, paste, slurry or, a wet semisolid. One of the main differences between this and milling/grinding dry is that a wet product is usually can more or less be treated like a liquid or semiliquid and thus, it can be pumped. This has several advantages in product handling, moving, pasteurization, manipulation, or movement and manipulation for further processing within a processing plant. Being able to flow a product through pipes/hoses using pressure allows it to move into and out of various vessels and machines efficiently and consistently. Disadvantages of wet products over dry forms include: they have potentially lower shelf life, they need to be kept cool, they need to be more carefully handled or processed to mitigate microbial growth, and they weigh more and have greater volume, which negatively impacts the economics of storing and shipping. Thus, shipping or storing a product in some dry format has several cost, quality, efficiency, and shelf life advantages. Additionally, wet grinding or milling tends to require at least slightly more technically advanced equipment or techniques. For that reason, and for the need to keep the product cool or stored to limit microbial growth, dealing with insects in wet form may not be ideal for some applications in areas without access to sufficient facilities, equipment, or technology. However, cooking or pasteurization (even sterilization) can be done to insects used in wet form as a liquid or slurry and wet milling or grinding may still be useful in these situations if the material is to be utilized immediately. There is a wide variety of equipment and methodology available to achieve grinding of wet products (wet grinding or wet milling) including but not limited to: grinders, mills, stone mills, mixers, peanut butter grinders, colloid mills, bead mills, dispersers, homogenizers, choppers, rotor stator devices, press grinders, mashers, macerators, food processors, rollers, or juicers of various sorts (centrifugal, masticating, etc.). The wet milling process can be applied by batch (using things like rotor–stator apparatus, etc.) or continuously (using in-line machines). Additionally, for insects, removal of chitin may be desired. This will almost certainly require the product to be wet and the insects to have not been previously roasted or dried. In a simple step, raw insects can be ground/milled and filtered to remove the chitin away using an apparatus such as cheese cloth or other filter presses.

Dry Milling

The most technologically simple forms of grinding, milling, or particle reduction are typically done with dry products. Even from ancient times, stone mills and mortar and pestles have been used to grind many products such as grains (wheat, corn, etc.), beans, seeds, nuts, and so on. Some insect-eating cultures (in Southeast Asia, Africa, South America, etc.) even pound their insects into pastes to use in dips, sauces, and other foodstuffs. The grinding of dried insects can create a dried powder or course meal useful similarly to powders produced using methods described in this chapter (although, as described earlier, drying insects prior to grinding almost always results in a lower-quality product with less utility and functionality, and potentially with a shorter shelf life). Since dry milling is old and often simple, a very wide variety of tools and methods exist to grind dry products into fine powders or course meals including but not limited to: grinders, mills, stone mills, peanut butter grinders, pin mills, bead mills, hammer mills, food processors, rollers, and Wiley mills. In some cases, the dry grinding or milling apparatus may be cheaper to utilize and useful at smaller scales and in lower technology regions than those used to wet grind/wet mill products. Nonetheless, in industrialized areas of the world, the optimal processes for grinding and processing insects into industrial food ingredients such as high-quality, functional powders, meals, pastes, liquids, slurries, or semisolids will almost always involve wet grinding and doing much of the processing and manipulation of the product in its wet state.

THE REAL PIONEERS: ENTREPRENEURS IN THE INSECT-BASED FOOD SPACE

The current edible insect food industry is in many ways a product of an interconnected age where food preferences and choices flow across distances and cultures. In an age of increased ease of international travel, people have accessed the idea of insects as food through exposure, both virtual and in person, to diverse cultures and new concepts of what can be food. While the insect-based food and insect farm industry to date has been noticeably secretive, we have been able to gather some information as to how things started (Table 5.4). For example, due to the Western world's increased exposure to other culture's use of insects as food, several companies have been able to use crowdfunding sites to raise capital. Prominent examples o f US-based companies include: Chapul, who raised just over $16,000 in 2012; Exo, whose 2013 campaign for its cricket bar generated nearly $55,000 (USD); Six Foods, who raised nearly $71,000 from almost 1300 people; Hopper Foods, who got almost $35,000 from 479 investors; and Crickers crackers, whose 406 backers pledged $33,000. Additionally, a group called LIVIN Farms by members of a design firm called Livin Studio raised $145,429 (USD) (the largest crowdfunding amount raised by an insect-based–food-related company to date). Their campaign featured and collected preorder sales for their small countertop personal mealworm rearing apparatus called the Hive.

Another way that small startup companies in this innovative new industry have been able to raise funds, particularly for more innovation and research and development (R&D) work has been through research grants and business or entrepreneurial competitions (many of the latter for students). For example, the company Aspire Food Group won the $1 million Hult Prize for student entrepreneurs in 2013. Other startups in the insect-based food industry have also won smaller yet significant business competitions for students as well as nonstudent entrepreneurs, particularly in North America and Europe. Government and philanthropic research grants also provide substantial potential for much needed funding in the insect-based food/agriculture industry, yet only one company to date, All Things Bugs, LLC, has been able to receive research grant funding from government and major funding organizations (over $750,000 to date). The company was first founded in order to qualify for receiving a Phase I Bill & Melinda Gates Foundation Grand Challenges research grant in 2012 for $100,000 to develop a Ready-to-Use Therapeutic Food (RUTF) to alleviate malnutrition in children. The application was submitted in 2011, but the founder (Dr Aaron T. Dossey) was not employed in a position eligible to receive the grant, so he founded the company All Thins Bugs, LLC to satisfy the Gates Foundation's requirements. Subsequently, All Thins Bugs, LLC received several SBIR grants from the USDA including a Phase I ($100,000) grant in 2013 for further RUTF and insect-based food ingredient process development (commercialized in Phase I, which is rare and speaks to the potential of the insect-based food market; technology from this grant now patent pending) (Dossey, 2015). Phase II of that project ($450,000) was received for further development of insect processing technology and evaluation of insect powders for functionality, shelf life, safety, quality, nutrient content, and other properties. Another Phase I ($100,000) was received for technology and innovation development of improved cricket-farming methods and equipment to lower costs and improve watering, harvesting, and feed, with technologies developed during its research projects funded by the USDA (Crabbe, 2012; Mermelstein, 2015; Day, 2015; Tarkan, 2015). The company has gone on to produce and sell over 10,000 pounds of cricket powder since it started manufacturing in 2014, which is the largest reported production scale of any insect-based food company to date which we can verify.

The connection between this industry and the internet goes deeper still, given that the majority of retail sales continue to be direct to consumer via website stores. A tremendous amount of marketing has been done through Facebook and

TABLE 5.4 List of Insect-Based Food Companies, Insect Farms, Insect-Based Animal Feed Companies, and Insect Industry Advocacy Organizations

Number	Name	Website URL	Location	Continent	Food	Feed	Consumer Products	Farming	Advocacy
1	3ento	www.4ento.com	Switzerland	Europe	X	X			X
2	All Things Bugs LLC and Griopro	www.allthingsbugs.com and www.cricketpowder.com	Athens, GA, US	North America	X	X	X	X	X
3	ArthroFarms	www.arthrofarm.com	Cypress, TX, US	North America	X			X	
4	Arthrofood		Madrid, Spain	Europe	X				
5	Aspire Food Group USA	www.aspirefg.com/usa/	Austin, TX, US	North America	X			X	
6	Ben's Bugs	www.bensbugs.be	Berigen, Belgium	Europe	X		X		
7	Big Cricket Farms	www.bigcricketfarms.com	Yougstown, OH, USA	North America	X			X	
8	Binpro	www.Binpro.de	Moers, Nordrhein-Westfalen, Germany	Europe		X		X	
9	Bitty Foods	www.bittyfoods.com	San Francisco, CA, US	North America	X		X		
10	Bodhi Protein	www.bodhiprotein.com	London, UK	Europe	X		X		
11	Bug Bites	www.bugbitesfood.com	Auburn, ME, US	North America		X			
12	Bug Box	www.bugbox.supply	Chicago, IL, US	North America	X		X		
13	Bug Boys	www.bugboys.co.uk	Brighton, UK	Europe					X
14	Bug Corp	www.bug.com.vn/	Vietnam	Asia	X		X		
15	Bug Eater Foods	www.bugeaterfoods.com	Lincoln, NE, US	North America	X		X		
16	Bug Foundation	www.bugfoundation.com	Osnabrück, Germany	Europe	X		X		
17	Bug Grub	www.buggrub.com	Norfolk, UK	Europe	X		X		
18	Bug Muscle	www.bugmuscle.com	Redlands, CA, US	North America	X		X		
19	Bugzz	www.bugzz.nl	Amsterdam, Netherlands	Europe	X		X		
20	Bush Grub	www.bush-grub.co.uk	Kent, UK	Europe	X		X		
21	C-fu Foods	www.c-fufoods.com	Ithaca, NY, US/Toronto, ON, Canada	North America	X		X		
22	Chapul	www.chapul.com	Salt Lake City, UT, US	North America	X		X		
23	Chloe's Treats		New York, NY, US	North America		X	X		
24	Chucky's Crickets		San Antonio, TX, US	North America	X				
25	Coalo Valley Farms	www.coalo.farm	Van Nuys, CA, US	North America	X			X	
26	Crickeatz	www.crickeatz.co.uk	Birmingham, UK	Europe	X		X	X	
27	Crickers	www.crickerscrackers.com	Austin, TX, US	North America	X		X		
28	Crickers Crackers	www.crickerscrackers.com	Austin, TX, US	North America	X		X		
29	Cricket Flours	www.cricketflours.com	Eugene, OR, US	North America	X		X		

Modern Insect-Based Food Industry **Chapter | 5** **137**

	Name	Website	Location	Region	C1	C2	C3	C4
30	Cricket Foods	www.cricketfoods.com	Redwood City, CA, US	North America	X			
31	Crik Nutrition	www.criknutrition.com	Winnipeg, MB, Canada	North America	X			
32	CRISPIzZ		Lyon, France	Europe	X			
33	Crobar	www.croprotein.com	London, UK	Europe	X			
34	Crunchy Critters	www.crunchycritters.com	Derby, UK	Europe	X			
35	Damhert	www.damhert.be	Belgium	Europe	X			
36	De Insectenbar	www.deinsectenbar.nl	Amsterdam, Netherlands	Europe	X			
37	Deli Bugs	www.delibugs.nl	Lelystad, Netherlands	Europe	X			X
38	Dimini Cricket	www.diminicricket.com	Branches, France	Europe	X			
39	Don Bugito	www.donbugito.com	San Francisco, CA, US	North America	X			
40	Eat Grub	www.eatgrub.co.uk	UK	Europe	X			
41	Edible Bug Farm	www.ediblebugfarm.com	Birmingham, UK	Europe	X		X	
42	Edible Bug Shop	www.ediblebugshop.com.au/	Sydney, Australia	Australasia	X			
43	Edible Inc	www.edible-bug.co	Seoul, Korea	Asia	X			
44	Edible Unique	www.edibleunique.com	UK	Europe	X			
45	EIF	www.eif-th.com	Chiang Mai, Thailand	Asia	X	X		
46	Ento	www.eat-ento.co.uk	London, UK	Europe	X			
47	EntoBento	www.launch.entobento.com/san-diego/	San Diego, CA, US	North America		X		
48	EntoCoopa		Languedoc, France	Europe	X	X	X	
49	Entocube	www.entocube.com	Espoo, Finland	Europe	X	X	X	
50	EntoFood	www.entofood.com	Kuala Lumpur, Malaysia	Asia		X	X	
51	Entologics	www.entologics.com	São Paulo, Brazil	North America	X	X	X	
52	Entomo Farm	www.entomo.farm	Orsay, France	Europe	X	X	X	
53	Entomo Foodafrica		Katanga, DRC	Africa	X			X
54	Entomochef (BCDE catering)	www.entomochef.be/en	Brussels, Belgium	Europe	X			X
55	EnviroFlight	www.enviroflight.net	Yellow Springs, OH, US	North America		X	X	
56	Essento	www.essento.ch	Berne, Switzerland	Europe	X		X	
57	Europe-Entomophagie	www.europe-entomophagie.com/en/#	Rouen, France	Europe	X			
58	Exo	www.exoprotein.com	New York, NY, US	North America	X			
59	Get Sharp	www.getsharp.eu	London, UK	Europe	X			
60	Goffard Sisters	www.goffardsisters.com	Liège, Belgium	Europe	X			
61	Gran Mitla	www.granmitla.us	Oaxaca, Mexico	North America	X			
62	Greenkow	www.greenkow.be/EN	Brussels, Belgium	Europe	X			
63	Grub	www.eatgrub.co.uk	London, UK	Europe	X			

(*Continued*)

138 Insects as Sustainable Food Ingredients

TABLE 5.4 List of Insect-Based Food Companies, Insect Farms, Insect-Based Animal Feed Companies, and Insect Industry Advocacy Organizations (cont.)

Number	Name	Website URL	Location	Continent	Food	Feed	Consumer Products	Farming	Advocacy
64	Grub Kitchen (restaurant)	www.grubkitchen.co.uk	Pembrokeshire, UK	Europe	X				
65	Gryö	www.gryobars.com	Paris, France	Europe	X		X		
66	Hakuna Mat	www.hakunamat.se	Göteborg, Sweden	Europe	X		X		
67	Hao Cheng Mealworm Inc.	www.hcmealworm.com	China	Asia	X	X	X	X	
68	Hexa foods	www.hexafoods.com	Montreal, QC, Canada	North America	X		X		
69	Hopper Foods	www.hopperatx.com	Austin, TX, US	North America	X		X		
70	Hotlix	www.hotlix.com/candy/	Grover Beach, CA, US	North America	X		X		
71	Insectement Votre	www.insectement-votre.com/	Rouen, France	Europe	X		X		
72	Insecteo	www.insecteo.com	Brussels, Belgium	Europe	X		X		
73	Insectes au Menu	www.insectesaumenu.fr	La côte d'Aime, France	Europe	X		X		
74	Insectes Comestibles	www.insectescomestibles.fr/en/	France	Europe	X		X		
75	Insectitos	www.insectitos.com	Aurora, IL, US	North America	X		X		
76	International Society for Food and Feed Insects	www.facebook.com/ISFFI	Athens, GA, US	North America	X	X			X
77	Invertebrate Studies Institute	www.facebook.com/InvertebrateStudiesInstitute	Athens, GA, US	North America	X	X			X
78	Jimini's	www.jiminis.com	Paris, France	Europe	X		X		
79	Jumping Jack Snacks	www.jumpingjacksnack.co	Wageningen, Netherlands	Europe	X		X		
80	Kaki 6		Kuala Lampur, Malaysia	Asia	X		X		
81	Khepri	www.khepri.eu	Paris, France	Europe	X		X	X	X
82	Kreca	www.kreca.eu	Ermelo, Netherlands	Europe	X	X		X	
83	LaViewEye/The Cricket Girl	www.lavieweye.com	Asheville, NC, US	North America	X		X		
84	Lazybone	www.lazyboneuk.com/categories/Edible-Insects-/	UK	Europe	X		X		
85	Little Herds	www.littleherds.org	Austin, TX, US	North America					X
86	Livin Studios	www.livinstudio.com	Vienna, Austria/ Hong Kong	Asia	X		X	X	
87	Manataka/WorldEnto	www.manataka.org/page2807.html	US	North America	X		X		
88	Marxfoods/Marx Pantry	www.marxpantry.com/Entomophagy	US	North America	X		X		
89	Mealfood Europe	www.mealfoodeurope.com	Madrid, Spain	Europe		X		X	
90	Meatmaniac	www.meatmaniac.com/SearchResults.asp?Cat=1824	Richardson, TX, US	North America	X		X		

Modern Insect-Based Food Industry **Chapter | 5** **139**

#	Name	URL	Location	Region					
91	Micronutris	www.micronutris.com	Toulouse, France	Europe	X				X
92	Multivores	www.en.multivores.com/20-insects-natures	Bordeaux, France	Europe	X			X	X
93	Next Millennium Farms	www.nextmillenniumfarm	Lakefield, ON, Canada	North America	X	X		X	X
94	NextProtein	www.nextprotein.co	Paris, France	Europe	X	X			X
95	Nordic Food Lab	http://nordicfoodlab.org/	Frederiksberg, Denmark	Europe	X		X		
96	Nutribug	www.nutribug.com	UK	Europe	X			X	
97	Omni Protein		Des Moines, IA, US	Europe	X			X	
98	Osgrow	www.osgrow.com/productsdetail.php?ProductID=44	UK	Europe	X			X	
99	Ozark Fiddler Farms	www.ozarkfiddlerfarms.co	Fayetteville, AR, US	North America	X				X
100	Proti-Farm	www.protifarm.com/welcome.html	Ermelo, Netherlands	Europe		X			X
101	Protix Biosystems	www.protix.eu	Dongen, Netherlands	Europe		X			X
102	Pupa Planet	www.pupaplanet.com	Madrid, Spain	Europe		X			X
103	Qvicket	www.qvicket.se	Stockholm, Sweden	Europe	X			X	
104	Rainbow Mealworms	www.rainbowmealworms.net	Compton, CA, US	North America	X	X			X
105	Rocky Mountain Micro Ranch	www.rmmr.co	Denver, CO, US	North America	X				X
106	Sahakhun Bug Farm	www.plus.google.com/+SahakhunBugFarm	Rayong, Thailand	Asia	X			X	X
107	Sahakhun Bug Farm Ltd.	www.senzufoods.com	Chiang Mai, Thailand	Asia	X			X	X
108	SENZU Foods	www.senzufoods.com	Miami, FL, US	North America	X			X	
109	Sexy Food	www.sexyfood.fr/en/	France	Europe	X			X	
110	Six Foods	www.sixfoods.com	Boston, MA, US	North America	X			X	
111	Snickets	www.snickets.co	London, UK	Europe	X			X	
112	Tarzan Nutrition	www.tarzannutrition.ca	Montreal, QC, Canada	North America	X			X	
113	Thailand Unique	www.thailandunique.com	Udon Thani, Thailand	Asia	X			X	
114	The Bug Shack	www.thebugshack.co.uk	UK	Europe	X			X	
115	The Gourmet Cricket Co.	www.gourmetcricket.bigca	Cornwall, UK	Europe	X			X	
116	The Grasshopper Suppliers	www.thegrasshoppersuppl	Amsterdam, Netherlands	Europe	X			X	
117	The Green Scorpion	www.thegreenscorpion.com.au/shopshow.toy?catnid=600438	Australia	Australasia	X	X		X	X
118	Thinksect	www.thinksect.com	Portland, Oregon, US	North America	X			X	
119	Tiny Farms	www.tiny-farms.com	Oakland, CA, US	North America	X				X
120	Tsukahara Chinmi	www.tsukahara-chinmi.com/shopinfo.html	Ina, Nagano, Japan	Asia	X	X			
121	Vfur farmed insects	www.fodurskordyr.is	Ísafjörður, Iceland	Europe	X	X			X
122	Ynsect	www.ynsect.com	Évry, France	Europe	X	X			X

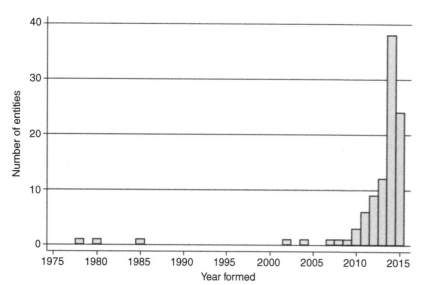

FIGURE 5.1 **Growth of the insect-based food industry in recent years.** Data showing the increase in number of insect-based food companies, edible insect farms, and nonprofit organizations working in the area of insects as a food ingredient/commodity, as represented by entity founding date. The data represented in this figure are derived from Table 5.4. Note that data collection ended during 2015, which is likely the reason for a lower value for 2015 compared with 2014. In general the increased number of companies and entities entering the insect-based food industry and field continue to rise rapidly as evidenced largely by new social media (Twitter accounts, Facebook pages, etc.) and press articles. However, due to the dispersed and informal nature of these enterprises, it is generally difficult to maintain updated lists of new entities in this space.

especially Twitter. Almost on a weekly basis, new Twitter accounts, Facebook pages, and companies they represent appear on these social media venues, making it difficult to collect complete lists of all new companies to date (hence the artificially low number for new companies in 2015 represented in Fig. 5.1). Additionally, the large majority of insect-based food product sales to date appear to be direct to consumers from the respective websites of the insect-based food companies themselves. Some of these companies are also utilizing various online venues such as Amazon, Etsy, and a wide variety of large as well as small online retailers and distributors. With the exception of a couple US companies, including Chapul (which sells its cricket protein bars in several health food grocery stores in the western United States), and Bug Eater Foods (which sells its flavored cricket protein shake powder mix in a Midwestern grocery chain), the vast majority of consumer sales have been through online sales via websites and generally outside traditional brick-and-mortar stores.

Summary of Current Companies, Farms, and Other Organizations

There is a growing number of companies producing insect-based food, particularly in places where people do not commonly consume insects yet (ie, Europe and North America) (Table 5.4) (Fig. 5.1). These range from insect farmers, mostly raising crickets and mealworms, to those producing consumer goods, particularly from milled and ground crickets (ie, cricket powder, which is sometimes erroneously referred to as cricket flour as previously discussed in this chapter). In many cases, cricket farms are also producing cricket powder or cricket meal, as well as selling whole insects. In Europe, a significant number of companies are raising mealworms.

These Western industries are marketed largely by entrepreneurial companies, many of which have acquired financial support through crowdfunding campaigns that focus on the benefit to the environment and human health from insect-based foods. Operating in a largely gray area in terms of regulations and industry standards in both North America and Europe, this vibrant and growing industry is estimated to be worth $20 million (Hoffman, 2014), increasing to $360 million by 2020 (Robinson, 2014).

The largest and perhaps most vibrant insect farming industries, in terms of number of producers but not necessarily scale of production, operates in Southeast and East Asia, where edible insects are a mundane part of the population's diet. In Thailand, an estimated 20,000 cricket farms are in operation. These are mostly small to medium enterprises. The low care time needed for crickets relative to the revenue from selling the crickets is a huge benefit to these communities. The Lao People's Democratic Republic also has many small to medium edible insect farming enterprises. In 2014 the government there proposed a set of national regulations to formalize the industry; the regulations are not yet ratified at the time of publication.

China is similarly a leading place for food and feed insect farming, as well as businesses raising insect species used in "traditional medicine" applications, like the aforementioned 100 farms raising cockroaches for medicinal and cosmetic uses.

Overall, worldwide, the interest in the emerging insect industry has increased greatly. Interestingly, the FAO maintains a global edible insect stakeholder directory online, listing insect-based food companies, civil societies, and farmers (FAO, 2015) (Fig. 5.2).

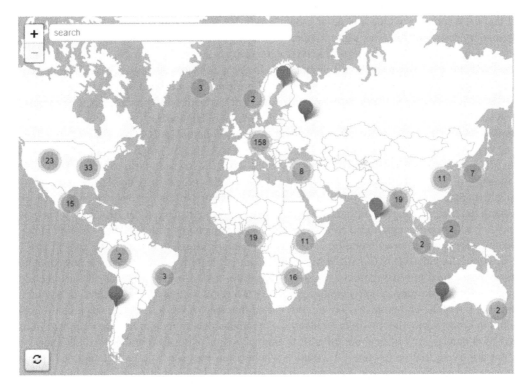

FIGURE 5.2 Screen shot of UN FAO's global stakeholder directory (version 2.0) on edible insects (map). This database was first established and published in Nov. 2013 and is continually updated to date (current database and map can be accessed at: http://www.fao.org/forestry/edibleinsects/stakeholder-directory/en/). It shows all types of practitioners who email UN FAO to be added. These include companies, universities, nonprofits and NGOs (nongovernmental organizations), farms, individual researchers, museums, institutions, and other types of entities. A wide variety of practitioners in the field of insects as food and feed are represented worldwide to date. *(Taken on January 22, 2016 at 11:16 PM US Eastern Time; last updated July 3, 2015)*

Outside of these aforementioned areas, insect food industries exist largely within the informal sector, particularly in parts of Africa where eating local insect species is common (Kelemu et al., 2015). Edible insects are generally gathered from the wild, sometimes via semicultivation, which is the practice of modifying natural habitats to enhance the quantity and/or quality of harvested insects (Van Itterbeeck and van Huis, 2012). Little postharvesting processing is done beyond killing and cooking the insects, rendering a product that has a limited shelf life. Overall, insects are sold to customers ranging from seasonal collectors to food markets, sometimes with intermediaries buying quantities from multiple collectors to sell directly to market vendors. Countries with these informal edible insect sectors typically do not have regulations concerning the sale of or standards for insects. Thus, concerns remain around safety, economic impact, and other issues in the informal sector. There is an increasing call for exploration of insect farming at the household and industrial levels of local species in places where these informal edible insect industries operate, particularly given the existence of food and nutritional insecurity that generally exists in the same areas and the potential for insect farming to address these problems.

Subindustry Niches: Importance of Finding One's Focused Role in the Industry and Product Selection

As the insect-based food and commodity industry grows, it will have to match the scale, pricing, and efficiencies of other commodities. The industry will have to adapt from a small niche industry into a large mainstream one. Addressing concerns of scale, pricing, and efficiency proactively will put the industry in the best position for success. Indeed, it has long been recognized that there are major differences in how large businesses and industries can and must do things versus smaller ones. For example, in Adam Smith's *Wealth of Nations'* (c.1776) (Smith, 1776), Smith describes the importance of "separation of labor" in larger more complex businesses versus smaller "trifling manufactures." There are a number of reasons why larger industries benefit from specialization and division of labor. Smith provides in one example, three reasons why division of labor increases productivity: (1) "increase of dexterity," likely related to increased skill with practice, (2) "saving of the time which is commonly lost in passing from one species of work to another," and (3) "the invention of a great number of machines which facilitate and abridge labor." Thus, even in the beginning of the industrial revolution, Smith and others realized that each company within a large or growing industry would do best to focus on a single aspect of the industry rather than to try doing all within the same company. Doing so creates greater efficiency and productivity within that company. Efficient and productive companies also make for a more efficient and productive industry overall. Likewise, the insect-based food and commodity industry, as a new emerging industry at its nascent stages, has the opportunity to organize itself with an eye toward the future, allowing it to scale faster. This means that it is ideal for emerging companies focused on insect-based products to specialize rather than try to do all aspects of the business themselves. Agriculture traditionally has been seasonal and thus less able to improve efficiency through division of labor and specialization (Smith, 1776). However, the indoor nature of insect farming certainly allows for both of these economic tools to be applied.

As this is a very new industry, particularly for North America and Europe, scale is small, efficiency is suboptimal, many processes are done manually and, likewise, costs of production are high (from farm to table). Thus, to insure a bright future for the industry, a rapid and efficient growth is critical. Costs of farmed insects need to be reduced to at least 25% to be competitive with other animal based protein commodities and production scale needs to increase many fold to supply the larger players in the food industry. Additionally, if the industry does not reach critical scale benchmarks within reasonable time frames, it risks being marginalized as a "passing fad" or something with fundamental limitations that prevent insect from becoming a mainstream commodity. Two critical factors in accelerating industry growth are: (1) efficiency and (2) synergy within the industry. A few recommendations can be made to current and new entrepreneurs and others entering the insect-based food industry, particularly small companies and farms. First select one of three segments of the industry and focus on it: (1) farming, (2) processing (providing insects as an ingredient, powder, or whole insect or other form), or (3) making consumer end products. Trying to do more than one of these would be too much for one startup and would run counter to synergy with the rest of the industry. For example, during cooperation between two companies within the insect-based food industry, it will be more difficult to find an equitable position within a collaboration or partnership if both companies are trying to do everything. A situation where companies are "vertically integrated" means that all are in competition with one another since each does the same as the other. It is better for one of the companies to focus on farming to become highly skilled and effective at mass producing their insects of choice at large scale, quickly with high quality, and at a low price. Specialization will allow companies to focus on optimizing their technological, innovative, and skill prowess, but also improve their overall business efficiency and productivity. If the farmer decides to also become an insect processor, a whole new set of skills and business challenges are added to those required for farming. The need for these new skills and capabilities will divert the attention of the company's executives and employees, making each aspect of the business (farming and processing) less likely to succeed. It also can greatly hamper attention to detail and thus hurt overall product quality (and in the worst cases, safety!).

Second, it is important that the insect-based food industry (particularly farmers and processors) reach certain scale benchmarks as quickly as possible for reasons discussed previously, and to attain economies of scale. For example, a new cricket or mealworm farm should be prepared to sell loads of 15,000–25,000 pounds of raw frozen crickets or mealworms at a time. This means renting or investing in freezer capacity and quickly scaling up capacity. A typical refrigerated trailer holds about 24 standard sized pallets. Thus, shipping frozen insects in loads of approximately 24 pallets maximizes the economies of scale related to transportation. There are many other benefits to both the farmer as well as their processor customer in larger-scale production, such as overall larger business, lower perunit (ie, perpound) pricing on processing at contract manufacturers (or the processor's own facility), maximization of profits from labor and raw material (feed), and maximizing efficiencies of facility and equipment investments. In fact, companies in general can be more innovative and efficient at larger scales based on their ability to invest in larger and higher tech equipment, which is acquired from growing in overall business scale/output.

Pricing is another aspect of the insect-based food industry which needs particular attention. Currently, the most popular edible insects in the United States, crickets and mealworms, sell at around $4 (USD) per pound frozen. Thus, farms should be prepared to sell their raw frozen insects at this price at a profit or with a reasonable "profit horizon timeline" (date at which that price will become profitable and begin to repay investments in facilities, labor, equipment, feed, production, etc.). While this price is currently marketable, it is not sustainable. Thus, for the industry to grow and thrive, it is critical that farms and other sections of the industry seek to utilize innovation, technology, mechanization, automation, and overall business efficiency and economies of scale to lower the price of raw frozen insects to under $1 (USD) per pound. This will allow their dry weight cost (dry powder being the most commonly utilized and useful formats for edible insects currently and for the foreseeable future) to be competitive with other protein sources such as whey.

Another consideration, as discussed earlier, will be for edible insect farmers and insect-based food processors and producers to meet market demands for more organic and gluten-free products while also maintaining and lowering product prices. This starts with the insect farmer and the need to utilize certified organic insect feeds which are also gluten free (ie, use materials such as corn in place of wheat). Fortunately, there are already companies offering specially formulated certified organic and gluten-free cricket feeds in North America. Some of these include: Coyote Creek Farm (an organic farm and feed mill) (Elgin, TX, USA), Nature's Grown Organics (Premier Cooperative, Westby, WI, USA), and Homestead Organics (Berwick, Ontario, Canada).

Given the recommendation that businesses in the insect-based food industry specialize, what follows are several suggestions and product ideas that will hopefully be beneficial to these businesses in their journey to find their role or niche and to grow rapidly and thrive.

Consider One's Product: Potential Products and By-Products From Mass-Produced Food or Feed Insects

It is already established that insects can be used as pollinators, biological control agents, vehicles for education and outreach programs, objects of art, pets (particularly in Europe and Asia), and feeder insects for animals (Yi et al., 2010). There is an increasing market for edible insects and insect-based food products worldwide, particularly outside the United States. Some US restaurants, particularly Latin American and Asian cuisines, are now offering insects on their menus (Gahukar, 2011; Durst et al., 2008). A number of companies, as previously mentioned in this chapter, are preparing to capitalize on this emerging market in its early stages by being the first food product developers using insects as a primary ingredient in the United States. However, promising technologies for addressing food security, sustainability, and meeting new and novel market demands take time to develop, perfect, and bring up to industrial scale. Thus, it is important for the insect-based food communities within academia, government, and industry to begin contemplating and developing the most feasible products on which this emerging industry can launch. This chapter has identified prominent examples to date. Once protocols are developed to produce various insect-based food ingredients, they can then be incorporated into numerous consumer items such as: meat substitutes, protein fortified bars, and nutritional powders, as well as numerous types of snack foods. The following are a few examples which merit examination.

Alternative "Meats"

Replacing vertebrate animal meat (muscle and other tissue) in our diet with protein-rich meat-like products derived from insects, the coup de grâce of insects replacing vertebrate livestock, means that our diets do not need to change drastically. We do not necessarily need to do without the delicious meaty products that most of us enjoy. Examples of common place foods in Western diets that could be changed to insect-based versions include tacos, hot dogs, and breaded meat nuggets (currently made of chicken and fish). With a small amount of innovation, most if not all of these can be made from insects. This can be achieved in part utilizing processes very similar to those used to make vegetarian or vegan meat substitutes from plant protein (tofu, tempeh, etc.). The internal protein of insects often appears to behave very similar to other proteins and meats when cooked, with similar textures, flavors, and odors. Even many popular vertebrate livestock-derived products such as hot dogs, sausages, ground meats, chicken nuggets, and others often contain substantial amounts of fillers and other nonmeat ingredients as well as seasoning to improve their palatability. With research and development, existing meat and meat substitute production methods can be applied to mass produced/farmed insects as the starting raw material. Many products currently made from vertebrate animals could quite easily be made from insects. Some companies in the United States are already developing strategies to make these types of insect-derived alternative meat products feasible. Further, the widespread acceptance of plant-based versions of these foods have arguably paved the way for acceptance of insect proteins in familiar food formats.

Snack Food, Protein, and Nutritional Supplements

In the short-term, protein supplements (bars, shakes, powders, cereals, etc.) for athletes and others wishing to increase their protein consumption also present a simple and ripe opportunity for incorporation of insect-derived protein. Much of the protein in these products comes from peanuts or soy. As mentioned, animal-derived protein is superior for human nutrition to plant protein, so the best protein supplements also must include some animal protein. Many of these products contain whey protein derived from milk, the production of which has a much larger environmental footprint than that of insects. These types of products have a relatively low barrier to entry, since they are very simple to produce and are typically sold to nutrition-conscience or environmentally conscience consumers, and the protein they contain is not different in look, flavor, or texture from just about any other protein source. As a result, simple strategies to produce powders and pastes from insects can constitute high-quality protein ingredients for high-end protein supplement foods and beverages. Some of the aforementioned companies are already exploring some of these types of products in the United States.

Chitin: Opportunities for New Products From Insects

For many insect-based food products, the chitinous exoskeleton must be removed. This is particularly true for products such as those already discussed to alleviate malnutrition (RUTF, etc.) and others from which high nutrient absorption efficiency is desired, as chitin can reduce absorption of some nutrients. However, the leftover chitin from industrial-scale insect-based food producing operations can itself be a desirable high-value product with a number of additional applications such as a nutraceutical for reduction of fat or cholesterol, as a drug carrier, in agricultural pest control, in water purification, as biodegradable materials and plastic-alternatives, as an antimicrobial ingredient in food and other perishable materials, to aid in wound healing, in cosmetics, and in a host of other applications (Dossey, 2013; Shockley and Dossey, 2014). Indeed, this topic has gained the attention of others who have reviewed the potential applied/industrial value of chitin in greater detail (Tharanathan and Kittur, 2003; Je and Kim, 2012). Marketing chitin as a high-value commodity can help subsidize the budding insect-based food industry, particularly in the United States and Europe. The vast majority of chitin currently used for these applications comes from unsustainable harvesting of shrimp which, like much of the ocean's other resources, are constantly being overharvested. As a consequence, chitin from farmed/mass-produced insects, particularly those fed with agricultural or food by-products, can present a much more sustainable alternative source of this chitin.

Insect Culinary Use: Restaurants, Chefs, Fairs, Outreach Events, Shows, and the Like

While a majority of insect-based food product companies are using insect powder in products, there are uses for whole insects in the culinary restaurant industry. The species marketed this way can vary widely, from the more common crickets and mealworms to grasshoppers and larger, more exotic species. The restaurant industry could be quite lucrative for the raw frozen insect market in particular, and allow beginning edible insect farmers to sell their product at a smaller scale and at a premium price until they are large enough to begin supplying frozen insects to the more substantial insect-based ingredient processing companies. Restaurants also employ insect powders in different food dishes and these are often produced by the restaurants and chefs themselves rather than being purchased from companies making cricket powder or similar products.

There are several high-end restaurants using insects in their "haute cuisine" dishes, including the Michelin-star rated Noma in Copenhagen and the Washington DC-based Oyamel (Oyamel is owned by the famous chef Jose Andrés). Some restaurants serving cuisine that traditionally include insects also serve dishes with whole insects or powders, including Mexican and various Asian restaurants. Additionally, a few chefs specialize in insect cuisine, including David George Gordon, author of the *Eat a Bug Cookbook*. Another very popular insect recipe cookbook over the years has been *Creepy Crawly Cuisine: The Gourmet Guide to Edible Insects*, written by Professor Julieta Ramos-Eludory from Mexico. Since these foundational works, particularly in the past 2–3 years, there have been many more recipe books published with insect-based food and edible insect recipes—many more than can be listed here. The same is true for restaurants serving insects around the world, and even in North America: the list is growing too rapidly and the information is too scattered for a comprehensive list in this chapter.

In addition to restaurants and cookbooks, insect-based foods are often showcased at public events such as fairs and conferences put on by nonprofit organizations, universities, students, professors, museums, and also government organizations. Both companies and nonprofit or nongovernmental organizations attend such events to provide insect-based food samples and information as an outreach to people who may not be familiar with the practice. Examples include company tables and presentations at food industry events like Western Natural Foods Expo and the Institute of Food Technologist's Annual Meeting and Food Expo, as well as public events such as the USA Science and Engineering Festival (insect section organized by the Invertebrate Studies Institute and All Things Bugs, LLC) and Slow Food USA's Slow Meat Fair. There are also regional and local events focusing on edible insects such as Bugfest in Raleigh, North Carolina, the Grasshopper Festival in Washington state, or the Bug Fest in Austin, Texas. See Chapter 1 for a more comprehensive list of insect-related fairs and outreach events promoting insects as a food source.

Global Malnutrition and How Insects Can Help

Malnutrition affects 178–195 million children worldwide (Black et al., 2008). The World Health Organization (WHO) estimates that malnutrition accounts for 54% of child mortality worldwide (Duggan et al., 2008). Malnutrition results in wasting, reduced ability to fight infection, and impaired cognitive and other developmental disorders, which can become permanent (Victora et al., 2008). About 60 million children suffer from moderate acute, and 13 million from severe acute malnutrition (Collins et al., 2006). Malnutrition is a contributing factor in over 50% of the 10–11 million children under the age of 5 years who die from preventable causes each year (Rice et al., 2000; Pelletier and Frongillo, 2003; Caulfield et al. 2004; Black et al. 2008). Fatality rates from severe acute malnutrition have averaged 20–30% since the 1950s (Schofield and Ashworth, 1996). It is estimated that 9% of sub-Saharan African and 15% of Asian children suffer from moderate acute malnutrition (Collins et al., 2006).

As mentioned previously, the UN has established 8 Millennium Development Goals (MDG). The first MDG is to "eradicate extreme poverty and hunger" and the fourth is to "reduce child mortality rates." Specifically, the goals involve reducing childhood mortality by two-thirds and worldwide hunger by half. Indeed, the topic of the potential for insects to contribute to sustainable human food security has taken the notice of multiple organizations. For example, as mentioned earlier in this chapter, the UN FAO has taken initiative and proposed a program of feeding people with alternative food sources, including insects (Gahukar, 2011). Two major meetings on the feasibility and benefits of insects as a food source have resulted from this initiative: (1) a Feb. 2008 workshop in Thailand which produced an important book in the field called *Forest Insects as Food: Humans Bite Back* (Durst et al., 2008), and (2) a technical consultation held Jan. 2012 at UN FAO headquarters in Rome, Italy (Vantomme et al., 2012). At the 2012 meeting, a global summit on insects and food security for 2014 was also proposed. Those in attendance at the 2012 meeting appeared to come away quite optimistic that insect-based food products are indeed an important part of our future. It was also generally recognized that government and industry backing is necessary to support the widespread implementation of insect-based diets.

Malnutrition in developing countries is to a significant degree a problem of protein and calorie deficiency. While content of particular nutrients and nutrient categories varies widely among the hundreds of species analyzed, insects are in general a very good source of protein, fat and other nutrients valuable to the human diet (see also the earlier section on the nutrient content of insects). Additionally, the problem of protein deficiency is one that affects low-income people worldwide. The consumption of some insects can provide both high-protein and caloric intake. For example, the nutrient content of African palm weevil larvae (*Rhynchophorus phoenicis*) was described in a Nigerian study and they were found to be very high in both carbohydrate and protein (Okaraonye and Ikewuchi, 2008). Because insects are a robust source of animal protein, it is important to improve production and preservation techniques, to market them, and make them available to all populations (Melo et al., 2011).

Given the fact that insects compare quite favorably in their nutrient content to vertebrate livestock, increasing their consumption in places where vertebrate livestock is unavailable, or their growth is not feasible, can greatly improve the health and nutritional status of a great number of people worldwide. Access to edible insects can be achieved by different methods including wild harvesting, efficient/sustainable farming, or even household rearing/gardening. Insects are much less resource-intensive and much more resistant to drought and disease (typically) than cattle and most other vertebrate livestock commonly utilized by humans (see the section that follows on insects as a sustainable source of human food). Research and education programs on insect farming/preparation and safe-selection/preparation of wild harvested insects are two activities which can greatly improve nutrition and food security worldwide. However, the primary methods of preparing, cooking, or processing insects in many areas of the world, particularly in places where poverty and malnutrition are highest, involve roasting, searing, or frying whole insects. Unfortunately, these methods of processing can remove fat (an important source of calories) and other nutrients, while destroying others or making them indigestible. Education programs on methods of cooking, processing, or preparing insects (either wild harvested or farmed) for human consumption in places where malnutrition rates are high might dramatically improve the nutritional status and overall health of the local people.

Relief foods such as RUTF products present another opportunity to begin utilizing insects as a more sustainable source of food of animal origin. One reason for this is that it is often highly desirable for organizations or companies to grow these products in the same country where they will be consumed and from ingredients produced locally. Whereas milk powder is almost always imported and is rarely locally available in places where malnutrition is prevalent, many types of insects can be locally farmed or harvested from the wild. There is already at least one US company working on just this sort of product (Table 5.3) (Crabbe, 2012). That firm was awarded a grant from the Bill & Melinda Gates Foundation in 2012 to develop an insect-based RUTF (Crabbe, 2012). If this sort of endeavor is successful, it could substantially improve the prospects of the insect-based food industry by drastically increasing demand for insects reared as human food ingredients for the insect-based RUTF.

Another potentially impactful way in which insects can improve food security is strategic utilization of pest species and/or species of high seasonal or regular abundance. This can not only expand the use of insects (currently an extremely underutilized resource), but also be a safer, cleaner, and more sustainable method of pest control, thus turning the proverbial "lemons into lemonade." For example, insects such as Lepidoptera larvae (caterpillars), locusts, grasshoppers (Orthoptera), and others tend to be of extremely high abundance in places where effective pesticides are not available or feasible to use. Cerritos and Cano-Santana (2008) found that manual harvest of grasshopper pests reduces the density of grasshoppers and suggests that implementation of this mechanical method of control may be an effective substitute for chemical control (Cerritos and Cano-Santana, 2008). Mechanical control provides general advantages that include (1) a second profitable product; (2) savings realized from reduced cost of insecticides; and (3) reduced risk of soil and water contamination by insecticides. This sort of dual-function pest control might be highly amenable to farming in developing nations as well as organic farms in the United States. Even in more modern agricultural settings, if insect pests can be allowed to consume inedible parts of crop plants, or crops which are otherwise deemed no longer useful as a harvest, they can potentially be an additional crop with value-added to the already ongoing farming practice. Other species, while not actual pests, can also be highly abundant at certain times of the year or in periodic outbreaks. Some examples of this include lake flies in Africa

(Ayieko and Oriaro, 2008; Ayieko et al., 2010a,b). Some walking stick insects (Phasmatodea), defoliate sections of forest in Australia from time to time (Graham, 1937; Campbell, 1960, 1961; Hennemann and Conle, 2008). Thus, more efficient and increased use of abundant insect resources result in the reduction of pesticides, as well as creating new economic opportunities for indigenous people (DeFoliart, 1992).

Some specific studies have already identified cases whereby pest insects can have a substantial positive health impact in areas where malnutrition is high, as well as economic benefits. For example, a grasshopper in Mexico known as Chapulines (*Sphenarium purpurascens Charpentier*) is controlled by harvesting them as a food item in addition to the crops they attack. This activity provides an annual profit of about $3000 per family and a total of 100 metric tons of edible insect mass (Premalatha et al., 2011). Studies on the nutritional value of edible insects from Mexico and Nigeria revealed that some of the insects, which are considered pests, also have high nutritional qualities (Ramos-Elorduy, 1997; Banjo et al., 2006). Christensen et al. (2006) identified edible insects as a mineral source in Kenya. Deficiencies of important minerals such as iron, zinc, and calcium are often widespread among people in developing countries. These deficiencies are caused by the low availability of these nutrients in staple foods such as cereals and legumes due to the considerable amount of phytic acid and other antinutrients present in these foods and to the lack of animal foods with higher content of these nutrients (Christensen et al., 2006). Iron and zinc deficiency is widespread in developing countries, especially in children and women of reproductive age. Iron deficiency leads to anemia, reduced physical activity, and increased maternal morbidity and mortality. As a result, use of insects as food ingredients and edible insects could prove to be a valuable measure to combat iron and zinc deficiency in developing countries. Some caterpillars have been found to be a rich source of iron, copper, zinc, thiamin (vitamin B1), and riboflavin (B2); 100 g of cooked insect provided >100% of the daily requirement of each of these minerals and vitamins. Iron deficiency is a major problem in women's diets in the developing world, particularly among pregnant women, and especially in Africa (DeFoliart, 1992).

Novelty Products

Among insect-containing foods, particularly in the United States and Europe, there are still various exotic or novelty items which contain only a small amount of whole insects. These types of products can be sold at relatively high prices relative to ingredient costs but are not generally considered to be a meal item (more of a candy or snack). Thus, they are a way for the emerging insect-derived food product industry to generate revenue while developing cheaper and more mainstream products and allowing cultural acceptance for insect-based foods to increase. On the other hand, as long as novelty or exotic food items containing whole visible insects remain available on the market, both cultural acceptability and overall market feasibility on a larger scale of insects as a primary food ingredient may continue to be severely marginalized. Thus, we propose that an increased focus on developing more standard processed insect-containing foods similar to those which people are already familiar with and enjoy (described throughout this chapter) is essential to truly realizing all of the benefits of humans eating insects.

SUPPLY CHAIN NEEDS: FEED, FARMS, INSECTS, TRANSPORTATION, PROCESSING, AND MANUFACTURING

Feed

Insect farmers producing for the pet food and fishing bait industry often use grain/corn-based commercially formulated feeds that are similar in nutrient content to chicken feed. Some of these companies are raising insects on certified organic feed, but there is no current official certification for organic insects. Some farms raising insects specifically for human consumption advertise that crickets are fed diets composed of human-grade food, including milled grains and in some cases, postconsumer food waste. Similarly, there are not yet standards, legal definitions, or certifications outlining what constitutes food-grade insects and, as such, standards are primarily being defined by the companies themselves, based on their marketing interests. However, it is not clear if this is necessary from a legal point of view or in general to maintain safety and quality of insects farmed for human consumption. It is possible, and even likely, that the indoor farming and feeding standards utilized by older farms raising insects for as live pet/reptile feed or fishing bait are perfectly fine for insects intended for human consumption. See Chapter 7 for a deeper discussion on regulatory considerations for insects as food.

As discussed earlier in this chapter, given the demand for gluten-free food products and the suitability of insect powders or meal to provide a high-protein substitution for glutenous flours, gluten-free feed will be important for insect farmers to utilize whenever practical to do so. Similarly, high demand for organically raised protein provides a corresponding demand for edible insects that are raised on organic feed. There are not yet certified organic edible insect farms, although some companies make claims that the edible insects they sell were fed a certified organic diet.

The burgeoning industry of insect farming to produce livestock feed, particularly for chicken, pig, and farmed fish feed, can most likely rely upon much lower cost biomaterials such as food waste, lower quality agricultural by-products, animal parts (fish offal, etc.) or animal waste as its feed, particularly for black soldier fly and other fly farms. Many insects, and flies in particular, are able to convert waste into nutritious feed. There is some discussion on the ability of fly larvae reared in mass to reduce the overall levels of microbes, including some pathogens such as $E.\ coli$, in the waste they are given as feed, thus also improving sanitation when fly farms are combined with other livestock farms or meat processing plants.

A largely unexplored environmentally beneficial source of insect feed could be the use of pre- and postconsumer organic waste, algae or nonfood crop biomass such as high growth grasses and other plants. While many insect growers at the home level regularly use vegetable, fruit, and/or grain food waste that could otherwise be composted, no insect farm has yet devised a system to obtain the quality and quantity of food waste streams that would allow high nutrition and low food-borne illness risk in feeding edible insect species at the industrial scale level. Once edible insect farms begin to safely utilize these sources of insect feed ingredients on a large scale, the costs of edible insects will likely go down and this will further improve their already low environmental impact.

Farms and Farmed Species

In general, simply put, we need many more and much larger scale edible insect farms. A pressing issue for industrial-scale insect farms is the need for innovation that allows farmers to fully exploit the expected financial benefits of raising animals that coverts feed to food up to 10 times more efficiently than cattle, as well as require much less space, both for raising them and for growing feed for them. As previously mentioned, current insect farming practices are primarily based on manual labor, which in developed countries is costly and a primary driver for continued high prices of edible insects in food. Additional research is greatly needed to improve practices that will maintain good health and quality of insects while also increasing automation and potentially optimizing and or reducing life cycles.

Particularly in North America and Europe, regulations that address land use, particularly for farms in urban areas, and other agriculture rules, are nearly completely lacking for all aspects of insect cultivation, as well as regulations about insects as food. Additionally, in these areas of the world there are no subsidies or tax or other governmental financial support systems in place for the insect industry in general, edible or otherwise. If such laws and incentives are developed that favor insect farming, including tax structures to support this food production, the industry will be further boosted towards long-term viability toward insects being the lowest cost animal-based protein commodity available on the market.

In addition to the need for more insect farms in general, we need much larger insect farms as well as a greater number and diversity of insects being produced by those farms worldwide. While the insect production industry is currently rising rapidly based almost entirely on crickets and mealworms, some limitations of the industry are exacerbated due to the lack of scale and insect diversity representation amid what is commercially available from indoor, industrial insect farms. Economies of scale are currently working against this industry in ways that do not affect larger, less efficient/sustainable and more environmentally damaging high-protein commodities such as beef and dairy. Insect-based food production, from the farm to the table (particularly in insect processing and insect-based food production) suffers from high perunit costs compared with other products since the entire industry is currently made up of small farms and small companies processing insects or using the processed insects in their end products. Part of the solution is simply the establishment of more and much larger insect farms offering a larger diversity of species beyond crickets, mealworms, and waxworms. More diverse offerings can provide farms with additional revenue streams of possibly higher-end products to supply a wider variety of tastes. For example, many restaurants may wish to offer grasshoppers, giant water bugs, or other less common dishes along with their cricket- and mealworm-based dishes. With an increased (and thus lower cost per unit) and more reliably available supply, the current startups in this industry will be able to grow more effectively and larger mainstream food companies will be more attracted to utilizing this emerging commodity in their own products.

Transportation, Storage, and Distribution

In general, the insect-based food and edible insect industry will likely initially be highly dependent on existing infrastructure available in the mainstream food industry. Fortunately, a very good infrastructure and systems already exist for the transpiration, storage, and distribution of raw and processed foods, so this is a resource that the insect-based food industry may be able to utilize indefinitely with minor modifications as it grows. An important question for an insect farmer or processor to ask might be: When does it make sense to ship frozen versus live, wet versus dried, raw versus processed? The coup de grâs for most efficient transport, shipping, and storage of insect-based food ingredients, as with most other food ingredient commodities, is dry and in powder form. Dry products overall benefit from light weight, long shelf life, and overall simplicity and low cost of their storage needs. Powders provide the least wasted airspace of any product format as the material is tightly packed most efficiently in powder form in bags, in boxes, on pallets—and indeed even for end-consumer use by taking up minimal space in food manufacturer warehouses, kitchen cupboards, and restaurant pantries. Whole dried insect products also benefit from shelf life and low tech storage given the low moisture level, but with whole insects the wasted airspace for storage and transport is much greater than for insect powders or meals. Additionally, whole dried insects are very fragile and may not be desirable by the end consumer or restaurant chef if greatly damaged or crushed in transport. To date there seems to be a consensus among modern insect farms that live is the most expensive because of the space it takes, size of boxes, costs of boxes, wasted space not filled with insects, and potential for the insects to die and then decompose in transit (ie, low shelf life of the product). Frozen shipping for raw insects is generally the next best/cheapest option in industrialized countries where frozen shipping is available using refrigerated trucks or trains. Transoceanic shipping of frozen products, on the other hand, can pose both shelf-life and cost challenges. The insect-based food and edible insect industry, or even individual companies within it, may also at some point consider stockpiling their products at centralized freezer warehouses for raw, or noncold storage facilities for dry or shelf-stable products (similar to how other food commodities are handled, such as at grain elevators, etc.). A sort of coop model may even be worth considering for the insect-based food industry. Centralized distribution can benefit farmers, processors, distributors, and buyers and can generally help the industry scale. Another option worth considering for farms and/or centralized distribution facilities might be locating farms near rail lines to maximize utilizations of trains, since trains are extremely reliable and efficient (low energy use, etc.) forms of bulk transportation.

Processing and Manufacturing Infrastructure

Insect-based food producers are currently using two primary methods to manufacture products:

1. Work with food production companies to find the best processes and thus outsource manufacturing once the desired product outcome is determined or;
2. Make the food products themselves, using commercial grade equipment.

Each of these prospective methods has advantages. In the first outsourcing scenario, companies are able to rely upon industry knowledge to produce high-quality products. In the second scenario, while the time and learning uptake could be higher to achieve a high-quality product to sell, the financial costs could be initially lower which is often advantageous for startup businesses with little or no capital. Indeed, several startups in the existing insect-based food industry are already utilizing small commercial kitchen facilities to make their own products by hand.

As with transportation and storage, the insect-based food industry, as a small nascent industry in the shadow of the larger, older, and more advanced mainstream food industry, will heavily depend on existing food processing equipment, facilities, and contract manufactures for the foreseeable future, at least for insect-based food businesses at significant scale. The good news is that, at least in the industrialized world, an excellent infrastructure of contract manufactures, equipment, and facilities already exist for those fortunate enough to have the financial resources required to access them. As small food companies grow, the utilization of contract manufacturers, copackers, and similar fee-based processing and packaging businesses becomes necessary. As a company grows, it will become too large and a commercial kitchen may not be able to supply the product demand, yet the company may not be large or well financed enough to build its own processing plant which can cost many millions of dollars (USD). Many such contract manufactures exist; however, many of these enterprises do not yet accept insects into their facility for a variety of reasons. Part of these are based on firm restrictions such as kosher status, allergens, or that certain facilities only process a specific and narrow range of products (ie, some only process plant material, some only process starches and no proteins, etc.), while others are simply concerned about having insects in their facility based on what their other clients might think about the unknown risks. As the insect-based food industry grows, gains more attention, is taken more seriously, and concerns about safety and consumer aversion are alleviated through research and communication, it will be ever easier for insect-based food companies to find and work with contract manufacturers and copackers.

INTRIGUING THE LARGER FOOD INDUSTRY: USES OF INSECTS AS INDUSTRIAL FOOD INGREDIENTS

While insect-based foods are beginning as a niche industry, there is potential for wider use in mainstream food products in the Western world and beyond. Given edible insect advantages as a sustainable resource and efficiencies in production, insect farming for food and feed shows great promise as a climate-smart agricultural practice. They can also be farmed very cleanly without use of steroids, hormones, antibiotics, or other products or methods which a growing segment of consumers (and thus food companies) find acceptable in their food supply. The nutritional profiles of edible insects can be added to these advantages plus the ability of industrial farming techniques to produce large quantities of sustainable protein to meet growing demand for both food and meat products. Additionally, it is yet unknown what this entirely unexplored yet vast resource has to offer as far as functionality and overall utility as an industrial food ingredient. While the potential of insects as a food source has not yet been met in terms of cost to produce and resulting insect-based food prices, many parties are researching innovations that will allow producers and farmers to test the premises of this ancient and yet new food commodity. Thus, it is highly advisable that the insect-based food industry engage regularly with larger food companies to develop a dialogue about what it will take for these companies to buy into the concept by actually buying insect-based food ingredients for their own products.

CONCLUSIONS

In summary, this an exciting and historic time for the food and agricultural industries. While we face a great many challenges with human population growth, climate change, and environmental destruction, these also offer opportunities to explore new solutions for an overall better world. We believe strongly that insects have a substantial role to play in this endeavor on many levels. With the recent emergence of the insect-based food industry, particularly in industrialized nations, it is an exciting time to be part of this revolution to provide insects as a *transformative* resource for food, agriculture, and other industries. In cultures that eat insects regularly, they are considered beneficial and are viewed as nutritious, medicinal, environmental, and as a sustainable and secure food item. Once insects become more widely accepted as a respectable food item in industrialized countries, the implications will have a profound and positive impact on businesses, industry, governments, and research. Replacing vertebrate livestock-derived foods and food ingredients with those derived from insects will also substantially improve the health of the earth's natural environment. Currently, insects are still not widely available in Western societies and industrialized nations and come at a high cost as compared to other animal-based food ingredients. However, as the production of insects as human food and animal feed increases worldwide, particularly in the United States, many aspects related to feasibility of insect-based foods, such as cost, safety, efficiency of insect mass production, and availability will improve. The resulting increase in demand will synergistically drive the status of insects as a standard mainstream food ingredient. A multidisciplinary approach is needed to realize the substantial potential benefits of human consumption of insects on a global scale. Engineers are needed to develop appropriate rearing systems for different environments and insects. Food scientists are needed to study the nature of insect foods and nutritional content, functionality, and organoleptic properties insect-based ingredients impart on various foodstuffs, the principles underlying insect food processing, and the improvement of insect-based foods. Nutritionists are needed to advise consumers on insect foods and nutritional impacts of insect foods on human health. Marketing, promotion, and advertising specialists are needed to promote the numerous benefits of insect-based foods and ingredients.

In order to realize their potential as a major suite/family of commodities, there are several constraints which must be overcome. These include: production cost and efficiency, commercialization, technology, regulation, and social change of food habits and food choices. The need for development of multiple-product food-insect systems is pressing (Gahukar, 2011). Efforts addressing all of these limitations and opportunities are already underway in industrialized areas of the world, particularly in North America, Europe, and Asia. Indeed, the call for using insects to improve human food security has become more prominent in the past couple of years as well (Crabbe, 2012; Dossey, 2013; Dzamba, 2010; Durst et al., 2008; Gahukar, 2011; Katayama et al., 2008; Nonaka, 2009; Premalatha et al., 2011; Ramos-Elorduy, 2009; Shockley and Dossey, 2014; Smith, 2012; Srivastava et al., 2009; van Huis, 2013; Vantomme et al., 2012; Vogel, 2010; Yen, 2009).

In order to realize the potential benefits that insects can provide to our food security as human food and as animal feed, we call for the following: (1) greater support and attention from government, agricultural, and regulatory agencies for funding and research on insect production and use as a human food ingredient, (2) support from industry to provide the means to move insect-based foods and other products from the laboratory to the market, and (3) the establishment of a formal international society (such as the International Society for Food and Feed Insects), industry association, and a journal for researchers, academics, industry partners, and other practitioners in the field of food and feed insect production. These will help tremendously in moving this emerging field forward. A formal international society would promote and present how insect-based food research could be beneficial to science, society, and industry and, thus, set the stage for what might be one of the most substantial revolutions in modern agriculture and food production: the human utilization of insects for food.

To close, we offer the following inspirational points of advice for all those interested in further exploration of the insect-based food industry:

1. Get *creative*! Imagine all the possibilities! Adopt this mantra: Anything can be made out of insects.
2. Remember that there are over a million insect species in the world, and food and feed are by far the proverbial tip of the iceberg.

3. Antibiotics, enzymes, industrial materials (chitin), and nutraceuticals are among many applications for mass-produced insect material—keep an open mind as to where your journey might lead as far as the most successful products your company can offer and from which species it makes the most sense. There is a lot of potential here beyond food and feed—it takes creativity, selecting the right insect, and a little R&D.
4. I (Dr Aaron T. Dossey) take a lot of inspiration for our industry from George Washington Carver and Percy Julian, seeing insects as the development of a whole new class and suite of commodities with wide-ranging uses. History can be a valuable and inspiring guide.
5. Remember in your marketing efforts that farm-raised insects are just another protein source and have a lot of benefits to consumers and the natural environment. They can be as clean; healthy; grown without antibiotics, steroids, hormones, or pesticides; and as safe as any animal, depending on how they are produced. We have never found *E. coli*, *Staphylococcus aureus*, *Listeria*, or *Salmonella* in any raw farmed cricket or mealworm, so they seem to be quite clean compared to other animals. Also, you should take caution with any new product and label it properly as an allergen, as necessary.
6. Both insect-based food startups and consumers may wish to start conservatively at first. Start with a bar, snack, baked good, or protein shake made with cricket or mealworm powder as your first experience. If people see that foods and beverages they are used to can contain insects and still taste as good or the same as they are used to, they'll see right away that there's nothing "icky" about foods and beverages that contain insects as an ingredient. The beverages and things like pastas or even some baked goods are excellent examples for those looking to really explore because these products are very sensitive to texture and flavor changes —so people can try these and really tell, "Hey, I can't detect the insect at all!"
7. Buy the pure cricket powder to use or cook with from companies specializing in insect processing, as they will be able to deliver the highest quality and most reliable supply. This is also the best way to help the industry grow the fastest, achieving critical levels and economies of scale for maximum benefit to the world.
8. Overall, have fun and remember that you are a pioneer working in a historic time for a transformative new industry for food and agriculture. You are utilizing the largest and most diverse group of organisms on the planet: Class Insecta.

REFERENCES

Amadi, E.N., Ogbalu, O.K., Barimalaa, I.S., Pius, M., 2005. Microbiology and nutritional composition of an edible larva (*Bunaea alcinoe* Stoll) of the Niger Delta. J. Food Safety 25, 193–197.

A Strategic Look at Protein, 2014. New hope 360 from supply to shelf, [blog] October 24, 2013. Available from: http://newhope360.com/print/news/strategic-look-protein

Ansc.purdue.edu, 2014. Purdue Food Animal Education Network. [online]. Available from: http://www.ansc.purdue.edu/faen/dairy%20facts.html.

Ayieko, M.A., Ndong'a, M.F., Tamale, A., 2010a. Climate change and the abundance of edible insects in the Lake Victoria region. J. Cell Anim. Biol. 4, 112–118.

Ayieko, M.A., Oriaro, V., 2008. Consumption, indigenous knowledge and cultural values of the lakefly species within the Lake Victoria region. Afr. J. Environ. Sci. Technol. 2, 282–286.

Ayieko, M.A., Oriaro, V., Nyambuga, I.A., 2010b. Processed products of termites and lake flies: improving entomophagy for food security within the Lake Victoria region. Afr. J. Food Agricult. Nutr. Dev. 10 (2), 2085–2098.

Banjo, A.D., Lawal, O.A., Songonuga, E.A., 2006. The nutritional value of fourteen species of edible insects in southwestern Nigeria. Afr. J. Biotechnol. 5, 298–301.

Beare-Rogers, J., Dieffenbacher, A., Holm, J.V., 2001. Lexicon of lipid nutrition. Pure Appl. Chem. 73, 685–744.

Black, R.E., Allen, L.H., Bhutta, Z.A., Caulfield, L.E., de Onis, M., Ezzati, M., et al., 2008. Maternal and child undernutrition: global and regional exposures and health consequences. Lancet 371, 243–260.

Bondari, K., Sheppard, D.C., 1981. Soldier fly larvae as feed in commercial fish production. Aquaculture 24, 103–109.

Calderone, A., 2015. Shailene Woodley Says Eating Ants and Pigs' Feet Tastes 'Great'. People Magazine, March 27, 2015. Available from: http://greatideas.people.com/2015/03/27/shailene-woodley-eats-insects-bugs-pigs-feet/

Campbell, K.G., 1960. Preliminary studies in population estimation of two species of stick insects (Phasmatidae:Phasmatodes) occurring in plague numbers in highland forest areas of south-eastern Australia. Proc. Linn. Soc. 85, 121–141.

Campbell, K.G., 1961. The effects of forest fires on three species of stick insects (Phasmatidae Phasmatodea) occurring in plagues in forest areas of south-eastern Australia. Proc. Linn. Soc. 85, 112–121.

Capinera, J.L., 2004. Encyclopedia of Entomology. Kluwer Academic, Dordrecht, London.

Caulfield, L.E., de Onis, M., Blossner, M., Black, R.E., 2004. Undernutrition as an underlying cause of child deaths associated with diarrhea, pneumonia, malaria, and measles. Am. J. Clin. Nutr. 80, 193–198.

Cdc.Gov. 2014. Nutrition for Everyone: Basics: Protein | DNPAO | CDC [online]. Available from: http://www.cdc.gov/nutrition/everyone/basics/protein.html.

Cerritos, R., Cano-Santana, Z., 2008. Harvesting grasshoppers Sphenarium purpurascens in Mexico for human consumption: a comparison with insecticidal control for managing pest outbreaks. Crop Protect. 27, 473–480.

Christensen, D.L., Orech, F.O., Mungai, M.N., Larsen, T., Friis, H., Aagaard-Hansen, J., 2006. Entomophagy among the Luo of Kenya: a potential mineral source? Int. J. Food Sci. Nutr. 57, 198–203.

Collins, S., Dent, N., Binns, P., Bahwere, P., Sadler, K., Hallam, A., 2006. Management of severe acute malnutrition in children. Lancet 368, 1992–2000.

Crabbe, N., 2012. Local expert gets funding to develop insect-based food for starving children. Gainesville Sun, pp. 1B–6A.

Day, AC., 2015. Insectpreneur Series: Interview with Dr Aaron T. Dossey of All Things Bugs, LLC. Available from: http://4ento.com/2015/08/05/insectpreneur-series-interview-aaron-dossey-all-things-bugs/

DeFoliart, G., 1992. Insects as human food. Crop Prot. 11, 395–399.

DeFoliart, G.R., 1995. Edible insects as minilivestock. Biodivers. Conserv. 4, 306–321.

Demick, B., 2013. Cockroach farms multiplying in China. Los Angeles Times. Available from: http://www.latimes.com/world/la-fg-c1-china-cockroach-20131015-dto-htmlstory.html

Dolson L., 2014. Protein info: how much protein do you need? [Online]. Available from: http://lowcarbdiets.about.com/od/nutrition/a/protein.htm

Dossey, A.T., 2010. Insects and their chemical weaponry: new potential for drug discovery. Nat. Prod. Rep. 27, 1737–1757.

Dossey, A.T., 2013. Why insects should be in your diet. The Scientist 27, 22–23.

Dossey, A.T., 2015. Insect products and methods of manufacture and use thereof. United States Patent Office Application Number: 14/537960, Publication Date: May 14, 2015, Filing Date: November 11, 2014, Priority Date: November 11, 2013.

Dowell, R., Worley, J., Gomes, P.V., Hendrichs, J., Robinson, A.S., 2005. Sterile insect supply, emergence, and release. In: Dyck, V.A., Hendrichs, J., Robinson, A.S. (Eds.), Sterile Insect Technique: Principles and Practice in Area-Wide Integrated Pest Management. Springer, Dordrecht, Netherlands, pp. 297–324.

Duggan, C., Watkins, J.B., Walker, W.A., 2008. Nutrition in pediatrics: basic science, clinical application.

Durst, P.B., DVJ, Leslie, R.N., Shono, K. (Eds), 2008. Forest Insects as Food: Humans Bite Back proceedings of a workshop on Asia-Pacific resources and their potential for development, pp. 1–4. Bangkok, FAO Regional Office for Asia and the Pacific.

Durst, P.B., Leslie, R.N., Shono, K., 2008. Forest Insects as Food: Humans Bite Back. FAO, Regional Office for Asia and the Pacific, Chiang Mai.

Dzamba J., 2010. Third Millennium Farming: Is it time for another farming revolution? Architecture, Landscape and Design. Toronto, CA. Available from: http://www.thirdmillenniumfarming.com/

Earth-Policy.Org. 2014. Bookstore—Plan B 2.0: Rescuing a planet under stress and a civilization in trouble | Chapter 9. Feeding Seven Billion Well: Producing Protein More Efficiently| EPI [online]. Available from: http://www.earth-policy.org/books/pb2/pb2ch9_ss4

FAO., 2015. Edible Insects Stakeholder Directory. Available from: www.fao.org/forestry/edibleinsects/stakeholder-directory/en/

Finke, M., 2013. Complete nutrient content of four species of feeder insects. Zoo Biol. 32, 27–36.

Gahukar, R.T., 2011. Entomophagy and human food security. Int. J. Trop. Insect Sci. 31, 129–144.

Giaccone, V., 2005. Hygiene and health features of minilivestock. In: Paoletti, M.G. (Ed.), Ecological Implications of Minilivestock: Potential of Insects, Rodents, Frogs and Snails. Science Publishers, Enfield, NH, pp. 579–598.

Gorham, J.R., 1979. Significance for human health of insects in food. Annu. Rev. Entomol. 24, 209–224.

Graham, S.A., 1937. The Walking Stick as a Forest Defoliator. University of Michigan School of Forestry and Conservation, Lansing, MI.

Hennemann, F.H., Conle, O.V., 2008. Revision of oriental phasmatodea: the tribe Pharnaciini Gunther, 1953, including the description of the world's longest insect, and a survey of the family Phasmatidae Gray, 1835 with keys to the subfamilies and tribes (Phasmatodea: "Anareolatae": Phasmatidae). Zootaxa 1906, 1–311.

Hoffman, A., 2014. Inside the Edible Insect Industrial Complex. Available from: http://www.fastcompany.com/3037716/inside-the-edible-insect-industrial-complex

Huang, C.B., Alimova, Y.M., Ebersole, J.L., 2011. Short- and medium-chain fatty acids exhibit antimicrobial activity for oral microorganisms. Arch. Oral Biol. 56, 650–654.

Je, J.Y., Kim, S.K., 2012. Chitosan as potential marine nutraceutical. Adv. Food Nutr. Res. 65, 121–135.

Jongema, Y., 2015. List of edible insects of the world. Available from: http://www.wageningenur.nl/en/Expertise-Services/Chair-groups/Plant-Sciences/Laboratory-of-Entomology/Edible-insects/Worldwide-species-list.htm.

Joseph, A., 2015. Angelina Jolie, Salma Hayek's weird eating habits, Times of India, Apr 30, 2015.Available from: http://timesofindia.indiatimes.com/entertainment/english/hollywood/news/Angelina-Jolie-Salma-Hayeks-weird-eating-habits/articleshow/47094839.cms

Katayama, N., Ishikawa, Y., Takaoki, M., Yamashita, M., Nakayama, S., Kiguchi, K., Kok, R., Wada, H., Mitsuhashi, J., 2008. Entomophagy: a key to space agriculture. Adv. Space Res. 41 (5), 701–705.

Kelemu, S., Niassy, S., Torto, K., Fiaboe, K., Affognon, H., Tonnang, N., et al., 2015. African edible insects for food and feed: inventory, diversity, commonalities and contribution to food security. J. Insect. Food Feed 1 (2), 103–119.

Klunder, H.C., Wolkers-Rooijackers, J., Korpela, J.M., Nout, M.J.R., 2012. Microbiological aspects of processing and storage of edible insects. Food Control 26 (2), 628–631.

Li, Q., Zheng, L., Cai, H., Garza, E., Yu, Z., Zhou, S., 2011. From organic waste to biodiesel: black soldier fly, *Hermetia illucens*, makes it feasible. Fuel 90, 1545–1548.

Melo, V., Garcia, M., Sandoval, H., Jimenez, H.D., Calvo, C., 2011. Quality proteins from edible indigenous insect food of Latin America and Asia. Emir. J. Food Agric. 23, 283–289.

Mermelstein, N., 2015. Crickets, Mealworms, and Locusts, Oh My! Food Technology Magazine, p. 69–73. Available from: http://www.ift.org/food-technology/past-issues/2015/october.aspx

Montanabeefcouncil.org, 2014. Montana Beef Council: Beef Trivia [online]. Available from: http://www.montanabeefcouncil.org/beeftrivia.aspx

Nakagaki, B.J., Defoliart, G.R., 1991. Comparison of diets for mass-rearing Acheta domesticus (Orthoptera: Gryllidae) as a novelty food, and comparison of food conversion efficiency with values reported for livestock. J. Econ. Entomol. 84 (3), 891–896.

News.cornell.edu, 2014. U.S. could feed 800 million people with grain that livestock eat, Cornell ecologist advises animal scientists. Cornell Chronicle [online]. Available from: http://www.news.cornell.edu/stories/1997/08/us-could-feed-800-million-people-grain-livestock-eat

Nonaka, K., 2009. Feasting on insects. Entomolog. Res. 39, 304–312.

Okaraonye, C.C., Ikewuchi, J.C., 2008. Rhynchophorus phoenicis (F) larva meal: nutritional value and health implications. J. Biol. Sci. 8, 1221–1225.

Oonincx, D.G., van Itterbeeck, J., Heetkamp, M.J., Van Den Brand, H., van Loon, J.J., van Huis, A., 2010. An exploration on greenhouse gas and ammonia production by insect species suitable for animal or human consumption. PLoS One 5, e14445.

Pelletier, D.L., Frongillo, E.A., 2003. Changes in child survival are strongly associated with changes in malnutrition in developing countries. J. Nutr., 107–119.

Pimentel, D., Pimentel, M., 2003. Sustainability of meat-based and plant-based diets and the environment. Am. J. Clin. Nutr. 78 (3), 660S–663S, Available from: http://ajcn.nutrition.org/content/78/3/660S.full.

Popa, R., Green, T.R., 2012. Using black soldier fly larvae for processing organic leachates. J. Econ. Entomol. 105, 374–378. http://w2.vatican.va/content/francesco/en/encyclicals/documents/papa-francesco_20150524_enciclica-laudato-si.html.

Pope Francis. 2015. Laudato Si': On care for our common home. Encyclical Letter.

Premalatha, M., Abbasi, T., Abbasi, T., Abbasi, S.A., 2011. Energy-efficient food production to reduce global warming and ecodegradation: The use of edible insects. Renewable Sust. Energy Rev. 15, 4357–4360.

Ramos-Elorduy, J., 1997. Insects: a sustainable source of food? Ecol. Food Nutr. 36, 247–276.

Ramos-Elorduy, J., 2009. Anthropo-entomophagy: cultures, evolution and sustainability. Entomolog. Res. 39, 271–288.

Rice, A.L., Sacco, L., Hyder, A., Black, R.E., 2000. Malnutrition as an underlying cause of childhood deaths associated with infectious diseases in developing countries. Bull. World Health Organ. 78, 1207–1221.

Robinson, N., 2014. Insect Proteins Will Take Off. Food Manufacturer, UK. Available from: http://www.foodmanufacture.co.uk/Ingredients/Insect-foods-could-be-worth-230M

Safina, C., 2011. Why Are We Using Up the Earth? CNN Opinion: Carbon Dioxide. CNN, New York, NY.

Schofield, C., Ashworth, A., 1996. Why have mortality rates for severe malnutrition remained so high? Bull. World Health Organ., 223–229.

Shackleton, C.S., 2004. The importance of non-timber forest products in rural livelihood security and as safety nets: a review of evidence from South Africa. S. Afr. J. Sci. 100, 658–664.

Sheppard, D.C., Tomberlin, J.K., Joyce, J.A., Kiser, B.C., Sumner, S.M., 2002. Rearing methods for the black soldier fly (Diptera: Stratiomyidae). J. Med. Entomol. 39, 695–698.

Shira, D., Oh, E., 2010. Salma Hayek: I Love Eating Bugs! Available from: http://www.people.com/people/article/0,20396060,00.html

Shockley, M., Dossey, A.T., 2014. Insects for human consumption. In: Morales-Ramos, J.A. (Ed.), Mass Production of Beneficial Organisms. Academic Press, London, pp. 617–652.

Singh, P., 1977. Artificial diets for insects, mites, and spiders. Plenum Publishing Co., New York, NY, 594 p.

Singh, P., Moore, R.F., 1985. Handbook of Insect Rearingvol. IIElsevier Science Publishers, Amsterdam, The Netherlands, 514 p.

Smith, A., 1776, Wealth of Nations.

Smith, A., 2012. Get Ready to Pay More for Your Steak. CNN Money, New York, NY.

Srivastava, S.K., Babu, N., Pandey, H., 2009. Traditional insect bioprospecting: as human food and medicine. Ind. J. Traditional Knowl. 8, 485–494.

Steinfeld, H., Gerber, P., Wassenaar, T.D., Castel, V., Rosales, M.M., Haan, C.D., Food and Agriculture Organization of the United Nations, Livestock Environment and Development (FIRM), 2006. Livestock's long shadow: environmental issues and options, vol. xxiv, Food and Agriculture Organization of the United Nations, Rome, p. 390.

Tarkan, L., 2015. Why these startups want you to eat bugs. Fortune Magazine, August 25, 2015. Available from: http://fortune.com/2015/08/25/edible-insects-bug-startups/

Tharanathan, R.N., Kittur, F.S., 2003. Chitin: the undisputed biomolecule of great potential. Crit. Rev. Food Sci. Nutr. 43, 61–87.

van Huis, A., 2013. Potential of insects as food and feed in assuring food security. Ann. Rev. Entomol. 58, 563–583.

van Huis, A., Van Itterbeeck, J., Klunder, H., Mertens, E., Halloran, A., Muir, G., et al., 2013. Edible Insects: Future Prospects for Food and Feed Security. Food and Agriculture Organization of the United Nations, Rome.

Van Itterbeeck, J., van Huis, A., 2012. Environmental manipulation for edible insect procurement: a historical perspective. J. Ethnobiol. Ethnomed. 8, 3.

Vantomme, P., Mertens, E., van Huis, A., Klunder, H., 2012. Assessing the Potential of Insects as Food and Feed in Assuring Food Security. United Nations Food and Agricultural Organization, Rome.

Victora, C.G., Adair, L., Fall, C., Hallal, P.C., Martorell, R., Richter, L., Sachdev, H.S., 2008. Maternal and child undernutrition: consequences for adult health and human capital. Lancet 371, 340–357.

Vogel, G., 2010. For more protein, filet of cricket. Science 327, 811–1811.

Walker, T.J., Masaki, S., 1989. Natural history of crickets. In: Huber, F., Loher, W., Moore, T.E. (Eds.), Cricket Behavior and Neurobiology. Cornell University Press, Ithaca, NY, pp. 1–43.

Wardlaw, G.M., 2006. Contemporary Nutrition. [S.l.]. McGraw-Hill.

Waterfootprint.org, 2014. Water footprint and virtual water [online]. Available from: http://www.waterfootprint.org/?page=files/Animal-products

Weissman, D.B., Gray, D.A., Pham, H.T., Tussen, P., 2012. Billions and billions sold: pet-feeder crickets (Orthoptera: Gryllidae), commercial cricket farms, an epizootic densovirus, and government regulations make for a potential disaster. Zootaxa 3504, 67–88.

Yen, A.L., 2009. Entomophagy and insect conservation: some thoughts for digestion. J. Insect Conserv. 13, 667–670.

Yi, C., He, Q., Wang, L., Kuang, R., 2010. The utilization of insect-resources in chinese rural area. J. Agricul. Sci. 2, 146–154.

Zheng, L., Hou, Y., Li, W., Yang, S., Li, Q., Yu, Z., 2012a. Biodiesel production from rice straw and restaurant waste employing black soldier fly assisted by microbes. Energy 47, 225–229.

Zheng, L., Li, Q., Zhang, J., Yu, Z., 2012b. Double the biodiesel yield: rearing black soldier fly larvae, *Hermetia illucens*, on solid residual fraction of restaurant waste after grease extraction for biodiesel production. Renew. Energ. 41, 75–79.

Chapter 6

Insect Mass Production Technologies

J.A. Cortes Ortiz*, A.T. Ruiz*, J.A. Morales-Ramos**, M. Thomas[†], M.G. Rojas**, J.K. Tomberlin[‡], L. Yi[†], R. Han[§], L. Giroud[¶], R.L. Jullien[††]

*Entomotech, S.L. PITA, Agroingroindustrial Technological Park of Almería, Almería, Spain; **USDA-ARS National Biological Control Laboratory, Stoneville, MS, United States; [†]Zetadec B.V., Wageningen, The Netherlands; [‡]Department of Entomology, Texas A&M University, College Station, TX, United States; [§]Guangdong Entomological Institute, Guangzhou, China; [¶]Insagri, S.L. Coin, Málaga, Spain; [††]Khepri, and FFPIDI, French Federation of Producers Importers and Distributors of insect, France

Chapter Outline

Introduction	153
Mass-Produced Insect Species and Their Respective Applications	154
Insects for Food and Feed	154
Insects for Medicinal Use	155
Insects for Other Applications	155
Potential of Using Conventional Feedstock for Rearing Insects	156
Principals of Feed Production for Insect Mass Production	156
Manufacture of Insect Feed	157
Nutritional Requirements for Farmed Insects	158
Macronutrients	158
Micronutrients	160
Considerations for Insect Mass Rearing Equipment and Mechanization	162
Production and Operation Management	162
Rearing Area	163
Feeding and Watering	164
Separation and Sorting Room and Product Traceability	165
Cleaning Room	165
Compost Area	166
Production Techniques by Species	166
Mealworm Production Technologies	166
Black Soldier Fly, *Hermetia illucens*, Production	173
Housefly, *Musca domestica*, Production	176
Cricket Production in the United States	179
Waxworm Production	182
Environmental Control for Efficient Production of Insects in General	184
Estimating Optimal Conditions Required for Design for Insects in General	184
Air Flow Design	186
Equipment for Climate Control	191
Automatic Control and Artificial Intelligence	194
Concluding Remarks	196
References	196

INTRODUCTION

Insects have been used for food since the dawn of humanity. Documented uses have been spotted throughout history, with insect traces found in coprolites in the Ozark Mountains. In Spain, cave paintings depicting the collection of wild bees during prehistory were also uncovered. In China, ancient silkworm species were probably consumed more than 4000 years ago (see also Chapter 2 for more on historic and cultural uses of insects as food around the world).

While agriculture developed, only two species were domesticated: *Apis mellifera* L. and *Bombyx mori* L. Both those species were used for food but also had other applications (pollination, silk production). Honey was in fact the most common source of simple sugars before the 18th century. Sericulture was invented as early as the Yangshao period (5000–3000 BC) due to the ease of rearing methods for *B. mori* and other Saturniidae, while apiculture could have appeared as early as 3000 years ago in the Middle East.

Apart from those two species, insects were mostly collected in the wild as the techniques and materials available for large-scale production were inefficient. The Bible, for example, mentions locusts as a food source for John the Baptist in the desert. Entomophagy remained a habit based on opportunity because the cost of collection was higher than the calories gained by it. However, in tropical regions, insects tend to reproduce faster and grow bigger thanks to the combination of heat and humidity all year round. Unsurprisingly, the most enduring entomophagous countries are in tropical and subtropical regions. On the other hands, insects grow and reproduce easily, have high feed conversion efficiency (since they are cold blooded) and can be reared on biowaste streams. On average, 1 kg of insect biomass can be produced from 2 kg of feed biomass (Collavo et al., 2005) (see Chapter 4 for more on insect farming sustainability and insect production efficiencies based on inputs compared with other livestock). Insects can feed on waste biomass and can transform this into high-value food and feed resource. A desk study (Veldkamp et al., 2012) has demonstrated that it is technically feasible to produce insects on a large scale and to use them as an alternative sustainable protein-rich ingredient in pig and poultry diets, particularly if they are reared on substrates of biowaste and organic sidestreams.

Mass production of insects for feed and food is not currently a common enterprise, little information is available, and the state of the art in the production is still primitive as compared with production technology for other farms animals. As an example, a small egg-producing farm with 30,000 hens will produce around 1,500 kg of eggs a day while only few insect farms in the world are currently at this level of production. Agriprotein, of South Africa, one of the biggest companies in the world offering insect meals as feeds, claimed to produce 10 tons of insect meal a day in 2015, which could satisfy the nutritional demands of several small hen farms. These data show the need to increase the number of insect farms and their efficiency in order for the insect-based meals to be competitive with other feed raw materials, as soy is nowadays. Process mechanization, cheap insect feed production, building design, and temperature-controlled rooms have to be adapted for insect production and all these elements have to be blended to be economically profitable and sustainable.

As much as 80% of the world's population consumes insects as an intentional part of their diet (see Chapter 2 for more information on this). For example, China has a long history of human consumption and animal feed of some insect species (Feng and Chen, 2002; Wei and Liu, 2001; Chen, 1999). According to the records, eating insects dates back 3000 years in China (Feng and Chen, 2002). In many regions of China, people eat insects habitually (Chen, 1999) and there are more than 100 species of edible insects. Studies on the suitability of insects as animal and bird feed have been conducted in China since the 1950s (Wei and Liu, 2001). Indeed, the Western world is experiencing a renewed interest in insects as food for a number of reasons. See Chapter 5 for more on the emerging insect-based food industry.

Insect farming for animal feed is a relatively novel activity, yet the practice is rapidly being pursued by both researchers and commercial producers (see Chapter 10 for more on how insects can be utilized in animal feed). In the past, the most used technique was to hang carcasses above aquaculture ponds, where flies would lay their eggs. At hatching time, the maggots would fall and be eaten by the fish. Other species have been raised as a way to feed reptiles and other animals raised as pets or in zoos. However, as mentioned before, insect mass production is a very blooming industry which is just starting, and there is not much information available regarding mass production. The earlier pioneers in insect mass production were Singh and Moore (1985) who edited the *Handbook of Insect Rearing*, which has become the basis for the industrialization of insects. Several other handbooks have appeared since those times addressing insect production to feed pets or as a hobby. More recently, the book *Principles and Procedures for Rearing High Quality Insects* (Schneider, 2009) has given a more modern vision of insect mass production, presenting the author's experience from insect large-scale production for sterile insect release technique (SIT), probably the largest insect production operation known.

In this chapter we attempt to summarize what is scientifically known about the production of some of the main insect species used as food and feed, in combination with the insect farmer's production knowledge, to provide a picture of the state-of-the-art of this dynamic insect rearing industry.

MASS-PRODUCED INSECT SPECIES AND THEIR RESPECTIVE APPLICATIONS

More than 1900 insect species have been found to be edible around the world (van Huis et al., 2013; Shockley and Dossey, 2014) (see also Chapter 2). However, only a handful are bred in sufficient quantity to qualify as farming.

Insects for Food and Feed

Some of the most common insects that have been grown with the purpose to produce food and feed include crickets like *Acheta domesticus* (L.), *Gryllodes sigillatus* (Walker), *Gryllus assimilis* (Fab.), *G. bimaculatus* De Geer, and *G. locorojo* Weissman and Gray; the greater wax moth, *Galleria mellonella* L.; mealworms like *Tenebrio molitor* L., *Zophobas atratus* Fab., *Z. morio* Fab., and *Alphitobius diaperinus* Panzer; the housefly, *Musca domestica* L.; and the black soldier fly, *Hermetia illucens* (L.) (Table 6.1).

TABLE 6.1 Insect Species for Feed and Food

Name	Latin Name	Food	Feed
House cricket	*Acheta domesticus*	x	
Mealworm	*Tenebrio molitor*	x	x
Mediterranean field cricket	*Gryllus bimaculatus*		x
Silkworm	*Bombyx mori*	x	x
Wax worm	*Galleria mellonella*	x	
European honeybee	*Apis mellifera*	x	
Common housefly	*Musca domestica*		x
Common green bottle fly	*Lucilia sericata*		x
Green rose chafer	*Cetonia aurata*		x
Jamaican field cricket	*Gryllus assimilis*	x	
Migratory locust	*Locusta migratoria*	x	
Palm weevil	*Rhynchophorus ferrugineus*	x	
Palm weevil	*Rhynchophorus phoenicis*	x	
Sun beetle	*Pachnoda marginata*		x

Insects for Medicinal Use

Many species of insects have been utilized by various cultures around the world medicinally (Dossey, 2010, 2011; Costa Neto, 2005) (see also Chapter 2). One of the only species known to be mass produced for medicinal purpose is the American cockroach *Periplaneta americana* (L.) (Blattodea: Blattidae), which is sold in China and Korea to treat baldness, among other things. However, other insects or insect by products are used in many cultures to cure various ailments. *Lucilia sericata* (Meigen) (Diptera: Calliphoridae) has been used for maggot therapy, and the larvae secretes allantoin which is used as a treatment for osteomyelitis. Honeybee venom has many applications against inflammation, pain, and asthma, among others (Kampmeier and Irwin, 2009). Fleas have been investigated as a way to propagate antibodies and vaccinate their hosts (rabbits, in that particular case). One of the most interesting applications of insects in medicine is the production of vaccines and some other useful proteins, using baculovirus as a vector. The baculovirus is inoculated in a moth larva and, after a period of incubation, the vaccine or protein could be harvested from the insect hemolymph (Madhan et al., 2010). However, insects could be used extensively in medicine for their mediating effect in antimicrobial and antifungal peptides, insulin production, angiotensin converting enzyme (ACE), and antioxidant enzymes (Mlcek et al., 2014).

Insects for Other Applications

Other species are currently raised for various purposes: cochineal is used to extract a food colorant and sometimes a dye. Their mass production is mainly done in a natural way, were the cochineal *Dactylopius* spp. Is allow to grow feeding on prickly pear *Opuntia ficus-indica* (L.) Miller and collected manually, then sundry (Barriga-Ruiz, 1994). *Dactylopius* spp has been also used as biological control of the cacti *Opuntia* spp. (Zimmerman, 1979).

Insects are also farmed for biological pest control and pollination. Ladybugs, bumblebees, and lacewings are all raised on a medium scale. Some species are also bred to feed those insects, namely aphids. It should be noted that some of those species (ladybugs) are not edible due to toxins.

Additionally, many species are raised for research purposes, such as *Manduca sexta* (L.) (Lepidoptera: Sphingidae) (tobacco hornworm moth), often used in biomedical experiments and *Drosophila melanogaster* Meigen (common fruit fly), a classic model organism for many research purposes, particularly genetics research (Yoon, 1985). But the most surprising application of insect could be biodegradation and mineralization of Polystyrene by *T. molitor* (Yang et al., 2015a,b), which have been demonstrated that is capable of developing feeding on this polymer assisted by symbiotic gut bacteria.

POTENTIAL OF USING CONVENTIONAL FEEDSTOCK FOR REARING INSECTS

Currently in most of the developing countries, livestock is one of the fastest growing agricultural subsectors and the demand for livestock products is rapidly increasing (Makkar et al., 2014). However, in many developing countries there are deficits in the supply of feed demand to raise livestock. New unconventional alternate feed resources could play an important role in meeting this deficit. For instance, fruit and vegetable waste could be used for animal feed and insect could also provide additional feed sources. The FAO published a report on utilization of fruit and vegetable wastes as livestock feed and as substrates for generation of other value-added products (Wadhwa and Bakshi, 2013).

We do realize that the present regulations and legislation could be an obstacle for using insects as an alternative protein source in feeds for terrestrial animals. However, in aquaculture, rising demand and cost of fishmeal-based diets have created an urgency for alternative protein sources to use in aquaculture feeds. Therefore, research on development of insect proteins for aquaculture is needed, which could possibly be used as supplement to fishmeal. FAO estimated that insects have a similar potential market as fishmeal and they could be employed as feed in aquaculture and livestock and also used in the pet industry (van Huis et al., 2014).

Feeding trials on trout diets in which 25 and 50% of fishmeal was replaced by black soldier fly, showed that the results from sensory perspective were as good as the control diet with 100% fishmeal (Sealey et al., 2011). With regard to sustainable production and protein nutritive properties, Barroso et al. (2014) and Sánchez-Muros et al. (2014) concluded that the use of insect meals as an alternative source of animal protein may be an option in comparison to fishmeal and soymeal. Plus, insects, like the mealworms, were able to transform the low-nutritive waste products into a high-protein diet. Assuming competitive pricing, mealworms could be used as a replacement of soymeal in animal feed. In the short term, insects as alternative and sustainable proteins to animal feed seems to be promising.

Needs for developing guidelines in producing insects on an industrial scale were formulated at the first international conference "Insects to Feed the World," FAO and Wageningen University, 2014. The European commission asked the European food safety authority (EFSA) to provide a scientific opinion on the risks associated with consumption of insect proteins. Results of this scientific opinion were positive as long as insects are produced properly and procedures are continuously revised using safe feedstock.

The utilization of insect meals in animal feed requires the mass production of insects. At present, the information required to upscale insect production is not well established, particularly for insect rearing on biowaste and organic sidestream substrates. Currently, the larvae of black soldier fly (BSF) can feed on a wide range of organic materials, such as rotting fruit and vegetables and coffee bean pulp, as well as animal manure and human excreta (Makkar et al., 2014). Organic wastes products from livestock production facilities could be used to produce insects as an added value product. The larval stage of black soldier fly can last from 2–4 months, depending on feed availability. The duration of the pupal stage extends from 14 days to 5 months. Adults do not feed, they survive on fat accumulated during the larval stage. According to Makkar et al. (2014), there are three methods for rearing black soldier flies: pig manure, poultry manure, and food wastes. The drawback of rearing black soldier flies is that they require a warm environment. Maintaining optimal temperature and humidity add to the cost of mass production (Makkar et al., 2014).

Principals of Feed Production for Insect Mass Production

It can be anticipated that the need for improved feeds and raw materials for insects will increase with increasing production levels. Balancing the nutrient requirements of the farmed insects with the nutrients offered from these feedstocks may require additional (high) value additives, feed manufacturing operations, and different feed presentation forms (liquid, powders, or mash; semimoist; pellets; extrudes; simple or complex coacervates). Additionally, the need to explore feed formulations utilizing lower cost and less environmentally damaging or resource intensive to produce ingredients, cheap or free/waste stream materials and/or unused, underutilized or novel biomass not yet used as animal feed (possibly some algaes, grasses, etc.) will become more relevant as the insect rearing industry leaves its infancy stage. An excellent review on the nutritional aspects for insects in animal feeding has been written by Makkar et al. (2014). This section will focus on prospective feed manufacturing technologies to produce half-products, concentrates and complete feeds for insects. In addition, a variety of possible feed forms that are suitable for use in insect feeding are discussed.

Insects can be classified according to their feeding adaptations as liquid feeders, possessing sucking mouthparts; solid feeders, possessing biting mouthparts enabling them to chew animal or plant materials; and piercing–sucking insects, which possess modified sucking mouthparts adapted to piercing the host and sucking its internal liquefied tissues from animal or plant origin (Chapman, 2003). Liquid or liquefied feeds are more suitable for insects with sucking mouthparts, although most liquid feeders are capable to liquefying solid food materials by extra oral digestion (Cohen, 2004). Solid feeders are capable of chewing large solid food particles and reducing them to ingestible pieces. Piercing insects penetrate their host to take up nutritious fluid and they may require a barrier or encapsulating media for a liquid or semiliquid diet (Morales-Ramos et al., 2014). The mouth structures of insect are adapted to specific feeding needs according to the species; thus, artificial feeds offered to insects must take into account these specific feeding requirements.

Liquid and Semisolid Feeds

Liquid feeds are not produced as such. In most cases where animals are fed on liquid type of feeds (eg, young and suckling animals), milk feeds or milk feed replacers are used. These consist of powders that can easily be dissolved or dispersed in (hot) water. Ingredients used are either soluble or ground to fine powder for reducing settling when offered to the subjects. Specific feeds can be made by spray drying liquid solutions or slurries to obtain free flowing powders with complete nutritionally balanced feeds or in a concentrate that can be blended with other ingredients, possibly on the farm.

The powdered feeds are premanufactured by grinding and blending of the appropriate ingredients and stored in pouches or bulks. For some insects, the range of raw materials may include animal or insect proteins. These need to be water soluble or water dispersible to facilitate dissolving and distribution of the feed to the designated insect species (Yi et al., 2013).

Semimoist is a feed-form in which the moisture content ranges are variable but usually well above 15%. More important is the water activity content of these feeds, which may range between 0.65 and 0.85. The water activity content determines the shelf life of the product. Shelf life is moderated by the use of sugars, salts, or glycols with the objective of reducing microbial or fungal degradation of the feed. Semimoist feed texture is softer than solid feeds (semisolid), as produced by pelleting and extrusion. To ingest this feed the insect needs to chew or bite small fragments. On the other hand, insects with sucking mouthparts are able to feed on semisolid diets easily. Many insects have extra-oral digestion capabilities. In the US Department of Agriculture, Agricultural Research Service (USDA-ARS) semisolid diets have been produced for several species of Heteropterans showing good results, but they can also feed on solids like peanuts and sunflower seeds (Morales-Ramos, personal communication).

Solid Feed Presentations

Feed mash is the simplest solid feed form that can be manufactured. It consists of grinding and mixing all raw materials into the correct proportions to meet nutritional requirements of the insect. No additional heat- or compaction treatments are conducted on the feed; hence, energy expenditure to prepare the feed is low compared to pellets and extruded feed. On the other hand, segregation of raw materials often occurs due to transport and handling. This may impair the nutritional quality of the feed, especially individual insect feed intake, where absolute feed intake per individual is low.

Pelleting is the most common feed manufacturing operation. Raw feed materials are dosed according to a certain feed recipe, ground, mixed, preconditioned (usually with steam), shaped in the pelletizer to small rod-like agglomerates (pellets) and cooled (Thomas and van der Poel, 1996; Thomas et al., 1997, 1998; Abdollahi et al., 2013).

Solid feeds also can be extruded. Extrusion processing consists of fine grinding of the ingredients in the recipe, mixing, preconditioning with steam and other liquids, followed by cooking and conditioning the mash, and shaped using a single- or twin-screw extruder. Advantages of the extrusion process include a high degree of cooking of the ingredients, inactivation of certain antinutrient factors (ANF's), sterilization of the feed, and shaping of the feed material into intricate forms. Overviews on the extrusion process are given in Guy (2001) and Mosicki (2011).

Complex coacervation is an encapsulation technique in which two oppositely charged polymers are used to create a particle consisting of a (soft) inner liquid core and a hard outer shell. In food applications, the polymers used are usually proteins and polysaccharides (Xiao et al., 2014). Proteins can be either from animal, insect, or plant origin. Polysaccharides may include Arabic gum, pectin, chitosan, agar, alginates, carrageenans, or carboxymethylcellulose to name but a few (Xiao et al., 2014). Coacervates are interesting in that they deliver a liquid inner product with a hard outer shell that needs to be pierced by the insect to gain access to the (nutritious) fluid. General manufacturing processes in relation to feed texture, size, and moisture content can be seen in Table 6.2.

Manufacture of Insect Feed

Manufacturing processes for insect feeds heretofore has not been discussed in the literature. Manufacturing feeds for insects using current food processing technology procedures is the obvious starting point. Size, shape, and texture of feed forms for use in insect feeding represent a knowledge gap in relation to feeding requirements of insects. Food processing technology significantly influences feed functional properties. It may lead to reduced digestibility or loss of feed nutritional value, as well as reduced acceptability to the insect due to changes in flavor, aroma, texture, color, or other properties.

TABLE 6.2 Various Manufacturing Processes for Insect Feed and Feed Forms Derived Thereof

Manufacturing Process	Starting Materials	Feed Form and Texture	Approximate Size (Typical) (mm)	Moisture Content at Feeding (%)	Approximate Capacity (t/h)
Spray drying	Slurries and solutions	Instant powder to be dissolved in water	—	>80	0.1–2
Semimoist	Finely ground solid materials, plasticizers	Soft solid feed form	0.5–20	15–30	0.05–2
Mash	(Finely) ground solid materials	Powder	0.5–2	<13.5	0.1–100
Pellets	(Finely) ground solid materials	Pellets, solid, and dry	2–10 diameter rods	<13.5	0.1–100
Extrudates	(Finely) ground solid materials	Extrudates, hard, brittle, and porous	0.5–20	<13.5	0.1–5
Coacervates	Dissolved nutrients, alginates, cross-linkers	Spheres: soft inside; hard outside	0.4–2	<90	0.01–0.1

Insect feed processing on an industrial scale should includes the use of sustainable procedures, for example, the use of water/energy and economic feasibility (cost, etc.).

NUTRITIONAL REQUIREMENTS FOR FARMED INSECTS

Most insect species being farmed for food or feed around the world are omnivores, often scavengers in their natural environment. Omnivores display a high level of nutritional flexibility and can consume food from a variety of origins. Nevertheless, their nutritional requirements are complex and difficult to determine. Because they utilize a variety of food sources with different nutritional characteristics, it is difficult to determine the correct proportion in which each item is consumed in nature (Morales-Ramos et al., 2014). On the other hand, the nutritional flexibility of omnivorous insects allow them to develop and reproduce for multiple generations feeding on suboptimal diets. This nutritional flexibility can make them easy to rear using low-value food sources, which make them ideal for large-scale farming (van Huis et al., 2013). Nevertheless, the productivity of farmed insects can be improved by providing a correctly balanced diet (Patton, 1967; Pickens and Lorenzen, 1983; Nakagaki and DeFoliart, 1991; Hogsette, 1992; Won and Kwon, 2009; Morales-Ramos et al., 2011; van Broekhoven et al., 2015).

Macronutrients

The three major nutrient groups include protein, lipid, and carbohydrate (P, L, and C). Morales-Ramos et al. (2014) used the absolute ratios (PLC ratios) of these three major nutrient groups as a basis for artificial diet development. The PLC ratios provide information on the relative amount consumed of each of the three major nutrient groups. Every animal species has specific requirements of these three nutrient ratios for optimal food utilization (Morales-Ramos et al., 2014). For instance, the efficiency of conversion of ingested food (ECI) was significantly lower in diets with low protein and high starch content as compared with diets of high protein and low starch content in *T. molitor*, *Z. atratus*, and *A. diaperinus* (van Broekhoven et al., 2015). To calculate PLC ratios, each major nutrient [protein (P), lipid (L), and carbohydrate (C)] percentage is divided by the sum of the percentages of protein, lipid, and digestible carbohydrate (total carbohydrate − fiber). For instance, protein ratio of a given diet is calculated as %P/(%P + %L + %C). The sum of the three ratios (P, L, and C) should always be = 1; however PLC ratios can be transformed to base 100 or 1000 by multiplying each ratio by the respective base. The content of major nutrient groups and PLC ratios of food ingredients commonly used in diets for farmed insects are presented in Table 6.3. Effective oligidic diet formulations developed for *T. molitor* (Morales-Ramos et al., 2013), *A. domesticus* (Patton, 1967; Nakagaki and DeFoliart, 1991), and *Musca domestica* L. (Hogsette, 1992) possess similar nutritional characteristics. PLC ratios for the most successful diets were calculated by the matrix method reported by Morales-Ramos et al. (2014) using formulations reported by the authors and nutrient contents of the ingredients from US Department of Agriculture, Agricultural Research Service (2015). Despite the taxonomic difference among these three species, the PLC ratios of the best oligidic diets showed similarities in the relatively low content of lipid (Fig. 6.1).

TABLE 6.3 Content of Major Nutrient Groups as Percentages and Calculated PLC Ratios of Food Ingredients Commonly Used in Diet Formulations for Farmed Insects

Food Ingredient	Total Protein	Total Lipid	Total Carbohydrate	Crude Fiber	Water	PLC Ratio[a]
Whole soy flour	34.5	20.6	35.2	9.6	5.2	428:255:317
Whole yellow corn flour	6.9	3.9	76.8	7.3	10.9	86:49:865
Crude wheat bran	15.5	4.3	64.5	42.8	9.9	373:104:523
White wheat	11.3	1.7	75.9	12.2	9.6	147:22:831
Alfalfa pellets	16.0	1.5	64.0	27.0	12.0	294:28:679
Dry potato flour	8.3	0.4	81.2	6.6	6.6	100:5:895
Carrot, dehydrated	8.1	1.5	79.6	23.6	4.0	123:23:854
Dry milk, skim	36.2	0.8	52.0	0	3.2	407:9:584
Baker's yeast	38.3	4.6	38.2	21.0	7.6	637:77:286
Brewer's yeast	53.3	0	43.3	20.0	7.0	696:0:304
Dry egg yolk	32.2	55.8	3.6	0	3.0	352:609:39
Dry beef liver	68.0	12.0	13.0	0	0.8	731:129:140

P, protein; L, lipid; C, digestible carbohydrate = total carbohydrate − fiber.
[a]Base 1000.
Source: Data from US Department of Agriculture, Agricultural Research Service, USDA National Nutrient Database for Standard Reference, Release 28 (2015).

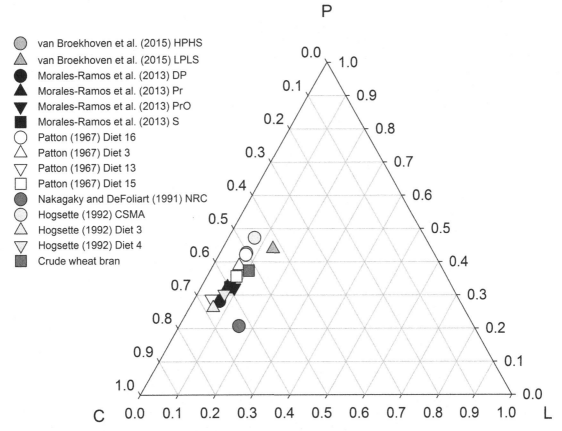

FIGURE 6.1 Ternary plot of protein, lipid, and carbohydrate (PLC) ratios of oligidic diet formulations for *T. molitor* (reference from 1st to 6th), *A. domesticus* (reference from 6th to 11th), and *M. domestica* (reference from 12th to 14th) as compared to crude wheat bran (reference 15th). (Ratios calculated from data published in van Broekhoven et al. (2015) (T. molitor, Z. atratus, and A. diaperinus.), Morales-Ramos et al. (2013) (T. molitor), Patton (1967) (A. domesticus), Nakagaki and DeFoliart (1991) (A. domesticus), Hogsette (1992) (M. domestica), and US Department of Agriculture, Agricultural Research Service (2015) (wheat bran))

Carbohydrates are an important energy source for insects as with all organisms, and insects also require them to produce chitin, which is an amino polysaccharide found on the exoskeleton of arthropods (Chippendale, 1978). The most common sugars involved in arthropod nutrition are monosaccharides (glucose, fructose, and galactose) and disaccharides (maltose and sucrose) (Singh, 1977). Most insects are able to absorb and metabolize fructose and glucose, but some monosaccharides such as arabinose, ribose, xylose, and galactose, while readily absorbed, are not metabolized (Chippendale, 1978). The farmed insect species which are generalist feeders are able to digest disaccharides, such as sucrose and maltose, while specialist feeders are not (Singh, 1977; Cohen, 2004). Insect amylase can also digest starch which is present in many food ingredients used in farmed insect diets (Chippendale, 1978).

Lipids are essential structural components of the cell membrane, provide an efficient way to store and provide metabolic energy during sustained demands, and serve as a barrier for water conservation in arthropod cuticle. Most lipids exist as triglycerides consisting of three fatty acid molecules linked by a molecule of glycerin. In general polyunsaturated fatty acids such as linoleic and linolenic acids are essential for insect nutrition because they are either unable to synthesize them altogether or incapable of synthesizing them in sufficient quantities. The inability of insects to synthesize polyunsaturated fatty acids has been confirmed in some species and limited capacity has been observed in other species such as mosquitoes, aphids, and cockroaches and derivatives of polyunsaturated fatty acids, known as eicosanoids, stimulate oviposition in crickets (Downer, 1978; Chapman, 1998).

Phospholipids are an important group of complex lipids, which play an important role in lipid transfer and in the synthesis of vitellin and other lipoproteins (Agosin, 1978; Shapiro, 1988). Phospholipids are easily digested and absorbed by insects and are a good source of polyunsaturated fatty acids (Chapman, 1998). Feed ingredients such as egg yolk and soybeans are good sources of phospholipids.

Protein is a fundamental component of the tissues of all organisms (particularly animals such as insects) and is required for growth and development. Increasing protein content in the diet of the yellow mealworm, *T. molitor*, shortens development time (Morales-Ramos et al., 2010) and increases weight gain rate (van Broekhoven et al., 2015). Similar results were obtained on the house cricket, *A. domesticus*, when the protein content was increased on diet formulations (Patton, 1967). Proteins, in addition of being structural components of tissues (such as muscle, including intracellular protoplasm and extracellular matrix proteins), also play an important role in the metabolism and regulation of physiological functions by being essential components of enzymes and hormones. Proteins are classified according to their solubility and function as globulins, nucleoproteins, lipoproteins, and insoluble protein. Globulins include enzymes, antibodies, and protein hormones. Nucleoproteins are associated with nucleic acids and ribosomes and lipoproteins often serve as transport of lipids and storage proteins like vitelline. Structural proteins, also known as fibrous proteins, are insoluble because they consist of rigid molecular structures containing a high degree of inter- and intrapolypeptide-chain cross-links (Agosin, 1978).

The chemical structure of proteins consists of a chain of amino acid molecules linked together. In some proteins, these chains are folded into specific shapes for specific proteins [to give them their specific function(s)], whereas other protein chains (otherwise known as "polypeptide" chains) tend to be loosely structured or unstructured, properties also intrinsic to the function of the specific protein in question. There are 20 amino acids known to be components of proteins in living organisms, including alanine, arginine, asparagine, aspartic acid, cysteine, glutamic acid, glutamine, glycine, histidine, isoleucine, leucine, lysine, methionine, phenylalanine, proline, serine, threonine, tryptophan, tyrosine, and valine (Stryer, 1988). Arthropod diets should contain at least 10 amino acids considered essential which include leucine, isoleucine, valine, threonine, lysine, arginine, methionine, histidine, phenylalanine, and tryptophan (Chapman, 1998; Cohen, 2004; Lundgren, 2009) because insects are unable to synthesize them. Other amino acids, such as tyrosine, can be synthesized only in insufficient quantities and require high energy consumption for their synthesis (House, 1961; Chapman, 1998;Lundgren, 2009). Tyrosine is a major component of sclerotin and is required in large quantities during molting (Richards, 1978; Hopkins, 1992).

Micronutrients

Micronutrients are required in small quantities, but are essential for insect nutrition and lack or deficiency of a micronutrient have a significant negative impact on insect biology, even if the major nutrient groups are present in adequate quantities and ratios in their diet. Micronutrients mostly include complex lipids like sterols, vitamins, and some minerals.

Sterols are essential nutrients for insects in general because they are unable to synthesize them (House, 1961; Chapman, 1998). Sterols play a variety of important roles in insect physiology as components of subcellular membranes, precursors of hormones, constituents of surface wax of cuticle, and constituents of lipoprotein carrier molecules (Downer, 1978; Chapman, 1998). Insects obtain sterols from cholesterol, but other important sources of sterols include plant phytosterols

TABLE 6.4 Sterol Content in Common Food Ingredients Used to Formulate Diets for Farmed Insects

Food Ingredient	Cholesterol	Phytosterols[a]	Ergosterol
Whole soy flour	0	59	0
Whole yellow corn flour	0	38	0
Soybean oil	0	293	0
Peanut oil	0	207	0
Corn oil	0	968	0
Canola oil	0	657	0
Carrot, dehydrated	0	94.0	0
Baker's yeast	0	0	800
Brewer's yeast	0	0	1200
Dry milk, skim	20	0	0
Dry egg yolk	2335	0	0
Dry beef liver	917	0	0

aIncluding beta-sitosterol, campesterol, and stigmasterol.
Source: Data on phytosterols and cholesterol obtained from US Department of Agriculture, Agricultural Research Service, USDA National Nutrient Database for Standard Reference, Release 28 (2015). Data on ergosterol obtained from Bills et al. (1930).

and ergosterol from fungi (Downer, 1978; Cohen, 2004). Sterol needs of farmed insects may be satisfied by supplementing diets with sources of cholesterol (such as egg yolk, milk, fish oil), phytosterol (such as vegetable oil, soybeans, or corn), or ergosterol (such as yeast) (Table 6.4).

Vitamins are usually required for growth in insects, which are unable to synthesize certain ones (Chapman, 1998). Water soluble vitamins include vitamin C and the B complex, which is comprised of thiamin (B_1), riboflavin (B_2), niacin = nicotinamide (B_3), pantothenic acid (B_5), pyridoxine (B_6), biotin (B_7), folic acid (B_9), and cobalamins (B_{12}). The B vitamins function as cofactors for enzymes and are required in the diet of all insects (Chapman, 1998), except for vitamin B12, which is not universally required (House, 1961; Cohen, 2004). Vitamin C (ascorbic acid) is known to play a role in the molting process (molting is the shedding of the outer exoskeleton cuticle of insects and other arthropods during growth or in their transition from one life stage to another) (Chapman, 1998; Cohen, 2004). Inositol and choline are constituents of some phospholipids. Choline plays a role in spermatogenesis (the manufacture of sperm) and oogenesis (the manufacture of eggs) in addition to its structural role in phospholipids and it is probably required in all insects. Inositol is known to be required in Coleoptera (beetles) and plays a role in the nervous system (Chapman, 1998). Lipid soluble vitamins include retinol, carotenoids (A), tocoferols (E), calciferol (D), and phyloquinone (K). Only vitamins A and E are known to be required in insects, where they play a role in the synthesis of pigments and in reproduction, respectively (Chapman, 1998). Vitamins C, A, and E are also antioxidants and may play an important role in the detoxification processes and protection against microbial infection (Cohen, 2004). The content of vitamins present in food ingredients commonly used in diet formulations for farmed insects are presented in Tables 6.5 and 6.6.

While some elements like nitrogen, sulfur, iron, and phosphorus can be obtained from organic sources other essential elements for growth and reproduction must be obtained from inorganic sources (minerals) which cannot be biosynthesized. The 24 elements known to be essential for living matter, in order of importance, are: hydrogen, oxygen, carbon, nitrogen, calcium, phosphorus, chlorine, potassium, sulfur, sodium, magnesium, iron, copper, zinc, silicone, iodine, cobalt, manganese, molybdenum, fluorine, tin, chromium, selenium, and vanadium (Hammond, 1996). Elements can be divided according to their ionic charge into cation (+) and anion (−). Cations include metals like iron, sodium, potassium, magnesium, manganese, calcium, zinc, and copper. Anions include chloride, sulfur, fluoride, phosphorus, and iodine. Minerals are compounds that consist of combinations of cations and anions. House (1961) mentions potassium, sodium, phosphorus, magnesium, manganese, zinc, and copper as important in insect growth. Iron is important in several enzyme pathways including in the synthesis of DNA (Cohen, 2004). Dunphy et al. (2002) determined that iron plays an important role in the immunity functions of the hemolymph of *Galleria mellonella* L. (the greater waxworm moth) calcium is required in a lesser extent in arthropods than in vertebrates, but still plays an important role in muscular excitation (Cohen, 2004).

TABLE 6.5 Content of Vitamins A, E, D, and C in Food Ingredients Commonly Used to Formulate Diets for Farmed Insects (Quantities in mg/100 g)

Food Ingredient	Vitamin			
	A	E	D	C
Whole soy flour	0.072	1.9	0	0
Whole yellow corn flour	0.171	0.4	0	0
Crude wheat bran	0.006	1.5	0	0
White wheat	0.005	1	0	0
Carrot, dehydrated	51.6	5.5	0	14.6
Dry potato flour	0.007	0	0	81.0
Dry milk, skim	1.354	0	0	6.8
Dry milk, whole	0.343	0.5	7.8	8.6
Baker's yeast	0	0	0	0.3
Brewer's yeast	0	0	0	0
Dry egg yolk	1.793	5.4	0	0
Dry beef liver	33.87	1.3	1.33	4.3

Vitamin A = retinol + alpha and beta carotene; vitamin E = alpha tocopherol; vitamin D = cholecalciferol and ergocalciferol; vitamin C = ascorbic acid.
Source: Data obtained from US Department of Agriculture, Agricultural Research Service, USDA National Nutrient Database for Standard Reference, Release 28 (2015).

TABLE 6.6 Content of Complex B Vitamins in Food Ingredients Commonly Used to Formulate Diets for Farmed Insects (Quantities in mg/100 g)

Food Ingredient	Vitamin						
	B1	B2	B3	B4	B5	B6	B12
Whole soy flour	0.6	1.2	4.3	191.0	1.6	0.5	0
Whole yellow corn flour	0.2	0.1	1.9	21.6	0.7	0.4	0
Crude wheat bran	0.5	0.6	13.6	74.4	2.2	1.3	0
White wheat	0.4	0.1	4.4	0	1.0	0.4	0
Carrot, dehydrated	0.5	0.4	6.6	72.1	1.5	1.0	0
Dry potato flour	1.0	0.1	6.3	54.9	2.1	0.7	0
Dry milk, skim	0.4	1.5	1.0	169.0	3.6	0.4	0.004
Dry milk, whole	0.3	1.2	0.6	119.0	2.3	0.3	0.003
Baker's yeast	2.4	5.5	39.8	32.0	11.3	1.5	0
Brewer's yeast	4.0	5.1	33.3	0	2.0	2.7	0.001
Dry egg yolk	0.3	1.9	0.1	1388.0	7.8	0.7	0.005
Dry beef liver	0.7	9.3	44.0	1110.0	24.0	3.7	0.198

Vitamin B1 = thiamin, B2 = riboflavin, B3 = nicotinamide, B4 = choline, B5 = pantothenic acid, B6 = pyridoxine, B12 = cobalamin.
Source: Data obtained from US Department of Agriculture, Agricultural Research Service, USDA National Nutrient Database for Standard Reference, Release 28 (2015).

CONSIDERATIONS FOR INSECT MASS REARING EQUIPMENT AND MECHANIZATION

Production and Operation Management

One of the main challenges in mass rearing of edible insects is to find the right balance between the levels of mechanization, automation, labor, investment, and productivity. The elements involved with insect production and related quality and production costs such as feeding, watering, handling, harvesting, cleaning systems, processing, packaging and storage can be improved significantly by adding technology.

Insect Mass Production Technologies Chapter | 6 163

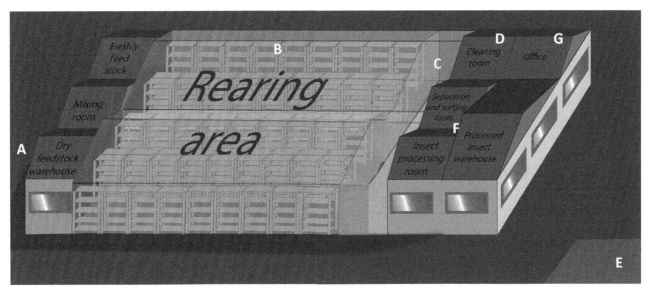

FIGURE 6.2 **Farm example, distribution of the operations.** (A) firstly the raw material is stacked at the silo, (B) then feed is introduced in the rearing room, (C) when insects are ready to be harvested they are cleaned and separated from frass in the harvesting room, (D) the insect containers are cleaned in the cleaning room, (E) then the frass goes to the composting area, (F) while the harvested insect are processed in the feed processing area, and (G) offices and management is set at control room.

Current edible-insect farms do not reach a high level of mechanization with two consequences: products available on the market are overpriced and can often be of low or widely varying quality. Insects could be a protein substitute to meat and fishmeal for animal feed only if farmers can offer large volumes of biomass, reliable supply, high and consistent quality, and at a competitive price. Reaching a level of at least 80% mechanization in the production processes, edible-insect farming could offer high enough volumes with costs per kilo sufficiently low to be competitive with any other animal-derived commodity. Mechanization can facilitate the partial replacement of manual farm labor, increase the productivity, and provide better efficiency and consistency in product quality and supply.

An integrated rearing system with mechanization, information, and automation can be applied competitively and effectively for large-scale insect farming. Monitoring systems can be integrated with control systems to provide historical performance data including feeding, mating, oviposition rates, environment and microbiological control, and life cycle.

As with traditional agricultural mechanization, the choice of equipment depends on the edible insect species being reared, the location and size of the farm, the economic status of the region among other factors. The following sections will describe Fig. 6.2 which shows an example of a farm designed and located in the south of Spain with the objective of industrial production of the yellow mealworm, *T. molitor*.

Rearing Area

For an efficient insect production it is necessary to use the rearing space efficiently (Fig. 6.2). For some insect species which can crawl vertically, jump, or fly, three-dimensional crawling space (such as cardboard dividers, egg crate material, or a more permanent lattice of some material or another) can allow for greater density versus depending only on two-dimensional flat spaces. For those limited to the bottom of a flat surface (most larvae such as mealworms and fly larvae, etc.), it is important to minimize space between trays with insects and feed/substrate while balancing against the need to deliver air and regulate temperature. Additionally, depth of feed/substrate in the trays of such species is critical as to not add too much so that it is wasted, as some insect species will only utilize feed and living space at a certain depth. To further minimize space needed to farm a maximum quantity of insects, rearing boxes can be held in multilevel shelves, which are filled with as many rearing boxes as possible to minimize the space usage per kilogram of insect produced. In some cases, stackable boxes can be used or the boxes can be set on a wagon or a pallet to allow free movement around the rearing area. An example of a rearing room with trolley to carry the rearing boxes of a moth species is shown in Fig. 6.3. The climate control system must be capable of maintaining adequate environmental conditions in the rearing area. Air conditioners equipment suitable to for environmental control will be described further in this chapter.

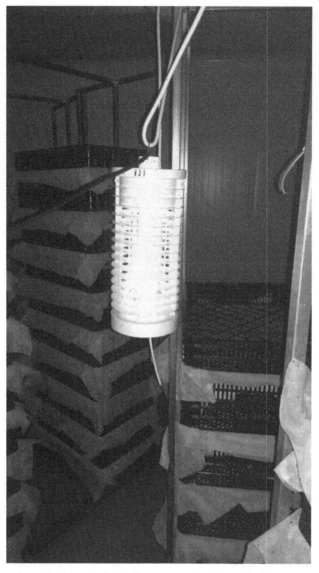

FIGURE 6.3 Rearing room holding *Ephestia kuehniella* Zell. (Lepidoptera: Pyralidae). Rearing boxes are stacked one top of the other and placed in a trolley to allow movement around the room. Walls are constructed with smooth insulating metal panels coated with resin (resistant and energy efficient) the corners are rounded to allow easy cleaning. Ceiling fans push the air down to distribute heat uniformly.

Feeding and Watering

Feed handling for any kind of farm is an integral part of the overall system. Likewise, this is also true for industrialized edible-insect farming (Fig. 6.2). Due to the perishable nature of some feeds used, especially for those with high water content such as those formulated from nonmarketable produce products, insect farmers have to serve-in-time (pickup/delivery: once or twice a week) to reduce feed spoilage. When feeds are formulated from dry material such as spent grain and yeast, wheat flour, dry oilseed cakes, spoilage is less of an issue but it is important to deliver only the quantity of feed needed and when it is needed to reduce unnecessary wasteful feed use, especially since feed delivered often becomes contaminated with insect waste, saliva, secretions and/or further environmental contaminants such as mold and moisture. It is unusual when feed can be recycled or reused after it is delivered to farmed insects, though this may be the case in some well-designed and very efficient systems for some species requiring very dry environments.

In one example (Fig. 6.2), tower silos with mechanical downloaders can be used, which fill mixer wagons equipped with several distribution units with dealing arms to reach every shell level, the wagons are robotic controlled by a central computer that will determine the position where the arms will go. For such a system, a computer program uses an array for every position on the shell guiding the wagon to the position to which it will deliver the feed. As the type, quantity, format and rate of delivery of feed depends largely on the species being farmed as well as its life stage, it is important to use an identification and feeding system that provides the right diet at the right stage at the right time. Data from a computerized version of such systems can be supplemented by observations of technical production using wireless transmission systems through tablets or similar devices and through an enterprise resource planning (ERP). This data would not be subjective, but it would be limited to a number of parameters with a value scale, which would help in objective decision making.

In a computerized insect rearing system, data obtained through various probes (temperature, humidity, etc.), could be transmitted to a computer, which could be analyzed by comparing the inputs, microclimate, development of the population, as well as the analysis of incidents and historical data. Thus, a PID system could make decisions independently, and move backward lots to warmer rooms, which would accelerate development production, and vice versa; and collect biomass in fresh condition or state of serious changes, such as decreased performance or occurrence of diseases.

In general, one of the best advantages of producing insect species is that they have reduced water needs compared with vertebrate livestock. For some species, water can be easily provided directly in their feed through fresh vegetables and fruits. Others require a small amount of water provided independently from the feed (if their feed itself is a dry feed). For those requiring water for drinking separately from their food, a low pressure water system has to be installed to provide the quantity of water required for each rearing container and, in most cases, not much more than is needed so the insects do not drown in it, the water does not experience excessive microbial growth or insect waste contamination or does not flood the insect habitat.

Separation and Sorting Room and Product Traceability

Traceability of edible insects, as with all agricultural products, from farm to table is essential to maintain safety and trace any hazards or disease outbreaks to their source. Each new production batch of insects harvested must be properly catalogued in a way which can be traced to the raw material used to feed and water that batch (Fig. 6.2). This information should follow the insects through each of the subsequent steps taken during the production process, including the results of the quality controls to be implemented at each critical point of the system. This traceability should cover those preset and self-sufficient procedures that provide insight into the history, location and path of a product or batch of products along the supply chain at any given time. All packages or batches can be characterized using RFID tags (allowing remote reading), indicating at least: date and time of the start of the batch, lots of raw material used, parental precedence, stage of life, density, separation and cleaning date, and feed components. An integrated management system of traceability allows individual control for each batch of production, stock control and final product, an absolute control of the evolution of the lots, and incident detection and critical points in the system. In this way, we can guarantee the quality and safety of products. The same management system can receive production orders; control packaging lines; and control production remotely by issuing, validating, and reading standard barcodes. These barcodes are essential for traceability in the food industry and use standards.

One example of this type of machine (Fig. 6.3) can separate up to 500 kg/h of product (such as mealworm larvae of commercial size). Using such equipment it is conceivable to reach a capacity of 2 t/h of separated product.

For such a machine used to handle cricket species such as the *A. domesticus* or *G. assimilis*, each box (Fig. 6.2) placed on a conveyor belt is manipulated by a mechanical arm, first to remove the rearing substrate material (hideaways) from inside the boxes and second to gently collect the crickets reaching the harvesting area where the crickets are gently brushed down with a soft circular brush from the dispersal material, and vacuum from the box. The vacuum is designed to harvest the cricket without producing any damage. Then the crickets are collected in a net recipient where they could be managed, weighed, and transported to other processes. During the separation phase, the feed conversion rate is automatically calculated by measuring the amount of feed consumed by the insects, the insect weight gained and the frass generated. Simple electronic logger cells can be used to record pre- and postfeeding insect weights after the separation phase.

Cleaning Room

In some cases, cleaning or rinsing of the insects is desired, and may be desired postharvest as well (although in many cases this will be unnecessary, particularly if they will be immediately frozen or heat processes upon harvesting). For other species, the insects may be reared their entire lives in the same container from hatching through harvest, so insect removal from those containers will not be necessary until harvest. Nonetheless, in other cases it may be desirable to rinse or clean the harvested insects, or even to transfer them between containers during their life cycle prior to harvesting. In such cases, after the separation phase, all boxes can be sent to a high pressure cleaner, filled with new substrate, and sent back to the separation area (Fig. 6.2). All rearing boxes must be designed for easy handling and made of structurally resistant materials that can be washed multiple times. Cleaned rearing boxes with new rearing substrate are filled with cleaned insects. Edible insects separated by age and size are either sent back to the rearing area or packaged for shipment to the process area. Dead insects and waste are stored and collected weekly by an authorized company for destruction.

Compost Area

Frass separated from live insects and feed residue can be collected and processed or packaged for other applications (such as fertilizer or feed for other animals) in a separate area from the rearing or cleaning room (Fig. 6.2). For example, for some fertilizer applications, frass can be composted by adding some carbon rich feedstock as sawdust or straw (amount of carbon depending on the insect species frass). It can then be moistened and pH adjusted to be around 7.5, then piled and turned every 48 h until the temperature of the pile start to decrease then let it rest until the pile does not heat up anymore. The processed frass or the compost are prepared for shipment using a packaging flour machine with identification on big bags.

PRODUCTION TECHNIQUES BY SPECIES

Mealworm Production Technologies

The most commonly produced mealworm species in the United States are *T. molitor* (yellow mealworm) and *Zophobas morio* F. (superworm or "super mealworm"). Commercial production of mealworms started in the 1950s in the United States and it has expanded considerably within the last four decades. The market for mealworms in the United States was initially for fishing bait, but during the late 1970s the pet food market opened, increasing demand considerably for live feeder insects such as mealworms (particularly for pet reptiles). Another important market for mealworms in the United States is for wild songbird feed. Mealworms are sold either live, fresh, roasted, or dried and canned or bag-packed. Mealworms are also commercially available as dry powders processed in different forms. At present, it is unknown the total volume of sales and the number of commercial producers in the United States.

Larvae and Adult Rearing

Conventional mealworm production is based on trays to hold larvae during their development and adults for colony reproduction. The most common tray size used is a standard 65 L × 50 W × 15 H cm, which are easy to handle and have sufficient depth to prevent larvae or adults from escaping. Trays can be made of wood, high density polyethylene, or fiberglass. Trays are usually arranged in specially constructed furniture with multilevel racks or shelves (Fig. 6.4).

Rearing systems based on trays have several disadvantages. Because trays are closed systems, there is a continuous accumulation of frass in the larvae trays, which can favor mite proliferation in the colony. Also, the adult oviposition in the bottom of the tray cannot be recovered safely and must remain in the tray. Adults must be separated from the food and place in new trays for every oviposition period. Increasing production scale using a tray system results in increasing difficulties and demands of labor as the number of trays required increase in number.

In recent years in some facilities, the tray system has been improved by replacing the tray bottoms with screens to allow frass particles to fall through. Similarly, bottom screens can be used in the adult holding trays to allow first instars to fall to a shallow tray at the bottom for collection. This system has been described by Morales-Ramos et al. (2012). A screened rearing system is not limited by size and can be adapted to different type of holding containers. Morales-Ramos et al. (2012) describe a system for rearing *T. molitor* consisting of stackable containers with nylon screens at the bottom. Containers hold larvae with diet consisting of wheat bran and supplements. Container stacks can consist of five or more containers sitting on a wheeled dolly (Fig. 6.5A). The screen openings have a dimension of 500 μm (0.5 mm), which is designed to allow frass particles to pass through with minimal loss of food. The continuous movement of larvae facilitate the movement of frass particles, which continuously fall from one container to the next until they end on a collection container at the bottom of the stack. Eliminating frass particles from the containers helps to control mite infestations and allows a more precise monitoring of larval food consumption. Changes in food consumption rates can be used to optimize production or to detect potential problems in the colony. Based on head capsule measurements (Morales-Ramos et al., 2015), the screen opening dimensions will hold larvae of the fifth instar or older. Smaller larvae must be temporarily (4–5 weeks at 25–28°C) held in containers with solid bottoms before transferring them to the screened containers.

Screened bottoms can also be used to collect progeny from adult holdings. *T. molitor* females normally oviposit on the bottom of trays gluing the eggs with a sticky substance. Eggs are commonly glued to food particles and are almost impossible to separate without damaging them. In a tray system this forces a continuous change of trays to separate adults from the oviposited substrate to prevent egg cannibalism. Morales-Ramos et al. (2012) describe a system consisting of a

FIGURE 6.4 **Conventional mealworm trays.** Commercial production of *T. molitor*. (A) Larvae growing trays. (B) Adult reproductive trays.

container with screen on the bottom, to hold *T. molitor* adults, sitting on top of a shallow tray (Fig. 6.5B). The screen has opening dimensions of 850 µm, which allow the passage of small larvae to the collection tray at the bottom. The screen opening dimensions allow in theory the passage of first to sixth instars (Morales-Ramos et al., 2015) with limited loss of food. Small food particles falling through the screen provide food for the early instars collected at the bottom tray. The continuous movement of adults encourage early instars to seek the bottom to escape cannibalism, inducing them to fall through the screen openings. In a screened container system adults do not need to be separated and removed from the container every week. Progeny can be continuously collected from the bottom tray.

Different variations of screen bottom container systems have been implemented in some commercial mealworm production facilities in the United States (Morales-Ramos, unpublished). In theory, this system can be used for any species of tenebrionid with some modifications.

Diets and Feeds

Although *T. molitor* can grow and reproduce by feeding exclusively in wheat bran, significant improvements in development time, larval survival, efficiency of food conversion, and adult fecundity can be obtained by the addition of simple food supplements (Morales-Ramos et al., 2010, 2011a, 2013; van Broekhoven et al., 2015). Most commercial producers in the

FIGURE 6.5 **Mealworm screened bottom tray system.** (A) Larvae growing tray system consisting of stackable boxes with screened bottoms. Mealworms had fully consumed the food and frass has fallen through the bottom screens. (B) Adult reproductive rearing system. Picture showing the top screened container with adult and the bottom tray with small food particles, frass, and early larval instars.

United States supplement the diet of *T. molitor* by adding sliced vegetables such as potato, carrot, and cabbage. Vegetable supplements can add important nutrients that may be missing in wheat bran such as vitamins, essential fatty acids, and sterols (van Broekhoven et al., 2015). Adding raw vegetables can also provide larvae with a source of water. Balancing the nutrient ratios of *T. molitor* diets is important. Increasing the protein content can increase survival and shorten development time in *T. molitor* (Morales-Ramos et al., 2011a, 2013; van Broekhoven et al., 2015), but protein supplements can increase the price of the feed. Using high protein by-products as supplements is a more viable option (van Broekhoven et al., 2015). Also, increasing the calorie content of the food mix increases growth rate and pupal mass (Morales-Ramos et al., 2011a, 2013), but high calorie diets may reduce larval survival (van Broekhoven et al., 2015). High calorie diets impact the ability of insects to respond to disease (Krams et al., 2015). For instance, increasing the diet lipid content make *T. molitor* more susceptible to entomopathogenic nematodes (Shapiro-Ilan et al., 2008, 2012).

Watering

Water requirements of *T. molitor* larvae are minimal because they have the ability to take up water dissolved in subsaturated air at a relative humidity of 90% or higher (Dunbar and Winston, 1975; Machin, 1976). For this reason it is possible to mass produce *T. molitor* without providing water to the larvae even at a relative humidity as low as 75% (Morales-Ramos et al., 2012). However, maintaining a high relative humidity in the growth chamber or rearing room can be expensive. Additionally, even at conditions of high relative humidity, *T. molitor* larvae grow faster when they are provided with water. This is probably because active water vapor absorption requires metabolic energy (Hansen et al., 2006). Also, adult beetles do not uptake atmospheric water and must be provided with water at least twice a week. Providing a continuous source of water is difficult under current production systems because any water lick favors fungal and mite growth. Most mealworm producers in the United States provide water by spraying or dispensing small amounts every few days. This method is labor-intensive since it requires the servicing of a large number of trays or rearing units. Another method of providing water consists of the use of water-absorbing polymers like polyacrylamide, which are dispensed fully saturated into the trays or rearing units. However, the long-term impact of these type of polymers to the colony is unknown as well as the consequences of using the insect product for human consumption. Developing better watering systems is one of the most important requirements of the industry at present time.

Rearing Density

Some factors can impact productivity in mealworm rearing systems and require the development of methods to recognize them and control them. One of these factors is larval densities maintained during production. Increased larval densities are known to delay or inhibited pupation in many tenebrionid species (Tschinkel and Willson, 1971; Nakakita, 1982; Botella and Ménsua, 1986). Increasing larval densities can also impact growth rates and reduce pupal mass (Weaver and McFarlane, 1990; Parween and Begum, 2001), and cause delayed development and increased mortality and cannibalism (Savvidou and Bell, 1994) in tenebrionids. Increasing crowding may be impacting the physiological state of larvae. For instance, Hirashima et al. (1995) observed a delay in the release of ecdysone in *Tribolium freemani* (Hinton) larvae under crowded conditions. More importantly, increasing larval densities are known to reduce the efficiency of food conversion in *T. molitor* (Morales-Ramos and Rojas, 2015). From the economic point of view, reducing efficiency of food conversion also reduces productivity and increases production costs. On the other hand, reducing larval density to improve conversion efficiency will impact the rearing space requirements. The number of rearing units must be increased to reduce larval densities; however, time is an important factor to consider for optimization of the use of rearing space. Improvements in food conversion efficiency (Morales-Ramos and Rojas, 2015) and reduction in development time (Savvidou and Bell, 1994), resulting from reducing larval densities in *T. molitor* production systems, may outweigh the temporary losses of rearing space. Future research should focus on determining the optimal larval densities required for *T. molitor* to maximize productivity as biomass weight produced per area during a determined period of time (eg, kg/m^2 per month).

Reproduction can also be negatively impacted by high rearing densities (Morales-Ramos et al., 2012). Increasing adult densities in *T. molitor* not only reduces oviposition rate, but also increases egg cannibalism and reduces adult longevity (Morales-Ramos et al., 2012). Reproduction can be optimized by maintaining a low adult density and by replacing reproductive colonies every 56–60 days.

Separation

Another biological aspect of *T. molitor* that impacts mass production is its developmental plasticity. The larval development of *T. molitor* transits through a variable number of instars (Cotton and St George, 1929; Esperk et al., 2007). The number of instars in *T. molitor* can be impacted by temperature (Ludwig, 1956), humidity (Murray, 1968; Urs and Hopkins, 1973), photoperiod (Tyshchenko and Sheyk Ba, 1986), oxygen concentration (Loudon, 1988; Greenberg and Ar, 1996), larval density (Connat et al., 1991), parental age (Ludwig, 1956; Ludwig and Fiore, 1960), and nutrition (Stellwaag-Kittler, 1954; Morales-Ramos et al., 2010). As a consequence, larval development time in *T. molitor* displays a high level of variability (Morales-Ramos et al., 2010, 2015) making it impossible to synchronize its life cycle in mass-production systems. Because commercial mealworm products are classified on standardized sizes for feeding specific types of pet animals, separation is an important part of the commercialization process. Industry has solved this problem by using separation methods based on larval size. These methods consist of passing the whole larval production through screens of different sizes. In most of the production facilities in the United States, separation procedures are performed manually by shaking large framed screens, which must be changed to different meshes to obtain different larval sizes one at the time. A mechanized procedure was developed by Morales-Ramos et al. (2011b), which consists of a three-screen circular separator fed by a conveyor (Fig. 6.6). This system does not require changing screen sizes and automatically separates four different sizes (large, medium, and small larvae, and frass particles) in a continuous manner. Two of the screens (first and second) have rectangular openings, instead of the conventional square, in order to facilitate the passage of elongated-shaped larvae. The third screen is a conventional standard No. 35 (500 µm) with square openings to allow the passage of frass particles. At least one such system is in operation in a commercial facility in the United States.

Size separation is important for obtaining the reproductive stock of the colony. Cannibalization of pupae by larvae is well documented in *T. molitor* (Martin et al., 1976; Weaver and McFarlane, 1990; Morales-Ramos et al., 2012, 2015). Breeding stock is commonly selected from the larger larvae group during the separation process. However, even after larvae have been separated by size, pupation occurs at different times among larvae creating opportunities for pupal cannibalism.

FIGURE 6.6 Mealworm larvae separation system. Separation system consists of a conveyor, which discharges mealworm larvae with diet into a three-screen circular separation system producing four discharge (only one discharge visible in picture) groups of different sizes.

In commercial production facilities, the breeding stock is just allowed to develop completely in a tray (or group of trays) and losses by pupal cannibalism are normally accepted and tolerated. Pupal losses from cannibalism cannot be completely eliminated because last instar larvae can have a development time range of over 30 days (Morales-Ramos, unpublished). Larvae are continuously pupating in the breeding stock tray and cannibalism constantly occurs in the tray (Morales-Ramos et al., 2015). Removing pupae daily from the breeding stock tray can reduce cannibalism, but increases the amount of labor required for colony maintenance. Pupal separation can potentially be mechanized and this is a good subject for future technological development.

Mealworm Production in China

The mealworm is the most commonly reared insect in China. They are produced industrially, for both domestic use and export, as well as food for people, pets, and zoo animals, including birds, reptiles, small mammals, amphibians, and fish and also for outdoor wild bird feeders (Wang et al., 1996; Ng et al., 2001; Ramos-Elorduy et al., 2002; Giannone, 2003; Wu et al., 2005; Gasco et al., 2014a,b; Piccolo et al., 2014; Schiavone et al., 2014). Thousands of tons of dry mealworm larvae are currently produced for export every year (Data of Chinese Customs, 2014).

The media used in China for mealworm production (larvae and adults) consists of wheat bran and a variety of vegetables. The optimum water content of mealworm feed (wheat bran and variable vegetables) is about 18% (Wu, 2009). The vegetables or their residues should be washed clean with tap water and dried in the air to remove the possible pesticide residues, soil, dust, and other unwanted contaminants.

The rearing rooms for mealworm are kept at the temperature of 25–30°C and relative humidity (RH) of 50–75% (Manojlovie, 1988; Chen and Liu, 1992). Adults are maintained and fed in cardboard containers (Fig. 6.7), in contrast with the plastic or fiberglass trays commonly used in European and North American farms (Fig. 6.8). The optimal rearing density of adults is about 0.94 larva/cm^2. The adults usually begin to oviposit from the fourth day after emergence. A screen (14 meshes) containing wheat bran as bedding is used as oviposition substrate (Fig. 6.9). To collect semisynchronized eggs, the adults need to be removed from the oviposition container and placed into to another container every 3–5 days during spawning period which can last for 40–50 days. The larvae usually hatch 7 days after the eggs are laid. The best rearing density for commercial larvae production is 1.18 larva/cm^2 (Wu, 2009). One hundred larvae need approximately 6–10 g of wheat bran and 0.6–0.1 g of fresh vegetables (carrot, cabbage, pumpkin, etc.) every day, which should be added separately at regular intervals (2 times per day). Rotten or uneaten vegetables should be removed immediately. Larval frass should be removed from rearing containers every 3–5 days.

FIGURE 6.7 **Large scale mealworm rearing in paper containers (China).** *(Pictures from Liu Yusheng, Shandong Agricultural University)*

FIGURE 6.8 **Mealworms in plastic trays (Spain).** Distance between trays allow good air circulation. *(Pictures from Liu Yusheng, Shandong Agricultural University)*

172 Insects as Sustainable Food Ingredients

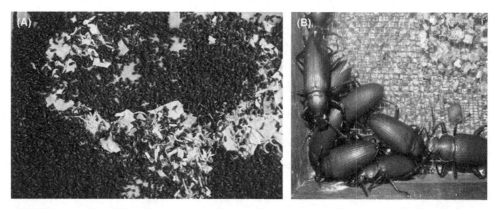

FIGURE 6.9   Adults of *T. molitor* (A) and *Z. morio* (B) for egg collection. *(Pictures from Liu Yusheng, Shandong Agricultural University.)*

Larval developmental period to full grown larva or pupa is between 45 and 60 days. Larval separation and collection is mainly accomplished through the use of a 60 mesh screen, a process which in some facilities has been automated by using an electric vibrating screen. When a large number of larvae have become pupae, the pupae should be collected and placed in a new tray filled with wheat bran for adult emergence. The collection of pupae currently still requires manual work, making it a possible target for improvement, mechanization, or automation to reduce cost as well as increase scale and efficiency of mealworm production. Fresh mealworms can be dried using numerous methods. One popular method utilized in China involves use of a microwave drying machine (Fig. 6.10) for production for sale and long storage.

The super worm *Z. morio*, a native of Southern and Central Africa, is also one species with great potential as a resource of insect protein. Thousands of tons are produced every year in China for animal feed at the present time. They are produced industrially mainly as feed for pets and zoo animals (Zhu, 2013; Zhao et al., 2011). This insect has the potential to convert garbage into feed and food protein, and the residues can also be used as organic fertilizer. The life cycle and breeding process of the super mealworm are similar to the mealworm (Cai et al., 2008). One of the key differences between the

FIGURE 6.10   Microwave drying machine. *(Pictures from Liu Yusheng, Shandong Agricultural University.)*

superworms and yellow mealworms is that for the superworm (*Z. morio*) to pupate, the mature larvae must be isolated from one another individually, presumably because disturbances from other superworms moving around prevent the mature superworm from relaxing long enough to go into pupation. Yellow mealworms, on the other hand, will pupate in their larval rearing trays amid their feed, frass, and other moving mealworms. This is a critical difference to consider when designing automated and mechanized systems for rearing these two species.

The cost of mealworm product depends on the diet used, rearing scale, mechanization, and labor costs. In China, based on retail cost on Sep. 15, 2015, the cost estimate of dry larvae is about 6.3 USD/kg. In Europe, the price of dry larvae dedicated to human consumption is about 50 USD/kg. The cost of dry super mealworm in China is about 15.76 USD/kg.

Black Soldier Fly, *Hermetia illucens*, Production

The black soldier fly, *H. illucens* (Diptera: Stratiomyidae), possesses several unique features that have resulted in it being recognized as a model system for reducing waste as a means to produce protein. Black soldier fly prepupae can be used as feed for a variety of animals including but not limited to catfish (Bondari and Sheppard, 1987), rainbow trout (St-Hilaire et al., 2007b; Sealey et al., 2011), poultry (Hale, 1973), and swine (Newton et al., 1977) (see Chapter 9). This species is found on all major continents, excluding Antarctica, and is found in major abundance predominately in temperate and tropic regions (Sheppard et al., 1994). The black soldier fly is holometabolous with an egg, larva, prepupa, pupa, and adult stage (May, 1961). The larvae are white, while the prepupae and pupae are black, which leads to easy identification (Sheppard et al., 2002). The resulting adults are wasp-like in appearance and will live for approximately 2 weeks (Tomberlin et al., 2002).

The life-history of the black soldier fly is simplistic in some regards. The black soldier fly is not synanthropic and thus is not a pest of humanity (Sheppard et al., 1994). Although cases of facultative myiasis have been documented (Lee et al., 1995), the evidence would suggest these interactions were accidental. Adults live solely to mate and lay eggs (Sheppard et al., 2002), while the presence of black soldier fly larvae in waste often results in the exclusion of other pest species, such as the house fly, *M. domestica* (Ditpera: Muscidae) (Sheppard, 1983). Furthermore, black soldier fly larvae are able to reduce pathogens in waste, such as *Escherichia coli* (Liu et al., 2008), and *Salmonella* spp. (Erickson et al., 2004). There is also some evidence that females, given the opportunity, may prefer to lay eggs in areas where other black soldier fly larvae have been feeding (suggesting a possible larva to female adult pheromone activity) (personal conversations with black soldier fly researchers and producers). Furthermore, larvae, which are high in protein and fat, are amendable for self-harvesting which allow for methods to be implemented that are low cost (Sheppard et al., 1994). However, it should be noted that development and ability to reduce varied wastes is population specific (Zhou et al., 2013). If interested in industrializing the black soldier fly, establishing a colony from a local population is recommended. This approach could also reduce the likelihood of using a strain of black soldier fly from a different region of the world that might be susceptible to environmental conditions in a given area.

Production of the black soldier fly falls within two realms—adult and larval colonies. Of course, adults are essential for producing the offspring that can be used for either, maintaining the colony and digesting waste. The colony is also responsible for mass production of the larvae that will be used for digesting waste and producing prepupae for harvesting and processing for protein and fat.

The purpose of this section will be to review the life-history of the black soldier fly and its use for mass production of protein and fat. Adult and larval colony maintenance will be reviewed as well as methods for mass production at the industry scale. Furthermore, current hurdles that need to be overcome before implementation will be discussed as well. Unfortunately, research to date has been conducted at the lab scale. Thus, data presented here should be taken with caution, as their translation to industry level is not known at this time.

Adult Colony

Very little is known about adult black soldier fly behavior. The primary reason is due to their short lifespan and propensity to spend a majority of their adult life in less developed settings (ie, more rural conditions). Males rarely return to sites of development (where substrates full of black soldier fly larvae exist), and females are only present at these sites to oviposit, as adults of this species are not known to eat or drink. This also makes them possibly more hygienic to produce than other flies, since the adults are less likely to spread microbes from various substrates they visit to feed. Once eggs are deposited, females die shortly thereafter. Thus, in colony, dead adults will accumulate in the bottom of a colony cage (see section "Mating Behavior" for description). These adults will need to be removed periodically to increase oviposition levels in desirable locations (see section "Oviposition" for description).

Adult Management

Limited information is available on the adult life-history of the black soldier fly. However, what is known about adult biology is critical for its mass production in colony. Males typically emerge first with females following 2 days later (Tomberlin et al., 2002). After an additional 2 days, mating will occur, and oviposition 2 days after mating (Tomberlin and Sheppard, 2002). This life-history trait is important to recognize when managing an adult colony as the release of multiple generations into a cage could potentially lead to disruptive mating attempts. Males will attempt to mate with females not receptive to copulation. Old males attempt to mate and prevent younger males from such opportunities. And, males seeking mates are accosting females that are attempting to oviposit. In each of these examples, mating and oviposition are being inhibited, which could have devastating consequences on egg production. As for the adults and their maintenance, they do not need to be fed or receive water as they rely on fat body reserves for energy; however, providing moisture through a misting system 2–3 times per day during the summer is highly recommended, as they are susceptible to heat.

Mating Behavior

Adults have a unique lekking (ie, aggregation) behavior that is critical for mating (Tomberlin and Sheppard, 2001). Adults are known to form aggregation sites where males will attempt to secure a female in flight and mate with her. Because of such a behavior, colonies are to be maintained in cages at minimum 1 m³ (Fig. 6.11A). Furthermore, mating behavior is regulated by sunlight (Tomberlin and Sheppard, 2002). This factor could be quite limiting during the winter when sunlight is less frequently encountered in some places and intensity is reduced. However, efforts have been made to develop artificial light systems to overcome this limitation. To date, efforts have produced systems that generate approximately 60% of mating that occur under natural light conditions (Zhang et al., 2010). Other factors to consider are temperature and humidity (Holmes et al., 2012) as temperatures below approximately 27°C result in reduced adult activity, which will translate into lower mating pairs and oviposition rates (Tomberlin and Sheppard, 2002).

Males compete for lekking sites meaning that males that land at a given location within a lekking site will aggressively chase other males (ie, avoid competition) or females (ie, mate) that enter these sites. And, personal observations indicate

FIGURE 6.11 *H. Illucens* (A) Adult cage, (B)larva, and (C) eggs.

male size could impact their ability to secure the sites and possible mates. Furthermore, mating appears to be limited with adults mating typically once. However, males will continue to attempt to mate with females multiple times. This behavior is a concern for colony maintenance (see previous discussion with regards to maintaining overlapping generations in a cage).

Oviposition

Females deposit one clutch during their life. They typically seek out sites indicative of appropriate larval resources. Previous research has determined that microbes associated with conspecific eggs play a role in attraction of ovipositing adults (Zheng et al., 2013). Furthermore, this research determined the response was dose dependent. In some instances, high concentrations of microbes increased oviposition at a site, while in other cases high concentrations decreased oviposition preference. Another interesting aspect of this work was that the response was strain specific. Strains of bacteria species from one species did not necessarily equate the same level of oviposition as strains isolated from the black soldier fly.

Oviposition sites should be reduced to specific locations within the colony (Sheppard et al., 2002). In most cases, researchers use containers inoculated with decomposing grain that is saturated with water. Corrugated cardboard blocks (2–3 cm thick, 3–5 cm long, 2–3 cm wide) are attached to the container directly above the medium. Adults will oviposit in the cardboard allowing for easy quantification based on weight. Cardboard can be removed as needed; however, it is recommended that cardboard remain in the oviposition site less than 4 days as eggs will begin to hatch. Eggs collected can be partitioned into two groups, those to be used to maintain the colony and those for mass production.

Larval Maintenance

As previously mentioned, eggs tend to hatch in approximately 4 days (Tomberlin et al., 2002). But, low temperatures can significantly impact hatch time and viability (Holmes et al., 2010). Typically, cardboard containing eggs should be placed in a small container (500 mL clear plastic cup), covered with a paper towel, which is secured with a rubber band. The cub can be placed on a shelf in a room designated as the colony larval space. This space is to be maintained under stable conditions approximately 27°C, 60–70% relative humidity, and 14:10 L:D. Temperature is a vital parameter that must be monitored closely as it can impact larval development (Tomberlin et al., 2009).

Cups containing cardboard should be checked daily for egg hatch. When observed, a small amount (~50 g) of a standard diet at 70% moisture should be provided to the neonate larvae. Previous research has used a Gainesville diet (Hogsette, 1992) for colony maintenance as it can be produced in high volume, standardized, and at a low cost (Sheppard et al., 2002). The larvae typically feed on this small amount for 2–4 days depending on the number of larvae. Once the time interval from placement of the cardboard in the colony to its remove has past, it can be removed from the cup as it is assumed all eggs have hatched. Once the larvae have digested the initial feeding allotment(s), they can be poured into a larger container—usually shoebox size (24 cm long, 12 cm deep, 13 cm wide)—and fed ad libitum until the box is approximately half full. At that time, the larvae can then be moved to a full-size container (76 cm long, 12 cm deep, 45 cm wide) and fed until pupation. Previous research has found that feeding can be terminated when approximately 40% of the larvae have entered the prepupal stage (black appearance and mobile) (Tomberlin et al., 2002). At this stage, containers should be covered with gauze, which will need to be secured. This step is taken so as to prevent emerging adults from escaping from the pan. Adults can be released into the colony as needed; however, care should be taken to partition emerging adults using a prescribed manner where initial emergence over the first 2 days is recognized as typically male and those emerging on days 3 and 4 are female. Emergence after day 4 has not been examined in detail and therefore cannot be commented on at this time. This approach is important so that mating is synchronized. Females typically will mate 2 days after emergence. If pools of emerging adults are mixed, optimization of oviposition of fertilized eggs can be impacted (see previous discussion on lekking behavior).

Costs

Costs associated with maintaining an adult colony are limited. Adults should be watered as needed; however, an automatic misting system can alleviate any associated personnel costs. True costs are with colony maintenance as related to cleaning out dead adults, monitoring oviposition, and releasing adults into the colony.

Larval Production

Black soldier fly larvae are able to consume a wide variety of waste. Research has determined they can successfully develop on vertebrate remains (Tomberlin et al., 2005), kitchen waste, fruits and vegetables, raw liver (Nguyen et al., 2013, 2015), fish offal (St-Hilaire et al., 2007a), municipal waste (Diener et al., 2011), human waste (Banks et al., 2014), and dairy cattle manure (Myers et al., 2008). This plasticity obviously makes them an ideal insect model for mass production of protein and oil; however, while black soldier fly larvae are able to consume a wide range of waste streams, the type of waste fed to them can impact their development time. Therefore, caution will need to be exercised when formulating waste streams to larvae in order to maximize production and efficiency. A potential solution could be the use of probiotics. Past research determined that select microbes, when applied to waste as a pretreatment, enhance black soldier fly digestion of waste, larval development, and prepupal mass (Yu et al., 2011).

Current systems for mass production of black soldier fly larvae fall into two categories. Initial efforts under field conditions focused on continuous production as observed within poultry operations (Sheppard et al., 1994). This system relies on large-scale troughs with inclined walls inoculated with soldier fly larvae daily. Larvae are fed a steady stream of waste at low volume, which results in continuous feeding of the larvae. Resulting prepupae disperse from the waste by climbing up the inclined wall of the troughs and are collected in pipes arranged at the lip of the ridge. Prepupae then move through the pipe to collection bins positions located under holes along the length of the pipe. However, there are concerns with this system. Any sort of infection could result in massive die-off in the trough. Determining the efficiency of production can be difficult as all age classes of larvae are mixed together. And, a large amount of space is needed for maintaining such a facility. This limitation can also be a concern with regards to heating and cooling expenses.

The second approach that has received considerable attention over the past 15 years relies on batches of larvae in containers, much like the methods described for colony maintenance (Sheppard et al., 2002). Each container is a given unit that can be fed daily and monitored for 40% of the larvae to become prepupae. This approach is labor intensive as it relies on individuals feeding each container daily, checking for prepupae, and finally moving the container to a separation area. Methods for separation have not been published at this time; however, mechanical separation would be most appropriate in order to maximize efficiency.

There are benefits to the batch approach over the continuous system for larval production. Concerns over mass die-off of larvae are avoided. Close monitoring of production by each unit (eg, how many eggs introduced and number of prepupae produced can be determined) can lead to optimization as related to waste streams that have the greatest survival, quickest larval development, and highest levels of protein.

Storage of Prepupae

Unfortunately, nothing has been published on the proper storage of harvested prepupae. Current practices range from freezing to immediate protein and lipid extraction. Determining an appropriate practice will depend on resource availability and level of production.

Quality Assurance

The production of black soldier fly prepupae has been studied in detail. However, there are still areas of production that need additional research prior to utilization of the raw material.

Heavy Metal Contamination. Because the larvae can feed on a variety of wastes, there is potential for heavy metal accumulation (Diener et al., 2011). Some research has indicated that heavy metals might accumulate in the exoskeleton of prepupae (Diener et al., 2015). This physiological response could be a benefit for mass production if the proteins and lipids are extracted, thus leaving the heavy metals in the chitinous exoskeleton.

Microbial Contamination. Although research has indicated that black soldier fly larvae reduce some pathogens (Erickson et al., 2004; Liu et al., 2008), efforts should be in place for proper processing and microbial mitigation (to reduce pathogens as well as spoilage microbes) as well as screening product, especially if used in raw form.

Nutritional Balance. Black soldier fly larvae can feed on a wide variety of wastes; however, the quality of the prepupae produced is still in question. In order to mass-produce this insect, efforts are needed to determine the impact of these different wastes on protein and lipid levels in resulting prepupae. Such data would be essential for quality assurance and standardization.

Housefly, *Musca domestica*, Production

Housefly is the most common fly (Diptera) species and one of important farmed insects in China. The ability of housefly maggots to grow on a large range of substrates can make them useful to turn wastes into a valuable biomass rich in protein and fat useful as animal feed, and the residues can be used as organic fertilizer.

Feed and Feed Ingredients

Housefly adults are usually reared in a diet containing 50% glucose and 50% milk powder (Chang et al., 2007); however, there are several different media recipes for larval rearing, including wheat bran, pig manure, and chicken manure (Richardson, 1932; Wu et al., 2001; Yang et al., 2004; Wang et al., 2010). The water content of the common rearing media for fly larvae is often kept at 65–70%. Due to the high water content of livestock manures, sometimes there is a need to add wheat bran or rice bran into the manures to adjust the water content. Usually, the livestock manures are pretreated by fermentation to kill heat-sensitive pathogens that may be present (Zhu, 2012; Chen et al., 2014). Hogsette (1992) described diet formulations for housefly larvae that do not contain manure. A commercial diet (CSMA) produced by Ralston Purina (St. Louis, Missouri) was commonly used in the United States during the 1990s, but this formulation is no longer commercially available. This formulation consisted of 33%, wheat bran, 27% alfalfa meal, and 40% brewer's yeast granules (Hogsette, 1992). A cheaper and more efficient (89.5% yield) diet was obtained by adding corn meal, while eliminating the yeast granules: 50% wheat bran, 30% alfalfa meal, and 20% corn meal. These fly diet formulations are mixed with water at 1:1 ratio before introducing the fly eggs (Hogsette, 1992). As with other insect species, diet formulation development in the housefly is still an open field for research. There is great potential for the use of agricultural and industrial by products, as houseflies have great flexibility of food consumption and utilization.

The mature larvae are separated from the medium by lowering oxygen content conditions (described later) for their use as fresh maggots, or as dry powder feed for animals such as chickens or shrimp. Larvae are stored at 4°C for 4–5 days to be used fresh. Larvae are converted to dry powder using the same process used to produce fishmeal.

Production Process

The flow process scheme is presented in Fig. 6.12 as one example of a type of housefly rearing system. The rearing rooms (25–32°C, RH = 65–75%) for adult houseflies are designed to prevent escape, provide aeration, and provide 8-h light (Fig. 6.13). Food and water are supplied simply with hand controlled equipment. A long pipe is inserted inside fly cages and water is introduced into the water reservoir, where the adult flies can drink. Another option for watering is to use an inverted cup (filled with water) on a plate covered with filter paper. The paper gets moist with water continuously. Optimal adult density is approximately 5625 adults (1:1 sex ratio) per cubic meter of rearing room space. When the adults become 3 days old, a piece of cotton cloth is placed onto wet wheat bran (water content 60–70%) as oviposition substrate for egg collection. Adults are attracted to these oviposition sites by the aroma of fermented wheat bran. Egg collection is carried out between 0900 (9:00 am) and 1600 (4:00 pm) each day, and for 5–10 days (Fig. 6.14). The adult oviposition period can be prolonged and egg yield increased by supplementing the adult diet with chicken eggs.

Once collected, the fly eggs are introduced onto the diet medium surface in plastic or metal containers at a proportion of approximately 1 g of eggs per kg of diet medium. At conditions ranging from 25–32°C temperature and RH 65–75% humidity, the larvae emerge from the eggs and become mature after 5 or 6 days. The larvae are separated from the diet medium by several methods, such as self-escape, negative phototaxy (Jia, 2007), screen mesh (Jia, 2007; Li et al., 1998),

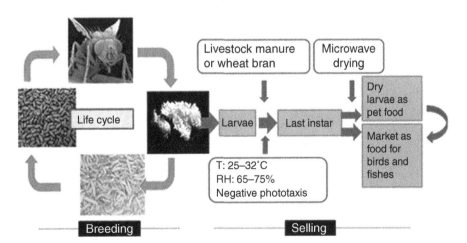

FIGURE 6.12 Flow chart process scheme for the production of housefly.

FIGURE 6.13 Rooms for house fly adult rearing. *(Pictures from Zhao Guoyu, Guandong Entomological Institute)*

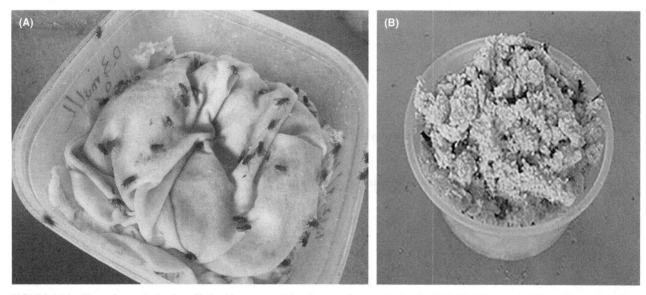

FIGURE 6.14 House fly egg collection. Cloth with attractant (A) and group of eggs covering the cloth (B). *(Pictures from Zhao Guoyu, Guandong Entomological Institute.)*

FIGURE 6.15 **Room for larval separation with low oxygen.** Larva move out of the boxes when the level of oxygen is low. *(Picture from Zhao Guoyu, Guandong Entomological Institute.)*

or lowering oxygen concentration (Fig. 6.15) (Han and Chen, 2010). Mature larvae can also be mixed with dry wheat bran for pupation in the rearing container. After 1 or 2 days at 25–30°C and 65–75% RH, they become pupae. The pupae are moved into the rooms for adult rearing.

When harvested for use in some application, larvae separated from the media are can be processed via any desired method such as using microwaves for drying to produce larval powder. Dry maggots can be stored seal hermetically at room temperature (25°C) for about 4–6 months. The cost estimate of rearing housefly larvae depends on the composition of the diet media, rearing scale, machines required, labor cost, and so forth. Based on the retail cost on Sep. 15, 2015, the estimated price of dry maggots is about 9.42 USD/kg.

Cricket Production in the United States

The most common cricket species produced commercially in the United States are the house cricket, *A. domesticus*, and the Indian cricket, *G. sigillatus* (also called the banded cricket) (Weissman et al., 2012). In recent years, many cricket farmers have adopted the Jamaican field cricket, *Grillus assimilis* (Fab.) to replace colonies of *A. domesticus* decimated by infections of the densovirus (AdDNV) (Weissman et al., 2012). Crickets have been produced commercially in the United States for at least 65 years. They are mostly used as fish bait or lures, but also as food for exotic pets including amphibians, reptiles, birds, and mammals. In the last decade, a new market has been developing in the United States for crickets as a food ingredient for human consumption. Crickets are dried and ground to produce a fine powder (cricket powder) or a course meal (the course meal sometimes erroneously referred to as "cricket flour"), which is then added to different manufactured food products. Although crickets have been grown commercially in the United States for decades, cricket production techniques are still primitive requiring a great deal of labor. Most of the rearing procedures are done manually and no significant mechanization has been developed to simplify or speed the procedures.

Rearing Units

Crickets are grown in large boxes ranging from 36–60 cm deep filled with different materials, most commonly cardboard, such as egg cartons or packing dividers for shipping to increase surface space. Rearing boxes can be made of different materials such as wood, cardboard, metal, high density polyethylene, or fiberglass. Smooth surfaces are preferred because this makes difficult for crickets to climb the walls. Boxes made of wood or cardboard should be lined with some hydrophobic material such as a plastic bag so that moisture from cricket waste or watering units does not destroy the box, especially during use. The top edges of boxes made out of wood or cardboard must be lined with Teflon (Fig. 6.16C), cello tape, or other

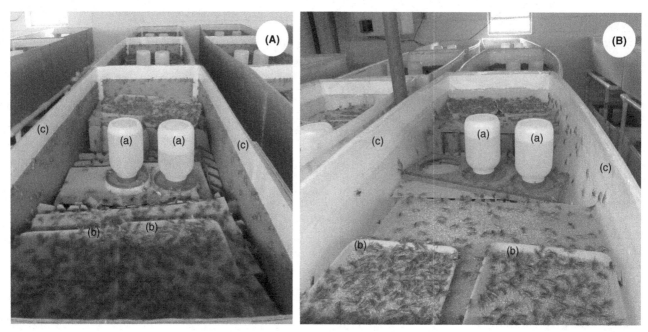

FIGURE 6.16 **Two examples of cricket rearing units.** (A) Boxes constructed of cardboard and (B) boxes made of high density polyethylene. (a) Watering units, (b) feeding trays with diet, and (c) smooth tape for preventing escapes.

smooth material close to the top to prevent escapes. Dimensions range from 50 to 160 cm long and from 40 to 100 cm wide (Fig. 6.16). Larger boxes provide more rearing space, reducing the number of units to service, but handling larger boxes requires more effort. On the other hand, smaller boxes are easier to handle, but a larger number of units are needed and more servicing labor is required. Water is provided with commercial chicken watering systems modified with sponges to prevent drowning (Fig. 6.16A). Some producers have developed their own watering systems, which require less maintenance. However, automatic watering systems have known to generate leaks, filling the rearing unit with water and drowning the crickets. Current rearing units, with solid bottoms, are closed systems not designed to allow excess water to flow out. Future automatic watering system designs should take this factor into account and consider redesigning the rearing boxes as well.

Diets and Feeds

Food is provided in shallow trays or paper plates placed at the top of the egg carton or filling material stack (Fig. 6.16B). Some commercial cricket feed mixes are available (such as the commonly used "Cricket Chow" from the feed company Purina), but many producers prepare their own proprietary feed mixes. Patton (1967) reported that diets containing from 20 to 30% protein, 32–47% carbohydrate, and from 3.2 to 5.2% lipid were the most satisfactory to rear *A. domesticus*. Yet, most commercial cricket feed mixes report contents of crude protein less than 20%. Ingredient cost is often a limitation on developing commercially viable diet formulations. Nakagaki and DeFoliart (1991) determined that commercial chicken feeds produced cricket biomass at a much lower price than Patton's (1967) diet 16 despite nutritional advantages of the later. However, protein content of the chicken feed used by Nakagaki and DeFoliart (1991) was 22%. Going forward, as the demand for crickets as food or feed increases, cheaper and more efficient and sustainable diets will be, and are already being, developed. These should seek to utilize underutilized biomass such as spent grains and yeast from brewing or ethanol operations, unused preconsumer produce or food industry by-products, algae, and/or currently unused biomass such as various grasses (like switchgrass).

Environmental Conditions

Crickets require higher rearing temperatures than other commonly produced insects. The optimal range of temperature for growth and development of *A. domesticus* is between 32 and 35°C (Patton, 1978). Cricket farmers report temperature ranges between 29.5 and 33°C in their rearing areas. Egg incubation period is from 10 to 14 days and nymphal development lasts from 6 to 8 weeks in *A. domesticus* at 32°C (Patton, 1978). Information on the optimal relative humidity conditions for growth and development of *A. domesticus* is lacking, but Clifford et al. (1977) reports a range between 25 and 50% as adequate. Some commercial growers report using between 33 and 45% RH in their cultures. McFarlane (1985) states that a relative humidity of 55% is adequate to prevent crickets from clustering around the water source, while at the same time ensures the food will remain dry and unsuitable for growth of microorganisms.

Reproduction

Adult crickets are transferred to new boxes for reproduction. Increased rearing space for adult crickets is obtained by using cardboard box dividers instead of egg cartons. Cardboard box dividers provide bigger open spaces for the larger adults to move. Food and water sources are placed at the top of the dividers. Oviposition takes place in pans or trays filled with a moist oviposition substrate which are placed at the top of the carton dividers. Although females will oviposit in any wet surface, they prefer to oviposit in moist oviposition substrates with some depth to bury the eggs. Oviposition substrates can consist of peat moss, coconut husk, vermiculite, sand, cotton, or tuff. These components can also be mixed in different ratios to obtain better oviposition substrates. A good cricket oviposition substrate must be soft enough for female crickets to penetrate with the ovipositor and must retain moisture. In the United States, peat moss is the most commonly used oviposition substrate in cricket farms. Shallow containers (5–7.5 cm deep), which can be pans or trays (Fig. 6.17), are filled with the oviposition substrate and exposed to the adult crickets for 48–72 h. As mentioned earlier, eggs develop in 10–14 days (Clifford et al., 1977). When eggs are close to hatching, oviposition trays are moved inside new rearing boxes provided with food and water and kept at the same environmental conditions until most of the eggs had hatch. Oviposition trays, water, and food sources are removed, leaving first instars alone in the box.

Batches of first instars are separated to start new rearing units. The number of first instar per batch is estimated by volume (34 mL is equivalent to between 10,000 and 12,000 first instars). The number of first instars can also be estimated by weight. The mean weight of a first instar is approximately 500 µg (Morales-Ramos, unpublished), so 2000 first instars weigh approximately 1 g. The rearing density varies greatly among commercial cricket farms. Some farmers report densities of between 500 and 750 crickets in a rearing space of 28.3 L (1 ft^3). This volume of rearing space is obtained by stacking six egg cartons (30 × 30 × 5 cm). The total surface provided by the egg cartons in this volume or rearing space

FIGURE 6.17 **Rearing unit with oviposition trays (a) filled with peat moss as oviposition substrate.** Gravid adult crickets will be transferred to this box.

is equivalent to 10,800 cm² according to the estimates by Lundy and Parrella (2015) of 1,800 cm² space available on each 30 × 30 cm egg carton. Using these estimates, the densities of cricket nymphs per area of rearing space are between 4 and 7 nymphs/dm² (= 100 cm²) in commercial cricket farms in the United States. Other farms utilize the cardboard dividers in place of egg carton material, which may help with more rapid harvesting. Adult crickets that have been used for colony reproduction are commonly sold as fish bait. Once the reproductive output of females decline, adult crickets are collected and packed for sale.

In general, for harvesting adults or younger cricket nymphs, all feeding, watering, and oviposition devices are removed from the rearing box. The cardboard box dividers are removed by shaking the adult crickets into the rearing box. Adult cricket counts are done by volume as described for the first instars, after establishing the average volume occupied by 1000 adult crickets. Crickets are collected from the boxes by allowing them to climb cardboard or paper cylinders, which are later shaken to induce crickets to fall into a funnel connected to a volume measuring container. Crickets are shipped alive in specially designed boxes filled with paper packing material and typically sold by the thousand. Crickets (or any insect) for the insect-based food or feed ingredient industry to be processed should be offered frozen and priced on a per-pound basis. If harvested for the insect-based food market, the crickets are typically frozen and packaged in food-grade plastic bags and stacked into cardboard boxes which are then stored in freezers, such as walk-in freezers. For shipping, those boxes are typically stacked on standard sized pallets and then shrink-wrapped in place to be hauled via refrigerated truck to the processing location. An ideal farm for servicing the insect-based food market should also have sufficient walk-in freezer space to store full pallets (ideally a truckload, or 25 pallets at any given time, to maximize economies of scale involved with shipping, etc.).

Waxworm Production

The insects commonly known as waxworms are the larval stage of the greater wax moth, *G. mellonella* The greater wax moth is a major pest of honeybee, *Apis mellifera* L., hives and feeds on wax, pollen, honey, and larvae in honeycombs of active bee colonies (Milam, 1970); however, wax moths do not cause death of the bee colony (Sanford, 1987; Somerville, 2007). *Galleria mellonella* has been reared for more than 80 years for different purposes including experimental bioassays; production of biological control agents, such as insect predators, egg parasitoids, tachinid parasitoids, entomopathogenic nematodes, and other entomopathogens; as food for pet reptiles, such as lizards, turtles, snakes, and so forth; and for fishing bait or lure. Waxworms have been commercially produced in the United States for at least 60 years and sold as food for exotic pets and for fishing bait.

Larval development of *G. mellonella* requires between 28 and 30 days at 29–30°C and pupal development requires 6–10 days at the same temperature range (Marston and Campbell, 1973; Krams et al., 2015). Adult females start ovipositing immediately after mating (no preovipositional period) and a single female can produce between 1450 and 1950 eggs during its lifetime, depending on temperature, relative humidity, adult density, and larval diet (Marston et al., 1975). Adults do not feed and can live for up to 12 days; however, most of the females' oviposition potential is realized within 7 days after emergence (King et al., 1979).

Larval Development and Diets

The natural food of *G. mellonella* consists of honey combs containing bee wax, honey, pollen, and bee brood. Although *G. mellonella* can develop exclusively on bee's wax (Krams et al., 2015), larvae grow faster and attain a larger mass when feeding on bee brood and pollen (Somerville, 2007). Bee's wax is not required for development in *G. mellonella*, but its inclusion in the diet reduces development time and improves growth rate, mostly due to metabolic water production (Haydak, 1936; Young, 1961). Since bee's wax is expensive, efforts have been made to eliminate if from *G. mellonella* diet formulations (Marston and Campbell, 1973). The oldest artificial diet formulation devoid of bee's wax was developed by Haydak (1936) and consisted of honey (273 g), glycerin (226 g), corn flour (182 g), baker's yeast (45 g), and equal amounts (91 g) of whole wheat flour, wheat bran, and powdered skim milk. A better larval survival was obtained by Balázs (1958) using a diet consisting of corn flour (220 g), bee's wax (175 g), baker's yeast (75 g), and equal amounts (110 g) of wheat flour, powdered milk, honey, glycerin, and wheat grain. Beck (1960) was able to reduce the amount of bees' wax (47 g), while obtaining a higher larval survival with a diet consisting of baby's cereal (Pablum) (321 g), honey (236 g), glycerin (208 g), brewer's yeast (94 g), and water (94 mL). Dutky et al. (1962) eliminated the use of honey in a formulation consisting of baby's cereal (Pablum) (440 g), glycerin (210 g), sucrose (180 g), a vitamin solution (1 mL), and water (170 mL). Marston and Campbell (1973) were able to substitute the baby's cereal formula, which was expensive, with CSMA housefly formulation (Hogsette, 1992) while eliminating the bee's wax with no significant loss in larval survival with a diet consisting of CSMA formulation (337 g), honey (248 g), glycerin (218 g), brewer's yeast (99 g), and water (99 mL). The earlier large-scale mass production systems of *G. mellonella* were aimed toward the mass production of biological control agents and relied on slight modifications of Dutky et al. (1962) diet (Marston et al., 1975; King et al., 1979; King and Hartley, 1985; Gross, 1994).

Early instar larvae are commonly grown in cylindrical containers or jars, which can be made of glass, polystyrene, polypropylene, or high density polyethylene. The size of the containers can range from 1 to 4 L and are commonly modified with screened tops, bottoms, or both. Rearing containers are filled halfway with diet and *G. mellonella* eggs at a ratio of 1 mg of eggs per 12–24 g of diet (King et al., 1979; Gross, 1994). Rearing conditions for the larvae are between 28 and 30°C and between 60–75% RH, and continuous darkness. Larvae are allowed to develop in these type of containers for 2 weeks and then transferred to large pans or trays with at least 3 times the volume of the cylindrical containers. Diet is added if needed at this point (Marston et al., 1975). Larvae develop in the trays for another 2 weeks to reach full size and then are separated from the diet using sifters or a large screen with standard numbers 10 or 12 (1.4–2-mm openings). At this point, larvae can be harvested for sale or frozen for storage.

In one example of a manual harvesting/separation apparatus, King et al. (1979) used a long-screened frame (approximately 1.5 m long × 600 cm wide). The mixture of larvae, diet, and frass was dumped at one end of the frame. A 150 W lamp was placed on that end of the frame to drive negatively phototaxic larvae to move to the opposite extreme of the frame away from the light. An electric resistance (coil or heating element) inside of the frame heated its inner side to 96.6°C (lower temperatures may still be effective) to prevent larvae from escaping as they were repelled by the heat. Fully grown larvae weighing between 200 and 300 mg each were manually collected free of diet, frass, and debris at the dark end of the frame, where they accumulated (Hartley et al., 1977; King et al., 1979).

Adult Rearing and Reproduction

Some pans with fully grown larvae are allowed to complete development and pupate to supply the reproductive colony with adults. Larvae spin their cocoons on the sides of the pan, or alternatively, sheets of cardboard can be placed vertically in the center of the tray. Larvae will get into the cardboard spaces and pupate. Pans with cocoons are placed inside emergence boxes constructed from ply wood or other opaque material to maintain darkness. Emergence box dimensions can vary, but an example can be 121.9 cm high, 101.6 cm wide, and 71.1 cm deep (Marston et al., 1975). Emergence boxes can have multiple shelves inside to accommodate several pans. Emergence cages are sealed with a door lined with insulation sealing material. Translucent tubes or hoses (2.54 cm diam.) are connected from the top of the emergence cage to the rearing cages to allow emerging adults to move into them. Moths are attracted by light passing through the translucent tube (Marston et al., 1975).

Alternatively, larvae can be allowed to complete development inside cylindrical cages of jars as described earlier. Larval density in the cylinders used for reproduction should be lower than that used for production (1 mg of eggs per 10–12 g of diet). Larvae migrate to the top of the container and spin cocoons to pupate. Cylinders set for reproduction must have additional ventilation to allow evaporation of moisture accumulating from the emptying of larval gut tracts before pupation. Cocoons can be easily collected from the top of the container and placed inside emergence boxes or cages (King et al., 1979; King and Hartley, 1985).

Adults of *G. mellonella* are nocturnal and they become active during darkness. The rearing conditions for the adults are 24°C, 78–80% RH, and 14 h photophase (14:10 h light:dark). Lighting is accomplished by fluorescent cool white lighting of medium intensity. Moths lose scales continuously and they become airborne due to their small size and light weight. High levels of scales in the environment can be hazardous to humans (Reinecke, 2009) and must be controlled. A scale collector is described by Hartley et al. (1977) consisting of a network of flexible tubes connected from a main PVC tube to the upper sides of each of the oviposition cages (or containers). Air is moved from the main tube through a filter using an air blower or a vacuum pump. Each of the tubes connected to the oviposition cages has a metal screen to prevent moths from being suck into the air circulation network (Hartley et al., 1977). Davis and Jenkins (1995) designed a filtration system to manage scales in the whole rearing room consisting of two continuously operating filtration units, each drawing air in five points at low points along opposite walls. Air passes through three filters: the outer filter is a polyester pad, followed by a DQP pleated panel filter, which serves to eliminate larger particles, and the main filter is a Viledon MF-90 pocket filter capable of removing 1 μm-sized particles. The filtered air is expulsed through a damper at the top of the units to generate a continuous air current in the room (Davis and Jenkins, 1995).

The design of the oviposition boxes vary in size and complexity, but mostly consist of square boxes constructed of polyvinyl chloride (PVC) tubing, high-density polyethylene framing, metal framing, or wood with screens on two or more sides (all sides can be screened). The screen material can be polypropylene, nylon, polyester, or metal, and the opening size can be from 0.4 to 300 mm. Oviposition substrates can also differ. In one design, the oviposition substrate is introduced from the top of the cage through slits lined with weatherstripping and hangs into the interior of the cage (Marston et al., 1975). The oviposition substrate consisted of plastic sheeting (45.7 × 61 cm), which had been sprayed with a fine mist of 10% sucrose solution on water, and coated in both sides with granulated sugar. After the sugar dried completely, the sheets were introduced through the cage slits for exposure to the moths. Eggs were collected by washing the sheets with water and biodegradable detergent and passing the water through a standard No. 60 sieve (Marston et al., 1975). King et al. (1979) used waxed paper as oviposition substrate, which was introduced into the oviposition cages as accordion-folded strips through slots cut on the top of the cage, as described previously. Eggs were collected by scraping them from the sheets (King et al., 1979).

Davis (1982, 2009) describes oviposition cages designed for rearing the southwestern corn borer, *Diatraea grandiosella* Dyar; however, the system is highly adaptable for many lepidopteran species including *G. mellonella*. Cages are constructed of steel angle siding (but could be replaced by aluminum or other material) and walls are covered with 320 mm mesh hardware cloth. The top of the cage presents slots similar to those described earlier to insert waxed paper sheets as oviposition substrate. This cage also includes an opening at the bottom to introduce pupae for emergence of adults inside the cage. However, this system could be modified to include emergence boxes as described earlier. The cage dimensions are 64 × 64 × 64 cm and can house up to 1500 adults. This system has been implemented for mass rearing *Spodoptera frugiperda* (Smith), *Helicoverpa zea* (Boddie), and *Heliothis virescens* Fab. in different programs (Davis, 2009).

ENVIRONMENTAL CONTROL FOR EFFICIENT PRODUCTION OF INSECTS IN GENERAL

There is no doubt about the efficiency of insects in terms of feed conversion, but for their efficient growth the right environmental temperature must be maintained, since as poikilothermic organisms (ectotherm), insect development speed is correlated to temperature as they are unable to regulate their own body temperatures metabolically. Lower temperatures than the optimum will lead in a delay in the development and higher temperatures than the optimum will produce stress and both extremes lead to increased mortality and lower overall farm productivity. Furthermore, for healthy insect growth, humidity plays an important role. When the humidity is too high, there are increased chances of diseases and food spoilage, while humidity that is too low will produce some physiological problems and feed may dry out too quickly (for insects who depend on moisture in their feed). At the same time, insects produce CO_2 during respiration, which can accumulate in a closed environment to toxic levels. This CO_2, together with other toxic gasses, must be removed by circulating and refreshing the enclosed atmosphere with clean air. Since the air conditioning process requires energy, considerable amount of energy can be saved if the system is well implemented through the design of the climatic rooms and by the right choice of equipment and automation system. Insects are produced in many different ways according to species and farming locations go through different environmental conditions around the year; therefore, needs of equipment to ensure an efficient and economical acclimatization vary. The most important factor in environmental control or air conditioning is uniformity through space (rearing rooms) and time (through the day). In the following sections, we will address these issues and provide recommendations for designs and decision making for present and future insect farming entrepreneurs.

Estimating Optimal Conditions Required for Design for Insects in General

Since insects are unable to regulate their temperature to maintain their biological activity, insect production efficiency is directly related to environmental temperature. Every insect species has a minimum temperature threshold at which development stops, a range of temperatures at which activity and development rates increase in linear relation to the increment of the temperature, a high temperature threshold at which development does not increase substantially (critical thermal maximum), that is, at temperatures beyond this threshold, activity and development rate decrease and mortality increases (Wigglesworth, 1972). Additionally, as with any animal, insects do produce some amount of metabolic heat. Thus, it is important for this and other reasons to optimize population density and avoid overcrowding. Optimal density is highly variable and dependent on the species being reared and design of the rearing habitat. For any insect species, the closer the temperature is to the optimum level, the higher the production. Thus, for the correct design of the production facilities, information about the optimum temperature for development and reproduction of the produced insect species must be determined. In addition, the climate record from the facility location area including information on temperature ranges, relative humidity (RH), and dominant winds will be required for deciding on the election of adequate equipment and ventilation openings.

Thermal Requirements for Production

Insect production industry, as a fast growing business, could be started in many locations and often within cities. Locations in temperate countries, with cold winters and warm summers, will require equipment for heating during winter and cooling during summer. Once the optimum temperature for insect growth and reproduction is determined and knowledge of the climate conditions of the area are obtained, estimations energy requirements could be calculated by determining the gap between the extreme (maximum and minimum) recorded external temperatures and the optimal for insect rearing. Other consideration in thermal requirements must be taken into account; for instance, insects do not regulate their temperature, but they actually produce heat from metabolic activity or by mechanical friction (Heinrich, 1981). In some species like crickets, metabolic heat production can be neglected because of the relatively low densities of their rearing conditions and their relatively low metabolic activity. But gregarious species with high metabolic activity and high density production systems like mealworms and flies, produce significant levels of heat as reported by Slone and Gruner (2007). Production systems for *H. illucens* and *T. molitor* experience measurable and significant levels of metabolic heat (Hansen et al., 2006) and the heat contribution from the insects to the rearing environment cannot be neglected. In addition to insect heat production, microorganisms present in the rearing media of flies like *H. illucens* also produce heat. As a result, the sum of both insects and microorganisms heat contribution to the rearing system should be estimated and taken into account in the system design.

Vital Measurements Assay

The bibliography search and estimation of the insect thermal requirements provide just an approximation of functional requirements of the system, and in many cases, insect farm machinery designs will be overestimated to ensure that the equipment will meet the year-long rearing needs. But one of the ways to empirically determine what equipment will be needed for the rearing of each species is by using the vital measurement assay, a method based on measurements of air parameters (temperature, RH, O_2, CO_2, and NH_4) for each species using a breathing chamber (Storm et al., 2012) and microcalorimeters (Hansen et al., 2006). This method has been modified to be able to estimate production of CO_2, consumption of O_2, and production of heat and humidity by a determined number of insects. The data could be expressed as gas concentration per kilogram of larvae per minute. Two types of assay could be done in the breading chamber: a continuous air flow assay and nonair exchange assay. Each of them provides a different type of information and is described next.

Nonair Exchange Assay

To carry out the assay, a certain number of insects are weighted and introduced into an insulated box with tight lid. The box is fitted in the inside with probes to measure RH, T^a, O_2, CO_2, and NH_4 (Fig. 6.18), but other parameters could be measured if needed. Data are recorded in a computer for later analysis, but real-time graphics could also be produced. Once the lid is

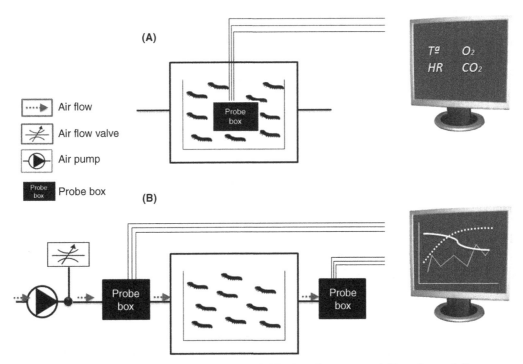

FIGURE 6.18 Scheme of air parameters measurement assay set up. (A) Nonair exchange assay and (B) continuous air flow assay.

tightly closed, the assay starts and is extended until any of the parameters reach levels that are lethal to the insects studied (normally maximum CO_2). Carbon dioxide levels higher than 6000 ppm become excessive even if the insect could survive them, because at these concentration levels, the CO_2 is dangerous for humans and should not be reached in a rearing facility. The same assay should be performed including the rearing substrate materials, since the materials can also contribute to the air parameter values. In addition air samples from the box could be extracted for mass spectrometry and combustion analysis to detect the possible presence of toxic gases produced and accumulated. The information obtained from the assay provides the amount of metabolic heat (Kcal/kg h) and water produced by respiration (g/h). Oxygen is consumed at a fix rate and when it declines to a certain minimum level, the curve start to flatten, which means oxygen consumption is decreasing. That point is the minimum oxygen the species could stand and, below this concentration, death will occur. Thus, the minimum oxygen level should be one of the criteria for the alarm system design.

Continuous Air Flow Assay

The principle of the device is to measure the different parameters (CO_2, O_2, NH_4, T^a, RH) or any other volatile solids in the air inlet and outlet of one insulated box (the breathing chamber). The airflow has to be accurately regulated and measured to know the precise amount of air passing through the breathing chamber. Precision electronic probes measure all the parameters continuously and send the information to a computer to register all the data, for later analysis and graphic construction.

All the data collected empirically can be used in the system design. Ventilation needs are calculated for the most limiting factor impacting the rearing; also, the setting points of the algorithms could be set to respond to excess of temperature, CO_2, humidity control, or any other of the measured parameters.

Light Requirements

Light could influence the development of some insect species, but most of the insect species produced for feed and food, like *T. molitor* and *Z. morio*, do not depend of light for correct development, thus their rearing could be done fully under dark conditions. Similarly, the larval stage of many fly species do not require light and have photophobic behavior. However, adult flies are often dependant on light for mating and oviposition. Fluorescent lights are sufficient for some fly species like *M. domestica*, but other fly species like *H. illucens* are impacted by light intensity and wavelength (Zhang et al., 2010), having the highest fertility when exposed to sunlight. Thus, a mating area for the adults with abundant sunlight and artificial light during low-light days has to be considered in the design, as well as a dark area for larval development. However, for some species, the day length has an impact in adult fertility and longevity and light could be controlled with an automatic timer to keep a constant photoperiod. The only lighting required in the larval rearing area are for workers to be able to see. Light emitting diode (LED) lights are compact, generate little heat, and have low energy consumption.

Space and Location

Location is important in insect production and facility design. Normally in places in which there are crowded cities, the land and space is expensive. Therefore, production systems should be very intensive, and precise equipment is needed. Since very intensive systems could be crushed easily and losses could be rather high, air purification is needed to prevent toxic substances from entering the system, as well as pests which normally are more resistant to pesticides. When space is not constrained and weather conditions are favorable (as happens in tropical countries), room acclimatization is not that important, and the equipment used is generally to mitigate the extreme conditions or to improve a certain part of the cycle.

Air Flow Design

Air flow is an important factor in the room design for mass rearing insects. Air flow is important as a way to exchange and refresh the room air maintaining the CO_2 at safe levels and eliminating other toxic gases out of the system. Air flow is also important to maintain the air movement inside the room to avoid areas of air stagnation. Stagnant air is one of the enemies of the insect farmer; stuffy air favors diseases, problems, and food degradation, resulting in unhealthy insect development, and also negatively affects the health of the workers. Because insects are normally enclosed in containers which somehow possess their own microenvironment, when we talk about optimal climate parameters for insect growth, we are actually referring to the levels of T^a, RH, CO_2, and O_2 inside the rearing boxes and not necessarily in the room per se. Conditions in the room may be similar but not exactly the same. For instance, T^a and RH in the climate room could be 23°C and 65%, while inside the rearing box, they may be 23.5°C and 75%. So, the correct relation between the set point in the room's acclimatization device and the conditions in the rearing boxes must be determined in order to maintain the insects at optimal conditions.

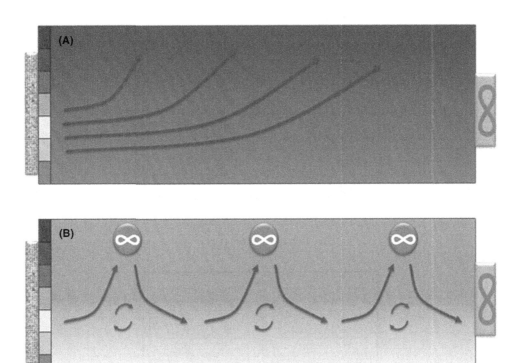

FIGURE 6.19 **Air movement tendency due to temperature-induced density gradients.** Air stratification without fans (A) and air stratification after using fans hanging down from the ceiling (B).

Probably the best air flow system is one based on laminar flow design. Bell et al. (1981) built a laminar flow chamber with excellent results attaining air uniformity and distribution throughout the room; however, the system used high-efficiency particulate arresting (HEPA) filters, which are too expensive for commercial facilities. Evaporative pads can be used in one of the walls instead HEPA filters and the air can exit through the opposite wall of the room for a more economical design (Fig. 6.22). In addition, conventional rearing rooms have a tendency to create a gradient of temperatures from the top of the room (warmer) to the bottom (colder). The stratification could produce a temperature gap of 5°C or more between the top and the bottom of the room, which can segregate the insect rearing boxes into different developmental speeds, with faster development occurring in the boxes located higher in the room. This developmental speed segregation results in a decreased production efficiency and increased complexity for management and control. Air stratification can be reverted by situating fans at the ceiling to blow air down, mixing the air from the top in a turbulent flow (Fig. 6.19).

However, boxes located in front of the air inlet or fans are normally dryer and have air parameters that are more similar to those in the room, while the less ventilated boxes have higher humidity than the average in the room. The lack of the uniformity also produces uneven development, which affects other activities of the production such as planning, selection, and so forth. All these problems could be prevented with a right airflow or laminar flow design. If installing a laminar flow design is not possible, a distribution pipe located in the middle of the main axis of the room and with exit (outlet) at the lower side of the lateral walls is a cheaper alternative and it is easy to install (Figs. 6.20 and 6.21).

In general, the design the airflow depends mainly on air inlet and outlet positioning and the caudal as well as the distribution of the shelves along the room. Laminar air flow is probably one of the most desirable airflows, as stated earlier. To achieve the laminar flow, one of the walls, with holes uniformly distributed on the surface, will act as air inlet and the opposite wall, with the same perforation pattern distribution, will act as air outlet. Thus the air will enter uniformly from one side, cross the room, and exit through the opposite side. If air demand for air renovation is high in the rearing area, the same effect could be achieved using an evaporative cooling system with evaporative pads. In this system one of the walls is constructed from evaporative panels (normally made of cardboard with evenly distributed pores for the air circulation) and in the opposite wall, heavy-duty fans act as air exhaust. Another advantage of this system is its ability to decrease air temperature while increasing humidity, which is very useful for the rearing of certain species. As a general rule, and especially in a laminar air flow design, shelves should be oriented in parallel to the airflow, or the shelves will act as a barrier for the air movement. When the air flows in parallel with the shelves, only the first boxes from the corridors are exposed directly

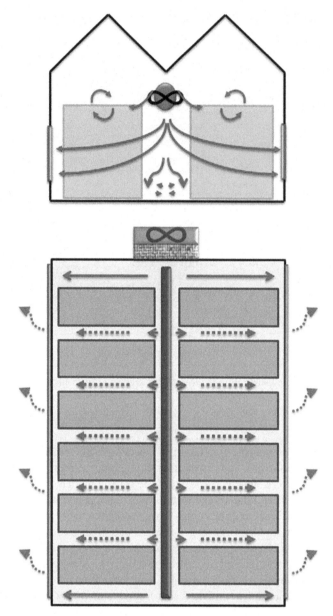

FIGURE 6.20 Turbulent air flow system scheme.

to the air inlet making them drier, but the rest of the boxes are exposed to uniform air conditions resulting in uniform insect development. When shelves are perpendicular to the air flow, the first shelf is exposed to the air inlet, acting as a wall, and the rest of the shelves will have very poor air conditions, especially in the lower parts (Fig. 6.22).

The design possibilities on air flow are infinite depending on the distribution of the air inlet and outlet, but as discussed before, the best results are achieved when the air flows between the shelves evenly. Special care should be taken not to relegate inlet and outlet to the ceiling because the airflow will short circuit, and the air distribution in the room will be uneven. One important consideration is the distance between air inlet and outlet. In intensive production operations, insects release high levels of metabolic heat and water, increasing both the temperature and humidity of the air as it crosses over the rearing boxes. The gap of temperature and humidity becomes higher with the distance to the outlet (Fig. 6.23).

Renovating and circulating air from the exterior creates contamination risks, thus it is recommended to adapt the air renovation system in such a way that the air is filtered at the point it enters and before it flows across the rearing room and exits the system. However, if air renovation requirements are excessive, like in regions of extreme climates, the cost of conditioning the air may be too high. Thus, implementing a complete internal filtration system becomes essential under these conditions.

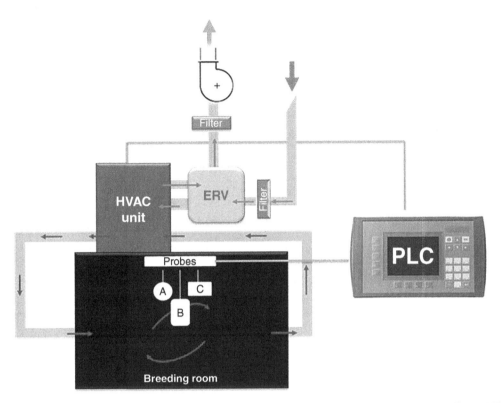

FIGURE 6.21 Central HVAC system scheme. Air enters the system, is filtered, and passes through the enthalpy recovery ventilator (ERV), where temperature is exchanged with the exiting, conditioned air to the set point in the HVAC. It is then introduced in the rearing rooms, after passing the room return to the HVAC, where a portion is rejected passing the ERV and the remaining air returns to the room after conditioned. The whole process is controlled with the PLC.

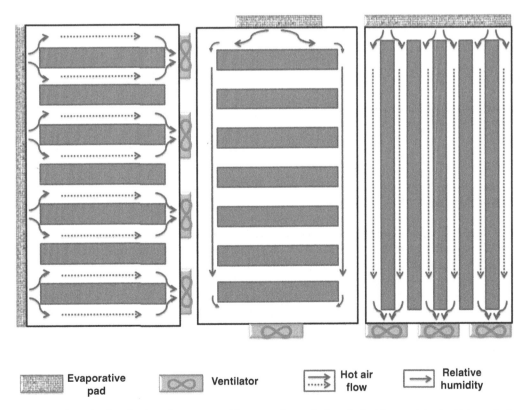

FIGURE 6.22 Shelves orientation in a laminar flow system with evaporative pads. Longitudinal (upper picture), which require lower investment but produces a higher difference in air parameters between inlet and outlet. Perpendicular air flow (lower picture), which require a higher investment but is capable of producing a higher level of air uniformity.

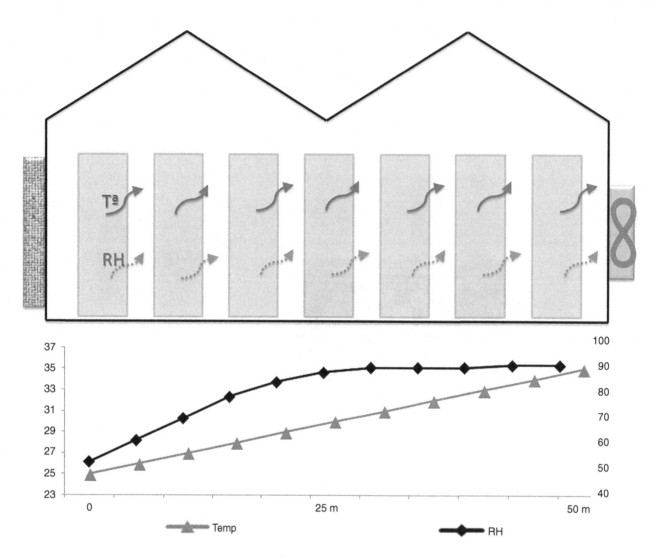

FIGURE 6.23 **Temperature and humidity evolution along the room.** Inlet air has similar air parameters than the set point in the HVAC systems but increase temperature and humidity though the rearing room.

Air flow design might require complex computer assisted programs, the most sophisticated program for the air flow design is computational fluid dynamics (CFD), which is a branch of fluid mechanics that uses numerical analysis and algorithms to solve and analyze problems that involve fluid flows, heat transfer systems, and other physical processes. The CFD operation is based on solving complex equations of fluid flows in a region whose characteristics are known. CFD is commonly applied by engineers and scientists in a wide range of areas such as: industrial processes (chemical reactors), building construction (design of ventilation, energy efficiency), health (effect of toxic gases), electronic (heat transfer circuits), power and energy (optimization of combustion processes), and motor industry (modeling of combustion systems and vehicle aerodynamics).

The CFD is a very powerful tool with applications in the design of insect production systems. It allows the planning of the construction to be adapted to the local characteristics. It is also capable of designing optimal airflow and energy consumption systems to provide the best conditions for the development of insects in a production facility. Using this method, it is possible to reduce the time required for design and construction of the facilities through simulation, visualization, calculation, and determination of the fluid dynamics in the system. The end result of the processed data, including the distributions of pressure, speed, flow, and turbulence (as well as the calculation of moments, forces, etc.), can be graphically shown. It is important to collect local information on the prevailing wind orientation and variations, temperature ranges, and other abiotic parameter information in order to obtain realistic simulations.

Equipment for Climate Control

Again, there are several elements for acclimatization which have to be considered during the production design setting. All the elements are somehow linked in such a way that the efficiency of the system could fail if one of the elements was badly selected. It is not uncommon to find designs of a mass rearing room incorporating a very efficient and well dimensioned inverter air conditioner for the facility, even overestimated for the room size, but with an inadequate humidifier. For example, assuming that the humidity set point is 65%, the following set of events are set in motion: (1) when the humidity probes register 60% the programmable logic controller (PLC) will activate the humidifiers; (2) if the humidifier overshoots the desired humidity mark by evaporating more water than needed, (3) in a very short time, the humidity could rise up to 75% due to the system's slow reaction time (time between reading and acting); (4) the system would start refreshing the air in the room to correct the parameter and the new air would cool down the room; (5) the PLC will receive the order to heat up again, dropping the humidity in the process, and finally; (6) the system will start over in continuous repetition. This could double the energy consumption as compared with a system in which the humidifiers could increase the humidity in the room in a short time. All the elements are equally important, since even a simple valve failure could make the system collapse. The next sections discuss the adequacy of different parts of the mass rearing acclimatization system, and important information regarding their use will be provided.

Filters

Filters are mainly used as a barrier for volatile solid removal. Filters are commonly neglected in many production systems, but they are important in order to maintain a barrier between the system and the external environment. It must be taken into consideration what type of contaminants could potentially enter the rearing system from the outside. The main concern are contaminants that can threaten the health and viability of the insect production, such as pesticides and biologicals, including entomophagous and entomopathogens. Insecticides may come from any nearby farm, garden, or just from any household insecticide treatment. Parasitoids, predators, or competing species are not highly deleterious for the integrity of the rearing, but could grow and spread up to the point of producing economical losses. Infestations with undesirable insect species are difficult to control or eliminate due to the fact that no insecticides can be sprayed inside the rearing rooms and only a limited number of chemicals can even be introduced within baiting systems or lures. The external air usually contains microorganisms that can include entomopathogenic species, such as fungi, bacteria, or viruses, which, once they enter the rearing system, could produce serious damage or the complete loss of the production. One of the most dramatic examples is the cricket densovirus, which has decimated the cricket industry in the United States and Europe (Weissman et al., 2012). In addition, insects, and especially insect exuviate, are known to be allergens to many people and must be removed from the facility environment to prevent or reduce allergic reactions to personnel or to the population living near the production facility. Filters are one of the cheapest preventive measures for all the potential problems arising in an insect mass-rearing operation as a consequence of external air circulation.

The choice of filter type depends on the potential hazards present in the adjacent area of the facility, for instance, activated carbon filters are recommended in urban or agricultural areas where the use of insecticides is high (in the latter case, relocation would probably be a better to choice). It would make less economic sense to install activated charcoal filters in areas considered clean from pesticides. The type of filters available and commonly used in insect rearing are described next.

Mechanical Filters

Mechanical filters are the cheapest to be incorporated in the system and are used to prevent the entrance of most of the contaminations. They are among the simplest filters consisting of nets that could be constructed from different materials (stainless steel, nylon, or polypropylene) and are useful to prevent rodents, birds, or any other strong animals to enter though the ventilation system. The majority of nets are made out of plastic materials, which are cheap, easy to clean, and simple to replace. Nets sizes range from 10 micron pores to several millimeters. Larger pore size could be used as prefilters while the smaller pore sizes are used as filters. Other types of mechanical filters are made of sponge-like materials or fibrous fabrics which create a three-dimensional structure where particles are retained. There are many choices available in the market and choice decision depends on facility needs, cost, particle size to retain, simplicity of cleaning, and replacement. Some of those filters also incorporate other characteristics such as activated carbon particles and electrostatic properties. One disadvantage of mechanical filters is that they produce load loses in the air circulation due to friction with the air, increasing dirt accumulation in a direct relation. To prevent the energy loses, the best option is to increase filter area, which allows the air to move slowly through a larger number of pores, decreasing friction.

Carbon Filter

This type of filters is useful to remove toxic elements from the air, mainly by absorption. The activated carbon is able to take out pesticides, organic solvents, and toxics from the air (Ao and Lee, 2004). The filter life is limited since the activated carbon absorbs particles, reducing its activity level over time, and it must be changed after certain number of operational hours. There are some filters that have two layers of porous material, which are filled in between with activated carbon. This type of filter will retain solid particles and volatile materials, but particle accumulation will reduce the operational life of the filter, so, in this case, it is recommended to be used in combination with a prefilter.

Electrostatic Precipitator

This type of filter negatively charges the incoming particles, which then get attracted to a positively charged plaque. They are highly efficient at capturing particles from the air, including microorganisms and some other pollutants. They are more expensive and difficult to clean than the filter types reviewed earlier, but do not get clogged with the accumulation of particles as mechanical and activated charcoal filter types do. Electrostatic precipitator filters are especially recommended for purifying the air inside the rearing area. One advantage over mechanical filters is that they do not produce load losses on the exhaust.

High-Efficiency Particle Arresting

This is a very sophisticated type of filter composed of randomly arranged fibers (typically fiberglass) creating very fine porous material with three-dimensional structure, which can retain 99.97% of airborne particles 0.3 μm in diameter, including bacteria and viruses. HEPA filters work better in precleaned environments (prefiltered air) because a large number of even small particle sizes can jam it. This type of filter has a high cost and its use is not common in mass production of insects, especially within the industry of insect production for food and feed, and this kind of filter is more often used in the United States.

Ultraviolet Filter

More than a filter the ultraviolet (UV) filter is a sanitation device which eliminates microorganisms from the atmosphere. UV filters do not remove particles, but can reduce the amount of entomopathogens that may be transmitted by air, including viruses. The UV light activity is inversely proportional to the distance, thus it should be located where the airflow will pass close to the UV bulb. For efficient pathogen removal, lamp power must be calculated taking airflow speed into account, or the UV exposure may not have the desired effect on the pathogens if the organisms pass too quickly through the exposure area.

Temperature Conditioners

To keep the temperature at the required level, several elements could be used depending the requirements of the insect species, and systems would be more focused on heating or cooling, depending of the local climate. In conventional animal farming, heating or cooling is normally provided just to prevent exposure to extreme conditions; in insect farming, it is a requirement to ensure productivity. The most commonly implemented devices available for heating in insect mass rearing are presented next.

Water Heating System

Hot water produced in a heater is distributed via pipes to the heated room. There are many types of water heating systems offered, which use a variety of energy sources such as fossil fuel, coal, biomass, geothermic, or solar. Equipment types include: storage tank water heaters, on demand water heaters, heat pump water heaters, hybrid heat pump water heaters, hot-water supply boilers, and the combination of them. The high variety of potential choices of equipment for heat delivery could create a "barrage" of possibilities. If heating is provided from the floor (eg, through hot water piping), the heat will distribute uniformly in the room by lifting the hot air from the base to the top. But if heat is to be distributed by air, this can be done using hot-water heat exchangers. Construction of this type of system is costly, but when heating requirements are high, this is one of the most commonly used systems, especially recommended for areas with cold weather. This system only provides heating, so if cooling is also required, it must be combined with cooling equipment.

Air Conditioners

This type of machine simply moves heat from one side to another, where the heat destination is named the heat sink. Air conditioners or heat pumps could be used either to heat or to cool down rooms. Heat pumps are several times more efficient than simple electric resistance heaters at utilizing electricity to heat a space. Heat pumps are often used in insect mass rearing facilities because, in addition to being energy efficient, they can be used for cooling and heating using one single device and are easy to implement in climate automatization systems. Implementation cost is high but they are very flexible and have a fast reaction to changes. One of the latest innovations in heat pumps is the use of geothermal energy (geothermal heat pumps). They save energy but are more expensive due to ground perforations required for installation of the geothermic heat exchanger. Geothermal heat pumps have not been implemented in insect farming facilities, but the technology is ready to be implemented and it is currently in use in conventional farms and greenhouses.

Evaporative Panels (Pads)

This system is extensively used for cooling in conventional animal farms and greenhouses. The system works by circulating air through a wall cover with evaporative pads (normally cardboard) continuously moistened by running water. In the opposite wall, fans act as exhaust and extract the air from the room forcing a continuous air flow through the moistened panels. The air cools down by transferring energy for water evaporation (Redding, 1981), and in the process, the air absorbs water increasing RH. The maximum theoretical temperature decrease attainable is 10°C from the original air temperature, but more commonly a decrease of between 4 and 5°C is achieved (Redding, 1981). Mineral content of the water source could be high since the solids will precipitate in the pad and remain in it until it is removed for washing without negatively affecting the system performance. The systems have some limitations; as RH increases to the point of saturation and water evaporation speed declines, the cooling process becomes inefficient. The best performance of the cooling pad system is attained when the external air is very dry, thus it is recommended for arid and semiarid areas.

Humidifiers

As a general rule, humidifiers work by evaporating water and mixing it with the air in different ways. As mentioned before, humidity is an important factor for insect production because an adequate level of air humidity is a requirement for healthy development and for preventing diseases and deformities. But a system to increase humidity also cools down the air. The most common systems to increase humidity are listed next.

Evaporative Pads

This is a very commonly used system to humidify and simultaneously cool the air before entry into the rearing area because its geometry allows a laminar flow configuration that facilitates air circulation. The system works exactly as an evaporative panel.

Nebulization With High and Low Pressure Nozzle

This system is based on water pulverization into small droplets, which evaporate easily, increasing humidity and cooling down the air in the process. The efficiency of the system increases as the droplets decrease in size, evaporating before falling and maintaining dryness of the surrounding surfaces. Low-pressure nozzles produce relatively big drops which normally wet the surface below. High-pressure nozzles produce smaller drops reducing surface wetting, but need the required pipes, nozzles, and pumps, which are more expensive. The best water pulverization is achieved with a combination water–high-air-pressure nozzle, which can produce very small droplets. Because the small openings of the nozzle could be clogged by solid particles, the system should use a water filter and water should be soft (low electrolyte concentrations). The system could be installed high in the ceiling along the rearing room to allow droplet evaporation before contacting any surface.

Ultrasonic Nebulization

This system is new on the market and it is capable of producing very small droplets using ultrasonic energy. The droplets are so small that they normally do not wet any material located in the vicinity. The required investment is higher than the conventional nozzles, but installation is easier and the sonic system can tolerate water of lower quality (but will last longer with high-quality water), making water filtration optional. This system can be recommended for smaller facilities with lower production outputs.

Steam Humidifiers

This system uses a water heating source to produce steam, which can be injected either directly in the rearing room or into the air conditioning system before entering the room. This system is capable of injecting large amounts of water in the rearing room atmosphere, while at the same time providing heat input when warming is needed. An alarm or cutoff system is necessary to avoid overheating in the event of steam malfunction. This system is recommended for facilities located in areas of cold weather, which can use water boilers to produce vapor to be used for heating and humidifying the rooms.

Centrifugal Atomization

This system is based on rotating blades (normally a fan) that breaks water drops into small droplets, which become smaller with each blade revolution and are propelled into the air by centrifugal force. These systems do not work well in dirty environments, requiring high maintenance. In addition, they produce a bad distribution of the water, which can create wet spots in the room.

Dehumidifiers

In some cases it is necessary to decrease humidity in the rearing room environment, but the operation of dehumidifiers is expensive and is probably used only in the production of insects for food. The excess humidity should be controlled by air ventilation as much as possible. Fans are an important element to force air movement inside the rearing room to create a turbulent flow that breaks down air stratification, allowing an even distribution of fresh air and preventing air stagnation. Air conditioners also can act as dehumidifiers because the water condenses in the evaporator, and is thus removed from the air. Dehumidification from air conditioners only works when the air exchange ratio with the outside is low but at high air exchange, the effect could be negligent.

Automatic Control and Artificial Intelligence

Automation in the Insect Production Industry

Industrial system automation is well developed in many areas, such as in the car industry, bottling and packaging, chemical plants, power generation, and so on. In these manufacturing industries, automation is mainly used for solving problems in the machinery sequencing and serialized production processes. Their goals are to obtain higher quality products (more accurate and homogeneous), in a more efficient way, and at reduced costs.

In the current insect production industry, much time, effort, and resources are invested in designing and building the facilities for insect mass rearing so that a certain professional level is reached. The design of these facilities takes into account many aspects such as building orientation, tightness, calculation of refrigeration and ventilation, and the like. However, the criteria followed by the automation of production processes are usually just orientated to control abiotic conditions in the facilities, such as temperature, humidity, CO_2, and lighting. As a result, numerous important parameters for insect production are ignored, and directly impact the productivity when attempting to monitor and automate the insect production system. Some of the biotic factors that are still out of the control of automatization could make an impact in the ventilation flow, dynamic changes in the feeding substrate, mobility of the supporting structures, measures to control entomopathogens, separation of biological material by size, and so on. However, the monitoring of abiotic factors such as oxygen consumption, production of CO_2, metabolic heat, food consumption, weight gain and so forth could be used as an indirect measurement of the biotic factors and could provide an idea of growth rates and development stage, and that data can be exchanged with the ERP.

Automation Control Platforms

Sensors allow monitoring and registering of environmental conditions and internal functioning of the productive system. They are able to generate responses to certain strategies and priorities previously programmed. This automation is possible thanks to massive data management capabilities controlled through PLC using proportional integral derivatives (PID) as heuristic layers to improve the control system.

Programmable Logic Controllers

The use of programmable logic controllers began in 1968 in the automotive industry, replacing complex and inefficient systems that required thousands of relays, timers, or sequencers. They are electronic devices that can be programmed to control machines or sequential processes (automatons) in industrial environments. PLC functions allow them to detect or receive signal inputs from the industrial process, produce and send commands to the system, receive operator settings, send reports, and alter its programming. PLCs are commonly used to replace the relays in machines, turbines, motors, and the like. They can also act as an intermediary between a personal computer (PC) and the manufacturing process, identifying faults and alarms, remote controlling machinery, and so forth.

PLCs require a lower level of wiring utilization, have great flexibility in programming and configuration of the automation process, and are easy to maintain. They can be adapted to any changes in production processes. Another important feature is their ability to communicate through a variety of protocols, allowing them to control different processes by sharing information between several PLCs, diagnose remotely, transfer files, send alarms, and remotely diagnose. There are many brands and models from different manufacturers, with different characteristics, providing a variety of PLC choices for the specific needs of each industrial process.

Programmable Automation Controllers

The evolution of PLCs has led to the development of programmable automation controllers (PACs). In fact, in the current field of automation, the use of PACs is more common than PLCs. The advances in the automation of processes have led to the use of multiple existing technologies, which have arisen from the routine use of personal computers. Therefore, modern controller drivers are more like computers than automatons. They add a great capacity for calculation and monitoring options. These platforms allow cost reductions and greater control over the automation process.

Automation Control Mechanisms

Along with the operation of the PLC, proportional integral derivative controllers (PID) drivers have been developed. These drivers include a series of algorithms to build a control mechanism. The PID calculates the deviation between a measured value and the value to be obtained, so that corrective action is taken to adjust the process. These algorithms represent a link between the automation programming processes, and the "reality" happening during the automated actions taking place according to the parameters collected by the PLC. The PID allows the automation system to "learn" how the system responds to the automated actions. To do this, the algorithm takes into account the current state (proportional), the past state (whole), and a forecast of the future (derivative) of the deviation given by the desired behavior or action. Learning improves response in the automation system, so that greater accuracy is achieved in the process, which lowers energy consumption and costs. PIDs are quite complex, and can be divided into three components that are expressed in the final PID algorithm formula:

$$y(t) = \text{MV}(t) = K_p e(t) + Ki \int_0^t e(\tau) d\tau + K_d \frac{de}{dt}$$

Proportional Effect

It depends on the error, but it is simply the difference between the set point value and the value of the variable that the process shows. The proportional gain (K_c) determines the ratio of the output response with respect to the error or deviation received.

Integral Effect

This component aims to integrate the error over time to minimize it. This means that the integral response will increase over time unless it is zero. Integral action can cause oscillation or instability problems if the integral gain (T_i) is too small.

Derivative Effect

The derivative component of the PID aims to anticipate the behavior that the error levels will have, since the response of this component is proportional to the rate of error change, thus eliminating the oscillations. It is important to note that most control systems use a little derivative gain (T_d), so, the derivative component is very sensitive to noise in the signal of the process variable. The PID must be adapted to each of the parameters controlled in the automation system, so it is necessary to have sufficient data and continuous monitoring to obtain an accurate result.

Artificial Intelligence

The purest application of artificial intelligence (AI) in the industry has been sought since the very beginning of the development of computing. That is, not only the automation of processes, but the development of ways to emulate the human information processing with electronic computers. Direct applications in solving problems in the production industry are not yet implemented, but facilitates are capable of data processing. As discussed in the section covering massive data processing, a large amount of information is constantly being received by the control systems, making it difficult to identify relevant information. An interesting approach would be to discover how the combination of AI and machine learning can optimize and facilitate the work of automation systems, decision making, and information processing to identify patterns in the data and provide quality and accurate knowledge.

CONCLUDING REMARKS

The field of insect mass rearing began with the mass production and release of sterile males for autocidal control of flies like the screwworm during the 1950s and 1960s. Some of the technologies presented in this chapter can be traced back to that time. Mechanization and environmental controlled room technology flourished with the efforts to produce large numbers of insect hosts and prey to mass-produce insect natural enemies for biological control during the 1970s, 1980s, and 1990s. Rearing technologies resulting from those efforts are still in use today, but many have been lost as government-funded mass production efforts had closed after sterile male releases succeeded in eradicating pests, and biological control production moved from government to commercial. Earlier insect mass-production programs had in common that they focus on pest control and success depended on the numbers of individuals produced. Future developments in insect mass rearing will likely be directed to the production of food and feed, with the main focus on the volume of biomass produced, rather than the number of individuals. As attitudes toward eating insects or including insect ingredients in food change, the demand for insect biomass will increase. At the present time, the state-of-the-art technology for insect mass production is inadequate to supply the potential demand for insect biomass. The technologies presented in this chapter, which represent the state-of-the-art, are still primitive and can be greatly improved. This presents an extraordinary opportunity for the scientific, food processing, and engineering communities for research and development of novel, meaningful technologies for the mass production and processing of insect biomass as food ingredients. We hope that the information provided in this chapter will inspire a new generation of professionals to pursue the development of such advances.

REFERENCES

Abdollahi, M.R., Ravindran, V., Svihus, B., 2013. Pelleting of broiler diets: an overview with emphasis on pellet quality and nutritional value. Anim. Feed Sci. Technol. 179, 1–23.

Agosin, M.R. (Ed.) 1978. Funtional role of proteins. In: Biochemistry of Insects. Academic Press, New York, pp. 93–144.

Althouse, A.D., Turnquis, C.H., Bracciano, A.F., 2004. Modern Refrigeration and Air Conditioning. The Coodheart-Willcox Company, Inc., Tinley Park, IL.

Ao, C.H., Lee, S.C., 2004. Combination effect of activated carbon with TiO_2 for the photodegradation of binary pollutants at typical indoor air level. J. Photochem. Photobiol. A 161, 131.

ASHRAE, 1980. American Society of Heating, Refrigeration and Airconditioning Engineers, Handbook of Fundamentals, New York, NY.

Balázs, A., 1958. Nutritional and nervous factors in the adaptation of *Galleria mellonella* to artificial diet. Acta Biol. Acad. Sci. Hung. 9, 47–69.

Banks, I.J., Gibson, W.T., Cameron, M.M., 2014. Growth rates of black soldier fly larvae fed on fresh human faeces and their implication for improving sanitation. Trop. Med. Int. Health 19, 14–22.

Barriga-Ruiz, C.A., 1994. Cochineal production in Peru. Paper presented at the FAO Expert Consultation Meeting on Non-wood Forest Products in Latin America, held in Santiago, Chile in July 1994.

Barroso, F.G., de Haro, C., Sánchez-Muros, M.J., Venegas, E., Martínez-Sánchez, A., Pérez-Bañón, C., 2014. The potential of various insect species for use as food for fish. Aquaculture 422, 193–201.

Beck, S.D., 1960. Growth and development of the greater wax moth *Galleria mellonella* (L.) (Lcpidoptera: Galleriidae). Trans. Wis. Acad. Sci. Arts Lett. 49, 137–148.

Bell, R.A., Owens, C.D, Shapiro, M., Tardif, J.R., 1981. Mass rearing and virus production, pp. 599–655. In: Doane, C.C., McManus, M.L. (Eds.), The Gypsy Moth: Research Toward Integrated Pest Management. Forest service, SEA, APHIS Tech. Bull. 1576.

Bills, C.E., Massingale, O.N., Prickett, P.S., 1930. Factors determining the ergosterol content of yeast. I. Species. J. Biol. Chem. 87, 259–264.

Bondari, K., Sheppard, D.C., 1987. Soldier fly *Hermetia illucens* L., as feed for channel catfish, *Ictalurus punctatus* (Rafinesque), and blue tilapia, *Oreochromis aureus* (Steindachner). Aquacult. Fish. Manage. 18, 209–220.

Botella, L.M., Ménsua, J.L., 1986. Larval arrest in development of *Tribolium castaneum* (Coleoptera: Tenebrionidae). Environ. Entomol. 15, 1264–1267.

Cai, K.J., Zhang, L.Q., Liu, L., 2008. Breeding technology of super mealworm. Mod. Agric. Sci. 15, 38–39.

Chang, B., Han, R.C., Cao, L., Liu, X.L., Liu, X.F., 2007. Effect of Musca domestica maggot and pupae as feed additives on the quality and flavor of Qingyuan chickens. Chinese Bul. Entomol. 44, 882–886.

Chapman, R.F., 1998. The insect: Structure and function, fourth ed. Cambridge University Press, Cambridge.

Chapman, R.F., 2003. In: Resh, V.H., Cardé, R.T. (Eds.), Encyclopedia of Insects. Academic Press; San Diego. 1266 p.

Chen, G.F., Liu, T.J., 1992. Studies on the blonomics and breeding of yellow mealworm, *Tenebrio molitor* L. J. Fujian Normal Univ. 8, 66–74.

Chen, J., Zhao, G., Han, R., 2014. Mass production of housefly (*Musca domestica*) for waste reutilization and as feed for chickens and shrimps. International Conference "Insects to Feed the World," Wageningen University, Wageningen, Netherlands, pp. 14–17.

Chen, X.O., 1999. The present situation and prospects of utilization of resource insects in China. World Forest. Res. 1, 46–52.

Chippendale, G.M., 1978. The functions of carbohydrates in the insect life process. In: Rockstein, M. (Ed.), Biochemistry of Insects. Academic Press, New York, NY, pp. 1–55.

Clifford, C.W., Roe, R.M., Woodring, J.P., 1977. Rearing methods for obtaining house crickets, *Acheta domesticus*, of known age, sex, and instar. Ann. Entomol. Soc. Am. 70, 69–74.

Cohen, A.C., 2004. Insect Diets Science and Technology. CRC Press, Boca Raton, FL.

Collavo, A., Glew, R.H., Huang, Y.S., Chuang, L.T., Bosse, R., Paoletti, M.G., 2005. House cricket small-scale farming. In: Paoletti, M.G. (Ed.), Ecological Implications of Minilivestock: Potential of Insects, Rodents, Frogs and Snails. Science Publishers, New Hampshire, pp. 519–544.

Data of Chinese Customs, 2014. Available from: http://www.haiguan.info/Complex/List.aspx?key=%u9EC4%u7C89%u866B.

Connat, J.L., Delbecque, J.P., Glitho, I., DelaChambre, J., 1991. The onset of metamorphosis on *Tenebrio molitor* larvae (Insecta, Coleoptera) under grouped, isolated, and starved conditions. J. Insect Physiol. 37, 653–662.

Costa Neto, E.M., 2005. Entomotherapy, or the medicinal use of insects. J. Ethnobiol. 25, 93–114.

Cotton, R.T., St George, R.A., 1929. The meal worms. Tech. Bull. US Dept. Agric. 95, 1–37.

Davis, F.M., 1982. Southwestern corn borer: oviposition cage for mass production. J. Econ. Entomol. 75, 61–63.

Davis, F.M., 2009. Insect rearing production systems, a case study: the South-western corn borer. In: Schneider, J.C. (Ed.), Principles and Procedures for Rearing High Quality Insects, Mississippi State University, Starkville, MS, pp. 307–333.

Davis, F.M., Jenkins, J.N., 1995. Management of scales and other insect debris: occupational health hazard in a lepidopterous rearing facility. J. Econ. Entomol. 88, 185–191.

Diener, S., Studt Solano, N., Roa Gutiérrez, F., Zurbrügg, C., Tockner, K., 2011. Biological treatment of municipal organic waste using black soldier fly larvae. Waste Biomass Valor. 2, 357–363.

Diener, S., Zurbrügg, C., Tockner, K., 2015. Bioaccumulation of heavy metals in the black soldier fly, *Hermetia illucens* and effects on its life cycle. J. Insects Food Feed 1, 261–270.

Dossey, A.T., 2011. Chemical defenses of insects: a rich resource for chemical biology in the Tropics. In: Vivanco, J., Weir, T. (Eds.), Chemical Biology of the Tropics: An Interdisciplinary Approach. Springer, New York, NY, pp. 27–57.

Dossey, A.T., 2010. Insects and their chemical weaponry: new potential for drug discovery. Nat. Prod. Rep. 27 (12), 1737–1757.

Downer, R.G., 1978. Functional role of lipids in insects. In: Rockstein, M. (Ed.), Biochemistry of Insects. Academic Press, New York, NY, pp. 57–92.

Dunbar, B.S., Winston, P.W., 1975. The site of active uptake of atmospheric water in larvae of *Tenebrio molitor*. J. Insect Physiol. 21, 495–500.

Dunphy, G.B., Niven, D.F., Chadwick, J.S., 2002. Iron contributes to the antibacterial functions of the haemolymph of *Galleria mellonella*. J. Insect Physiol. 48, 903–914.

Dutky, S.R., Thomsom, J.V., Catwell, G.E., 1962. A technique for mass rearing of greater wax moth. (Lepidoptera: Galleridae). Proc. Entomol. Soc. Wash. 64, 56–58.

Erickson, M.C., Islam, M., Sheppard, C., Liao, J., Doyle, M.P., 2004. Reduction of *Escherichia coli* O157:H7 and *Salmonella enterica* Serovar enteritidis in chicken manure by larvae of the black soldier fly. J. Food Protect. 67, 685–690.

Esperk, T., Tammaru, T., Nylin, S., 2007. Intraspecific variability in number of larval instars in insects. J. Econ. Entomol. 100, 627–645.

Feng, Y., Chen, X.M., 2002. Resource value of edible insects and utilizable ways. Forest Res. 15, 105–110.

Gasco, L., Belforti, M., Rotolo, L., Lussiana, C., Parisi, G., Terova, G., Roncarati, A., Gai F, 2014a. Mealworm (*Tenebrio molitor*) as a potential ingredient in practical diets for rainbow trout (*Oncorhynchus mykiss*). Abstract Book Conference Insects to Feed The World, The Netherlands, p. 78.

Gasco, L., Gai, F., Piccolo, G., Rotolo, L., Lussiana, C., Molla, P., Chatzifotis, S., 2014b. Substitution of fish meal by *Tenebrio molitor* meal in the diet of *Dicentrarchus labrax* juveniles. In: Abstract Book Conference Insects to Feed The World, The Netherlands, p. 80.

Giannone, M., 2003. A natural supplement made of insect larvae. Rivista di Avicoltura 72, 40–41.

Greenberg, S., Ar, A., 1996. Effects of chronic hypoxia, normoxia and hyperoxia on larval development in the beetle *Tenebrio molitor*. J. Insect Physiol. 42, 991–996.

Gross, H.R., 1994. Mass propagation of *Archytas marmoratus* (Diptera: Tachinidae). Environ. Entomol. 23, 183–189.

Guy, R., 2001. Extrusion Cooking: Technologies and Applications. Woodhead Publishing, Cambridge, 206.

Hale, O.M., 1973. Dried *Hermetia illucens* larvae (Stratiomyidae) as a feed additive for poultry. J. Georgia Entomol. Soc. 8, 16–20.

Hammond, C.R., 1996. The elements. In: Lide, D.R. (Ed.), CRC Handbook of Chemistry and Physics, 77th edition 1996–1997, Section 4. CRC Press, Boca Raton, FL, pp. 1–34.

Han, R.C., Chen, J.H., 2010. A method for rapid separation of housefly larvae. Chinese Patent CN 101317559 B.

Hansen, L.L., Ramløv, H., Westh, P., 2006. Metabolic activity and water vapour absorption in the mealworm *Tenebrio molitor* L. (Coleoptera Tenebrionidae): real-time measurements by two-channel microcalorimetry. J. Exp. Biol. 207, 545–552.

Hartley, G.G., Gantt, C.W., King, E.G., Martin, D.F., 1977. Equipment for the mass rearing of the greater wax moth and the parasite *Lixophaga diatraeae*. US Department of Agriculture, Agricultural Research Service, ARS-S-164, 4 p.

Haydak, M.H., 1936. Is wax a necessary constituent of the diet of wax moth larvae? Ann. Entomol. Soc. Am. 29, 581–588.

Heinrich, B., 1981. Insect Thermoregulation. John Wiley & Sons, Inc., New York, NY.

Hirashima, A., Takeya, R., Taniguchi, E.M., Eto, M., 1995. Metamorphosis, activity of juvenile-hormone esterase and alteration of ecdysteroid titres: effects of larval density and various stress on the red flour beetle, Tribolium freemani Hinton (Coleoptera: Tenebrionidae). J. Insect Physiol. 41, 383–388.

Hogsette, J.A., 1992. New diets for production of house flies and stable flies (Diptera: Muscidae) in the laboratory. J. Econ. Entomol. 85, 2291–2294.

Holmes, L.A., Van Laerhoven, S.L., Tomberlin, J.K., 2010. Lower temperature threshold for black soldier fly (Diptera: Stratiomyidae) egg and adult eclosion. American Academy of Forensic Sciences, Seattle, Washington, p. 310.

Holmes, L.A., VanLaerhoven, S.L., Tomberlin, J.K., 2012. Relative humidity effects on the life history of *Hermetia illucens* (Diptera: Stratiomyidae). Environ. Entomol. 41, 971–978.

Hopkins, T.L., 1992. Insect cuticle sclerotization. Annu. Rev. Entomol. 37, 273–302.

House, H.L., 1961. Insect nutrition. Annu. Rev. Entomol. 6, 13–26.

Jia, S.F., 2007. Study on maggot separation, processing and preservation. Feed Res. 7, 36–37.

Kampmeier, G.E., Irwin, B.E., 2009. Commercialization of insects and their products. In: Resh, V.H., Cardé, R.T. (Eds.), Encyclopedia of Insects. second ed. Academic Press, Burlington, pp. 220–227.

King, E.G., Hartley, G.G., 1985. Galleria mellonella. Singh, P., Moore, R.F. (Eds.), Handbook of Insect Rearing, vol. II, Elsevier, New York, NY, pp. 301–305.

King, E.G., Hartley, G.G., Martin, D.F., Smith, J.W., Summers, T.E., Jackson, R.D., 1979. Production of the tachinid *Lixophaga diatraeae* on its natural host, the sugar cane borer, and on an unnatural host, the greater wax moth. US Department of Agriculture, Science and Education Administration, AAT-S-3, 15 p.

Krams, I., Kecko, S., Kangassalo, K., Moore, F.R., Jankevics, E., Inashkina, I., Krama, T., Lietuvietis, V., Meija, L., Rantala, M.J., 2015. Effects of food quality on trade-offs among growth, immunity and survival in the greater wax moth *Galleria mellonella*. Insect Sci. 22, 431–439.

Lee, H.L., Chandrawathani, P., Wong, W.Y., Tharam, S., Lim, W.Y., 1995. A case of human enteric myiasis due to larvae of *Hermetia illucens* (Family: Stratiomyiadae): first report in Malaysia. Malaysian J. Pathol. 17, 109–111.

Li, G.S., Qi, F., Cui, M.X., 1998. Breeding technology of mealworm and housefly. Chinese Agric. Sci. Bull. 14, 92–93.

Liu, Q., Tomberlin, J.K., Brady, J.A., Sanford, M.R., Yu, Z., 2008. Black soldier fly (Diptera: Stratiomyidae) larvae reduce *Escherichia coli* in dairy manure. Environ. Entomol. 37, 1525–1530.

Loudon, C., 1988. Development of *Tenebrio molitor* in low oxygen levels. J. Insect Physiol. 34, 97–103.

Ludwig, D., 1956. Effect of temperature and parental age in the life cycle of the mealworm, *Tenebrio molitor* Linnaeus (Coleoptera Tenebrionidae). Ann. Entomol. Soc. Am. 49, 12–15.

Ludwig, D., Fiore, C., 1960. Further studies on the relationship between parental age and the life cycle of the mealworm, *Tenebrio molitor*. Ann. Entomol. Soc. Am. 53, 595–600.

Lundgren, J.G., 2009. Nutritional aspects of non-prey foods in the life histories of predaceous Coccinellidae. Biol. Control 51, 294–305.

Lundy, M.E., Parrella, M.P., 2015. Crickets are not a free lunch: protein capture from scalable organic side-streams via high-density populations of *Acheta domesticus*. PLoS One 10, e0118785.

Machin, J., 1976. Passive exchanges during water vapour absorption in mealworms (*Tenebrio molitor*): a new approach to studying the phenomenon. J. Exp. Biol. 65, 603–615.

Madhan, S., Prabakaran, M., Kwang, J., 2010. Baculovirus as vaccine vectors. Curr. Gene Ther. 10, 201–213.

Makkar, H.P., Tran, G., Heuzé, V., Ankers, P., 2014. State-of-the-art on use of insects as animal feed. Anim. Feed Sci. Technol. 197, 1–33.

Manojlovie, B., 1988. Influence of food and temperature on post–embryonal survival of the *Tenebrio molitor* L. Zastita–bilja (Yugoslavia) 39 (l), 43–53.

Marston, N., Campbell, B., 1973. Comparison of nine diets for rearing *Galleria mellonella*. Ann. Entomol. Soc. Am. 66, 132–136.

Marston, N., Campbell, B., Boldt, P.E., 1975. Mass producing eggs of the greater wax moth, *Galleria mellonella* (L.). US Department of Agriculture, Technical Bulletin 1510, 15 p.

Martin, R.D., Rivers, J.P., Cowgill, U.M., 1976. Culturing mealworms as food for animals in captivity. Int. Zoo Yearb. 16, 63–70.

May, B.M., 1961. The occurrence in New Zealand and the life-history of the soldier fly *Hermetia illucens* (L.) (Diptera: Stratiomyidae). N. Z. J. Sci. 4, 55–65.

McFarlane, J.E., 1985. *Acheta domesticus*. Singh, P., Moore, R.F. (Eds.), Handbook of Insect Rearing, vol. I, Elsevier Science Publishing, Amsterdam, The Netherlands, pp. 427–434.

Milam, V.G., 1970. Moth pests of honey bee combs. Glean. Bee Cult. 68, 424–428.

Mlcek, J., Rop, O., Borkovcova, M., Bednarova, M., 2014. A comprehensive look at the possibilities of edible insects as food in Europe—a review. Polish J. Food Nutr. Sci. 64, 147–157.

Morales-Ramos, J.A., Kay, S., Rojas, M.G., Shapiro-Ilan, D.I., Tedders, W.L., 2015. Morphometric analysis of instar variation in *Tenebrio molitor* (Coleoptera: Tenebrionidae). Ann. Entomol. Soc. Am. 108, 146–159.

Morales-Ramos, J.A., Rojas, M.G., 2015. Effect of larval density on food utilization efficiency of *Tenebrio molitor* (Coleoptera: Tenebrionidae). J. Econ. Entomol. 108, 2259–2267.

Morales-Ramos, J.A., Rojas, M.G., Coudron, T.A., 2014. Artificial diet development for entomophagous arthropods. In: Morales-Ramos, J.A., Rojas, M.G., Shapiro-Ilan, D.I. (Eds.), Mass Production of Beneficial Organisms, Invertebrates and Entomopathogens. Academic Press, Waltham, MA, pp. 203–240.

Morales-Ramos, J.A., Rojas, M.G., Kay, S., Shapiro-Ilan, W.L., Tedders, W.L., 2012. Impact of adult weight, density, and age on reproduction of *Tenebrio molitor* (Coleoptera: Tenebrionidae). J. Entomol. Sci. 47, 208–220.

Morales-Ramos, J.A., Rojas, M.G., Shapiro-Ilan, D.I., Tedders, W.L., 2010. Developmental plasticity in *Tenebrio molitor* (Coleoptera: Tenebrionidae): analysis of instar variation in number and development time under different diets. J. Entomol. Sci. 45, 75–90.

Morales-Ramos, J.A., Rojas, M.G., Shapiro-Ilan, D.I., Tedders, W.L., 2011a. Self-selection of two diet components by *Tenebrio molitor* (Coleoptera: Tenebrionidae) larvae and its impact on fitness. Environ. Entomol. 40, 1285–1294.

Morales-Ramos, J.A., Rojas, M.G., Shapiro-Ilan, D.I., Tedders, W.L., 2011b. Automated insect separation system. US Patent No. US 8,025,027 B1.

Morales-Ramos, J.A., Rojas, M.G., Shapiro Ilan, D.I., Tedders, W.L., 2013. Use of nutrient self-selection as a diet refining tool in *Tenebrio molitor* (Coleoptera: Tenebrionidae). J. Entomol. Sci. 48, 206–221.

Moscicki, L., 2011. Extusion-Cooking Techniques: Applications, Theory and Sustainability. Wiley-VCH Verlag GmbH & Co. KGaA, Weinheim, Germany, 236 p.

Murray, D.R.P., 1968. The importance of water in the normal growth of the larvae of *Tenebrio molitor*. Entomol. Exp. Appl. 11, 149–168.

Myers, H.M., Tomberlin, J.K., Lambert, B.D., Kattes, D., 2008. Development of black soldier fly (Diptera: Stratiomyidae) larvae fed dairy manure. Environ. Entomol. 37, 11–15.

Nakagaki, B.J., DeFoliart, G.R., 1991. Comparison of diets for mass-rearing *Acheta domesticus* (Orthoptera: Gryllidae) as a novelty food, and comparison of food conversion efficiency with values reported for livestock. J. Econ. Entomol. 84, 891–896.

Nakakita, H., 1982. Effect of larval density on pupation of *Tribolium freemani* (Coleoptera: Tenebrionidae). Appl. Entomol. Zool. 17, 269–276.

Nash, W.J., Chapman, T., 2014. Effect of dietary components on larval life history characteristics in the medfly (Ceratitis capitata: Diptera, Tephritidae). PLoS One 9 (1), e86029.

Newton, G.L., Booram, C.V., Barker, R.W., Hale, O.M., 1977. Dried *Hermetia illucens* larvae meal as a supplement for swine. J. Anim. Sci. 44, 395–400.

Ng, W.K., Liew, F.L., Ang, L.P., Wong, K.W., 2001. Potential of mealworm (*Tenebrio molitor*) as an alternative protein source in practical diets for African catfish *Clarias gariepinus*. Aquacult. Res. 32 (Suppl. 1), 273–280.

Nguyen, T.T.X., Tomberlin, J.K., VanLaerhoven, S.L., 2013. Influences of resources on *Hermetia illucens* (Diptera: Stratiomyidae) larval development. J. Med. Entomol. 50, 898–906.

Nguyen, T.T.X., Tomberlin, J.K., VanLaerhoven, S.L., 2015. Ability of black soldier fly (Diptera: Stratiomyidae) larvae to recycle food waste. Environ. Entomol. 44, 406–410.

Oonincx, D.G.A.B., van Itterbeeck, J., Heetkamp, M.J.W., van den Brand, H., van Loon, J.J.A., van Huis, A., 2010. Anexploration on greenhouse gas and ammonia production by insect species suitable for animal or human consumption. PLoS One 5, e14445.

Parween, S., Begum, M., 2001. Effect of larval density on the development of the lesser mealworm, Alphitobius diaperinus Panzer (Coleoptera: Tenebrionidae). Int. Pest Control 43, 205–207.

Patton, R.L., 1967. Oligidic diets for *Acheta domesticus* (Orthoptera: Gryllidae). Ann. Entomol. Soc. Am. 60, 1238–1242.

Patton, R.L., 1978. Growth and development parameters for *Acheta domesticus*. Ann. Entomol. Soc. Am. 71, 40–42.

Piccolo, G., Marono, S., Gasco, L., Iannaccone, F., Bovera, F., Nizza, A., 2014. Use of *Tenebrio molitor* larvae meal in diets for gilthead sea bream *Sparus aurata* juveniles. In: Abstract Book Conference Insects to Feed The World, The Netherlands, p. 76.

Pickens, L.G., Lorenzen, K.J., 1983. A new larval diet for *Musca domestica* (Diptera: Miscidae). J. Med. Entomol. 5, 572–573.

Ramos-Elorduy, J., 1997. Insects: a sustainable source of food? Ecol. Food Nutr. 36, 247–276.

Ramos–Elorduy, J., Avila Gonzalez, E., Rocha Hernandez, A., Pino, J.M., 2002. Use of *Tenebrio molitor* (Coleoptera: Tenebrionidae) to recycle organic wastes and as feed for broiler chickens. J. Econ. Entomol. 95, 214–220.

Ramos-Elorduy, J., 2009. Anthropo-entomophagy: cultures, evolution and sustainability. Entomol. Res. 39, 271–288.

Redding, G.J., 1981. Functional Design Handbook for Australian Farm Buildings. Agricultural Engineering Section, University of Melbourne, Melbourne.

Reinecke, J.P., 2009. Health and safety issues in the rearing of arthropods. In: Schneider, J.C. (Ed.), Principles and Procedures for Rearing High-Quality Insects. Mississippi State University, Starkville, MS, pp. 71–85.

Richards, A.G., 1978. The chemistry of insect cuticle. In: Rockstein, M. (Ed.), Biochemistry of Insects. Academic Press, New York, NY, pp. 1–55.

Richardson, H.H., 1932. An efficient medium for rearing houseflies throughout the year. Science 14, 350–351.

Rueda, L.M., Axtell, R.C., 1996. Temperature dependent development and survival of the lesser mealworm Alphitobius diaperinus. Med. Vet. Entomol. 10, 80–86.

Sánchez-Muros, M.J., Barroso, F.G., Manzano-Agugliaro, F., 2014. Insect meal as renewable source of food for animal feeding: a review. J. Clean. Prod. 65, 16–27.

Sanford, M.T., 1987. Diseases and Pests of the Honey Bee. University of Florida IFAS Extension Series, Document CIR766, 13 p.

Savvidou, N., Bell, C.H., 1994. The effect of larval density, photoperiod and food change on development of *Gnatocerus cornutus* (F.) (Coleoptera: Tenebrionidae). J. Stored Prod. Res. 30, 17–21.

Schneider, J.C., 2009. Principles and Procedures for Rearing High Quality Insects. Mississippi State University, Starkville, MS, 352 p.

Schiavone, A., De Marco, M., Rotolo, L., Belforti, M., Martinez Mirò, S., Madrid Sanchez, J., Hernandez Ruiperez, F., Bianchi, C., Sterpone, L., Malfatto, V., Katz, H., Zoccarato, I., Gai, F., Gasco, L., 2014. Nutrient digestibility of *Hermetia illucens* and *Tenebrio molitor* meal in broiler chickens. Abstract Book Conference Insects to Feed The World 2014, The Netherlands, p. 84.

Sealey, W.M., Gaylord, T.G., Barrows, F.T., Tomberlin, J.K., McGuire, M.A., Ross, C., St-Hilaire, S., 2011. Sensory analysis of rainbow trout, *Oncorhynchus mykiss*, fed enriched black soldier fly prepupae, *Hermetia illucens*. J. World Aquacult. Soc. 42, 34–45.

Shapiro, J.P., 1988. Lipid transport in insects. Annu. Rev. Entomol. 33, 297–318.

Shapiro-Ilan, D., Rojas, M.G., Morales-Ramos, J.A., Lewis, E.E., Tedders, W.L., 2008. Effects of host nutrition on virulence and fitness of entomopathogenic nematodes: lipid- and protein-based supplements in *Tenebrio molitor* diets. J. Nematol. 40, 13–19.

Shapiro-Ilan, D., Rojas, M.G., Morales-Ramos, J.A., Tedders, W.L., 2012. Optimization of a host for *in vivo* production of entomopathogenic nematodes. J. Nematol. 44, 264–273.

Sheppard, D.C., 1983. House fly and lesser fly control utilizing the black soldier fly in manure management systems for caged laying hens. Environ. Entomol. 12, 1439–1442.

Sheppard, D.C., Newton, G.L., Thompson, S.A., Savage, S., 1994. A value added manure management system using the black soldier fly. Bioresour. Technol. 50, 275–279.

Sheppard, D.C., Tomberlin, J.K., Joyce, J.A., Kiser, B., Sumner, S.M., 2002. Rearing methods for the black soldier fly (Diptera: Stratiomyidae). J. Med. Entomol. 39, 695–698.

Shockley, M., Dossey, A.T., 2014. Insects for human consumption. In: Morales-Ramos, J., Rojas, G., Shapiro-Ilan, D.I. (Eds.), Mass Production of Beneficial Organisms: Invertebrates and Entomopathogens. Academic Press, New York, NY, pp. 617–652.

Singh, P., 1977. Artificial Diets for Insects, Mites, and Spiders. Plenum Press, New York, NY.

Singh, P., Moore, R.F., 1985. Handbook of Insect Rearingvol. I and vol. IIElsevier, New York, NY.

Slone, D.H., Gruner, S.V., 2007. Thermoregulation in larval aggregations of carrion-feeding blow flies (Diptera; Calliphoridae). J. Med. Entomol. 44, 516–523.

Somerville, 2007. Wax Moth. NSW Department of Primary Industries, Primefact 658, 4 p.

St-Hilaire, S., Cranfill, K., McGuire, M.A., Mosley, E.E., Tomberlin, J.K., Newton, L., Sealey, W., Sheppard, C., Irving, S., 2007a. Fish offal recycling by the black soldier fly produces a foodstuff high in omega-3 fatty acids. J. World Aquacult. Soc. 38, 309–313.

St-Hilaire, S., Sheppard, C., Tomberlin, J.K., Irving, S., McGuire, M.A., Mosley, E.E., Hardy, R.W., Sealey, W., 2007b. Fly prepupae as a feedstuff for rainbow trout, *Oncorhynchus mykiss*. J. World Aquacult. Soc. 38, 59–67.

Stellwaag-Kittler, F., 1954. Zur physiologie der käferhäutung untersuchungen am Meklkäfer *Tenebrio molitor* L. Biol. Zentralbl. 73, 12–49.

Storm, I.M.L.D., Hellwing, A.L.F., Nielsen, N.I., Madsen, J., 2012. Methods for measuring and estimating methane emission from ruminants. Animals 2, 160–183.

Stryer, L., 1988. Biochemistry, third ed. W.H. Freeman and Company, New York, NY.

Thomas, M., van der Poel, A.F.B., 1996. Physcial quality of pelleted animal feed. 1. Criteria for pellet quality. Anim. Feed Sci. Technol. 61, 89–112.

Thomas, M., van Vliet, T., van der Poel, A.F.B., 1998. Physical quality of pelleted animal feed 3. Contribution of feedstuff components. Anim. Feed Sci. Technol. 70, 59–78.

Thomas, M., van Zuilichem, D.J., van der Poel, A.F.B., 1997. Physical quality of pelleted animal feed. 2. Contribution of processes and its conditions. Anim. Feed Sci. Technol. 64, 173–192.

Tomberlin, J.K., Adler, P.H., Myers, H.M., 2009. Development of the black soldier fly (Diptera: Stratiomyidae) in relation to temperature. Environ. Entomol. 38, 930–934.

Tomberlin, J.K., Sheppard, D.C., 2001. Lekking behavior of the black soldier fly (Diptera: Stratiomyidae). Fla. Entomol. 84, 729–730.

Tomberlin, J.K., Sheppard, D.C., 2002. Factors influencing mating and oviposition of black soldier flies (Diptera: Stratiomyidae) in a colony. J. Entomol. Sci. 37, 345–352.

Tomberlin, J.K., Sheppard, D.C., Joyce, J.A., 2002. Selected life-history traits of black soldier flies (Diptera: Stratiomyidae) reared on three artificial diets. Ann. Entomol. Soc. Am. 95, 379–386.

Tomberlin, J.K., Sheppard, D.C., Joyce, J.A., 2005. Black soldier fly (Diptera: Stratiomyidae) colonization of pig carrion in south Georgia. J. Forensic Sci. 50, 152–153.

Tschinkel, W.R., Willson, C.D., 1971. Inhibition of pupation due to crowding in some tenebrionid beetles. J. Exp. Zool. 176, 137–146.

Tyshchenko, V.P., Sheyk Ba, A., 1986. Photoperiodic regulation of larval growth and pupation of *Tenebrio molitor* L. (Coleoptera: Tenebrionidae). Entomol. Rev. 66, 35–46.

Urs, K.C.D., Hopkins, T.L., 1973. Effect of moisture on growth rate and development of two strains of *Tenebrio molitor* L. (Coleoptera Tenebrionidae). J. Stored Prod. Res. 8, 291–297.

US Department of Agriculture, Agricultural Research Service, Nutrient Data Laboratory, 2015. USDA National Nutrient Database for Standard Reference, Release 28. Available from: http://www.ars.usda.gov/ba/bhnrc/ndl

van Broekhoven, S., Oonincx, D.G.A.B., van Huis, A., van Loon, J.J.A., 2015. Growth performance and feed conversion efficiency of three edible mealworm species (Coleoptera: Tenebrionidae) on diets composed of organic by-products. J. Insect Physiol. 73, 1–10.

Van Huis, A., Van Itterbeeck, J., Klunder, H., Mertens, E., Halloran, A., Muir, G., Vantomme, P., 2013. Edible insects–future prospects for food and feed security. FAO Forestry Paper, p. 171.

van Huis, A., Van Itterbeeck, J., Klunder, H., Mertens, E., Halloran, A., Muir, G., Vantomme, P., 2014. Edible Insects: Future Prospects for Food and Feed Security. FAO, ONU, Rome,, 187 p.

Veldkamp, T., Van Duinkerken, G., Van Huis, A., Lakemond, C.M.M., Ottevanger, E., Bosch, G., Van Boekel, M.A.J.S., 2012. Insects as a sustainable feed ingredient in pig and poultry diets–a feasibility study. Rapport 638–Wageningen Livestock Research. Available from: http://www.wageningenur.nl/uploadmm/2/8/0/f26765b9-98b2-49a7-ae43-5251c5b694f6234247%5B1%5D.

Wadhwa, M., Bakshi, M.P.S., 2013. Utilization of Fruit and Vegetable Wastes as Livestock Feed and as Substrates for Generation of Other Value-Added Products. RAP Publication, Bangkok, p. 4.

Wang, F., Zhu, F., Lei, C.L., 2010. Animal manure breeds housefly and its application. Chinese Bull. Entomol. 47, 657–664.

Wang, Y.C., Chen, Y.T., Li, X.R., Xia, J.M., Du, Q., Sheng, Z.C., 1996. Study on rearing the larvae of *Tenebrio molitor* Linne and the effects of its processing and utilization. Acta Agriculturae Universitatis Henanensis 30, 288–292.

Weaver, D.K., McFarlane, J.E., 1990. The effect of larval density on growth and development of *Tenebrio molitor*. J. Insect Physiol. 36, 531–536.

Wei, M.C., Liu, G.Q., 2001. The research and exploitation of insect protein. J. Central South Forest. Univ. 21, 86–90.

Weissman, D.B., Gray, D.A., Pham, H.T., Tijssen, P., 2012. Billions and billions sold: pet-feeder crickets (Orthoptera: Gryllidae), commercial cricket farms, an epizootic densovirus, and government regulations make for a potential disaster. Zootaxa 3504, 67–88.

Whitaker, J.H., 1979. Agricultural Buildings and Structures. Reston Publishing Co., Reston, VA.

Wigglesworth, V.B., 1972. The Principles of Insect Physiology, seventh ed. Chapman and Hall, London, 827 p.

Won, R., Kwon, O., 2009. Relationship between diet composition and the fecundity of *Musca domestica*. Entomol. Res. 39, 412–415.

Wu. S.X.. 2009. Studies on optimization of rearing condition and nutriment content of larvae of *Tenebrio molitor* L. Anhui Agricultural University, Master Degree.

Wu, J.W., Chen, M., Peng, W.F., 2001. Study on the nutritional value of the housefly larva fed with pig manure. J. Guiyang Med. Coll. 26, 377–379.

Wu, F.Z., Lin, H.F., Liu, Z.H., Hu, C., 2005. The situation and tactics of the utilization of *Tenebrio molitor* production in China. Chinese Agric. Sci. Bull. 21, 72–75.

Xiao, Z., Liu, W., Zhu, G., Zhou, R., Niu, Y., 2014. A review of the preparation and application of flavour and essential oils microcapsules based on complex coacervation technology. J. Sci. Food Agric. 94, 1482–1494.

Yang, H.P., Xu, D.G., Wu, J.H., Xue, C.L., 2004. Effect of ecological treatment by *Musca domestica* larvae to pig manure on the oviposition and larvae hatching rate. Chin. J. Parasitol Parasit Dis. 22, 9–10.

Yang, Y., Yang, J., Wu, W.M., Zhao, J., Song, Y., Gao, L., Yang, R., Jiang, L., 2015a. Biodegradation and mineralization of polystyrene by plastic-eating mealworms: Part 1. Chemical and physical characterization and isotopic tests. Environ. Sci. Technol. 49, 12080–12086.

Yang, Y., Yang, J., Wu, W.M., Zhao, J., Song, Y., Gao, L., Yang, R., Jiang, L., 2015b. Biodegradation and mineralization of polystyrene by plastic-eating mealworms: Part 2. Role of gut microorganisms. Environ. Sci. Technol. 49, 12087–12093.

Yi, L., Lakemond, C.M.M., Sagis, L.M.C., Eisner-Schadler, V., van Huis, A., van Boekel, M.A.J.S., 2013. Extraction and characterization of protein fractions from five insect species. Food Chem. 141, 3341–3348.

Yoon, J.S., 1985. *Drosophila melanogaster*. Singh, P., Moore, R.F. (Eds.), Handbook of Insect Rearing, vol. II, Elsevier, New York, NY, pp. 75–85.

Young, R.G., 1961. The effects of dietary beeswax and wax components on the larvae of the greater wax moth. Galleria mellonella (L.). Ann. Entomol. Soc. Am. 54, 657–659.

Yu, G., Cheng, P., Chen, Y., Li, Y., Yang, Z., Chen, Y., Tomberlin, J.K., 2011. Inoculating poultry manure with companion bacteria influences growth and development of black soldier fly (Diptera: Stratiomyidae) larvae. Environ. Entomol. 40, 30–35.

Zhang, J., Huang, L., He, J., Tomberlin, J.K., Li, J., Lei, C., Sun, M., Liu, Z., Yu, Z., 2010. An artificial light source influences mating and oviposition of black soldier flies, *Hermetia illucens*. J. Insect Sci. 10, 202.

Zhao, H.Y., Zhao, B., Zhang, L., Su, Y.J., 2011. Application of *Zophobas morio* protein powder to penaeus vannamei larva feed. J. Tianjin Agric. Univ. 18, 20–23, 30.

Zheng, L., Crippen, T.L., Holmes, L., Singh, B., Pimsler, M.L., Benbow, M.E., Tarone, A.M., Dowd, S., Yu, Z., VanLaerhoven, S.L., Wood, T.K., Tomberlin, J.K., 2013. Bacteria mediate oviposition by the black soldier fly, *Hermetia illucens* (L.), (Diptera: Stratiomyidae). Nat. Sci. Rep. 3, 2563.

Zhou, F., Tomberlin, J.K., Zheng, L., Yu, Z., Zhang, J., 2013. Developmental and waste reduction plasticity of three black soldier fly strains (Diptera: Stratiomyidae) raised on different livestock manures. J. Med. Entomol. 50, 1224–1230.

Zhu, F.X., 2012. Rapid production of maggots as feed supplement and organic fertilizer by two-stage composting of fresh pig manure. Zhejiang University, Doctoral Dissertation.

Zhu, H.J., 2013. Study on the protein extraction from *Zophobas Morio*. XinJiang Agriculture University, Master Degree.

Zimmerman, H.G., 1979. Herbicidal control in relation to *Opuntia aurantiaca* Lindley and effects on cochineal populations (Dactylopius austrinus). Weed Res. 19, 89–93.

Chapter 7

Food Safety and Regulatory Concerns

P.A. Marone
Department of Pharmacology and Toxicology, Virginia Commonwealth University, Toxicology and Pathology Associates, LLC, Richmond, VA, United States

Chapter Outline

Introduction	203
Regulatory Considerations for Insects-as-Food Ingredient	204
Present History of Use and Regulations	204
Labelling Regulation and Health Claims Applicable to Insects	207
Safety Considerations for Insects as Food	207
Species Identity and Characterization of the Insect-Based Product for Whole Body or Processed Feed	209
Toxicological Assessment	210
Clinical Evaluation of Safety	211
Toxicological Hazards of Insect-Based Foods and Food Ingredients	211
Chemical	211
Physical	213
Farming and Novel Considerations Driving Insect Food/Feed Safety	214
Processing, Preparation, Packaging, and Transport of Insect-Based Foods and Food Ingredients	216
Conclusions on the Use of the Insects as Food and Feed	217
References	218

INTRODUCTION

According to the Food and Agriculture Organization (FAO) of the United Nations (FAO, IFAD, and WFP, 2015), by targeting economic progress and growth of the family farm in addressing needs of food security and nutrition, great strides have been made toward alleviating hunger and undernourishment since 1990–92, particularly in developing nations. Despite reductions of greater than 10% in the prevalence of undernourished populations, and more than half of developing countries reaching the United Nations Millennium Development Goal (UN, 2015a,b) of reductions in extreme poverty and hunger, much work remains to be done to assure food security for a growing world population—an estimated 70% increase in food production by 2050 is needed to feed a population of over 9 billion. In industrialized countries, where food security is of less concern, dramatic increases in the globalization and process complexity of the food supply raises questions of food safety, sustainability, and traceability. In the ongoing balance to maximize nutritional quality with the preservation of natural resources while continuing to increase productivity, insects, with a ratio of 200 million to every one human (Smithsonian Encyclopedia, 2015), have the potential to provide a low-cost, high-quality protein, fat, mineral, and micronutrient alternative to traditional meat, poultry, and aquaculture farming. Insects are and historically have been a traditional acceptable source of nourishment in many largely developing societies, with over 1700 species edible by an estimated 80% of the world's population in over 80 countries across the globe (Michels, 2012; Bidau, 2015) (see also Chapter 2). Despite this, consumers in industrialized countries have progressively contracted their food acceptances and subsequent choices over time to where little is known, available, or even considered regarding insect-as-food safety and use in the present day. Currently, the most common edible insects worldwide include various species of grasshoppers, crickets, locusts, silkworms, ants, mealworms, and beetles, while in the United States and in other developed nations, crickets and cricket powder (flour/paste) are the most popular and marketable items. As such, regulatory standards of insects as a human food and animal feed product, though largely unavailable to this point, are now beginning to be addressed as nations prioritize and coordinate their health and sustainable nutrition needs. For example, following an initial "Expert Consultation" at UN FAO in 2012 (Vantomme et al., 2012) international conference of "Insects to Feed the World" organized by FAO and Wageningen UR in the Netherlands in May, 2014 drew

450 participants from 45 nations with the objective of promoting insects as food and feed (FAO, Wageningen University and Research Centre UR (WUR), 2014; van Huis et al., 2013d). Conclusions of the meeting highlighted the socioeconomic opportunities, consumer acceptance, and regulatory/data challenges coordinate with minimizing the insect-as-food gap activities, and align goals between developing and industrialized countries for the standards necessary to harmonization efforts. In light of such enthusiastic meeting attendance and increasing awareness, interest in the use of insects as food is growing. The time and resources devoted by prestigious organizations such as FAO to initiating the dialogue of this potential nutritional source has captured the attention of consumers, entrepreneurs, the food industry, and international regulatory agencies worldwide to consider more fully the potential of insects. Furthermore, as upscale industrialized consumers demand ever greater variety and quality in their food choices, particularly when they coordinate with ecological responsibility and sustainability, the timing may be acutely suitable for broadening attitudes of Western diet preferences to include insects. The present review attempts to detail the current regulatory and safety issues associated with this potential nutrient source.

REGULATORY CONSIDERATIONS FOR INSECTS-AS-FOOD INGREDIENT

Safe, large-scale industrial production of insects is necessary to the acceptance of entomophagy by Western cultures. Industrial production allows for incorporation of controlled high standards of food quality and safety while ensuring a sustained economically and environmentally feasible means to consumer confidence. Industrialized production avoids the growing scarcity of wild-caught favored species that have been currently reported in some cultures (see Thailand, Section "Present History of Use and Regulations") while concurrently maintaining industry-wide economic stability for optional safe varieties. Whether by native custom or preference, or the possibility of farm-raised stock inconsistencies, common global safety guidelines for both wild-caught and commercial cultivation of insects as food are needed. Ironically, the barriers to this process are a two-way interdependence among the confidence of entrepreneurs (known as "entopreneurs") investing in this industry and the need to sanction it by regulatory oversight, often in the context of existing regulations and recommendations. An impasse is created when the uncertainties of regulatory framework governing the type, production, labeling and sale of novel foods serves as deterrent to the monetary start-up capital necessary to instill and grow assurance and prosperity in the industry (Burdock and Matulka, 2014). Further, Premalatha et al. (2011) from India aver the "supreme irony" that "all over the world monies worth billions of rupees are spent every year to save crops that contain no more than 14% of plant protein by killing another food source (insects) that may contain up to 75% of high-quality animal protein."

Indeed, for most of the industrialized world, there is a weighty imbalance vastly favoring guidelines for insects as agricultural pests from those of insects as a food source in and of themselves. As a result, little structured, standardized, or harmonized regulations exist to guide the emerging insect-based food industry.

Present History of Use and Regulations

Whether by either traditional acceptance or current necessity, many developing countries have led, encouraged, and benefitted from the economic cultivation of insects as food even as regulatory standards and the governmental bodies responsible for them remain ambiguous or undetermined.

Thailand is among the most progressive countries for insect farming. The United Nation's Food and Agriculture Organization (FAO) noted Thailand as "one of the few countries to have developed a viable and thriving insect farming sector" with "more than 20,000 insect farming enterprises … registered in the country," (Hanboonsong et al., 2013). Insect farming in Thailand, with more than 200 species comprised largely of crickets, grasshoppers, palm weevil larvae, and bamboo caterpillars, is outpacing governmental regulatory oversight. The country's large and profitable commercial edible insect trade has elevated the standards of living for the small farmer. Generally, in Asia as well as across the globe in East Africa and other developing areas, FAO has promulgated insect food and feed guidelines according to its Codex Alimentarius, the 186 member country/215 observer consortium responsible for the internationally recognized collection of standards for food production and safety. These voluntary standards, including recommendations for hygiene, labeling, contaminants, and residues, and presently considered comparable to other foods of animal origin, serve as a basis for multinational food regulations, now regarded for paving the way for Western awareness and consideration of insects as food/feed (CAC, 2010; Halloran, 2014; van Huis et al., 2013d). Serving as the basis for hygienic practices of food production is the Hazard Analysis and Critical Control Points (HACCP) system, designed to mitigate food safety risks by identifying and controlling potential hazards along the food processing chain (FAO/WHO, 2001).

The fundamental economic, environmental, and nutritional potential of insects as a sustainable food resource has captured the attention of Western cultures, particularly the European Union (EU), Canada, and Australia/New Zealand as early regulatory players. Within the European Union, a consortium of 28 member countries, the European Food Safety Authority

(EFSA), monitored by the European Commission (EC), is responsible for regulatory oversight of food and feed safety. Here, the Scientific Committee on Food has established strict health regulations throughout the production process on animal by-products for human food and livestock feed, specifically that of processed animal proteins (PAP). Presently, insect production is considered under the jurisdiction of "novel foods," a food or food ingredient that has not been traditionally used for human consumption before May, 1997, and which requires that a risk assessment be performed prior to marketing (EC regulation 258/1997) (Belluco et al., 2013). Considering that insects have been a traditionally consumed food from a defined "third country" prior to this date, a detailed and scientific history of safe use for *the given species* would remain to be demonstrated from established history of use and scientific data. Risk assessment dossiers include, but are not limited to, information as to the type of food, including analytic identity, its history and intended usage, nutritional, microbial, and toxicological information, and production processes. While insects are not specifically mentioned in the new European Council (EC) regulation on novel foods (European Council, 2015), the European Union does not prevent the feeding of insects to livestock or its PAPs in aquaculture despite its limitation of PAPs in farm animals and insects would qualify under the definition as "a food that is eaten elsewhere but has never been traditionally eaten before in the EU" (Day, 2015). As if to further highlight ambiguity in the individual laws and regulations of member states, insects are marketed in the United Kingdom by interpretation of regulations barring foods *isolated from animals*, but not *consisting of animals* thereby offering the potential to establish history of use for all EU members (Belluco et al., 2013) as insects as whole animals are considered under traditionally consumptive fare over human existence. Yet to be determined is the means of slaughter for insects as questions remain regarding their applicability to humane treatment apart from those of warm-blooded species (see later). The EC, in considering standards of insects for food, has petitioned EFSA for scientific opinion, which was most recently issued (EFSA, 2015). The report acknowledges that for farmed products, the substrate, stage of harvest, insect species, its developmental stage, and postproduction processing will all impact the possible contaminants present. Nonetheless, the microbiologic hazards associated with insects as food and feed are "expected to be comparable to other nonprocessed sources of protein of animal origin" when currently approved feedstock is used. Assessment of the biological, chemical, and allergenic concerns stemming from other growth substrate transfer need further evaluation, particularly for protein derived from human or livestock (prion risk) [Chapter 8 for a discussion on the microbiological considerations and prion risk (eg, mad cow disease and others), and Chapter 9 for more details on the allergenicity of insects]. Further, while the hazards related to the environment are expected to be similar to that of other animal production systems, methodical human data is needed to evaluate the occurrence of hazardous chemicals in farmed insects and the general uncertainty related to the possible variable risk associated with insects as food. Previous to this advisory, the EC, along with Belgian authorities, published a preliminary advisory report for the safe cultivation, distribution, and consumption of insects applying the same health and sanitary rules as traditional foods for microbial, chemical, allergenic, and physical hazards along the production path (Halloran, 2014; SciCom, 2014). In this report, Belgium's Scientific Committee of the Federal Agency for the Safety of the Food Chain of the Superior Health Council (SHC) describes the hazards associated with the specific ingestion of crickets, mealworms, and locusts, as appropriate consideration for human consumption with recommendations for further risk analysis. Most recently, Swiss authorities of the Federal Food Safety and Veterinary Office have agreed to follow Belgium's lead by finding these particular species "suitable" for sale and consumption in early 2016 (Raaflaub, 2015). In addition, individual government-funded research grants are exploring the potential of insects as food and feed. PROteinINSECT [through the Food and Environmental Research Agency (FERA)] and GREEiNSECT [through the Danish International Development Agency (DANIDA)] research consortiums funded by EU and Danish development agencies, respectively, are two recent examples of coordination between European, African (Kenya) and Asian feed production companies' and the exchange of technological expertise for "adoption and adaption" of insects as food (Halloran, 2014). Additionally, in the United States, All Things Bugs, LLC has received grant support from national (USDA, 2014) and private foundations (Bill & Melinda Gates Foundation) to explore various aspects of insects as food ingredients, including certain allergy and safety concerns (Crabbe, 2012; Mermelstein, 2015).

The Canadian Food Inspection Agency of Health Canada is the regulatory agency responsible for food safety standards in Canada. With this agency as a guide, local provincial health authorities may independently regulate production and usage of various foodstuffs, including those served in restaurants. Traditionally, the larvae of certain species (eg, the Mezcal worm, the larval stage of the agave snout weevil species *Scyphophorus acupunctatus*) have been permitted for Mexican alcoholic beverages, but insect-as-food practice is considered under "novel food," where safe history of use is generally unavailable—though it would be permitted following a risk assessment from the Novel Food Section of the Bureau of Microbial Hazards, Food Directorate, Health Canada.

Similar standards exist for Food Standards Australia/New Zealand, where insects fall under the purview of novel or, more distinctively, nontraditional, food for some species [super mealworm (*Zophobas morio*), mealworm beetle, and house crickets (*Acheta domesticus*); Halloran, 2014].

While some member countries have been more tolerant of insects as food, allowing production and marketing through private companies, Finland, in suit with most other EU member countries, follows the prevailing regulation of the EU commission in that the "importation, selling, marketing or growing of whole or processed insects for use as food is, at present, forbidden until the history of use of the species within EU has been established or novel food authorization has been granted to the species" (FFSA, 2015).

In China, among the most populous and economically prosperous countries in the world, where many species of insects, as well as other arthropods such as spiders and scorpions have been consumed for centuries, there is no coordinated industrial or governmental effort to bring entomophagy to mass market. While large-scale insect breeding, with many local farms, is based in Shandong province, investment in the industry is not forthcoming. Experts from Kunming Institute of Zoology in Yunnan province, as the nation's major research institute in the field, cite the lack of validated scientific research on the potential of insects as food for little industry advancement in that country (Yao, 2013). In China, as well as for many other developing countries, awareness from current "wait and see" attitudes may quickly transform to business and regulatory action as heightened interest in this fast-changing industry continues.

In the United States, it is the FDA, in coordination with the United States Department of Agriculture (USDA) (for regulation on meat, poultry, and eggs), through its Animal and Plant Health Inspection Service (APHIS) agency for safeguarding insect farming, which serves to govern the use and safety assessment of insects as food. Fundamentally, little, if any, regulation exists for insects as food. Present regulations largely consider insects as contaminants of food, specifically, under the Environmental Protection Agency (EPA), to distinguish tolerance levels for insecticides in crop protection through its Toxic Substances Control Act (TSCA). In addition, with the understanding that "it is economically impractical to grow, harvest, or process raw products that are totally free of nonhazardous, naturally occurring, unavoidable defects" the FDA, through the Center for Food Safety and Applied Nutrition (CFSAN), established food defect action levels that would present no hazards to human health. This directorate, binding upon the manufacturer to comply via acceptable hygienic practices, lists acceptable *maximum* levels from such contaminants as rodent hair and feces, insects and their fragments, mold and other "foreign matter." Presumably (ironically), these directives would also apply to insect-based foods and their production.

Going forward, regulation for the safety of insects for the processing and marketability as food would fall under the jurisdiction of FDA's Federal Food, Drug and Cosmetic Act (FFDCA, 1908), revised in 1938, and again in a 1958 amendment, out of concern for new food additives, the ingredient category to which processed food insects are considered to belong (Duhaime-Ross, 2014). Briefly, the FDA considers food to be inherently safe, without the need for premarket approval, [Sections 201(f); 401(a)(1)] presuming that the food is uncontaminated. The applicable category of "generally recognized as safe" (GRAS) may qualify food additive safety when sanctioned by scientific experts, qualified by experience and training, based on history of product use before 1958, or scientific studies and condition of use. At the same time, the "Delaney clause" stated that a substance known to cause cancer in man or animal could not be used as a food additive. Both directives would pertain to insects as food/feed. Interestingly, while a case could be made for (long) history of use, at the same time, an examination of the potential toxins harbored within or consumed by insects themselves stands to make safety assessment unique under existing guidelines. Since the time of its inception, the GRAS listing of approved food ingredients has moved from an active FDA affirmation process (1972) to one of notification (1997), thus alleviating the potential backlog of prospective substances from FDA dockets (Kotsonis and Burdock, 2013; Kruger et al., 2014; Marone and Birkenbach, 2015 for detailed discussion). Nonetheless, the law places the responsibility of safety directly on the manufacturer while FDA maintains oversight of product market approval. Further clarification of the law was made in 1994 with the Dietary Supplement Health and Education Act (DSHEA) which distinguishes and regulates dietary supplements as food apart from drugs promulgating current good manufacturing practices (cGMP) under Section 21 CFR Part 111 (Rehnquist, 2003). These laws, along with expanding HACCP regulations presently in force throughout the food industry (Zuraw, 2015), serve to standardize and safeguard food "farm to fork" and by extension, insect-based food/feed production. Most recently, the Food Safety Modernization Act (FSMA) of 2011 has mandated sweeping legislation intended for securing the confidence and safety of the food supply, particularly for the prevention of foodborne illness, by making the FDA directly responsible for both US and foreign manufacturers. Final rules for preventative controls in the manufacture of human and animal food have recently been published (http://www.fda.gov/downloads/Food/GuidanceRegulation/FSMA/UCM461834.pdf). Surveillance actions include directives for facility hazard analysis and risk-based preventative controls, cGMP's, and inspections of foreign suppliers and third-party verification accreditation for high-quality food production, packaging, and distribution guidance with technical assistance for process, reporting, and traceability accounting along an expanding global food chain network. The most recent rules to the law include model standards for the accreditation of third-party auditors/certification bodies for which user fees are applied, a legislation to become particularly important to the insect-as-food industry as import control of potentially exotic, nonnative species (the anticipated consequence of consumer delicacy demand) would place foreign suppliers directly under FDA jurisdiction (FDA, 2015).

LABELLING REGULATION AND HEALTH CLAIMS APPLICABLE TO INSECTS

Following upon the enactment of the original Federal Food Drug Act stemming from the horrendous unsanitary conditions in the meat-packing industry detailed in Upton Sinclair's "The Jungle" (1924), the Supreme Court formally condemned any statement on a product label that attempted to mislead or deceive the consumer. Since then, product labels have undergone numerous modifications and updates for additives, packaging, and nutrient content, all with the purpose of properly informing the consumer for safety and quality, particularly elevating awareness of possible intolerance to individual ingredients (such as dairy, lactose, and certain allergens). In keeping with the manufacturer as the responsibility party of product safety, by 1990 the Nutrition Labeling and Education Act (NLEA) required "all packaged foods to bear nutrition labeling and all health claims for foods to be consistent with terms defined by the Secretary of Health and Human Services" (Weingarten, 1990). The 20-year-old nutrition label as we know it today, revised from the original 1990 act, including a listing of the most important nutrients, serving size, calories, and % daily value, is currently in the process of being further revised. Proposed changes by FDA are expected "to expand and highlight the information they (consumers) most need when making food choices." Both for current labelling and those expected with the upcoming proposed changes (within 2 years; http://www.fda.gov/Food/GuidanceRegulation/GuidanceDocumentsRegulatoryInformation/LabelingNutrition/ucm387533.htm), listed nutrients include separate line items for protein and mineral content, particularly applicable to insects. Nutrient content and subsequent labeling, important for informing on potential health-related issues such as allergenicity/bioactivity, has the capacity to vary greatly depending on insect species, including how and where they are cultivated, harvested, and processed. In this regard, standardization throughout the industry remains to be determined and would be expected to be reflected in the label (Fig. 7.1). When insect-developed powders and pastes are included as ingredients in other processed foods such as protein bars, snacks, bake mixes, baked goods, pastas, meat alternatives, sauces, protein shakes and powders, beverages, and other derivatives, labeling of ingredients would be expected to reflect their order of prominence, a statement of clear insect-derived identification and any regulations limiting their use. Finally and most importantly, distribution of the edible insects and insect based foods must comply with federal regulations for labeling with regard to their identity, which includes common and scientific name of the insect (Section 403 of FFDC Act), and any associated health claims. (http://www.fda.gov/Food/GuidanceRegulation/GuidanceDocumentsRegulatoryInformation/LabelingNutrition/ucm2006828.htm). To date, numerous reports have been written on the health benefits and ecologic responsibility of insect protein (Carter, 2014; McCall, 2014). However, the FDA has been largely silent on insects as food, including qualification of health claims, citing only its reference to Section 201(f) of the FFDC Act (insects as adulterants) and stating its firm position that a food should be hygienically derived and wholesome (Ziobro, 2015).

SAFETY CONSIDERATIONS FOR INSECTS AS FOOD

It is generally considered that insects, as members of the segmented arthropoda phylum, display a high degree of sequence homology for tropomysin cross-reactivity with crustaceans/shellfish, thereby creating potential common allergies in susceptible individuals (Panzani and Ariano, 2001; Ayuso, 2011; Belluco et al., 2013). Therefore, specific regulatory recommendations for the raising, harvesting, processing, and labeling of insects for food/feed would reasonably derive from existing FDA/EPA seafood guidance and is so considered (http://www.epa.gov/agriculture/tfsy.html).

In keeping with the FDA's Federal Food, Drug, and Cosmetic Act which requires a manufacturer wishing to market a new dietary ingredient not available before October 15, 1994 (FDA [Section 413(d) of the Federal Food, Drug, and Cosmetic Act, 21 U.S.C. 350b(d)]), sufficient information must be provided by the manufacturer attesting that the dietary supplement containing a new dietary ingredient is reasonably expected to be safe under the conditions of use. Under the existing guidance, insects as new dietary ingredient would generally also conform to the requirement for the ingredient as a supplement since it is expected that their availability as food would not be the sole source of human or animal nutrition. This directorate would be coordinate with FDA's definition of supplement as "a product (other than tobacco) intended to supplement the diet that contains one or more dietary ingredients. A dietary supplement is limited to products that are intended for ingestion in tablet, capsule, powder, softgel, gelcap, liquid, or other form, that are *not represented as conventional food or as the sole item of a meal or of the diet*, and that are labeled as dietary supplements" (http://www.fda.gov/Food/DietarySupplements/ucm109764.htm).

Depending on the history of use, regulatorily-approved routes for the qualification of safety applicable to insects, likely each individual species, with specific feeding, methods, and environments of rearing and methods of processing for a particular insect based ingredient, would be either as a premarket approved new dietary ingredient (NDI) or as a premarket exempt GRAS determination. The FDA has distinguished between these determinations by providing guidance to the industry, fundamentally requiring a 75-day premarket evidentiary notification of safety by the manufacturer for an NDI,

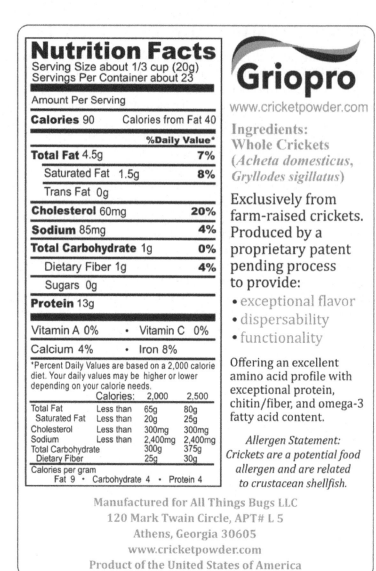

FIGURE 7.1 *(Reproduced with permission from: www.cricketpowder.com)*

and a published manuscript indicative of consensual scientific expert safety approval for GRAS determination. Nonetheless, the overall product requirements remain the same for both NDI and GRAS determinations: that any and all available information on the test product be thoroughly evaluated for general recognition of safety as addressed in 21 CFR 571 under the FFDCA, including identity and composition, intended use, analytic characterization using validated test methods, assessment of risk/tolerance based upon exposure and concern levels, available or determined clinical and preclinical safety studies under approved protocols, and environmental impact (http://www.fda.gov/Food/GuidanceRegulation/GuidanceDocumentsRegulatoryInformation/ucm257563.htm). Safety evaluation through carefully planned toxicological testing programs are designed to attain a no-observed-adverse-affect-level (NOAEL) with a dosage high enough to insure safety. Similar to that for substances added to animal and veterinary foods promulgated through the Center of Veterinary Medicine (CVM, 2010), insect-as-food ingredients may be considered under the GRAS proposal given that it is safe under its intended use and in accordance with federal rules and regulations.

The FDA, in harmonization with international regulatory bodies, has approved and promulgated standardized toxicological and pharmacological preclinical safety studies under Good Laboratory Practices (GLP), first adopted from Europe in 1978, to ensure reliable, reproducible results supported by precise record-keeping for the qualification of safety. These practices were later endorsed and coordinated under the Organization of Economic Cooperative Development (OECD), a consortium of 34 member countries created in 1960, dedicated to global economic development.

Since that time, OECD guidelines for the testing of chemicals (Sections 1–4; http://www.oecd.org/chemicalsafety/testing/oecdguidelinesforthetestingofchemicals.htm), instituted, adopted, and refined according to appropriate technological and animal usage recommendations, have served to standardize product safety testing through accepted protocols by the member states and as universally accepted. In the United States, coordinated safety assessment with OECD through FDA's *Toxicological Principles for the Safety Assessment of Direct Food Additives and Color Additives Used in Food* through its Center for Food Safety and Applied Nutrition (1982) (Kokoski, 1992), and since revised (known as the Redbook II; FDA, 2000; http://www.fda.gov/Food/GuidanceRegulation/GuidanceDocumentsRegulatoryInformation/IngredientsAdditivesGRASPackaging/ucm078044.htm) and currently in the process of again being amended, has provided the acceptable standards for test product assessment, both clinically and preclinically.

While the rigor of testing is assured through GLP (FDA, 1981), the degree of testing is determined by the toxicological concern level appropriate to the given food. In the case of insects, the prospective hazard level would be dependent on the given (so-called edible) derived taxonomic species in question, their life stage of appropriate consumption, and the level of processing, from whole animal to processed powders, pastes, or other ingredient formats. Determination of safety is such that the expected dietary intake level (EDI) be lower than the acceptable daily intake (ADI) (derived from long-term preclinical safety studies) from all sources of a given ingredient (Alger et al., 2013). To this end, known physical and chemical relationships of the ingredient have been used to determine bioactivity for levels of toxicological potential and concern, which, in turn, will determine the level of safety testing applicable to the additive (insect ingredient). Categories of low, intermediate, and high concern may require increasing complexity of toxicological testing from in vitro/in vivo short-term acute (<1 month) studies to longer-term (≥6 mos) studies requiring rodent and nonrodent species, including multigenerational reproductive assessment. Newer, in silico methods have the capacity to use quantitative structure–activity relationships (QSAR) and high throughput screening for identifying bioactivity and toxicological potential of chemicals in foods and packaging (Hartung and Hoffmann, 2009; Price and Chaudhry, 2014; Schilter et al., 2014). These methods may be used to characterize insect species in the same way that chemicals are characterized by their functional groups. Current general recommendations for GRAS and NDI safety determinations include (1) thorough characterization of the test product with determination of exposure level, (2) performance and assessment of preclinical toxicology and pharmacology studies, including absorption, distribution, metabolism, and elimination studies (pharmaco-/toxico-kinetic), and (3) clinical assessment, when warranted or available by use. These assessments traditionally include review and integration of the existing published and unpublished literature and history of use (Kruger et al., 2014).

Species Identity and Characterization of the Insect-Based Product for Whole Body or Processed Feed

Like other food additives and coordinate with their level of processing (whole animal or powder), insects destined for food applications would be characterized based on their purity, stability/shelf-life, and potency for the duration of specified usage in coordination with cGMP's for large scale-up and reliably producing the same product over time with as few impurities as reasonably achievable. Important to the safety characterization of insect-based food and feed is the requirement for strict species identification and traceability, the ability to identify geographic location of origin and third-party food chain management. Many laboratories now perform taxonomic DNA identification as a quality control measure for other food types (seafood, for example), serving to not only identify the product but also ensure the public health, supply chain security, environmental sustainability, and economic stability of the market through purity validation and avoidance of mislabeling or other forms of food fraud or adulteration (eg, dilution, substitution, addition). Establishment of sanctioned methods of analysis through the Association of Official Analytic Chemists or other recognized organization and used in testing laboratories certified for this purpose are often based on state-of-the-art mass spectroscopies and chromatographies, coordinated with verification of dose level (quantitation), homogeneity, and stability at the time of toxicological testing (Physical/Chemical Properties; http://www.oecd-ilibrary.org/environment/oecd-guidelines-for-the-testing-of-chemicals-section-1-physical-chemical-properties_20745753). A vendor quality process validation scheme, one element in the enforcement of FSMA, ensures the rigorous means of identifying, evaluating, approving, and maintaining standards of supply in the industry. An industry mandated certificate of analysis (COA) for each batch of product by the supplier provides purchasers with independent verification of product quality, safety, and compliance with regulations. Vendor quality controls such as these also apply to product processing and process validation. By assuring raw material security through the institution of reliable methods of analysis and validation, processors, sellers, and consumers can be confident of what they are getting to keenly sanction a fledgling industry.

Once standardized methods for basic product identity are established, ongoing assessment for product variability is similar to that of any functional food additive. Insects can be subject to the ecologic variabilities of the feed they are given, soils, their cultivation/environmental conditions, the treatment of them, and the climate that influences them. The plethora of possible impurities that this may introduce in a natural, wild-caught setting would presume to be reduced in controlled farming. Like all other living organisms, insects have the potential to concentrate contaminants from dietary exposure; and similar to that of animals raised for their meat, their bodies can bioaccumulate toxic compounds following exposures and metabolize toxic by-products, in turn, exposing consumers. Nonetheless, the ability to define, assess, and mitigate impurities from residues in the raw material through the production and packaging process is paramount to manufacturer brand, particularly in this novel, impressionable, and competitive food-market landscape. To this end, it is the insect powder (flour/paste) form that will be (1) most aesthetically appealing and useful to the broadest range of consumers, and (2) considered more easily regulated for quality and safety standardization, in cooperation with regulatory authorities for acceptable risk. That being said, it is expected that whole animal products would remain, to the greatest extent possible, subject to rigorous safety standards based on species identification and supplier confidence. The Food Chemicals Codex (FCC) and the EU Regulation No 231/2012 food additives listed in Annexes II and III to Regulation (EC) No 1333/2008 (food additives) along with the Joint FAO/WHO Expert Committee on Food Additives (JECFA), specify acceptable levels of quality and purity of food-grade ingredients in the United States and European Union, respectively (http://www.fda.gov/Food/GuidanceComplianceRegulatoryInformation/GuidanceDocuments/ChemicalContaminantsandPesticides/default.htm).

Toxicological Assessment

Determined from FDA's draft guidance on dietary supplements (July, 2011 http://www.fda.gov/Food/GuidanceRegulation/GuidanceDocumentsRegulatoryInformation/ucm257563.htm) promulgated from the 2011 FSMA directorate to ensure a safe food supply, preclinical evaluation remains the industry standard for reliable toxicological and pharmacological safety prediction in humans and is generally derived from designs of the OECD and FDA Redbook guidelines mentioned earlier (http://www.oecd-ilibrary.org/environment/oecd-guidelines-for-the-testing-of-chemicals-section-4-health-effects_20745788 and http://www.fda.gov/Food/GuidanceRegulation/GuidanceDocumentsRegulatoryInformation/IngredientsAdditivesGRASPackaging/ucm2006826.htm.).

Genotoxic assessment remains among the initial safety standard in food additive petitions, scientifically supported for its wide-ranging predictive benefit of low through high toxicological levels of concern. Through their combined in vitro/in vivo-tiered battery approach, these studies provide the first screen to carcinogenicity by detecting agents which may induce genetic effects, either directly or indirectly, for point mutations, structural chromosomal aberrations, recombinations, and/or numerical changes (clastogenicity) (Ames et al., 1973; Kirsch-Volders et al., 2010; Kirkland et al., 2005, 2011; EFSA, 2011). With study designs easily adaptable for use with insect powders, these studies are useful for their low cost, good accuracy, and expeditious turn-around times.

Progressing further to increasing levels of concern for toxicological prediction are short-term acute studies (1 month or less) which may be performed in vitro (often with reconstructed human tissue) or in vivo, (most often using rodent or rabbit models), employing oftentimes high dose levels for target organ specificity and lethality. Similar to that of their genotoxicity counterparts, these studies offer low cost, and, due to their generally small quantity of test substance requirement, offer a good alternative to longer term studies when product availability is limited.

The most robust studies for the greatest levels of toxicological concern are those of repeated-dose studies of longer duration. Within the constructs of standard OECD and FDA Redbook recommendations and intended use, the selection and tailoring of design (objective) of these studies depends on product-specific anticipated target population (sensitive: infants/children, elderly, pregnant women), the duration and administration of use, and the expected exposure effects. They may also provide a sound scientific platform for any future clinical studies based on health claims. Preclinical studies may vary from short- (<1 month) to long- (2 years) term duration and may target various physiologic states for changes in reproduction and development, neurologic, immunologic effects, and pharmacodynamics, among others. In general, these studies are performed in rodents, with the option for a second sensitive species (dogs, rabbits, monkeys) as the concern level rises. They consist of three dose levels plus a control at approximately 90%, 50%, and 10–25% of maximal tolerated dose, in a test vehicle according to the expected human exposure route, oftentimes the diet (Kotsonis and Burdock, 2013; Marone and Birkenbach, 2015) for determination of a no-observed-adverse-effect-level, no-observed-affect-level or lowest-observed-adverse-effect-level. Pathologic endpoints of clinical chemistry, hematology, urinalysis, and coagulation as well as histopathology of selected organs and available toxicokinetic analysis of blood and target organs accompany in-life parameters of clinical observation, body weight, and food consumption to qualify a safety determination.

As commercial production and distribution of insect powder becomes more widely accepted and common, safety studies will become more frequent in the scientific literature adding to the fund of knowledge from which to further guide regulatory standards. Among the few toxicological studies published to date are the investigations of: raw and processed larva of *Cirina forda* (Emporer moth) administered orally to mice and rats showing neurotoxicity in mice (LD50 value of 7000 mg/kg bw; Akinnawo et al., 2002); the protein of *Antheraea pernyi* silkworm pupae tested in acute and subacute toxicological studies where the oral acute maximum tolerated dose was >15.0 g/kg bw in mice, and a maximum dose of 1.50 g/kg per day in rats with no mutagenicity, (Zhou and Han, 2006); and freeze-dried powder of *Allomyrinna dichotoma* (Japanese rhinocerous beetle) larvae gavaged to rats producing no treatment-related signs for a NOAEL of >2500 mg/kg per day (Noh et al., 2015).

For now, similar in requirements as that for other food-additive GRAS safety petitions, it is anticipated that insect powder would generally, at minimum, require test product derivation, characterization, and analysis, a standard genotoxicity battery, and tailored toxicology study(ies) for compliance in the United States, Europe, and most other countries.

Clinical Evaluation of Safety

Edible insects (and, by corollary, insect-based food ingredients) are unique in that it can be reliably assumed that many species have been consumed by humans for all of hominid existence, however, the history of use *for a given edible species* is not well documented. Therefore, while traditional consumption can be relatively undisputed, modern-day safety standards dictate, along with the strict requirement for analytic product characterization, satisfactory preclinical safety history, with clinical trials to support safety claims. Clinical trials are not a general requirement of regulatory GRAS dossiers, but under good clinical practice guidelines (ICH, 1996), are mandated by the use of function and disease mitigation health claims as applicable for insect-derived foods. Performance of these studies are costly, requiring extensive planning and test-subject regulation and monitoring; and with thorough and precise reporting, can enable sound safety assessment for specific populations, particularly those of potentially sensitive individuals (Kotsonis and Burdock, 2013).

In general, the ultimate design of a comprehensive safety evaluation plan for insect-based food would follow that of any other novel food including individual product consultation with the FDA or appropriate regulatory authority prior to product development and marketing.

TOXICOLOGICAL HAZARDS OF INSECT-BASED FOODS AND FOOD INGREDIENTS

Worldwide regulatory insufficiencies have largely reflected a lack of consensus of which insect species are suitable for consumption. In addition, the capacity of insects to harbor toxins, chemical (pesticide) residues, heavy metals, and bacteria, among others, both natural and otherwise, without standardized cultivation, processing, and/or preparation techniques to mitigate potential risk, provide obstacles to mass consumption of insects, be they wild-caught or farmed.

Most recently, ANSES, the French Agency for Food, Environmental and Occupational Health & Safety, conducted a safety assessment in the European Union of the hazards associated with edible insects and insect-based food consumption. They agree that despite uncertainty and insufficient information, the industry is "to be controlled through the establishment of specific standards to reduce any potential risks of their consumption," with hazards based upon (1) chemical substances; (2) physical agents; (3) allergens to all arthropods; (4) microbiologic agents—parasites, viruses, bacteria, fungi, and their toxins; and (5) farming and production conditions, processes, and storage (ANSES, 2015). Using these categories as a guide, the risks associated with these vulnerabilities and the extent with which they are managed will depend on the level of processing from raw to whole animal to ground powder.

Chemical

Although most regulatory associations of insects with food have been historically concentrated on their agricultural mitigation as pests, a number of considerations for chemical exposure associated with their use as food, largely having to do with their inherent nature, are potentially adverse. With the recognition that insects specifically cultivated/grown for food should have a reduced risk for all potential hazards, in general, one of the primary considerations concerning insects, particularly those wild-caught, is that of their bioaccumulation of pesticides. Any pesticide/fungicide used for agriculture and those used specifically against insects is potentially dangerous to consumers. Several reports are available attesting to the adverse health effects to consumers when insects are brought to market following their exposure, such as the high concentrations of organophosphate residues in locusts collected for food in Kuwait or the numerous generalized instances of lost (pesticide-induced) host environments and biodiversity causing unknown changes to the natural habitat subsequently affecting highly adaptive insect species (Saeed et al., 1993; DeFoliart, 1999).

Exogenous Chemical Contaminants and Residues

Despite chemical contaminants as the greatest single threat, both in the case of wild-caught insects and those cultivated under controlled conditions, the general consensus is that growth and harvesting conditions can be largely controlled. This is to indicate that the incorporation of chemical residues in insects may be considered directly related to their presence in the exogenous feed and/or nutrient media. Residues for natural plant and insect feed stock can, theoretically, be monitored (and eventually controlled) for pesticides, herbicides, mycotoxins, veterinary residues, polychlorinated biphenyls (PCBs), phytochemicals, and xenoestrogens (for endocrine disruptor capability), organophosphates/neuroactives, antibiotics, processing contaminates, and others (van der Spiegel et al., 2013; SciCom, 2014). Nonetheless, the diversity of the natural habitat of insects in the wild (as opposed to in farms) makes monitoring challenging and possible bio-incorporation exponentially variable. For example, consumption of emergent aquatic insect prey who feed from polychlorinated biphenyl-contaminated water and sediment near shorelines (and contaminated pools) become toxin-concentrated vectors to terrestrial invertebrates like spiders and wasps in the area who feed on them (Raikow et al., 2011). By extension, information on the accumulation of various residues *in specific parts of the insect*, such as PCBs in fat, undoubtedly dependent upon species, remains to be elucidated. It is important to consider that as other novel foods are gathered from increasingly variable and unconventional sources (sea life/seaweed, for example), the opportunity and natural ease for insect adaptation and exploitation into those habitats prompts a greater vigilance for the insect-based food industry. These considerations may be expected to be mitigated for insects farmed indoors, which are likely to represent the vast majority of future uses of insects as food ingredients. Likewise, given the diversity of edible-insect species, it is unlikely that a single regulation or set of standards will sufficiently apply to all edible-insect species, though some regulations and sets of standards may be composed as to be applicable to more than a single species (eg, it may be possible to set standards that work well for both mealworms and crickets farmed indoors).

Natural Toxicants of Insects

While pesticides are considered the most commonly suggested dangers, ingestion of some insect species carries with it potential hidden threats. Among these are the natural toxicants well described by Belluco et al. (2013), and defined as "naturally present in insect feed or synthesized by insects," thereby underscoring the importance of identifying safe and unsafe insects (Warner, 2007; Dossey, 2010). Briefly, noxious substances innately produced by the insect may be further distinguished by their purpose, that is, they may function in guarding against predators (superficial) or in their maintenance of the metabolic processes of the organism itself (internal). Finally, while insects feeding on edible plants may be considered safe, sequestration and resulting tolerance of toxins from feed may change upon food preparation (Berenbaum, 1993).

Potentially toxic superficially present substances associated with consumer exposure include those associated with natural color mimicry/patterns and/or those of direct chemical production apparent as a defensive glandular secretion on the surface of the body (Belluco et al., 2013; SciCom, 2014). Biologically-derived carmine dye from cochineal insects often used in juices, yogurt, candy, and cosmetics has been linked to anaphylaxis-like reactions. Shellac used for confectionary glaze is derived from red dye and clear resin secreted from the body of the *Laccifer lacca* (lac) bug (Finn, 2011). More potentially overt adverse reactions distinguish defense mechanisms of various insect groups: phanerotoxic insects include those with specific venom-synthetic organs such as bees, ants, and assassin bugs, while cryptotoxic insects, such as blister beetles, present with toxins at high levels derived from their synthesis or accumulation in either specific structures or diffused throughout the body (poisonous blood plasma).

Other examples of insect-derived chemicals produced as a result of normal animal homeostasis presenting potential toxic hazards to sensitive humans include graded reactions resulting from metabolic steroids (testosterone family of beetles), inhibitory enzymes such as cyanotoxics associated with metabolic pathways (Coleoptera and Lepidoptera), and natural neurodepressants (longhorn beetles) (Blum, 1994; Belluco et al., 2013).

In addition, to the endemic chemicals mentioned, other unforeseen contaminants may enter into the human food chain when an insect's soil-dwelling feeding on solid waste is converted from discarded consumer products leaving excrement in farmed areas (Belluco et al., 2013). Alternatively, uses of organic waste streams (manure and water) developed as insect food may raise concerns for the presence of pathogenic organisms and toxins as they are sequestered in the insect body (van Huis et al., 2013b). It is worth noting here that, while many insects are well known for their endogenously produced, modified, or dietsequestered toxins, the majority of species likely to be mass produced for human consumption (typically crickets or mealworms) must be qualified to be devoid of such acute toxins or venoms. Nonetheless, toxicity evaluation is merited for all species, due to the lack of data on long-term consumption of any insect as a major part of the human diet or chronic nondietary exposure (such as occupational exposure at processing plants producing insect-based foods and food ingredients).

Heavy Metal Contamination

As with any naturally grown food source ultimately dependent on the changing conditions of their environment, insects are potentially vulnerable to heavy metal bioaccumulation, particularly for wild harvested insects but the risk may extend to farm-raised species depending on the feed and water sources as well as other farm conditions. Mercury in water and soil sediment not only appears to bioaccumulate, but may be empirically modeled to the taxonomy of some insect species and could be used to predict pollutant levels in them (Zhang et al., 2012). Reports for arsenic (Green et al., 2001), cadmium (Vijver et al., 2003; Diener et al., 2015), lead (Handley et al., 2007), and zinc (Diener et al., 2015) and those along the soil–plant–insect–chicken food chain (Zhuang et al., 2009) are among the many documenting heavy metal accumulation in insect organs. In addition, the implication for reuse of organic (animal and agricultural) waste in farming, particularly in countries where resources are limited, may serve to propagate and magnify this hazard in a food source that is highly productive for its environmental input (van Huis et al., 2013a). As such, heavy metals such as arsenic, cadmium, and lead will accumulate in the bodies of insects who will tolerate significantly higher levels of metals than mammals, in some cases surpassing acceptable EU levels (Charleton of PROteINSECT in Anthes, 2014; Charleton et al., 2015). Thus, it is prudent for farms utilizing agricultural or food by-products as insect feed ingredients to utilize food grade and preconsumer materials (such produce not attractive enough to sell, yeast, and spent grains from brewing operations, etc.). In a case particularly noteworthy for its adverse effects in sensitive populations, documented increased lead levels traced to dried grasshopper (chapulines), together with environmental and food preparation contaminants, were found in the blood of children and pregnant women in Oaxaca, Mexico (Handley et al., 2007). These cases point out the need for strict regulatory guidance when evaluating insects as feed for humans or livestock.

Insects as an Antinutrient Source

Indirectly associated with the physico-chemical construct of the animal, the reliance of insects as the sole source of nutrients is a potential metabolic health threat in persons and populations with food insufficiency and poor diet. In this regard, insects have been implicated for their antinutritional potential through interference of normal physiologic processes, thus resulting in nutritional deficiencies (Nishimune et al., 2000; Musundire et al., 2014). For example, determination of the presence of saponins, oxalates and tannins, phytates and phenolics, gossypol pigments, oxalates, and glucosinolates to interfere with various physiologic processes such as protein and carbohydrate digestion, mineral utilization, or immune stimulation requires further investigation and would be a limitation on insect use as food (Kiranmayi, 2014). It is expected that as appropriate edible-insect consumption becomes more widespread and accepted, production and preparation practices will become increasingly understood and advanced to address and mitigate these concerns.

Additionally, more recent focus on the human gut microbiome and the consideration of entomophagy for its gastrointestinal effects and the consequences for systemic health overall may become subjects of study for the near future.

Physical

The greatest physical hazards associated with consumption of insects, particularly in raw form, are those of choking. The inhomogeneous body construct of the insect, combined with the various hard appendages/exoskeleton components allow for the most unfamiliar of textures to the normal Western palate. Exposure to whole-body insects presents the unique possibility of physical hazards in the form of choking or asphyxia when the likes of appendicular spines, stingers, rostrums, course hairs, cuticles, or other unusual body parts, particularly hazardous to the elderly and children, are less familiar or accommodated to tastes given to adventurous gourmands. For this reason, ground insect powders, over raw or whole insects, are more suitable/useful, are considered safer to the general public, and afford the best opportunity for acceptance, both at the personal and commercial sale level.

Similar vulnerabilities may exist around the natural curiosity of children to ingest insects on the impulse to explore (Belluco et al., 2013). These tendencies could lead to unforeseen adverse effects associated with choking and swallowing, particularly when the species body construct or chemical make-up exacerbates the hazard. This has been documented by a number of authors for caterpillar ingestion producing symptoms of toxic reactions mimicking allergic reaction in several species of the genus *Lophocampa* hickory moth (Lee et al., 1999; Pitetti et al., 1999).

In summary, in light of the uncertainty and insufficient data associated with insect consumption, ANSES in coordination with the new Food Safety Guide from EU's Federal Agency for the Safety of the Food Chain (SciCom, 2014; Robinson, 2015) in its prescient recommendations may provide a good preliminary guide for the EU, and by extension, the international community, when it states the importance of: "concentrating research efforts on potential sources of hazards; establishing EU-level positive and negative lists for different insect species and stages of development that can and cannot be consumed; exploring the animal well-being issue for these categories of invertebrates; defining a specific regulatory framework for the farming and production conditions of insects and insect-based products which ensures the control of health risks; and establishing allergy risk prevention measures, both for consumers and in the occupational environment" (ANSES, 2015).

FARMING AND NOVEL CONSIDERATIONS DRIVING INSECT FOOD/FEED SAFETY

Of the many potential areas where the use of insects as food or as food ingredients possibilities are great, animal feed and aquaculture remain promising, though cultivation and processing standards for the industry remain largely undefined. While detailed to greater extent elsewhere in this book (see also Chapter 4), it is well-known that the feed conversion efficiency of insects is well below that of beef (van Huis et al., 2013b) and as such, the use and practice of insects for higher animal nutrition warrant their consideration. "Breeding trials conducted by the EU initiative PROteINSECT have found that one hectare of land could produce at least 150 tons of insect protein per year" as compared to less than a ton of soybean for the same area (Zanolli, 2014). Aside from insects being natural food for birds and fish, land use for insect farming is economical and environmentally sustainable. The recent EPA spill of toxic metals into the Animas River in Colorado, with its potential, still-to-be-determined effects on wildlife and cattle, are clear indications of the need to safeguard natural resources and food supply through the use of industrialized insect farming (Turkewitz, 2015). This and other environmental challenges may indicate that the time is right for controlled cultivation of insects as shrinking of natural habitats, the largest of which is the tropical rainforest biome, may deselect for desired species and render them unavailable or inadequate for wild-caught sale.

As regulatory standards for insect use become better defined and entomophagy become more accepted in global markets, the basic regulatory platform for these commodities in the context of animal feed use would be expected to fall under the United States' CVM's Food Additive Petitions for Animal Feed, including the GRAS petition procedures (http://www.fda.gov/AnimalVeterinary/SafetyHealth/AnimalFeedSafetySystemAFSS/default.htm) and the FSMA regulations for imported foods mentioned previously.

Depending on the extent of use and the growing acceptance of the industry, potential novel areas of development for insects include those commonly in use for other functional foods today including the use of insects as colorants, flavorants, texturizers, antioxidants, protein energy supplements, or other organoleptic enhancers. These considerations broaden the usage of insect-based food and feed ingredients beyond the basic need of nutritional maintenance and security to that of offering the possibility of health effects and a positive, sensational eating experience for a larger global population. As examples, insect powders can be purposed in a multitude of forms to fortify beverages/smoothies, and as ingredients in energy drinks, health bars, baking mixes, granolas, meat alternatives, and in a myriad of other uses, for both home and commercial consumption. Most recently, the use of home and commercial 3D food printers (McFarland, 2015) may allow the benefit of extrusion of high-quality insect protein paste for striking custom visual and textural forms (Spencer, 2014), while masking awareness of the basic component ingredient (insect derivation is not recognizable as such) (Fig. 7.2). This utility assumes that purity and safety for all ingredients, including the insect component, are as according to regulatory standards. These possibilities will surely heighten attitudes and use of insect high-nutrient protein powder/pastes, resulting in greater attention for FDA and international regulatory oversight (Fig. 7.3).

Of the more controversial issues surrounding ingredients and functional foods, genetically modified organisms (GMO), and nanotechnology may have potential ramifications for the insect-based food and feed market. While some may consider it premature or even inappropriate to contemplate the implication of insects to these technologies, these possibilities spur the necessity for regulation sooner rather than later. Furthermore, the present ubiquitous use of genetically engineered plants such as soybeans, corn, canola, and cotton, to name just few, as common components of sauces, sweeteners, thickeners, emollients, and their derivatives are a potential platform for the incorporation and enhancement of any natural substance into foods for their desired nutritional value. The safety of these modified foods for human and animal consumption must meet the same safety requirements as traditionally derived foods, according to EFSA and the FDA, EPA and USDA (US FDA, 2001; EFSA, 2010). Presently, with the exception of the introduction of transgenic genes for functional enhancement, the FDA considers most GMO foods to be "substantially equivalent to and historically similar to non-GM crops under a 1992 policy and, as such, these products are usually designated as GRAS without the need for premarket approval under the Federal Food, Drug, and Cosmetic Act" (Marone and Birkenbach, 2015). Like all other functional foods, insect-derived protein changes in the form of transgenic modifications would be the subject of careful scrutiny for possible allergens and toxins of unintended consequence. Unlike other foods, the living, whole-body nature of these insect-derived food sources necessitates a still higher level of vigilance particularly for a life form that is highly adaptable both cultivated and in the wild.

FIGURE 7.2 3D Printed Pizza. *(From: Foodini, a 3D food printer: Real Food, 3D printed. Reprinted with permission from www.naturalmachines.com)*

Like that of other novel foods developed for the industry, insect-derived food regulatory address must parallel advances in food technology while at the same time, debate for mandatory US GMO labeling continues. In addition to the cultivation of the insects themselves, since in the vast majority of cases insects are and will be utilized in their whole form, careful consideration must be taken for the GMO, organic, health, and safety nature of the feed they are given as feed components are likely to remain in the guts of insects at the time they are processed and consumed, be they consumed in whole or in powder/paste form. Nonetheless, the outcomes of the legal directorates and controversies now surrounding safety regulations and labeling for traditionally sourced GMO foods would be expected to have great impact for the inevitable consequential questions arising for insect-derived foods (McAuliff, 2015).

Similarly, the use of nano-sized insect powders and its packaging would be expected to fall under the same present regulations as those of traditional foods (OECD, 2010; FDA, 2014a). With the understanding that the formalized industry processing of insect powders is technolognically advanced enough to standardize (flour) particulates, it is reasonable to assume that the effect would be similar to that of other nanotechnogically produced foods: that reductions in particulate size via increased surface area-to-mass ratio would increase solubility, absorption, and bioavailability resulting in changes in

FIGURE 7.3 3D Chocolate Snowflake. *(From: Foodini, a 3D food printer: Real Food, 3D printed. Reprinted with permission from www.naturalmachines.com)*

potential tissue response and variable safety risks (Warheit and Donner, 2010; Burdock and Matulka, 2014). The address of these considerations, like that of other functional foods, awaits the insect-as-food industry in turn.

PROCESSING, PREPARATION, PACKAGING, AND TRANSPORT OF INSECT-BASED FOODS AND FOOD INGREDIENTS

Similar to that for other foods, the processing, cooking, storage, and packaging techniques for the insect-based food industry are expected to preserve the natural ingredients of the raw material to the greatest extent possible. For insects as with all other biological materials utilized as food, these procedures have a significant potential to affect the sensory visual, aroma, taste, and texture appeal to the potential consumer, particularly initially when market acceptability is most impressionable. Given the relatively novel status of insects as food ingredients, particularly in North America and Europe, extra care must be taken in processing, packaging, and transport as to show to the consumer the highest possible quality, safety, and care standards available in the food industry as to not add to an already skeptical public's concerns. As with any other food material, contaminants can be introduced into the insect-derived product anywhere throughout the breeding, manufacturing, and packaging process to result in undesirable changes which could greatly affect market appeal, critical to a fledgling company. Here, in coordination with regulatory agencies working to develop universal standards, manufacturers can maximize consumer confidence in the industry and in their brand through current responsible HACCP compliance and traditional food-safety distribution practices (FAO/WHO, 2001). Also see Chapter 5 of this book for more discussion on considerations for insect processing for use as an ingredient in food products.

Insects are a novel food in that they are live organisms that are generally sacrificed prior to what may be direct whole-animal consumption or in preparation for processing (Mlcek et al., 2014). Unlike that for the meat, fish, and poultry industry (USDA Humane Slaughter Act of 1978; Fed Reg, 2014), the process of slaughter for insects used as or in food is unique in that there exists no standard procedures or regulation, particularly for the enormity of biomass/biodiversity this involves. While no specific recommendations for humane treatment exist, it is becoming increasingly clear that, using crustaceans as a behavioral and physiological guide, invertebrates experience through nociceptors, what we would define as pain upon potentially harmful events (Wigglesworth, 1980; Tomasik, 2015). The use of anesthesia, analgesia, and chemical euthanasia for insects has been reported and is not without its own practical considerations and controversies (Cooper, 2011; Bennie et al., 2012). It stands to reason that an industry-wide consensus regarding these procedures must comply with universally accepted sanitary guidelines and practices *for each given species*. The Netherlands is among the first to pioneer hygienic whole-animal processing for "ready-to-eat" human consumption specific for the mealworm beetle, including steps consisting of: (1) one-day fasting (to ensure an empty gut), followed by (2) heat treatment, and (3) whole-body freeze-drying for extended 52-week shelf-life (van Huis et al., 2013c; NFCPSA, 2014). In North America and other areas of Europe, freezing appears to be the consensual best practice for humane slaughter and simultaneous preservation of the nutrients in raw insects as food. This is because insects, as poikilothermic (cold blooded) creatures become anesthetized at low temperatures prior to dying once frozen in most species. In fact, in the natural environments of locations which experience seasonal freezing winters, this is how many insects in the wild are naturally extinguished. As the insect food industry takes shape, a standardization of safe slaughter and commercialization procedures is expected to be forthcoming, particularly in areas of the world where animal rights activism is sensitive, even as the considered unattractiveness of the insect body (at least in Western cultures) lends no particular regard.

Processed and whole-animal dried insect-derived foods can be compact, easy to store, and transport. Their shelf life is considered advantageously longer than traditional meat/poultry/fish protein sources. This serves to heighten their practical safety appeal (van Huis et al., 2013c). Nonetheless, the insect-based food consumable is subject to the same processing, preparation, and packaging (leachate) potential to form toxic by-products similar to that of other foods (van der Spiegel et al., 2013; Marone and Birkenbach, 2015). Specifically, traditional thermal and nonthermal (acidifying) processing techniques and heat gradients specific to the species, to preserve and sterilize insect products for crushed, pulverized, raw, boiled, ground, or mashed final products may be used with newer technologies, such as microencapsulation, for core timed-release and to minimize microstructural changes (Wang and Bohn, 2012).

Little, if anything, is known regarding adverse processes resulting from cooking insects for use in food; however, it is presumed that the same general preparation precautions apply as for conventional foods. Awareness of species identification and body construct becomes ever more important when macromolecular composition, specific to the insect type, is a determinant in conventional cooking-related hazards associated with the formation of toxic acrylamides, polycyclic aromatic hydrocarbons, phenolic compounds, furans, heterocyclic amines, and/or lipid oxidation products, produced as a result of frying, baking, sautéing, broiling, roasting, grilling, and the like. The FDA and WHO have not determined if acrylamides pose a health threat to humans although the European Union has developed a toolbox on this preparation by-product (FoodDrinkEurope, 2014). As such, defining preparation changes and determination of shelf-life standards for a given species will further enhance consumer confidence as known ingredient bioactivity or potential toxicity-generating changes can be addressed and mitigated.

Product packaging primarily maintains an inert barrier against exogenous pathogens and other forms of contaminant agents as well as environmental elements (humidity, light, etc.), while also providing an attractive mechanism for displaying the product (either by showing it directly, concealing it while displaying marketing material, or some combination). For the insect-based food industry, given the newness of the insect-based food concept to the Western world, packaging has the heightened potential, depending on whole animal or powder, to not only be significantly marketing impactful (clear or opaque cover; design appeal), but to offer enhanced antimicrobial protection through creative synergistic flavoring capacities, while providing its role to preserve the quality and safety of the product (Sultanbawa, 2011). This is in keeping with the current understanding that as with any other food product, leachates from packaging have the potential to act as indirect food additives which can and should be controlled to the greatest extent possible. As with traditional foods, the use of nanomaterials in food formulation and packaging is growing and their use for insect product packaging would be expected to be regulated under current guidance in the United States and the European Union (Duncan, 2011; FDA, 2014b; Panagiotou and Fisher, 2013). Similarly, safe transport and sale of insect-derived foods, both statewide and internationally, would be expected to be subject to the same existing regulations (CFSAN and CVM) as those that are currently in existence for human and animal feed (http://www.fda.gov/Food/GuidanceRegulation/GuidanceDocumentsRegulatoryInformation/SanitationTransportation/default.htm), where the primary focus guides sanitary transport, defect action levels, and the minimization of potential microbial hazards.

CONCLUSIONS ON THE USE OF THE INSECTS AS FOOD AND FEED

It is a fortuitous and forgiving irony that the countries which suffer the greatest food insufficiency and security are those that are most traditionally accepting of insects as part of their diet. That basic acceptance allows a potentially great impact of utilizing insects as a food source upon global nutritional benefit for all peoples, through greater consumer awareness and education, compelling, most importantly, a responsible regulatory global standard. With the promise of an equitable model solution to the fundamental standards of human nutrition, the insect-based food industry makes less vulnerable both the reliance on extensive natural resources, and the tendency for external social/civil upheaval often threatening to the food security of a developing world. For this reason, it is incumbent upon manufacturers and consumers, in coordination with good globally harmonized scientific policy, to fairly and thoughtfully consider this abundant food source. Action to assure the safe application of this developing industry, beginning with species identification and thorough characterization for toxicological threat level, will safeguard an alternative food supply adequate to address the extensive population growth expected within the next generation.

REFERENCES

Akinnawo, O.O., Abatan, M.O., Ketiku, A.O., 2002. Toxicological study on the edible larva of Cirina forda (Westwood). Afr. J. Biomed. Res. 5, 43–46.

Alger, H.M., Maffini, M.V., Kulkarni, N., Bongard, E.D., Neltner, T., 2013. Perspectives on how FDA assesses exposure to food additives when evaluating their safety: workshop proceedings. Pew Charitable Trust Comp. Rev. Food Sci. Food Safety 12, 90–119.

Ames, B.N., Lee, F.D., Durston, W.E., 1973. An improved bacterial test system for the detection and classification of mutagens and carcinogens. PNAS 70 (3), 782–786.

ANSES—French Agency for Food, Environmental and Occupational Health and Safety. Insect-as-food: A review of potential hazards and research needs. April 9, 2015. Available from: https://www.anses.fr/en/content/insects-food-review-potential-hazards-and-research-needs.

Anthes, E., 2014. Could insects be the food of the future? BBC Available from: http://www.bbc.com/future/story/20141014-time-to-put-bugs-on-the-menu.

Ayuso, R., 2011. Update on the diagnosis and treatment of shellfish allergy. Curr. Allergy Asthma Rep. 11 (4), 309–316.

Belluco, S., Losasso, C., Maggioletti, M., Alonzi, C.C., Paoletti, M.G., Ricci, A., 2013. Edible insects in a food safety and nutritional perspective: a critical review. Comprehens. Rev. Food Sci. Food Safety 12, 296–313.

Bennie, N.A.C., Loaring, C.D., Bennie, M.G.M., Trim, S.A., 2012. An effective method for terrestrial arthropod euthanasia. J. Exp. Biol. 215, 4237–4241.

Berenbaum, M., 1993. Sequestering of plant toxins by insects. Food Insect Newslett. 6 (3), , Available from: http://labs.russell.wisc.edu/insectsasfood/files/2012/09/Volume_6_No_3.pdf.

Bidau, C.J., 2015. Bug delicacies: insects as a powerful food resource for a troubled world. Entomol. Ornithol. Herpetol. 4, 1–2.

Blum, M., 1994. The limits of entomophagy: a discretionary gourmand in a world of toxic insects. Food Insects Newslett. 7 (1.), , Available from: http://www.food-insects.com/Vol7%20no1.htm.

Burdock, G.A., Matulka, R.A., 2014. Nanotechology: The regulatory impact. The World of Food Ingredients pp. 56–59.

Codex Alimenarius Commission (CAC), 2010. Development of regional standard for Edible Crickets and their products. Joint FAO/WHO Food Standards Programme FAO/WHO Coordinating Committee for Asia, Seventeenth Session, Bali, Indonesia, 22–26 November 2010, pp. 1–9. Available from: ftp://ftp.fao.org/codex/Meetings/CCASIA/ccasia17/CRDS/AS17_CRD08x.pdf.

Carter, C., 2014. You Won't Believe This Shocking New Nutrition Trend. November 2014. Men's Health. Available from: http://www.menshealth.com/nutrition/crickets-perfect-protein.

Charleton, A.J., Dickinson, M., Wakefield, M.E., Fitches, E., Kenis, M., Han, R., Zhu, F., Kone, N., Grant, M., Devic, E., Bruggeman, G., Prior, R., Smith, R., 2015. Exploring the chemical safety of fly larvae as a source of protein for animal feed. J. Ins. Food Feed 1 (1), 7–16.

Cooper, J.E., 2011. Anesthesia, analgesia, and euthanasia of invertebrates. ILAR J. 52 (2), 196–204.

Crabbe, N., 2012. Local expert gets funding to develop insect-based food for starving children. Gainesville Sun, 1B–6A, 2012.

CVM., 2010. Available from: http://www.fda.gov/AnimalVeterinary/ResourcesforYou/ucm047111.htm.

Day, A.C., 2015. Europe agrees to allow insects as food. November 24, 2015. Available from: http://4ento.com/2015/11/24/europe-allow-insects-as-food/

DeFoliart, G.R., 1999. Insects as human food: why the Western attitude is important. Ann. Rev. Entomol. 44 (1), 21–50.

Diener, S., Zurbrügg, C., Tockner, K., 2015. Bioaccumulation of heavy metals in the black soldier fly, *Hermetia illucens* and effects on its life cycle. J. Ins. Food Feed, 1(1), Wageningen Academic Publishers, available on-line. Available from: http://www.wageningenacademic.com/doi/pdf/10.3920/JIFF2015.0030

Dossey, A.T., 2010. Insects and their chemical weaponry: new potential for drug discovery. Nat. Prod. Rep. 27, 1737–1757.

Duhaime-Ross, A., 2014. I ate crickets because they're future food. The Verge. Available from: http://www.theverge.com/2014/9/16/6096821/crunch-time-canadian-farm-wants-to-put-crickets-in-your-kitchen

Duncan, T.V., 2011. Applications of nanotechnology in food packaging and food safety: Barrier materials, antimicrobials and sensors. J. Coll. Inter. Sci. 363 (1), 1–24, doi: http://dx.doi.org/10.1016/j.jcis.2011.07.017.

EFSA, 2011. Scientific opinion on genotoxicity testing strategies applicable to food and food safety assessment. EFSA J. 9, 2379.

EFSA, 2015. Risk profile related to production and consumption of insects as food and feed. EFSA J. 13 (10), 4257, 1–60.

European Council (EC), 2015. Council of the European Union. New rules on novel foods get Council's approval. Press Release. Available from: http://www.consilium.europa.eu/en/press/press-releases/2015/11/11-novel-foods-new-rules/

European Food Safety Authority (EFSA) Panel on Genetically Modified Organisms (GMO), 2010. Scientific opinion on the assessment of potential impacts of genetically modified plants on non-target organisms. EFSA J. 8 (11), 1–72.

Food and Agriculture Organization, International Fund for Agriculture Development, World Food Program (FAO, IFAD, and WFP), 2015. The State of Food Insecurity in the World. Meeting the 2015 international hunger targets: Taking stock of uneven progress Food and Agriculture Organization of The United Nations, Rome, 2015. Available from: Available from: http://www.fao.org/3/a-i4646e/index.html

FAO, Wageningen University and Research Centre UR (WUR), 2014. Insects to feed the world. Summary Report. 1st International Conference 14–17 May 2014, Wageningen (Ede), The Netherlands. Available from: https://www.wageningenur.nl/en/show/Insects-to-feed-the-world.htm

FAO/WHO, 2001. Codex Alimentarius: Joint FAO/WHO Standards Programme, Rome, Italy. Available from: www.codexalimentarius.org

Food and Drug Administration, 1981. Guidance for Industry, Good Laboratory Practices, Questions and Answers. U.S. Department of Health and Human Services, Food and Drug administration, Office of Regulatory Affairs, June 1981, (Minor editorial and formatting changes made December 1999 & July 2007), pp. 1–25.

FDA, 2000. Guidance for Industry and Other Stakeholders Toxicological Principles for the Safety Assessment of Food Ingredients Redbook 2000. U.S. Department of Health and Human Services, Food and Drug Administration Center for Food Safety and Applied Nutrition, July 2000; Updated July 2007. Available from: http://www.fda.gov/downloads/Food/GuidanceRegulation/UCM222779.pdf

FDA, 2014a. Guidance for Industry: Considering whether an FDA-regulated product involves the application of nanotechnology. U.S. Department of Health and Human Services Food and Drug Administration, Office of the Commissioner. June 2014. Available from: http://www.fda.gov/RegulatoryInformation/Guidances/ucm257698.htm

FDA, 2014b. Guidance for Industry: Assessing the effects of significant manufacturing process changes, including emerging technologies on the safety and regulatory status of food ingredients and food contact substances, including food ingredients that are color additives, June 2014. http://www.fda.gov/Food/GuidanceRegulation/GuidanceDocumentsRegulatoryInformation/IngredientsAdditivesGRASPackaging/ucm300661.htm

FDA, 2015. User fee program to provide for accreditation of third-party auditors/certification bodies to conduct food safety audits and to issue certifications. 21 CFR Part 1 [Docket No. FDA–2011–N–0146] RIN 0910–AG66, Federal Register 80 (142) Friday, July 24, 2015.

Finnish Food Safety Authority (FFSA) EVIRA, 2015. Available from: http://www.evira.fi/portal/en/food/manufacture+and+sales/novel+foods/insects+as+food/

Finn, A., 2011. Shellac as food glaze. Gentle World August 15, 2011. Available from: http://gentleworld.org/shellac-food-glaze/.

FoodDrinkEurope, 2014. FoodDrinkEurope acrylamide toolbox 2013. Available from: http://ec.europa.eu/food/food/chemicalsafety/contaminants/toolbox_acrylamide_201401_en.pdf

Green, K., Broome, L., Heinze, D., Johnston, S., 2001. Long distance transport of arsenic by migrating Bogong moths from agricultural lowlands to mountain ecosystem. Victor. Nat. 118 (4), 112–116.

Halloran, A., 2014. Discussion paper: Regulatory frameworks influencing insects as food and feed. Preliminary Draft. Version December 5, 2014.

Hanboonsong, Y., Jamjanya, T., Durst, P.B., 2013. Six-legged livestock: Edible insect farming, collection and marketng in Thailand. Food and Agriculture Organization of The United Nations, Regional Office for Asia and the Pacific, Bangkok, Thailand, pp. 1–32.

Handley, M.A., Hall, C., Sanford, E., Diaz, E., Gonzalez-Mendez, E., Drace, K., Wilson, R., Villalobos, M., Croughan, M., 2007. Globalization, binational communities, and imported food risks: results of an outbreak investigation of lead poisoning in Monterey County. CA Am. J. Public Health 97 (5), 900–906.

Hartung, T., Hoffmann, S., 2009. Food for thought … on in silico methods in toxicology. Altex 26 (3), 155–166.

International Conference on Harmonization. (ICH), 1996. Guidance for Industry, E6 Good Clinical Practice: Consolidated Guidance. U.S. Department of Health and Human Services, Food and Drug Administration, Center for Drug Evaluation and Research (CDER), Center for Biologics Evaluation and Research (CBER)

Kiranmayi, P., 2014. Is bio active compounds in plants acts as anti-nutritional factors. Int. J. Curr. Pharm. Res. 6 (2), 36–38.

Kirkland, D., Aardema, M., Henderson, L., Muller, L., 2005. Evaluation of a battery of three in vitro genotoxicity tests to determine rodent carcinogens and non-carcinogens. I. Sensitivity, specificity and relative predictivity. Mutat. Res. 584, 1–256.

Kirkland, D., Reeve, L., Gatehouse, D., Vanparys, P., 2011. A core in vitro genotoxicity battery comprising the Ames test plus the in vitro micronucleus test is sufficient to detect rodent carcinogens and in vivo genotoxins. Mutat. Res. 271, 27–73.

Kirsch-Volders, M., Decordier, I., Eljhjouji, A., Plas, G., Aardema, M., Fenech, M., 2010. In vitro genotoxicity testing using the micronucleus cell assay in cell lines, human lymphocytes and 3D models. Mutagen 26, 177–184.

Kokoski, C.J., 1992. Overview of FDA's Redbook guidelines. Crit. Rev. Food Sci. Nutr. 32 (2), 161–163.

Kotsonis, F.N., Burdock, G.A., 2013. Food Toxicology. In: Klaassen, D. (Ed.), Casarett and Doull's Essentials of Toxicology. eighth ed. McGraw-Hill, New York, NY, USA, pp. 1306–1356.

Kruger, C.L., Reddy, C.S., Conze, D.B., Hayes, A.W., 2014. Food safety and foodborne toxicants. In: Hayes, A.W., Kruger, C.L. (Eds.), Hayes' Principles and Methods of Toxicology. sixth ed. CRC Press, Boca Raton, Florida, pp. 621–675.

Lee, D., Pitetti, R.D., Casselbrant, M.L., 1999. Oropharyngeal manifestations of lepidopterism. Arch. Otolaryngol. Head Neck Surg. 125 (1), 50–52.

Marone, P.A., Birkenbach, K.E., 2015. Safety of Functional Foods, Chapter 2. In: Bagchi, D., Preuss, H.G., A. Swaroop, A. (Eds.), Nutraceuticals and Functional Foods in Human Health and Disease Prevention. CRC Press, Taylor and Francis Group, Boca Raton, Florida.

McAuliff, M., 2015. House votes to ban states from labeling GMO foods. HuffPost Politics. Available from: http://www.huffingtonpost.com/entry/gmo-labels-food_55b12fabe4b08f57d5d3f393

McCall, A., 2014. Startups pitch cricket flour as the best protein you could eat. Natl. Pub. Radio. Available from: http://www.npr.org/sections/thesalt/2014/08/15/340653853/startups-pitch-cricket-flour-as-the-best-protein-you-could-eat

McFarland, M., 2015. 5 amazing ways 3-D-printed food will change the way we eat. Washington Post. Available from: http://www.washingtonpost.com/news/innovations/wp/2015/01/28/5-amazing-ways-3d-printed-food-will-change-the-way-we-eat/

Mermelstein, N., 2015, Crickets, Mealworms, and Locusts, Oh My! Food Technol., pp. 69–73. Available from: http://www.ift.org/food-technology/past-issues/2015/october.aspx.

Michels, S., 2012. Bugs for dinner? PBS News Hour Rundown. Available from: http://www.pbs.org/newshour/rundown/bugs-for-dinner/.

Mlcek, J., Rop, O., Borkovcova, M., Bednarova, M., 2014. A comprehensive look at the possibilities of edible insects as food in Europe: a review. Pol. J. Food Nutr. Sci. 64 (3), 147–157.

Musundire, R., Zvidzai, C.J., Chidewe, C., Samende, B.K., Manditsera, F.A., 2014. Nutrient and anti-nutrient composition of *Henicus whellani* (Orthoptera: Stenopelmatidae), an edible ground cricket, in south-eastern Zimbabwe. Int. J. Trop. Insect Sci. 34, 1–9.

Netherlands Food and Consumer Product Safety Authority (NFCPSA), 2014. Ministry of Economic Affairs. Advisory report on the risks associated with consumption of mass-reared insects. Advice from the director Office for Risk Assessment and Research, Bureau Risicobeoordeling & onderzoeksprogrammering, Utricht, The Netherlands, pp. 1–21.

Nishimune, T., Watanabe, Y., Okazaki, H., Akai, H., 2000. Thiamine is decomposed due to Anaphe spp. entomophagy in seasonal ataxia patients in Nigeria. J. Nutr. 130 (6), 1625–1628.

Noh, J.H., Yun, E.Y., Par, H., Jung, K.J., Hwang, J.A., Jeong, E.J., Moon, K.S., 2015. Subchronic oral dose toxicity of freeze-dried powder of Allomyrina dichotoma larvae. Toxicol. Res. 31 (1), 69–75.

Organization of Economic Cooperative Development (OECD), 2010. Guidance manual for the testing of manufactured nanomaterials: OECD's sponsorship programme; first revision. Environment directorate joint meeting of the chemicals committee and the working party on chemicals, pesticides and biotechnology. ENV/JM/MONO(2009)20/REV. Available from: http://search.oecd.org/officialdocuments/displaydocumentpdf/?cote=env/jm/mono(2009)20/rev&doclanguage=en

Panagiotou, T., Fisher, R.J., 2013. Producing micron- and nano-size formulations for functional foods applications. Funct. Foods Health Dis. 3 (7), 274–289.

Panzani, R.C., Ariano, R., 2001. Arthropods and invertebrates allergy (with the exclusion of mites): the concept of panallergy. Allergy 56 (Suppl. 69), 1–22.

Pitetti, R.D., Kuspis, D., Krenzelok, E.P., 1999. Caterpillars: an unusual source of ingestion. Pediatr. Emerg. Care 15 (1), 33–36.

Premalatha, M., Abbasi, T., Abbasi, T., Abbasi, S.A., 2011. Energy-efficient food production to reduce global warming and ecodegradation: the use of edible insects. Renew. Sustain. Energy Rev. 15 (9), 4357–4360.

Price, N., Chaudhry, Q., 2014. Application of in silico modelling to estimate toxicity of migrating substances from food packaging. Food Chem. Toxicol. 71, 136–141.

Raaflaub, C., 2015. Creepy-crawlies conquer Swiss menus. Available from: http://www.swissinfo.ch/eng/insects-on-your-plate_creepy-crawlies-conquer-swiss-menus/41597534

Raikow, D.F., Walters, D.M., Fritz, K.M., Mills, M.A., 2011. The distance that contaminated aquatic subsidies extend into lake riparian zones. Ecol. Appl. 21 (3), 983–990.

Rehnquist, J., 2003. Dietary Supplement Labels: Key Elements. Department of Health and Human Services. Office of Inspector General, 1–24. Available from: https://oig.hhs.gov/oei/reports/oei-01-01-00120.pdf

Robinson, N., 2015. Scientific committee of the federal agency for the safety of the food chain. Available from: http://www.foodmanufacture.co.uk/Food-Safety/Europe-s-first-insect-food-safety-guide

Saeed, T., Dagga, F.A., Saraf, M., 1993. Analysis of residual pesticides present in edible locusts captured in Kuwait. Arab Gulf J. Sci. Res. 11 (1), 1–5.

Schilter, B., Benigni, R., Boobis, A., Chiodini, A., Cockburn, A., Cronin, M.T., Worth, A., 2014. Establishing the level of safety concern for chemicals in food without the need for toxicity testing. Regul. Toxicol. Pharmacol. 68 (2), 275–296.

Scientific Committee of The Federal Agency for the Safety of the Food Chain (SciCom), 2014. Common Advice, SciCom 14-2014 and SHC Nr. 9160. Subject: Food Safety Aspects of Insects Intended For Human Consumption (SciCom dossier 2014/04; SHC dossier n(9160). September 12, 2014.

Smithsonian Encyclopedia, 2015. BugInfo. Numbers of insects (species and individuals). Available from: http://www.si.edu/encyclopedia_si/nmnh/buginfo/bugnos.htm

Spencer, K., 2014. Insects made into meals using 3D printers. Sky News. http://news.sky.com/story/1258692/insects-made-into-meals-using-3d-printers

Sultanbawa, Y., 2011. Plant antimicrobials in food applications: Minireview. In: Méndez-Vilas, A., (Ed.), Science against microbial pathogens: Communicating current research and technological advances. Formatex Microbiology Book Series No 3, Vol. 2, pp. 1084–1093.

Tomasik, B., 2015. Do bugs feel pain? Available from: http://reducing-suffering.org/do-bugs-feel-pain/

Turkewitz, J., 2015. Environmental agency uncorks its own toxic water spill at Colorado mine. NY Times. Available from: http://www.nytimes.com/2015/08/11/us/durango-colorado-mine-spill-environmental-protection-agency.html?_r=0

US Department of Agriculture (USDA), 2014. Food Safety and Inspection Service 9 CFR Parts 381 and 500 [Docket No. FSIS–2011-0012] RIN 0583–AD32 Modernization of Poultry Slaughter Inspection Federal Register 79 (162), pp. 49566–49637, 2014.

US FDA (US Food and Drug Administration), 2001. Draft guidance for industry: voluntary labeling indicating whether foods have or have not been developed using bioengineering. Available from: http://www.fda.gov/Food/GuidanceRegulation/GuidanceDocumentsRegulatoryInformation/ucm059098.htm.

United Nations (UN), 2015. Millennium Development Goals and Beyond 2015. Available from: http://www.un.org/millenniumgoals/poverty.shtml

United Nations (UN), 2015. The millenium development goals report. United Nations, New York, NY, 2015, pp. 1–72.

van der Spiegel, M., Noordam, M.Y., van der Fels-Klerx, H.J., 2013. Safety of novel protein sources (insects, microalgae, seaweed, duckweed, and rapeseed) and legislative aspects for their application in food and feed production. Comp. Rev. Food Sci. Food Safety 12, 662–678.

van Huis, A., Van Itterbeeck, J., Klunder, H., Mertens, E., Halloran, A., Muir, G., Vantomme, P., 2013. Environmental opportunities for insect rearing for food and feed. Chap 5. In: Edible Insects: Future prospects for Food and Feed Security. Food and Agriculture Organization Forestry Paper 171. Food and Agriculture Organization of the United Nations, Rome, Italy 2013, pp. 59–66.

van Huis, A., Van Itterbeeck, J., Klunder, H., Mertens, E., Halloran, A., Muir, G., Vantomme, P., 2013. Insects as animal feed. Chap 7. In: Edible Insects: Future prospects for Food and Feed Security. Food and Agriculture Organization Forestry Paper 171. Food and Agriculture Organization of the United Nations, Rome, Italy, 2013, pp. 89–97.

van Huis, A., Van Itterbeeck, J., Klunder, H., Mertens, E., Halloran, A., Muir, G., Vantomme, P., 2013. Food safety and preservation. Chap 10. In: Edible Insects: Future Prospects for Food and Feed Security. Food and Agriculture Organization Forestry Paper 171. Food and Agriculture Organization of the United Nations, Rome, Italy, 2013, pp. 117–124.

van Huis, A., Van Itterbeeck, J., Klunder, H., Mertens, E., Halloran, A., Muir, G., Vantomme, P., 2013. Regulatory frameworks governing the use of insects for food security. Chap 14. In: Edible Insects: Future prospects for Food and Feed Security. Food and Agriculture Organization Forestry Paper 171. Food and Agriculture Organization of the United Nations, Rome, Italy, 2013, pp. 153–159.

Vantomme, P., Mertens, E., van Huis, A., Klunder, H., 2012. Assessing the potential of insects as food and feed in assuring food security. United Nations Food and Agricultural Organization, Rome, Italy. Available from: http://www.fao.org/docrep/015/an233e/an233e00.pdf

Vijver, M., Jager, T., Posthuma, L., Peijnenburg, W., 2003. Metal uptake from soils and soil-sediment mixtures by larvae of *Tenebrio molitor* (L.) (Coleoptera). Ecotoxicol. Environ. Safety 54 (3), 277–289.

Wang, L., Bohn, T., 2012. Health-promoting food ingredients and functional food processing technology. In: Bouayed, J. (Ed.), Nutrition, Well-Being and Health, pp. 201–224. Available from: http://www.intechopen.com/books/nutrition-well-being-and-health/health-promoting-food-ingredientsdevelopment-and-processing

Warheit, D.B., Donner, E.M., 2010. Rationale of genotoxicity testing of nanomaterials: regulatory requirements and appropriateness of available OECD test guidelines. Nanotoxicology 4, 409–413.

Warner, M., 2007. From: Why not eat insects? Holt VM, 1885. Available from: http://bugsandbeasts.com/whynoteatinsects/

Weingarten, H., 1990. 1862–2014: A brief history of food and nutrition labeling. Available from: http://blog.fooducate.com/2008/10/25/1862-2008-a-brief-history-of-food-and-nutrition-labeling/.

Wigglesworth, V.B., 1980. Do insects feel pain? Antenna 1, 8–9.

Yao, Y., 2013. Nation not ready for insect diet, expert warns. Chinadaily.com. Available from: http://www.chinadaily.com.cn/china/2013-05/17/content_16505412.htm

Zanolli, L., 2014. Insect farming is taking shape as demand for animal feed rises. Available from: http://www.technologyreview.com/news/529756/insect-farming-is-taking-shape-as-demand-for-animal-feed-rises

Zhang, Z., Song, X., Wang, Q., Lu, X., 2012. Mercury bioaccumulation and prediction in terrestrial insects from soil in Huludao City. Northeast China Bull. Environ. Contam. Toxicol. 89, 107–112.

Zhou, J., Han, D., 2006. Safety evaluation of protein of silkworm (*Antheraea pernyi*) pupae. Food Chem. Tox. 44 (7), 1123–1130.

Zhuang, P., Zou, H., Shu, W., 2009. Biotransfer of heavy metals along a soil-plant-insect-chicken food chain: field study. J. Environ. Sci. 21 (6), 849–853.

Ziobro, G., 2015. FDA Center for Food Safety and Applied Nutrition Presentation at the Institute of Food Technologists (IFT) 2015: Regulatory Issues, Concerns, and Status of Insect Based Foods and Ingredients. Chicago, IL.

Zuraw, L., 2015. The Expanding Role of HACCP. Food Safety News. June 29, 2015. Available from: http://www.foodsafetynews.com/2015/06/the-expanding-role-of-haccp/#.VeiVN_lVhBc

Chapter 8

Ensuring Food Safety in Insect Based Foods: Mitigating Microbiological and Other Foodborne Hazards

D.L. Marshall*, J.S. Dickson**, N.H. Nguyen[†]
*Eurofins Microbiology Laboratories, Inc., Fort Collins, CO, United States; **Department of Animal Science, 215F Meat Laboratory, Ames, IA, United States; [†]Department of Environmental Science, Policy & Management, University of California, Berkeley, CA, United States

Chapter Outline

Introduction	223		Milk Sanitation	237
Microbes Associated with Insects	224		International Administration	237
Insects as a Vector of Foodborne Disease Hazards	225		Prerequisite Programs	238
Bacterial Infections	229		Good Manufacturing Practices	238
Salmonellosis	230		Training and Personal Hygiene	238
Shigellosis	230		Pest Control	238
Vibriosis	230		Sanitation	239
E. Coli	231		Sanitary Facility Design	239
Yersiniosis	231		Sanitary Equipment Design	239
Campylobacteriosis	231		Cleaning and Sanitizing Procedures	240
Listeriosis	232		**Hazard Analysis Critical Control Point System**	**241**
C. perfringens	232		HACCP Plan Development	242
Other Bacterial Foodborne Infections	232		**Food Safety Modernization Act**	**243**
Nonbacterial Foodborne Infections	232		**Validation**	**244**
Infectious Hepatitis	233		**Food Preservation**	**244**
Enteroviruses	233		Principles of Food Preservation	245
Multicellular Parasites	233		Asepsis/Removal	245
Prions	234		Modified Atmosphere Conditions	246
Foodborne Bacterial Intoxications	234		High-Temperature Preservation	246
Staphylococcus aureus Enterotoxin	234		Low-Temperature Preservation	247
B. cereus Enterotoxin	234		Drying	248
Botulism	235		Preservatives	248
Chemical Intoxications	235		Irradiation	249
Physical Hazards	236		Fermentation	249
Administrative Regulation	**236**		**Conclusions**	**250**
United States Department of Agriculture (USDA)	237		**References**	**250**
United States Food and Drug Administration (FDA)	237			

INTRODUCTION

The purpose of food processing and preparation is to provide safe, wholesome, and nutritious food to the consumer. Responsibilities for accomplishing this outcome collectively lie with every step in the food chain; from farm or wild harvest food production through processing, storage, distribution, retail sale, and consumption. Each party has obligations to meet reasonable expectations of the other parties involved in the process, with no group solely responsible for food safety. Although regulatory expectations may vary by jurisdiction, they hold similar desired outcomes that material presented for food consumption will be safe and wholesome.

Producers of edible insects and insect-derived ingredients must process (eg, blanching, chilling, and/or drying) such materials to ensure that microbial and other forms of contamination are minimized. Although it may not be practical to deliver fresh unprocessed food that is completely free of food safety hazards, control of the production environment is essential to reduce the presence of harmful materials by using good agricultural practices (GAPs) (Food and Drug Administration, 1998). As an example, insects treated with antibiotics for infectious disease control during production should be withheld from consumption until an appropriate withdrawal time is established to ensure absence of antibiotic residues in edible tissues (Hirose et al., 2006; Cappellozza et al., 2011). In contrast, insects reared or harvested from waste products may be contaminated with pathogens associated with their feed. In such cases, application of good management practices and implementation of postharvest controls is imperative to ensure the safety of resulting edible products.

Prudent producers of edible insect products must ensure that raw materials delivered to the processing facility are of reasonable quality and not contaminated with harmful levels of chemicals (natural toxins, drugs, or pesticides), pathogenic microorganisms, or injurious physical defects (Belluco et al., 2013). After processing, steps must be taken to ensure that food will be properly handled through the distribution and retail chain. Furthermore, care must be exercised to ensure that products will not be improperly prepared by the consumer. The latter is important because processors have responsibility over their products due to labeling instructions. Processors must manipulate raw foods in a manner that minimizes growth of existing microorganisms as well as minimizes additional contamination during processing. To achieve this outcome, a properly constructed and implemented food safety plan is essential (US FDA, 2015a).

Consumers and the regulatory community have reasonable expectations that foods have been produced and processed under hygienic conditions. Such expectation can include foods should not be adulterated by the addition of any biological, chemical, or physical hazards and that labels have information available on both composition and nutritional aspects of products. Because many consumers in developed countries may not have experience consuming insects or insect-containing products, the burden on the industry to meet demanding expectations is crucial for market acceptance and exploitation. To build trust, most governments establish regulations that govern production, processing, distribution, and retailing of foods.

A growing cadre of critical consumers is increasingly advocating greater food safety control expectations, many of which may not be based on sound risk assumptions and may be unreasonable. On the surface, the argument that raw foods should be free of infectious microorganisms may seem reasonable; however, technologies or processes may not exist in a legal or practical form to assure that such raw foods are not contaminated with infectious agents. Because production and processing practices for insect based foods may be immature, the industry must exercise due diligence to ensure the safety of products.

Consumers and food-service operators require education to not hold insect-based foods under unsanitary conditions prior to consumption and not adulterating such products with the addition of biological, chemical, or physical agents. Improper handling can increase foodborne illness risks by allowing populations of harmful bacteria to increase or by encouraging cross contamination between raw and cooked foods.

MICROBES ASSOCIATED WITH INSECTS

Insects, as any other organism, have an associated complex microbial community (fungi, bacteria, archaea, protozoa, and viruses) that span the symbiotic spectrum from mutualist to pathogenic. These microbes may be associated with insects in nature and in farming environments. Together, these microbes are collectively known as the microbiota of an organism. Most of the microbes present in or on insects are not considered harmful to human consumers and are not known to be common causes of food spoilage. It is important to recognize that many of the microbes associated with industrialized insect farming for human consumption and insects as food ingredients are likely to be found in other food production and processing environments. Nonetheless, it is important to review aspects of insect internal microbiome here as insects are somewhat unique in the food livestock space since the industry utilizes whole insects including their digestive system and its contents. The subject of insect microbiome is a vast and well studied field with yet much to be learned and as such is largely beyond the scope of this chapter. Nonetheless, here we will discuss microbes associated with the more common industrially produced insect species used for food production, such as crickets, mealworms, and waxworms.

The gut of *Acheta domestica* L. (house cricket) contains bacteria in the genera *Citrobacter*, *Klebsiella*, *Yersinia*, *Bacteroides*, and *Fusobacterium* (Ulrich et al., 1981). These microbes may have active and important roles for crickets in digestion of carbohydrates, proteins, and lipids (Kaufman et al., 1991). However the microbial community may not be stable as a change of diet causes a change in composition of the gut microbiota (Domingo et al., 1998). The yellow mealworm, *Tenebrio molitor* L. hosts bacteria in the genera *Actunobacillus*, *Propionibacterium*, *Citrobacter*, *Serratia*, *Bacillus*, *Dermabacter*, *Brachybacterium*, *Clavibacter,* and *Exiguobcacterium* (Liu et al., 2011). *Enterobacter asburiae* Brenner et al. and an unclassified *Bacillus* species have been found in the gut of the waxworm, *Galleria mellonella* L. (Yang et al., 2014).

The mopane caterpillar, *Imbrasia belina* Westwood, which is commonly consumed in African countries, has been found to contain several molds, including *Aspergillus*, *Cladosporium*, *Fusarium*, *Penicillium*, and *Phycomycetes*. Potential human pathogenic bacteria *Escherichia coli* Migula, *Klebstella pneumoniae* Schroeter, and *Bacillus cereus* Frankland and Frankland have been found in some samples of processed caterpillars, which suggests that handling or consumption of these caterpillars has potential health risks. Because many of these bacteria and fungi are common in the production and processing environment, finding such microbes in edible caterpillar products after processing (boiling and drying) may not be reflective of insect gut origin (Gashe et al., 1997).

Each insect species appears to host a different group of bacteria, with a few common genera such as *Enterobacter*, *Bacteriodes*, *Citrobacter*, and *Bacillus* being found (Belluco et al., 2013). These genera, especially the first two, also are found in the human digestive tract. Members of the genus *Yersinia* that had been detected once in the gut of the house cricket can be pathogenic to humans (Ulrich et al., 1981). This genus also occurs in insects (and other arthropods such as fleas), which serve as a natural reservoir. The genus-level classification isn't enough to connect these bacteria to those that might be harmful to humans, but overall, naturally occurring gut bacteria of insects do not appear to overlap with those found in farm environments or those that may be harmful to humans.

Definitive human foodborne pathogens *Salmonella* and *Listeria monocytogenes* Murray et al. were not isolated from the guts of farm-raised superworm (*Zophobas morio* Fab.), yellow mealworm (*T. molitor*), wax moth (*G. mellonella*), butterworm (*Chilecomadia moorei* Silva Figueroa), and house cricket (*A. domesticus*) (Giaccone, 2005). *E. coli*, *Citrobacter intermedius* Sledak, *Micrococcus* spp., *Streptococcus* spp., and *Bacillus subtilis* Ehrenberg have been isolated from lesser meal worms (*Alphitobius diaperinus* Panzer) collected from poultry houses (Harein et al., 1970). Simple food processing intervention, such as boiling insects in water for a few minutes, eliminated live bacterial cells, but spores were found to have survived this process and could germinate at a later time and cause food spoilage or human illness (Klunder et al., 2012).

It is important to note that microbial prevalence and populations associated with wild insects and those with cultivated insects may differ. These differences are largely determined from the source of food the insects are eating and production environment. Therefore, proper farming techniques should be adapted for control of pathogens that might harm humans.

INSECTS AS A VECTOR OF FOODBORNE DISEASE HAZARDS

Due to vast genetic, morphological, and biochemical differences between insects and humans (as compared with the relatively greater similarity between humans and other vertebrates we eat including cows, pigs, and chickens), the foodborne transmission of several types of narrow-host range foodborne pathogens from insects to humans likely to be relatively minor, particularly for viruses, prions (pathogenic proteins), and parasites. Such agents have very specific host ranges and specific life cycles that make them unlikely to be transmitted by consumption of insects (van Huis et al., 2013). This is particularly true for food made from insects raised in hygienically maintained farms and subsequently processed utilizing appropriate food-grade facilities and techniques with proper cool storage, heating, and drying to kill and otherwise mitigate the presence, cross contamination, and growth of such pathogens in those products. Nonetheless, given the vast diversity of insects in the world, and the many hundreds or more that are edible and may be mass produced as food, the nature of foodborne risk from insect consumption is a topic needing much more study to better understand. Thus, it is prudent to treat insects appropriately, as with any other livestock, to mitigate any risks from potential foodborne pathogens or other contaminants.

The European Food Safety Authority (EFSA, 2015) considers the risks of consuming insects as food is similar to other food protein sources. This risk profile includes biological hazards (bacteria, viruses, parasites, fungi, and prions), chemical hazards (heavy metals, toxins, veterinary drugs, hormones, and others) as well as allergens and hazards related to the environment. Factors influencing presence of biological and chemical hazards include specific production methods, cultivation substrate (such as feed and bedding) used, insect species, stage of insect development at harvest, and methods used for further processing. Conventional animal feed materials used as a substrate for insect production are assumed to have similar microbial hazards as those feeds used to grow other animals. Presence of prions in such feed material will depend on the source material used, such as human or ruminant sources. The occurrence of harmful chemicals in edible insects is generally not well characterized.

Contrary to popular consumer perception about the risk of chemicals in foods, major hazards associated with foodborne illness are clearly of biological origin (Mead et al., 1999). The US Centers for Disease Control and Prevention (CDC) has published summaries of foodborne diseases by etiology for the years 1993 through 1997 (Table 8.1) (Olsen et al., 2000). Foodborne disease agents are reported for four categories; bacterial, parasitic, viral, and chemical. Greater than 95% of all reported outbreaks foodborne illnesses are caused by microorganisms or their toxins. Fully 97% of reported cases are likewise linked to a microbial source. Only around 3% of the outbreaks and less than 1% of cases can be truly linked to chemical (heavy metals, monosodium glutamate, and other chemicals) contamination of foods. Furthermore, 97% of reported deaths are due to microbial sources. These data are from reported outbreaks. CDC estimates for the actual number of cases of foodborne disease caused by microbial agents is much higher due to under reporting (Table 8.2).

TABLE 8.1 Reported Foodborne Diseases in the United States, 1993–97

Etiologic Agent	Outbreaks No.	Outbreaks %	Cases No.	Cases %	Deaths No.	Deaths %
Bacterial						
B. cereus	14	0.5	691	0.8	0	0.0
Brucella	1	0.0	19	0.0	0	0.0
Campylobacter	25	0.9	539	0.6	1	3.4
Clostridium botulinum	13	0.5	56	0.1	1	3.4
Clostridium perfringens	57	2.1	2,772	3.2	0	0.0
E. coli	84	3.1	3,260	3.8	8	27.6
L. monocytogenes	3	0.1	100	0.1	2	6.9
Salmonella	357	13.0	32,610	37.9	13	44.8
Shigella	43	1.6	1,555	1.8	0	0.0
S. aureus	42	1.5	1,413	1.6	1	3.4
Streptococcus, Group A	1	0.0	122	0.1	0	0.0
Streptococcus, other	1	0.0	6	0.0	0	0.0
Vibrio cholerae	1	0.0	2	0.0	0	0.0
Vibrio parahaemolyticus	5	0.2	40	0.0	0	0.0
Yersinia enterocolitica	2	0.1	27	0.0	1	3.4
Other bacterial	6	0.2	609	0.7	1	3.4
Total bacterial	655	23.8	43,821	50.9	28	96.6
Parasitic						
Giardia lamblia	4	0.1	45	0.1	0	0.0
Trichinella spiralis	2	0.1	19	0.0	0	0.0
Other parasitic	13	0.5	2,261	2.6	0	0.0
Total parasitic	19	0.7	2,325	2.7	0	0.0
Viral						
Hepatitis A	23	0.8	729	0.8	0	0.0
Norwalk/Norwalk-like	9	0.3	1,233	1.4	0	0.0
Other viral	24	0.9	2,104	2.4	0	0.0
Total viral	56	2.0	4,066	4.7	0	0.0
Chemical						
Ciguatoxin	60	2.2	205	0.2	0	0.0
Heavy metals	4	0.1	17	0.0	0	0.0
Monosodium glutamate	1	0.0	2	0.0	0	0.0
Mushrooms	7	0.3	21	0.0	0	0.0
Scrombotoxin	69	2.5	297	0.3	0	0.0
Shellfish	1	0.0	3	0.0	0	0.0
Other chemical	6	0.2	31	0.0	0	0.0
Total chemical	148	5.4	576	0.7	0	0.0
Unknown Etiology	1,873	68.1	35,270	41	1	3.4
Grand Total	2,751	100.0	86,058	100.0	29	100

Soure: Data from Olsen et al. (2000).

TABLE 8.2 Reported and Estimated[a] Illnesses, Frequency of Foodborne Transmission, and Hospitalization and Case-Fatality Rates for Known Foodborne Pathogens, United States

Disease or Agent	Estimated Total Cases	Reported Cases by Surveillance Type			% Foodborne Transmission	Hospitalization Rate	Case Fatality Rate
		Active	Passive	Outbreak			
Bacterial							
B. cereus	27,360		720	72	100	0.006	0.0000
Botulism, foodborne	58		29		100	0.800	0.0769
Brucella spp.	1,554		111		50	0.550	0.0500
Campylobacter spp	2,453,926	64,577	37,496	146	80	0.102	0.0010
C. perfringens	248,520		6,540	654	100	0.003	0.0005
E. coli O157:H7	73,480	3,674	2,725	500	85	0.295	0.0083
E. coli, non-O157 STEC	36,740	1,837			85	0.295	0.0083
E. coli, enterotoxigenic	79,420		2,090	209	70	0.005	0.0001
E. coli, other diarrheogenic	79,420		2,090		30	0.005	0.0001
L. monocytogenes	2,518	1,259	373		99	0.922	0.2000
Salmonella typhi[b]	824		412		80	0.750	0.0040
Salmonella, nontyphoidal	1,412,498	37,171	37,842	3,640	95	0.221	0.0078
Shigella spp.	448,240	22,412	17,324	1,476	20	0.139	0.0016
Staphylococcus food poisoning	185,060		4,870	487	100	0.180	0.0002
Streptococcus, foodborne	50,920		1,340	134	100	0.133	0.0000
V. cholerae, toxigenic	54		27		90	0.340	0.0060
V. vulnificus	94		47		50	0.910	0.3900
Vibrio, other	7,880	393	112		65	0.126	0.0250
Y. enterocolitica	96,368	2,536			90	0.242	0.0005
Subtotal	5,204,934						
Parasitic							
Cryptosporidium parvum	300,000	6,630	2,788		10	0.150	0.005
Cyclospora cayetanensis	16,264	428	98		90	0.020	0.0005
G. lamblia	2,000,000	107,000	22,907		10	n/a	n/a
Toxoplasma gondii	225,000		15,000		50	n/a	n/a
T. spiralis	52		26		100	0.081	0.003
Subtotal	2,541,316						
Viral							
Norwalk-like viruses	23,000,000				40	n/a	n/a
Rotavirus	3,900,000				1	n/a	n/a
Astrovirus	3,900,000				1	n/a	n/a
Hepatitis A	83,391		27,797		5	0.130	0.0030
Subtotal	30,883,391						
Grand Total	38,629,641						

[a]Numbers in italics are estimates; others are measured.
[b]>70% of cases acquired abroad.
Source: Data from http://www.cdc.gov/ncidod/eid/vol5no5/mead.htm and http://www.cdc.gov/epo/mmwr/preview/mmwrhtml/ss4901a1.htm

Bacterial agents are by far the leading cause of illness, with total numbers estimated as high as 76 million cases per year and deaths as high as 5000 annually in the United States (Snowdon et al. 2002). Costs are estimated to be $9.7 billion annually in medical expenses and lost productivity in the United States (Snowdon et al., 2002). The high incidence of foodborne disease is paralleled in other developed countries (Todd, 1992). Enteric viruses are now recognized as the leading cause of foodborne infections, although the bacteria are better known. Predominant bacterial agents are *Campylobacter* spp., *Salmonella* spp., *Shigella* spp., and *C. perfringens* Veillon and Zuber. More recent assumptions show fewer estimated illnesses and deaths (Table 8.3) (Scallan et al., 2011).

TABLE 8.3 Estimated Number of Annual US Foodborne Illnesses, Hospitalizations, and Deaths, 2000–08

Pathogen Type	Pathogen	Illnesses	Hospitalizations	Deaths
Bacteria	*B. cereus*	63,000	20	0
	Brucella spp.	840	55	1
	Campylobacter spp.	850,000	8,500	76
	C. botulinum	55	43	9
	C. perfringens	970,000	440	26
	E. coli O157	63,000	2,100	20
	E. coli non O157	110,000	270	1
	Enterotoxigenic *E. coli*	18,000	12	0
	Diarrheagenic *E. coli*	12,000	8	0
	L. monocytogenes	1,600	1,500	250
	Mycobacterium bovis	60	31	3
	Salmonella spp.	1,000,000	19,000	380
	S. enterica serotype Typhi	1,800	200	0
	Shigella spp.	130,000	1,500	10
	S. aureus	240,000	1,100	6
	Streptococcus spp.	11,000	1	0
	Vibrio spp.	18,000	83	8
	V. cholerae	84	2	0
	Vibrio vulnificus	96	93	36
	V. parahaemolyticus	35,000	100	4
	Y. enterocolitica	98,000	530	29
Parasites	*Cryptosporidium* spp.	58,000	210	4
	C. cayetanensis	11,000	11	0
	Giardia intestinalis	77,000	230	2
	T. gondii	87,000	4,400	330
	Trichinella spp.	160	6	0
Viruses	Astrovirus	15,000	87	0
	Hepatitis A virus	1,600	99	8
	Norovirus	5,500,000	15,000	150
	Rotavirus	15,000	350	0
	Sapovirus	15,000	87	0
Known agent total		9,400,000	56,000	1,400
Unspecified agents		38,000,000	72,000	1,700
Grand total		47,000,000	128,000	3,100

Source: Scallan et al. (2011).

TABLE 8.4 Places Where Foodborne Outbreaks Occurred, 1993–97

Place	Number	Percentage
Home	582	21.3
Deli, café, restaurant	1185	43.1
School	91	3.3
Picnic	34	1.2
Church	63	2.3
Other	664	24.1
Unknown	99	3.6

Source: Data from Olsen et al. (2000).

TABLE 8.5 Contributing Factors Leading to Foodborne Outbreaks, 1993–97

Factors	Number	Percentage
Improper holding temperature	938	37.0
Inadequate cooking	274	10.8
Contaminated equipment	400	15.8
Food from unsafe source	153	6.0
Poor personal hygiene	490	19.3
Other	282	11.1

Source: Data from Olsen et al. (2000).

Foodborne bacterial hazards are classified based on their ability to cause infections or intoxications. Foodborne infections are usually the predominant type of foodborne illness reported. Foodborne outbreaks most often occur with foods prepared at food service establishments and at home (Table 8.4). Improper holding temperatures and poor personal hygiene are the leading factors contributing to reported outbreaks (Table 8.5).

Bacterial hazards are further classified based upon the severity of risk (Snowdon et al., 2002). Severe hazards are those capable of causing widespread epidemics. Moderate hazards can be those that have potential for extensive spread, with possible severe illnesses, complications, or sequelae in susceptible populations. Mild hazards can also cause outbreaks but have limited ability to spread. Those involved with food production, processing, and service should pay careful attention to controlling these biological hazards by: (1) destroying or minimizing the hazard, (2) preventing contamination of food with the hazard, or (3) inhibiting growth or preventing toxin production by the hazard. Control steps will follow in later sections of this chapter.

When investigating foodborne disease outbreaks, the most important factor is time (Bryan, 1988). Prompt reporting of an outbreak is essential to identifying implicated foods and stopping potentially widespread epidemics. Initial work in the investigation should be inspection of the premises where the outbreak occurred. Look for obvious sources, including sanitation and worker hygiene. Food preparation, storage, and serving should be carefully monitored. Interview those involved in the outbreak. Obtain case histories of victims and healthy individuals. Discuss health history and work habits of food handlers. Collect appropriate specimens for laboratory analysis, including stool samples, vomitus, and swabs of rectum, nose, and skin. Attempt to collect suspect foods, including leftovers or garbage if necessary. Specific tests for pathogens or toxins will depend on potential etiological agents and food type. Analysis of data should include case histories, illness specifics (incubation time, symptoms, and duration), lab results, and attack rates. All foodborne disease outbreaks should be reported to local and state health officers and to the CDC.

Bacterial Infections

Predominant human bacterial infections transmitted via foods are salmonellosis, campylobacteriosis, yersiniosis, vibriosis, and shigellosis (Cliver and Riemann, 2002). Most causative agents are Gram-negative, rod-shaped bacteria that are inhabitants of the intestinal tract of animals. Indeed, most regulatory agencies consider foods of animal origin (meat, poultry and eggs, fish and shellfish, and milk and dairy products) potentially hazardous foods. Because of life cycle and rearing conditions, insects also should fall in this category.

Insect microbiota consist of multiple common bacterial genera, including *Staphylococcus, Streptococcus, Bacillus, Proteus, Pseudomonas, Escherichia, Micrococcus, Lactobacillus,* and *Acinetobacter* (Agabou and Alloui, 2010; Amadi et al., 2005; Braide et al., 2011; Giaccone, 2005). Most of these are found naturally in production and processing environments and many species of these genera are not pathogenic to humans. Microbes found in the insect intestinal tract are thought to be mostly commensal and vary little among insect species (Klunder et al., 2012). Indicator counts (aerobic plate counts, anaerobic plate counts, and *Enterobacteriaceae* counts) of insects can be large populations ($>10^7$ CFU/g), and vary by insect species (Belgian Scientific Committee of the Federal Agency for the Safety of the Food Chain, 2014).

Human pathogen prevalence in insects are generally low (ranging from not detected to <100 CFU/g) (Netherlands Food and Consumer Product Safety Authority, 2014). Despite this finding, it is well-known that beetles, cockroaches, and flies in food production and processing environments can harbor definitive enteric human pathogens such as *Salmonella, Campylobacter,* and pathogenic *E. coli*. Such insects are thought to carry the pathogens rather than serve as vehicles that support bacterial multiplication. Such pathogen carriage is likely due to acquisition from outside environments associated with vertebrate animal feces (Leffer et al., 2010). It is expected that carefully controlled insect rearing and processing environments will not provide a significant source of human pathogens in ready-to-eat products. What follows is a general overview of the common foodborne microbial hazards with opportunities for control.

Salmonellosis

The genus *Salmonella* resides primarily in the intestinal tract of animals, including some insects in the wild (Tauxe, 1991). Many people are permanent, often asymptomatic carriers. Salmonellosis varies with serotype and strain, susceptibility of host, and total number of cells ingested. Several dozen serotypes cause foodborne outbreaks. Incubation time is 24–36 h, which may be longer or shorter. Symptoms include nausea, vomiting, abdominal pain, and diarrhea, which may be preceded by headache, fever, and chills. Weakness and prostration may occur. Duration is 1–4 days with a low mortality rate (0.1%). Individuals at high risk include the very young and the elderly; both may have a considerably higher mortality rate (3.8%) (Gray and Fedorka-Cray, 2002). The condition needed for an outbreak is the ingestion of live cells present in the food. For high-fat foods such as chocolate or cheese, 50 cells or lower may be a sufficient infectious dose due to protective enrobement of cells by fat allowing survival in highly acidic gastric fluid during intestinal transit. This is an important situation, for example, for manufacturers of chocolate covered insects. Foods primarily involved in outbreaks include meat, poultry, fish, eggs, and milk products. *Salmonella enterica* subsp. *enterica* Enteritidis Le Minor and Popoff is present in raw uncooked eggs even with sound shells (Humphrey et al., 1989). Most often the bacterium is transferred from a raw food to a processed food via cross-contamination. Control of *Salmonella* in insect-based foods can be accomplished in several ways. Avoidance of contamination by using only healthy food handlers and adequately cleaned and sanitized food contact surfaces, utensils, and equipment works best. Making sure that insect ingredients are free of *Salmonella* contamination also is essential. Most heat treatments of foods by cooking or pasteurization are sufficient to kill *Salmonella* as long as treatment temperatures and times have been validated for the matrix. Refrigeration temperatures at or below 5°C are sufficient to prevent *Salmonella* multiplication, as the minimum temperature for growth is 7–10°C. The prevalence of salmonellosis as a foodborne disease has prompted regulatory agencies to adopt a zero tolerance for the genus in ready-to-eat foods. Presence of the bacterium in these foods renders them unwholesome and unfit for consumptions. These foods must then be destroyed or reprocessed to eliminate the pathogen.

Shigellosis

Four *Shigella* species are associated with foodborne transmission of dysentery, *S. dysenteriae* Castellani and Chalmers, *S. flexneri* Castellani and Chalmers, *S. boydii* Ewing, and *S. sonnei* Sonne (Lampel and Maurelli, 2002). The disease is characterized with an incubation period of 1–7 days (usually less than 4 days). Symptoms include mild diarrhea to very severe with blood, mucus, and pus. Fever, chills, and vomiting also occur. Duration is long, typically 4 days–2 weeks. *Shigella* spp. have a very low infectious dose of around 10–200 cells. Foods most often associated with shigellosis are any that are contaminated with human fecal material, with salads frequently implicated. Control is best focused on worker hygiene and avoidance of human waste as insect growth material.

Vibriosis

Most *Vibrio* spp. are obligate halophiles (organisms that thrive in high salt concentrations) that are found in coastal waters and estuaries (Hackney and Dicharry, 1988). Consequently most foodborne outbreaks are associated with consumption of raw or undercooked shellfish (oysters, crabs, shrimp) and fish (sushi or sashimi) (Holmberg, 1992). *V. parahaemolyticus* Fujino causes most vibriosis outbreaks in developed countries and is primarily foodborne. *V. cholerae* Pacini is primarily waterborne, but has been associated with foods from aquatic origin (Popovic et al., 1993). Because *V. cholerae* is halotolerant, it can survive and grow in nonsalt foods. Hence, the bacterium has been spread through foods of terrestrial origin in addition to nonsaline fresh water. *V. vulnificus* Reicheit et al. is capable of causing very serious infections leading to septicemia and a high mortality rate (30–40%) (Tacket et al., 1984). This very high mortality rate is the highest of all foodborne infectious agents. Fortunately the incidence of *V. vulnificus* infections is extremely low. Consumption of raw oysters harvested from warm waters (US Gulf Coast) among high-risk individuals (chronic alcoholics, severely immunocompromised) are factors involved with fatalities (Sakazaki, 2002). Several other *Vibrio* species may be pathogenic (Sakazaki, 2002). Incubation period for vibriosis is 2–48 h, usually 12 h. Symptoms include abdominal pain, watery diarrhea, usually nausea and vomiting, mild fever, chills, headache, and prostration. Duration is usually 2–5 days. Cholera typically expresses profuse rice water stools as a characterizing symptom. *V. vulnificus* infections can include septicemia and extremity cellulitis. Vibrios have potential to be a food safety issue for insect-based foods if insects are cultivated in aquatic systems. Control is achieved by use of uncontaminated water sources, cooking, preventing cross contamination, and chilling foods to less than 10°C (National Advisory Committee on Microbiological Criteria for Foods, 1992).

E. coli

There are at least six pathogenic types of *E. coli* associated with foodborne illness (Fratamico et al., 2002). The infectious dose for most strains is high (10^6–10^8 cells), although enterohemorrhagic strains may be much lower (2–45 cells). Enteropathogenic (EPEC) strains are serious in developing countries but rare in the United States. These strains are a leading cause of neonatal diarrhea in hospitals. Likewise, diffusely adherent (DAEC) and enteroaggregative (EAEC) *E. coli* strains are associated with childhood diarrhea. Enteroinvasive (EIEC) strains have an incubation period of 8–24 h, with 11 h most often seen. Symptoms are similar to *Shigella* infections, with bloody diarrhea lasting for several days. Enterotoxigenic (ETEC) strains are a notable cause of traveler's diarrhea. Onset for illness by these strains is 8–44 h, 26 h normal. Symptoms are similar to cholera, with watery diarrhea, rice water stools, shock, and maybe vomiting lasting a short 24–30 h. Enterohemorrhagic or verotoxigenic strains (EHEC) are the most serious *E. coli* found in foods, especially in developed countries. *E. coli* O157:H7 is the predominant serotype among these shiga-like toxin producing bacteria, although other serotypes are found. EHEC cause three syndromes (Padhye and Doyle, 1992; Tarr, 1994). Hemorrhagic colitis (red, bloody stools) is the first symptom usually seen. Hemolytic uremic syndrome (HUS), which is the leading cause of renal failure in children, is characterized by blood clots in kidneys leading to death or coma in children and the elderly. Rarely, individuals may acquire thrombotic thrombocytopenic purpura (TTP), which is similar to HUS but causes brain damage and has a very high mortality rate. Verotoxic strains have an incubation period of 3–4 days. Symptoms include bloody diarrhea, severe abdominal pain, and no fever. Duration ranges from 2–9 days. Vehicles of transmission for *E. coli* include untreated water, cheese, salads, raw vegetables, and water. For *E. coli* 0157:H7, ground beef, raw milk, and raw apple juice or cider are common vehicles. Prevention of *E. coli* outbreaks associated with insects includes treatment of rearing water and ruminant-based feed ingredient supplies, applying a sufficient heat kill step during processing of raw insects as food ingredients, and proper cooking of insect-containing food by consumers or restaurants working with raw insects or food containing raw insects.

Yersiniosis

Most environmental *Y. enterocolitica* Schleifstein and Coleman strains are avirulent; however, pathogenic strains are often isolated from porcine or bovine foods (Kapperud, 2002). The disease is predominantly serious to the very young or the elderly and is more common in Europe and Canada compared to the United States. Incubation period for the disease is 24 h to several days with symptoms including severe abdominal pain similar to acute appendicitis, fever, headache, diarrhea, malaise, nausea, vomiting, and chills. It is not uncommon for children involved in outbreaks to experience unnecessary appendectomies. Duration is usually long, 1 week to perhaps several months. The majority of foods involved in yersiniosis outbreaks are pork and other meats. Milk, seafood, poultry, and water may also serve as vehicles. Control of this bacterium in insects is achieved by adequate pasteurization of animal-sourced feed material, proper cooking of finished products, and avoiding cross-contamination. Refrigeration is not adequate because the bacterium is psychrotrophic and will grow at cool temperatures.

Campylobacteriosis

Three *Campylobacter* species are linked to foodborne diseases, *C. jejuni* Jones et al., *C. coli* Doyle, and *C. laridis* Benjamin et al, (Altekruse and Swerdlow, 2002). *Campylobacter jejuni* is most often associated with poultry, *C. coli* with swine, and

C. laridis with shellfish. *Campylobacter jejuni* gastroenteritis is the most frequent infection among the bacterial agents of foodborne disease (Table 8.1). Campylobacters and related pathogens *Arcobacter* spp. and *Helicobacter pylori* Marshall et al. are microaerophilic and are thus sensitive to normal atmospheric oxygen concentrations (21% O_2) and very low oxygen concentrations (less than 3%). Growth is favored by 5%O_2. Disease characteristics are an incubation period of 1–10 days, 3–5 days normal. Symptoms include fever, abdominal pain, vomiting, bloody diarrhea, and headache, which last for 1 day to several weeks. Relapses are common. The infectious dose is low, 10–500 cells. Foods linked to outbreaks include raw milk, animal foods, raw meat, and fresh mushrooms. Although gastroenteritis is the predominant clinical presentation of campylobacteriosis, chronic sequelae may occur. Guillian-Barré syndrome, which is a severe neurological condition, and Reiter's syndrome, which is reactive arthritis are rare but serious consequences of campylobacteriosis. *H. pylori* is associated with chronic peptic ulcers. Control of these bacteria is achieved by adequate cooking, pasteurization, and cooling and by avoiding cross-contamination. Further control can be achieved by treating animal-based feed material to eliminate the bacteria prior to feeding to insects.

Listeriosis

L. monocytogenes emerged as a cause of foodborne disease in 1981 (Farber and Peterkin, 1991; Smith and Fratamico, 1995). Susceptible humans include pregnant women and their fetuses, newborn infants, the elderly, and immunocompromised individuals due to cancer, chemotherapy, and AIDS. The disease has a high 30% mortality rate. Incubation period is variable, ranging from 1 day to a few weeks (Harris, 2002). In healthy individuals symptoms are mild fever, chills, headache, and diarrhea. In serious cases septicemia, meningitis, encephalitis, and abortion may occur. The duration is variable. The infectious dose is unknown, but for susceptible individuals it may be as low as 100–1000 cells. Foods associated with listeriosis are milk, soft cheeses, meats, and vegetables. Like *Y. enterocolitica*, the bacterium is psychrotrophic and will grow at refrigeration temperatures, though slowly. This bacterium is widespread in wet food processing environments so it is significant risk in foods that can support multiplication during extended shelf life chill storage. Control is best done by avoiding cross-contamination and adequately processing/cooking food.

C. perfringens

C. perfringens Veillon and Zuber is a moderate thermophile showing optimal growth at 43–47°C, with a maximum of 55°C (Labbe and Juneja, 2002). Large numbers of viable cells ($>10^8$) must be consumed, which then pass through the stomach into the intestine. The abrupt change in pH from stomach to intestine causes sporulation to occur, which releases an enterotoxin. Furthermore, the bacterium can grow in the intestine leading to a toxicoinfection. The illness is characterized by an incubation period of 8–24 h. Symptoms are abdominal pain, diarrhea, and gas. A cardinal symptom is explosive diarrhea. Fever, nausea, and vomiting are rare. Duration is short, 12–24 h. Because of the large infectious dose, foods often associated with outbreaks are cooked meats and poultry that have been poorly cooked, such as gravy (anaerobic environment at bottom of pot), stews, and sauces. Outbreaks frequently occur in food service establishments where large quantities of food are made and poorly cooled. Control is best achieved by rapidly cooling cooked food to less than 7°C, holding hot foods at greater than 60°C, and reheating leftovers to greater than 71°C.

Other Bacterial Foodborne Infections

Many other bacteria have been linked to foodborne diseases including *Plesiomonas shigelloides* Bader (raw seafood), *Aeromonas hydrophila* Chester (raw seafood), *Streptococcus pyogenes* Rosenbach (milk, eggs), and perhaps *Enterococcus faecalis* Schleifer and Kilpper-Balz (Cliver, 2002a). Their contribution to foodborne illness appears to be minimal but they may contribute to opportunistic infections.

Nonbacterial Foodborne Infections

Numerous infectious viruses and parasitic worms are capable of causing foodborne illness. All are easily controlled by proper heat treatment of foods. Difficulty with laboratory confirmation of viral agents as causes of foodborne illness leads to probable under reporting (Cliver, 1994a, 1994b). Although insects are well-known carriers and vectors of numerous viruses that infect humans, such viruses are transmitted by arthropods through direct blood injection. There is scant evidence of circumstances that such virus particles can be transmitted by insect ingestion. There remains the possibility of ingestion transmission if insects carry environmentally acquired human viruses (Hepatitis A or norovirus) from human feces or animal manure exposure (Wei et al., 2010). In such circumstances, the virus does not replicate in the insect but simply catches a ride. Evidence of possible foodborne insect transmission of parasites (trematodes) belonging to the families Lecithodendridae and Plagiorchidae is known (Chai et al., 2009). Chagas disease has been linked with the accidental ingestion of insects or consumption of contaminated food (Pereira et al., 2000). *Dicrocoelium dendriticum* Rudolphi is a trematode passed to humans during ingestion of ants containing Metacercariae. The parasite also can be transmitted by consumption of infected animal liver (Jeandron et al., 2011). Many protozoa (*Entamoeba histolytica* Schaudinn, *G. lamblia* Lambl, *Toxoplasma* spp., and *Sarcocystis* spp.) have been found in cockroaches (Graczyk et al., 2005). An overview of common foodborne viruses and parasites follows.

Infectious Hepatitis

Hepatitis A virus is a fairly common infectious agent having an incubation period of 10–50 days, mean of 4 weeks (Cliver, 2002b). Symptoms include loss of appetite, fever, malaise, nausea, anorexia, and abdominal distress. Approximately 50% of cases develop jaundice that may lead to serious liver damage. The duration is several weeks to months. The infectious dose is quite low, less than 100 particles. The long incubation period and duration of the disease means that affected individuals will shed virus for a prolonged period. Foods handled by an infected worker or those that come in contact with human feces are likely vehicles (raw shellfish, salads, sandwiches, and fruits). Filter feeding mollusks concentrate virus particles from polluted waters. Control for insect-based foods is achieved by cooking, stressing personal hygiene by employees in insect farms and in processing facilities of ingredient and food manufacturers, and by avoiding use of human waste as feed material.

Enteroviruses

Noroviruses in the calicivirus family (Coxsackie, ECHO, Norwalk, Rotavirus, Astrovirus, Calicivirus, Parvovirus, and Adenovirus) are now considered the leading cause of foodborne gastroenteritis in the United States (Mead et al., 1999). Other viruses most certainly are involved but our ability to isolate them from infected consumers and foods is limited. Incubation period is typical for infectious organisms, 27–72 h (Cliver, 2002b). Symptoms are usually mild and self limiting and include fever, headache, abdominal pain, vomiting, and diarrhea. Duration is from 1–6 days. The infectious dose for these agents is thought to be very low, 1–10 particles. Foods associated with transmission of viral agents are raw shellfish, vegetables, fruits, and salads. Control is primarily achieved by cooking, proper personal hygiene by employees in insect farms and in processing facilities of ingredient and food manufacturers, and not using human waste as insect growth material.

Multicellular Parasites

Nematodes (roundworms) linked to foodborne illness in humans include *T. spiralis* Owen, *Ascaris lumbricoides* L., *Trichuris trichiura* L., *Enterobius vermicularis* L., *Anisakis* spp., and *Pseudoterranova* spp. (Dubey et al., 2002). *T. spiralis* can invade skeletal muscle and cause damage to vital organs leading to fatalities. Incubation period of trichinosis is 2–28 days, usually 9 days. Symptoms include nausea, vomiting, diarrhea, muscle pains, and fever. Several days duration is common. Foods linked to the disease are raw or undercooked pork and wild game meat (beaver, bear, and boar). Control in pork is accomplished by: (1) cooking to 60°C for 1 min, (2) frozen storage at −15°C for 20 days, −23°C for 10 days, or −30°C for 6 days, or (3) following USDA recommendations for salting, drying, and smoking sausages or other cured pork products. *Anisakis simplex* Rudolphi and *Pseudoterranova decipiens* Krabbe are found in fish and are potential problems for consumers of raw fish. The incubation period is several days with irritation of throat and digestive tract as primary symptoms. Control of these nematodes is by thoroughly cooking food or by freezing food prior to presenting for raw consumption. *Ascaris lumbricoides* L. is commonly transmitted by use of improperly treated water or sewage fertilizer on crops.

Cestoda (tapeworms) are common in developing countries. Examples include *Taenia saginata* Goeze (raw beef), *Taenia solium* L. (raw pork), and *Diphyllobothrium latum* L. (raw fish) (Dubey et al., 2002). Incubation period is 10 days to several weeks with usually mild symptoms including abdominal cramps, flatulence, and diarrhea. In severe cases weight loss can be extreme. Control methods are limited to cooking and freezing. Salting has been suggested as an additional control technique. Use of untreated animal products as feed material for insects is discouraged.

Protozoa cause a large number of foodborne and waterborne outbreaks each year. *Entamoeba histolytica*, *T. gondii* Nicolie and Manceaux, *C. cayetanensis* Ortega et al., *Crytosporidium parvum* Tyzzer, and *G. lamblia* cause dysentery-like illness that can be fatal (Dubey et al., 2002). Incubation period is a few days to weeks leading to diarrhea. Duration can be several weeks, with chronic infections lasting months to years. Those foods that contacted feces or contaminated water are common vehicles. Control is best achieved by proper personal hygiene and water and sewage treatment. Clearly, use of untreated human feces for insect cultivation is unwise.

Prions

Prions are small proteins found in animal nervous tissues (brain, spinal cord) (Cliver, 2002b). They are capable of forming holes in brains of affected animals leading to neurological deficits. In cattle, prions are associated with bovine spongiform encephalopathy (BSE) and consumers of beef from affected animals are at risk of obtaining the human form of the disease called variant Creutzfeldt–Jakob disease. Although this link is tenuous, a few human cases in Europe are thought to be based on consumption of contaminating nervous tissue in beef. The disease is characterized by progressive brain dysfunction ultimately leading to death. Little is known about the incubation period or the infectious dose, as this is a newly emerged condition. Meat and milk from affected animals are not considered a significant transmission risk. Insects are genetically quite dissimilar from ruminants and other mammals, which are the origin of infectious prions. Thus, insects themselves as sources of prion diseases in humans may be unlikely. However, as mentioned previously, the vast genetic diversity of edible insects and the relatively nascent nature of the insect based food industry require prudence, vigilance, and additional research to more fully define such risks. Likewise, proper risk mitigation strategies similar to those employed with other animal-based food products is necessary. For example, insects feeding on prion-containing animal carcass material can harbor infectious prion proteins (Post et al., 1999). Such insects have been shown to transmit prion-related diseases to animals after feeding. Therefore, use of infected ruminant animals and their by-products or human waste as feed is strongly discouraged.

Foodborne Bacterial Intoxications

Foodborne microbial intoxications are caused by a toxin in the food or production of a toxin in the intestinal tract of one who consumes the toxin-producing microbe. Normally the microorganism grows in the food prior to consumption. There are several differences between foodborne infections and intoxications. Intoxicating organisms normally grow in the food prior to consumption, which is not always true for infectious microorganisms. Microorganisms causing intoxications may be dead or nonviable in the food when consumed; only the toxin need be present. Microorganisms causing infections must be alive and viable when food is consumed. Thus, heat treatments such as pasteurization or cooking may not be sufficient to mitigate microbe derived (or other types of) toxins, so materials contaminated with these toxins cannot be simply heated or reprocessed for use and must be discarded or used for a nonfood application. Infection-causing microorganisms invade host tissues and symptoms usually include headache and fever. Toxins usually do not cause fever and toxins act by widely different mechanisms. There may be opportunities for production of bacterial toxins in insect-based foods if such foods or ingredients can support multiplication of toxin-producing bacteria. What follows is an overview of the most common intoxicating bacterial agents.

Staphylococcus aureus Enterotoxin

Certain strains of *S. aureus* Rosenbach produce a heat stable enterotoxin that is resistant to denaturation during thermal processing (cooking, canning, and pasteurization) (Wong and Bergdoll, 2002). The bacterium is salt (10–20% NaCl) and nitrite tolerant, which enables survival in cured meat products (luncheon meats, hams, sausages, etc.). Conditions that favor optimum growth favor toxin production, that is, high protein and starch foods. *S. aureus* competes poorly with other microorganisms, so if competitors are removed by cooking and *S. aureus* is introduced, noncompetitive proliferation is possible. The toxin affects the vagus nerve in the stomach causing uncontrolled vomiting shortly after consumption (1–6 h). Other symptoms include nausea, retching, severe abdominal cramps, and diarrhea, which clear in 12–48 h. Fortunately, fatalities are rare. Sources of the bacterium are usually from nasal passages, skin, and wound infections of food handlers. Hence, suspect foods are those rich in nutrients, high in salt, and those that are handled, with ham, salami, cream-filled pastries, and cooked poultry common vehicles. Control is accomplished by preventing contamination, personal hygiene, and no hand-food contact. Refrigeration below 5°C prevents multiplication, and heating foods to greater than 60°C will not destroy the toxin but will kill the bacterium. Prolific growth of the bacterium is possible in the 5–40°C range. Problems with the bacterium occur most frequently with foods prepared at home or at food service establishments, where gross temperature abuse has occurred. Insects or insect-based ingredients that are properly processed and sufficiently dried should not be a major issue with this bacterium.

B. cereus Enterotoxin

This spore-forming bacterium produces a cell-associated endotoxin that is released when cells lyse upon entering the digestive tract (Griffiths and Schraft, 2002). There are two distinct types of disease syndromes seen with this bacterium. The diarrheal syndrome occurs 8–16 h after consumption. Symptoms include abdominal pain, watery diarrhea, with vomiting and nausea rarely seen. Duration is a short 12–24 h. Foods linked to transmission of this syndrome are pudding, sauces, custards, soups, meat loaf, and gravy. The second, emetic syndrome, is similar to *S. aureus* intoxication. The incubation period is very short, 1–5 h. Symptoms commonly are nausea and vomiting, with rare occurrence of diarrhea. Duration again is short, less than 1 day. This syndrome is commonly linked to consumption of fried rice in Asian restaurants. Other foods include mashed potatoes and pasta. The infectious dose for both is thought to be at least 500,000 bacterial cells. Because the bacterium forms spores, prevention of outbreaks is by proper temperature control. Hot foods should be held at greater than 65°C, leftovers should be reheated to greater than 72°C, and chilled foods should be quickly cooled to less than 10°C.

Botulism

This rare disease is caused by consumption of neurotoxins produced by *C. botulinum* (Parkinson and Ito, 2002). This spore-forming bacterium grows anaerobically and sometimes produces gas that can swell improperly processed canned foods. The bacterium produces several types of neurotoxins that are differentiated serologically. The toxins are heat-labile exotoxins. Two main food poisoning groups (proteolytic and nonproteolytic) are found in nature. Nonproteolytic strains can be psychrotrophic and grow at refrigeration temperatures without the food showing obvious signs of spoilage (no swollen cans or off odor). Incubation period is 12–48 h, but may be shorter or longer. Early symptoms, which may be absent, include nausea, vomiting, and occasionally diarrhea. Other symptoms are dizziness, fatigue, headache, constipation, blurred vision, double vision, difficulty in swallowing, breathing, and speaking, dry mouth and throat, and swollen tongue. Later, paralysis of muscles followed by the heart and respiratory system can lead to death due to respiratory failure. Duration is 3–6 days for fatal cases, several months for nonfatal cases. Treatment of suspect cases is by immediate administration of antisera, which can be useful if given early. Respiratory assistance is usually required.

Foods frequently linked to botulism are inadequately heated or processed home canned foods, primarily low-acid vegetables, preserved meats, and fish (more common in Europe), cooked onions and leftover baked potatoes. The bacterium generally will not grow at a pH of less than 4.6 or at a water activity below 0.85. Thus, high-acid foods, like tomatoes and some fruits, generally are safer than low-acid foods, like corn, green beans, peas, muscle foods, and so forth. Control is by applying a minimum botulinum cook (also known as a "12 D" process) to all thermally processed foods held in hermetically sealed containers. Each particle of food must reach 120°C and be held at that temperature for 3 min to reach a 12 D process. Consumers should reject swollen or putrid cans of food. Properly cured meats hams, bacon, and luncheon meats should not support growth and toxin production by the bacterium. Properly dried insects will not support growth of the bacterium but can be a source of spores when used as ingredients.

A related illness caused by *C. botulinum* is infant botulism. The bacterium can colonize and grow in the intestinal tract of some newborn infants who have not developed a desirable competing microflora. The toxin is then slowly released in the intestines leading to weakness, lack of sucking, and limpness. Evidence suggests that infant botulism may be associated with sudden infant death syndrome. Consumption of honey by young infants has been linked to this type of disease. There is potential risk of this conditions should contaminated insects be incorporated into infant diets.

Chemical Intoxications

Chemical hazards are minimally important as etiological agents of foodborne disease (Table 8.1). It should be noted that a number of chemicals, whether naturally occurring or intentionally added, have tolerance limits in foods. These limits are published in the Code of Federal Regulations, Title 21. Informal limits are available through FDA Compliance Policy Guidelines (Center for Food Safety and Applied Nutrition, Washington, DC). Prohibited substances (CFR 21, Part 189) are not allowed in human foods either because they have been shown to be a public health risk or they have not been shown to be safe using sound scientific data (Biehl and Buck, 1989). Safe food additives are oftentimes referred to as generally recognized as safe (GRAS) substances. There are no documented occurrences of foodborne disease associated with the proper use of such industrial chemicals as insecticides, herbicides, fungicides, fertilizers, food additives, or package material migration chemicals.

Most human-made chemicals associated with foodborne disease find their way into foods by unintentional means. Accidental or inadvertent contamination with heavy metals, detergents, or sanitizers can occur (Taylor, 2002). Although infrequently reported to CDC, most chemical intoxications are likely to be short in duration with mild symptoms. CDC does not attempt to link exposure to these chemicals with chronic diseases. There are measurable levels of pesticides, herbicides, fungicides, fertilizers, and veterinary drugs and antibiotics in most foods. For the vast majority of instances where these residues are found, levels are well below tolerance. Heavy metal poisonings have occurred primarily due to leaching of lead, copper, tin, zinc, or cadmium from containers or utensils in contact with acidic foods. Although usually considered minor contributors to human illness, toxic chemicals in foods may be significant contributors to morbidity and mortality of consumers.

A number of toxic chemicals found in foods are of microbial origin. For example, mycotoxins are secondary metabolites produced by fungi (Chu, 2002). The aflatoxins were the first fungal metabolite in foods regulated by the US government. Grains and nut products are common carriers of these and other mold toxins. Other fungal toxins not associated with microscopic molds include toxic alkaloids associated with certain mushrooms. In this case, direct consumption of wild mushrooms that are frequently confused with edible domesticated species can lead to acute toxicity (Gecan and Cichowicz, 1993). There are no current food processing or sanitation methods that can render these mushrooms acceptable as human food.

Large populations of fungi (yeasts and molds) can be found in fresh, freeze-dried, and frozen insects (Belgian Scientific Committee of the Federal Agency for the Safety of the Food Chain, 2014). Mycotoxigenic molds (*Aspergillus* spp. and *Penicillium* spp.) have potential to grow on harvested insects post mortem if water activity is sufficient (Simpanya et al., 2000). If such populations reach secondary metabolism, many potent toxins can be elaborated. Indeed, aflatoxins have been found in commercial lots of mopane worms (Schabel, 2010).

A number of seafood toxins are naturally associated with shellfish and some predatory reef fish (Johnson and Schantz, 2002). Again, the ultimate cause of these intoxications is traced to the presence of microorganisms. Under favorable environmental conditions, populations of planktonic algae (dinoflagellates) are high (algal bloom) in shellfish growing waters. The algae are removed from the water column during filter feeding of molluscan shellfish (oysters, clams, mussels, cockles, and scallops). The shellfish then concentrate the algae and associated toxins in their edible flesh. Four primary shellfish intoxications have been identified: amnesic shellfish poisoning (ASP), diarrhetic shellfish poisoning (DSP), neurotoxic shellfish poisoning (NSP), and paralytic shellfish poisoning (PSP). ASP has been linked to mussels, DSP with mussels, oysters, and scallops, NSP with oysters and clams, and PSP with all mentioned shellfish. Control of shellfish toxins is best accomplished by monitoring harvest waters for the toxic algae. Post harvest control is not presently possible; however, depuration or relaying may be of some use.

Some marine fish harvested from temperate or tropical climates may contain toxic chemicals. Scombroid fish (anchovy, herring, marlin, sardine, tuna, bonito, mahi mahi, tuna, mackerel, bluefish, and amberjack) under time/temperature abuse during storage can support growth of bacteria that produce histidine decarboxylase (Johnson and Schantz, 2002). This enzyme releases free histamine from the fish tissues. High histamine levels leads to an allergic response among susceptible consumers. Prompt and continued refrigeration of these fish after harvesting will limit microbial growth and enzyme activity. Fish most often associated with histamine scombrotoxicity are mahi mahi, tuna, mackerel, bluefish, and amberjack. Another form of naturally occurring chemical food poisoning found in tropical and subtropical fish is ciguatera. Like shellfish toxicity, ciguatera results when fish bioconcentrate dinoflagellate toxins though the food chain. Thus, large predatory fish at the top of the food chain can accumulate enough toxins to give a paralysis-type response among consumers. Fish associated with ciguatera poisoning are grouper, barracuda, snapper, jack, mackerel, and triggerfish. Again, monitoring of harvest waters is the essential control step to avoid human illness. It should be assumed that insects produced using contaminated seafood materials or contaminated aquatic water sources may have potential to transmit these types of biotoxins.

Physical Hazards

Consumers frequently report physical defects with foods, of which presence of foreign objects predominate (Pierson and Corlett, 1992). Glass is the leading object consumers report and is evidence of manufacturing or distribution error. Most physical hazards are not particularly dangerous to the consumer, but their obvious presence in a food is disconcerting. Most injuries are cuts, choking, and broken teeth. Control of physical hazards in foods is often difficult, especially when these hazards are a normal constituent of the food, such as insect exoskeletons, bones, and shells. Good manufacturing practices and employee awareness are the best measures to prevent physical hazards. Metal detectors and X-ray machines may be installed where appropriate. Consumer education will be important to minimize the risk of potentially harmful physical components normally associated with insect consumption. Some insects are consumed whole, which can itself be a choking hazard.

ADMINISTRATIVE REGULATION

Several regulatory groups are involved in the regulation of food safety and quality standards, from local and state agencies to international agencies. Since there is tremendous variation within and between local and state agencies, this discussion will be confined to the national and international agencies that regulate food. At the US national level, two federal agencies regulate the vast majority of food produced and consumed (Department of Agriculture, Food Safety and Inspection Service, 2003a; Food and Drug Administration, 2003a,b).

United States Department of Agriculture (USDA)

The USDA has responsibility for certification, grading, and inspection of all agricultural products. All federally inspected meat and meat products, including animals, facilities, and procedures, are covered under a series of meat inspection laws that began in 1906 and have been modified on several different occasions, culminating in the latest revisions in 1996 (Department of Agriculture, Food Safety and Inspection Service, 1996). These laws cover only meat that is in interstate commerce, leaving the legal jurisdiction of intrastate meats to individual states. In the states that do have state inspected meats, in addition to federally inspected meats, the regulations require that the state inspection program be "equivalent" to the federal program. Key elements in meat inspection are examination of live animals for obvious signs of clinical illness and examination of gross pathology of carcasses and viscera for evidence of transmissible diseases. The newest regulations also require the implementation of a HACCP system and microbiological testing of carcasses after chilling. Eggs and egg products are also covered by USDA inspection under the Egg Products Inspection Act of 1970 (Department of Agriculture, Food Safety and Inspection Service, 2003b). This act mandates inspection of egg products at all phases of production and processing. USDA inspection of meat processing is continuous; that is, products cannot be processed without an inspector or inspectors present to verify the operation.

United States Food and Drug Administration (FDA)

The FDA has responsibility for ensuring that foods are wholesome, safe, and have been stored under sanitary conditions, as outlined by the Food Drug and Cosmetic Act of 1938. This act has been amended to include food additives, packaging, and labeling. The last two issues relate not only to product safety and wholesomeness, but also to nutritional labeling and economic fraud. FDA is also empowered to act if pesticide residues exceed tolerances set by the US Environmental Protection Agency. Unlike USDA inspection, FDA inspection is discontinuous, with food processing plants being required to maintain their own quality control records while inspectors themselves make random visits to facilities.

Edible insects will likely fall under the jurisdiction of the FDA. The FDA is responsible for ensuring that foods sold or imported into the United States are safe and wholesome. However, the FDA has a risk-based approach to foods. Foods which are considered the highest risk are those which are intended to be consumed without further processing by the consumer, such as cheese or yogurt. Foods that would be considered lower risk would be those which would be processed by the consumer, such as dry pasta or flour. While the presence of *Salmonella* in foods is not desirable, it is far less of a concern in flour, which will be cooked or baked before it is consumed, than in cheese, which will be consumed directly. Some insects may be sold for direct consumption by the consumer, that is, pasteurized and processed into a powder ingredient, or whole insects previously cooked, boiled, or fried. Other insect products may require further preparation by the consumer, such as using dried or frozen insects for cooking.

Milk Sanitation

Perhaps one of the greatest public health success stories of the 20th century has been the pasteurization of milk. The US Public Health Service drafted a model milk ordinance in 1924, which has been adopted by most local and state regulatory authorities and has become known as the Grade A PMO (Pasteurized Grade A Milk Ordinance) (Food and Drug Administration, 2002). This ordinance covers all phases of milk production, including but not limited to animal health, design and construction of milk processing facilities, equipment, and most importantly, the pasteurization process itself. The PMO sets quality standards for both raw and processed milk, in the form of cooling requirements and bacteriological populations. The PMO also standardizes the pasteurization requirements for fluid milk, which insures that bacteria of public health significance will not survive in the finished product. From a historical perspective, it is interesting to note that neither the public nor the industry initially embraced pasteurization, but that constant pressure from public health officials finally succeeded in making this important advance in public health almost universal.

International Administration

The Codex Alimentarius Commission, created by the Food and Agriculture Organization and the World Health Organization, has the daunting task of implementing food standards on an international scale (Food and Agriculture Organization, 2003). These standards apply to both general and specific food categories and also set limits for pesticide residues in foods. Acceptance of these standards is voluntary and at the discretion of individual governments, but acceptance of the standards requires that the country apply them equally to both domestically produced and imported products. The importance of international standards is growing daily as international trade in food expands. Many countries find that they are both importing and exporting foods, and a common set of standards is critical in establishing trade without the presence of nontariff trade barriers. The European Union recently enacted a Novel Foods law that treats edible insects as a novel food that requires demonstration that such products are safe to eat.

Prerequisite Programs

In order to achieve the goal of producing a safe food product, food processors should have in place a variety of fundamental programs covering the general operation of the process and the processing facility. These programs are considered "prerequisites," as without these basic programs in place, it is impossible to produce safe and wholesome foods, irrespective of the available technology, inspection process, or microbiological testing. These prerequisite programs fall generally under the term "good manufacturing practices," but also include sanitation, equipment and facility design, personal hygiene issues, and pest control.

Good Manufacturing Practices

GMPs cover a broad range of activities with the food-processing establishment. Although there is general guidance in the Code of Federal Regulations (Food and Drug Administration, 2003a,b), GMPs are established by the food processor, and are specific to their operation. There is also general guidance on GMPs available from a variety of organizations representing specific commodities or trades. Specific applications of GMPs are discussed in the following sections ("Training and Personal Hygiene," "Pest Control," "Sanitation," "Sanitary Facility Design," "Sanitary Equipment Design," and "Cleaning and Sanitizing Procedures"), but GMPs also apply activities that affect not only the safety of the product, but also the quality. As an example, a refrigerated holding or storage temperature may be set by a GMP at a point below that which is actually required for product safety, but is set at that point for product quality reasons. Conversely, if a raw material or partially manufactured product, which under normal circumstances would be kept refrigerated, were subsequently found to be at a higher temperature, it would be deemed to be out of compliance with the GMP.

GMPs may also focus on the actual production processes and controls within those processes. GMPs may be viewed as rules that assure fitness of raw materials and ingredients, rules that maintain the integrity of processed foods, and rules to protect the finished product (foods) from deterioration during storage and distribution. Other GMPs may address the presence of foreign materials in the processing area, such as tramp metal from equipment maintenance or broken glass from a shattered light bulb. These GMPs are established to provide employees with specific guidance as to the company's procedures for addressing certain uncommon but unavoidable issues.

While GMPs by their nature cover broad areas of operation, the individual GMP is usually quite specific, presenting complete information in a logical, step-wise fashion. An employee should be able to retrieve a written GMP from a file, and should be able to perform the required GMP function with little or no interpretation of the written material.

Training and Personal Hygiene

Personnel who are actually involved in food processing operations should also understand the necessity for proper cleaning and sanitation, and not simply rely on the sanitation crew to take care of all issues (Marriott, 1999a). In addition, all employees must be aware of basic issues of personal hygiene, especially when they are in direct contact with food or food processing equipment. Some key elements, such as hand washing and clean clothing and gloves, should be re-emphasized on a periodic basis. An important aspect of this is an emphasis on no "bare handed" contact with the edible product, using utensils or gloves to prevent this from occurring. This information has been outlined by the US Food and Drug Administration in the Good Manufacturing Practices section of the Code of Federal Regulations.

Pest Control

Pests, such as insects and rodents, present both physical as well as biological hazards (Marriott, 1997). While the consumer would undoubtedly object to the proverbial "fly in the soup," the concerns with the introduction of biological hazards into the foods by pests are even greater. Integrated pest management (IPM) includes the physical and mechanical methods of controlling pests within the food processing environment and the surrounding premises. At a minimum, the processing environment and the area surrounding the processing plant should be evaluated by a competent inspector for both the types of pests likely to be present, and the potential harboring of such pests. A comprehensive program should be established that addresses flying insects, crawling insects, and rodents, the objective of which is to prevent access to the processing environment. These of course apply to food processing, packaging, and handling facilities and, in this case, the reference to insects applies to those from the environment and not the farm-raised insects intentionally incorporated into insect-based foods. In the case of insect-based foods and edible-insect farms, the farmed insect or the key ingredient may be referred to as the livestock and distinguished from unwanted insects from the environment, deemed to be pests. Given that it is impossible to completely deny pest access to the processing environment, internal measures should be taken to reduce the numbers of any pests that enter the processing area. Since it is undesirable to have poisonous chemicals in areas surrounding actual food production, active pest reduction methods should be mechanical in nature (traps, insect electrocuting devices, etc).

Record keeping is an important aspect of pest management. Documentation of pest management activities should include maps and maintenance schedules for rodent stations, bait stations, insect electrocuters, an inventory of pesticides on the premises, and reports of inspections and corrective actions. There should be standard operating procedures for applying pesticides, and they should only be applied by properly trained individuals. Many food-processing establishments contract with external pest control operators to address their pest control needs.

Sanitation

The World Health Organization defines *sanitation* as "all precautions and measures which are necessary in the production, processing, storage and distribution of food, in order to assure an unobjectionable, sound and palatable product which is fit for human consumption." Sanitation is the fundamental program for all food processing operations, irrespective of whether they are converting raw products into processed food or preparing food for final consumption. Sanitation impacts all attributes of processed foods, from organoleptic properties of the food to the safety and quality of the food itself. From a food processors perspective, an effective sanitation program is essential to producing quality foods with reasonable shelf lives. Without an effective program, even the best operational management and technology will ultimately fail to deliver the quality product that consumers demand.

Sanitation programs are all encompassing, focusing not only on the details of soil types and chemicals, but the broader environmental issues of equipment and processing plant design. Many foodborne microorganisms, both spoilage organisms and bacteria of public health significance, can be transferred from the plant environment to the food itself (FDA/MIF/IICA, 1988). Perhaps one of the most serious of these microorganisms came to national and international attention in the mid 1980s (and even in more recent cases), when *L. monocytogenes* was found in processed dairy products. The genus *Listeria* was considered to be a relatively minor veterinary pathogen until that time, and not even considered a potential foodborne agent. However, subsequent research demonstrated that *L. monocytogenes* was a serious human health concern, and more importantly was found to be widely distributed in nature. In many food processing plants, *Listeria* spp. were found to be in the general plant environment, and subsequently efforts have been made to improve plant sanitation, through facility and equipment design as well as focusing more attention on basic cleaning and sanitation. *Salmonella* also can be a foodborne pathogen spread by environmental contamination, especially in the case of low moisture foods.

Sanitary Facility Design

Some of the basic considerations of food processing facility design include the physical separation of raw and processed products, adequate storage areas for nonfood items (such as packaging materials), and a physical layout that minimizes employee traffic between raw and processed areas. While these considerations are easily addressed in newly constructed facilities, they may present challenges in older facilities that have been renovated or added on to. Exposed surfaces, such as floors, walls, and ceilings, in the processing area should be constructed of material that allows for thorough cleaning. Although these surfaces are not direct food contact surfaces, they contribute to overall environmental contamination in the processing area. These surfaces are particularly important in areas where food is open to the environment, and the potential for contamination is greater when temperature differences in the environment result in condensation (Gabis and Faust, 1988). As an example, a large open cooking kettle will generate some steam that may condense on surfaces above the kettle. This condensate may, without proper design and sanitation, drip back down into the product carrying any dirt and dust from overhead surfaces back into the food. Other obvious considerations are basic facility maintenance as well as insect and rodent control programs, as all of these factors may contribute to contamination of food.

Sanitary Equipment Design

Many of the same considerations for sanitary plant design also apply to the design of food processing equipment. Irrespective of its function, processing equipment must protect food from external contamination and from undue conditions that will allow existing bacteria to grow. The issue of condensate as a form of external contamination has already been raised. Opportunities for existing bacteria to reproduce may be found in the so-called "dead spaces" within some equipment. These areas can allow food to accumulate over time under conditions that allow bacteria to grow. These areas then become a constant inoculation source for additional product as it moves through the equipment, increasing the bacteriological population within the food. Other considerations of food equipment design include avoiding construction techniques that may allow product to become trapped within small areas of the equipment, creating the same situation that occurs in the larger dead spaces within the equipment. As an example, lap seams that are tack welded provide ample space for product to become trapped. Not only does this create a location for bacteria to grow and contaminate the food product, it also creates a point on the equipment that is difficult if not impossible to clean.

Cleaning and Sanitizing Procedures

Cleaning and sanitizing processes can be generically divided into five separate steps that apply to any sanitation task (Ingham et al., 1996). The first step is removal of residual food, waste materials, and debris. This is frequently referred to as a "dry" clean up. The dry clean up is followed by a rinse with warm (48–55°C) water, to remove material that is only loosely attached to surfaces and to hydrate material that is more firmly attached to surfaces. Actual cleaning follows the warm water rinse, which usually involves the application of cleaning chemicals and some form of scrubbing force, either with mechanical brushes or with high-pressure hoses. The nature of the residual food material will determine the type of cleaning compound applied. After this, surfaces are rinsed and inspected for visual cleanliness. The inspection process is a critical part of sanitation, as the effectiveness of most chemical sanitizers is greatly reduced in the presence of residual soil or cleaning solutions. Visual inspection would include but not be limited to evidence of remaining soil, or a cloudy or hazy liquid which could indicate the presence of cleaning chemical residues.

At this point, the cleaning process is repeated on any areas that require further attention. Carbohydrates and lipids can generally be removed with warm to hot water and sufficient mechanical scrubbing. Proteins require the use of alkaline cleaners, while mineral deposits can be removed with acid cleaners. Commercially available cleaning compounds generally contain materials to clean the specific type of food residue of concern, as well as surfactants and, as necessary, sequesterants that allow cleaners to function more effectively in hard water (Marriott, 1999b).

When surfaces are visually clean, a sanitizer is applied to reduce or eliminate remaining bacteriological contamination. Inadequately cleaned equipment cannot be sanitized, as the residual food material will protect bacteria from the sanitizer. One of the most common sanitizing agents, widely used in small- and medium-sized processing facilities, is hot water. Most regulatory agencies require that when hot water is used as the sole method of sanitization, the temperature must be at or above 85°C. As with other thermal processes, the effectiveness of hot water is a result of the surface temperature of the equipment, as well as the time at that temperature. While heat sanitization is effective, it is not as economical as chemical sanitizers because of the energy costs required to maintain the appropriate temperature. Chlorine containing sanitizers are economical and effective against a wide range of bacterial species, and are widely used in the food industry (Marriott, 1999c). Typically, the concentrations of chlorine applied to equipment and surfaces are in the 150–200 ppm range. Chlorine sanitizers are corrosive and can, if improperly handled, release chlorine gas into the environment.

Iodine containing sanitizers are less corrosive than chlorine sanitizers, but are also somewhat less effective. These sanitizers must be used at slightly acidic pH values to allow for the release of free iodine. The amber color of iodine sanitizers can give an approximate indication of concentration, but can also leave residual stains on treated surfaces. Quaternary ammonium compounds (QACs) are noncorrosive and demonstrate effective biocidal action against a wide range of microorganisms. These sanitizers are generally more costly and are not as effective as chlorine compounds, but they are stable and provide residual antimicrobial activity on sanitized surfaces. Food processing plants will frequently alternate between chlorine and QAC sanitizers to prevent development of resistant microbial populations or will use chlorine sanitizers on regular production days and then apply QACs during periods when the facility is not operating (ie, over a weekend).

Another element in food plant sanitation programs is the personnel who perform the sanitation operations as well as the employees who work in the processing area. Sanitation personnel should be adequately trained to understand the importance of their function in the overall processing operation in addition to the training necessary to properly use the chemicals and equipment necessary for them to perform their duties. This is important not only from the standpoint of assuring that the food processing environment is properly cleaned and sanitized, but also as a matter of employee safety. Many of the procedures, processes, and chemicals used in sanitation have the potential to cause serious bodily injury to the employees, and proper training is important to minimize the risk to worker health.

HAZARD ANALYSIS CRITICAL CONTROL POINT SYSTEM

The basic concept of (hazard analysis critical control point system) HACCP was developed in the late 1950s and early 1960s as a joint effort to produce food for the manned space program. The US Air Force Space Laboratory Project Group, the US Army Natick Laboratories, and the National Aeronautics and Space Administration (NASA) contributed to the development of the process, as did the Pillsbury Company, which had a major role in developing and producing the actual food products for manned space missions. Since that time, the HACCP system has evolved and been refined, but still focuses on the original goal of producing food that is safe for consumption (Pierson and Corlett, 1992).

Since their development, HACCP principles have been used in many different ways. However, recent interest in the system has been driven by changes in the regulatory agencies, specifically the US Department of Agriculture Food Safety and Inspection Service (USDA-FSIS), and the US Food and Drug Administration (FDA). USDA-FSIS revised the regulations that govern meat inspection to move all federally inspected meat plants to a HACCP-based system of production and inspection (Department of Agriculture, Food Safety and Inspection Service, 1996). The FDA has also changed the regulations for fish and seafood, as well as for juices, again moving this to a HACCP-based system for production (Department of Health and Human Services, Food and Drug Administration, 1995). With the publication of the final rule preventative controls for human foods, most foods produced or imported into the United States will be under an HACCP, or preventative controls (similar in concept to HACCP) system by 2020 (US FDA, 2015a).

The goal of an HACCP system is to produce foods that are free of biological, chemical, and physical hazards (Stevenson and Bernard, 1995). HACCP is a preventative system, designed to prevent problems before they occur, rather than trying to fix problems after they occur. Biological hazards fall into two distinct categories, those that can potentially cause infection and those that can potentially cause intoxications. Infectious agents require the presence of viable organisms in the food and may not, depending on the organisms and the circumstances, require that the organism actually reproduce in the food. As an example, *E. coli* O157:H7 has an extremely low infectious dose for humans (possibly less than 100 viable cells), and as such, the mere presence of the bacterium in foods is a cause for concern. In contrast, organisms involved in intoxications usually require higher numbers of the organism in the food to produce sufficient amounts of toxin to cause clinical illness in humans. However, some of the toxins involved in foodborne diseases are heat stable, so that absence of viable organisms in the food is not necessarily an indication of the relative safety of the food. *S. aureus* is a good example, where it typically requires greater than 1,000,000–10,000,000 cells per gram of food to produce sufficient toxin to cause illness in humans (Noleto and Bergdoll, 1982). However, because the toxin itself is extremely heat stable, cooking the food will eliminate the bacterium but not the toxin, and the food can still potentially cause an outbreak of foodborne illness.

Chemical hazards include chemicals that are specifically prohibited in foods, such as cleaning agents, as well as food additives that are allowed in foods but only at regulated concentrations. Foods containing prohibited chemicals or food additives in levels higher than allowed are considered adulterated. Adulterated foods are not allowed for human consumption and are subject to regulatory action by the appropriate agency (USDA or FDA). Chemical hazards can be minimized by assuring that raw materials (foods and packaging materials) are acquired from reliable sources that provide written assurances that the products do not contain illegal chemical contaminants or additives. During processing, adequate process controls should be in place to minimize the possibility that an approved additive will be used at levels not exceeding maximum legal limits for both the additive and the food product. Other process controls and GMPs should also insure that industrial chemicals, such as cleaners or lubricants, will not contaminate food during production or storage (Food and Drug Administration, 2003a,b).

In a traditional HACCP system, allergens are usually considered as chemical hazards. However, they are separate and distinct form how we traditionally view chemical hazards, and there is a growing trend to consider them as a separate hazard category. The primary human food allergens are milk, eggs, fish, crustacean shellfish, tree nuts, peanuts, wheat, and soybeans (US FDA, 2015b). While these are normal and expected ingredients in many foods (eg, wheat flour in bread), the presence of even trace amounts of these food items may trigger severe reactions in sensitive individuals. Because of this, foods containing these products as ingredients must be adequately labeled. Based on historical trends, approximately one-third of the product recalls in the United States over the last five years are attributable to undeclared allergens, those that are in the food as an ingredient but not represented on the label (see Chapter 9).

Physical hazards are extraneous material or foreign objects that are not normally found in foods. For example, wood, glass, or metal fragments are extraneous materials that are not normally found in foods. Physical hazards typically affect only a single individual or a very small group of individuals, but because they are easily recognized by the consumer, are sources of many complaints. Physical hazards can originate from food processing equipment, packaging materials, the environment, and from employees. While some physical hazards can be detected during food processing (eg, metal by the use of metal detectors), many nonferrous materials are virtually impossible to detect by any means and so control often resides with employees.

HACCP Plan Development

Prior to the implementation of HACCP, a review should be conducted of all existing prerequisite programs. These would include but would not be limited to GMPs, employee GMPs and training, sanitation, pest control, and facility design. Deficiencies in these programs should be addressed prior to the implementation of HACCP because an HACCP plan presumes that these basic programs are fully functional and effective. Without properly functional prerequisite programs, the HACCP system will not achieve the desired goal of improving the safety of the food products. Development of an HACCP plan begins with the formation of an HACCP team (American Meat Institute Foundation, 1994). Individuals on this team should represent diverse sections within a given operation, from purchasing to sanitation. The team is then responsible for development of the plan. Initial tasks that the team must accomplish are to identify the food and method of distribution, and to identify the consumer and intended use of the food. Having done this, the HACCP team should construct a flow diagram of the process and verify that this diagram is accurate.

The development of an HACCP plan is based on seven principles or steps in logical order (Table 8.6) (National Advisory Committee on Microbiological Criteria for Foods, 1998). With the flow diagram as a reference point, the first principle or step is to conduct a hazard analysis of the process. The HACCP team identifies all biological, chemical, and physical hazards that may occur at each step during the process. Once the list is completed, it is reviewed to determine the relative risk of each potential hazard, which helps identify significant hazards. Risk is the interaction of "likelihood of occurrence" with "severity of occurrence." As an extreme example, a sudden structural failure in the building could potentially contaminate any exposed food with foreign material. However, the likelihood of the occurrence of such an event is small. In contrast, if exposed food is held directly below surfaces that are frequently covered with condensate, then the likelihood of condensate dripping on exposed food is considerably higher. An important point in the determination of significant hazards is a written explanation by the HACCP team regarding how the determination of "significant" was made. This documentation can provide a valuable reference in the future, when processing methods change or when new equipment is added to the production line.

The second principle in the development of an HACCP plan is the identification of critical control points (CCPs) within the system. A CCP is a point, step, or procedure where control can be applied and a food safety hazard can be prevented, eliminated, or reduced to acceptable levels. An example of a CCP is the terminal heat process applied to canned foods after cans have been filled and sealed. This process, when properly conducted according to FDA guidelines, effectively eliminates a potential food safety hazard, *C. botulinum*. Once CCPs have been identified, the third principle in the development of an HACCP plan is to establish critical limits for each CCP. These limits are not necessarily the ideal processing parameters, but the minimum acceptable levels required to maintain the safety of the product. Again, in the example of a canned food, the critical limit is the minimum time and temperature relationship to insure that each can has met the appropriate standards required by FDA.

The fourth principle, following in logical order, is to establish appropriate monitoring requirements for each CCP. The intent of monitoring is to insure that critical limits are being met at each CCP. Monitoring may be on a continuous or discontinuous basis. The presence of a physical hazard, such as metal, can be monitored continuously by passing all of the food produced through a metal detector. Alternately, the presence of foreign material can be monitored on a continuous basis by visual inspection. Discontinuous inspection may involve taking analytical measurements, such as temperature or pH, at designated intervals during the production day. Some analytical measurements can be made on a continuous basis by the use of data recording equipment, but it is essential that continuous measures be checked periodically by production personnel to assure that the critical limits are being met.

The fifth principle in the development of an HACCP plan is to establish appropriate corrective actions for occasions when critical limits are not met. Corrective actions must address the necessary steps to correct the process that is out of

TABLE 8.6 Seven HACCP Principles

Hazard analysis
Identify critical control points (CCP)
Establish critical limits for each CCP
Monitor CCP
Establish corrective action
Record keeping
Verification

control (such as increasing the temperature on an oven) as well as addressing disposition of the product that was made while the process was out of control. A literal interpretation of the HACCP system and a CCP is that when a CCP fails to meet the critical limits, then the food product is potentially unsafe for human consumption. As a result, food produced while the CCP was not under control cannot be put into the normal distribution chain without corrective actions being taken to that product. Typically this means that the product must be either re-worked or destroyed, depending on the nature of the process and the volume of product that was produced while the CCP was out of control. This argues for frequent monitoring, so that the actual volume of product produced during each monitoring interval is relatively small.

The sixth principle in the development of an HACCP plan is verification. Verification can take many forms. At its' most basic level, verification simply means that the monitoring procedures and corrective actions described in the HACCP plan are in fact being performed as described. Microbiological tests of finished products can be performed to evaluate the effectiveness of an HACCP plan. Alternately, external auditors can be used to evaluate all parts of the HACCP plan, to insure that the stated goals and objectives are being met. An HACCP plan must also be periodically reviewed and updated, to reflect changes in production methods and use of different equipment. Another critical aspect of verification is education of new employees on the HACCP plan itself. As HACCP is phased in to many food-processing environments, many employees who are unfamiliar with the concepts and goals of HACCP will have to be educated on the necessity of following the plan. In one sense, USDA-FSIS regulations have guaranteed that meat processors will follow HACCP plans, as the penalty for not following the HACCP plan can be as severe as the loss of inspection at an establishment resulting in termination of production. However, HACCP is an excellent system for monitoring and improving production of food products, and many food processors will discover that HACCP plans offer many benefits, well above and beyond the legal requirements of the regulatory agencies.

The seventh principle in the development of an HACCP plan is the establishment of effective record keeping procedures. In many respects, an HACCP plan is an elaborate record-keeping program. Records should document what was monitored, when it was monitored and by whom, and what was done in the event of a deviation. Reliable records are essential from both a business and regulatory perspective. From the business perspective, HACCP records allow a processor to develop an accurate longitudinal record of production practices and deviations. Reviewing HACCP records may provide insight on a variety of issues, from an individual raw material supplier whose product frequently results in production deviations, to an indication of an equipment or environmental problem within a processing plant. From a regulatory perspective, records allow inspectors to determine if a food processor has been fulfilling commitments made in the HACCP plan. If a processor has designated a particular step in the process as a CCP, then they should have records to indicate that the CCP has been monitored on a frequent basis and should also indicate corrective actions taken in the event of a deviation.

FOOD SAFETY MODERNIZATION ACT

In 2015, the FDA published the final rules for Preventative Controls for Human Foods, as part of the Food Safety Modernization Act (US FDA, 2015a). This is a comprehensive regulation intended to address food safety issues. It requires the development of a Food Safety Plan, which includes prerequisite programs (GMPs and sanitation), hazard identification, process controls, verification, validation, and record keeping programs (Fig. 8.1). Although the terminology is somewhat different, the concepts are very similar to HACCP. Perhaps the biggest difference if the emphasis on allergen control, which is far more prominent in preventative controls.

The focus of the Preventative Controls for Human Foods rules, as with HACCP, is prevention. The approach follows a logical progression of identifying hazards, identifying control measures, and then designing a systematic approach to

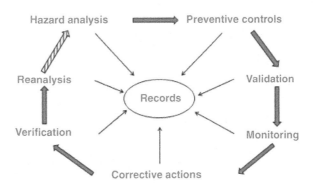

FIGURE 8.1 Food Safety Modernization Act food safety plan elements.

ensuring that the controls are followed. The Preventative Controls for Human Foods publication applies these well-established principles to all foods, whereas previously HACCP was only applied to low-acid canned foods, seafood, and juices. Since HACCP was introduced into the meat industry in the 1990s, the overall incidence of foodborne pathogens detected on meat and poultry products has decreased substantially. While this decrease cannot be solely attributed to HACCP, there is little doubt that HACCP has contributed to this decline. It is hoped that the implementation of the Preventative Controls for Human Foods will contribute to similar reductions in other foods.

VALIDATION

Validation is an important concept in both HACCP and Preventative Controls for Human Foods. Validation is the collection and evaluation of scientific, technical, and observational information to demonstrate that a process is effective in controlling the identified hazard. In its most basic form, validation assures that the control measures actually control the identified hazard. Validation is commonly thought of as a two-part process. The first is the scientific and technical justification for the control measure. This is the fundamental knowledge that indicates that a given set of conditions, for example a temperature and time combination, are capable of eliminating a certain population of bacteria. The second part of validation is the in-plant demonstration that the process, as implemented in the processing establishment, can in fact control the hazard. Validation varies by different food processes, but a basic protocol for in-plant validation has been described (European Food Safety Authority, 2015).

One of the main concerns of validation is the level of expected hazard. This can be the frequency of occurrence, or in the case of biological hazards, the population. This is important because if a process control is designed to eliminate a hazard which occurs 1% of the time, or a bacterial population of 10 cells per gram, it may fail if the incidence is at 7% or the population is 100 cells per gram. It is incumbent upon the processing establishment to have some concept of the extent of the hazard. This can be derived from historical data within the specific establishment, or by a trade association relevant to the specific food. In other cases, the relevant regulatory agency may have published survey data which provides a basis for the necessary level of control of a hazard. It is important for an establishment to know not only the typical level of a hazard, but also the maximum expected level of the hazard. Ideally, a process control should be sufficient to control the maximum expected level of the hazard, with an additional margin of safety.

FOOD PRESERVATION

Normal microflora of foods are characterized by food type and growing/handling practices. Foods of plant origin have flora on outer surfaces. Animals too, including insects, have flora on surfaces, but they also have intestinal flora and secretion flora. Outside sources, such as soil, dust, water, humans, and equipment, can be significant sources of disease causing microbes. Use of diseased animals for foods is dangerous because they often carry human pathogens. It should be noted that the inner tissues of plants and animals are generally sterile; however, cabbage inner leaves have lactobacilli and animal intestinal tracts have numerous microbes. Pathogens found on fruits and vegetables are from soil origin (*Clostridium, Bacillus*) or from contaminated water, fertilizer, or food handlers. Some grain and nut products are naturally contaminated in the field with mycotoxin producing molds. Soil is also a source of contamination of foods from animal origin. Animal feces can harbor coliforms, *C. perfringens*, enterococci, and enteric pathogens. Milk from infected udders (mastitis) can carry disease causing *S. pyogenes* and *S. aureus*. Nonmastitic udders can shed *Brucella*, Rickettsia, and viruses.

Outside sources of contamination that are not normally associated with food can be important in terms of food safety. Soil and dust contain very large numbers and large variety of microbes. Many microorganisms responsible for food spoilage come from these sources. Contamination is by direct contact with soil, water, or by airborne dust particles. Air can carry microorganisms from other sources such as sneezing, coughing, dust, and aerosols. Pathogens, mold spores, yeasts, and spoilage bacteria can be then disseminated. Organic debris from plants or animals is an excellent source. Microorganisms can grow on walls, floors, and other surfaces and act as a source of contamination during food processing and preparation. Airborne particles can be removed by filtration or by electrostatic precipitation.

Treated sewage may be used for fertilizer and might be considered a good substrate for growing insects, although due to large amounts of toxic compounds like heavy metals and human pathogens, it is not used often for this purpose (Cliver, 1994a). Sewage can be an excellent source of pathogens including all enteric gram negative bacteria, enterococci, *Clostridium*, viruses, and parasites. Sewage that contaminates lakes, streams, and estuaries has been linked to many seafood outbreaks. In addition, water used for food must be safe for drinking and must be treated and free of pathogens. Furthermore, water must not contain toxic wastes. Water in food processing is typically used for washing, cooling, chilling, heating, ice, or as an ingredient. Stored water (reservoirs) and underground water (wells) are usually self-purifying.

TABLE 8.7 Methods of Food Preservation

Methods	Description
Asepsis	Keeping microorganisms out of food, "aseptic packaging"
Removal	Limited applications, difficult to do, filtration
Anaerobiosis	Sealed, evacuated container, vacuum packaging
High temperatures	Sterilization, canning, pasteurization
Low temperatures	Refrigeration, freezing
Dehydration	Drying or tying-up water by solutes and hydrophilic colloids, lower water activity (a_w)
Chemical preservatives	Natural, developed, or added (propionic acid, nisin, spices), acids lower pH
Irradiation	γ-Rays (ionizing) or UV (nonionizing)
Mechanical destruction	Grinding, high pressures, not widely used
Combinations	Most frequently employed, multiple hurdle concept

Numbers and types of microorganisms found in foods depends on: 1) the general environment from which the food was obtained, 2) the quality of raw food, 3) the sanitary conditions under which the food was processed or handled, and 4) the adequacy of packaging, handling, and storage of foods. General methods of food preservation are shown in Table 8.7. The hurdle concept uses multiple methods (multibarrier approach) to food preservation and is the most common. Examples include pasteurized milk (heat, refrigeration, and packaging) or canned beans (heat, anaerobiosis, and packaging). The most common antimicrobial hurdles used in insect processing include drying and heat processing.

Principles of Food Preservation

Principles of food preservation rely on preventing or delaying microbial decomposition (Potter and Hotchkiss, 1995; Jay, 1996). This can be accomplished by using asepsis or removal. Preventing growth or activity of microbes with low temperatures, drying, anaerobic conditions, or preservatives can also be done. Killing or injuring microbes with heat, irradiation, or some preservatives is certainly effective. A second principle is to prevent or delay self-decomposition, which is done by destruction or inactivation of enzymes (blanching) or by preventing or delaying autoxidation (antioxidants). The third principle is to prevent physical damage caused by insects, animals, and mechanical forces, which prevents entry of microorganisms into food. Physical barriers (packaging) are the primary means of protection. To control microorganisms in foods, many methods of food preservation depend not on the destruction or removal of microbes but rather on delaying the initiation of growth or hindering growth once it has begun.

For food preservation to succeed, one must be able to manipulate the microbial growth curve. Many steps can be done to lengthen lag phase or positive acceleration phase of a population. These steps include: (1) prevent introduction of microbes by reducing contamination (fewer numbers gives a longer lag phase), (2) avoid addition of actively growing microorganisms that may be found on unclean containers, equipment, and utensils, and (3) create unfavorable environmental conditions for microbial growth. The last step is the most important in food preservation and can be done by low water activity, extremes of temperature, irradiation, low pH, adverse redox potential, and by adding inhibitors and preservatives (Table 8.6). Some of these steps may only damage or injure microorganisms without killing them; hence, the need for multiple barriers becomes essential (Potter and Hotchkiss, 1995). For each of these steps to be effective, other factors should be considered. For example, the number of organisms present determines kill rate. Smaller numbers give faster kill rates. Vegetative cells are most resistant to lethal treatments when in late lag or stationary phase and least resistant when in log phase of growth.

Asepsis/Removal

Keeping microorganisms out of food is often difficult during food production. Processing and postprocessing are much easier places to apply asepsis. Protective covering of foods such as skin, shells, and hides are often removed during processing, thereby exposing previously sterile foods to contaminating microbes. Raw agricultural commodities normally carry a natural bioburden upon entering the processing plant. Packaging is the most widely used form of asepsis and includes wraps, packages, cans, etc.

Removal of microorganisms from foods is not very effective. Washing of fruits and vegetables can remove some surface microorganisms. However, if wash water becomes dirty, it can add microbes to the food. Trimming is an effective way to remove spoiled or damaged parts. Filtration is good for clear liquids (juices, beer, soft drinks, wine, and water) but is of little value for solid foods. Centrifugation, such as used in sedimentation/clarification steps, is not useful for removal of bacteria or viruses.

Modified Atmosphere Conditions

Altering the atmosphere surrounding a food can be a useful way to control microbes. Examples include packaging with vacuum, CO_2, N_2, or combinations of inert gases with or without oxygen. Some CO_2 accumulation is possible during fermentations or vegetable respiration. It is important to note that vacuum packaging can lead to favorable environments for proliferation of anaerobic pathogens such as *C. botulinum*.

High-Temperature Preservation

Use of high-temperature processing is based on destroying microbes, but may also injure certain thermoduric microbes. Not all microorganisms are killed, that is, spore formers usually survive (Jay, 1996). Other barriers are combined with a thermal process to achieve adequate safety and product shelf life. Commercial sterilization used in the canning process usually destroys all viable microbes that can spoil the product. Thermophilic spores may survive but will not grow under normal storage conditions.

Several factors affect heat resistance of microorganisms in foods (Jay, 1996). Species variability and the ability to form spores plus condition of the microbial population can affect heat resistance. Environmental factors such as food variability and presence of other preservative measures employed also dictate thermal resistance. For example, heat resistance increases with decreasing water activity. Hence, moist air heating is better than dry heating. High fat foods tend to increase the heat resistance of cells. The larger the initial number of microorganisms present means a higher heat resistance. Older (stationary phase) cells are more resistant to heat than younger cells. Resistance increases as growth temperature increases. A microbe with a high optimum temperature for growth will generally have a high heat resistance. Addition of other inhibitors, such as nitrite, will decrease heat resistance. Likewise, high-acid foods (pH less than 4.6) will not generally support growth of pathogens. There is a time–temperature relationship that is a very important factor governing heat resistance of a microbial population. As temperature increases, the time needed for a given kill decreases. The relationship is dependent on type and size of food container. Larger containers require longer process times. Metal conducts heat better than glass, which can lower process times.

Microorganisms are killed by heat at a rate nearly proportional to the numbers present. This is a log order of death, which means that under a constant temperature, the same percentage of a population will die at a given time interval, regardless of the population size (Fig. 8.2). For example, 90% die in 30 s, 90% of those remaining die in the next 30 s, and so on. Thus, as the initial number of microorganisms increases, then the time required for the reduction of all organisms at a given temperature also increases. Food microbiologists express this time–temperature relationship by calculating a number of constants. D value is the time required to reduce a population by one log cycle (90% population reduction) at a given

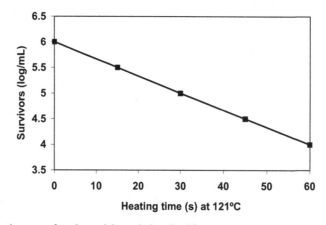

FIGURE 8.2 **Typical heat inactivation curve for a bacterial population.** $D = 30$ s.

temperature. Thermal death time (TDT) is the time needed to kill a given number of organisms at a given temperature. Thermal death point (TDP) is the temperature needed to kill a given number of organisms at a set time (usually 10 min; D_{10}).

In food canning, the time–temperature profile must be calculated for each size container, for each food type, and for each retort used. When done correctly, these time–temperature conditions provide a large margin of safety since one rarely knows the numbers and types of microbes in a given container, but one must assume that *C. botulinum* is present. To insure safety, inoculated pack studies are done using *Clostridium sporogenes* Metchnikoff PA 3679, which is 6 times more heat-resistant than *C. botulinum*. A known number of PA 3679 is added to cans fitted with thermocouples. Cans are then processed to 120°C (250°F) and held for various time periods. Survivors are enumerated to construct a TDC for that particular food and a D-value is calculated. For canned foods, a 12D margin of safety is used. Thus, heat at a given temperature is applied for a time equal to D × 12 log cycle reductions of PA 3679. Therefore, if a can had 10^9 spores, only 1 in 1000 cans would have the theoretical possibility of a single viable remaining spore. Thus, the probability of survival for *C. botulinum* would be 1 in 10^{12} if a can is heated at 120°C for 3 min. A minimum botulinum cook is one where every particle of food in a container reaches 120°C and remains at that temperature for at least 3 min.

Several factors affect heat transfer and penetration into food packages. Food type (liquids, solids, size, and shape) determine mixing effects during heating. Heat transfer by conduction occurs in solid foods (such as pumpkin) and results in slow heat transfer because there is no mixing of contents. Convection gives liquids (juice) faster heat transfer due to mixing by currents or mechanical agitation. Combination of conduction and convection is observed with particles suspended in liquid (such as peas), though heating is primarily by convection and depends on the viscosity of the liquid component. Container size, shape, and composition are important. Tall thin cans transfer heat faster than short round cans. Large cans require more time than small cans. Metal (tin, steel, and aluminum) containers transfer heat faster than glass, resulting in shorter process times. Plastics can have rapid heat transfer due to thinness. Retort pouches, which are laminates of foil and plastic, have rapid heat transfer; however, pinhole problems can occur. Preheating foods prior to filling containers and preheating retorts will shorten the process time. Rotation or agitation of cans during processing increases convection, providing faster heating.

Canning is the preservation of foods in hermetically sealed containers, usually by heat treatments. The typical sequence in canning is as follows. Freshly harvested good quality foods are washed to remove soils. Next, a blanch, or mild heat treatment, is applied to set color of fruits and vegetables, inactivate enzymes, purge dissolved gases, and to kill some microorganisms. Clean containers are then partially filled, leaving some head space. Hot packing is filling of preheated food to provide faster processing, although cold packing can be done. Containers are sealed under vacuum then placed into a retort. The retort is sealed and heated with pressurized steam. After heating, cans should be rapidly cooled to avoid overcooking and to prevent growth of thermophiles. Cooling is done by submerging cans in a sanitized water bath, which can cause problems if pinhole leaks are present in the cans, allowing water to enter containers.

Less severe heat processing is pasteurization, which usually involves heating to less than 100°C. Pasteurization has two purposes, to destroy all pathogens normally present in a product and to reduce numbers of spoilage microorganisms. This thermal process kills some but not all species of microorganisms present in the food. Pasteurization is used when more rigorous heat treatments might alter food quality. For example, overheated milk will coagulate, brown, and burn. Pasteurization should kill all pathogens normally associated with the product. This is useful when spoilage microorganisms are not heat resistant and when surviving microbes can be controlled by other methods. Another reason for pasteurization is to kill competing microorganisms to allow for a desirable fermentation with starter cultures. Pasteurization is used to manufacture cheeses, wines, and beers. Milk pasteurization may use three equivalent treatments. Low-temperature long time (LTLT) treatment uses 63°C (145°F) for 30 min. High-temperature short time (HTST) uses 72°C (161°F) for 15 s. Ultra high temperature or ultrapasteurized (UHT) uses 138°C for only 2 s. UHT processes are used for shelf-stable products.

Cooking temperatures involve heating of products to at or below 100°C, even though heating media may be much hotter. Baking, roasting, simmering, boiling, and frying are examples of cooking methods. All pathogens are usually killed except sporeformers. Microwaving does not exceed 100°C and can result in uneven heating. Microwave cooking should allow an equilibration time after removal from the oven for more even heating (Fruin and Guthertz, 1982; Sawyer et al., 1983).

Low-Temperature Preservation

Low temperatures retard chemical reactions and refrigeration slows microbial growth rates. Freezing prevents growth of most microorganisms by lowering water activity. Several psychrotrophic pathogens (*L. monocytogenes, Y. enterocolitica,* and nonproteolytic *C. botulinum*) are able to multiply at refrigeration temperatures (Lechowich, 1988). Among factors influencing chill storage, temperature of the compartment is critical (Scott, 1989). Temperature of food products should be held as low as possible. Relative humidity should be high enough to prevent dehydration but not too high to favor growth of microorganisms. Air velocity in coolers helps to remove odors, control humidity, and maintain uniform temperatures.

Atmosphere surrounding food during chill storage can affect microbial growth. Modified atmosphere packaging can help ensure the safety of chill-stored foods. Some plant foods respire resulting in removal of O_2 and release of CO_2. Ultraviolet irradiation can be used to kill microorganisms on surfaces and in the air during chill storage of foods.

For chill storage to be effective in controlling microorganisms, the rate of cooling should be rapid. Temperature should be maintained as low as possible for refrigerated foods (less than 4.5°C). Thawing of frozen foods presents special problems because drip loss provides ample nutrients for microorganisms. In addition, thawing should be done as rapidly as possible and the food used as quickly as possible to avoid opportunity for microbial growth. Often, thawing is done at room temperature over many hours, which can lead to exposure of surfaces to ambient temperatures for extended periods. Another problem is incomplete thawing of large food items (turkeys). By cooking a large item that is not completely thawed, the internal temperature may not reach lethal levels to kill even the most heat-ensitive enteric pathogen. In fact, a spike in the number of salmonellosis and camplyobacteriosis outbreaks occurs every Thanksgiving and Christmas holidays because of consumption of undercooked turkey and stuffing.

Drying

Foods can be preserved by removing or binding water. Any treatment that lowers water activity can reduce or eliminate growth of microorganisms. Some examples include sun drying, heating, freeze drying, and addition of humectants. Humectants act not by removing water but rather by binding to water making it unavailable to act as a solvent. Humectants in common use are salt, sugars, and sugar alcohols (sorbitol). Intermediate moisture foods are those that have 20–40% moisture and a water activity of 0.75–0.85. Examples include soft candies, jams, jellies, honey, pepperoni, and country ham. These foods often require antifungal agents for complete stability.

Preservatives

Food preservatives can be extrinsic (intentionally added), intrinsic (normal constituent of food), or developed (produced during fermentation) (Potter and Hotchkiss, 1995; Jay, 1996). Factors affecting preservative effectiveness include: (1) concentration of inhibitor, (2) kind, number, and age of microorganisms (older cells more resistant), (3) temperature, (4) time of exposure (if long enough. some microbes can adapt and overcome inhibition), and (5) chemical and physical characteristics of food (water activity, pH, solutes, etc.). Preservatives that are cidal are able to kill microorganisms when large concentrations of the substances are used. Static activity results when sublethal concentrations inhibit microbial growth.

Some examples of inorganic preservatives are sodium chloride (NaCl), nitrate and nitrite salts, sulfites, and sulfur dioxide (SO_2). NaCl lowers water activity and causes plasmolysis by withdrawing water from cells. Nitrites and nitrates are curing agents for meats (hams, bacons, sausages, etc.) to inhibit *C. botulinum* under vacuum packaging conditions. Sulfur dioxide (SO_2), sulfites (SO_3), bisulfite (HSO_3), and metabisulfites (S_2O_5) form sulfurous acid in aqueous solutions, which is the antimicrobial agent. Sulfites are widely used in the wine industry to sanitize equipment and reduce competing microorganisms. Wine yeasts are resistant to sulfites. Sulfites are also used in dried fruits and some fruit juices. Sulfites have been used to prevent enzymatic and nonenzymatic browning in some fruits and vegetables (cut potatoes).

Nitrites can react with secondary and tertiary amines to form potentially carcinogenic nitrosamines during cooking; however, current formulations greatly reduce this risk. Nitrates in high concentrations can result in red blood cell functional impairment; however, at approved usage levels they are safe (Nitrite Safety Council, 1980; Hotchkiss and Cassens, 1987). Sulfiting agents likewise can cause adverse respiratory effects to susceptible consumers, particularly asthmatics (Stevenson and Simon, 1981; Schwartz, 1983). Therefore, use of these two classes of agents is strictly regulated.

A number of organic acids and their salts are used as preservatives. These include lactic acid and lactates, propionic acid and propionates, citric acid, acetic acid, sorbic acid, and sorbates, benzoic acid and benzoates, and methyl and propyl parabens (benzoic acid derivatives). Benzoates are most effective when undissociated; therefore, they require low pH values for activity (2.5–4.0). The sodium salt of benzoate is used to improve solubility in foods. When esterified as parabens, benzoates are active at higher pH values. Benzoates are primarily used in high-acid foods (jams, jellies, juices, soft drinks, ketchup, salad dressings, and margarine). They are active against yeast and molds, but minimally so against bacteria. They can be used at levels up to 0.1%.

Sorbic acid and sorbate salts (potassium most effective) are effective at pH values less than 6.5 but at a higher pH than benzoates. Sorbates are used in cheeses, baked or nonyeast goods, beverages, jellies, jams, salad dressings, dried fruits, pickles, and margarine. They inhibit yeasts and molds, but few bacteria except *C. botulinum*. They prevent yeast growth during vegetable fermentations and can be used at levels up to 0.3%.

Propionic acid and propionate salts (calcium most common) are active against molds at pH values less than 6. They have limited activity against yeasts and bacteria. They are widely used in baked products and cheeses. Propionic acid is found naturally in Swiss cheese at levels up to 1%. Propionates can be added to foods at levels up to 0.3%.

Acetic acid is found in vinegar at levels up to 4–5%. It is used in mayonnaise, pickles, and ketchup, primarily as a flavoring agent. Acetic acid is most active against bacteria, but has some yeast and mold activity, though less active than sorbates or propionates. Lactic acid, citric acid, and their salts can be added as preservatives, to lower pH, and as flavorants. They are also developed during fermentation. These organic acids are most effective against bacteria.

Some antibiotics may be found in foods, although medical compounds are not allowed in human food, trace amounts used for animal therapy may occasionally be found. Bacteriocins, which are antimicrobial peptides produced by microorganisms, can be found in foods. An example of an approved bacteriocin is nisin, which is allowed in process cheese food as an additive. Some naturally occurring enzymes (lysozyme and lactoferrin) can be used as preservatives in limited applications where denaturation is not an issue. Some spices, herbs, and essential oils have antimicrobial activity, but such high levels are needed that the food becomes unpalatable. Ethanol has excellent preservative ability but is underutilized because of social stigma. Wood smoke, whether natural or added in liquid form, contains several phenolic antimicrobial compounds in addition to formaldehyde. Wood smoke is most active against vegetative bacteria and some fungi. Bacterial endospores are resistant. Activity is correlated with phenolic content. Carbon dioxide gas can dissolve in food tissues to lower pH and inhibit microbes. Developed preservatives produced during fermentation include organic acids (primarily lactic, acetic, and propionic), ethanol, and bacteriocins. All added preservatives must meet government standards for direct addition to foods. All preservatives added to foods are GRAS.

Irradiation

Foods can be processed or preserved with a number of types of radiation. Nonionizing radiations used include ultraviolet, microwave, and infrared. These function by exciting molecules. Ionizing radiations include gamma, X-rays, β-rays, protons, neutrons, and α-particles. Neutrons make food radioactive, while β-rays (low-energy electrons), protons, and α-particles have little penetrating ability and are of little practical use in foods. Ionizing gamma, X-rays, and high-energy electrons produce ions by breaking molecules and can be lethal to microorganisms.

Ultraviolet (260 nm) lamps are used to disinfect water, meat surfaces, utensils, air, walls, ceilings, and floors. UV can control film yeasts in brines during vegetable fermentations. UV effectiveness is dose dependant. Longer exposure time increases effectiveness. UV intensity depends on lamp power, distance to object, and amount of interfering material in path. For example, humidity greater than 60% reduces intensity. UV will not penetrate opaque materials and is good only for surface decontamination. Infrared heats products, but has little penetrating power. Microwaves cause rapid oscillation of dipole molecules (water) and results in the production of heat. Microwaves have excellent penetrating power. However, there are problems with the time–temperature relationship because microwaves cause foods to reach hot temperatures too quickly. Also, microwave-treated foods rarely exceed 100°C. Thus, instances of microbial survival in these foods has been reported (Fruin and Guthertz, 1982; Sawyer et al., 1983).

X-rays have excellent penetrating ability but are quite expensive. They are not widely used in the food industry. Gamma rays from radioactive sources (Cs^{135} and Co^{150}) have good penetration and are widely used to pasteurize and sterilize foods. Electron-beam generators also are gaining appeal as ionizing sources of radiation to process foods. Food irradiation is much more widespread in countries other than the United States. There is much untapped potential to use ionizing radiations to reduce or eliminate microbial pathogens in foods (Ingram and Roberts, 1980; Radomyski et al., 1994). This technology remains underexploited due to consumer weariness about the safety of the technology (WHO, 1981; Institute of Food Technologists, 1983; Skala et al., 1987).

Fermentation

A number of foods use beneficial microorganisms in the course of their processing (Jay, 1996). Bread, cheeses, pickles, sauerkraut, some sausages, and alcoholic beverages are made by the conversion of sugar to organic acids, ethanol, or carbon dioxide. These three by-products not only serve as desirable flavors but also provide a significant antimicrobial barrier to pathogens. There have been instances where poorly fermented foods have been linked to foodborne illness. Furthermore, cheese made from unpasteurized milk has a distinctly higher risk of carrying pathogens than cheese made from pasteurized milk. Proper acid development and avoidance of cross-contamination are essential control steps in manufacturing fermented foods. Alcoholic beverages have not been linked to foodborne disease other than excess consumption leading to ethanol toxicity.

CONCLUSIONS

The intent of food processing is to deliver safe and wholesome products to the consumer. Basic food safety programs, including GMPs and sanitation, are the minimum requirements to achieve this goal. A comprehensive food-safety plan that includes HACCP is a logical extension of these programs, and focuses on the prevention of hazards before they occur, rather than waiting for a failure to occur, and then addressing the problem. Preventative controls provide the most comprehensive approach to food safety in the processing environment, but they are not fool-proof. Perhaps the most challenging aspect is that, even with the best designed and implemented food safety plan, it may not always be possible to "prevent, eliminate or reduce to acceptable levels" the pathogen of concern. This is particularly true with foods that are purchased by the consumer in their raw state, and then cooked. A specific example is *E. coli* O157:H7 in ground beef. Irrespective of the preventative efforts of the processor, it is not possible to assure that the product is free of the bacterium, and there is no "acceptable level" of this organism in ground beef.

Because of the predominance of hazardous biological contaminants found in raw foods, most food processing unit operations are designed to reduce or eliminate these hazards. Successful implementation of these processing steps can greatly minimize the risk of foodborne disease transmission associated with insect based foods. Unsuccessful implementation or failure to recognize the need for interventions sets the stage for production of potentially dangerous products. Due to the varied nature of insects and foods that contain insect-based ingredients, it is imperative that prudent processors understand the inherent risks of their products and ensure the proper application of interventions to reduce these risks. This fundamentally sound recommendation will help move processed insects as a competitive product in the marketplace and will help maintain and enhance consumer confidence in the safety of their food supply.

REFERENCES

Agabou, A., Alloui, N., 2010. Importance of *Alphitobius diaperinus* (Panzer) as a reservoir for pathogenic bacteria in Algerian broiler houses. Vet. World 3, 71–73.

Altekruse, S.F., Swerdlow, D.L., 2002. *Campylobacter jejuni* and related organisms. In: Cliver, D.O., Riemann, H.P. (Eds.), Foodborne Diseases. 2nd Ed. Elsevier Science Ltd., London, UK, pp. 103–112.

Amadi, E.N., Ogbalu, O.K., Barimalaa, I.S., Pius, M., 2005. Microbiology and nutritional composition of an edible larva (*Bunaea alcinoe* Stoll) of the Niger Delta. J. Food Safety 25, 193–197.

American Meat Institute Foundation, 1994. HACCP. The Hazard Analysis Critical Control Point System in the Meat and Poultry Industry. American Meat Institute Foundation, Washington, DC, USA.

Belgian Scientific Committee of the Federal Agency for the Safety of the Food Chain, 2014. Food safety aspects of insects intended for human consumption. Common advice of the Belgian Scientific Committee of the Federal Agency for the Safety of the Food Chain (FASFC) and of the Superior Health Council (SHC). Available from: http://www.favv-afsca.fgov.be/scientificcommittee/advices/_documents/ADVICE14-2014_ENG_DOSSIER2014-04.pdf

Belluco, S., Losasso, C., Maggioletti, M., Alonzi, C.C., Paoletti, M.G., Ricci, A., 2013. Edible insects in a food safety and nutritional perspective: a critical review. Comp. Rev. Food Sci. Food Safety 12, 296–313.

Biehl, M.L., Buck, W.B., 1989. Chemical contaminants: their metabolism and their residues. J. Food Prot. 50, 1058–1073.

Braide, W., Oranusi, S., Udegbunam, L.I., Oguoma, O., Akobondu, C., Nwaoguikpe, R.N., 2011. Microbiological quality of an edible caterpillar of an emperor moth, *Bunaea alcinoe*. J. Ecol. Nat. Environ. 3, 176–180.

Bryan, F.L., 1988. Risks of practices, procedures and processes that lead to outbreaks of foodborne diseases. J. Food Prot. 51, 663–673.

Cappellozza, S., Saviane, A., Tettamanti, G., Squadrin, M., Vendramin, E., Paolucci, P., Franzetti, E., Squartini, A., 2011. Identification of *Enterococcus mundtii* as a pathogenic agent involved in the "flacherie" disease in *Bombyx mori* L. larvae reared on artificial diet. J. Invert. Pathol. 106, 386–393.

Chai, J.Y., Shin, E.H., Lee, S.H., Rim, H.J., 2009. Foodborne intestinal flukes in South-east Asia. Korean J. Parasitol. 47, 69–102.

Chu, F.S., 2002. Mycotoxins. In: Cliver, D.O., Riemann, H.P. (Eds.), Foodborne Diseases. 2nd Ed. Elsevier Science Ltd., London, UK, pp. 271–304.

Cliver, D.O., 1994a. Viral foodborne disease agents of concern. J. Food Prot. 57, 176–178.

Cliver, D.O., 1994b. Epidemiology of viral foodborne diseases. J. Food Prot. 57, 263–266.

Cliver, D.O., 2002a. Infrequent microbial infections. In: Cliver, D.O., Riemann, H.P. (Eds.), Foodborne Diseases. 2nd Ed. Elsevier Science Ltd., London, UK, pp. 151–159.

Cliver, D.O., 2002b. Viruses. In: Cliver, D.O., Riemann, H.P. (Eds.), Foodborne Diseases. 2nd Ed. Elsevier Science Ltd., London, UK, pp. 161–175.

Cliver, D.O., Riemann, H.P., 2002. Foodborne Diseases, 2nd ed. Elsevier Science Ltd., London, UK.

Department of Agriculture, Food Safety and Inspection Service: Pathogen reduction; hazard analysis and critical control point (HACCP) systems; action: Final rule, July 25, 1996. 9 CFR Parts 304, 308, 310, 320, 327, 381, 416, and 417. Federal Register: vol. 61, Number 144, 38805.

Department of Agriculture, Food Safety and Inspection Service: Agency Mission and Organization. Code of Federal Regulations, Title 9, Animals and Animal Products, 2003a. Part 300, 2003.

Department of Agriculture, Food Safety and Inspection Service: Inspection of eggs and egg products (Egg Products Inspection Act). Code of Federal Regulations, Title 9, Animals and Animal Products, 2003b. Part 590, 2003.

Department of Health and Human Services, Food and Drug Administration: Procedures for the safe and sanitary processing and importing of fish and fishery products; final rule, December 18, 1995. 21 CFR Parts 123 and 1240. Federal Register: Vol 60, No. 242, 65096.

Domingo, J.W.S., Kaufman, M.G., Klug, M.J., Tiedje, J.M., 1998. Characterization of the cricket hindgut microbiota with fluorescently labeled rRNA-targeted oligonucleotide probes. Appl. Environ. Microbiol. 64, 752–755.

Dubey, J.P., Murrell, K.D., Cross, J.H., 2002. Parasites. In: Cliver, D.O., Riemann, H.P. (Eds.), Foodborne Diseases. 2nd Ed. Elsevier Science Ltd., London, UK, pp. 177–190.

European Food Safety Authority, 2015. In: Risk profile related to production and consumption of insects as food and feed. EFSA J. 13 (10), 4257.

Farber, J.M., Peterkin, P.I., 1991. *Listeria monocytogenes*, a food-borne pathogen. Microbiol. Rev. 55, 476–511.

FDA/MIF/IICA, 1988. Recommended guidelines for controlling environmental contamination in dairy plants. Dairy Food Environ. Sanitation 8, 52–56.

Food and Agriculture Organization: Understanding the Codex Alimetarius. Available from: http://www.fao.org/docrep/w9114e/w9114e00.htm, 2003

Food and Drug Administration: Guide to minimize microbial food safety hazards for fresh fruits and vegetables, 1998. Available from: http://www.cfsan.fda.gov/~dms/prodguid.html

Food and Drug Administration: Grade "A" Pasteurized Milk Ordinance 2001 Revision, 2002. Available from: http://vm.cfsan.fda.gov/~ear/pmo01toc.html/

Food and Drug Administration, Department of Health and Human Services: Product Jurisdiction. Code of Federal Regulations, Title 21, Food and Drugs, Part 3, 2003a.

Food and Drug Administration, Department of Health and Human Services: Current Good Manufacturing Practice in manufacturing, packing, or holding human food. Code of Federal Regulations, Title 21, Food and Drugs, Part 110, 2003b.

Fratamico, P.M., Smith, J.L., Buchanan, R.L., 2002. Escherichia coli. In: Cliver, D.O., Riemann, H.P. (Eds.), Foodborne Diseases. 2nd Ed. Elsevier Science Ltd., London, UK, pp. 79–101.

Fruin, J.T., Guthertz, L.S., 1982. Survival of bacteria in food cooked by microwave oven, conventional oven, and slow cookers. J. Food Prot. 45, 695–698.

Gabis, D., Faust, R.E., 1988. Controlling microbial growth in food processing environments. Food Technol. 42 (12), 81–83.

Gashe, Berhanu A., et al., 1997. The microbiology of phane, an edible caterpillar of the emperor moth, Imbrasia belina. J. Food Prot. 60 (1), 1376–1380.

Gecan, J.S., Cichowicz, S.M., 1993. Toxic mushroom contamination of wild mushrooms in commercial distribution. J. Food Prot. 56, 730–734.

Giaccone, V., 2005. Hygiene and health features of mini livestock. In: Paoletti, M.G. (Ed.), Ecological implications of minilivestock: role of rodents, frogs, snails and insects for sustainable development. Science Publishers Inc., Enfield, New Hampshire, pp. 579–598.

Graczyk, T.K., Knight, R., Tamang, L., 2005. Mechanical transmission of humanprotozoan parasites by insects. Clin. Microbiol. Rev. 18, 128–132.

Gray, J.T., Fedorka-Cray, P.J., 2002. Salmonella. In: Cliver, D.O., Riemann, H.P. (Eds.), Foodborne Diseases. second ed. Elsevier Science Ltd., London, UK, pp. 53–68.

Griffiths, M.W., Schraft, H., 2002. *Bacillus cereus* food poisoning. In: Cliver, D.O., Riemann, H.P. (Eds.), Foodborne Diseases. second ed. Elsevier Science Ltd., London, UK, pp. 261–270.

Hackney, C.R., Dicharry, A., 1988. Seafood-borne bacterial pathogens of marine origin. Food Technol. 42 (3), 104–109.

Harein, Philip K., et al., 1970. Salmonella spp. and serotypes of Escherichia coli isolated from the lesser mealworm collected in poultry brooder houses. J. Econ. Entomol. 63 (1), 80–82.

Harris, L.J., 2002. Listeria monocytogenes. In: Cliver, D.O., Riemann, H.P. (Eds.), Foodborne Diseases. second ed. Elsevier Science Ltd., London, UK, pp. 137–150.

Hirose, E., Panizzi, A.R., Cattelan, A.J., 2006. Potential use of antibiotic to improve performance of laboratory-reared *Nezara viridula* (L.) (Heteroptera: Pentatomidae). Neotrop. Entomol. 35, 279–281.

Holmberg, S.D., 1992. Cholera and related illnesses caused by Vibrio species and *Aeromonas*. In: Gorbach, S.L., Bartlett, J.G., Blacklow, N.R. (Eds.), Infectious Disease. WB Saunders Co., Philadelphia, PA, USA, pp. 605–611.

Hotchkiss, J.H., Cassens, R.G., 1987. Nitrate, nitrite, and nitroso compounds in foods. Food Technol. 41 (4), 127–134.

Humphrey, J.J., Baskerville, A., Mawer, S., Rowe, B., Hopper, S., 1989. *Salmonella enteritidis* phage type 4 from the contents of intact eggs: a study involving naturally infected hens. Epidemiol. Infect. 103, 415–423.

Ingham, S.C., Ingham, B.H., Buege, D.R., 1996. Sanitation Programs and Standard Operating Procedures for Meat and Poultry Plants. American Association of Meat Processors, Elizabethtown, PA, USA.

Ingram, M., Roberts, T.A., 1980. Ionizing irradiation. Microbial Ecology of Foods, vol. I, Academic Press, New York, NY, USA, pp. 46–47.

Institute of Food Technologists, 1983. Radiation preservation of foods. Food Technol. 37, 55–60.

Jay, J.M., 1996. Modern Food Microbiology, fifth ed. Chapman and Hall, New York, NY, USA.

Jeandron, A., Rinaldi, L., Abdyldaieva, G., Usubalieva, J., Steinmann, P., Cringoli, G., Utzinger, J., 2011. Human infections with *Dicrocoelium dendriticum* in Kyrgyzstan: the tip of the iceberg? Int. J. Parasitol. 97, 1170–1172.

Johnson, E.A., Schantz, E.J., 2002. Seafood toxins. In: Cliver, D.O., Riemann, H.P. (Eds.), Foodborne Diseases. 2nd Ed. Elsevier Science Ltd., London, UK, pp. 211–229.

Kapperud, G., 2002. Yersinia enterocolitica. In: Cliver, D.O., Riemann, H.P. (Eds.), Foodborne Diseases. 2nd Ed. Elsevier Science Ltd., London, UK, pp. 113–118.

Kaufman, Michael, G., Michael, J. Klug., 1991. The contribution of hindgut bacteria to dietary carbohydrate utilization by crickets (Orthoptera: Gryllidae). Comp. Biochem. Physiol. Part A Physiol. 98.1, 117–123.

Klunder, H.C., Wolkers-Rooijackers, J., Korpela, J.M., Nout, M.J.R., 2012. Microbiological aspects of processing and storage of edible insects. Food Control 26, 628–631.

Labbe, R.G., Juneja, V.K., 2002. Clostridium perfringens. In: Cliver, D.O., Riemann, H.P. (Eds.), Foodborne Diseases. second ed. Elsevier Science Ltd., London, UK, pp. 119–126.

Lampel, K.A., Maurelli, A.T., 2002. Shigella. In: Cliver, D.O., Riemann, H.P. (Eds.), Foodborne Diseases. second ed. Elsevier Science Ltd., London, UK, pp. 69–77.

Lechowich, R.V., 1988. Microbiological challenges of refrigerated foods. Food Technol. 42 (12), 84–89.

Leffer, A.M., Kuttel, J., Martins, L.M., Pedroso, A.C., Astolfi-Ferreira, C.S., Ferreira, F., Ferreira, A.J., 2010. Vectorial competence of larvae and adults of *Alphitobius diaperinus* in the transmission of *Salmonella* Enteritidis in poultry. Vector-borne Zoonot. 10, 481–487.

Liu, Yu-Sheng, et al., 2011. Isolation and identification of intestinal bacterial flora of yellow mealworm. Chinese J. Microecol. 23 (10), 894–896.

Marriott, N.G., 1997. Pest control. In: Marriott, N.G. (Ed.), Essentials of Food Sanitation. Chapman and Hall, New York, NY, USA, pp. 129–149.

Marriott, N.G., 1999a. Personal hygiene and sanitary food handling. In: Marriott, N.G. (Ed.), Principles of Food Sanitation. fourth ed. Aspen, Gaithersburg, MD, USA, pp. 60–74.

Marriott, N.G., 1999b. Cleaning compounds. Principles of Food Sanitation. fourth ed. Aspen, Gaithersburg, MD, USA, pp. 114–138.

Marriott, N.G., 1999c. Sanitizers. In: Marriott, N.G. (Ed.), Principles of Food Sanitation. fourth ed. Aspen, Gaithersburg, MD, USA, pp. 139–157.

Mead, P.S., Slutsker, L., Dietz, V., McCaig, L.F., Bresee, J.S., Shapiro, C., Griffin, P.M., Tauxe, R.V., 1999. Food-related illness and death in the United States. Emerg. Infect. Dis. 5, 607–625.

National Advisory Committee on Microbiological Criteria for Foods, 1998. Hazard analysis and critical control point principles and applications guidelines. J. Food Prot. 61, 1246–1259.

NVWA (Netherlands Food and Consumer Product Safety Authority), 2014. Advisory report on the risks associated with the consumption of mass-reared insects. Available from: http://www.nvwa.nl/actueel/risicobeoordelingen/bestand/2207475/consumptie-gekweekte-insecten-advies-buro

Nitrite Safety Council, 1980. A survey of nitrosamines in sausages and dry-cured meat products. Food Technol. 34, 45–53.

Noleto, A.L., Bergdoll, M.S., 1982. Production of enterotoxin by a *Staphylococcus aureus* strain that produces three identifiable enterotoxins. J. Food Prot. 45, 1096–1097.

Olsen S.J., MacKinon L.C., Goulding J.S., Bean N.H., Slutsker, L., 2000. Surveillance for foodborne disease outbreaks—United States, 1993–1997. MMWR 49:No. SS01:1-51.

Padhye, N.V., Doyle, M.P., 1992. *Escherichia coli* O157:H7: Epidemiology, pathogenesis, and methods for detection in food. J. Food Prot. 55, 555–565.

Parkinson, H., Ito, K., 2002. Botulism. In: Cliver, D.O., Riemann, H.P. (Eds.), Foodborne Diseases. 2nd Ed. Elsevier Science Ltd., London, UK, pp. 249–259.

Pereira, N.R., Tarley, C.R.T., Matsushita, M., de Souza, N.E., 2000. Proximate composition and fatty acid profile in Brazilian poultry sausages. J. Food Comp. Anal. 13, 915–920.

Pierson, M.D., Corlett, D.A., 1992. HACCP: Principles and Applications. Chapman and Hall, New York, NY, USA.

Popovic, T., Olsvik, O., Blake, P.A., Wachsmuth, K., 1993. Cholera in the Americas: foodborne aspects. J. Food Prot. 56, 811–821.

Post, K., Riesner, D., Walldorf, V., Mehlhorn, H., 1999. Fly larvae and pupae as vectors for scrapie. Lancet 354, 1969–1970.

Potter, N.N., Hotchkiss, J.H., 1995. Food Science, fifth ed. Chapman and Hall, New York, NY, USA.

Radomyski, T., Murano, E.A., Olson, D.G., Murano, P.S., 1994. Elimination of pathogens of significance in food by low-dose irradiation: a review. J. Food Prot. 57, 73–86.

Recommendations by the National Advisory Committee on Microbiological Criteria for Foods, 1992. Microbiological criteria for raw molluscan shellfish. J. Food Prot. 55, 463–480.

Sakazaki, R., 2002. Vibrio. In: Cliver, D.O., Riemann, H.P. (Eds.), Foodborne Diseases. second ed. Elsevier Science Ltd., London, UK, pp. 127–136.

Sawyer, C.A., Naidu, Y.M., Thompson, S., 1983. Cook/chill foodservice systems: microbiological quality and endpoint temperature of beef loaf, peas and potatoes after reheating by conduction, convection and microwave radiation. J. Food Prot. 46, 1036–1043.

Scallan, E., Hoekstra, R.M., Angulo, F.J., Tauxe, R.V., Widdowson, M.-A., Roy, S.L., et al., 2011. Foodborne illness acquired in the United States—major pathogens. Emerg. Infect. Dis. 17, 7–15.

Schabel HG, 2010. Forest insects as food: A global review. In: *Forest insects as food: humans bite back*. FAO Regional Office for Asia and the Pacific, Bangkok, Thailand. Proceedings of a workshop on Asia-Pacific resources and their potential for development. 19–21, Chiang Mai, Thailand, pp. 37–64.

Schwartz, H.J., 1983. Sensitivity to ingested metabisulfite: variations in clinical presentation. J. Allergy Clin. Immunol. 71, 487–489.

Scott, V.N., 1989. Interaction of factors to control microbial spoilage of refrigerated foods. J. Food Prot. 52, 431–435.

Simpanya, M.F., Allotey, J., Mpuchane, S.F., 2000. A mycological investigation of phane, an edible caterpillar of an emperor moth, *Imbrasia belina*. J. Food Prot. 63, 137–140.

Skala, J.H., McGown, E.L., Waring, P.P., 1987. Wholesomeness of irradiated foods. J. Food Prot. 50, 150–160.

Smith, J.L., Fratamico, P.M., 1995. Factors involved in the emergence and persistence of food-borne diseases. J. Food Prot. 58, 696–716.

Snowdon, J.A., Buzby, J.C., Roberts, T.A., 2002. Epidemiology, cost, and risk of foodborne disease. In: Cliver, D.O., Riemann, H.P. (Eds.), Foodborne Diseases. second ed. Elsevier Science Ltd., London, UK, pp. 31–51.

Stevenson, K.E., Bernard, D.T., 1995. HACCP: Establishing Hazard Analysis Critical Control Point Programs. The Food Processors Institute, Washington, DC, USA.

Stevenson, D.D., Simon, R.A., 1981. Sensitivity to ingested metabisulfites in asthmatic subjects. J. Allergy Clin. Immunol. 68, 26.

Tacket, C.O., Brenner, F., Blake, P.A., 1984. Clinical features and an epidemiological study of *Vibrio vulnificus* infections. J. Infect. Dis. 149, 558–561.

Tarr, P.I., 1994. *Escherichia coli* O157:H7: overview of clinical and epidemiological issues. J. Food Prot. 57, 632–637.

Tauxe, R.V., 1991. *Salmonella*: a postmodern pathogen. J. Food Prot. 54, 563–568.

Taylor, S.L., 2002. Chemical intoxications. In: Cliver, D.O., Riemann, H.P. (Eds.), Foodborne Diseases. second ed. Elsevier Science Ltd., London, UK, pp. 305–316.

Todd, E.C.D., 1992. Foodborne disease in Canada: a 10-year summary from 1975–1984. J. Food Prot. 55, 123–132.

Ulrich, R.G., Buthala, D.A., Klug, M.J., 1981. Microbiota associated with the gastrointestinal tract of the common house cricket, *Acheta domestica*. Appl. Envir. Microbiol. 41, 246–254.

US FDA, 2015a. FSMA Final Rule for Preventive Controls for Human Food. Available from: http://www.fda.gov/Food/GuidanceRegulation/FSMA/ucm334115.htm

US FDA, 2015b. Food Allergies: What You Need to Know. Available from: http://www.fda.gov/Food/IngredientsPackagingLabeling/FoodAllergens/ucm079311.htm

van Huis, A., Itterbeeck, J.V., Klunder, H., Mertens, E., Halloran, A., Muir, G.,Vantomme, P., 2013. Food and Agriculture Organization of the United Nations. Edible Insects: Future Prospects for Food and Feed Security, Food and Agriculture Organization of the United Nations, Rome, Italy.

Wei, J., Jin, Y., Sims, T., Kniel, K.E., 2010. Survival of murine norovirus and hepatitis A virus in different manure and biosolids. Foodborne Pathog. Dis. 7, 901–906.

WHO: Wholesomeness of Irradiated Food, 1981. World Health Organization Technical Report Series, No. 659. Geneva, Switzerland.

Wong, A.C.L., Bergdoll, M.S., 2002. Staphylococcal food poisoning. In: Cliver, D.O., Riemann, H.P. (Eds.), Foodborne Diseases. second ed. Elsevier Science Ltd., London, UK, pp. 231–248.

Yang, Jun, et al., 2014. Evidence of polyethylene biodegradation by bacterial strains from the guts of plastic-eating waxworms. Environ. Sci. Tech. 48.23, 13776–13784.

Chapter 9

Insects and Their Connection to Food Allergy

M. Downs, P. Johnson, M. Zeece
Department of Food Science and Technology, Food Allergy Research and Resource Program, University of Nebraska-Lincoln, Food Innovation Center, Lincoln, NE, United States

Chapter Outline

Introduction	255	Hemocyanin	263
Food Allergy	255	Phospholipase	264
Insects and Food Allergy	257	Other Allergens	264
Insect Allergens	259	Known Aero-Allergens	264
Tropomyosin	259	Novel Allergens	264
Arginine Kinase	260	Effects of Processing	264
Sarcoplasmic Calcium Binding Protein	261	Methods of Allergen Detection	266
Myosin Light Chain	262	Conclusions	267
Troponin C	263	References	268
Sarcoplasmic Endoreticulum Calcium ATPase	263		

INTRODUCTION

The concept of insects as a sustainable source of protein and other nutrients has gained considerable attention in the recent past. Commonly consumed insects such as grasshoppers, crickets, mealworm larvae, and moth larvae contain high levels of good quality protein that is comparable to traditional animal sources such as poultry and beef (Rumpold and Schluter, 2013). On a fresh weight basis, most insects contain as much or more protein as beef but require only a small fraction of the inputs needed to produce it. Insect protein quality is demonstrated in the balance of essential and nonessential amino acids. Surveyed edible insects scored between 46 and 96% for an ideal protein essential amino acid balance (Ramos-Elorduy et al., 1997; Bukkens, 2005; Belluco et al., 2013). While the potential for insect protein as a source of food protein is high, there are some safety issues that have not been widely investigated. Specifically, insect protein allergenicity represents a major hurdle to its widespread use as a human food or food ingredient. While all novel food protein sources have the potential to be allergenic, the structural similarities of insect proteins to known allergens in more widely consumed arthropods presents an additional risk of allergic reactions due to cross-reactivity in individuals with preexisting allergies. For example, individuals with a shrimp allergy could potentially have an allergic reaction upon consumption of other arthropod species such as mealworm (Broekman et al., 2015). Therefore, the purpose of this chapter is to provide an overview of insect protein allergy risk as a food ingredient.

FOOD ALLERGY

Food allergies are serious and potentially life-threatening conditions that affect an estimated 5% of adults and 8% of children (Sicherer and Sampson, 2014). Allergic reactions to foods are responsible for approximately 125,000 emergency department visits, 2000 hospitalizations, and 150–200 deaths each year in the United States alone (Ross et al., 2008; Sampson, 2003; Yocum et al., 1999). While 90% of food allergies are caused by 8 foods (milk, eggs, fish, crustacean

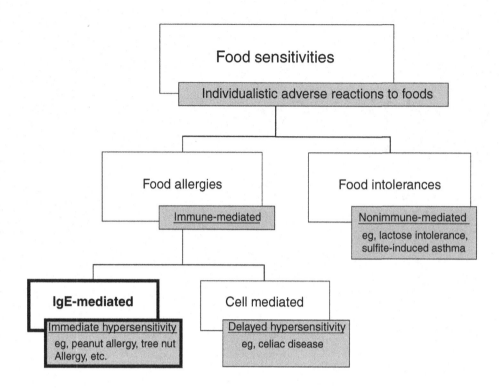

FIGURE 9.1 Food sensitivity classification.

shellfish, peanuts, tree nuts, soybean, and wheat), almost any food can be allergenic, and more than 160 foods have been reported in the literature as causing food allergies (Food and Agriculture Organization, 1995; Hefle et al., 1996). The prevalence of allergies to individual foods can vary geographically, based largely on differences in consumption patterns. As novel protein sources such as insects are introduced into the food supply, issues of allergenicity are critical considerations in order to protect the health of consumers. Individual reactions to foods can be broadly classified as either food allergies or food intolerances, as illustrated in Fig. 9.1 (Taylor and Hefle, 2006).

Food allergies are adverse reactions resulting from immune-mediated responses to a given food and occurring reproducibly upon exposure to that food (Boyce et al., 2010; Schneider Chafen et al., 2010). Food intolerances (eg, lactose intolerance), on the other hand, are nonimmune-mediated sensitivities. Food allergies, and specifically Immunoglobulin E (IgE)-mediated allergies, are the most relevant concern for the potential allergenicity of insect foods. For most individuals, the consumption of foods, even commonly allergenic foods, does not result in allergy due to the development of oral tolerance. Oral tolerance describes the active suppression of systemic immune responses to an antigen when the initial exposure to that antigen occurs via the oral route (Chase, 1946; Chehade and Mayer, 2005). In many respects, the development of food hypersensitivities can be viewed as a breakdown of the oral tolerance mechanisms that function in most individuals for the vast majority of food antigens (Chehade and Mayer, 2005). Interestingly, early (4–11 months) exposure of infants to peanuts reduced the risk of peanut allergy development in later life (Du Toit et al., 2015), indicating a potential route for food allergy prevention.

Upon an initial exposure to an allergenic food, the immune system responds to a perceived risk and produces IgE antibodies specific for the food allergens, which are almost exclusively naturally occurring proteins in foods (Taylor and Hefle, 2006). The allergen-specific IgE then binds to the high affinity Fc receptors on mast cells and basophils (Gould and Sutton, 2008). Upon a subsequent exposure, such as ingestion or contact with the food, the allergens cross-link IgE bound on the mast cells and basophils, which triggers the release of mediator molecules (eg, histamine), and an allergic reaction ensues (Gould and Sutton, 2008; Taylor and Hefle, 2006). A simplified representation of sensitization and elicitation is shown in Fig. 9.2. Symptoms of an allergic reaction to foods can be quite varied and range from localized cutaneous reactions (eg, hives) to systemic responses, including anaphylactic shock. Food-induced anaphylaxis is very serious with 14,000–30,000 reported events annually in the United States (Ross et al., 2008; Yocum et al., 1999). Unfortunately, the only currently approved treatment for food allergies is adherence to strict avoidance diets (Sampson, 2013). If an

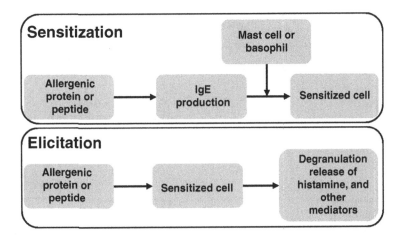

FIGURE 9.2 Basic mechanism of the sensitization and elicitation steps necessary for an IgE-mediated reaction to occur.

unintended exposure occurs, epinephrine can be used to mitigate symptoms and slow the progression of a reaction until additional medical attention can be received (Taylor and Hefle, 2006).

The prevalence and severity of food allergies, along with the necessity for strict avoidance, has led a number of regulatory authorities worldwide to implement food allergen labeling regulations. In the United States, the Food Allergen Labeling and Consumer Protection Act of 2004 (FALCPA) mandates that when any of the following eight foods/food groups or any ingredients derived thereof are intentionally added to a product formulation, they must be declared in plain language in the ingredients list or in a separate contains statement: milk, egg, fish, Crustacean shellfish, peanuts, tree nuts, soybeans, and wheat (United States Congress, 2004). In the case of the food groups (eg, tree nuts), the species of the allergenic food must also be identified on the label. Similar labeling requirements have been adopted by a number of other regulatory bodies, including those in the European Union, Canada, Japan, and Australia (Gendel, 2012). In many cases, the list of priority allergens in these countries are modified to reflect the prevalent food allergies in the specific region.

INSECTS AND FOOD ALLERGY

The introduction of insect-based food ingredients poses two different allergenicity risks. The first source of risk is the potential for cross-reactivity between ingested insects and taxonomically related species to which an individual has an existing allergy. In this case, clinical symptoms of cross-reactivity can occur with homologous proteins between closely related species. This type of cross-reactivity can be seen, for example, among related species of tree nuts (eg, pecans and walnuts) or among related species of crustacean shellfish (eg, shrimp, crab, and lobster). Food allergic symptoms can also occur in individuals sensitized to pollen allergens who consume fruits and vegetables that contain homologous allergens (Vieths et al., 2002).

With respect to insects and foods, much of the relevant cross-reactivity has been documented between insects and crustacean shellfish. Shellfish allergy is prevalent, with self-reported rates of 2 % in the United States (Sicherer et al., 2014). Both insects (Class Insecta) and crustacean shellfish (Subphylum Crustacea) are closely related taxonomically as illustrated in Fig. 9.3, leading to the possibility that crustacean shellfish-allergic individuals may exhibit reactions to similar insect proteins.

In some cases, cross-reactivity can occur between sources that sensitize an individual via different routes, for example inhalation and ingestion. One frequently reported occurrence of this type involves cross-reactivity between house dust mites [HDM, *Dermatophagoides pteronyssinus* (Trouessart)], which are frequently inhalant allergens, and molluscan and crustacean shellfish, which are food allergens (Antonicelli et al., 1992; Ayuso et al., 2002; Carrillo et al., 1992; De Maat-Bleeker et al., 1995; Fernandes et al., 2002; Mistrello et al., 1992; Petrus et al., 1999; Rame et al., 2002; van Ree et al., 1996a, 1996b; Vuitton et al., 1998; Witteman et al., 1994). For example, a number of clinical reports have indicated that allergy to snail can occur in individuals with HDM allergy (Antonicelli et al., 1992; Banzet et al., 1992; De Maat-Bleeker et al., 1995; Rame et al., 2002; van Ree et al., 1996a, 1996b; Vuitton et al., 1998). One well-documented study examined a population of individuals with combined HDM and snail allergy. Radioallergosorbent test (RAST) inhibition analysis indicated that the HDM was the primary sensitizer and demonstrated cross-reactivity between the HDM and snail allergens (van Ree et al., 1996b). Other authors have reported IgE cross-reactivity between HDM, shrimp, cockroach, and

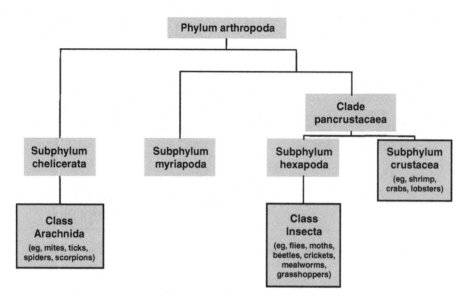

FIGURE 9.3 Taxonomic relationship between arthropods with notable allergen content.

squid (Crespo et al., 1995; Fernandes et al., 2002; Petrus et al., 1999). Some evidence indicates that the cross-reactive allergen is tropomyosin, however, other reports suggest there may be additional allergens responsible for the cross-reactivity between insects and shellfish (Ayuso et al., 2002; van Ree et al., 1996b; Witteman et al., 1994). Additionally, Gamez et al. (2014) reported cross-reacting IgE from both shrimp- and HDM-allergic patients to proteins identified as α-actinin and ubiquitin. While much of the clinical evidence for insect–shellfish cross-reactivity has focused on the shellfish as the ingestion allergen, it is foreseeable that similar reactivity could be observed upon the ingestion of insects with proteins homologous to HDM or other inhalant allergens.

With increased interest in insects as a protein source, more recent studies have focused on the possibility of allergic reactions occurring after ingestion of insects in individuals with existing HDM or shellfish allergy. One recent study evaluated sera from seven subjects allergic to crustacean shellfish and HDM (Verhoeckx et al., 2014). Six of the seven subject sera demonstrated IgE cross-reactivity yellow mealworm (*Tenebrio molitor* L.) extract upon immunoblotting. Five of the subject sera were also used for basophil activation test (BAT), and all five demonstrated activation with mealworm. The authors also evaluated the gastric stability of the mealworm proteins with a pepsin digestion assay and found them to be moderately stable, with incomplete digestion after 60 min and retention of IgE binding. Based on all of these results, the authors concluded a reasonable possibility exists that individuals with HDM and crustacean shellfish allergy may react to foods containing yellow mealworm.

Another recent study investigated potential cross-reactivity between giant freshwater prawns (*Macrobrachium rosenbergii* De Man) and field crickets (*Gryllus bimaculatus* De Geer) (Srinroch et al., 2015). The authors used pooled sera from 16 prawn-allergic subjects and found IgE reactivity to both prawn and cricket proteins by immunoblotting. The major cricket protein recognized by sera from the prawn-allergic sera was arginine kinase. Additionally, hexamerin 1B, an insect storage protein, was identified as a novel but minor cricket allergen. As with the mealworm study, cross-reactivity between prawn and crickets indicates that prawn-allergic subjects may experience an allergic reaction upon consumption of cricket. Given the homology among allergens from insects and shellfish, it could be expected that more examples of this type of cross-reactivity could be found with further study. If unsuspecting shellfish or HDM-allergic individuals were to consume products containing insect ingredients, they may be at risk for quite severe reactions due to the potentially very high dose of allergen ingested. This type of risk due to cross-reactivity should be a concern for processors using these ingredients.

The second source of allergenicity risk related to the introduction of insects as foods or food ingredients is the ability of the ingredient to sensitize individuals themselves and cause food allergy. In regions where edible insects are common, food allergic reactions have been reported. In China, for example, consumption of silkworm (*Bombyx mori* L.) pupa has been estimated to cause over 1000 anaphylactic reactions and result in 50 emergency department visits annually (Ji et al., 2008). While some of these cases are indicative of cross-reactivity, with reactions occurring on the first known ingestion of silkworm, others may be due to sensitization to the silkworm pupa itself. A review of the Chinese allergy literature investigating the causes of food-induced anaphylaxis indicated that insects were the third most common category for causative foods, behind fruits and aquatic products (Ji et al., 2009). Out of the 358 cases of food-induced anaphylaxis found in the literature,

27 were attributed to locusts, 27 to grasshoppers, 5 to silkworm pupa, and 1 each to cicada pupa, bee pupa, bee larva, and the moth species *Clanis bilineata* (Walker).

Severe allergic reactions to insects have also been reported in Thailand (Jirapongsananuruk et al., 2007; Piromrat et al., 2008). Researchers at a university hospital in Thailand reviewed inpatient anaphylaxis cases over a 6-year period and found 24 food-induced cases, with one patient reporting fried insects as the causative food. Another study of emergency department visits at a tertiary care hospital in Thailand found that over 2 years, 36 cases of food-induced anaphylaxis were seen, with 7 of those reactions due to fried insects. In parts of the world where insects are commonly consumed, they do seem to be allergenic and capable of causing severe reactions.

While not consumed directly, the cochineal insect (*Dactylopius coccus* L.) has been linked to allergic reactions through indirect exposure in foods that contain its extracted colorant, carmine. Carmine is a highly stable red pigment widely used in food, textiles, and cosmetics. In food applications, carmine is classified as a natural dye by the Food and Drug Administration in the United States (Natural Red No.4) and by the European Commission (E120). The industrial process of extracting and purifying the dye from the insect is a noted cause of occupational allergy. In particular, allergic reactions such as asthma and/or rhinitis have been reported for individuals working in various aspects of carmine production (Lizaso et al., 2000; Tarbar-Purroy et al., 2003). It is presumed that workers developed carmine allergy via inhalation exposure in the work place. The proteins responsible for development of occupational allergy have not yet been identified. There are also reports of allergic reactions occurring in individuals consuming food containing carmine. Drinking the alcoholic beverage Campari reportedly resulted in an allergic reaction (rhinoconjunctivitis and uticaria) (Wuthrich et al., 1997). Proteins in the carmine colorant used in this drink were proposed as the cause. Similarly, allergic reactions to carmine-containing yogurt and carmine used as coating colorant for azithromycin tablets have been reported (DiCello et al., 1999; Greenhawt et al., 2009).

INSECT ALLERGENS

Most food allergens tend to be abundant in consumed tissues, whether this is a cereal grain, nut, seed or in the muscle tissues of animals. Individual proteins that bind IgE from allergic patients can be identified by standard laboratory methodology such as Western blotting. Proteins thus identified can then be partially sequenced by techniques such as Edman degradation, or, as is becoming increasingly common, by protein mass spectrometry. As proteins capable of binding IgE from food-allergic individuals are identified and presented in the scientific literature, they are catalogued in repositories such as AllergenOnline (www.allergenonline.org). Nomenclature of allergens is assigned by the WHO/IUIS Allergen Nomenclature Sub-Committee and follows the format of the first three letters of the Genus followed by the first letter of the species and a number (eg, shrimp, *Penaeus monodon* Fab., myosin light chain = Pen m 3).

The prediction of allergens in novel foods such as insects is problematic, due largely to the fact that the particular traits of a protein that enable it to sensitize individuals and elicit allergic reaction are unknown. In addition, appropriate animal models for food allergies do not currently exist. The ability of insect proteins to both sensitize and elicit reactions in exposed individuals will only truly become apparent by monitoring allergic reactions in newly exposed populations, that is, through occupational exposure and, indeed, through consumption. Novel food allergens must be identified using the sera, and IgE therein, from food-allergic individuals. The process is therefore retrospective in nature and relies upon the exposure of a human population to the food source in question. However, the close taxonomic relationship between insects and other arthropods, some of which are already well-characterized sources of allergens, allows us to predict those proteins which may cause reactions in those already sensitized to shellfish and other closely related species. Another potential source of cross-reactivity with existing known allergens comes from sensitization to insect bites, stings, and aero-allergens such as those typified by a house dust mite allergy. The likelihood of cross-reaction is governed primarily by the degree of structural similarity between individual allergens and by their abundance. In this section, we will discuss the major allergens of crustacean shellfish and evidence of cross-reaction with similar proteins occurring in insects for human consumption.

Tropomyosin

Tropomyosin is an abundant protein found in both vertebrate and invertebrate animals and plays essential roles in muscle contraction and the cytoskeleton of nonmuscle cells. The sequence of amino acids in tropomyosin is highly conserved throughout the animal kingdom. Vertebrate and invertebrate tropomyosin sequences in fish and shellfish for example, are approximately 50% identical in sequence (Reese et al., 1997; Reese et al., 1999). The native protein is composed of two polypeptide chains arranged in a coiled-coil, rod-like, three-dimensional structure (Perry, 2001). Tropomyosin has a subunit molecular weight of approximately 33 kDa and is almost completely alpha helical in conformation. Short segments of

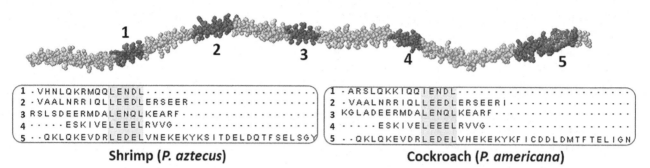

FIGURE 9.4 **Repeat regions of shrimp (*Penaeus aztecus*) and American cockroach (*Penaeus americana*) tropomyosin.** Five regions of similar amino acid sequence were identified in shrimp tropomyosin (Ayuso et al., 2002) and are highlighted in red on the structure of tropomyosin. The repeated regions are thought to be essential for the structural role of tropomyosin. These regions are also present in the cockroach tropomyosin. The amino acid sequences present in these regions are shown, with the repeat regions highlighted. Within each repeat region, three amino acid residues with overall negative charge, for example, glutamic acid (E) or aspartic acid (D) are contained within a pair of hydrophobic residues [here isoleucine (I) or leucine (L)].

random coil occurring principally at the N- and C-terminals are notable exceptions to its alpha helical conformation. End region sequences are responsible for its head-to-tail polymerization in vivo and the ionic strength-dependent polymerization of the purified protein (Spudich and Watt, 1971). In muscle, tropomyosin and troponin form a complex that is bound to the thin filament of myofibrils. This complex enables a calcium-dependent association with myosin cross-bridges on thick filaments during contraction (Perry, 2001). The molecular structure of tropomyosin contains a unique amino acid repeat that is responsible for its structure and function (Fig. 9.4). Repeat sequences provide sites for binding actin, tropomyosin, and the troponin complex within a precisely defined periodicity (Spudich and Watt, 1971; Greenfield and Hitchcock-DeGregori, 1995; Hitchcoock-DeGregori et al., 2002).

Tropomyosin was first identified as a crustacean allergen and is thought to be responsible for the majority of crustacean shellfish allergies (Daul et al., 1994; Shanti et al., 1993; Leung et al., 1994). The high degree of similarity between tropomyosin of different crustaceans means shellfish-allergic individuals are often sensitive to many if not all edible crustacea. Reaction to molluscan shellfish is also a possibility, though some crustacean-allergic individuals can tolerate molluscan shellfish (Leung et al., 1996). Similarly to arginine kinase, cross-reactions with vertebrate tropomyosin (including those from fish) are unlikely due to the limited degree of similarity between invertebrate and vertebrate proteins (Leung et al., 1996; Reese et al., 1996). Tropomyosins from shellfish and insects share a high degree of similarity. There is 75–80% sequence identity between shrimp, house dust mite, and American cockroach (inhalant allergen). This similarity raises the possibility of reactivity to shellfish tropomyosin due to sensitization from inhaled insect sources (Witteman et al., 1994). Less well understood is the role of insect tropomyosin as a primary food allergen. A recent study by Broekman et al. (2015) demonstrated that of 13 shrimp-allergic patients, 11 reacted to mealworm under food challenge. Cross-reactivity to tropomyosin appeared to be a key factor in these reactions. Cross-reactivity is more likely where IgE binding epitopes are conserved between crustacean and insect tropomyosin protein sequences, as is often apparently the case (Fig. 9.5).

Arginine Kinase

Arginine kinase is a highly abundant phosphagen, an ATP phosphotransferase found in invertebrate animals that catalyzes the phosphorylation of arginine residues (Kang et al., 2011). The invertebrate enzyme is a monomeric protein with a molecular weight of approximately 40 kDa and is fairly heat stable. Arginine kinase purified from cockroach was found to retain 50% of its activity after heating to 50°C for 10 min (Brown et al., 2004). It has also been reported that arginine kinase is a glycoprotein as a result of a positive PAS stain and putative N-glycosylation site in its sequence (Mao et al., 2013; Chen et al., 2013). Invertebrate arginine kinase performs the same biochemical function as creatine kinase (a dimeric molecule) in vertebrate animals, that is, coupling energy production with cellular function (Kang et al., 2011; France et al., 1997). An interesting property of arginine kinase is that 1 mM Zn^{2+} induces unfolding of the protein and subsequent formation of aggregates that are enzymatically inactive (Liu and Wang 2010). An open but intriguing question is whether Zn-induced unfolding might prove useful as a means to reduce arginine kinase allergenicity. Arginine kinase has been identified as a food allergen in variety of crustacean sources such as crab (Abdel Rahman et al., 2010; Rosmilah et al., 2012; Mao et al., 2013), various shrimp species (García-Orozco et al., 2007) and lobster (Binder et al., 2001). The enzyme is highly conserved both structurally and functionally within Arthropoda and is a significant source of cross-reactivity both within the crustacean species and between crustacea and insects. Yu et al. (2003) found that sera from shrimp-allergic individuals

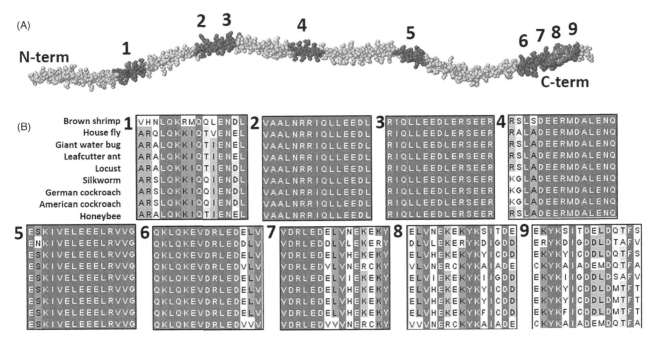

FIGURE 9.5 Alignment of insect tropomyosins and shrimp tropomyosin showing conservation of 9 IgE epitopes identified by Ayuso et al. (2002). Conservation of IgE epitopes is important in determining possible cross-reactivity between proteins. (A) Shows how identified epitopes map onto the structure of Pen a 1 (SDAP Model 284). In some cases identified epitopes overlap, (B) shows the amino acid alignment of insect tropomyosin sequences with brown shrimp within these epitopes. Species and NCBI accessions used were as follows: Brown shrimp (*P. aztecus*, DQ151457), House fly (*Musca domestica*, XP_005181083), Giant water bug (*Lethocerus indicus*, CAQ19275), Leafcutter ant (*Acromyrmex. Echinatior*, XP_011058810), Locust (*Locusta migratoria*, P31816), Silkworm (*B. mori*, NP_001103782), German cockroach (*Blatella germanica*, Q9NG56), American cockroach (*Periplaneta Americana*, AAD19606), and Honeybee (*Apis mellifera*, XP_006571122).

recognized other crustacean arginine kinase proteins such as those from crawfish and crab, with 11–12 out of 13 individual serum samples displaying cross-reactivity. Arginine kinase from snow crab (*Chionoecetes opilio* Fab.) is a major allergen and has the IUIS designation Chi o 3. Notably, snow crab has been associated as a cause of asthma and fatal anaphylaxis in fishermen and processing plant workers (Lopata et al., 2010). Investigation of snow crab allergens found IgE binding to arginine kinase and four other proteins including sarcoplasmic calcium binding protein, troponin, α actin, and sarcoplasmic endoreticulum Ca^{2+} ATPase (Abdel Rahman et al., 2011). Using a protein mass spectrometry approach, these authors identified a signature peptide from arginine kinase (AK 246-255) and proposed it as an analytical tool for the detection of snow crab. Arginine kinase was also reported to be a major allergen in mud crab, *Scylla paramamosain* Estampador, which is commonly consumed in China (Mao et al., 2013). This study found substantial cross-reactivity on Western blots to arginine kinase using sera from individuals allergic to crayfish, white shrimp, and octopus. Additionally, it was found that IgE binding to arginine kinase in these eight allergic sera differed if the protein was denatured. Only three of the eight allergic sera bound arginine kinase on Western blots of SDS-PAGE separations, whereas all gave positive results with the enzyme-linked immunosorbent assay (ELISA). The cross-reactivity between crustacean (king prawn, lobster) and insect (moth, dust mite, cockroach) arginine kinases has also been demonstrated serologically and with skin prick testing (Binder et al., 2001). Conversely, the likelihood of sera from shellfish-allergic individuals reacting to shrimp arginine kinase in less related species such as fish or mammals is low because of structural differences between arginine kinase and its analogous protein creatine kinase, found in vertebrate animals.

Sarcoplasmic Calcium Binding Protein

Sarcoplasmic calcium binding protein (SCP) is a member of the intracellular calcium binding protein family found in invertebrate but not vertebrate animals. The family includes calmodulin, troponin C, and others that function in calcium homeostasis (Cook et al., 1991; Herman and Cox, 1995). SCP's function in muscle is coupled to sarcoplasmic endoreticulum calcium ATPase (SERCA), the ATP-driven pump that lowers intracellular calcium concentration and causes muscle relaxation. Invertebrate SCP is the functional equivalent of parvalbumin in vertebrate muscle. SCP acts as a calcium buffer and contributes to the control of calcium-activated processes. The native SCP molecule exits as a dimer containing two polypeptide chains designated α and β, of which there are several isotypes. Each SCP subunit contains four EF hand calcium binding sites (dissociation constant 10^{-7} M) however, only three are functional (Cook et al., 1991). SCP is an acidic protein with a subunit molecular weight of approximately 21.8 kDa and a pI of 4.8–5.0 (Herman and Cox, 1995). Comparison of cloned DNA sequences from abdominal crayfish muscle [*Procambarus clarkii* (Girard)] revealed that SCP was 78.6–81.2% identical to shrimp muscle SCP (*Pemaeus sp.*) (Gao et al., 2006). White et al. (2011) cloned and expressed transcripts from the same freshwater crayfish (*P. clarkii*) and identified three SCP variants whose sequences differed between the second and third EF hand calcium binding domains. The authors also reported that the expression patterns of SCP in various tissues or in response to cold exposure was different than that for vertebrate muscle parvalbumin, suggesting that the two proteins may not have equivalent function in their respective tissues.

Shiomi et al. (2007) reported SCP as a significant component of shellfish allergy. In the study, the authors purified SCP from shrimp abdominal muscle (*Penaeus monodon* F.) and identified 2 enzymatic fragments that bound IgE from the sera in 8 out of 16 crustacean-allergic individuals. Further examination of individual sera positive for SCP found IgE binding to the 20 kDa band (SCP) in extracts of kuruma shrimp (*Penaeus japonicus* [Spence Bate]), pink shrimp (*Pandalus eous* Makarov) and American lobster (*Homarus americanus* Mine-Edwards) but not to king or snow crab [*Paralithodes camtschaticus* (Tilesius) and *C. opilio*, respectively]. These authors suggested that the distribution of IgE reactivity to SCP was probably limited to shrimp and crayfish. A study by Ayuso et al. (2009) identified SCP as a major shrimp allergen. Using immunoblots to a boiled shrimp extract, the authors found IgE-binding to a 20 kDa protein in 31 of 52 sera from shellfish-allergic individuals. The 20 kDa IgE-binding protein was identified as SCP (both α and β chains) using mass spectrometry analysis of in gel trypsin digested samples. The authors further examined IgE binding to a recombinant form of SCP with the 32 SCP-positive individual sera and found that 20 of the 32 subject sera also recognized cloned and expressed SCP. It is of note that 17 of those 20 sera (85%) came from shrimp-allergic children. The authors suggested the SCP may be a more significant allergen than tropomyosin for children. Shrimp SCP has been designated as Lit v 4.0101 in allergen nomenclature.

Myosin Light Chain

Myosin is the major protein of animal muscle. It is a calcium-dependent ATPase enzyme responsible for generating contractile force. Myosin molecules are composed of 6 polypeptide chains including two heavy chains with M_r of approximately 200 kDa, 2 essential light chains (ELC) with M_r of 18–20 kDa, and 2 regulatory light chains (RLC) with M_r of 18 kDa. ELC are calcium-binding proteins and play a role in the motor function of myosin. The calcium binding properties of ELC are contributed by four EF hand domains. Calcium binding to specific domain(s) alters the force and velocity of contraction. RLC are reversibly phosphorylated, a modification that alters the calcium level required for activation and the force of contraction (Fromherz and Szent-Gyorgyi, 1995; Sweeney, 1995). Substantial variation in myosin light chain isoforms exists based on developmental stage and muscle fiber type and between vertebrate and invertebrate animals (Fromherz and Szent-Gyorgyi, 1995; Hooper and Thuma, 2005). Invertebrates possess a wide variety of biological functions such as flight and catch muscles that are reflected in the composition of corresponding myosin motors performing those functions. Specific variations in myosin light chains include differences in stoichiometry, additional ELC variants (α and β), and switching of the EF hand domain responsible for calcium binding (Hooper and Thuma, 2005). A study by Ayuso et al. (2008) reported myosin light chain (MLC) as a new food allergen in shrimp and crab. This study used sera from a group of 38 shrimp-allergic individuals to examine IgE-binding proteins in extracts of raw and boiled shrimp [*Litopenaeus vannamei* (Boone)]. IgE binding was found to tropomyosin and a novel 20 kDa protein. Specifically, IgE binding to this protein was found in 21 of 38 patients. Additionally, IgE binding to the 20 kDa protein was greater in extracts of boiled shrimp compared to raw shrimp. The 20 kDa protein was identified as an MLC by mass spectrometry analysis of samples taken from electrophoretic separations. Subsequently, a cloned and expressed form of this protein was found to bind IgE in the sera of 19 patients that previously recognized the 20 kDa band. While most (17 of 19) patient sera recognized the recombinant MLC, the signal intensity was weaker compared to extracted protein. MLC was thus identified by the authors as a new shrimp allergen, designated as Lit v 3.0101. Recently, an 18 kDa protein purified from crayfish (*P. clarkii*) was identified as an MLC allergen (Zhang et al., 2015). The purified MLC protein bound IgE from crustacean-allergic individuals and had some unusual physicochemical properties. Specifically, it was glycosylated (approximately 4.9% by weight), thermostable to 100°C, and exhibited some digestive stability when treated individually with pepsin and pancreatin. Interestingly, neither thermal nor enzymatic treatment were found to completely eliminate IgE binding. The authors cloned and purified two isoforms of crayfish MLC (MLC1 and MLC2), which were found to correspond to ECL and RLC, respectively. They also found MCL1 to be the major allergenic form.

Troponin C

In skeletal muscle, troponin exists as a complex of three polypeptide chains: troponin I, T, and C. Troponin I is the inhibitory component, troponin T is the tropomyosin binding component, and troponin C is the calcium-binding component. Troponin I inhibits the interaction between actin and myosin when calcium is absent from its binding partner, troponin C. Troponin T binds tropomyosin, which in turn anchors the complex to the thin filament at precisely repeating intervals of 365 Angstroms (Zot and Potter, 1987; Ohtsuki et al., 1986; Brown and Cohen, 2005). Together, the troponin complex and tropomyosin regulate the interaction between myosin (thick filament) and actin (thin filament) in the sliding filament model of contraction. Contraction is initiated by a nerve impulse causing calcium release from the sarcolemma to the sarcoplasm. Binding of calcium ions to troponin C results in a conformational shift in the troponin complex. This structural change removes the troponin's steric hindrance of interaction between actin and myosin. Calcium also activates myosin's ATPase located in its heavy chain, providing energy required for movement when the nucleotide is hydrolyzed. Troponin C (TnC) is a member of the EF hand family of calcium-binding proteins. TnC exists as a single polypeptide chain with a molecular weight of approximately 18 kDa and a pI of 4.0. Troponin C has four calcium binding sites (I–IV) organized into two domains, one each in the C-terminal and N-terminal. C-terminal domains III and IV represent the high affinity site principally responsible for troponin's regulatory role in vertebrate muscle. However, invertebrate muscle proteins can be substantially different from their vertebrate counterparts, given the wide diversity of invertebrate organisms. A comprehensive review of invertebrate muscle-specific genes and proteins is given by Hooper and Thuma (2005). Troponin C from the American lobster (*H. americanus*) contains 11 isoforms encoded by 7 genes (Chao et al., 2010). There are also notable differences in calcium binding sites of invertebrate troponin C. A study of calcium binding sites in lobster muscle troponin C found that sites II and IV were necessary for full regulation of contraction (Tanaka et al., 2013). A recent investigation of the allergenicity of troponin C from shrimp (*P. monodon*) found IgE binding, albeit weak, to a recombinant form of troponin C in 8 of 35 sera from shrimp-allergic patients (Kalyanasundram and Santiago, 2015).

Sarcoplasmic Endoreticulum Calcium ATPase

Sarcoplasmic Endoreticulum Calcium ATPase (SERCA) is an enzyme that functions in coupling the energy from hydrolysis of ATP to the active transport of key ions across membranes. In muscle, SERCA is responsible for lowering intracellular calcium concentration to 10^{-7} M and thus bringing about relaxation of a contractile event. SERCA is a P-type ion channel enzyme. Calcium ions are exported to the luminal side of the sarcoplasmic reticulum membrane and protons are imported to the sarcoplasm. Two calcium ions are transported for each ATP hydrolyzed. SERCA is a single polypeptide chain folded into multiple domains. Its trans-membrane domain anchors the protein to the sarcoplasmic reticulum. SERCA is found in both plants and animals in various isoforms (more than 10), all with a molecular weight of approximately 110 kDa (Bublitz et al., 2013; Periswamy and Kalyanasundaram, 2007). The SERCA molecule is composed of four domains, each having individual and cooperative functions. The M domain is composed of 10 transmembrane alpha helices and is responsible for binding calcium ions. The three other domains (P, N, and A) are cytosolic. Conformational changes initiated by the hydrolysis of ATP predominate in the alpha helices of the M domain and represent a major gate-keeping function for ion mobility (Periswamy and Kalyanasundaram, 2007). Because SERCA performs an essential ion transport function in muscle and other cells, various isoforms are found in vertebrates and invertebrates. SERCA has been most studied in vertebrate muscle (particularly cardiac) because of the importance of abnormalities that are linked to disease states. The isoforms and respective tissue location of SERCA are SERCA1 (vertebrate fast twitch skeletal muscle), SERCA2 (cardiac and slow twitch skeletal muscle), and SERCA3 (cell signaling in nonmuscle cells). A major difference between vertebrate and invertebrate SERCA is that the latter originates from only one gene that is spliced to form two protein isoforms, SERCA2a and SERCA2b. Specifically, SERCA2b is the predominant form in invertebrate muscle (Wuytack et al., 2002). A study by Abdel Rahman et al. (2011) found IgE binding to several snow crab (*C. opilio*) muscle proteins including SERCA. In this study, immunoblotting of crab leg muscle extracts using a pool of sera from 16 snow–crab-allergic patients found IgE binding to several proteins including one with an M_r of approximately 110 kDa. Subsequent analysis using mass spectrometry identified the protein as a smooth muscle form of SERCA. While this is a single report linking SERCA to crustacean allergy, it was felt that the molecular similarities of SERCA to other calcium binding proteins (sarcoplasmic calcium binding protein, myosin light chain, and troponin C) warranted its description in this chapter.

Hemocyanin

Hemocyanin is an oxygen-transport protein found only in some invertebrates including many shellfish and insects. Hemocyanin has been identified as an allergen from shrimp (Piboonpocanun et al., 2011) and as a cross-reactive allergen of crustacean, cockroach, and dust mites (Ayuso et al., 2011). Hemocyanin has additionally been suggested to be clinically relevant as a crustacean allergen due to relatively high numbers of allergic individuals having IgE which recognizes the protein, a significant number of which have histories of anaphylactic reactions (Giuffrida et al., 2014).

Phospholipase

Phospholipase is a ubiquitous enzyme among eukaryotes which catalyzes the hydrolysis of phospholipids into fatty acids. Phospholipases are a bioactive venom component and have been implicated in allergic reactions to insect bites, for example, from the fire ant (Hoffman et al., 2005) (known as Sol I 1), and stings, as from the honeybee (Sobotka et al., 1976) (known as Api m 1). It is regarded as a major allergen due to preponderance of IgE to phospholipase in allergic individuals. Interestingly, this major allergen first identified as being a nonfood allergen also seems to be responsible for food-related reactions, such as those described from cochineal extracts derived from *D. coccus* (Ohgiya et al., 2009), where the authors showed binding of recombinant phospholipase to IgE from three individuals with histories allergic reactions to carmine-containing drinks. Reactions to carmine red have also been described from dietary supplements and over-the-counter drugs (Voltolini et al., 2014). The observation that phospholipases can function as both aero and ingestion allergens (Tarbar-Purroy et al., 2003; Gultekin and Doguc, 2013) raises the possibility of multiple routes of sensitization and elicitation of reactions by this enzyme from insects.

Other Allergens

In addition to the major shellfish allergens tropomyosin and arginine kinase, other shellfish proteins have been shown to bind IgE from the sera of shellfish-sensitized individuals. It is likely that these allergens are clinically less important, at least in terms of the number of individuals sensitized. These allergens, together with their identified sources and references, are summarized in Table 9.1. It should be noted that although these allergens appear to be of less significance in terms of shellfish allergy, their role, if any, in insect food allergy has yet to be elucidated and is likely to be unclear for some time.

Known Aero-Allergens

Inhalation is a recognized route of exposure and sensitization and can lead to food allergy (Ramirez and Bahna, 2009). Occupational exposure through dust and particulates has particularly been identified as problematic in the development of allergic disease. Dust allergies involving arthropods are most frequently associated with cockroaches and dust mites. Although the relationship between aero (primarily lung inhalation) and food allergy (gastrointenstinal tract involvement) is not clearly understood, there is the potential for those with existing sensitization to insect aero-allergens to be susceptible to insect-mediated food allergies. House dust mites in particular possess multiple proteins known to elicit allergic responses, with 30 allergens currently recognized (www.allergenonline.org), including tropomyosin (Der p 10) and arginine kinase (Der p 20). Tropomyosin has been identified as an inhalant allergen of cockroach (Asturias et al., 2015). Many of these allergens display significant homology to and IgE cross-reactivity with shellfish allergens (Sidenius et al., 2001), raising the possibility that these allergens may also be capable of eliciting responses from edible insects. However, the difference in route of exposure makes predicting whether food-related responses are likely in those already sensitized to airborne allergens difficult.

Novel Allergens

In addition to allergic reactions elicited by cross-reactivity in shellfish-allergic individuals, insect proteins also have the potential to themselves sensitize individuals. There is limited information available on such de novo sensitization, due largely to the lack of large-scale intentional consumption of insects in locales where food allergy is an issue and is studied.

EFFECTS OF PROCESSING

Food processing, particularly protein hydrolysis (Walker-Smith, 2003) and various heat-treatments have been demonstrated to reduce elicitation of allergic reactions in allergenic foods such as milk. Relatively little information exists concerning the effects of processing technologies on the allergenic properties of proteins from insects such as crickets grasshoppers, mealworm, and silkworm moth larva. However, insight regarding the effects of various technologies on insect proteins can be inferred based on investigations of closely related arthropod species, and the overview of existing reports is included here. A study reported by Kamath et al. (2013) used monoclonal antibodies to detect and measure tropomyosin in heated

TABLE 9.1 Other Notable Shellfish Allergens

Allergen	Source Organism	IUIS Name	Key References
Sarcoplasmic reticulum Ca-binding protein	Black tiger shrimp White shrimp	Pen m 4 Lit v 4	Shiomi et al. (2008) Ayuso et al. (2009) Abdel Rahman et al. (2011)
Sarcoplasmic Ca-binding protein (SCP)			White et al. (2011)
Myosin light chain	White shrimp Black tiger prawn	Lit v 3 Pen m 3	White et al. (2011) Abdel Rahman et al. (2010a)
Myosin heavy chain	Garden snail		Martins et al. (2004)
Troponin C	Abalone		Juji et al. (1990)
Actin	Mussel, Oyster		Koshte et al. (1989)
Hemocyanin	Keyhole limpet		Maeda et al. (1991)

Abdel Rahman, A., Kamath, S., Lopata, A. & Helleur, R. (2010a). Analysis of the allergenic proteins in black tiger prawn (*Penaeus monodon*) and characterization of the major allergen tropomyosin using mass spectrometry. Rapid communications in mass spectrometry 24(16), 2462–2470.
Ayuso, R., Grishina, G., Ibáñez, M., Blanco, C., Carrillo, T., Bencharitiwong, R., Sánchez, S., Nowak-Wegrzyn, A., Sampson, H., 2009. Sarcoplasmic calcium-binding protein is an EF-hand-type protein identified as a new shrimp allergen, J. Allergy Clin. Imm. 124(1), 114–120.
Juji, F., Takashima, H., Suko, M., Doi, M., Takaishi, T., Okudaira, H., Ito, K. & Miyamoto, T., 1990. A case of food-dependent-exercise induced anaphylaxis possibly induced by shellfish (*Sulculus supertexta* and *Turbo cornutus*), Arerugi [Allergy] 39(11), 1515–1522.
Koshte, V., Kagen, S. & Aalberse, R. 1989. Cross-reactivity of ige antibodies to caddis fly with arthropoda and mollusca, J. Allergy Clin. Imm. 84(2), 174–183.
Maeda, S., Morikawa, A., Kato, M., Motegi, Y., Shigeta, M., Tokuyama, K., Kuroume, T., Naritomi, Y., Suehiro, K., Kusaba, K. 1991. 11 cases of anaphylaxis caused by grand keyhole limpet (abalone like shellfish), Arerugi [Allergy] 40(11): 1415–1420.
Martins, L., Peltre, G., da Costa Faro, C., Vieira Pires, E. & da Cruz Inácio, F., 2004. The *Helix aspersa* (brown garden snail) allergen repertoire, Int. Arch. Allergy Immunol. 136(1): 7–15.
Shiomi, K., Sato, Y., Hamamoto, S., Mita, H., Shimakura, K., 2008. Sarcoplasmic calcium-binding protein: Identification as a new allergen of the black tiger shrimp (*Penaeus monodon*), Int. Arch. Allergy Immunol. 146(2), 91–98.
White, A., Northcutt, M., Rohrback, S., Carpenter, R., Niehaus-Sauter, M., Gao, Y., Wheatly, M., Gillen, C., 2011. Characterization of sarcoplasmic calcium binding protein (scp) variants from freshwater crayfish *Procambarus clarkii*, Comp. Biochem. Physiol. B: Biochem. Mol. Bio. 160(1), 8–14.
Source: Abdel Rahman, A., Kamath, S., Lopata, A., Robinson, J., Helleur, R., 2011. Biomolecular characterization of allergenic proteins in snow crab (*Chionoecetes opilio*) and de novo sequencing of the second allergen arginine kinase using tandem mass spectrometry. J. Proteomics 74(2), 231–241.

shellfish. This study included 11 crustacean species and seven mollusk species. Crustacea species examined included prawn [*P. monodon*, *Melicertus latisulcatus* Kishinouye, *Litopenaeus vannamei* (Boone), *Fenneropenaeus merguiensis* De Man, *Penaeus semisulcatus* De Haan], crab [*Portunus pelagicus* (L.), *Ovalipes australiensis* Stephenson & Rees, *C. opilio*], and lobster [*Thenus orientalis* (Lund), *Jasus edwardsii* (Hutton), *Cherax destructor* Clark]. The Mollusca species examined included bivalve (*Perna viridis* L., *Mytilus edulis* L., *Pecten fumatus* Reeve, *Crassostrea gigas* Thunberg) gastropod (*Turbo cornutus* Lightfoot), and cephlopod (*Octopus vulgaris* Cuvier, *Sepioteuthis lessoniana* Férussac). In this study, whole specimens were heated at 100°C for 20 min and the soluble fraction examined for tropomyosin by Western blotting and ELISA using a tropomyosin-specific monoclonal antibody. This study found that heat treatment actually increased tropomyosin antibody binding in Western blots of both crustacean and mollusk samples. It is interesting to note that no tropomyosin binding was found in unheated mollusk species. Additionally, a competitive ELISA using recombinant prawn tropomyosin as the standard showed greater inhibition for heat-treated crustacean samples compared to unheated ones. The proposed rationale for increased antibody response in heat-treated samples was that thermal treatment altered tropomyosin conformation in a way that increased availability of the epitope region(s) (Kamath et al., 2013).

A distinguishing characteristic of tropomyosin between vertebrate and invertebrate animals is the greater allergenicity of the latter form (Reese et al., 1999). A contributing factor to the greater allergenicity of invertebrate tropomyosin may be its higher thermostability. Ozawa et al. (2011) examined the thermostability of tropomyosin purified from three marine species; Japanese squid (*Todarodes pacificus* Steenstrup), tokobushi abalone (*Haliotis diversicolor* Reeve), and kuruma prawn (*Marsupenaeus japonicus* Spence Bate) using circular dichroism and differential scanning calorimetry. This work reported that the protein melting temperatures from the species examined ranged from 43.5–50.2°C and proposed that thermal stability of tropomyosin in these species followed the order: prawn > abalone > squid. In contrast, the melting temperature of vertebrate tropomyosin is approximately 40°C (Holtzer et al., 1986). It is also of note that thermal treatment can result in an increase of IgE binding to proteins. A recent investigation of IgE binding using a pool of sera from shellfish-allergic subjects to tropomyosin from blue swimmer crab (*P. pelagicus*) and black tiger shrimp (*P. monodon*) reported markedly increased IgE binding to cooked (100°C for 20 min) versus raw samples (Abramovitch et al., 2013). The authors proposed that the observed increase in IgE binding to tropomyosin in heated crab and shrimp may have resulted from Maillard reaction that can cause chemical modification of proteins. Maillard reaction between proteins and reducing carbohydrates can produce advanced glycation end products (AGE).

A study by Long et al. (2015) examined the effect of high pressure and/or heat treatment to alter the allergenic properties of tropomyosin from shrimp (*L. vannamei*). This study of processing effects was conducted on the myofibrillar fraction from shrimp. Samples were prepared by extracting shrimp muscle several times with low ionic strength buffer. The low-speed pellet fraction (myofibrillar) obtained after removal of soluble proteins was used for subsequent processing experiments. Shrimp samples were treated with high pressure ranging from 100 to 600 MPa in combination with temperatures ranging from 25°–75°C, for up to 30 min. The effects of processing treatments on these shrimp samples were assessed by competitive ELISA using pooled sera from eight shrimp-allergic patients and with a mouse model of shrimp allergy. Competitive ELISA analysis of these treatments showed the most effective method for reducing IgE binding to shrimp tropomyosin resulted from combined high pressure (500 MPa) and heat (55°C) treatment. The authors further examined allergenicity of tropomyosin in heat and high-pressure-treated samples using a mouse allergy model. In this experiment, control and tropomyosin-sensitized BALB/c mice were fed shrimp samples processed using the most effective method, 500 MPa and 55°C. Mice fed samples prepared from combined high pressure and heat treatment showed little or no tropomyosin allergic symptoms as judged by cytokine and IgE levels. Based on these results, the authors suggested that combined treatment using high pressure and heat may represent a way to lower shrimp allergenicity.

Yu et al. (2011) investigated the effect of several processing methods, boiling, boiling plus ultrasound, and high-pressure steam (0.14 MPa), as means to increase digestibility of tropomyosin in crab extracts (*S. paramamosain*). The effect of processing treatments on extracts was assessed using separate enzymatic digestions with pepsin, trypsin, and chymotrypsin. Western blotting with polyclonal antitropomyosin antibody and sera from crab-allergic individuals showed that all three processing methods increased enzymatic digestion of tropomyosin. While high-pressure steam (autoclaving) was the most effective for increasing enzymatic digestibility, only boiling was able to decrease IgE binding to enzymatically digested tropomyosin.

As previously described, there is good evidence for cross-reactivity between house dust mite and crustacean allergens in allergic individuals. Specifically, there is a possibility that HDM-allergic individuals may also react to yellow mealworm (*T. molitor*) proteins (Verhoeckx et al., 2014). Additionally, while thermal treatment including blanching, boiling, and baking affected the solubility of proteins, these treatments had no effect on the mealworm allergenicity as assessed by IgE binding (Broekman et al., 2014). In particular, the authors noted that tropomyosin is known for its extreme stability and is able to survive thermal processing common in food preparation and cooking, which also holds true for tropomyosin derived from insects. This thermal stability indicates that consumption of even highly heated insect material may still cause allergic reactions in sensitized individuals. Broekman et al., concluded that thermal processing did not lower allergenicity but clearly changed solubility of mealworm allergens.

Thermal treatment of crayfish (*P. clarkii*), a species that is native to a large part of the world, was investigated for the ability of processing to alter its allergenicity (Chen et al., 2013). This investigation found that thermal treatment at temperatures less than 44°C resulted in little effect. However, at temperatures in the range of 44–70°C, increased aggregation and precipitation was observed along with an increase in IgE binding. Similarly, increased IgE binding was observed for acidic (pH 1–3) treatment.

METHODS OF ALLERGEN DETECTION

The ability to analyze for trace amounts of food allergens is desirable for the validation of procedures to control for their presence in food manufacturing facilities handling multiple food ingredients (Jackson et al., 2008). Analytical methods will therefore be crucially important for the upscaling of insect protein manufacture for use beyond niche markets. Given the relative novelty of the use of insects as food, and therefore their presence in food manufacturing environments, few such methods currently exist, with those that do focused on supporting grain quality law (Kitto et al., 1994) and not allergen detection. It is likely, however, that such detection methods will be developed as the need for them arises. Current analytical methods for food allergens focus on three major types of analysis: immunological, polymerase chain reaction (PCR), and the newly developing field of mass spectrometry (MS). These are discussed later.

Immunological methods: Immunological methods rely on the detection of proteins from the allergenic food source using specific antibodies, typically IgG. These antibodies may detect one or more proteins of the source of allergen, often the allergenic proteins themselves, or indeed may be raised to whole extracts of the allergenic foodstuff. The format of immunological methods can be adapted according to need. Two commonly used formats are lateral flow devices (LFDs) and enzyme-linked immunosorbent assays (ELISAs). LFDs are commonly used in conjunction with swabbing for rapid analysis of potential contamination on surfaces and can be used to validate cleaning protocols. However, they are also

TABLE 9.2 Sequence Information Available for Selected Species of Edible Insects

Common Name	Taxonomic Name	Selected Species	Number of Protein Sequences	Tropomyosin Sequence (NCBI)
Mealworm	T. molitor	T. molitor	322	Unknown
Grasshopper	Suborder Caelifera	Schistocerca americana	59	Unknown
Cockroach	Order Blattodea	P. americana	345	Q9UB83
Silk worm	B. mori	B. mori	18,028	NP_001103782
Cochineal	D. coccus	D. coccus	4	Unknown
House fly/larva	M. domestica	M. domestica	6273	XP_005181083
Agave worm	H. agavis	H. agavis	0	Unknown
Honeybee	Genus Apis	A. mellifera	16,585	XP_003689891

The availability of sequence information for the major cross-reactive allergen tropomyosin is also indicated. Numbers of available protein sequences were obtained from UniProtKB (www.uniprot.org) as of Aug. 25, 2015.

sometimes used to analyze foodstuffs or ingredients for contamination. LFDs are typically rapid and easy to use, and yield largely qualitative results.

The enzyme-linked immunosorbent assay (ELISA) is commonly used by off-site testing laboratories. The testing procedure is more involved and typically takes longer than comparable lateral flow devices. However, ELISA is typically more sensitive (low ppm) and can yield quantitative information, which can be useful in analyzing potential risk of any allergen cross-contact. Although action levels for allergen cross-contact are largely undefined, ELISA methods are thought to be capable of the levels of sensitivity required to protect the majority of food-allergic consumers (Taylor et al., 2009). ELISA is currently the most frequently used method for detecting allergens in foods.

Polymerase chain reaction (PCR): PCR methods detect the presence of DNA specific to the food being tested for. It should be emphasized that, as PCR does not detect protein, it is unable to directly analyze the protein molecules in food that cause allergic reactions (Poms et al., 2004). As such, it may be less suitable for use in heavily processed foodstuffs where proteins and DNA may react very differently to the processing techniques used (Iniesto et al., 2013). PCR, however, can be extremely sensitive and can be multiplexed—developed into assays that detect multiple allergenic foods in one analysis. Proper use of PCR does require careful consideration of the allergenic food tested for, and the matrix in which it is present.

Mass spectrometry methods: Similarly to the immunological methods, mass spectrometry methods for allergen detection use proteins as analytes. Typically these proteins will be broken down to peptides by the action of enzymes such as trypsin, allowing analysis by conventional proteomic methods. The selection of peptide targets for analysis is of crucial importance, as such peptides need to represent the food to be analyzed for (including processed forms) and be specific to that food only to prevent false positives (Johnson et al., 2011). As MS methods rely on the preselection of peptide targets, it is crucial that there is sufficient nucleotide (DNA) information available for peptide design. For some allergenic food materials, particularly tree nuts, there are few available DNA, and therefore protein sequences, currently available (Johnson et al., 2014a). A summary of the publicly available protein sequences for common edible insects is shown in Table 9.2. It is clear that for many edible insects there is scant information on which to base MS methods for allergen detection.

The use of well-defined materials as reference standards is essential for method comparison and quality control, but there is notable lack of availability of such materials for food allergens (Taylor et al., 2009). This likely reflects the difficulty of preparing, characterizing, and maintaining shelf-life of protein standards in food materials (Johnson et al., 2014b). Although the lack of incurred allergen materials is currently being addressed by initiatives such as the MoniQA Food Allergen Reference Material Task Force, there are few commercially available materials currently available to assist in method harmonization.

CONCLUSIONS

The dynamics of increased world population and fixed reserves of arable land will contribute to substantial change in food and agricultural practices in the coming decades. Future food is likely to include foods that previously were not widely consumed. Examples of novel food and food technologies that are currently being proposed include: nonanimal milk and dairy products, cultured meat, vegetarian egg replacements, algae and algal oil from bio-fermentation, and insects. While the ability to feed the world's growing population is of utmost importance, other factors such as the safety of novel foods warrant consideration. In particular, food allergy represents a significant food safety concern for novel foods, especially insects. At present, the number of documented allergic reactions to insects/insect-containing foods are few. There is however, cause for concern because of the phenomena of cross-reactivity. Numerous examples of allergic reactions resulting from cross-reactivity are cited in this work. Thus, it is logical to assume that some individuals with shellfish allergy will develop allergic reactions when consuming insects. Additionally, with the exception of Broekman et al. (2014, 2015), little data exists concerning insect proteins as food allergens. Although comparisons of shellfish with insects with respect to reaction thresholds should be treated with caution, the ED_{10} (amount of protein causing a response in the most sensitive 10% of shellfish-allergic patients studied) was 2.5 g, significantly higher amount than other food allergens studied (Ballmer-Weber et al., 2015). Food challenge data with insects is currently lacking and should form a part of future allergen safety assessments.

Based on the number of reports, the most important invertebrate allergens are tropomyosin and arginine kinase. Additional proteins with potential for being cross-reactive allergens include sarcoplasmic calcium binding protein, troponin C, and myosin light chain 1. It is of interest to note that these proteins share several molecular and functional properties. They are all muscle proteins, often with highly conserved primary sequences and calcium-binding domains. They are thermostable and generally resistant to pepsin digestion. Use of conventional food processing methods has been largely ineffective in reducing IgE binding by these proteins. Contrary to expectation, thermal processing of shrimp resulted in increased IgE binding to tropomyosin. An exception to the lack of success with processing technology is the work reported by Long et al. (2015). They demonstrated the potential for combined application of high-pressure and thermal treatment (500 MPa and 55°C) to significantly reduce tropomyosin-induced elicitation of allergenicity using a mouse model. High-pressure processing holds potential for altering protein digestibility and perhaps allergenicity because of its ability to bring about changes in protein conformation (Hoppe et al., 2013).

Occupational allergy is a significant problem for individuals involved in the production (eg, farming) and processing of shellfish and insects. Issues predominantly occur where there is a risk of aerosolization of protein-containing material (eg, powder handling, spray drying). Allergic reactions ranging in severity from rhinitis to anaphylaxis and death have been reported for workers involved in crab and cochineal processing. It is assumed that workers became sensitized via inhalation of aerosols and developed allergic symptoms upon repeated exposure. Occupational allergy is a serious cause of illness and death resulting from allergic reaction and warrants research and intervention to protect workers.

Food applications of insects are more likely to be as ingredients that are added for nutritional or functional purposes. Given the potential for shellfish-allergic individuals to experience an allergic reaction after consuming a food containing insect protein, methods for determining insect protein in ingredients and processed foods are a high priority. ELISA methods using antibodies for specific target proteins have been a reliable technology for this purpose. However, recent advances in mass spectrometry and proteomics have made it possible to quantitatively detect multiple target proteins based on specific peptides. Proteomics also offer the capability to determine whether proteins have been modified by chemical or enzymatic reactions occurring in foods as a result of processing. For example, Maillard reaction can modify protein lysine and asparagine residues resulting in glycation and acrylamide products. It is possible that these posttranslational modifications are linked to protein allergenicity. Thus, proteomics represent an important methodology for the detection and study of protein allergenicity. The lack of sequence information for genes encoding proteins in consumed insects is a significant barrier to development of MS methods for both allergen research and detection sequence information would also help inform nutritional and bioactives research to identify potential benefits of insects over more traditional sources of protein. While there are potential allergenicity risks associated with insects as foods, adequate safety considerations require information on the use of insects as ingredients and levels of consumption as well as the important allergenic proteins. It is clear that more research is needed to address these information gaps.

REFERENCES

Abdel Rahman, A.M., Lopata, A.L., O'Hehir, R.E., Robinson, J.J., Banoub, J.H., 2010. Characterization and denovo sequencing of snow crab tropomyosin enzymatic peptides by electrospray ionization and matrix-assisted laser desorption ionization QqTof tandem mass spectrometry. J. Mass Spectrom. 45, 372–381.

Abdel Rahman, A.M., Kamath, S.D., Lopata, A.L.R.E., Robinson, J.J., Helleur, R.J., 2011. Biomoelcular characterization of allergenic proteins in snow crab (*Chionoecetes opilio*) and de novo sequencing of the second allergen arginine kinase using tandem mass spectrometry. J. Proteomics 74, 231–241.

Abramovitch, J.B., Kamath, S., Varses, N., Zubrinich, C., Lopata, A.L., O'Hehir, R.E., Rolland, J.M., 2013. IgE reactivity of blue swimmer crab *(Portunus pelagicus)* tropomyosin, Por 1 and other allergens: cross-reactivity with black tiger prawn and effects of heating. PLoS One 8, 1–13.

Antonicelli, L., Falagiani, P., Pucci, S., Garritani, M.S., Bilò, M.B., Bonifazi, F., 1992. Is there a cross-reacting allergen between mollusca gastropoda and mite? Allergy 47, 244.

Asturias, J.A., Gómez-Bayón, Carmen Arilla, N.M., Martínez, A., Palacios, R., Sánchez-Gascón, F., Martínez, J., 2015. Molecular characterization of American cockroach tropomyosin (*Periplaneta americana* allergen 7), a cross-reactive allergen. J. Immunol. 162 (7), 4342–4348.

Ayuso, R., Reese, G., Leong-Kee, S., Plante, M., Lehrer, S.B., 2002. Molecular basis of arthropod cross-reactivity: IgE-binding cross-reactive epitopes of shrimp, house dust mite and cockroach tropomyosins. Int. Arch. Allergy Imm. 129, 38–48.

Ayuso, E., Grishina, G., Bardina, L., Carrillo, T., Blanco, C., Ibanez, M.D., Sampson, H.A., Beyer, K., 2008. Myosin light chain is a novel shrimp allergen, Lit v 3. J. Allergy Clin. Immunol. 122, 795–802.

Ayuso, E., Grishina, G., Ibanez, M.D., Blanco, C., Carrillo, T., Bencharitiwong, R., Sanchez, S., Nowak-Wegrzyn, A., Sampson, H.A., 2009. Sarcoplasmic calcium binding protein is an EF hand-type protein identified as a new allergen. J. Allergy Clin. Imm. 124, 114–120.

Ayuso, R., Grishina, G., Pascal, M., Sanchez-Garcia, S., Towle, D., Smith, C., Ibáñez, M., Sampson, H.A., 2011. Hemocyanin, troponin C and fatty acid-binding protein (FABP) may be cross-reactive allergens between crustaceans, cockroach and dust mites. J. Allergy Clin. Imm. 127 (2), 235.

Ballmer-Weber, B.K., Fernandez-Rivas, M., Beyer, K., Defernez, M., Sperrin, M., Mackie, A.R., Salt, L.J., Hourihane, J.O'B., Asero, R., Belohlavkova, S., Kowalski, M., de Blay, F., Papadopoulos, N.G., Clausen, M., Knulst, A.C., Roberts, G., Popov, T., Sprikkelman, A.B., Dubakiene, R., Vieths, S., van Ree, R., Crevel, R., Mills, E.N., 2015. How much is too much? Threshold dose distributions for 5 food allergens. J. Allergy Clin. Imm. 135 (4), 964–971.

Banzet, M.L., Adessi, B., Vuitton, D.A., Amecoforcal, 1992. Allergic manifestations following eating snails in 12 patients with acaris allergy: a new crossover allergy? Manifestations allergiques après ingestion d'escargots chez 12 malades allergiques aux acariens: une nouvelle allergie croisée? Revue Francaise d'Allergologie et d'Immunologie Clinique 32, 198–202.

Belluco, S., Losasso, C., Maggioletti, M., Alonzi, C.C., Paoletti, M.G., Ricci, A., 2013. Edilble insects in a food safety and nutritional perspective: a critical review. Compr. Rev. Food Sci. F. 12, 296–313.

Binder, M., Mahler, V., Hayek, B., Sperr, W.R., Scholler, M., Prozell, S., Wiedermann, G., Valent, P., Valenta, R., Duchene, M., 2001. Molecular and immunological characterization of arginine kinase from the indianmeal moth *Plodia interpunctella*, a novel cross-reactive pan-allergen. J. Immunol. 167, 5470–5477.

Boyce, J.A., Assa'ad, A., Burks, A.W., Jones, S.M., Sampson, H.A., Wood, R.A., Plaut, M., Cooper, S.F., Fenton, M.J., 2010. Guidelines for the diagnosis and management of food allergy in the United States: summary of the NIAID-sponsored expert panel report. J. Allergy Clin. Imm. 126, 1105–1118.

Broekman, H., Knulst, A., den Hartog, J.S., Gaspari, M., de Jong, G.A., Houben, G., Verhoeckx, K.C., 2014. The effect of processing on the allergenicity of mealworm proteins. Allergy 69 (Suppl. 99), 283.

Broekman, H., Knulst, A., Jager, S.H., Monteleone, F., Gaspari, M., de Jong, G., Houben, G., Verhoeckx, K., 2015. Effect of thermal processing on mealworm allergenicity. Mol. Nutr. Food Res. 59 (9), 1855–1864.

Brown, J.H., Cohen, C., 2005. Regulation of muscle contraction by tropomyosin and tropomyosin: how structure illuminates function. Adv. Protein Chem. 71, 121–159.

Brown, A.E., France, R.M., Grossman, S.H., 2004. Purification and characterization of arginine kinase from the American cockroach (*Periplanta Americana*). Arch. Ins. Biochem. Physiol. 56, 51–60.

Bublitz, M., Musgaard, M., Poulsen, H., Thorgerson, L., Olesen, C., Schiott, B., Morth, J.P., Moller, J.V., Nissen, P., 2013. Ion pathways in the sarcoplasmic reticulum Ca + 2 –ATPase. J. Biol. Chem. 288 (15), 10759–10765.

Bukkens, S.G.F., 2005. Insects in the diet: nutritional aspects. In: Paoletti, M.G. (Ed.), Ecological Implications of Minilivestock: Potential of Insects, Rodents, Frogs, and Snails. Science Publishers, Enfield, NH, USA, pp. 545–577.

Carrillo, T., Castillo, R., Caminero, J., Cuevas, M., Rodriguez, J.C., Acosta, O., Rodríguez de Castro, F., 1992. Squid hypersensitivity: a clinical and immunologic study. Ann. Allergy 68, 483–487.

Chao, E., Kim, H.W., Mykles, D.L., 2010. Cloning and expression of eleven troponin-c isoforms in the American lobster *Homarus americanus*. Comp. Biochem. Physiol. B 157, 88–101.

Chase, M.W., 1946. Inhibition of experimental drug allergy by prior feeding of the sensitizing agent. Exp. Bio. Med. 61, 257–259.

Chehade, M., Mayer, L., 2005. Oral tolerance and its relation to food hypersensitivities. J. Allergy Clin. Imm. 115, 3–12.

Chen, H.L., Mao, H.Y., Cao, M.J., Cai, Q.F., Su, W.j, Zhang, Y.X., Liu, G.M., 2013. Purification, physiochemical properties and immunological characterization of arginine kinase, an allergen of crayfish (*Procambarus clarkii*). Food Chem. Toxicol. 62, 475–484.

Cook, W.J., Ealick, S.E., Babu, S.E., Cox, J.A., Vijay-Kumar, S., 1991. Three-dimensional structure of a sarcoplasmic calcium-binding protein from *Nereis diversicolor*. J. Biol. Chem. 266, 652–656.

Crespo, J.F., Pascual, C., Helm, R., Sanchez-Pastor, S., Ojeda, I., Romualdo, L., Martín-Esteban, M., Ojeda, J.A., 1995. Cross-reactivity of IgE-binding components between boiled Atlantic shrimp and German cockroach. Allergy 50, 918–924.

Daul, C.B., Slattery, M., Reese, G., Lehrer, S.B., 1994. Identification of the major brown shrimp (Penaeus aztecus) allergen as the muscle protein tropomyosin. Int. Arch. Allergy Immunol. 105 (1), 49–55.

De Maat-Bleeker, F., Akkerdaas, J.H., van Ree, R., Aalberse, R.C., 1995. Vineyard snail allergy possible induced by sensitization to house-dust mite (*Dermatophagoides pteronyssinus*). Allergy 50, 438–440.

DiCello, M.C., Myc, A., Baker, J.R., Baldwin, J.L., 1999. Anaphylaxis after ingestion of carmine colored foods: two case reports and a review of the literature. Allergy Asthma Proc. 20, 377–382.

Du Toit, G., Roberts, G., Sayre, P.H., Bahnson, H.T., Radulovic, S., Santos, A.F., Brough, H.A., Phippard, D., Basting, M., Feeney, M., Turcanu, V., Sever, M.L., Gomez Lorenzo, M., Plaut, M., Lack, G., 2015. LEAP study team: randomized trial of peanut consumption in infants at risk for peanut allergy. New Engl. J. Med. 372 (9), 803–813.

Fernandes, J., Reshef, A., Ayuso, R., Patton, L., Reese, G., Lehrer, S.B., 2002. IgE antibody reactivity to the major shrimp allergen in unexposed Orthodox Jews. J. Allergy Clin. Imm. 109, S217.

Food and Agriculture Organization, 1995. In: Food, Agriculture Organization of the United Nations, (Ed.), Report of the FAO Technical Consultation on Food Allergies. FAO, Rome, Italy.

France, R.M., Sellers, D.S., Grossman, S.H., 1997. Purification, characterization, and hydrodynamic properties of arginine kinase from Gulf shrimp (*Penaeus aztececus*). Arch. Biochem. Biophys. 1, 73–78.

Fromherz, S., Szent-Gyorgyi, A.G., 1995. Role of essential light chain EF hand domains in calcium binding and regulation of scallop myosin. Proc. Natl. Acad. Sci. 92, 7652–7656.

Gao, Y., Gillen, C.M., Wheatley, M.G., 2006. Molecular characterization of the sarcoplasmic calcium binding protein (SCP) form crayfish (*Procambarus clarkii*). Comp. Biochem. Physiol. B 144, 478–487.

Gamez, C., Zafra, M.P., Boquete, M., Sanz, V., Mazzeo, C., Ibanez, M.D., Sanchez-Garcia, S., Sastre, J., Pozo, V., 2014. New shrimp IgE-binding proteins involved in mite-seafood cross-reactivity. Mol. Nutr. Food Res. 58, 1915–1925.

García-Orozco, K.D., Aispuro-Hernández, E., Yepiz-Plascencia, G., Calderón-de-la-Barca, A.M., Sotelo-Mundo, R.R., 2007. Molecular characterization of arginine kinase, an allergen from the shrimp *Litopenaeus vannamei*. Int. Arch. Allergy Immunol. 144 (1), 23–28.

Gendel, S.M., 2012. Comparison of international food allergen labeling regulations. Regul. Toxicol. Pharm. 63, 279–285.

Giuffrida, M.G., Villalta, D., Mistrello, G., Amato, S., Asero, R., 2014. Shrimp allergy beyond tropomyosin in Italy: clinical relevance of arginine kinase. Sarcoplasmic calcium binding protein and hemocyanin. European Ann. Allergy Clin. Immunol. 46, 172–177.

Gould, H.J., Sutton, B.J., 2008. IgE in allergy and asthma today. Nat. Rev. Immunol. 8, 205–217.

Greenfield, N.J., Hitchcock-DeGregori, S.E., 1995. The stability of tropomyosin, a two-stranded coiled-coil protein, is primarily a function of the hydrophobicity of residues at the helix–helix interface. Biochemistry 34, 16797–16805.

Greenhawt, M., McMorris, M., Baldwin, J., 2009. Carmine hypersensitivity masquerading as azithromycin hypersensitivity. Allergy Asthma Proc. 30, 95–101.

Gultekin, F., Doguc, D.K., 2013. Allergic and immunologic reactions to food additives. Clin. Rev. Allergy Immunol. 45, 6–29.

Hefle, S.L., Nordlee, J.A., Taylor, S.L., 1996. Allergenic foods. Crit. Rev. Food Sci. Nutr. 36, 69S–89S.

Herman, A., Cox, J.A., 1995. Sarcoplasmic calcium-binding protein. Comp. Biochem. Physiol. B 111, 337–345.

Hitchcoock-DeGregori, S.E., Song, Y., Greenfield, N.J., 2002. Functions of tropomyosin's periodic repeats. Biochemistry 41, 15036–15044.

Hoffman, D.R., Sakell, R.H., Schmidt, M., 2005. Sol i 1, the phospholipase allergen of imported fire ant venom. J. Allergy Clin. Imm. 115 (3), 611–616.

Holtzer, M.E., Askins, K., Holtzer, A., 1986. Alpha-helix to random-coil transition of two-chain, coiled coils: experiments on the thermal denaturation of doubly cross-linked beta beta tropomyosin. Biochemistry 25 (7), 1688–1692.

Hooper, S.L., Thuma, J., 2005. Invertebrate muscles: muscle-specific genes and proteins. Physiol. Rev. 85, 1001–1060.

Hoppe, A., Jung, S., Patnaik, A., Zeece, M.G., 2013. Effect of high pressure treatment on egg white protein digestibility and peptide products. Innov. Food Sci. Emerg.Innov. Food Sci. Emerg. 17, 54–62.

Iniesto, E., Jiménez, A., Prieto, N., Cabanillas, B., Burbano, C., Pedrosa, M.M., Rodríguez, J., Muzquiz, M., Crespo, J.F., Cuadrado, C., Linacero, R., 2013. Real time PCR to detect hazelnut allergen coding sequences in processed foods. Food Chem. 138 (2), 1976–1981.

Jackson, L.S., Al-Taher, F.M., Moorman, M., DeVries, J.W., Tippett, R., Swanson, K.M., Fu, T.J., Salter, R., Dunaif, G., Estes, S., Albillos, S., Gendel, S.M., 2008. Cleaning and other control and validation strategies to prevent allergen cross-contact in food-processing operations. J. Food Proteins 71 (2), 445–458.

Ji, K.M., Zhan, Z.K., Chen, J.J., Liu, Z.G., 2008. Anaphylactic shock caused by silkworm pupa consumption in China. Allergy 63, 1407–1408.

Ji, K.M., Chen, J., Li, M., Liu, Z., Wang, C., Zhan, Z., Wu, X., Xia, Q., 2009. Anaphylactic shock and lethal anaphylaxis caused by food consumption in China. Trends Food Sci. Tech. 20, 227–231.

Jirapongsananuruk, O., Bunsawansong, W., Piyaphanee, N., Visitsunthorn, N., Thongngarm, T., Vichyanond, P., 2007. Features of patients with anaphylaxis admitted to a university hospital. Ann. Allergy Asthma Immunol. 98, 157–162.

Johnson, P.E., Aldick, T., Giosafatto, C.V.L., Watson, A., Baumgartner, S., Bessant, C., Heick, J., Mamone, G., O'Connor, G., Poms, R., Popping, B., Reuter, A., Ulberth, F., Monaci, L., Mills, E.N.C., 2011. Current perspectives and recommendations for the development of mass spectrometry methods for the determination of allergens in foods. J. AOAC Int. 94, 1026–1033.

Johnson, P.E., Marsh, J.T., Mills, E.N.C., 2014a. Mass Spectrometry-based Quantification of Proteins and Peptides in Food. In: Eyers, C.E., Gaskell, S. (Eds.), Quantitative Proteomics. Royal Society of Chemistry Books, Burlington House, London, UK, pp. 329–348.

Johnson, P.E., Rigby, N.M., Dainty, J.R., Mackie, A.R., Immer, U., Rogers, A., Titchener, P., Shoji, M., Ryan, A., Mata, L., Brown, H., Holzhauser, T., Dumont, V., Wykes, J.A., Walker, M., Griffin, J., White, J., Taylor, G., Popping, B., Crevel, R., Miguel, S., Lutter, P., Gaskin, F., Koerner, T.B., Clarke, D., Sherlock, R., Flanagan, A., Chan, C.-H., Mills, E.N.C., 2014b. A multi-laboratory evaluation of a clinically-validated incurred quality control material for analysis of allergens in food. Food Chem. 148, 30–36.

Kalyanasundram, A., Santiago, T.C., 2015. Identification and characterization of a new allergen troponin C (Pen m 6 0101) from Indian black tiger shrimp *Penaeus monodon*. European Food Res. Tech. 240, 509–515.

Kamath, S.D., Abdel Rahman, A.M., Komoda, T., Lopats, A.L., 2013. Impact of heat processing on detection of the major shellfish allergen tropomyosin in crustaceans and molluscs using specific monoclonal antibodies. Food Chem. 141, 4031–4039.

Kang, L., Shi, H., Liu, X., Zhang, C., Yao, Q., Wang, Y., Chang, C., Shi, J., Cao, J., Kong, J., Chen, 2011. Arginine kinase is highly expressed in a resistant strain of silkworm (*Bombyx mori*, Lepidoptera): Implication of its role in resistance to *Bombyx mori*, nucleopolyhedrovirus. Comp. Biochem. Biophys. B 158, 230–234.

Kitto, G.B., Quinn, F.A., Burkholder, W.E., 1994. Development of immunoassays for quantitative detection of insects in stored products. In Highley, E., Wright, E.J., Banks, H.J., Champ, B.R., (Eds.), Stored Product Protection: Proceedings of the 6th International Working Conference on Stored-product Protection, CAB International, Wallingford, UK, pp. 415–420.

Leung, P.S., Chu, K.H., Chow, W.K., Ansari, A., Bandea, C.I., Kwan, H.S., Nagy, S.M., Gershwin, M.E., 1994. Cloning, expression, and primary structure of *Metapenaeusensis* tropomyosin, the major heat-stable shrimp allergen. J. Allergy Clin. Imm. 94 (5), 882–890.

Leung, P.S., Chow, W.K., Duffey, S., Kwan, H.S., Gershwin, M.E., Chu, K.H., 1996. IgE reactivity against a cross-reactive allergen in crustacea and mollusca: evidence for tropomyosin as the common allergen. J. Allergy Clin. Imm. 98 (5), 954–961.

Liu, T., Wang, X., 2010. Zinc induces unfolding and aggregation of dimeric arginine kinase by trapping reversible unfolding intermediate. Acta Biochem. Biophys. Sinica 10, 1–8.

Lizaso, M.T., Moneo, I., Garcia, B.E., Acero, S., Quirce, S., Tabar, A.I., 2000. Identification of allergens involved in occupational asthma due to carmine dye. Ann. Allergy Asthma Immunol. 84, 549–552.

Long, F., Yang, X., Wang, R., Hu, X., Chen, F., 2015. Effects of combined high pressure and thermal treatments on the allergic potential of shrimp (*Litopenaeus vannamei*) tropomyosin in mouse model of allergy. Innov. Food Sci. Emerg. 29, 119–124.

Lopata, A.L., O'Hehir, R.E., Lehrer, S.B., 2010. Shellfish allergy. Clin. Exp. Allergy 40, 850–858.

Mao, H.Y., Cao, M.J., Maleki, S.J., Cai, O.F., Su, W.J., Yang, Y., Liu, G.M., 2013. Structural characterization and IgE epitope analysis of arginine kinase from Scylla paramamosain. Mol. Immunol. 56 (4), 463–470.

Mistrello, G., Falagiani, P., Riva, G., Gentili, M., Antonicelli, L., 1992. Cross-reactions between shelfish and house-dust mite. Allergy 47, 287.

Ohgiya, Y., Arakawa, F., Akiyama, H., Yoshioka, Y., Hayashi, Y., Sakai, S., Ito, S., Yamakawa, Y., Ohgiya, S., Ikezawa, Z., Teshima, R., 2009. Molecular cloning, expression, and characterization of a major 38-kd cochineal allergen. J. Allergy Clin. Imm. 123, 1157–1162.

Ohtsuki, I., Maruyama, K., Ebashi, S., 1986. Regulatory and cytoskeletal proteins of vertebrate skeletal muscle. Adv. Protein Chem. 38, 1–67.

Ozawa, H., Watabe, S., Ochiai, Y., 2011. Thermodynamic characterization of muscle tropomyosins from marine invertebrates. Comp. Biochem. Physiol. B 160, 64–71.

Periswamy, M., Kalyanasundaram, A., 2007. SERCA isoforms: their role in calcium transport and disease. Muscle Nerve 35, 430–442.

Perry, S.V., 2001. Vertebrate tropomyosin: distribution, properties, and function. J. Muscle Res. Cell Mot. 22, 5–49.

Petrus, M., Nyunga, M., Causse, E., Chung, E., Cossarizza, G., 1999. (Squid and house dust mite allergy in a child) Allergie au calmar et aux acariens chez l'enfant. Archives de Pediatrie 6, 1075–1076.

Piboonpocanun, S., Jirapongsananuruk, O., Tipayanon, T., Boonchoo, S., Goodman, R.E., 2011. Identification of hemocyanin as a novel non-cross-reactive allergen from the giant freshwater shrimp Macrobrachium rosenbergii. Mol. Nutr. Food Res. 55 (10), 1492–1498.

Piromrat, K., Chinratanapisit, S., Trathong, S., 2008. Anaphylaxis in an emergency department: a 2-year study in a tertiary-care hospital. Asian Pacific J. Allergy Immunol. 26, 121–128.

Poms, R.E., Anklam, E., Matthias, K., 2004. Polymerase Chain Reaction (PCR) Techniques for Foor Allergen Detection. J. AOAC Int. 87, 1391–1397.

Rame, J.-M., Lavaud, F., Staevska, M., Dubiez, A., Adessi, B., Vigan, M., Girardin, P., Vuitton, D.A., 2002. Prevalence of the sensitization to snails and shrimps in patients allergic to house dust mites (HDM) a prospective European multicenter study. J. Allergy Clin. Imm. 109, S218.

Ramirez, Jr., D.A., Bahna, S.L., 2009. Food hypersensitivity by inhalation. Clin. Mol. Allergy 7, 4–9.

Ramos-Elorduy, J., Moreno, J.M., Prado, E., Otero, J., De Guevara, O., 1997. Nutritional value of edible insects from the state of Oaxaca, Mexico. J. Food Comp. Anal. 10, 142–157.

Reese, G., Tracey, D., Daul, C.B., Lehrer, S.B., 1996. IgE and monoclonal antibody reactivities to the major shrimp allergen Pen a 1 (tropomyosin) and vertebrate tropomyosins. Adv. Exp. Med. Bio. 409, 225–230.

Reese, G., Jeung, B.J., Daul, C.B., Lehrer, S.B., 1997. Characterization of recombinant shrimp allergen Pen a 1 (tropomyosin). Int. Arch. Allergy Immunol. 113, 240–242.

Reese, G., Ayuso, R., Lehrer, S.B., 1999. Tropomyosin: an invertebrate pan-allergen. Arch. Allergy Immunol. 119, 247–258.

Rosmilah, M., Shahnaz, M., Zailatul, H.M., Noormalin, A., Normilah, I., 2012. Identification of tropomyosin and arginine kinase as major allergens of *Portunus pelagicus* (blue swimming crab). Trop. Biomed. 29 (3), 467–478.

Ross, M.P., Ferguson, M., Street, D., Klontz, K., Schroeder, T., Luccioli, S., 2008. Analysis of food-allergic and anaphylactic events in the National Electronic Injury Surveillance System. J. Allergy Clin. Imm. 121, 166–171.

Rumpold, B.A., Schluter, O.K., 2013. Potential challenges of insects as an innovative source for food and feed production. Innov. Food Sci. Emerg. 17, 1–11.

Sampson, H.A., 2003. Anaphylaxis and emergency treatment. Pediatrics 111, 1601–1608.

Sampson, H.A., 2013. Peanut oral immunotherapy: is it ready for clinical practice? J. Allergy Clin. Imm. 1, 15–21.

Schneider Chafen, J.J., Newberry, S.J., Riedl, M.A., Bravata, D.M., Maglione, M., Suttorp, M.J., Sundaram, V., Paige, N.M., Towfigh, A., Hulley, B.J., Shekelle, P.G., 2010. Diagnosing and managing common food allergies: a systematic review. J Amer. Med. Assoc. 303, 1848–1856.

Shanti, K.N., Martin, B.M., Nagpal, S., Metcalfe, D.D., Rao, P.V., 1993. Identification of tropomyosin as the major shrimp allergen and characterization of its IgE-binding epitopes. J. Immunol. 151 (10), 5354–5363.

Shiomi, K., Sato, Y., Hamamoto, S., Mita, H., 2007. Sarcoplasmic calcium binding protein: Identification as a new allergen of the black tiger *shrimp Penaeus monodon*. Int. Arch. Allergy Immunol. 146, 91–98.

Sicherer, S.H., Sampson, H.A., 2014. Food allergy: Epidemiology, pathogenesis, diagnosis, and treatment. J. Allergy Clin. Imm. 133, 291–307, e295.

Sicherer, S.H., Muñoz-Furlong, A., Sampson, H.A., 2014. Prevalence of seafood allergy in the United States determined by a random telephone survey. J. Allergy Clin. Imm. 114 (1), 159–165.

Sidenius, K.E., Hallas, T.E., Poulsen, L.K., Mosbech, H., 2001. Allergen cross-reactivity between house-dust mites and other invertebrates. Allergy 56 (8), 723–733.

Sobotka, A.K., Franklin, R.M., Adkinson, Jr., N.F., Valentine, M., Baer, H., Lawrence, M., 1976. Allergy to insect stings: II. Phospholipase A: the major allergen in honeybee venom. J. Allergy Clin. Imm. 57, 29–40.

Spudich, J.A., Watt, S., 1971. The regulation of rabbit skeletal muscle contraction. I. Biochemical studies of the tropomyosin-troponin complex with actin and the proteomlytic fragments of myosin. J. Biochem. 246 (15), 4866–4871.

Srinroch, C., Srisomsap, C., Chokchaichamnankit, D., Punyarit, P., Phiriyangkul, P., 2015. Identification of novel allergen in edible insect, *Gryllus bimaculatus*, and its cross-reactivity with Macrobrachium spp. allergens. Food Chem. 184, 160–166.

Sweeney, H.L., 1995. Function of the N-terminus of the myosin essential light chain of vertebrate muscle. Biophys. J. 68, 118S–1118S.

Tanaka, H., Takahashi, H., Ojima, T., 2013. Ca2 + -binding properties and regulatory roles of lobster troponin site II and IV. FESBS Lett. 587, 2612–2616.

Tarbar-Purroy, A.M.J., Alvarez-Puebla, S.A., Acero-Sainz, B.E., Garcia-Figueroa, S., Echechipia-Madoz, J.M., Olaguibel-Rivera, Quirce-Gancedo, S., 2003. Carmine (E-120)-induced occupational asthma revisited. J. Allergy Clin. Imm. 111, 415–419.

Taylor, S.L., Hefle, S.L., 2006. Food allergies and intolerances. In: Shils, M., Shike, M., Ross, A.C., Caballero, B., Cousins, R.J. (Eds.), Modern Nutrition in Health and Disease. 10th Ed. Lippincott Williams & Wilkins, Philadelphia, PA, USA, pp. 1512–1530.

Taylor, S.L., Nordlee, J.A., Niemann, L.M., Lambrecht, D.M., 2009. Allergen immunoassays: considerations for use of naturally incurred standards. Anal. Bioanal. Chem. 395, 83–92.

United States Congress, 2004. Food Allergen Labeling and Consumer Protection Act of 2004, 21 USC 301, US Code of Federal Regulations.

van Ree, R., Antonicelli, L., Akkerdaas, J.H., Garritani, M.S., Aalberse, R.C., Bonifazi, F., 1996a. Possible induction of food allergy during mite immunotherapy. Allergy 51, 108–113.

van Ree, R., Antonicelli, L., Akkerdaas, J.H., Pajno, G.B., Barberio, G., Corbetta, L., Ferro, G., Zambito, M., Garritani, M.S., Aalberse, R.C., Bonifazi, F., 1996b. Asthma after consumption of snails in house-dust-mite-allergic patients: a case of IgE cross-reactivity. Allergy 51, 387–393.

Verhoeckx, K.C.M., van Broekhoven, S., den Hartog-Jager, C.F., Gaspari, M., de Jong, G.A.H., Wichers, H.J., van Hoffen, E., Houben, G.F., Knulst, A.C., 2014. House dust mite (Der 10) and crustacean allergic patients may react to food containing yellow mealworm proteins. Food Chem. Toxicol. 65, 364–373.

Vieths, S., Scheurer, S., Ballmer-Weber, B., 2002. Current understanding of cross-reactivity of food allergens and pollen. Ann. NY Acad. Sci. 964, 47–68.

Voltolini, S., Pellegrini, S., Contatore, M., Bignardi, D., Minale, P., 2014. New risks from ancient food dyes: cochineal red allergy. European Ann. Allergy Clin. Immunol. 46, 232–233.

Vuitton, D.-A., Rancé, F., Paquin, M.-L., Adessi, B., Vigan, M., Gomot, A., Dutau, G., 1998. Cross-reactivity between terrestrial snails (*Helix* species) and house-dust mite (*Dermatophagoides pteronyssinus*) I. In vivo study. Allergy 53, 144–150.

Walker-Smith, J., 2003. Hypoallergenic formulas: are they really hypoallergenic? Ann. Asthma Allergy Immunol. 90, 112–114.

White, A.J., Northcutt, M.J., Rohrback, S.E., Carpenter, R.O., Niehaus-Sauter, M.M., Gillen, C.M., 2011. Characterization of sarcoplasmic calcium binding protein from freshwater crayfish *Procambarus clarkii*. Comp. Biochem. Physiol. B 160, 8–14.

Witteman, A.M., Akkerdaas, J.H., van Leeuwen, J., van der Zee, J.S., Aalberse, R.C., 1994. Identification of a cross-reactive allergen (presumably tropomyosin) in shrimp, mites and insects. Int. Arch. Allergy Immunol. 105 (1), 56–61.

Wuthrich, B., Kagi, M.K., Stucker, W., 1997. Anaphylaxis reactions to ingested carmine (E120). Allergy 52, 1133–1137.

Wuytack, F., Raeymaekers, L., Missiaen, L., 2002. Molecular physiology of the SERCA and SPCA pumps. Cell Calcium 32, 279–305.

Yocum, M.W., Butterfield, J.H., Klein, J.S., Volcheck, G.W., Schroeder, D.R., Silverstein, M.D., 1999. Epidemiology of anaphylaxis in Olmsted County: a population-based study. J. Allergy Clin. Imm. 104, 452–456.

Yu, C.J., Lin, Y.F., Chaing, B., Chow, L.P., 2003. Proteomics and immunological analysis of a novel shrimp allergen, Pen m 2. J. Immunol. 170, 445–453.

Yu, H.L., Cao, M.J., Cai, Q.F., Weng, W.Y., Su, W.J., Liu, G.M., 2011. Effects of different processing methods on digestibility of Scylla paramamosain allergen (tropomyosin). Food Chem. Toxicol. 49, 791–798.

Zhang, Y.X., Chen, H.L., Maleki, S.J., Cao, M.J., Zhang, L.J., Su, W.J., Liu, Guang, M.L., 2015. Purification, characterization, and analysis of the allergenic properties of myosin light chain in *Procambarus clarkii*. J. Agr. Food Chem. 63, 6271–6280.

Zot, A.S., Potter, J.D., 1987. Structural aspects of troponin-tropomyosin regulation of skeletal muscle contraction. Ann. Rev. Biophys. Chem. 16, 5353–5559.

Chapter 10

Brief Summary of Insect Usage as an Industrial Animal Feed/Feed Ingredient

M.J. Sánchez-Muros, F.G. Barroso, C. de Haro
Department of Biology and Geology, University of Almeria, Carretera de Sacramento, Almería, Spain

Chapter Outline

Overview	273
Justification of Using Insects in Animal Feed	273
Current Overview of the Use of Insects in Animal Feeding	274
Examples of Livestock Fed With Insects as Feed Ingredients	277
Poultry	277
Pigs (*Sus* sp.)	282
Fish	283
Hybrid Fish	290
Polyculture	290
Crustaceans (Shrimp, Crabs, Lobsters and Their Relatives)	291
Mollusks (Clams, Oysters, Snails and Their Relatives)	292
Overview	292
Other Animals	292
Benefits and Constraints Associated with Using Insects as Livestock Feed Ingredients	293
Nutritional	293
Feed Security and Safety	295
Animal Welfare	297
Promising Opportunities for Research and Technological Advancement	298
Ecological Aspects and Sustainability	298
Environmental Enrichment for Livestock Animals	299
Chitin	299
Insect Nutritive Value Improvement Using Different Rearing Systems	299
Physical and/or Chemical Treatments of Insect Meals to Improve Their Assimilation	300
Conclusions	300
References	301

OVERVIEW

Justification of Using Insects in Animal Feed

The production of food from animal origins has become increasingly expensive in economic and environmental terms. This situation is mainly provoked by increased demand because of increasing human populations and changes in human diets, both of which lead to increases in the demand for and ingestion of animal-derived products. Global demand for meat products will increase by 58% between 1995 and 2020, and meat consumption will rise from 233 million t (metric tons) in 2000 to a possible 300 million t by 2020. Milk consumption will increase from 568 to 700 million t by 2020, and there will be an estimated 30% increase in the demand for egg production (FAO, 2004a).

Animal feeding is one of the most expensive aspects of animal production, and it is very damaging from an environmental point of view. The global production of animal feed is estimated at about 1000 million t/year, including 600 million t of compound feed (FAO, 2004a). Moreover, it was identified as a major contributor to land occupation, primary production (the net amount of biomass produced each year by plants) use, acidification of soil, climate change, energy use, and water dependence (Mungkung et al., 2013).

The nutritive needs of monogastric species include high quality and high quantities of protein in the diet. From a nutritional point of view, in addition to stable quantity and quality production, protein sources must have high protein content, an adequate amino acid profile, high digestibility, good palatability, and no antinutritional factors (Barrows et al., 2008). Currently, the principal protein sources for animal feed are fishmeal (fish meal) and soy meal, and both products are linked to environmental problems. Soy cultivation causes the deforestation of areas with high biological value (Carvalho, 1999; Osava, 1999), high water consumption (Steinfeld et al., 2006), pesticide and fertilizer utilization (Carvalho, 1999), and transgenic variety usage (Garcia and Altieri, 2005), which cause significant environmental deterioration (Osava, 1999). On the other hand, fish meal is a resource that depends on the catch. The deterioration of the marine environment and the stripping of fisheries have resulted in decreased fish meal production, and its production is therefore quantitatively and qualitatively variable (AFRIS, 2015). In addition, an increase in demand has led to higher prices, including an increase from $600USD/t ton in 2005 to $2000USD/t to in June 2010. This trend of increasing prices is likely to continue (International Monetary Fund, 2010), because the prohibitive costs of feed (eg, meat meal, fish meal, and soybean meal) are major constraints to further development.

Change and innovation are required in many livestock production systems if they are to meet the present and future demands for animal products. In this context, research on and commercial implementation of new feeds (especially those rich in protein) for animal feeding is needed for sustainable animal production.

Several by-products and wastes from different industries have been examined for their potential as animal feed options with variable results, and they allow different inclusion levels of ingredients, thus saving traditional feed. Nevertheless, the variable quality and limited production are two important constraints.

Currently, there is great interest in the role of insects in animal feeding. Nutritive composition studies showed that insect protein values in most species have high protein quality and quantities (Ladrón de Guevara et al., 1995; Ramos-Elorduy et al., 1981, 1982, 1984, 1997). Studies of the protein percentages of numerous insect species revealed many species with higher protein levels than fish meal or soy meal. The highest values were found in the coleopteran [cactus weevil (*Metamasius spinolae* Gyllenhal, 69.1%) and *Rhantus atricolor* Aubé, 71.1%], dipterous [common fruit fly (*Drosophila melanogaster* Meigen, 70.1%)], and orthopteran orders (*Boopedon flaviventris* Bruner, 76.0%), (*Melanoplus mexicanus* Saussure, 77.1%), and (*Sphenarium histrio* Gerstaecker, 74.8%) (Sánchez-Muros et al., 2014). In addition to their nutritive qualities, the utilization of insects as feed implies certain environmental benefits such as high food conversion. Insects also feed on organic wastes, which could aid in the recycling of organic matter. Furthermore, compared to livestock, the use of insects could lead to the reduction of released greenhouse gasses (Oonincx and de Boer, 2012) and ammonia, and it could lead to decreased land occupation and water consumption (van Huis et al., 2013).

We can find many studies in the scientific literature that evaluate insect ingredients in animal feed. There is also a growing commercial interest in insects as animal feed, and the research is conducted in parallel with the increasing costs of traditional raw materials (López-Vergé et al., 2013). However, if we want to determine the nutritional potential of insects, research cannot only focus on their chemical composition (percentage of fat or protein and amino acid or fatty acid), it is also necessary to evaluate the nutritive and physiological utilization. Feeding trials in different species are needed to determine their future role as protein components in animal feed. As Finke et al. (1985) argued, the evaluation of protein quality through bioassay techniques is a more precise indicator of limiting amino acids than amino acid analyses. It is incorrect to think that only the proportion of each type of amino acid in each insect species can indicate its protein quality.

On the other hand, the potential of insect in animal feeding is not only defined by the nutritive characteristics. Although there are several companies reaching impressive scale and low prices for black soldier fly larvae, the need for mass production of various species of insects cheaply and with lower environmental impact is probably one in major need of research.

Current Overview of the Use of Insects in Animal Feeding

It is reasonable to assume that the number of scientific publications on the use of insects as animal feed has increased in recent years, and the numbers of papers found on this topic indicate that this is the case (Fig. 10.1). For instance, the number of publications in the last 15 years (2000–15) has tripled relative to the previous 30 years (1969–99). During the first 8 years of the 21st century (2000–07), there were 34 publications, and there were 56 publications during the next 8 years (2008–15). This trend suggests that the production of publications will continue to increase.

Another important aspect is that scientific interest is higher for some types of animal production, particularly in aquaculture production. Therefore, while the number of livestock feeding experiments has doubled in the past 15 years, the number of feeding experiments in aquaculture has quadrupled.

The increased interest in insects as an alternative protein source is probably due to the increased cost and limited availability of fish meal, which is the ideal protein ingredient in animal feed. The role of fish meal in the formulation of commercial aquaculture feed is higher than in feed for livestock, and this could explain why the number of feeding studies in aquaculture is much higher than in livestock.

The nutritional potential of up to 24 different species of insects belonging to 6 different orders (Blattodea, Coleoptera, Diptera, Isoptera, Lepidoptera, and Orthoptera) has been evaluated. However, most publications have tested species from

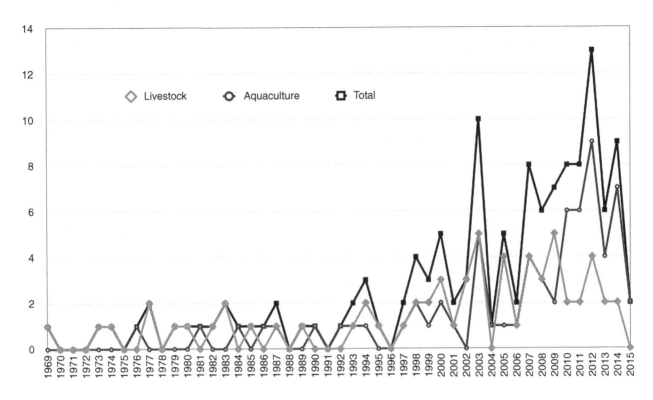

FIGURE 10.1 **Papers per year.** A literature review was conducted using the number of web publications in the ISI (ISI Web of Knowledge) as a reference, but as there were many articles in nonindexed journals, we had to search in Google academic and other websites in general. The keywords used were as follows: insect meal; animal feeding; livestock; aquaculture; Bombyx; Hermetia; Musca; Tenebrio.

the Diptera (48%) and Lepidoptera (29%) orders (Fig. 10.2). Clear differences between the insects evaluated in the feeding trials for livestock and aquaculture have been observed. In livestock, there is a predominance of experiments with house fly (*Musca domestica* L.; 43%), followed by silkworm/silkmoth (*Bombyx mori* L.;15%). However, that predominance is not so clear in aquaculture. Although experiments with the house fly (34%) are the most abundant, a similar number (29%) have evaluated silkworm. Furthermore, studies on other species (eg, black soldier fly, *Hermetia illucens* L., and yellow mealworm, *Tenebrio molitor* L.) are also common.

Lastly, we want to focus on the location of research studies (Fig. 10.3). Although experiments have been conducted in 27 countries, most of the studies are conducted in 10 Asian (40%) and 7 African countries (34%). In Asia, countries such as India (15%) and China (12%) are highlighted, but most of the feeding trials in Africa have been conducted in Nigeria (29%). It seems clear that the potential of insects as animal feed has been further investigated in this African country, because one-third of studies worldwide were conducted in Nigeria.

It should also be noted that researchers in the United States have conducted 12% of the feeding studies, but there is a downward trajectory to research conducted in this country. For instance, 10 studies were carried out before 2000, but only 4 were conducted after that date. In contrast, the interest observed in European countries seems to be increasing in recent years, and 12 (out of 13 total studies) have been published since 2000.

Besides gains in scientific development, there has been an increase in the number of companies that produce insects as pet feed and also products derived from insects. Nodaways, the major number of stakeholders is located in Europe (157, mainly research), followed by the United States (56). A growing interest in insects as an alternative protein source has also led to the establishment of companies in the Netherlands, Spain, South Africa, and the United States. More complete information can be found at the following website: http://www.fao.org/forestry/edibleinsects/stakeholder-directory/en/.

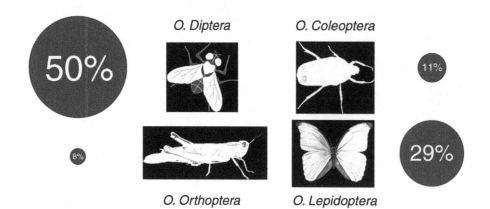

FIGURE 10.2 Insects used in animal food (% papers).

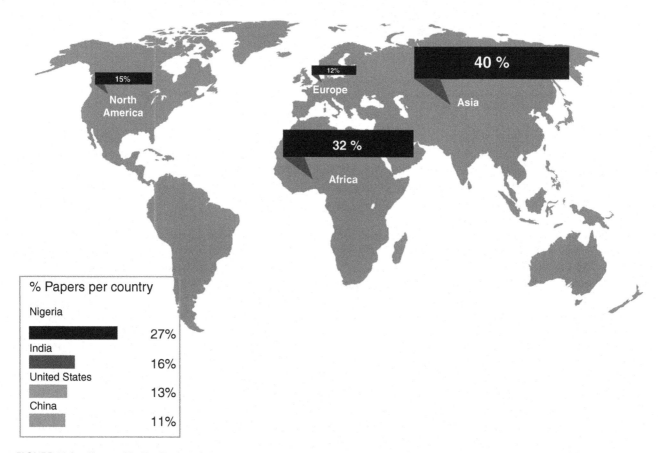

FIGURE 10.3 Geographic distribution of papers.

EXAMPLES OF LIVESTOCK FED WITH INSECTS AS FEED INGREDIENTS

Poultry

In contrast to ruminant livestock production, the major problem facing the poultry industry is a food supply that will contain all diet components needed by birds for rapid growth within a short period of time (Oyegoke et al., 2006). This is because birds are monogastric, and hence they lack the complex digestive anatomy for the synthesis of proteins and vitamins that is found in ruminants (Adeniji, 2007). Therefore, soybean meal is the major protein source of their diet, and it is supplied together with fish meal, which covers any amino acid deficiency associated with vegetable proteins (Miles and Jacobs, 1997). Thus, fish meal is a very important feed ingredient in poultry production (Ijaiya and Eko, 2009).

The animal protein source is the most costly ingredient for the formulation of poultry diets compared to other nutrient sources (Khatun et al., 2003). Soybean meal and fish meal are expensive, and our ability to replace them is limited. Therefore, the search for alternative sources used for total or partial replacement in the future is an important task (Ramos-Elorduy et al., 2002). The search for alternative and sustainable proteins is an issue of major importance that needs viable solutions in the short term, and this makes insect meal an increasingly attractive feed option for poultry (Makinde, 2015). In contrast to fish, wild birds and free-ranging poultry naturally consume many insects (Zuidhof et al., 2003), especially during their early life stages. For example, chickens can be found picking worms and larvae from the topsoil and litter where they walk (van Huis et al., 2013).

Although there are some very interesting reviews on insect use in poultry feeds (Khusro et al., 2012; Makkar et al., 2014; Veldkamp & Bosch, 2015), we have attempted to update the maximum number of publications on this subject in the next section. We believe that the most important contributions to this subject are presented here, but a number of articles have been difficult to access because of their local interest and distribution.

*Broiler Chickens (*Gallus gallus domesticus L.*)*

Order Blattodea (Cockroaches)

American Cockroach (Periplaneta americana L.) The only feeding trial with cockroaches was conducted by Aigbodion et al. (2012) who fed broilers insect diets enhanced with American cockroach adults. These researchers obtained significantly higher growth at 8 weeks with a diet containing cockroaches.

Order Coleoptera (Beetles)

Unlike the studies on maggots or silkworms, there are limited feeding trials on the use of mealworms or other beetle species in poultry diets.

Maize Weevil (*Sitophilus zeamais* Motschulsky)

López-Vergé et al. (2013) studied the effects of adding *maize weevil* larvae to the diet on performance parameters. The animals were divided into two treatment groups (insect-infested diet and an untreated diet as the control). The broiler chickens fed the insect-infested diet had a higher final body weight and a higher average daily feed intake than did animals fed the control diet. Furthermore, they did not detect pathogens in the insect (neither *Salmonella* sp. nor *Listeria monocytogenes* [Murray]), and hence concluded that maize weevil could be a safe ingredient.

Yellow Mealworm (*T. molitor*)

Ramos-Elorduy et al. (2002) found that yellow mealworms had potential usage as a protein source for raising broilers. Feeds with three percentage levels of mealworms (0, 5, and 10% dry weight) were used in a 19% protein content sorghum-soybeanmeal-based diet. They found no significant differences in feed intake, weight gain, or feed efficiency among the treatments.

Ballitoc and Sun (2013) conducted other experiments to determine the growth performance and carcass characteristics of broiler chicks fed with feed containing different percentage levels of ground yellow mealworm larvae. The study used five treatments at inclusion levels of 0, 0.5, 1, 2, and 10% yellow mealworm to replace commercial feed. They concluded that supplementation with ground yellow mealworms produced an increase in feed intake, body weight, and the efficiency of feed consumed due to a lower feed conversion rate (FCR). However, the best performing inclusion level was low (2% ground yellow mealworms).

Order Diptera (Flies)

Black Soldier Fly (H. illucens) Oluokun (2000) examined the effects of treatments with either fish meal or black soldier fly larvae meal regarding a full-fat soybean meal diet as a control. The average live weight gains of broilers fed with fish meal or black soldier fly larvae meal were higher than those fed the control diet. Oluokun also reported that the diet upgraded with the larvae meal did not affect the rate of gain, feed consumption, or the feed/gain ratio regarding fish meal. Moreover, there were improvements in the carcass yield, internal organ measurements (kidney, gizzard, and liver), and abdominal fat in animal feed with larvae diet regarding fish meal or control diet. Then the author concludes that maggot meal could replace fish meal in the broiler rations without any adverse effect on zootechnic indices.

Elwert et al. (2010) conducted a study during the starter and grower phases of broilers. In the starter phase, all-vegetable (wheat-corn-soybean meal) and fish meal diets (3% fish meal) served as negative and positive controls, respectively. In the test diets, different black soldier fly larvae meal proportions, depending of different level of defatting, were supplemented: 6.6% black soldier fly larvae meal with a fat content of 37%, 5.4% black soldier fly larvae meal with a fat content of 22%, and 4.7% black soldier fly larvae meal with a fat content of 15%. During the starter period, the full fat *Hermetia* meal (crude fat 37) yielded similar high body weights as compared to the fish meal diet. However, during the grower phase, the results comparing all-vegetable diets and a partially defatted diet (5% black soldier fly larvae meal, crude fat 22) did not lead to clear conclusions.

House Fly (*M. domestica*)

House fly maggots (*M. domestica* larvae) are most commonly used as feed for poultry. In many rural areas of the world, maggots have always constituted part of the daily diet of scavenging poultry (Téguia et al., 2002). Thus, maggot meal has been included in broiler diets as a replacement for conventional protein sources (Makinde, 2015). Calvert et al. (1969) found that a corn-house fly diet led to a slight (but not significant) improvement in the growth rate of chicks compared to a conventional corn-soybeanmeal diet.

In general, most of the studies using maggot meal have described similar or improved growth rates in chicks as compared to that resulting from vegetable meals. Teotia and Miller (1973) found no significant differences in final body weights or feed/gain ratios between chicks fed house fly pupae diets and those fed soybean meal control diets. Hwangbo et al. (2009) used a corn-soybean diet as basic feed (control), and diets supplemented with 5.0, 10.0, 15.0, or 20.0% maggot meal were used as experimental feed. They also found that diets containing 10–15% maggot meal improved the carcass quality and growth performance of broiler chickens.

Pro et al. (1999) used two sorghum diet treatments with maggot meal (24 and 19% for starter and grower phases, respectively), and soybean meal (38 and 32% for starter and grower phases, respectively) was used as a control. Although the maggot meal inclusion rate was high, there were no differences in body weight gain, feed consumption, and feed efficiency between the two treatments.

Adeniji (2007) investigated the replacement value of maggot meal for groundnut cake (GNC) in broiler diets. Several experiment diets were used in this study with increasing maggot meal inclusions (up to 22% of the diet). The results of this experiment indicated that broilers tolerated the 100% maggot meal replacement in their diet without adverse effects on performance. However, there were no significant differences in weight gain, feed intake, feed/gain ratio, or nutrient retention.

Adesina et al. (2011) evaluated the performance of broilers fed cassava peel-maggot meal mixture as a partial or total replacement for maize. In particular, they used a 4:1 ratio of dry cassava peel to maggot meal. The weight gain to FCRs were similar in the broilers fed the control diet (0% peel-maggot meal mixture) and the 50% peel-maggot meal mixture (29% diet). Regarding the cost of production, they also observed a significant reduction in the total feed costs when the peel-maggot meal mixture mixture was included (up to a 50% replacement).

Maggot meal has also been used to replace meat meal in broiler diets. Bamgbose (1999) used a control diet with meat meal (8% diet) and maggot meal with 0, 50, and 100% replacement levels with or without methionine. He concluded that maggot meal could completely replace meat meal without any adverse effects on performance and nutrient utilization. Moreover, nutrient utilization and performance were significantly enhanced with methionine supplementation (0.20%).

Several feeding trials have been conducted to evaluate the performance of broiler chickens fed maggot meal as a replacement of fish meal. In a yellow maize-soy meal diet, part of the soy was replaced by either maggot meal (4% of the diet) or fish meal (3% of the diet) (Ocio et al., 1979). No significant differences were found for weight increase or FCE between the three experimental groups. In a similar experiment, Ren et al. (2011) divided chickens into three groups: control, basal diet supplemented with 4% fish meal, and basal diet supplemented with 4.44% maggot meal. The results showed that the average daily weight gain and daily feed intake for the house fly larvae meal (maggot meal) group were significantly higher than those of the fish meal and control groups at the small broiler stage.

On the other hand, Djordjevic et al. (2008) used a 5% fish meal diet as a control. Experimental diets included those with a 50% fish meal substitution (2.5% fish meal-3% MGM), a 100% fish meal substitution (0% fish meal-6% maggot meal), and a group supplemented with washed fresh housefly larvae without fish meal or maggot meal. The results indicated that the substitution of fish meal with housefly larvae had no negative effects on body mass, daily weight gain, or food conversion. However, the most interesting result was that broilers from the experimental group supplemented with washed fresh housefly larvae had the highest feed intake. As Djordjevic et al. (2008) pointed out, this result is probably associated the physiological and biological characteristics of birds.

Okah and Onwujiariri (2012) formulated five diets where fish meal (4% of the diet) was replaced with 0 (control diet), 20, 30, 40, and 50% maggot meal. They observed that the highest substitution used (2% fish meal and 2% maggot meal) showed superior performance characteristics compared to the control diet (4% fish meal), and it was also proven to be a more economical option. The cost of diets decreased with increased levels of maggot meal. A live weight kilogram of the birds fed the 50% dietary maggot meal was 34.22% cheaper than those fed the control diet (4% fish meal). Similarly, Pieterse et al. (2014) investigated the effects of three diets containing 10% fish meal, 10% maggot meal, or a control diet with soybean meal as the protein source on broilers. Chicks that received the control diet had significantly lighter carcasses and a lower breast meat yield than either the 10% maggot meal or 10% fish meal chicks.

Conversely, neither Atteh and Ologbenla (1993), Awoniyi et al. (2003), nor Okubanjo et al. (2014) obtained such satisfactory results. They observed that maggots only provided a partial replacement for fish meal in broiler diets. For instance, Atteh and Ologbenla (1993) examined the effects of replacing 0, 33, 68, or 100% of dietary fish meal (9% of diet) with maggot meal. They noted that increased dietary levels of maggot meal reduced weight gain, and that maggot meal could only replace 33% of dietary fish meal to obtain comparable results to those of the control diet with fish meal. In a similar experiment, Awoniyi et al. (2003) used a control diet with 4% fish meal, and the treatment diets were formulated with maggot meal replacing 0, 25, 50, 75, and 100% of the fish meal. They reported that production rates tended to decrease with increasing levels of maggot meal replacement, and the 25% level of fish meal replacement with maggot meal (3% fish meal and 1.17% maggot meal) was the best replacement level. Moreover, Okubanjo et al. (2014) used maggot meal to replace fish meal at 0, 25, 50, 75, and 100% levels, which constituted 0, 1.4, 2.7, 4.1, and 5.4% of the diet, respectively. As in previous experiments, the final live weight decreased with the higher levels of dietary maggot meal, and only a 25% substitution of fish meal (1.4% maggot meal) showed no significant difference compared to the control diet.

There are several studies that examined the effects of maggot meal on broiler meat quality and carcass characteristics. Teotia and Miller (1973) incorporated houseflies into broiler diets, but did not detect any differences between chicks fed the control and experimental diets based on an informal taste panel. Ren et al. (2011) also observed no negative effects on the slaughter performance and meat quality associated with the incorporation of maggot meal in the diets.

Awoniyi et al. (2003) also observed that maggot meal (maggot meal) supplementation had no significant influence on dressing percentage, leg muscle yield, or breast muscle yield. These results were in agreement with the findings of Téguia et al. (2002), but they differed from the findings of Hwangbo et al. (2009). Hwangbo et al. (2009) found that the use of maggot meal had no effect on breast meat color, but it improved the dressing percentage compared to a corn-soy diet. According to Makinde (2015), these contradictory reports could also be attributed to the trial design where Hwangbo et al. (2009) had 30 replicates per treatment relative to the 6 and 4 replicates of Awoniyi et al. (2003) and Téguia et al. (2002), respectively.

Moreover, Okubanjo et al. (2014) also found no significant effects on organoleptic flavor, color, overall acceptability scores, and meat/bone ratios in broilers when maggot meal replaced fish meal (4.5% of diet), but sensory tenderness and juiciness were significantly higher with diets containing maggot meal. Pieterse et al. (2014) compared diets containing either 10% fish meal, 10% maggot meal, or a control diet with soybean meal, and reported no differences in breast and thigh muscle color, pH, water holding capacity, or cooking losses among the treatments. However, they observed significant differences in drip loss with the highest fish meal-fed samples (followed by the control diet), and the lowest drip loss was reported for the maggot meal-fed samples. Based on these results, it appears that the inclusion of larvae meal into the diets of broilers could have positive, rather than detrimental, effects on most carcass, meat, and sensory characteristics.

Order Lepidoptera (Butterflies, Moths and Their Caterpillar Larval and Chrystalis/Pupal Stages)

In a feeding trial for broilers, Ijaiya and Eko (2009) replaced 100% fish meal (8.75% of the diet) with *Anaphe infracta* (Walsingham) meal, and they observed no significant differences in feed intake, body weight gain, FCE, or the protein efficiency ratio. Furthermore, they also observed a decline in the costs associated with increased dietary levels of insect meal.

Muga silkworm, *Antheraea assamensis* Helfer, pupae meal has already been established as an effective protein supplement in broiler rations. The effects of replacing fish meal with muga silkworm pupae meal was examined in several experiments performed at the College of Veterinary Science Khanapara (India), and Sapcota et al. (2003) reported that there were no adverse effects on body weight gain when fish meal (5% level in the broiler diet) was completely replaced with muga silkworm meal. They also observed no adverse effects on various carcass qualities (Sheikh et al., 2005).

Furthermore, according to Sheikh and Sapcota (2007), the production per live weight kilogram of broilers was more economical with diets containing 100% muga silkworm pupae meal supplements. On the other hand, Chaudhary et al. (1998) noted that broiler chicks reared on rations fortified with antibiotics exhibited significantly higher growth rates and improved feed efficiency, and the antibiotic ration was found to be the most economical of the rations studied.

In a study by Sinha et al. (2009), the objective was to compare the growth performance of commercial broiler chicks fed different levels of giant silkworm pupae, *Antheraea mylitta* Drury, meal as an economic substitute for fish meal. The best animal performance was observed in the group fed the 50% silkworm pupae meal/50% fish meal diet.

Silkworm/Silkmoth (*B. mori*)

Within Lepidoptera, most experimental diets were based on silkworm. Silkworm pupae (silkworm pupae meal) are the byproduct after the silk thread has been removed from the cocoon. These pupae also commonly serve as human food sources (Jintasataporn, 2012). It is also one of the most studied insect species in recent years regarding its potential use in broiler feeding. Although Fagoonee (1983) suggested that the substitution of fish meal with silkworm pupae meal in broiler chicks depressed consumption due to the high oil and fiber content of silkworms, different feeding trials seemed to indicate that silkworm pupae meal was a good protein source.

The first evidence of a positive effect of the replacement of fish meal with silkworm pupae meal was found by Wijayasinghe and Rajaguru (1977). They conducted three feeding trials to examine the effects of various silkworm pupae meal replacement levels on the performance of broiler and laying hens. Up to 12% silkworm pupae meal was added to the diet, and it was generally observed that fish meal could be successfully replaced by silkworm pupae meal.

In the experiment of Khatun et al. (2003), the chicks were fed four dietary treatments (6% fish meal + 0% silkworm pupae meal; 4% fish meal + 2% silkworm pupae meal; 2% fish meal + 4% silkworm pupae meal; and 0% fish meal + 6% silkworm pupae meal). They observed that the growth rate, feed conversion, liability, meat yield, and profitability increased almost linearly based on increasing silkworm pupae meal levels. Moreover, profits (Tk/broiler and Tk/Kg broiler) were significantly higher as the level of dietary silkworm pupae meal increased.

Dutta et al. (2012) found that broilers fed a mixed diet (50% fish meal/50% silkworm pupae meal) showed improved productive indices. However, the degree of protein fish meal inclusion in the diet was unknown, because they did not include that data in their publication. When they evaluated the economics of feed costs and broiler production, they found that profits increased significantly as the level of dietary silkworm pupae meal increased, and these results were also found in the two previous studies. Similarly, Konwar et al. (2008) pointed out that although silkworm pupae meal can be incorporated into broiler diets by replacing up to 100% fish meal, improved performance was observed in broilers fed 50% fish meal/50% silkworm pupae meal diets with enzyme supplementation.

Using fish meal as reference material, Jintasataporn (2012) evaluated the quality of silkworm pupae from the spun silk industry and from the silk yarn reeling industry on the growth performance, carcass yield, and sensory evaluation of broilers. Although they included up to 20% spun silkworm pupae meal in an experimental diet (0% fish meal/20% spun silkworm pupae meal/0% reeling silkworm pupae meal), treatment with 0% fish meal/5% spun silkworm pupae meal/5% reeling silkworm pupae meal showed no difference in daily weight gain compared to the control diet (10% fish meal/0% spun silkworm pupae meal/0% reeling silkworm pupae meal). However, the results indicated a higher FCR than the control, and the carcass muscle of the control treatment was higher. On the other hand, no significant differences were observed in sensory evaluation.

Another interesting study was conducted by Venkatachalam et al. (1997) on the presence of certain antinutritional substances in defatted silkworm pupae. High amounts of phenols (2%) were found in the pupae, so they were detanned and defatted. In a feeding trial using fish meal (10% of the diet) as the control, detanned silkworm pupae meal and defatted silkworm pupae meal were experimentally tested at 2.5 and 5% levels (of the total diet), respectively. The results indicated that the production rates of chicks were better on diets containing detanned silkworm pupae meal than those containing defatted silkworm pupae meal, and the same results were observed compared to the control.

Ojewola et al. (2005) conducted an experiment comparing a control diet (4% fish meal) with two experimental diets containing pallid emperor moth (*Cirina forda* Westwood) (2% fish meal/2% pallid emperor moth, and 0% fish meal/4% pallid emperor moth). They found no significant differences between the growth performances of the broilers fed on the compound larvae diets compared to those fed fish meal.

Order Orthoptera (Crickets, Grasshoppers, Locusts, and Katydids)

House Cricket (Acheta domesticus) The replacement of soybean meal as the major source of protein by dried house cricket (*A. domesticus*) meal in practical diets doesn't show significant differences in weight gain between chicks fed corn-soybean meal diet and those fed corn-cricket diets. Feed:gain ratios improved significantly when diets were supplemented with methionine and arginine (Nakagaki et al., 1987).

Mormon Cricket (*Anabrus simplex* Haldeman)

High inclusion levels of Mormon cricket have been tested with good results. Thus, DeFoliart et al. (1982) found that corn-cricket-based diets (30% cricket in the diet) produced significantly better broiler chicken growth than that produced by a conventional corn-soybean-based diet. Finke et al. (1985) used crickets directly collected from crops, and also found no significant differences in weight gain in a feeding trial where a corn-cricket diet (including ground crickets at 18–28% of total feed; decreasing during chicken growth) was compared to a corn-soybean meal diet. Moreover, they concluded that, even at high levels, the incorporation of insect protein into the diet had no effect on the carcass quality and the taste of meat. Similar to Nakagaki et al. (1987), Finke et al. (1985) also used this same cricket species, and noted that methionine and arginine were probably colimiting in this insect.

Acrida cinerea (*Thunberg*)

Wang et al. (2007) found that the amino acid content of the grasshopper species *Acrida cinerea* was comparable to that found in fish meal. When maize-insect-soybean meal diets were formulated, the inclusion of 15% of this grasshopper meal did not affect broiler weight gain, feed intake, or the gain/feed ratio. Moreover, Liu and Lian (2003) replaced 20 and 40% fish meal in broiler diets, and observed similar growth rates and feed consumption to those seen in the control diet.

Similar to the results observed in *A. cinerea* experiments, Wang et al. (2005) found that feed containing up to 15% of the cricket *Teleogryllus mitratus* Burmeister could be included in corn-soybean meal diets without any adverse effects on broiler weight gain, feed intake, or the gain/feed ratio. Moreover, their results indicated that the field cricket had considerable amounts of digestible amino acids.

Migratory Locust (*Locusta migratoria* L.)

This locust (grasshopper) species is famous for its seasonal abundance in vast swarms in the wild and occurs in Africa, Asia, Australia, and New Zealand. In a study using increasing replacements of fish meal with the migratory locust (0, 1.7, 3.4, and 6.8% in the diet), the best production rate results were found at 1.7% substitution trials (Adeyemo et al., 2008). Nevertheless, because of the lack of statistical analysis, it is difficult to draw reliable conclusions from this study.

Laying Hens (Gallus gallus domesticus L.)
Order Diptera (Flies)

House Fy (M. domestica) Parshikova et al. (1981) reported an increase in egg yield when housefly larvae replaced fish meal in the diet of hens. Akpodiete et al. (1998) investigated the replacement value of fish meal with maggot meal (house fly) in the diets of laying chickens using performance indices, egg quality characteristics, and egg yolk biochemistry. A control diet (4% fish meal) was replaced with diets containing 25, 50, 75, and 100% maggot meal. They observed that the maggot meal substitution in the diets of birds did not show adverse consequences on performance and egg quality characteristics for all parameters measured, with the exception of albumen weight. Furthermore, egg yolk cholesterol and calcium concentrations decreased significantly with increased maggot meal inclusion. According to the authors, this may be of nutritional interest regarding the dietetic treatment of patients with atherosclerosis and other cardiovascular diseases.

In a similar experiment, Agunbiade et al. (2007) used fish meal and maggot meal as animal protein sources to supply the remaining 25% of the total dietary protein: 25% fish meal/0% maggot meal, 18.75% fish meal/6.25% maggot meal, 12.5% fish meal/12.5% maggot meal, 6.25% fish meal/18.75% maggot meal, and 0% fish meal/25% maggot meal. Although the average daily feed intake, weight gain, and FCR were not significantly affected, hen-day egg production was significantly influenced by the dietary treatments. The 12.5% fish meal/12.5% maggot meal diet showed the highest hen-day production, and the authors thought this could be a result of the complementary effects of the maggot meal and fish meal amino acid profiles. Moreover, regarding egg quality, they found that eggs from birds on diets containing maggot meal exhibited a linear decrease in shell thickness and shell weight. They concluded that the lower calcium content of maggot meal compared to fish meal was a feasible explanation.

Although the work of Dankwa et al. (2002) was conducted using rural poultry instead of laying hens, we are including it here because they studied the effects of supplementing the diet of scavenging chickens with live maggots on productivity (eg, body weight, age at first lay, egg weight, number of eggs hatched, and weight of chicks hatched). The experiment lasted for 14 months, and the experimental group of birds was supplemented with 30–50 g of live housefly larvae. They reported that clutch size, egg weight, number of eggs hatched, and chick weight were significantly higher in supplemented birds than in control birds. Previously, Ekoue and Hadzi (2000) found that the movements of live larvae stimulated consumption by chickens.

Order Lepidoptera (Butterflies, Moths, and Their Caterpillar Larval and Chrystalis/Pupal Stages)

Silkworm/Silkmoth (B. mori) To determine the effects of silkworm pupae meal on the growth and egg production performance of Rhode Island Red chickens, Khatun et al. (2005) compared layer chicks fed three dietary treatments (6% protein concentrate (PC) + 0% silkworm pupae meal; 0% PC + 6% silkworm pupae meal; and 0% PC + 8% silkworm pupae meal). In this study, the values of the most significant parameters (ie, profitability, growth, and egg production performance) were significantly higher in the 0% PC + 6% silkworm pupae meal diet. Therefore, according to the authors, cheaper silkworm pupae meal could be an excellent substitute for the costly protein concentrate used to formulate diets for layers, which could lead to increased profitability.

Turkeys (Meleagris gallopavo L.)
Order Coleoptera (Beetles)

An experiment conducted by Despins and Axtell (1994) demonstrated that lesser mealworm (*Alphitobius diaperinus* Panzer) larvae can be used as selective feed for turkey poults. Insect meal was not used in this feeding trial. Instead, larvae were placed directly on the ground so the birds could freely forage for them. No significant difference was found between the body weight of poults that fed on larvae and starter feed compared to that of poults that fed on starter feed only.

On the contrary, the same authors (Despins and Axtell, 1995) used a similar design in a later study, and they reported that the body weight of chicks feeding on starter feed and larvae was significantly greater than the weight of chicks consuming feed only. Furthermore, these experiments showed that broiler chicks feed readily on darkling beetle larvae in the litter, and they consume large numbers during the first days of life.

Overview

From these studies using insects as food ingredient for poultry, we can draw the following conclusions:

- Although there are many studies, most are not comparable to each other. The diversity of insect species, insect breeding systems, varied control diets, number of chickens, environmental conditions, and so forth do not allow definitive conclusions to be drawn. Thus, larger scale and more systematic studies utilizing more standardized and industrially relevant methodologies are needed.
- Most of the experiments were conducted in developing countries. As noted by Khatun et al. (2003), a limited number of feed ingredients are available for the formulation of balanced diets. Moreover, fish meal (a conventional animal protein source) is scarce and expensive, and it may even contain lethal pesticides that are detrimental deleterious to the poultry industry (Khatun et al., 2003). Therefore, if we want to increase the profitability of poultry production in these countries, the search for unconventional feed ingredients is especially relevant.
- Native insect species in each area are often studied. Furthermore, some species have been traditionally consumed by humans (Nigeria) or are by-products (eg, silkworm pupae), but have not been previously tested in broilers.
- The available literature confirms the feasibility of total or partial replacement of fish meal (fish meal) with insect meal (insect meal). The highest inclusion percentages were reported in some Orthoptera (up to 30 of the diet) (DeFoliart et al., 1982; Finke et al., 1985). Regarding house fly, most trials indicate that partial or even total replacement of fish meal is possible, although the optimal inclusion rate is generally lower than 10% (Makkar et al., 2014). Higher rates have resulted in lower intake and performance, which may be due to a decrease in palatability. For instance, the darker color of the meal may be less appealing to chickens (Atteh and Ologbenla, 1993; Bamgbose, 1999).
- Several studies analyzed economic valuation. Dutta et al. (2012), Khatun et al. (2003, 2005), Ijaiya and Eko (2009), and Sheikh and Sapcota (2007) estimated that increased benefits are obtained by replacing the fish meal or protein concentrate with silkworm pupae in poultry production. Therefore, silkworm meal could be a good substitute for scarce and expensive fish meal in broiler diets, which could increase economic gains (Ijaiya and Eko, 2009). Similarly, from an economic point of view, Akpodiete and Inoni (2000), Awoniyi et al. (2003), and Téguia et al. (2002) concluded that maggot meal could replace fish meal or maize (Adesina et al., 2011). Furthermore, it has no negative effect on the performance of the birds. According to Atteh and Ologbenla (1993), the cost of maggot harvesting and processing was about 83.3% lower than the costs associated with an equivalent weight of fish meal.

Pigs (*Sus* sp.)

There is limited information on the use of insects in pig feeding, but the results of a few tested species are promising. The use of silkworm pupae has been examined in growing and/or finishing pigs. The replacement of 100% soymeal (Coll et al., 1992) or fish meal (Medhi, 2011; Dankwa et al., 2000; Medhi et al., 2009a,b) with nondefatted silkworm meal resulted in no adverse effects on growth performance and carcass characteristics. However, at substitution levels higher than 50%, there was a reduction in intake, and it was compensated by a better FCR that did not alter carcass quality, meat quality, or blood parameters (Coll et al., 1992).

Regarding breeding pigs, only data for Diptera (specifically house fly and black soldier fly) are available. These studies indicate that the administration of insects in pig feed does not have adverse effects on reproduction. The sows and their offspring were fed a diet containing processed house fly meal (maggot meal), and there were no adverse effects on piglet performance, health (Bayandina, 1979; Bayandina and Inkina, 1980; Poluektova et al., 1980), or organoleptic properties or on the physiology and breeding performance of the sows (Bayandina and Inkina, 1980).

Supplementation with 10% house fly meal did not have a negative effect on body weight gain or FCE (Viroje and Malin, 1989), and the complete replacement of fish meal with house fly did not compromise growth performance or weaner pig economics (Dankwa et al., 2000).

Black soldier fly is especially valuable for pig feeding because of its lipid and calcium content, and it was as palatable to pigs as a soybean meal-based diet. However, it is deficient in methionine, cysteine, and threonine, and has high ash content (Newton et al., 1977). Black soldier fly prepupae meal allowed a 50% replacement of dried plasma in diets of early-weaned pigs, which resulted in increased weight gain and feed efficiency (Newton et al., 2005).

Fish

Fish are cultured animals that need high protein qualities and quantities. Aquaculture feeding is dependent on fish meal because of its particular nutritive characteristics. This dependence on fish meal and the high need for fish protein makes finding protein sources for fish meal replacement particularly important, and insects are good candidates. A sign of this interest is evident in the number of fish feeding trials using insects that has increased since the early century, and the increasing numbers seen in recent years are presented in Fig. 10.1.

Family Bagridae (Bagrid Catfishes)
Pelteobagrus vachellii Richardson

A 6-week growth trial was conducted to compare the effects of dietary supplementation with house fly and soybean meal on the growth performance and antioxidant responses of *P. vachellii*. The results showed that the 71% fish meal replacement reduced the growth rates and antioxidant capacity in both the house fly and soybean meal supplementation trials compared to a control diet (Dong et al., 2013).

Family Cichlidae (Cichlids)
Nile Tilapia (Oreochromis niloticus L.)
Order Coleoptera (Beetles)

Asiatic Rhinoceros Beetle (Oryctes rhinoceros L.). The results of experiments in Nile tilapia with Asiatic rhinoceros beetle culture under laboratory conditions did not seem to give promising results. The 16% fish meal substitution decreased the weight and survival. However, a negligible weight increase was also observed in the controls (2 g in 10 weeks) (Omoyinmi and Olaoye, 2012), which could be due to poor-quality fish meal in the diets used (35% protein), stress conditions, and/or fish disease (Henry et al., 2015).

Yellow Mealworm (T. molitor L.). The partial (25–50%) or total fish meal substitution of soybean meal with yellow mealworm larvae meal was examined. The results indicated that it did not affect the intake levels in fish, biometric indices, or the balance of essential and nonessential amino acids in fish muscle. However, the inclusion of yellow mealworm meal at any of the tested levels decreased the growth of Nile tilapia, worsened nutritional parameters, and affected the lipid profile of the fish muscle (Sánchez-Muros et al., 2015). The authors related the worse nutritive indices to the chitin content, which represents 1.4 and 2.8% for 25 and 50% fish meal substitutions in feeds, respectively. Moreover, the administration of a yellow mealworm larvae meal-based diet at early ages of development (with 25% fish meal substitution) did not cause irreversible effects on zootechnical parameters and biometric indices in Nile tilapia. Additionally, these parameters could recover when they were subsequently provided with a control diet (without insect meal), and the same was true for high unsaturated fatty acid (HUFA) muscle composition, which was altered by ingestion of the insect-based diet (de Haro, 2015).

Superworm (Zophobas morio Fab.). Utilizing superworm as a fish feed suppliment, Jabir et al. (2012b) achieved a 50% replacement of fish meal with superworm in Nile tilapia diets that resulted in optimal growth. Moreover, even a 75% replacement was well tolerated in terms of growth, FCE, and protein digestibility. Nevertheless, the apparent digestibility coefficients of protein and lipids were less for superworm meal than fish meal, and this was reflected in low lipid percentages found in muscles at all substitution levels tested as compared to the fish fed fish meal. This result occurred despite the major lipid content in diets formulated with superworm. Therefore, the authors suggested that slight improvements to the superworm containing fish diets are necessary before they can entirely replace fish meal as feed for Nile tilapia (Jabir et al., 2012a).

Order Diptera (Flies)

House Fly (M. domestica). The substitution percentage of house fly in place of fish meal has been established at 50% (Ajani et al., 2004) and 34% (Ogunji et al., 2007, 2008a,b,c). Nevertheless, a resulting weight gain similar to fish meal-fed fish is possible with a 100% substitution (Ajani et al., 2004; Ogunji et al., 2007, 2008b,c). Moreover, results suggest that feeding Nile tilapia fingerlings maggot meal diets did not cause physiological stress (Ogunji et al., 2007), but did affect the fatty acid profile in fish. Therefore, adequate sources of n-6 and n-3 fatty acids should be included in the diet (Ogunji et al., 2008b).

On the other hand, the substitution percentage seemed dependent on the minimum fish meal levels. Nile tilapia growth decreased significantly compared to that of the control groups when fish meal was reduced by 17.31% in the feed, and when 15% house fly was subsequently introduced. The results indicated that the protein efficiency ratio was not affected, although the inclusions were even higher (Ogunji et al., 2008c). This may be related to the composition and the dietary protein/energy ratio of the different diets. In addition, with higher substitutions, the apparent digestibility of crude protein worsened in Nile tilapia, and this was probably due to higher ash content in the diet (Ogunji et al., 2008c).

The combination of wheat bran and live house fly added to Nile tilapia feed has also been studied. At a 4:1 ratio of wheat bran: live house fly, the results showed better growth performance, specific growth rate, FCR, and survival than when wheat offal was fed singly (Ebenso and Udo, 2003). The results also revealed that diets with house fly protein were as effective in terms of growth and survival as fish meal for that species (Ogunji et al., 2007, 2008a,b,c; Omoyinmi and Olaoye 2012).

Order Orthoptera (Crickets, Grasshoppers, Locusts, and Katydids)

Migratory Locust (L. migratoria) The replacement of fish meal with migratory locust meal in Nile tilapia fingerling diets showed that 25, 50, and 75% substitutions were not significantly different from the control diet regarding protein digestibility. Furthermore, 25 and 50% substitutions did not differ significantly in lipid digestibility compared to the control. However, considering growth performance and hematological parameters, migratory locust meal could replace fish meal up to 25% without any adverse effects on growth performance and hematological parameters (Abanikannda, 2012; Emehinaiye, 2012 in Makkar et al., 2014).

Order Lepidoptera (Butterflies, Moths, and Their Caterpillar Larval and Chrystalis/Pupal Stages)

Defatted and nondefatted *silkworm* meal led to very good digestibility values in Nile tilapia (Hossain et al., 1992; Boscolo et al., 2001). However, a low inclusion level (5%) significantly reduced fish growth compared to fish meal-fed fish in control diets formulated with 5% fish meal (Boscolo et al., 2001).

Family Clariidae (Airbreathing Catfishes)

African Catfish (*Heterobranchus bidorsalis* Geoffroy Saint-Hilaire)

Order Isoptera (Termites; Macrotermes nigeriensis) Solomon et al. (2007) conducted a trial with African catfish fingerlings fed fish meal-based diets supplemented with *M. nigeriensis* and soybean meal at different ratios for 56 days. They found better growth performance and nutrient utilization when *M. nigeriensis* was included in the blend at a 75:25 ratio (*M. nigeriensis*: soybean meal). However, the variation in soybean meal and fish meal between the experimental diets made it difficult to establish whether the improved fish growth was due to dietary insects or to the increased dietary fish meal. On the other hand, there were marked significant differences in the ash, lipid, and crude protein of the carcasses of fish fed the different experimental diets (Solomon et al., 2007).

*African Sharptooth Catfish (*Clarias gariepinus *Burchell)*

Order Coleoptera (Beetles)

Yellow Mealworm (T. molitor) Yellow mealworm, whether used unaltered or transformed into a dry meal, is an acceptable alternative protein source that is highly palatable for African sharptooth catfish (Ng et al., 2001). During a 7-week feeding trial, Ng et al. (2001) showed that yellow mealworm could replace up to 40% fish meal without affecting growth performance and feed utilization efficiency. Diets with up to 80% replacement of fish meal with yellow mealworm still displayed good growth and feed utilization efficiency, but the values were lower than fish fed the control diet. Feeding solely with mealworms displayed a slight depression in growth performance, but when fed in combination with commercial pellets, African sharptooth catfish grew better than those fed the commercial pellets only (Ng et al., 2001). Regarding body composition, the ingestion of yellow mealworm significantly increased the lipids in the carcass (Ng et al., 2001).

Order Diptera (Flies)

House Fly (M. domestica) The results of house fly studies as feed ingredient for African sharptooth catfish are contradictory. In an 8-week study, house fly meal was detrimental to the growth performance at all levels of diet inclusion (12.5%, 50%, and particularly at 100%). However, nutrient utilization was less affected, and the best FCR and PER results were found in diets with 25 and 50% inclusion, respectively (Idowu et al., 2003). In contrast, African sharptooth catfish fed diets with 50 or 100% fish meal replacement with house fly meal for 10 weeks grew well (Nsofor et al., 2008; Aniebo et al., 2009). Oyelese (2007) found that African sharptooth catfish fed 50% live house fly in conjunction with a 50% artificial diet that was poor in fish meal (3.5%) grew better than fish fed the artificial diet alone over the 7-week experimental period. Moreover, no differences in organoleptic properties were found in African sharptooth catfish fed house fly meal (Aniebo et al., 2011). Therefore, it has been concluded that house fly meal is a viable alternative protein source to fish meal in the diet of African sharptooth catfish, (Aniebo et al., 2009; Oyelese, 2007).

On the other hand, the combination of hydrolyzed chicken feather meal, chicken offal meal, and house fly meal at a ratio of 4:3:2 as alternative animal protein mixture was also evaluated as a replacement for fish meal in African sharptooth catfish fingerling diets. The results indicated no significant differences in weight gain, specific growth rate, FCR, or protein efficiency ratio in fish fed the diet with 25–50% substitution and those fed the control diet (Adewolu et al., 2010).

In one study, processing methods, such as defatting and drying, influenced the nutrient concentration of house fly meal. However, the growth performance and nutrient utilization of African sharptooth catfish fed a diet with 100% fish meal replaced with defatted, oven dried and defatted, sun-dried maggot meals were not significantly different from each other, and the parameters were similar to those obtained in fish fed the fish meal-based diet. Generally, African sharptooth catfish performed better when fed diets containing defatted maggot meals than full-fat maggot meals, and this compared favorably with fish fed the fish meal-based diet (Fasakin et al., 2003).

Order Orthoptera (Crickets, Grasshoppers, Locusts, and Katydids)

Desert Locust (Schistocerca gregaria Forsskål) This insect is famous for its vast seasonal swarms which can ravage most plants in their path across Africa, the Middle East, and Asia. It is best known for devastating crops in Africa over many centuries and mentioned in the Bible as an agricultural plague even in that time. It remains an extremely abundant insect and an important agricultural pest today. For its utility as a livestock feed, one study to date has demonstrated that the replacement of up to 25% fish meal with desert locust in African sharptooth catfish juvenile diets did not affect growth (Balogun, 2011).

Variegated Grasshopper (Zonocerus variegatus L.) Using an 8-week trial, Alegbeleye et al. (2012) showed that adult variegated grasshopper meal could replace up to 25% of the fish meal in African sharptooth catfish fingerling diets without any adverse effects on growth and nutrient utilization. The results indicated increases in the inclusion level, performance, and carcass lipids, and decreases in apparent protein and lipid digestibility were observed. No statistical differences in the FCR and protein efficiency ratio were detected.

Mudfish (*Clarias anguillaris* L.)
Order Diptera (Flies)

House Fly (M. domestica). The use of frozen or live house fly larvae (maggot meal) to supplement artificial diets has been recommended for mudfish production, because it promoted fast growth rates in the fingerlings than did commercial feed (Achionye-Nzeh and Ngwudo, 2003; Madu and Ufodike, 2003). Mudfish juveniles fed unconventional diets exhibited the greatest increase in body weight, specific growth rate, and condition factors when fish were fed the commercial diet supplemented with live house fly, followed by an exclusive house fly diet (Madu and Ufodike, 2003). The combination of house fly and the commercial diet might have formed a better balanced diet for juvenile mudfish (Madu and Ufodike, 2003).

Order Isoptera (Termites)

M. nigeriensis. The growth response of mudfish fingerlings fed diets formulated with 10–40% *M. nigeriensis* Sjostedt without fish meal for 42 days showed that the growth rate of fish increased with increased protein content of the feed. The best daily weight gain and total weight gain were observed in fish fed a diet with 40% *M. nigeriensis* (Achionye-Nzeh et al., 2004).

Vundu (*Heterobranchus longifilis* Valienciennes)
Order Diptera (Flies)

House Fly (M. domestica). Comparisons between soymeal, cattle brain meal, and house fly meal included at 80% showed that house fly meal supplemented with amino acids results in better performance in vundu than soymeal. Moreover, these conditions resulted in lower performance than fish that were fed cattle brain meal during a 7-week feeding trial (Ossey et al., 2012), but no fish meal control diet was used to compare results.

Order Isoptera (Termites)

Macrotermes sp. In a 12-week experiment focusing on *Macrotermes* sp. meal as substitute for fish meal diets (0–100%), Sogbesan and Ugwumba (2008) used broken-line analysis and found that 50% inclusion levels of *Macrotermes* sp. meal yielded the best results. For instance, fish fed this diet exhibited the highest mean weight gain, relative growth rate, specific growth rate, and protein efficiency ratios as well as the lowest FCRs.

Walking Catfish (*Clarias batrachus* L.)

Order Lepidoptera (Butterflies, Moths, and Their Caterpillar Larval and Chrystalis/Pupal Stages)

Silkworm pupae meal was found to be a suitable fish meal substitute in walking catfish diets (Venkatesh et al., 1986; Borthakur and Sarma, 1998a,b). The results indicated that walking catfish fed silkworm meal had slightly lower specific growth rates, minor protein efficiency ratios, and poorer FCRs than fish fed fish meal diets. However, the digestibility of crude protein (Borthakur and Sarma, 1998a) and intestinal protease activity (Venkatesh et al., 1986) were similar to fish fed fish meal diets.

Order Orthoptera (Crickets, Grasshoppers, Locusts, and Katydids)

Although research suggests the suitability of utilizing the grasshopper *Poekilocerus pictus* (Fab.) as a partial substitute for fish meal in formulated walking catfish diets, results of studies using this insect have not been promising; the fish feed with grasshopper (Johri et al., 2010; Johri et al., 2011b) grew lower than fish feed fish meal diets. The 100% substitution of fish meal in a 91-day trial had no effect on hematological parameters. However, minor shrinkage of the gills was observed in fish fed with grasshopper diet and a reduction in ovarian steroidogenesis was noted, which may be associated with reduced fertility (Johri et al., 2011a,c).

Family Cyprinidae (Carps and Minnows)

Black Carp (*Mylopharyngodon piceus* Richardson)

Order Diptera (Flies)

House Fly (M. domestica). Supplementation with 25 g/kg house fly meal promoted growth in black carp, which improved weight gain and specific growth rates without changes in muscle composition. The growth enhancement seemed to result from an improvement in immunological status (Ming et al., 2013).

Common Carp (*Cyprinus carpio* L.)

Order Lepidoptera (Butterflies, Moths, and Their Caterpillar Larval and Chrystalis/Pupal Stages)

Silkworm/Silkmoth (B. mori). Nondefatted *silkworm* pupae meal could replace fish meal in the diet of common carp (Jeyachandran and Raj, 1976; Nandeesha et al., 1990, 2000) without affecting growth performance, FCE, and quality in terms of color, odor, texture, and flavor (Nandeesha et al., 1990, 2000).

Defatted silkworm pupae were less digestible than fish meal (Kim, 1974), and the protein and fat digestibility of nondefatted silkworm pupae were better than that of fish meal. This difference is due to higher protease enzyme activity and a significantly higher deposition of protein in fish fed defatted insect diets (Nandeesha et al., 1990, 2000). However, a 100% substitution of fish meal with defatted silkworm pupa could be utilized (Kim, 1974). Finally, the usefulness of silkworm pupa oil as an energy source in comparison to sardine oil for common carp has been examined with acceptable results (Nandeesha et al., 1999).

Deccan Mahseer (*Tor khudree* Sykes)

The use of silkworm as feed was examined in Deccan mahseer fingerlings with promising results. Better growth and survival were obtained in fish fed a diet containing 50% defatted silkworm pupae (Shyama and Keshavanath, 1993 in Makkar et al., 2014).

Prussian Carp (*Carassius gibelio* Bloch)

Order Diptera (Flies)

House Fly (M. domestica). A 6-week growth trial was conducted to investigate the effects of dietary supplementation with house fly and soybean meal (with a 71% fish meal substitution in the control diet) on the growth performance and antioxidant responses of Prussian carp. The results indicated a significant decrease in the specific growth rate in fish fed the soybean meal diet compared to those fed the control diet (Dong et al., 2013).

Putitor Mahseer (*Tor putitora* Hamilton)

Order Lepidoptera (Butterflies, Moths, and Their Caterpillar Larval and Chrystalis/Pupal Stages) An experiment was conducted to study the growth performance of Putitor Mahseer fingerlings fed during a 60-day trial, and the results indicated that acceptable growth was observed in fish feed a diet with a 70% substitution of fish meal with silkworm pupae (Sawhney, 2014).

Rohu (*Labeo rohita* Hamilton)

Order Lepidoptera (Butterflies, Moths, and Their Caterpillar Larval and Chrystalis/Pupal Stages)

Silkworm/Silkmoth (B. mori). A 90-day experiment that focused on the effects of a diet with a 50% replacement of fish meal with silkworm pupae and clam meat on rohu fingerlings revealed a significantly better utilization of the experimental diet (Begun et al., 1994).

Regarding the processing of the meal, silkworm that did not undergo the deffating process were better than defatted silkworm, but the differences were not significant (Hossain et al., 1997). Nevertheless, both silkworm and defatted silkworm pupae that were used as the sole source of protein produce significantly better apparent and true protein digestibility values than those fed the fish meal diet (Hossain et al., 1997).

Silver Barb (*Barbonymus gonionotus* Bleeker)

Order (Butterflies, Moths, and Their Caterpillar Larval and Chrystalis/Pupal Stages)

Silkworm/Silkmoth (B. mori). In silver barb fingerlings, the substitution of fish meal with silkworm pupae was examined. The highest growth rate, FCR, protein efficiency ratio, apparent net protein utilization, apparent protein digestibility, and growth performance were observed in fish fed a diet in which about 38% of the total dietary protein was replaced silkworm pupae meal. However, the diet containing lower levels of silkworm pupae (19.1%) and higher levels of mustard oilcake (19.8%) led to reduced fish growth. Economic analyses of the diets suggested the possibility of using silkworm pupae as an alternative source of protein in silver barb feed (Mahata et al., 1994).

Family Heteropneustidae (Airsac Catfishes)

Stinging Catfish (*Heteropneustes fossilis* Bloch)

Order Lepidoptera (Butterflies, Moths, and Their Caterpillar Larval and Chrystalis/Pupal Stages)

Silkworm/Silkmoth (B. mori). Researchers managed to successfully achieve a 75% substitution of fish meal with silkworm pupae meal in stinging catfish diets without adverse effect on growth (Hossain et al., 1993 in Makkar et al., 2014).

Family Ictaluridae (Freshwater Catfishes)

Black Bullhead (*Ameiurus melas* Rafinesque)

Order Coleoptera (Beetles)

Yellow Mealworm (T. molitor). In a 90-day trial, the 50% substitution of fish meal with yellow mealworm meal showed good growth performance in black bullhead. However, fish in the insect meal group reached a final mean body weight that was significantly lower than that of the fish meal group, and the survival rate of the fish meal group was significantly higher (Roncarati et al., 2014). Fish in the fish meal group were fed a control diet (51.6% protein and 18.1% lipid), whereas those in the insect meal group received an experimental diet (50.8% protein and 22.1% lipid). However, the differences in the macronutrient contents of the diets could have affected the results.

Channel Catfish (*Ictalurus punctatus* Rafinesque)

Order Diptera (Flies)

Black Soldier Fly (H. illucens). The level of positive results for replacement of fish meal with black soldier fly depends on the culture system. For example, the replacement of 10% fish meal with 10% dried black soldier fly prepupae (whole or chopped and reared on poultry manure) over a 15-week period for subadult channel catfish grown in cages resulted in slower growth rates (weight gain and total body weight) and decreased animal crude protein content. In contrast, if channel catfish was cultured in tanks at a slower growth rate, the replacement did not significantly reduce the growth rate (Bondari and Sheppard, 1987).

Regarding the administration method, results have indicated increased uneaten larval waste in the chopped larvae fed tanks compared to those in the whole larvae fed tanks. However, chopping improved weight gain and efficient utilization by the fish (Bondari and Sheppard, 1987).

Comparison between menhaden fish meal and prepupae black soldier fly meal showed that 25% fish meal replacement was possible without compromising weight gain. Moreover, a 100% replacement was possible if the diets were supplemented with soybean meal in order to obtain isoproteic diets (Newton et al., 2005). More recently, Zhang et al. (2014) found that the replacement of 25% fish meal with chicken manure conversion with microorganisms and black soldier fly larvae powder resulted in similar growth indices, immunity indices, and bodies compared to the control group. However, higher substitutions did not produce satisfactory results.

Family Lateolabracidae (Ray-Finned Fish)
Japanese Seabass (*Lateolabrax japonicus* Cuvier)
Order Lepidoptera (Butterflies, Moths, and Their Caterpillar Larval and Chrystalis/Pupal Stages)

Silkworm/Silkmoth (B. mori). In Japanese seabass that were fed a combination of commercial feed and various food sources at a 70:30 ratio, the energy and protein apparent digestibility coefficient of silkworm pupae that were not defatted was lower than poultry by-product meal, blood meal, and soybean meal. However, it was comparable to that of feather meal (Ji et al., 2010).

Family Moronidae (Temperate Basses)
European Seabass (*Dicentrarchus labrax* L.)
Order Coleoptera (Beetles)

Yellow Mealworm (T. molitor). The effects of yellow mealworm inclusion on the growth and feed efficiency of European seabass juveniles were studied by Gasco et al. (2014b), with a 25% substitution satisfactorily achieved. However, at a 50% substitution, yellow mealworm induced growth reduction and less favorable outcomes for both the specific growth rate and feed consumption ratio. The protein efficiency ratio and feed consumption were not affected by the inclusion of yellow mealworm, and the whole body proximate composition analysis did not show any differences between the treatments. On the other hand, yellow mealworm inclusion influenced the fatty acid composition of body lipids. In particular, a decrease in the contents of eicosapentaenoic acid (EPA) and docosahexaenoic acid (DHA) was observed with an increase to a 50% inclusion of yellow mealworm meal.

Family Osphronemidae (Giant Gouramis)
Snakeskin Gourami (*Trichogaster pectoralis* Regan)
Order Lepidoptera (Butterflies, Moths, and Their Caterpillar Larval and Chrystalis/Pupal Stages)

Silkworm/Silkmoth (B. mori). Jintasatapom et al. (2011) showed that silkworm could replace 50% of the protein from fish meal in snakeskin gourami broodstock diets without any adverse effects on egg quality (in terms of fry number from the first spawn and fingerling number). However, growth performance and egg fecundity were significantly decreased with the increased percentage of silkworm pupae in the diet. The results indicated that protein digestibility decreased according to the inclusion level of silkworm pupae in the diets. In contrast, the dry matter digestibility was similar to the control, and egg quality (in terms of fry number and fingerling number of the first spawn) was not significantly different ($P > 0.05$). The survival rate of 1-month nursing fish of the first spawn was significantly different ($P < 0.05$) at the 100% substitution. Hence, a 50% protein substitution of fish meal with silkworm pupae (14.57% by weight) could be used in snakeskin gourami broodstock diets without any adverse effects on egg quality.

Family Paralichthyidae (Sand Flounders)
Bastard Halibut (*Paralichthys olivaceus* Temminck and Schiegel)

According to a study by Lee et al. (2012), silkworm can replace 10% fish meal in bastard halibut diets. The replacements of fish meal with 10 and 20% silkworm pupae meal was tested in bastard halibut diets, and the best performance was achieved with the 10% silkworm pupae meal replacement, as compared to the control diet (Lee et al., 2012).

Family Salmonidae (trouts and Salmons)
Atlantic Salmon (*Salmo salar* L.)

Lock et al. (2014) tested the replacement of fish meal with black soldier fly larva meal (0, 25, 50, and 100%) in Atlantic salmon. The feed intake and FCR decreased with increasing black soldier fly inclusion, which increased the utilization of feed for growth. The protein and lipid digestibility, histology, and sensory testing of fillets did not show any differences among the treatments. The authors concluded that a favorable amino acid profile and high medium-chain fatty acid content are needed to make black soldier fly meal a promising ingredient for use in Atlantic salmon diets.

The nutrient isolation and processing methods of the insect meal have an important impact on the performance of the product. For instance, highly defatted black soldier fly meal that was dried at a conventional temperature resulted in low growth in fish compared to those fed lightly defatted black soldier fly meal that was dried at a low temperature. Moreover, the lightly defatted black soldier fly meal allowed a 100% substitution of fish meal, which had similar growth indices compared to fish fed the control diet (Henry et al., 2015).

Chum Salmon (Oncorhynchus keta Walbaum)
Order Lepidoptera (Butterflies, Moths, and Their Caterpillar Larval and Chrystalis/Pupal Stages)
Silkworm/Silkmoth (B. mori). In a 6-week feeding experiment with chum salmon fry, Akiyama et al. (1984) evaluated the replacement of fish meal with silkworm preferentially over other ingredients, but they did not obtain results better for insect meal over fish meal or any other protein meals. Fish were fed fish meal diets or diets supplemented with silkworm pupae powder (5%), dried beef liver (5%), krill meal (5%), or earthworm powder (5%) in place of fish meal. None of the dietary treatments appreciably increased the food intake of the fish over that of the control group, and no significant differences were noted in body protein and ash contents among all dietary treatments (Akiyama et al., 1984).

Rainbow Trout (Oncorhynchus mykiss Walbaum)
Order Coleoptera (Beetles)
Yellow Mealworm (T. molitor). Gasco et al. (2014a) reached an inclusion level of up to 50% yellow mealworm without a growth performance reduction in rainbow trout feedstuffs. In a 75-day feeding trial, there were no statistical differences for 0, 25, or 50% fish meal replacements in growth performance metrics and proximate composition of the resulting fish. The lowest hepatosomatic indices were observed in fish fed the 25 and 50% yellow mealworm replacement diets.

Order Diptera (Flies)
Common House Mosquito (Culex pipiens L.). The common house mosquito has been evaluated as a feed for rainbow trout fed exclusively with frozen mosquitos, and the results showed a decrease in growth compared with fish given control supplemented with amino acids (dipeptide glycine-lysine or free glycine and free lysine) but higher than control diet without suppletetation of amino acids (Ostaszewska et al., 2011).

Black Soldier Fly (H. illucens). In a 9-week study, researchers replaced 0, 25, or 50% of the fish meal protein in juvenile rainbow trout diets with black soldier fly prepupae meal (reared on swine manure). Compared to the control, no differences in total weight gain and feed conversion of fish fed at 25% were observed (St-Hilaire et al., 2007b). The protein content of fish after the end of the experiment did not differ significantly between treatments, and the lipid content was less in fish fed with black soldier fly at any level. Moreover, the inclusion of black soldier fly pupae in fish feed altered the fatty acid profile of rainbow trout, in that fish fed black soldier fly diets were low in fish oil had low levels of Omega-3 fatty acids in muscle (St-Hilaire et al., 2007b). Nevertheless, the fatty acid profile of insects could be modified by feeding methods that allow increased fish meal replacement, which could restore the levels of n-3 in fish muscle (Sealey et al., 2011).

House Fly (M. domestica). During a 9-week feeding trial, St-Hilaire et al. (2007a) evaluated the inclusion of 25% house fly meal (reared on cow manure) in place of fish meal in rainbow trout diets. In contrast to the results described for black soldier fly, the total weight gain was less for fish fed the 25% house fly diet compared to the control. However, compared to the controls, no differences were observed in the FCR, protein and lipid content, muscle fatty acid profile for linoleic acid, EPA, DHA, or ARA contents of the whole-body proximate composition.

Order Lepidoptera (Butterflies, Moths, and Their Caterpillar Larval and Chrystalis/Pupal Stages)
Silkworm pupae have been examined as a shrimp meal substitute in rainbow trout. A 100% replacement of shrimp meal indicated a significantly higher specific growth rate compared to fish fed a control diet (Dheke and Gubhaju, 2013).

Family Scophthalmidae (Turbots)
Turbot (Psetta maxima L.)
Kroeckel et al. (2012) showed that the incorporation of black soldier fly protein in juvenile turbot diets is possible, but it is limited to a 33% inclusion to avoid significant changes in feed intake, feed conversion, and protein retention. However, both indices, feed intake and specific growth rate, decreased with increasing black soldier fly incorporation, and this was likely due to the palatability and low apparent digestibility coefficients of organic matter, crude protein, crude lipid, and

gross energy. Regarding body composition, the whole-body protein content was not affected by the treatment, but body lipid content decreased with increasing black soldier fly inclusion levels.

Family Sparidae (Porgies, Sea Breams)

Gilt-Head Bream (*Sparus aurata* L.)

Order Coleoptera (Beetles) The substitution of fish meal with up to 25% yellow mealworm protein in gilt-head bream juvenile diets is feasible without adverse effects on growth performance and whole-body proximate composition (Piccolo et al., 2014). However, a slight depression was observed in the protein efficiency ratio and the FCR during a 60-day trial. Furthermore, the 50% substitution induced less favorable indices (Piccolo et al., 2014).

Order Diptera (Flies)

Common Green Bottle Fly (Lucilia sericata Meigen). Gilt-head bream fed during a 33-day trial with different substitution percentages (0, 25, and 50%) of fish meal with common green bottle fly larvae meal (cultured with pork liver) resulted in a final weight that was similar to fish fed the control diet (without common green bottle fly). Furthermore, with respect to body composition, the only significant difference was observed for regarding the muscle lipid percentage of fish fed the maximum replacement (de Haro et al., 2015).

Hybrid Fish

Vundu (H. longifilis) × African Sharptooth Catfish (C. gariepinus)

Order Diptera (Flies)

House Fly (*M. domestica*) The study of growth responses, FCRs, and cost benefits of hybrid catfish vundu × African sharptooth catfish fed different substitution percentages of fish meal with house fly maggots meal (0–100%) showed that a 25% substitution had the best growth performance and the highest mean growth rate. However, no significant differences were observed on specific growth rate (SGR) and final weight for diets at 0, 50, and 75% replacements of fish meal (Sogbesan et al., 2006).

Polyculture

Major Indian Carp (Catla catla Hamilton), Mrigal Carp (Cirrhinus mrigala Bloch), Rohu (L. rohita), and Silver Carp (Hypophthalmichthys molitrix Valenciennes)

Order Lepidoptera (Butterflies, Moths, and Their Caterpillar Larval and Chrystalis/Pupal Stages)

Silkworm/Silkmoth (*B. mori*) Silkworm has been studied as feed ican a polyculture system consisting of 30% major Indian carp, mrigal carp, rohu and 10% silver carp. Silkworm treatment was evaluated for silkworm pupae silage and untreated fresh pastes that were incorporated in fish feed formulations to replace fish meal. The results clearly indicated that the survival rate, FCR, and specific growth rate with fermented silkworm pupae silage were nutritionally superior to untreated silkworm pupae silage or fish meal. The dietary influence on the proximate composition of whole fish was marginal (Rangacharyulu et al., 2003).

African Sharptooth Catfish (C. gariepinus) and Vundu (H. longifilis)

Order Coleoptera (Beetles)

Asiatic Rhinoceros Beetle (*O. rhinoceros* L.) Asiatic rhinoceros beetle can be used to completely replace fish meal in fish diets in mix cultures of African sharptooth catfish and vundu. However, for optimal growth and nutrient utilization, a 25% level of fish meal replacement with Asiatic rhinoceros beetle meal is most suitable for the fingerlings of both fish species. The study also showed that Asiatic rhinoceros beetle meal is more suitable for African sharptooth catfish than vundu (Fakayode and Ugwumba, 2013).

Common Carp (C. carpio) and Nile Tilapia (Oreochromis niloticus)

Order Diptera (Flies)

House Fly (*M. domestica*) The digestibility of house fly compared to a reference diet (fish meal as a primary protein source) and a test diet (70% reference diet + 30% house fly meal) was evaluated in Nile tilapia and common carp. The apparent digestibility coefficients for dry matter, crude protein, crude fat, and gross energy of house fly were significantly lower for Nile tilapia than for common carp. However, the spawning activities of experimental Nile tilapia and the soft feces consistency of common carp may have affected the results (Ogunji et al., 2009).

Channel Catfish (I. punctatus) and Blue Tilapia (Oreochromis aureus Steindachner)
Order Diptera (Flies)

Black Soldier Fly (H. illucens) The substitution of a commercial diet with 50 or 75% black soldier fly larvae was tested during a 10-week trial for the polyculture of channel catfish and blue tilapia, and the results indicated that fish body weight and total length were not affected. Moreover, the taste test results regarding the aroma and texture of fish fed black soldier fly larvae were acceptable to the consumer (Bondari and Sheppard, 1981).

Chinese White Shrimp (Fenneropenaeus chinensis Osbeck) and Japanese Blue Crabs (Portunus trituberculatus Miers)
Order Diptera (Flies)

House Fly (M. domestica) In a polyculture of Chinese white shrimp and Japanese blue crabs that were fed a diet containing 30–50% housefly maggot meal, the yields and survival of the shrimp and crabs were higher than those fed control diets. Furthermore, shrimp and crabs fed the MGM diet had significantly higher body weight, especially in the mid- and late-culture periods (Zheng et al., 2010a).

Crustaceans (Shrimp, Crabs, Lobsters and Their Relatives)

Family Palaemonidae
Giant River Prawn (Macrobrachium rosenbergii De Man)

Feces from black soldier fly larvae (reared on dried distiller's grains) were used in commercial prawn farms to replace the regular commercial feed. The results showed a performance similar to that of the regular giant river prawn feed, and the economic returns were enhanced. The only notable difference was that the prawns fed the black soldier fly feces diet were slightly more pale than those fed the traditional diet, but there were no changes in flavor (Tiu, 2012).

Family Penaeidae (Penaeid Shrimps)
Chinese White Shrimp (F. chinensis Osbeck)

Juvenile Chinese white shrimp fed diets containing different proportions of house fly exhibited enhanced body length, body weight, specific growth rate, and survival, which corresponded with increased dietary house fly. On the other hand, the total n-3 HUFA and n-6 HUFA in the shrimp muscles increased with increased house fly proportions in the diets. However, the essential fatty acid content and higher polyunsaturated fatty acids in the muscles of shrimp fed house fly were significantly lower than those of the shrimp in the control group (Zheng et al., 2010a,b).

Speckled Shrimp (Metapenaeus monoceros Fab.)
Order Lepidoptera (Butterflies, Moths, and Their Caterpillar Larval and Chrystalis/Pupal Stages)

Silkworm/Silkmoth (B. mori). Shrimp growth trials showed that digestive efficiency was reduced when silkworm meal replaced fish meal (Sumitra-Vijayaraghavan Wafar and Royan, 1978 in Makkar et al., 2014)

Whiteleg Shrimp (Litopenaeus vannamei Boone)

According to growth performance parameters, antioxidant indices, and nonspecific immune indices of juvenile whiteleg shrimp, the best replacement ratio of fish meal with house fly meal was 40%, and the maximum level should not exceed 60% (Cao et al., 2012a). Specific growth rate tended to decrease with house fly inclusion, and the muscle nutritional composition showed a positive correlation between the crude protein and ash content of the muscle and the proportion of house fly meal in the diet. The crude protein content at a 100% substitution and the ash content at 80 and 100% replacements were significantly higher than that of the control group, and no effects on moisture and crude lipid content were observed. No significant effects on the nutritional composition, essential amino acid, delicious amino acid, or inosine monophosphate content in the muscle were observed. However, the arginine content increased with increasing fish meal substitution increments. Digestive enzymes were also studied, and no effects on activity associated with house fly inclusion were observed. However, pathological changes in the histological structure of the hepatopancreas were observed with replacement levels over 60% (Cao et al., 2012b).

Mollusks (Clams, Oysters, Snails and Their Relatives)

*Disk Abalone (*Haliotis discus *Reeve)*

Order Lepidoptera (Butterflies, Moths, and Their Caterpillar Larval and Chrystalis/Pupal Stages)

Silkworm/Silkmoth (*B. mori*) Cho (2010) studied the growth and survival of disk abalone juveniles that were fed the following diets: control, soybean meal, poultry meal, corn gluten meal, dehydrated silkworm pupae meal, meat and bone meal, soybean meal and corn gluten meal, soybean meal and dehydrated silkworm pupae meal, and corn gluten meal and dehydrated silkworm pupae meal. The combined soybean meal and corn gluten meal or the silkworm pupae meal and soy meal could completely replace fish meal in the abalone diet, resulting in improved performance.

Overview

It is difficult to obtain clear conclusions because of the variety of fish and insects species examined in studies of diet formulation, ingredients, and so forth. In general, the results showed that insects could play an important role in aquaculture feeding. Nevertheless, substantial knowledge is needed to successfully use insects as alternatives to fish meal. All of these studies indicate the need to expand upon the following key points:

- To determine the most adequate insects species for each fish species.
- To evaluate and develop various insect species and production methodologies which are most economical end practical as an animal feed commodity.
- To evaluate influence of insect meal on muscle quality.
- To study efficient administration techniques: chopped, whole, alive, and so on.
- To analyze the influence of culture systems on the nutritive utilization of insect-based diets.
- To assess the optimal fish meal substitution percentage.
- To test different strategies to improve the nutritive values of insect meals.

Other Animals

*American Alligator (*Alligator mississippiensis *Daudin)*

Order Diptera (Flies)

Black Soldier Fly (*H. illucens*) Black soldier fly larvae cannot be recommended as a complete replacement for commercial feed in young American alligators (Bodri and Cole, 2007). The feeding of American alligators with dried black soldier fly larvae (reared on restaurant food waste) reduced weight and increased the snout-vent length (Bodri and Cole, 2007).

*Giant Ditch Frog (*Leptodactylus fallax *Müller)*

Order Diptera (Flies)

Black Soldier Fly (*H. illucens*) Black soldier fly larvae had poor nutrient digestibility in giant ditch frogs. However, the digestibilities of almost all nutrients except Na and K were enhanced through processing (pureeing). After processing or chewing that broke the exoskeleton, black soldier fly larvae supplied high levels of dietary minerals without the need for additional external Ca. Moreover, Ca and P digestibilities were approximately two-fold higher than that for mashed larvae (about 90% versus 45–50%), and they were similar to values measured for supplemented cricket-based diets (Dierenfeld and King, 2008 in Makkar et al., 2014).

*Rat (*Rattus *sp.)*

Order Diptera (Flies)

House Fly (*M. domestica*) The feed value of dried house fly larva meal was compared in a nutritional study of growing rats (Bouafou et al., 2011a,b). The chemical composition of rats, feed intake and weight gain of subjects, feed and protein efficiency ratios, digestibility, net protein utilization, and biological value parameters were similar at 2.5, 5, and 7.5% inclusions of dried house fly meal. The regime at 5% inclusion had the best nutritional value (Bouafou et al., 2011b). Nevertheless, damage to kidney tissue and liver portal veins was found in rats fed dried house fly meal (Bouafou et al. 2011a,b).

BENEFITS AND CONSTRAINTS ASSOCIATED WITH USING INSECTS AS LIVESTOCK FEED INGREDIENTS

Nutritional

Source of Protein and Amino Acids

The use of insects in animal feed has been mainly focused on their value as a protein source. Studies of insect protein values indicated that most species had high protein quantities (Ladrón de Guevara et al., 1995; Ramos-Elorduy et al., 1981, 1982, 1984, 1997) and quality. Moreover, the protein included essential amino acids, such as lysine, methionine, and leucine, which are limited in protein sources of vegetal origin (Hall, 1992). Insect meal has protein levels and amino acid profiles that are better than soy meal, and it is even similar to fish meal in some species (Sánchez-Muros et al., 2014). Nevertheless, amino acid profiles vary among species (see Chapter 3). The comparison between the amino acid profile of fish meal and various insect species indicates that the amino acids are taxon-related. However, the amino acid profile of the flies (Order Diptera) is similar to that of fish meal (Barroso et al., 2014). The protein digestibility is another criterion to consider. Wang et al. (2007) studied the digestibility of amino acids in grasshopper meal and fish meal in broilers, and the results indicated higher values for grasshopper meal than fish meal, with the exception of serine, glutamic acid, and histidine. However, little information about the digestibility of insects is available. Fish studies indicate that the diet can include a percentage of insect meal without negative effects on digestibility, but this percentage depends on the insect species and the animal species that consumes it. For example, a 33% substitution with black soldier fly did not affect protein digestibility in turbot (Kroeckel et al., 2012), which is contrary to the results observed in Atlantic salmon fed a diet containing this insect (low protein digestibility was found) (Lock et al., 2014). On the other hand, the 75% substitution of fish meal with variegated grasshopper, migratory locust, superworm, or house fly in the diets of African sharptooth catfish and Nile tilapia did not affect digestibility (Ogunji et al., 2008c; Alegbeleye et al., 2012; Emehinaiye, 2012; Jabir et al., 2012b). Moreover, high digestibility *was* observed in diets containing silkworm for several fish species, including rohu (Begun et al., 1994), Mozambique tilapia (*Oreochromis mossambicus* Peters) (Hossain et al., 1992), Nile tilapia (Boscolo et al., 2001), black bullhead (Borthakur and Sarma, 1998b), snakeskin gourami (Jintasatapom et al., 2011), and walking catfish (with a 100% replacement of fish meal) (Habib et al., 1994). Nevertheless, the digestibility of silkworm varied based on the treatment of the meal, and the defatted meal exhibited better digestibility than the nondefatted meal (Hossain et al., 1997).

Furthermore, the nitrogen content of chitin molecules could provoke an overestimation of protein, however de cuticular chitin content is low, as Finke (2007) demonstrate the standard 6:25 conversion ratio between nitrogen and protein is adequate for insect CP.

The protein and amino acid availability especially for the amino acids from proteins that are either highly sclerotized or which may be bound to chitin (Finke, 2007). However, this nitrogen composes a low percentage of the total, so it does not affect the in vitro protein digestibility (Sánchez-Muros et al., 2015).

Source of Lipids and Fatty Acids

Another interesting aspect of insects as a food or feed ingredient is the level and quality of lipid content in some species, which can reach up to 77% of the body composition in some insects, such as *Phassus triangularis* Edwads larvae (the caterpillar of a moth species from Mexico) (Ramos-Elorduy et al., 1997). The lipid percentage in insects varies with the developmental phase, and, for holometabolous species which have larval and pupal stages, it is higher in larvae and pupae than in adults (Sánchez-Muros et al., 2014). Moreover, the lipid percentage in larvae and adults tends to be higher than fish meal (~8%) or soy meal (~3%). The high lipid content has some advantages, including as high-energy supply, and it is useful when high-energy diets are required (eg, broiler chickens). However, this elevated lipid content limits inclusion in livestock or fish diets, because the increased lipid percentage is above the required levels. Manzano-Agugliaro et al. (2012) proposed the use of insect fat in biodiesel production, and the resulting protein-rich paste could also be used in animal feeding (aquaculture or livestock).

Regarding fatty acid profiles, insect meal shows reduced levels of linoleic acid (18:2 n6) compared to soy meal. Furthermore, compared to fish meal, it also exhibits major levels of PUFAS n-6 (Omega-6 fatty acids) and low levels of HUFAS n-3 (Omega-3 fatty acids) (especially 20:5 n3 [EPA] and 22:6 n3 [DHA]). However, the levels of Omega-3 and 6 fatty acids are higher for many insects than for other animal livestock such as beef, dairy, pork, and chicken. The inclusion of HUFAS fatty acids in the diet is necessary because they affect important biological functions in vertebrates. Moreover, the Omega-3 fatty acids found in fish are considered a fatty acid source for humans. Terrestrial insects are clearly deficient in 20:5 n3 (EPA) and 22:6 n3 (DHA) compared with fish meal. However, 20:5 n3 is present in aquatic insects, and 22:6

n3 appears at a very low percentage in a few *Notonecta* species of order Ephemeroptera (the Mayflies) (Bell et al., 1994). The low Omega-3 fatty acid content imposes a limit on its inclusion in diet formulation, and this is particularly relevant to fish when fish meal is replaced by insect meal. In fact, the 25% replacement of fish meal with black soldier fly meal in trout diets affected the Omega-3 fatty acid levels in the fish (St-Hilaire et al., 2007b). Nevertheless, the fatty acid composition of insects varies with species, development stage, and feeding (Stanley-Samuelson et al., 1988; Ghioni et al., 1996; Bukkens, 2005; Raksakantong et al., 2010; Sánchez-Muros et al., 2014), which allows us to modify the insect fatty acid profile. For instance, St-Hilaire et al. (2007a) found increased Omega-3 fatty acids (EPA, DHA, and ALA) in black soldier fly prepupae when fish offal was included in the diet. These Omega-3 fatty-acid-enhanced prepupae may be a suitable fish meal and fish oil replacement for carnivorous fish and other animal diets. Black soldier fly meal obtained from flies fed fish offal allowed higher fish meal replacement percentages, and it also improved the Omega-3 and Omega-6 fatty acid content in fish muscles (Sealey et al. 2011). This ability to accumulate selected fatty acids from the diet allowed researchers to use the larvae as suitable vectors for the introduction of valuable Omega-6 fatty acids to animal muscle. de Haro et al. (2015) found an increase of arachidonic content in the muscle of gilt-head bream fed a diet containing common green bottle fly, which contains a high ARA percentage (10.6% total fatty acids when cultured with pig liver). The fish fed common green bottle fly meal enriched with ARA exhibited increased levels (80%) of this fatty acid in the muscle compared to fish fed the control diet (de Haro et al., 2015). Therefore, the use of this feed allowed fish to obtain a functional food that was rich in this fatty acid.

Chitin

In some studies, the presence of chitin might have affected the growth performance by influencing the feed intake, availability, and digestibility of the nutrients (Kroeckel et al., 2012). With a caloric content of 17.1 kJ g^{-1}, chitin could be a source of carbohydrates, and it could also constitute a substantial percentage of the total energy intake; nevertheless, its digestibility, which is discussed later, should also be considered. The hydrolysis of chitin requires the involvement of chitinase and chitobiase enzymes. High digestive chitinase levels were observed in cobia (*Rachycentron canadum* L.) (Fines and Holt, 2010) and in a broad range of marine teleost fish, and it was measured in the blood and lymphomyeloid tissues (in addition to digestive tissues) (Fange et al., 1976; Danulat and Kausch, 1984; Lindsay, 1986; Jeuniaux, 1993; Gutowska et al., 2004). Mammalian gut chitinases have been identified in humans, mice, cows, and chickens (Boot et al., 2001; Suzuki et al., 2001). However, despite high chitinase activity, low chitin digestibility was described in rainbow trout (Lindsay et al., 1984). On the other hand, chinolytic activity has also been described in gut bacteria, and it may play a role in the hydrolysis and apparent absorption of carbohydrates from chitin (Sugita et al., 1999).

However, chitin could affect the digestibility of other nutrients, including proteins (Longvah et al., 2011) or lipids (Kroeckel et al., 2012), leading to a reduction in growth. The results of these experiments that fed fish diets containing different chitin levels showed highly variable results, which could be dependent on the fish species or the chitin origin (crustacean or insect).

Moreover, chitin may have a positive effect on the function of the immune system. For instance, the use of insects to feed chickens may diminish the use of antibiotics in the poultry industry (van Huis et al., 2013), which is discussed in the section on antibiotic resistance.

Vitamins and Minerals

The vitamins and minerals needed for livestock varies based on species, age, physiological status, and the production quality and quantity. Rumpold and Schlüter (2013) published an exhaustive review on the mineral and vitamin contents found in different insect species. High variation was observed among species, but all analyzed insects were generally low in calcium, potassium, and sodium (with the exception of house fly larvae) (Hwangbo et al., 2009). Furthermore, most insects exhibited very high amounts of phosphorous. The magnesium content varied, and especially rich magnesium levels were found in the Order Hemiptera ("true bugs" such as stink bugs, etc.) and some species of the Order Orthoptera (eg, crickets, grasshoppers, locusts, and katydids). The micronutrient content (including copper, iron, magnesium, manganese, phosphorous, selenium, and zinc) in some species allowed the use of these species as sources of micronutrients for livestock (Rumpold and Schlüter, 2013, 2015)

Regarding vitamins, the content depends on the species, and some species are rich in some vitamins and deficient in others (see Chapter 3, and Rumpold and Schlüter, 2013). To a great extent, the insect vitamin profile also depends on the composition of the insect diet (Ramos-Elorduy et al., 2002).

In general, insects are rich in Vitamin B2 (0.11–8.9%), pantothenic acid, and biotin (FAO, 2004b; Bukkens, 2005). Some species showed high levels of Vitamin B12, (eg, yellow mealworms and house crickets) (Finke, 2002; Bukkens, 2005), folic acid (grasshoppers, crickets, locusts, and beetles) (FAO, 2004b), and Vitamin B1 (0.1– 4 mg of dry matter). On the other hand, insects are deficient in Vitamin A, Vitamin C, niacin, and Vitamin E (FAO, 2004b).

To include insects in animal feed, it is necessary to understand the contribution of minerals and vitamins to the diet, and it is important to specifically select the desired mineral and vitamin provisions. Moreover, the capacity to bioaccumulate minerals in insect exoskeletons, adipose cells, gonads, and digestive tracts must also be taken into account (van Huis et al., 2013). Therefore, the mineral content of the feed used to culture insects must be exhaustively controlled to avoid mineral toxicity or interaction between minerals. On the other hand, the capacity of insect of bioaccumulation of vitamins, mineral, and fatty acids could be used to engineer their nutrient content for feed.

See also Chapter 3 of this book for a more detailed discussion of insect nutrient content, particularly as it relates to human nutrition.

Feed Security and Safety

The feed used for livestock is required to meet standards of food safety, which includes the control of microorganisms (natural or contamination from handling), antinutritional factors, and toxic compounds from endogenous or exogenous origins. The food safety conditions of non-European insects has been studied (see Van der Spiegel et al., 2013), but much more work on this topic is needed. See Chapters 7 and 8 of this book for a detailed discussion on the food safety, regulatory, microbiological, and allergen issues related to insects as a human food ingredient.

Heavy Metals, Pesticides, and Contaminants

For insects and in general, there are several main subclasses of chemical contaminants which are monitored for animal feeds, human foods, and veterinary medicines. These include pesticides, heavy metals, dioxins and polychlorinated biphenyls, polyaromatic hydrocarbons, and mycotoxins of different species (Charlton et al., 2015). For instance, oriental leafworm moth larvae/caterpillars (*Spodoptera litura* Fab.) showed a higher capacity for the accumulation of cooper and zinc. However, chickens fed with insects containing high levels of these minerals exhibit lower levels of these minerals than the insect. This could be due to fecal excretion or low ingestion of insects (Zhuang et al., 2009). Further investigation of this field is required to avoid possible toxicity via insects, and this is needed even though toxicity is easy to prevent by rearing the insects in controlled conditions and feeding them safe feeds.

Pesticides are also bioaccumulated in some species of insects. For instance, high concentrations of organophosphorus pesticide residues were detected in locusts collected for food in Kuwait after the 1988/1989 outbreak (Saeed et al., 1993), the pesticides accumulated in the insects would be transferred to the food chain. Elevated glutathione S-transferase activity was observed when fish received higher dietary house fly larva meal concentrations, suggesting the presence of pesticides, drugs, or toxins in the hen waste used to feed the flies (Ogunji et al., 2007). Other repercussions of contaminated insect meal could also affect feed intake. For example, broiler chicks that were fed locusts that had been sprayed with an insecticide exhibited lower intake than those fed unsprayed locusts (Gibril, 1997). The capacity of insects to bioaccumulate drugs (eg, antibiotics) from the rearing substrate must also be considered. For instance, nicarbazin, which was used as an anticoccidial in poultry feed, was detected in house flies that were fed poultry manure (Charlton et al., 2015). Nonetheless, as most pesticides are designed to kill insects, insects are very sensitive to these compounds. Thus, it is unlikely that pesticides will be present or tolerated in farm-raised insects, as the pesticides woud likely limit productivity of the farm by killing the insects. Additionally, farm-raised insects and those mass produced indoors should be very easy to control in general, including eliminating their contamination by pesticides or any other chemicals or pathogens.

Microorganisms

Since insects are genetically very different than humans, vertebrates, and mammals, pathogens affecting insects are considered by many to be safe for humans (particularly viruses) (Banjo et al., 2006; Opara et al., 2012; van Huis et al., 2013) and probably for livestock. Nevertheless, insects could harbor some pathogens including microorganisms such as viruses, fungi, protozoa, or bacteria (Vega and Kaya, 2012), and in particular larvae could act as reservoirs of these pathogens (McAllister et al., 1994). Pathogens mainly associated with the cuticle (exterior of the insect) tend to be fungi (eg, *Aspergillus, Penicillium,* and *Fusarium*) (Mpuchane et al., 1996; van Huis et al., 2013), or bacteria (eg, *Staphylococcus aureus* Rosenbach, *Pseudomonas aeruginosa* Migula, and nonpathogenic *Bacillus* species) (Banjo et al., 2006; Opara et al., 2012), which produce toxins or cause the deterioration of the insect meal. For livestock, these pathogens hinder their conservation, and they are a risk to the health of the animal.

On the other hand, insects could be a transmission vector of some pathogens. For instance, house fly maggots were implicated in the transmission of *Escherichia coli* (Migula) from reservoir animals to other animals and humans (Moriya et al., 1999). *Cronobacter* spp., *Salmonella* spp., and *L. monocytogenes* (Murray) were also detected in house flies and blowflies that were collected from the dumpsters of urban restaurants (Pava-Ripoll et al., 2012). Despins et al. (1994) demonstrated the transmission of enteric viral pathogens to turkey poults that fed on infected lesser mealworm larvae. Another mode of contamination associated with insects is inadequate handling during insect meal manufacturing (van Huis et al., 2013). Regarding insects that were fed slaughterhouse waste, the possible transmission of prions has been discussed (Charlton et al., 2015). Therefore, this aspect must be carefully studied. Their presence may also be passed along to food products made from the animals fed with these insects, though many of these pathogens are common in such livestocks already fed on noninsect feeds. However, most of these pathogens, being associated with the exterior of the insects, likely come from their environment. Thus, a clean insect farm with good hygiene can likely mitigate or eliminate most if not many of these pathogens from the facility.

Insect Derived Toxins, Venoms, and Allergens

Toxic and venomous insects from the Orders Lepidoptera (butterflies and moths), Hemiptera (true bugs such as stink bugs), and Hymenoptera (ants, bees, wasps, and hornets) (Blum, 1981) are classified as neurotoxic, hemolytic, digestive, hemorrhagic, and allogeneic. The toxins chemically consist of alkaloids, terpenes, polysaccharides, biogenic amines (eg, histamine), organic acids (eg, formic acid), and amino acids, but the majority are peptides and proteins (Blum, 1981; Schmidt, 1986). Toxic/venomous insects typically use their chemical defense systemsto immobilize or kill prey and/or for defense against predator attacks.

Blum (1981) provided several examples of toxic species that should be avoided as human food, including cyanogenic species (eg, butterflies), vesicant species (eg, *Lonomia* moths), and those that produce steroids (eg, *Ilybius fenestratus* Fab.), corticosteroid hormones (eg, great diving beetle, *Dytiscus marginalis* L.), necrotoxic alkaloids (eg, fire ants, *Solenopsis* spp.), and toluene (eg, cerambicids). There are no studies on the toxins from these insects for livestock, but it is expected that toxins that affect humans also affect livestock. In chickens fed house flies, increased liver weight, which could indicate a toxic effect, was observed. In rats, the inclusion of 10% house fly meal resulted in histological changes in the liver and the kidneys (Bouafou et al., 2011a), which could indicated a certain level of toxicity.

The allergies provoked by insects have been mainly studied in humans (see Chapter 8 of this book for a detailed discussion on insect allergenicity in humans). However, it is also an important issue in animal feed, because the allergic response in farm animals could result in animal welfare issues, lower weight, and decreased meat performance (Charlton et al., 2015).

Digestive Enzyme Inhibitors and Antinutritional Compounds

Compared to other enzyme inhibitors, protease inhibitors have been studied the most, and they are widespread (Eguchi, 1993). In insects, protease inhibitors have a defensive function, but the effects of these when insects are included in animal diets have not been examined. Studies of protease genetic patterns in tse-tse fly (*Glossina morsitans* Westwood), common fruit fly, tobacco hornworm hawk moth (*Manduca sexta* L.), and silkworm indicated properties similar to those of serpins (Eguchi, 1993; Gubb et al., 2010).

Other antinutrients found in insect species, such as oxalate, phytate, tannin, and hydrocyanide, were detected, but the amounts were far below toxic levels (Omotoso, 2006; Ekop et al., 2010).

Preservation and Storage

The artificial breeding of insects may allow greater control over hygiene practices, and increased safety associated with edible-insect supplies might mitigate potential hazards. The greatest risk to humans and animals is the emergence of opportunistic disease-causing microbes in insect rearing systems, and hence insect diseases should also be avoided and controlled (Eilenberg et al., 2015).

Similar to other meat products, insects are rich in nutrients and moisture, and this provides a suitable environment for microbial growth and survival (Klunder et al., 2012). Studies indicated that the storage of dried housefly larvae with excessive moisture levels (23%) promoted the growth of bacteria and fungi, so processing to and storage at a maximum moisture level of 4–5% was recommended to minimize microbial activity (Makkar et al., 2014).

After washing, live insects are often transported in ice coolers shortly after collection. Furthermore, the preservation of meal of high-fat content insect species (especially those with high Omega-3 fatty-acid content) may require the addition of antioxidants and storage in dry areas, though more studies are needed to determine shelf-life of various insect-derived products overall.

Antibiotic Resistance

Antibiotics are poorly absorbed in the animal digestive tract, and they are released into the environment via animal feces (Binh et al., 2008; Looft et al., 2012; Zhu et al., 2013). Therefore, insects that are reared on manure for use as animal feed have a high likelihood of acquiring and carrying bacteria with antibiotic resistance. Furthermore, several studies demonstrated the proliferation of bacteria and the horizontal transfer of resistance genes via the insect digestive tract, and the transmission of resistant bacteria by insects to new substrates was also observed. Zurek and Ghosh (2014) proposed "that insect management should be an integral part of pre- and postharvest food safety strategies to minimize spread of zoonotic pathogens and antibiotic resistance traits from animal farms. Furthermore, the insect link between the agricultural and urban environment presents an additional argument for adopting prudent use of antibiotics in the food animal industry." On the other hand, adequate treatment methods, such as disinfection of insect meal, could help avoid the transmission of resistant bacteria from insects to livestock and good hygiene at insect production facilities might help mitigate the need for antibiotic use.

Animal Welfare

The role of insects in animal welfare could be an additional benefit of insect use in animal culture in several ways. For example, it has been shown that chitin-increased activity in the innate immune system of sea bream has been described by different authors (Sakai et al., 1992; Esteban et al., 2000; Kawakami et al., 1998). However, most studies used noninsect chitin or chitin derivatives, and the administration of chitin was via injection. In rainbow trout, chitin administration stimulated macrophage activity (Sakai et al., 1992) and also stimulated respiratory burst, phagocytic activity, and cytotoxic activity in sea bream (Esteban et al., 2000). Furthermore, some evidence suggests that chitin confers protection against infections. For instance, rainbow trout and yellow tail individuals injected with chitin exhibited increased resistance to *Vibrio anguillarum* Bergeman (Sakai et al., 1992) and *Pasteurella piscicida* (Kawakami et al., 1998), respectively.

A few studies have added chitin to diets, and the results indicated an improvement of the *Epinephelus bruneus* Bloch immune response (Harikrishnan et al., 2012). In addition, the observed nonspecific modulation of hemolytic complement activity, leucocyte respiratory burst activity, and cytotoxicity indicated the enhancement of sea bream immune activity (Esteban et al., 2001). It is important to note that these studies added pure chitin to the diet. Nevertheless, the inclusion of insects supports the results obtained with pure chitin. For instance, Ming et al. (2013) found that a 2.5% supplementation with house fly meal increased the lysozyme, serum alkaline phosphatase, and glutathione peroxidase in serum. Moreover, superoxide dismutase and catalase activities in the liver increased, and it also reduced the malondialdehyde levels in the serum and liver. Furthermore, *Aeromonas hydrophila* (Chester) Stanier contamination resulted in lower mortality in fish that were fed house flies (Ming et al., 2013). This result indicated that chitin from insects played a role in the immune system, which was similar to the results observed when pure chitin was administered via injection.

The inclusion of chitin derivatives in the diet (eg, chitosan) was studied in different species. The results indicated that the particle size has different effects. For instance, particles measuring between 10^3–10^4 Da modulated the immune response and reduced the establishment of pathogens (Wang et al., 2003). However, if the molecular weight was below 10^3 Da, it stimulated the growth of *Bifidobacteria*, which has a number of health-promoting properties (Shigehiro et al., 1990; Zhou and Lin, 2000; Hou and Gao, 2001). The antibiotic/prebiotic activity was described for chitin in rats and chickens (Chen and Chen, 1999; Chen et al., 2002), and the hypolipidemic properties were described in broilers (Hirano et al., 1990; Hossain and Blair, 2007), hens, and rabbits (Hirano et al., 1990) that were fed chitosan instead of chitin (Hirano et al., 1990). However, there are few studies that have examined the effects of ingested insect chitin in livestock, and thus this is a field that requires further study.

Oxidative stress is considered a welfare indicator. The administration of insect meal that resulted in decreased oxidative stress was described (Manzano-Agugliaro et al., 2012), and the authors related these results to lower long chain HUFAS n-3 values and low peroxidation levels. The 71% supplementation of house fly meal in place of fish meal (compared to the control diet) resulted in an enhancement of the antioxidant capacity of Prussian carp (Dong et al., 2013). On the contrary, silkworm pupae meal substitutions above 50% in common carp caused oxidative stress, which significantly reduced fish growth, decreased superoxide dismutase and intestinal protease activities, and increased the amount of heat shock protein (Ji et al., 2013). These contradictory results could be due to the possible bioaccumulation of contaminants in insects, so the substrates used to rear insects must be free of contaminants.

PROMISING OPPORTUNITIES FOR RESEARCH AND TECHNOLOGICAL ADVANCEMENT

We would like to conclude by reflecting on some aspects that we consider important if insects are to be used as common animal feed ingredients in the near future. Although there are a large number of publications that address this issue, fundamental knowledge gaps regarding the potential use of insect-based animal feed remain. The insect world is huge and diverse, and that leaves much to be studied and discovered. Therefore, it will take many years to optimize the exploitation of insects as a feeding resource. Indeed, insects are already being utilized commercially as livestock feed by a rapidly growing number of companies around the world. Thus, in our opinion, some particular areas of research must be addressed.

Ecological Aspects and Sustainability

Environmental Benefits

As mentioned in other chapters of this book on use of insects as human food, compared to livestock, insect cultures are associated with low greenhouse gas emissions, low land-based activity, low ammonia emissions, and efficient conversion of feed into protein. Furthermore, insects can feed on organic by-products and other biomass that does not compete directly with the human food supply. From the point of view of animal feeding, insect use has additional benefits. For instance, decreased use of the main protein sources in animal feeding (fish meal and soy meal) would have significant environmental repercussions. Soy cultivation causes the deforestation of areas with high biological value (Carvalho, 1999; Osava, 1999), high water consumption (Steinfeld et al., 2006), pesticide and fertilizer utilization (Carvalho, 1999), and transgenic variety use (Garcia and Altieri, 2005), which cause significant environmental deterioration (Osava, 1999). On the other hand, fish meal is a resource that depends on the catch, and an exhaustive stripping of fisheries promotes the deterioration of the marine environment resulting in extinction of many marine species.

Utilization of the Native Insects by Local Farms

Taking advantage of natural resources without causing environmental deterioration is a great challenge. Currently, natural resources are overused owing to overcrowding, and are therefore ceasing to be renewable or sustainable. The traditional use of insects by local communities was not dangerous for the environment because it utilized small amounts of insects for either human or animal consumption. However, current interest in the use of insects as feed could increase the demand and the price, which could generate environmental problems such as extinction or entomofauna (insects) degradation. Therefore, the exploitation of wild entomofauna must be controlled with strict programs that sustain a healthy environment. Additionally, it would be beneficial in most cases to replace the use of wild collected insects with more efficiently captive farmed insects. In many cases, due to the high levels of biodiversity in Class Insecta, these could still be local native species, thus preventing the need to move nonnative and possibly invasive species from one area to another.

Expansion of the Number of Selected Species for Mass Rearing

In contrast with livestock culture in extensive or traditional systems, insects are part of the natural diet of animals, and small amounts of insects are consumed without harmful environmental effects. However, to supply the volume of insects needed for animal feeding, it is necessary to produce tons of insects; this could only be accomplished via mass-rearing systems.

The risks associated with insect cultivation stems from the rearing of foreign species. The FAO (2013) recommended the rearing of local species, particularly in tropical countries, because they pose virtually no risk to the environment, and there is no need for climate control. On the other hand, the methods associated with the use of insects to control plant pests are well developed, and there are established systems of control over cultured insect populations, which could be applied to the mass rearing of insects as animal feed.

Despite the development of insect culture systems, it is essential to continue research initiatives aimed at developing the "micro-livestock" of numerous major species, different taxa, and varied nutritional habits. Until now, mass rearing has focused on very few species. From an ecological point of view, it would be ideal to produce local species that are adapted to the environmental conditions and feed on local wastes and by-products, which would make production more efficient. Moreover, this would subsequently promote greater sustainability in integrated production systems, and it would result in lower production costs and the economic development of rural communities through small-scale and family-run farms and fish farms.

However, the ecological footprints of the insects or the water requirements for these intensive culture systems are not well established. Therefore, future studies that examine the development of mass insect cultures are needed to determine the real impact of mass-rearing insect production.

Environmental Enrichment for Livestock Animals

Insects could play an important role in environmental enrichment. Animals have feeding behaviors that include foraging. Neuringer (1969) demonstrated that when given a choice between "working" for food and having the food provided ad libitum, many animals chose to work for their food. Searching for food is a natural behavior that animals must express, and denying an animal appetitive opportunities might be a source of frustration or stress (Hughes and Duncan, 1998; Shepherdson et al., 1993; Shepherdson, 1998). In addition, tossing insects on the ground currently provides the opportunity and motivation for ground foraging and investigation behaviors (Mellen and MacPhee, 2001).

Therefore, future studies should focus on determining the effects of providing whole insects on the welfare of fish, chickens, and pigs. Artificial and intensive livestock production systems are increasingly being implemented. Moreover, the environment of the animals is very stressful, which induces annoyance and abnormal behavior, and these conditions can lead to serious health problems and decreased productivity.

Health problems associated with the legs and meat quality of broiler chickens reared in intensive systems are an example of this issue. Leg problems are one of the most common causes of culling, early mortality, and late mortality in broiler stocks, and they are considered one of the more serious welfare problems facing the broiler industry (SCAHAW, 2002). Walking ability in poultry is considered an important welfare parameter (Mench, 2004), but it is also critical for obtaining optimum flock performance, maintaining good energy efficiency and feed utilization, reducing bone disorders, improving bone strength, reducing skeletal fractures, improving food safety, and reducing processing plant condemnation issues (Oviedo-Rondón, 2007). One way to improve leg condition is to increase the locomotive activity of birds (Bizeray et al., 2002). However, the environment in the intensive systems provides minimal stimulation to the animals. One purpose of the enrichment is to provide the animals with objects, sounds, or odors that are not directly linked with the performance of some behavior, but instead provide the animals with a more stimulating environment (Newberry, 1995). In this context, we believe that insects could provide a fundamental enriching element. In nature, chickens (and other Galliformes) mainly eat insects during the first week of life to cover their high protein needs, and the percentage of insects in the diet decreases during the second and third weeks of life. Therefore, we believe that providing whole insects (live, if possible) in addition to their food supply, would favor the natural feeding and harvest behavior of the birds. Moreover, this higher level of physical activity would improve the muscular and skeletal systems of chickens, and it would also increase poultry productivity.

Chitin

Determining the digestibility of chitin and its effect on the digestibility of other nutrients for each livestock or aquaculture species is the first step. It is also necessary to obtain the adequate or acceptable levels of chitin in the diet.

On the other hand, the removal of chitin during meal manufacturing should also be investigated. The chitin could be removed from insect meal via alkaline extraction (DeFoliart et al., 1982; Belluco et al., 2013), but it can also be done using mechanical methods.

The addition of hydrolytic enzymes to diets is a global practice in animal feed production. The addition of chitinase and its effect on digestibility has not been studied, and it could be a solution to increase chitin digestibility. Alternatively, chitin could be degraded by chemicals or enzymatic methods before being added to diets as product of hydrolysis (eg, chitooligosaccharides, acetylglucosamine, or chitosan) (Shiau and Yu, 1999; Se-Kwon and Niranjan, 2005; Lin et al., 2012a,b).

Chitin also being its own potentially high value by-product. Chitin and its derivates (chitosan) are used as antioxidant, anticancer, antiinflammatory, drug delivery, and plastics, Park and Kim (2010). If insect-based animal feeds take off and expand commercially, chitin as a by-product of animal feed production will be a much more sustainable and potentially larger scale source of chitin than current marine sources.

Insect Nutritive Value Improvement Using Different Rearing Systems

There are indications that food sources may affect the nutritional composition of insects. St-Hilaire et al. (2007b) observed that fatty acid composition in the black soldier fly could be manipulated by changing the substrate composition. On the other hand, the results of de Haro et al. (2015) indicated the accumulation capacity of essential fatty acids in common green bottle fly. However, several studies are needed to determine the bioaccumulation capacity of fatty acids or other compounds required for nutrient contribution or for the prevention of toxicity.

The capacity of insects to accumulate fatty acids, vitamins, minerals, and other nutrients is a very promising concept since the whole insects can be utilized to modulate/engineer the nutrient content of the resulting insect meal. Furthermore, knowledge of the effects of the substrate on lipid, protein, chitin, vitamins, or mineral levels should be improved.

Physical and/or Chemical Treatments of Insect Meals to Improve Their Assimilation

So far, few studies have examined the effects of different treatments to optimize the use of nutritious insect meals. Meal processing, including drying, hydrolyzing, ensiling, or defatting could improve the palatability, nutrient availability, digestibility, and composition of insect meals, which could make them more suitable for fish nutrition (Newton et al., 2005).

Drying

The drying of insect meal prevents the growth of bacteria and fungi. Regarding the house fly, a maximum 4–5% moisture content is recommended to minimize bacterial and fungal activity (Makkar et al., 2014). On the other hand, the drying method could modify the nutritive characteristics. For instance, Fasakin et al. (2003) found lower daily weight gain, protein efficiency ratios, and specific growth in fish fed a diet containing sun-dried MGM as compared to oven-dried meal. The sun-dried meal was richer in lipids and poorer in protein than the oven-dried meal (Aniebo and Owen, 2010), but oven drying increased the risk of lipid oxidation (Henry et al., 2015) and the loss of volatile fatty acids. Palatability was also affected by the drying method, and increased with oven and sun drying (Ng et al., 2001). Therefore, the recommended moisture percentage for the insect meal of each species, the drying methods, and the effects on nutritive and feeding parameters should be studied.

Silage

Ensiling is a method used to increase the shelf life of livestock feed. Few studies have examined the silage of insect feed, but promising results indicated the increased preservation of silkworm pupae meal and improved zootechnical indices. When fish meal was replaced with silkworm pupae or fermented silkworm pupae silage in a polyculture system for major Indian carp, mrigal carp, rohu, and silver carp, the fermented silkworm pupae silage resulted in better survival rates, FCRs, and specific growth rates compared to feeding with the untreated fresh silkworm pupae or fish meal, which had little influence on the body proximate composition (Rangacharyulu et al., 2003).

The addition of organic acids, enzymes, and the like increased the nutritive value of the feed. The inclusion of additives in insect ensiling is an area that should be investigated to optimize the use of insects in animal feed. Yashoda et al. (2008) obtained good-quality silage by ensiling with molasses or curd as a lactic acid culture. Therefore, ensiling with additives, such as chitinases, to promote the degradation of chitin, should be further studied.

Defatting

The quality and quantity of insect lipids limits the inclusion of insects in livestock diets, but the defatting of insects is one solution. The extracted lipids could be used as biofuel, and the resulting protein could also be used as a protein source in animal feed (Manzano-Agugliaro et al., 2012). The defatting of insect meal enabled the increased dietary inclusion of insects without affecting the growth of African catfish (Fasakin et al., 2003). However, defatting methods must be further studied. In Atlantic salmon, better results were obtained with meal that was lightly defatted and dried at a low temperature compared to meal that was highly defatted and dried at a conventional temperature (Henry et al., 2015). However, the insect lipid levels and the possible oxidation of lipids (Aniebo and Owen, 2010) could have influenced these results. Henry et al. (2015) proposed the addition of antioxidants to insect meal to reduce the negative effects of drying processess on lipids.

Disinfection

The microbiology of insects is significantly different from that of conventional food animals, and it deserves further study (Belluco et al., 2013). Moreover, other insect species should also be investigated as potential pathogen hosts to minimize disease transmission via the food chain (Khusro et al., 2012). In addition, the disinfection of insect meal should be considered, and the establishment of hygienic rules for insect processing is necessary.

CONCLUSIONS

There is currently a growing interest in insects as agricultural products, and the number of studies that consider the use of insects in animal feeding is also increasing. The studies reviewed in this chapter show the great potential that insects have as animal feed. Although they will not be able to match the nutritional characteristics of fish meal, insects could become a major animal feed source. In many cases, insect meals could partially replace fish meal, but they could completely replace some vegetable or soy meals found in many livestock and aquaculture feeds.

So far, the insect species tested in animal feeds are limited, so the number of evaluated and mass-reared insect species should be expanded. Moreover, it seems clear that more feeding trials in different species are needed to determine their future roles as protein components in animal feed.

Several lines of research show great promise in the future use of insects in animal feeding. The role of insect-based feeds in animal welfare, nutritional value improvement via physical means or breeding systems, feed security and safety, increased assimilation of insect meal via physical or chemical treatments, chitin modification, and so forth are all issues that should be addressed in future studies. Another challenge that must be met is the standardization of the nutritive composition of insect species with feeding interest in animals. This will allow us to optimize insects as renewable livestock feed resources.

REFERENCES

Abanikannda, M.F., 2012. Nutrient digestibility and haematology of Nile tilapia (*Oreochromis niloticus*) fed with varying levels of locust (*Locusta migratoria*) meal. Bachelor of Aquaculture and Fisheries Management, Federal University of Agriculture. Abeokuta, Ogun State.

Achionye-Nzeh, C.G., Ngwudo, O.S., 2003. Growth response of *Clarias anguillaris* fingerlings fed larvae of *Musca domestica* and soyabean diet in the laboratory. Biosci. Biotech. Res. Comm. 15, 221–223.

Achionye-Nzeh, C.G., Salami, S., Ogidiolu, O., 2004. Growth response of *Clarias anguillaris* fingerlings fed diets formulated with *Microtermis nigerensis*. Nigerian J. Appl. Sci. 19, 1570–1573.

Adeniji, A.A., 2007. Effect of replacing groundnut cake with maggot meal in the diet of broilers. Int. J. Poult. Sci. 6, 822–825.

Adesina, M.A., Adejinmi, O.O., Omole, A.J., Fayenuwo, J.A., Osunkeye, O., 2011. Performance of broilers finishers fed graded levels of cassava peel-maggot meal-based diet mixtures. J. Agric. Forest. Soc. Sci. 9, 226–231.

Adewolu, M.A., Ikenweiwe, N.B., Mulero, S.M., 2010. Evaluation of an animal protein mixture as a replacement for fishmeal in practical diets for fingerlings of *Clarias gariepinus* (Burchell, 1822). Israeli J. Aquaculture-Bamidgeh 62, 237–244.

Adeyemo, G.O., Longe, O.G., Lawal, H.A., 2008. Effects of feeding desert locust meal (*Schistocerca gregaria*) on performance and haematology of broilers. In: Tielkes, E. (Ed.), Tropentag 2008, International Research on Food Security, Natural Resource Management and Rural Development. University of Hohenheim, Centre for Agriculture in the Tropics and Subtropics, Hohenheim, Germany, p. 361.

AFRIS, 2015. Animal Feed Resource Information System. Food and Agriculture Organization of the United Nations, Rome, Italy.

Agunbiade, J.A., Adeyemi, O.A., Ashiru, O.M., Awojobi, H.A., Taiwo, A.A., Oke, D.B., Adekunmisi, A.A., 2007. Replacement of fish meal with maggot meal in cassava-based layers' diets. J. Poult. Sci. 44, 278–282.

Aigbodion, F.I., Egbon, I.N., Erukakpomren, E., 2012. A preliminary study on the entomophagous response of *Gallus gallus domesticus* (Galliformes: Phasianidae) to adult *Periplaneta americana* (Blattaria: Blattidae). Int. J. Trop. Insect Sci. 32, 123–125.

Ajani, E.K., Nwanna, L.C., Musa, B.O., 2004. Replacement of fishmeal with maggot meal in the diets of Nile tilapia, *Oreochromis niloticus*. World Aquaculture 35, 52–54.

Akiyama, T., Murai, T., Hirasawa, Y., Nose, T., 1984. Supplementation of various meals to fishmeal diet for chum salmon fry. Aquaculture 37, 217–222.

Akpodiete, O.J., Inoni, O.E., 2000. Economics of production of broiler chickens fed maggot meal as replacement for fish meal. Nigerian J. Anim. Prod. 27, 59–63.

Akpodiete, O.J., Ologhobo, A.D., Onifade, A.A., 1998. Maggot meal as a substitute for fish meal in laying chicken diet. Ghana J. Agric. Sci 31, 137–142.

Alegbeleye, W.O., Obasa, S.O., Olude, O.O., Otubu, K., Jimoh, W., 2012. Preliminary evaluation of the nutritive value of the variegated grasshopper (*Zonocerus variegatus* L.) for African catfish *Clarias gariepinus* (Burchell. 1822) fingerlings. Aquaculture Res. 43, 412–420.

Aniebo, A.O., Owen, O.J., 2010. Effects of age and method of drying on the proximate composition of housefly larvae (*Musca domestica* Linnaeus) meal (HFLM). Pakistan J. Nutr. 9, 485–487.

Aniebo, A.O., Erondu, E.S., Owen, O.J., 2009. Replacement of fish meal with maggot meal in African catfish (*Clarias gariepinus*) diets. Revista Cientifica UDO Agricola 9, 666–671.

Aniebo, A.O., Odukwe, C.A., Ebenebe, C.I., Ajuogu, P.K., Owen, O.J., Onu, P.N., 2011. Effect of housefly larvae (*Musca domestica*) Meal on the carcass and sensory qualities of the mud catfish, (*Clarias gariepinus*). Adv. Food Energ. Sec. 1, 24–28.

Atteh, J.O., Ologbenla, F.D., 1993. Replacement of fish meal with maggots in broiler diets: effects on performance and nutrient retention. Nigerian J. Anim. Prod. 20, 44–49.

Awoniyi, T.A.M., Aletor, V.A., Aina, J.M., 2003. Performance of broiler-chickens fed on maggot meal in place of fishmeal. Int. J. Poultry Sci. 2, 271–274.

Ballitoc, D.A., Sun, S., 2013. Ground yellow mealworms (*Tenebrio molitor* L.) feed supplementation improves growth performance and carcass yield characteristics in broilers. Open Science Repository Agriculture, Online (open-access), e23050425. doi:10.7392/openaccess.23050425.

Balogun, B.I., 2011. Growth performance and feed utilization of *Clarias gariepinus* (Teugels) fed different dietary levels of soaked *Bauhinia monandra* (Linn.) seed meal and sun-dried locust meal (*Schistocerca gregaria*). PhD, Dept Biological Sciences, Faculty of Science, Ahmadu Bello University, Zaria, Nigeria.

Bamgbose, A.M., 1999. Utilization of maggot-meal in cockerel diets. Indian J. Anim. Sci. 69, 1056–1058.

Banjo, A.D., Lawal, O.A., Adeyemi, A.I., 2006. The microbial fauna associated with the larvae of *Oryctes monocerus*. J. Appl. Sci. Res. 2, 837–843.

Barroso, F.G., de Haro, C., Sánchez-Muros, M.J., Venegas, E., Martínez-Sánchez, A., Pérez-Bañón, C., 2014. The potential of various insect species for use as food for fish. Aquacultres, 422–423, 193–201.

Barrows, F.T., Bellis, D., Krogdahl, A., Ashild, S., Jeffrey, T., Herman, E.M., 2008. Report of the plant products in aquafeed strategic planning workshop: an integrated, interdisciplinary research roadmap for increase utilization of plant feedstuffs in diets for carnivorous fish. Rev. Fisheries Sci. 16, 449–455.

Bayandina, G.V., 1979. Fattening of pigs with diets containing various amounts of essential amino acids. Nauchnye Trudy Novosibirskogo Sel'skokhozyaistvennogo Instituta 123, 100–106.

Bayandina, G.V., Inkina, Z.G., 1980. Effects of prolonged use of housefly larvae in the diet of sows and their offspring on fattening and meat quality of the young. Nauchnye Trudy Novosibirskogo Sel'skokhozyaistvennogo Instituta 134, 52–59.

Begun, N.N., Chakraborthy, S.C., Zaher, M., Abdul, M.M., Gupta, M.V., 1994. Replacement of fish meal by low cost animal protein as a quality fish feed ingredients for the Indian major carp, *Labeo rohita*, fingerlings. J. Sci. Food Agri. 64, 191–197.

Bell, J.G., Ghioni, C., Sargent, J.R., 1994. Fatty acid compositions of 10 freshwater invertebrates which are natural food organisms of Atlantic salmon parr (*Salmo salar*): a comparison with commercial diets. Aquaculture 128, 301–313.

Belluco, S., Losasso, C., Maggioletti, M., Alonzi, C., Paoletti, M., Ricci, A., 2013. Edible insects in a food safety and nutritional perspective: a critical review. Comp. Rev. Food Sci. 12, 296–313.

Binh, C.T.T., Heuer, H., Kaupenjohann, M., Smalla, K., 2008. Piggery manure used for soil fertilization is a reservoir for transferable antibiotic resistance plasmids. FEMS Microbiol. Ecol. 66, 25–37.

Bizeray, D., Estevez, I., Leterrier, C., Faure, J.M., 2002. Influence of increased environmental complexity on leg condition, performance, and level of fearfulness in broilers. Poult. Sci. 81, 767–773.

Blum, M.S., 1981. Chemical Defenses in Arthropods. Academic Press, New York, NY.

Bodri, M.S., Cole, E.R., 2007. Black soldier fly (*Hermetia illucens* Linnaeus) as feed for the American alligator (*Alligator mississippiensis* Daudin). Georgia J. Sci. 65, 82–88.

Bondari, K., Sheppard, D.C., 1981. Soldier fly larvae as feed in commercial fish production. Aquaculture 24, 103–109.

Bondari, K., Sheppard, D.C., 1987. Soldier fly, *Hermetia illucens* L., larvae as feed for channel catfish, *Ictalurus punctatus* (Rafinesque), and blue tilapia, *Oreochromis aureus* (Steindachner). Aquacult. Res. 18, 209–220.

Boot, R.G., Blommaart, E.F.C., Swart, E., Ghauharali–van der Vlugt, K., Bijl, N., Moe, C., Place, A., Aerts, J.M.F.G., 2001. Identification of a novel acidic mammalian chitinase distinct from chitotriosidase. J. Biol. Chem. 276, 6770–6778.

Borthakur, S., Sarma, K., 1998a. Protein and fat digestibility of some non conventional fish meal replacers incorporated in the diets of fish *Clarias batrachus* (Linn.). J. Environ. Ecol. 16, 368–371.

Borthakur, S., Sarma, K., 1998b. Effect of some non-conventional fish meal replacers on the growth, feed conversion and body composition of *Clarias batrachus* (Linn.) fingerlings. J. Environ. Ecol. 16, 694–698.

Boscolo, W.R., Hayashi, C., Meurer, F., 2001. Fish, meat and bone, poultry by-products and silkworm meals as attractive in diets for Nile tilapia (*Oreochromis niloticus*) fingerlings. Revista brasileira de zootecnia 30, 1397–1402.

Bouafou, K.G.M., Doukoure, B., Konan, B.A., Amonkan, K.A., Katy-Coulibally, S., 2011a. Substitution of fish meal with dried maggots' meal in growing rats' diet: histological and histopathological consequences. J. Appl. Biosci. 48, 3279–3283.

Bouafou, K.G.M., Konan, B.A., Meite, A., Kouame, K.G., Katy-Coulibally, S., 2011b. Determination of the nutritional value of dried maggots' meal in growing rats diet. J. Anim. Plant Sci. 12, 1553–1559.

Bukkens, S.G.F., 2005. Insects in the human diet: nutritional aspects. In: Paoletti, M.G. (Ed.), Ecological implications of minilivestock; role of rodents, frogs, snails, and insects for sustainable development. Science Publishers, Enfield, NH, USA, pp. 545–577.

Calvert, C.C., Martin, R.D., Morgan, N.O., 1969. House fly pupae as food for poultry. J. Econ. Entomol. 62, 938–939.

Cao, J.M., Yan, J., Huang, Y.H., Wang, G.X., Zhang, R.B., Chen, X.Y., Wen, Y.H., Zhou, T.T., 2012a. Effects of replacement of fish meal with housefly maggot meal on growth performance, antioxidant and non-specific immune indexes of juvenile *Litopenaeus vannamei*. J. Fishery Sci. China 36, 529–537.

Cao, J.M., Yan, J., Wang, G.X., Huang, Y.H., Zhang, R.B., Zhou, T.L., Liu, Q.F., Sun, Z.W., 2012b. Effects of replacement of fish meal with housefly maggot meal on digestive enzymes, transaminases activities and hepatopancreas histological structure of *Litopenaeus vannamei*. South China Fisheries Sci. 8, 72–79.

Carvalho, R., 1999. A Amazônia rumo ao 'ciclo da soja, Amazônia Papers, Programa Amazônia, Amigos da Terra, São Paulo, Brazil.

Charlton, A.J., Dickinson, M., Wakefield, M.E., Fitches, E., Kenis, M., Han, R., Zhu, F., Kone, N., Grant, M., Devic, E., Bruggeman, G., Prior, R., Smith, R., 2015. Exploring the chemical safety of fly larvae as a source of protein for animal feed. J. Insect. Food Feed 1, 7–16.

Chaudhary, K., Das, J., Saikia, S., Sengupta, S., Chaudhary, S.K., 1998. Supplementation of broiler diets with antibiotic and probiotic fed muga Silkworm pupae meal. Indian J. Poultry Sci. 33, 339–342.

Chen, S.H., Chen, H.C., 1999. Effect of oral ad- ministration of *Cellulomonas flavigena* NTOU 1-degraded chitin hydrolysate on physiological changes in rats. Food Sci. Agri. Chem. 3, 186–193.

Chen, H.C., Chang, C.C., Mau, W.J., Yen, L.S., 2002. Evaluation of N-acetylchitooligosaccharides as the main carbon sources for the growth of intestinal bacteria. FEMS Microbiol. Lett. 209, 53–56.

Cho, S.H., 2010. Effect of fishmeal substitution with various animal and/or plant protein sources in the diet of the abalone *Haliotis discus hannai* Ino. Aquacult. Res. 41, e587–e593.

Coll, J.F.C., Crespi, M.P.A.L., Itagiba, M.G.O.R., Souza, J.C.D., Gomes, A.V.C., Donatti, F.C., 1992. Utilization of silkworm pupae meal (*Bombyx mori* L.) as a mendhsource of protein in the diet of growing-finishing pigs. Rev. Soc. Bras. Zootec. 21, 378–383.

Dankwa, D., Oddoye, E.O.K., Mzamo, K.B., 2000. Preliminary studies on the complete replacement of fishmeal by house-fly-larvae-meal in weaner pig diets: effects on growth rate, carcass characteristics, and some blood constituents. Ghana J. Agric. Sci. 33, 223–227.

Dankwa, D., Nelson, F.S., Oddoye, E.O.K., Duncan, J.L., 2002. Housefly larvae as a feed supplement for rural poultry. Ghana J. Agri. Sci 35, 185–187.

Danulat, E., Kausch, H., 1984. Chitinase activity in the digestive tract of the cod *Gadus morhua* (L.). J. Fish Bio. 24, 125–133.

de Haro, C., 2015. Evaluación de la harina de insectos como fuente alternativa a la harina de pescado en dietas para peces. PhD, Biology and Geology. University of Almería, p. 211.

de Haro, C., Ramos-Bueno, R., Sánchez-Muros, M.J., Barroso, F.G., Rincón-Cervera, M.A., Guil-Guerrero, J.L., 2015. Insect larvae as feed ingredient selectively increase arachidonic acid content in farmed gilthead sea bream (*Sparus aurata* L.). Aquacult. Res.

DeFoliart, G.R., Finke, M.D., Sunde, M.L., 1982. Potential value of the Mormon cricket (Orthoptera: Tettigoniidae) harvested as a high-protein feed for poultry. J. Entomol. 75, 848–852.

Despins, J.L., Axtell, R.C., 1994. Feeding behavior and growth of turkey poults fed larvae of the darkling beetle, *Alphitobius diapernius*. Poultry Sci. 73, 1526–1533.

Despins, J.L., Axtell, R.C., 1995. Feeding behavior and growth of broiler chicks fed larvae of darkling beetle *Alphitobius diaperinus*. Poultry Sci. 74, 331–336.

Despins, J.L., Axtell, R.C., Rives, D.V., Ficken, M.D., 1994. Transmission of enteric pathogens of turkeys by darkling beetle larvae. J. Appl. Poult. Res. 3, 61–65.

Dheke, S., Gubhaju, R.S., 2013. Growth response of rainbow trout (*Oncorhynchus mykiss*) on substitution of shrimp meal by different protein sources. Nepalese J. Zool. 1, 24–29.

Dierenfeld, E.S., King, J., 2008. Digestibility and mineral availability of phoenix worms, *Hermetia illucens*, ingested by mountain chicken frogs, *Leptodactylus fallax*. J. Herpetological Med. Surg. 18, 100–105.

Djordjevic, M., Radenkovic-Damnjanovic, B., Vucinic, M., Baltic, M., Teodorovic, R., Jankovic, L., Vukasinovic, M., Rajkovic, M., 2008. Effects of substitution of fish meal with fresh and dehydrated larvae of the house fly (*Musca domestica* L.) on productive performance and health of broilers. Acta Veterinaria 58, 357–368.

Dong, G.F., Yang, Y.O., Song, X.M., Yu, L., Zhao, T.T., Huang, G.L., Hu, Z.J., Zhang, J.L., 2013. Comparative effects of dietary supplementation with maggot meal and soybean meal in gibel carp (*Carassius auratus gibelio*) and darkbarbel catfish (*Pelteobagrus vachelli*): growth performance and antioxidant responses. Aquaculture Nutr. 19, 543–554.

Dutta, A., Dutta, S., Kumari, S., 2012. Growth of poultry chicks fed on formulated feed containing silk worm pupae meal as protein supplement and commercial diet. J. Anim. Feed Res. 2, 303–307.

Ebenso, I.E., Udo, M.T., 2003. Effect of live maggot on growth of the Nile perch, *Oreochromis niloticus* (Cichlidae) in South Eastern Nigeria. Glob. J. Agric. Sci. 2, 72–73.

Eguchi, M., 1993. Protein protease inhibitors in insect and comparison with mammalian inghibitors. Comp. Biochem. Physiol. 105B, 449–456.

Eilenberg, J., Vlak, J.M., Nielsen-LeRoux, C., Capellozza, S., Jensen, A.B., 2015. Diseases in insects produced for food and feed. J. Insect. Food Feed 2, 87–102.

Ekop, E.A., Udoh, A.I., Akpan, P.E., 2010. Proximate and anti-nutrient composition of four edible insects in Akwa Ibom state. Nigeria World J. Appl. Sci. Technol. 2, 224–231.

Ekoue, S.K., Hadzi, Y.A., 2000. Maggot production as a protein source for young poultry in Togo: preliminary observations. Tropicultura 18, 212–214.

Elwert, C., Knips, I., Katz, P., 2010. A novel protein source: maggot meal from *Hermetia illucens*. In: Gierus, M., Kluth, H., Bulang, M., Kluge, H. (Eds): 11. Tagung Schweine und Geflügelernährung, November 23–25, 2010 Lutherstadt Wittenberg, Institut für Agrar und Ernährungswissenschaften, Universität Halle-Wittenberg.

Emehinaiye, P.A., 2012. Growth performance of *Oreochromis niloticus* fingerlings fed with varying levels of migratory locust (*Locusta migratoria*) meal, Bachelor of Aquaculture and Fisheries Management, Federal University of Agriculture. Abeokuta, Ogun state.

Esteban, M.A., Mulero, V., Cuesta, A., Ortuño, J., Meseguer, J., 2000. Efects of injecting chitin particles on the innate immune reponse of gilthead seabream (*Sparus aurata* L.). Fish Shellfish Immunol. 10, 543–554.

Esteban, M.A., Cuesta, A., Ortuño, J., Meseguer, J., 2001. Immunomodulatory effects of dietary intake of chitin on gilthead seabream (*Sparus aurata* L.) innate immune system. Fish Shellfish Immunol. 11, 303–315.

Fagoonee, I., 1983. Inclusion of silkworm pupae in poultry rations. Tropical Vet. J. 1, 91–96.

Fakayode, O.S., Ugwumba, A.A.A., 2013. Effects of replacement of fishmeal with palm grub (*Oryctes rhinoceros* [Linnaeus, 1758]) meal on the growth of *Clarias gariepinus* (Burchell, 1822) and *Heterobranchus longifilis* (Valenciennes, 1840) fingerlings. J. Fisheries Aquatic Sci. 8, 101–107.

Fange, R., Lundblad, G., Lind, J., 1976. Lysozyme and chitinase in the blood and lymphomyeloid tissues of marine fish. Marine Bio. 36, 277–282.

FAO, 2004a. Protein Sources for the Animal Feed Industry, FAO Animal Production and Health Proceedings. FAO, Bangkok, April 29–May 3, 2002.

FAO, 2004b. Report of a joint FAO/WHO expert consultation, B., Thailand, Sept. 21–30, 1998. Vitamin and mineral requirements in human nutrition: FAO/WHO Expert Consultation on Human Vitamin and Mineral Requirements, Bangkok, Thailand. Available from: http://www.who.int/nutrition/publications/micronutrients/9241546123/en/.

FAO, 2013. Edible insects Future prospects for food and feed security (ONLINE), FAO, Food and Agriculture Organization, Rome, Italy. Available from: http://www.fao.org/docrep/018/i3253e/i3253e.pdf.

Fasakin, E.A., Balogun, A.M., Ajayi, O.O., 2003. Evaluation of full-fat and defatted maggot meals in the feeding of clariid catfish *Clarias gariepinus* fingerlings. Aquacult. Res. 34, 733–738.

Fines, B.C., Holt, G.J., 2010. Chitinase and apparent digestibility of chitin in the digestive tract of juvenile cobia, *Rachycentron canadum*. Aquaculture 303, 34–39.

Finke, M.D., 2002. Complete nutrient composition of commercially raised invertebrates used as food for insectivores. Zoo Biol. 21, 269–285.

Finke, M.D., 2007. Estimate of chitin in raw whole insects. Zoo Biol. 26, 105–115.

Finke, M.D., Sunde, M.L., DeFoliart, G.R., 1985. An evaluation of the protein quality of Mormon cricket (*Anabrux simplex* H.) when used as a high protein feedstuff for poultry. Poultry Sci. 64, 708–712.

Garcia, M.A., Altieri, M.A., 2005. Transgenic crops: implications for biodiversity and sustainable agriculture. Bull. Sci. Technol. Soc. 25, 335–353.

Gasco, L., Belforti, M., Rotolo, L., Lussiana, C., Parisi, G., Terova, G., Roncarati, A., Gai, F., 2014a. Mealworm (*Tenebrio molitor*) as a potential ingredient in practical diets for rainbow trout (*Oncorhynchus mykiss*), In: Vantomme, P., Munke, C., van Huis, A., (Eds.), First International conference "Insects to Feed the World," Wageningen University, Ede-Wageningen, The Netherlands, p. 69.

Gasco, L., Gai, F., Piccolo, G., Rotolo, L., Lussiana, C., Molla, P., Chatzifotis, S., 2014b. Substitution of fish meal by *Tenebrio molitor* meal in the diet of *Dicentrarchus labrax* juveniles, In: Vantomme, P., Munke, C., van Huis, A., (Eds.), 1st International conference "Insects to Feed the World," Wageningen University, Ede-Wageningen, The Netherlands, p. 80.

Ghioni, C., Bell, J.G., Sargent, J.R., 1996. Polyunsaturated fatty acids in neutral lipids and phospholipids of some freshwater insects. Comp. Biochem. Physiol. B 114, 161–170.

Gibril, S., 1997. Utilization of locust meal in poultry diets. J. Nat. Res. Environ. Stud. 1, 19–23.

Gubb, D., Sanz-Parra, A., Barcena, L., Troxler, L., Fullaondo, A., 2010. Protease inhibitors and proteolytic signalling cascades in insects. Biochime 92, 1749–1759.

Gutowska, M.A., Drazen, J.C., Robison, B.H., 2004. Digestive chitinolytic activity in marine fishes of Monterey Bay, California. Comp. Biochem. Physiol. A 139, 351–358.

Habib, M.A.B., Hasan, M.R., Akand, A.M., Siddiqua, A., 1994. Evaluation of silkworm pupae meal as a dietary protein source for *Clarias batrachus* fingerlings. Aquaculture 124, 62.

Hall, G.M., 1992. Fish processing technology. In: Ockerman, H.W. (Ed.), Fishery By-Products. VCH Publishers, New York, NY, USA, pp. 155–192.

Harikrishnan, R., Kim, J.S., Balasundaram, C., Heo, M.S., 2012. Dietary supplementation with chitin and chitosan on haematology and innate immune response in *Epinephelus bruneus* against *Philasterides dicentrarchi*. Exp. Parasitol. 131, 116–124.

Henry, M., Gasco, L., Piccolo, G., Gountoulaki, E., 2015. Review on the use of insects in the diet of farmed fish: past and future. Anim. Feed Sci. Tech. 203, 1–22.

Hirano, S., Itakura, C., Seino, H., Akiyama, Y., Nonaka, I., Kanbara, N., Kawakami, T., 1990. Chitosan as an ingredient for domestic animal feeds. J Agric. Food Chem. 38, 1214–1217.

Hossain, S.M., Blair, R., 2007. Chitin utilisation by broilers and its effect on body composition and blood metabolites. British Poultry Sci. 48, 33–38.

Hossain, M.A., Nahar, N., Kamal, M., Islam, M.N., 1992. Nutrien digestibility coefficients of some plant and animal proteins for tilapia (*Oreochromis mossambicus*). J. Aqua. Trop. 7, 257–266.

Hossain, M.A., Islam, M.N., Alim, M.A., 1993. Evaluation of silkworm pupae meal as dietary protein source for catfish (*Heteropneustes fossilis* Bloch), Fish Nutrition in Practice: Fourth International Symposium on Fish Nutrition and Feeding, Biarritz, France, pp. 785–791.

Hossain, M.A., Nahar, N., Kamal, M., 1997. Nutrient digestibility coefficients of some plant and animal proteins for rohu (*Labeo rohita*). Aquaculture 151, 37–45.

Hou, Q.L., Gao, Q.S., 2001. Chitosan and Medicine. Shanghai Science and Technology Press, Shanghai, China.

Hughes, B., Duncan, I., 1998. The notion of ethological "need," models of motivation, and animal welfare. Anim. Behav. 36, 1696–1707.

Hwangbo, J., Hong, E.C., Jang, A., Kang, H.K., Oh, J.S., Kim, B.W., Park, B.S., 2009. Utilization of house fly-maggots, a feed supplement in the production of broiler chickens. J. Environ. Bio. 30, 609–614.

Idowu, A.B., Amusan, A.A.S., Oyediran, A.G., 2003. The response of *Clarias gariepinus* fingerlings (Burchell 1822) to the diet containing Housefly maggot (*Musca domestica*) (L). Nigerian J. Anim. Prod. 30, 139–144.

Ijaiya, A.T., Eko, E.O., 2009. Effect of replacing dietary fish meal with silkworm (*Anaphe infracta*) caterpillar meal on growth, digestibility and economics of production of starter broiler chickens. Pakistan J. Nutr. 8, 845–849.

International Monetary Fund, 2010. International monetary fund primary commodity prices. Available from: http://www.imf.org/external/np/res/commod/index.aspx.

Jabir, M.D.A.R.J., Razak, S.A., Sabaratnam, V., 2012a. Effect of mushroom supplementation as a prebiotic compound in super worm based diet on growth performance of red tilapia fingerlings. Sains Malaysiana 41, 1197–1203.

Jabir, M.D.A.R.J., Razak, S.A., Vikineswary, S., 2012b. Nutritive potential and utilization of super worm (*Zophobas morio*) meal in the diet of Nile tilapia (*Oreochromis niloticus*) juvenile. African J. Biotech. 11, 6592–6598.

Jeuniaux, C., 1993. Chitinolytic systems in the digestive tract of vertebrates: a review. In: Muzzarelli, R.A.A. (Ed.), Chitin Enzymology. European Chitin Society, Ancona, Italy, pp. 233–244.

Jeyachandran, P., Raj, S.P., 1976. Experiments with artificial feed son *Cyprinus carpio* fingerlings. J. Inland Fish. Soc. India 8, 33–37.

Ji, W.X., Wang, Y., Tang, J.Y., 2010. Apparent digestibility coefficients of selected feed ingredients for Japanese sea bass (*Lateolabrax japonicus*) reared in sea water. J. Fisheries China 34, 101–107.

Ji, H., Zhang, J.L., Huang, J.Q., Cheng, X.F., Liu, C., 2013. Effect of replacement of dietary fish meal with silkworm pupae meal on growth performance, body composition, intestinal protease activity and health status in juvenile Jian carp (*Cyprinus carpio* var Jian). Aquacult. Res. 46, 1209–1221.

Jintasatapom, O., Chumkam, S., Jintasatapom, O., 2011. Substitution of silkworm pupae (*Bombyx mori*) for fishmeal in broodstock diets for snakeskin gourami (*Trichogaster pectoralis*). J. Agric. Sci. Tech. A1, 1341–1344.

Jintasataporn, O., 2012. Production performance of broiler chickens fed with silkworm pupa (*Bombyx mori*). J. Agric. Sci. Technol. A2, 505–510.

Johri, R., Singh, R., Johri, P.K., 2010. Effect of different formulated plant and animal diet on hematology of *Clarias batrachus* Linn. under laboratory conditions. Biochem. Cell. Arch. 10, 283–291.

Johri, R., Singh, R., Johri, P.K., 2011a. Studies on ovarian activity in formulated feed treated *Clarias batrachus* Linn. J. Exp. Zool. India 14, 111–115.

Johri, R., Singh, R., Johri, P.K., 2011b. Impact of formulated plant and animal supplemented diets on nutritional efficiency, growth and body composition in juveniles of *Clarias batrachus* in experimental tanks. J. Exp. Zool. India 14, 59–68.

Johri, R., Singh, R., Johri, P.K., 2011c. Histopathological examination of the gill, liver, kidney, stomach, intestine, testis and ovary of *Clarias batrachus* Linn. during the feeding on different formulated feeds. J. Exp. Zool. India 14, 77–79.

Kawakami, H., Shinohara, N., Sakai, M., 1998. The non-specific immunostimulation and adjuvant effects of Vibrio anguillarum bacterin, M-glucan, chitin and Freund's complete adjuvant against Pasteurella piscicida infection in yellowtail. Fish Pathol. 33, 287–292.

Khatun, R., Howlider, M.A.R., Rahman, M.M., Hasanuzzama, M., 2003. Replacement of fish meal by silkworm pupae in broiler diets. Pak. J. Bio. Sci. 6, 758–955.

Khatun, R., Azmal, S.A., Sarker, M.S.K., Rasid, M.A., Hussain, M.A., Miah, M.Y., 2005. Effect of silkworm pupae on the growth and egg production performance of Rhode Island Red (RIR) pure line. Int. J. Poult. Sci. 4, 718–720.

Khusro, M., Andrew, N.R., Nicholas, A., 2012. Insects as poultry feed: a scoping study for poultry production systems in Australia. World. Poultry Sci. J. 68, 435–446.

Kim, Y.K., 1974. Determination of true digestibility of dietary proteins in carp with Cr_2O_3 containing-diet. Nippon Suisan Gakk 40, 651–653.

Klunder, H.C., Wolkers-Rooijackers, J., Korpela, J.M., Nout, M.J.R., 2012. Microbiological aspects of processing and storage of edible insects. Food Control 26, 628–631.

Konwar, P., Konwar, B.K., Ahmed, H.F., Nath, N.C., Ghosh, M.K., 2008. Effect of feeding silkworm pupae meal with enzyme supplementation on growth performance of broilers. Indian Vet. J. 85, 47–49.

Kroeckel, S., Harjes, A.G.E., Roth, I., Katz, H., Wuertz, S., Susenbeth, A., Schulz, C., 2012. When a turbot catches a fly: evaluation of a pre-pupae meal of the black soldier fly (*Hermetia illucens*) as fish meal substitute—Growth performance and chitin degradation in juvenile turbot (*Psetta maxima*). Aquaculture, 364–365, 345–352.

Ladrón de Guevara, O., Padilla, P., García, L., Pino, J.M., Ramos-Elorduy, J., 1995. Amino acid determination in some edible Mexican insects. Amino Acids 9, 161–173.

Lee, J., Choi, I.C., Kim, K.T., Cho, S.H., Yoo, J.Y., 2012. Response of dietary substitution of fishmeal with various protein sources on growth, body composition and blood chemistry of olive flounder (*Paralichthys olivaceus*, Temminck & Schlegel, 1846). Fish Physiol. Biochem. 38, 735–744.

Lin, S., Mao, S., Guan, Y., Lin, X., Luo, L., 2012a. Dietary administration of chito oligosaccharides to enhance growth, innateimmune response and disease resistance of Trachinotus ovatus. Fish Shellfish Immunol. 32, 909–913.

Lin, S., Mao, S., Guan, Y., Luo, L., Pan, Y., 2012b. Effects of dietary chitosan, oligosaccharides and Bacillus coagulans on the growth, innateimmunity and resistance of koi (*Cyprinus carpio koi*). Aquaculture, 342–343, 36–41.

Lindsay, G.J.H., 1986. The signifcance of chitinolytic enzymes and lysozyme in rainbow trout (*Salmo gairdneri*) defense. Aquaculture 51, 169–173.

Lindsay, G.J.H., Walton, M.J., Adron, J.W., Fletcher, T.C., Cho, C.Y., Cowey, C.B., 1984. The growth of rainbow trout *Salmo gairdneri* given diets containing chitin and its relationship to chitinolytic enzymes and chitin digestibility. Aquaculture 37, 315–334.

Liu, C.M., Lian, Z.M., 2003. Influence of *Acrida cinerea* replacing Peru fish meal on growth performance of broiler chickens. J. Econ. Anim. 7, 48–51.

Lock, E.J., Arsiwalla, T., Waagbø, R., 2014. Insect meal: a promising source of nutrients in the diet of Atlantic salmon (*Salmo salar*), In : Vantomme, P., Munke, C., van Huis, A., (Eds.), First International conference "Insects to Feed the World," Wageningen University, Ede-Wageningen, The Netherlands, 74.

Longvah, T., Mangthya, K., Ramulu, P., 2011. Nutrient composition and protein quality evaluation of eeri silkworm (*Samia ricinii*) prepupae and pupae. Food Chem. 128, 400–403.

Looft, T., Johnson, T.A., Allen, H.K., Bayles, D.O., Stedtfeld, R.D., 2012. In-feed antibiotic effects on the swine intestinal microbiome. Proc. Natl. Acad. Sci. USA 109, 1691–1696.

López-Vergé, S., Barroeta, A.C., Riudavets, J., Rodríguez-Jerez, J.J., 2013. Utilization of *Sitophilus zeamais* (Motschulsky) larvae as a dietary supplement for the production of broiler chickens. Proc. Nutr. Soc. 72, E315.

Madu, C.T., Ufodike, E.B.C., 2003. Growth and survival of catfish (*Clarias anguillaris*) juveniles fed live tilapia and maggot as unconventional diets. J. Aqua. Sci. 18, 47–51.

Mahata, S.C., Bhuiyan, A.K.M., Zaher, M., Hossain, M.A., Hasan, M.R., 1994. Evaluation of silkworm pupae as dietary protein source for Thai sharpunti *Puntius gonionotus*. J. Aquacul. Trop. 9, 77–85.

Makinde, O.J., 2015. Maggot meal: a sustainable protein source for livestock production,a review. Adv. Life Sci. Tech., 31. Available from: http://www.iiste.org/Journals/index.php/ALST/article/view/21273.

Makkar, H.P.S., Tran, G., Heuzé, V., Ankers, P., 2014. State-of-the-art on use of insects as animal feed. Anim. Feed Sci. Tech. 197, 1–33.

Manzano-Agugliaro, F., Sanchez-Muros, M.J., Barroso, F.G., Martínez-Sánchez, A., Rojo, S., Pérez-Bañón, C., 2012. Insects for biodiesel production. Renew. Sust. Energ. Rev. 16, 3744–3753.

McAllister, J.C., Steelman, C.D., Skeeles, J.K., 1994. Reservoir competence of the lesser mealworm (Coleoptera: Tenebrionidae) for *Salmonella typhimurium* (Eubacteriales: Enterobacteriaceae). J. Med. Entomol. 31, 369–372.

Medhi, D., 2011. Effects of enzyme supplemented diet on finishing crossbred pigs at different levels of silk worm pupae meal in diet. Indian J. Field Vet. 7, 24–26.

Medhi, D., Math, N.C., Sharma, D.N., 2009a. Effect of silk worm pupae meal and enzyme supplementation on blood constituents in pigs. Indian Vet. J. 86, 433–434.

Medhi, D., Nath, N.C., Gohain, A.K., Bhuyan, R., 2009b. Effect of silk worm pupae meal on carcass characteristics and composition of meat in pigs. Indian Vet. J. 86, 816–818.

Mellen, J., MacPhee, M.S., 2001. Philosophy of environmental enrichment: past, present, and future. Zoo Biol. 20, 211–226.

Mench, J., 2004. Lameness. In: Weeks, C., Butterworth, A. (Eds.), Measuring and Auditing Broiler Welfare. CAB International, Oxford, UK, pp. 3–17.

Miles, R.D., Jacobs, J.P., 1997. Fish meal understanding why this feed ingredient is so valuable in poultry diets, University of Florida Cooperative and Extension Service. Institute of Food and Agricultural Science, FL, USA.

Ming, J., Ye, J., Zhang, Y., Yang, X., Wu, C., Shao, X., Liu, P., 2013. The influence of maggot meal and l-carnitine on growth, immunity, antioxidant indices and disease resistance of black carp (*Mylopharyngodon piceus*). J. Chinese Cereals Oils Assoc. 28, 80–86.

Moriya, K., Fujibayashi, T., Yoshihara, T., Matsuda, A., Sumi, N., Umezaki, N., Kurahashi, H., Agui, N., Wada, A., Watanabe, H., 1999. Verotoxinproducing *Escherichia coli* O157:H7 carried by the housefly in Japan. Med. Vet. Entomol. 13, 214–216.

Mpuchane, S., Taligoola, H.K., Gashe, B.A., 1996. Fungi associates with *Imbrasia belina*, an edible grasshopper. Botswana Notes Rec. 28, 193–197.

Mungkung, R., Aubin, J., Prihadi, T.H., Slembrouck, J., van der Werf, H.M.G., Legendre, M., 2013. Life Cycle Assessment for environmentally sustainable aquaculture management: a case study of combined aquaculture systems for carp and tilapia. J. Cleaner Prod. 57, 249–256.

Nakagaki, B.J., Sunde, M.L., DeFoliart, G.R., 1987. Protein quality of the house cricket *Acheta domesticus* when fed to rooster chicks. Poultry Sci. 66, 1367–1371.

Nandeesha, M.C., Srikanth, G.K., Keshavanath, P., Varghese, T.J., Basavaraja, N., Das, S.K., 1990. Effects of non-defatted silkworm-pupae in diets on the growth of common carp, *Cyprinus carpio*. Biol. Wastes 33, 17–23.

Nandeesha, M.C., Gangadhara, B., Manissery, J.K., 1999. Silkworm pupa oil and sardine oil as an additional energy source in the diet of common carp, *Cyprinus carpio*. Asian Fisheries Sci. 12, 207–215.

Nandeesha, M.C., Gangadhara, B., Varghese, T.J., Keshavanath, P., 2000. Growth response and flesh quality of common carp, *Cyprinus carpio* fed with high levels of non-defatted silkworm pupae. Asian Fisheries Sci. 13, 235–242.

Neuringer, A., 1969. Animals respond for food in the presence of free food. Science 166, 339–341.

Newberry, R.C., 1995. Environmental enrichment; increasing the biological relevance of captive environments. Appl. Anim. Behav. Sci. 44, 229–243.

Newton, G.L., Booram, C.V., Barker, R.W., Hale, O.M., 1977. Dried *Hermetia illucens* larvae meal as a supplement for swine. J. Anim. Sci. 44, 395–400.

Newton, L., Sheppard, C., Watson, D.W., Burtle, G., Dove, R., 2005. Using the black soldier fly, *Hermetia illucens*, as a value-added tool for the management of swine manure. Rep. for Mike Williams, Dir. Anim. Poult. Waste Manag. Cent., North Carolina State Univ., Raleigh, NC, 1–17.

Ng, W.K., Liew, F.L., Ang, L.P.K.W. W., 2001. Potential of mealworm (*Tenebrio molitor*) as an alternative protein source in practical diets for African catfish, *Clarias gariepinus*. Aquacult. Res. 32, 273–280.

Nsofor, C.J., Osayamwen, E.M., Ewuim, S.C., Etaga, H.O., 2008. Effects of varying levels of maggot and fishmeal on food utilization and growth of *Clarias gariepinus* fingerlings reared in net hopas in concrete ponds. J. Appl. Nat. Sci. 9, 79–84.

Ocio, E., Viñaras, R., Rey, J.M., 1979. House fly larvae meal grown on municipal organic waste as a source of protein in poultry diets. Anim. Feed Sci. Tech. 4, 227–231.

Ogunji, J.O., Nimptsch, J., Wiegand, C., Schulz, C., 2007. Evaluation of the influence of housefly maggot meal (magmeal) diets on catalase, glutathione S-transferase and glycogen concentration in the liver of *Oreochromis niloticus* fingerling. Comp. Biochem. Physiol. A 147, 942–947.

Ogunji, J.O., Kloas, W., Wirth, M., Neumann, N., Pietsch, C., 2008a. Effect of housefly maggot meal (magmeal) diets on the performance, concentration of plasma glucose, cortisol and blood characteristics of *Oreochromis niloticus* fingerlings. J. Anim. Psys. Nutr. 92, 511–518.

Ogunji, J.O., Kloas, W., Wirth, M., Schulz, C., Rennert, B., 2008b. Housefly maggot meal (magmeal) as a protein source for *Oreochromis niloticus* (Linn.). Asian Fisheries Sci. 21, 319–331.

Ogunji, J.O., Toor, R., Schulz, C., Kloas, W., 2008c. Growth performance, nutrient utilization of Nile tilapia *Oreochromis niloticus* fed housefly maggot meal (magmeal) diets. Turkish J. Fisheries Aqua. Sci. 8, 141–147.

Ogunji, J., Pagel, T., Schulz, C., Kloas, W., 2009. Apparent digestibility coefficient of housefly maggot meal (magmeal) for Nile tilapia (*Oreochromis niloticus* L.) and carp (*Cyprinus carpio*). Asian Fisheries Sci. 22, 1095–1105.

Ojewola, G.S., Okoye, F.C., Ukoha, O.A., 2005. Comparative utilization of three animal protein sources by broiler chickens. Int. J. Poultry Sci. 4, 462–467.

Okah, U., Onwujiariri, E.B., 2012. Performance of finisher broiler chickens fed maggot meal as a replacement for fish meal. J. Agric. Tech. 8, 471–477.

Okubanjo, A.O., Apata, E.S., Babalola, O.O., 2014. Carcass and organoleptic qualities of chicken broilers fed maggot meal in replacement for dietary fish meal. Am. J. Res. Comm. 2, 147–156.

Oluokun, J.A., 2000. Upgrading the nutritive value of full-fat soyabeans meal for broiler production with either fishmeal or black soldier fly larvae meal (*Hermetia illucens*). Trop. J. Anim. Sci. 3, 51–61.

Omotoso, O.T., 2006. Nutritional quality, functional properties and anti-nutrient compositions of the larva of *Cirina forda* (Westwood) (Lepidoptera: Saturniidae). J. Zhejiang Univ. Sci. B 7 (1), 51–55.

Omoyinmi, G.A.K., Olaoye, O.J., 2012. Growth performance of Nile tilapia *Oreochromis niloticus* fed diets containing different sources of animal protein. Libyan Agric. Res. Ctr. J. Int. 3, 18–23.

Oonincx, D.G.A.B., de Boer, I.J.M., 2012. Environmental impact of the production of mealworms as a protein source for humans: a life cycle assessment. PLoS ONE 7, e51145.

Opara, M.N., Sanyigha, F.T., Ogbuewu, I.P., Okoli, I.C., 2012. Studies on the production trend and quality characteristics of palm grubs in the tropical rainforest zone of Nigeria. Int. J. Agric. Tech. 8, 851–860.

Osava, M., 1999. Soy production spreads, threatens Amazon in Brazil. InterPress Service. Available from: http://www.ipsnews.net/1999/09/environment-brazil-soy-production-spreads-threatens-amazon/.

Ossey, Y.B., Koumi, A.R., Koffi, K.M., Atse, B.C., Kouame, L.P., 2012. Use of soybean, bovine brain and maggot as sources of dietary protein in larval *Heterobranchus longifilis* (Valenciennes, 1840). J. Anim. Plant Sci. 15, 2099–2108.

Ostaszewska, T., Dabrowski, K., Kwasek, K., Verri, T., Kamaszewski, M., Sliwinski, J., Napora-Rutkowski, L., 2011. Effects of various diet formulations (experimental and commercial) on the morphology of the liver and intestine of rainbow trout (*Oncorhynchus mykiss*) juveniles. Aquacult. Res. 42, 1796–1806.

Oviedo-Rondón, E.O. 2007. Predisposing factors that affect walking ability in turkeys and broilers. Annual Carolina Poultry Nutrition Conference Research, Triangle Park, N.C. Available from: http://www.thepoultrysite.com/articles/1323/predisposing-factors-that-affect-walking-ability-in-turkeys-and-broilers.

Oyegoke, O.O., Akintola, A.J., Fasoranti, J.O., 2006. Dietary potentials of the edible larvae of *Cirina forda* (westwood) as a poultry feed. African J. Biotech. 5, 1799–1802.

Oyelese, O.A., 2007. Utilization of compounded ration and maggot in the diet of *Clarias gariepinus*. Res. J. Appl. Sci. 2, 301–306.

Park, B.K., Kim, M.M., 2010. Applications of chitin and its derivatives in biological medicine. Int. J. Mol. Sci. 11, 5152–5164.

Parshikova, O.A., Chaphinskaya, K.N., Molchanova, N.V., Romasko, U.N., 1981. Metabolism and productivity of hens given diets with ground housefly larvae. Nutr. Abs. Rev. 53, 361.

Pava-Ripoll, M., Pearson, R.E., Miller, A.K., Ziobro, G.C., 2012. Prevalence and relative risk of *Cronobacter spp.*, *Salmonella spp.*, and *Listeria monocytogenes* associated with the body surfaces and guts of individual filth flies. Appl. Environ. Microbiol. 78, 7891–7902.

Piccolo, G., Marono, S., Gasco, L., Lannaccone, F., Bovera, F., Nizza, A., 2014. Use of *Tenebrio molitor* larvae meal in diets for gilthead sea bream *Sparus aurata* juveniles, In: Vantomme, P., Munke, C., van Huis, A., (Eds.), First International conference "Insects to Feed the World," Wageningen University, Ede-Wageningen, The Netherlands, p. 76.

Pieterse, E., Pretorius, Q., Hoffish mealan, L.C., Drew, D.W., 2014. The carcass quality, meat quality and sensory characteristics of broilers raised on diets containing either *Musca domestica* larvae meal, fish meal or soya bean meal as the main protein source. Anim. Prod. Sci. 54, 622–628.

Poluektova, L.S., Chaplinskaya, K.N., Dement'eva, T.A., Kozlova, L.S., Ganenkova, N.M., 1980. Effect of adding into the diet of pigs a meal from house-fly larvae on metabolism, development, and meat quality of the pigs. Nauchnye Trudy Novosibirskogo Sel'skokhozyaistvennogo Instituta. 128, 24–27.

Pro, A., Cuca, M., Becerril, C., Bravo, H., Bixler, E., Pérez, A., 1999. Estimación de la energía metabolizable y utilización de larva de mosca (*Musca domestica* l.) en la alimentación de pollos de engorda. Arch. Latinoam. Prod. Anim. 7, 39–51.

Raksakantong, P., Meeso, N., Kubola, J., Siriamornpun, S., 2010. Fatty acids and proximate composition of eight Thai edible terricolous insects. Food Res. Int. 43, 350–355.

Ramos-Elorduy, J., Pino, J.M., González, O., 1981. Digestibilidad in vitro de algunos insectos comestibles de México. Folia Entomolológica Mexicana 49, 141–152.

Ramos-Elorduy, J., Bourges, H., Pino, J.M., 1982. Valor nutritivo y calidad de la proteína de algunos insectos comestibles de México. Folia Entomolológica Mexicana 53, 111–118.

Ramos-Elorduy, J., Pino-Moreno, J.M., Márquez-Mayaudon, C., 1984. Protein content of some edible insects in Mexico. J. Ethnobiol. 4, 61–72.

Ramos-Elorduy, J., Pino, J.M., Escamilla, E., Alvarado, M., Lagunez, J., Ladron de Guevara, O., 1997. Nutritional value of edible insects from the state of Oaxaca, Mexico. J. Food Comp. Anal. 10, 142–157.

Ramos-Elorduy, J., Avila, E., Rocha, A., Pino, J.M., 2002. Use of *Tenebrio molitor* (Coleoptera: Tenebrionidae) to recycle organic wastes and as feed for broiler chickens. J. Econ. Entomol. 95, 214–220.

Rangacharyulu, P.V., Giri, S.S., Paul, B.N., Yashoda, K.P., Rao, R.J., Mahendrakar, N.S., Mohanty, S.N., Mukhopadhyay, P.K., 2003. Utilization of fermented silkworm pupae silage in feed for carps. Bioresource Tech. 86, 29–32.

Ren, J., Wu, Y., Lin, J., 2011. Effect of house fly larvae meal on growth performance and slaughter performance of yellow dwarf chickens. J. China Poultry 33, 8–11.

Roncarati, A., Gasco, L., Parisi, G., Terova, G., 2014. Growth performance of common catfish (*Ameiurus melas* Raf.) fingerlings fed insect meal diets, In: Vantomme, P., Munke, C., van Huis, A., (Eds.), First International conference "Insects to Feed the World," Ede-Wageningen, The Netherlands, p. 162.

Rumpold, B.A., Schlüte, O., 2015. Insect-based protein sources and their potential for human consumption: nutritional composition and processing. Anim. Frontiers 5, 20–24.

Rumpold, B.A., Schlüter, O.K., 2013. Nutritional composition and safety aspects of edible insects. Nutr. Food Res. 57, 802–823.

Saeed, T., Dagga, F.A., Saraf, M., 1993. Analysis of residual pesticides present in edible locusts captured in Kuwait. Arab. Gulf. J. Sci. Res. 11, 1–5.

Sakai, M., Kamiya, H., Ishii, S., Atsuta, S., Kobayashi, M., 1992. The immunostimulating effects on chitin in rainbow trout, *Oncorhynchus mykiss*. Dis. Asian Aqua. 1, 413–417.

Sánchez-Muros, M.J., Barroso, F.G., Manzano-Agugliaro, F., 2014. Insect meal as renewable source of food for animal feeding: a review. J. Cleaner Prod. 65, 16–27.

Sánchez-Muros, M.J., de Haro, C., Sanz, A., Trenzado, C.E., Villareces, S., Barroso, F.G., 2015. Nutritional evaluation of *Tenebrio molitor* meal as fishmeal substitute for tilapia (*Oreochromis niloticus*) diet. Aqua. Nutr.

Sapcota, D., Sheikh, I.U., Dutta, K.K., Sarma, S., Goswami, R., 2003. Effect of dietary Muga silkworm supplementation on the performance of broilers. Indian Vet. J. 80, 19–22.

Sawhney, S., 2014. Effect of partial substitution of expensive ingredient i.e. fish meal on the growth of *Tor putitora* fed practical diets. Int. H. Multidiciplinary Res. 2, 482–489.

SCAHAW, 2002. Scientific Committee on Animal Health and Animal Welfare 2002. The welfare of chickens kept for meat production (broilers). Available from: http://ec.europa.eu/food/fs/sc/scah/out39_en.pdf. European Union.

Schmidt, J.O., 1986. Chemistry, pharma-cology and chemical ecology of ant venoms. In: Pyek, T. (Ed.), Venoms of the Hymenoptera. Academic Press, London, UK, pp. 425–508.

Sealey, W.M., Gaylord, T.G., Barrows, F.T., Tomberlin, J.K., McGuire, M.A., Ross, C., St-Hilaire, S., 2011. Sensory analysis of rainbow trout, *Oncorhynchus mykiss*, fed enriched black soldier fly prepupae, *Hermetia illucens*. J. World Aqua. Soc. 42, 34–45.

Se-Kwon, K., Niranjan, R., 2005. Enzymatic production and biological activities of chitosan oligosaccharides (COS): a review. Carbohydr. Polym. 62, 357–368.

Sheikh, I.U., Sapcota, D., 2007. Economy of feeding muga silk worm pupape meal in the diet of broilers. Indian Vet. J. 84, 722–724.

Sheikh, I.U., Sapcota, D., Dutta, K.K., Sarma, S., 2005. Effect of dietary silkworm pupae meal on the carcass characteristics of broilers. Indian Vet. J. 82, 752–755.

Shepherdson, D.J., 1998. Tracing the path of environmental enrichment in zoos. In: Shepherdson, D.J., Mellen, J.D., Hutchins, M., (Eds.), Second Nature: Environmental Enrichment for Captive Animals, Washington, DC, pp. 1–12.

Shepherdson, D.J., Carlstead, K., Mellen, J., Seidensticker, J., 1993. The influence of food presentation on the behavior of small cats in confined environments. Zoo Biol. 12, 203–216.

Shiau, S.-Y., Yu, Y.-P., 1999. Dietary supplementation of chitin and chitosan depresses growth in tilapia, *Oreochromis niloticus* × *O. aureus*. Aquaculture 179, 439–446.

Shigehiro, H., Chitoshi, I., Haruyoshi, S., 1990. Chitosan as an ingredient for domestic animal feeds. J. Agric. Food Chem. 87, 1214–1217.

Shyama, S., Keshavanath, P., 1993. Growth response of *Tor khudree* to silkworm pupa incorporated diets, Fish Nutrition in Practice, Fourth International Symposium on Fish Nutrition and Feeding, Biarritz, France, pp. 779–783.

Sinha, S., Dutta, A., Chattopadhyay, S., 2009. Effect of replacement of fish meal by de-oiled silkworm pupae of *Antheraea mylitta* (Drury) on the growth performance of broiler chickens. Bull. Indian Acad. Sericult. 13, 70–72.

Sogbesan, A.O., Ugwumba, A.A.A., 2008. Nutritional evaluation of termite (*Macrotermes subhyalinus*) meal as animal protein supplements in the diets of *Heterobranchus longifilis* (Valenciennes, 1840) fingerlings. Turkish J. Fisheries Aqua. Sci. 8, 149–157.

Sogbesan, A.O., Ajuonu, N., Musa, B.O., Adewole, A.M., 2006. Harvesting techniques and evaluation of maggot meal as animal dietary protein source for Heteroclarias in outdoor concrete tanks. World J. Agric. Sci. 2, 394–402.

Solomon, S.G., Sadiku, S.O.E., Tiamiyu, L.O., 2007. Wing reproductive termite (*Macrotremes nigeriensis*)–soybean (*Glyxine max*) meals blend as dietary protein source in the practical diets of *Heterobranchus bidorsalis* fingerlings. Pakistan J. Nutr. 6, 267–270.

Stanley-Samuelson, D.W., Jurenka, R.A., Cripps, C., Blomquist, G.J., Renovales, M., 1988. Fatty acids in insects: composition, metabolism, and biological significance. Arch. Ins. Biochem. Physiol. 9, 1–33.

Steinfeld, H., Gerber, P., Wassenaar, T., Castel, V., Rosales, M., De Haan, C.P.R.F., 2006. Livestock's Long Shadow: Environmental Issues and Options. FAO, Rome, Italy.

St-Hilaire, S., Cranfill, K., McGuire, M.A., Mosley, E.E., Tomberlin, J.K., Newton, L., Sealey, W., Sheppard, C., Irving, S., 2007a. Fish offal recycling by the black soldier fly produces a foodstuff high in omega-3 fatty acids. J. World Aqua. Soc. 38, 309–313.

St-Hilaire, S., Sheppard, C., Tomberlin, J.K., Irving, S., Newton, L., McGuire, M.A., Mosley, E.E., Hardy, R.W., Sealey, W., 2007b. Fly prepupae as a feedstuff for rainbow trout, *Oncorhynchus mykiss*. J. World Aqua. Soc. 38, 59–67.

Sugita, H., Yamada, S., Konagaya, Y., Deguchi, Y., 1999. Production of b-*N*-acetylglucosaminidase and chitinase by Aeromonas species isolated from river fish. Fish Sci. 65, 155–158.

Sumitra-Vijayaraghavan Wafar, M.V.M., Royan, J.P., 1978. Feeding experiments with the shrimp, *Metapenaeus monoceros* (Fabricius). Indian J. Marine Sci. 7, 195–197.

Suzuki, M., Morimatsu, M., Yamashita, T., Iwanaga, T., Syuto, B., 2001. A novel serum chitinase that is expressed in bovine liver. FEBS Lett. 506, 127–130.

Téguia, A., Mpoame, M., Okourou-Mba, J.A., 2002. The production performance of broiler birds as affected by the replacement of fish meal by maggot meal in the starter and finisher diets. Tropicultura 20, 187–192.

Teotia, J.S., Miller, B.F., 1973. Fly pupae as a dietary ingredient for starting chicks. Poultry Sci. 52, 1830–1835.

Tiu, L.G., 2012. Enhancing Sustainability of Freshwater Prawn Production in Ohio, In: Newsletter, O.S.U.S.C., (Ed.). Ohio State University South Centers Newsletter, Columbus, OH, USA. Fall, p. 4.

Van der Spiegel, M., Noordam, M.Y., van der Fels-Klerx, H.J., 2013. Safety of novel protein sources (insects, microalgae, seaweed, duckweed, and rapeseed) and legislative aspects for their application in food and feed production. Comp. Rev. Food Sci. Safety 12, 662–678.

van Huis, A., Itterbeek, J.V., Klunder, H.C., Mertens, E., Halloran, A., Muir, G., Vantomme, P., 2013. Edible Insects: Future Prospects for Food and Feed Security. Food and Agriculture Organization, Rome, Italy.

Vega, F.E., Kaya, H.K., 2012. Insect Pathology. Academic Press, San Diego, San Diego, USA, pp. 490.

Veldkamp, T., Bosch, G., 2015. Insects: a protein-rich feed ingredient in pig and poultry diets. Anim. Front. 5, 45–50.

Venkatachalam, M., Thangamani, R., Pandiyan, V., Shanmugsundram, S., 1997. Effect of reducing antinutritional factors in silkworm pupae meal on its feeding value of broilers. Indian J. Poultry Sci. 32, 182–184.

Venkatesh, B., Mukherji, A.P., Mukhopadhyay, P.K., Dehadrai, P.V., 1986. Growth and metabolism of the catfish *Clarias batrachus* fed with different experimental diets. Proc. Indian Acad. Sci. (Anim. Sci.) 95, 457–462.

Viroje, W., Malin, S., 1989. Effects of fly larval meal grown on pig manure as a source of protein in early weaned pig diets. Thurakit Ahan Sat. 6, 25–31.

Wang, X.W., Du, Y.G., Bai, X.F., Li, S.G., 2003. The effect of oligochitosan on broiler gut flora, microvilli density, immune function and growth performance. Acta Zoonutrim Sin. 15, 32–45.

Wang, D., Zhai, S.-W., Zhang, C.-X., Bai, Y.-Y., An, S.-H., Xu, Y.-N., 2005. Evaluation on nutritional value of field crickets as a poultry feedstuff. Asian-Aust. J. Anim. Sci. 18, 667–670.

Wang, D., Zhai, S.-W., Zhang, C.-X., Zhang, Q., Chen, H., 2007. Nutrition value of the Chinese grasshopper *Acrida cinerea* (Thunberg) for broilers. Anim.Feed Sci. Tech. 135, 66–74.

Wijayasinghe, M.S., Rajaguru, A.S., 1977. Use of silkworm (*Bombyx mori* L.) pupae as a protein supplement in poultry rations. J. Natn. Sci. Coun. Sri Lanka 5, 95–104.

Yashoda, K.P., Rao, R.J., Rao, D.N., Mahendrakar, N.S., 2008. Chemical and microbiological changes in silkworm pupae during fermentation with molasses and curd as lactic culture. Bull. Ind. Acad. Sericult. 12, 58–66.

Zhang, J.B., Zheng, L.Y., Jin, P., Zhang, D.N., Yu, Z.N., 2014. Fishmeal substituted by production of chicken manure conversion with microorganisms and black soldier fly, In: Vantomme, P., Munke, C., van Huis, A., (Eds.), First International Conference Insects to Feed The World, Wageningen University, Ede-Wageningen, The Netherlands, p. 153.

Zheng, W., Dong, Z.G., Li, X.Y., Cheng, H.L., Yan, B.L., Yang, S.H., 2010a. Effects of dietary fly maggot *Musca domestica* in polyculture of Chinese shrimp with swimming crab *Portunus trituberculatus*. Fisheries Sci. 29, 344–347.

Zheng, W., Dong, Z.G., Wang, X.Q., Cao, M., Yan, B.L., Li, S.H., 2010b. Effects of dietary fly maggot Musca domestica on growth and body compositions in Chinese shrimp *Fenneropenaeus chinensis* juveniles. Fisheries Sci. 29, 187–192.

Zhou, L., Lin, H., 2000. Chitin and chitosan. Chin. Feeds 23, 16–18.

Zhu, Y.G., Johnson, T.A., Su, J.Q., Qiao, M., Guo, G.X., Stedtfeld, R.D., Hashsham, S.A., Tiedje, J.M., 2013. Diverse and abundant antibiotic resistance genes in Chinese swine farms. Proc. Natl. Acad. Sci. USA 110, 3435–3440.

Zhuang, P., Zou, H., Shu, W., 2009. Biotransfer of heavy metals along a soil-plant-insect-chicken food chain: field study. J. Environ. Sci. 21, 849–853.

Zuidhof, M.J., Molnar, C.L., Morley, F.M., Wray, T.L., Robinson, F.E., Khan, B.A., Al-Ani, L., Goonewardene, L.A., 2003. Nutritive value of house fly (*Musca domestica*) larvae as a feed supplement for turkey poults. Anim. Feed Sci. Tech. 105, 225–230.

Zurek, L., Ghosh, A., 2014. Insects represent a link between food animal farms and the urban environment for antibiotic resistance trait. Appl. Environ. Microbiol. 80, 3562–3567.

Appendix

Documented Information for 1555 Species of Insects and Spiders

Table A1 Documented Information for 1555 Species of Insects and Spiders

Order	Family	Genus	Species	Common Name	Faunal	Distribution and References
Araneae	Nephilidae	*Nephila*	*clavipes* (L.)		Neo	Mexico (Vázquez-Dávila, 2008)
		Nephila	*inaurata* (Walckenaer)		Afr	Madagascar (Decary, 1937)
		Nephila	*madagascariensis* Vinson		Afr	Madagascar (Decary, 1937; Bergier, 1941)
		Nephila	*edulis* (Labil.)	Golden orb-weaver	Aus	New Caledonia (Bergier, 1941)
		Nephila	*antipodiana* (Walck.)		Or	Thailand (Bristowe, 1932)
		Nephila	*pilipes* (F.)		Or	Thailand (Bristowe, 1932)
	Sparassidae	*Heteropoda*	*venatoria* Latreille	Brown huntsman spider	Neo	Venezuela (Araújo and Beserra, 2007)
	Theraphosidae	*Holothele*	*waikoshiemi* Bertani and Araújo	Bird eater spider	Neo	Venezuela (Araújo and Beserra, 2007)
		Pseudotheraphosa	*apophysis* Tinter	False goliath tarantula	Neo	Venezuela (Schultz and Schultz, 2009)
		Haplopelma	*albostriatum* (Simon)	Thai zebra tarantula	Or	Laos (Bristowe, 1945), Cambodia (Yen et al., 2013)
		Theraphosa	*blondi* (Latreille)	Bird eater spider	Neo	Venezuela (Araújo and Beserra, 2007)
Coleoptera	Buprestidae	*Chalcophora*	*yunnana* Faim.		Pal	China (Chen and Feng, 1999)
		Coraebus	*sauteri* Kerremans		Pal	China (Chen and Feng, 1999)
		Coraebus	*sidae* Kerremans		Pal	China (Chen and Feng, 1999)
		Chrysobothris	*fatalis* Harold	Jewel beetle	Afr	Angola (DeFoliart, 2002)
		Chrysobothris	*femorata* (Olivier)	Flat headed apple borer	Or	Thailand (DeFoliart, 2002)
		Euchroma	*gigantea* L.	Ceiba borer beetle	Neo	Colombia (DeFoliart, 2002; Dufour, 1987), Mexico (Ramos-Elorduy and Pino Moreno, 2004), Brazil (Costa Neto and Ramos-Elorduy, 2006)
		Sphenoptera	*kozlovi* Jak.		Pal	China (Chen and Feng, 1999)
		Steraspis	*amplipennis* (Fahr.)	Jewel beetle	Afr	Angola (DeFoliart, 2002)
		Sternocera	*aequisignata* Saund.	Jewel beetle	Or	Thailand (DeFoliart, 2002), Laos (Yhoung-Aree and Viwatpanich, 2005)
		Sternocera	*ruficornis* Saund.	Jewel beetle	Or	Thailand (Hanboonsong, 2010)
	Sternocera	*Sternicornis* (L.)		Jewel beetle	Or	India (Singh et al., 2007)
		Sternocera	*castanea* (Olivier)	Jewel beetle	Afr	Benin (Riggi et al., 2013), Burkina Faso (Tchibozo, 2015)
		Sternocera	*feldspathica* White	Jewel beetle	Afr	Angola (DeFoliart, 2002)
		Sternocera	*funebris* Boheman	Jewel beetle	Afr	Zimbabwe (Chavanduka, 1976; Gelfand, 1971)
		Sternocera	*interrupta* (Olivier)	Jewel beetle	Afr	Benin (Riggi et al., 2013), Cameroun (Seignobos et al., 1996), Burkina Faso (Tchibozo, 2015)
	Carabidae	*Cicindela*	*curvata* Chevr.	Tiger beetle	Neo	Mexico (DeFoliart, 2002)
		Cicindela	*roseiventris* Chevr.	Tiger beetle	Neo	Mexico (DeFoliart, 2002)

Cerambycidae	Acanthophorus	confinis Laporte de Cast.	Long-horned beetle	Afr	Zambia (Mbata, 1995)
	Acanthophorus	capensis White	Long-horned beetle	Afr	Zambia (Mbata, 1995)
	Acanthophorus	serraticornis (Oliv.)		Or	India (Singh et al., 2007)
	Acanthophorus	maculatus (F.)	Long-horned beetle	Afr	Zambia (Mbata, 1995)
	Acrocinus	longimanus (L.)	Harlequin beetle	Neo	Colombia (DeFoliart, 2002; Dufour, 1987), Mexico (Ramos-Elorduy and Pino Moreno, 2004)
	Agrianome	spinicollis (McLeay)	Long-horned beetle	Aus	Australia (DeFoliart, 2002)
	Analeptes	trifasciata (F.)	Long-horned beetle	Afr	Nigeria (Banjo et al., 2006), D. R. Congo (Hoare, 2007), CAR (Bahuchet, 1985)
	Ancylonotus	tribulus (Fabr.)	Long-horned beetle	Afr	Gabon, Senegal (Netolitzky, 1919), West Africa (Bergier, 1941)
				Neo	South America (DeFoliart, 2002)
	Anoplophora	chinensis (Forster)		Pal	China (Chen and Feng, 1999)
	Anoplophora	glabripennis (Motsch.)	Asian longhorn beetle	Pal	China (Chen and Feng, 1999)
	Anomophysis	inscripta (Waterhouse)		Or	India (Singh et al., 2007)
	Aplagiognathus	spinosus (Newman)		Neo	Mexico (DeFoliart, 2002)
	Apriona	germani (Hope)	Mulberry longhorn beetle	Or	Thailand, Vietnam (Hanboonsong et al., 2000)
	Apriona	rugicollis Chevr.		Pal	China (Chen and Feng, 1999)
	Arhophalus	cf rusticus (L.)		Pal	Japan (Schimitschek, 1968)
				Neo	Mexico (Ramos-Elorduy et al., 1998; DeFoliart, 2002)
	Arhophalus	rusticus montanus (LeConte)		Neo	Mexico (Ramos-Elorduy and Pino Moreno, 2004)
	Aristobia	approximator (Thomson)		Or	Thailand (Hanboonsong, 2010)
	Aromia	bungii (Faldermann)		Pal	China (Chen and Feng, 1999)
	Bardistus	cibarius Newman	Long-horned beetle	Aus	Australia (DeFoliart, 2002; Yen, 2015)
	Batocera	gigas (Drap.)		Or	Indonesia (Meer Mohr, 1965)
	Batocera	lineolata Chevr.		Pal	Japan (Schimitschek, 1968)
	Batocera	numitor Newman	Mango-tree longhorn borer	Or	Philippines (DeFoliart, 2002)
	Batocera	parryi (Hope)		Or	India (Singh et al., 2007)
	Batocera	roylei Hope		Or	India (Chakravorty et al., 2011, 2013)
	Batocera	rubus (L.)		Or	Indonesia, India, Sri Lanka (DeFoliart, 2002), Laos (Yhoung-Aree and Viwatpanich, 2005)
	Batocera	tigris (Voet)		Or	Indonesia (Roepke, 1952)

(*Continued*)

Table A1 Documented Information for 1555 Species of Insects and Spiders (cont.)

Order	Family	Genus	Species	Common Name	Faunal	Distribution and References
		Batocera	wallacei Thomson		Or	Malaysia (Chung, 2010)
		Callipogon	barbatus (F.)		Neo	Mexico (Ramos-Elorduy et al., 1998; DeFoliart, 2002)
		Celosterna	scabrator (F.)	Babul-root boring longcorn	Or	India (DeFoliart, 2002)
		Ceroplesis	burgeoni Breuning	Long-horned beetle	Afr	Southern Afr (Malaisse, 1997)
		Cnemoplites	edulis Newman	Long-horned beetle	Aus	Australia (DeFoliart, 2002)
		Cnemoplites	flavilipis Thomson	Long-horned beetle	Aus	Australia (DeFoliart, 2002)
		Derobrachus	procerus Thomson		Neo	Mexico (Ramos-Elorduy and Pino Moreno, 2004)
		Diastocera	wallichii Hope		Or	India (Singh et al., 2007)
		Dorysthenes	buqueti (Guérin-M.)		Or	Thailand (Utsunomiya and Masumoto, 1999)
		Dorysthenes	forficatus Fabr.		Pal	North Africa (Ghesquière, 1947)
		Dorysthenes	granulosus (Thonsom)		Or	Thailand (Utsunomiya and Masumoto, 1999)
		Dorysthenes	montanus (Guérin-M.)		Or	India (Singh et al., 2007)
		Dorysthenes	walkeri (Waterhouse)		Or	Thailand (Utsunomiya and Masumoto, 1999)
		Eburia	stigmatica Chevr.		Neo	Mexico (Ramos-Elorduy and Pino Moreno, 2004)
		Ergates	spiculatus Lec.	Pine sawyer beetle	Nea	N. Am. (DeFoliart, 2002)
		Eurynassa	australis odewahni Pascoe	Long-horned beetle	Aus	Australia (DeFoliart, 2002)
		Glenea	cantor obesa (Thons.)		Or	India (Singh et al., 2007)
		Hevorodon	maxillosum (Drury)		Neo	Mexico (Ramos-Elorduy, 2006)
		Hoplocerambyx	spinicornis (Newman)	Sal heartwood borer	Or	India (Singh et al., 2007), Malaysia (DeFoliart, 2002)
		Hoplocerambyx	severus Pascoe	Long-horned beetle	Aus	Papua New Guinea (DeFoliart, 2002)
		Lagocheirus	rogersi (Bates)		Neo	Mexico (Ramos-Elorduy and Pino Moreno, 2004)
		Macrodontia	cervicornis (L.)		Neo	Brazil, Guyana, Paraguay, West Indies, Jamaica (DeFoliart, 2002), Ecuador (Onore, 2005)
		Macrotoma	edulis Karsch	Long-horned beetle	Afr	Sao Tomé and Principe (Netolitzky, 1919)
		Macrotoma	fisheri (Waterhouse)		Or	Thailand (Utsunomiya and Masumoto, 1999)
		Macrotoma	natala Thomson	Long-horned beetle	Afr	Botswana (Roodt, 1993)
		Mallodon	costatus Montrouzier	Long-horned beetle	Aus	New Caledonia (DeFoliart, 2002)
		Mallodon	downesi Hope	Long-horned beetle	Afr	Central Afr (Bergier, 1941), south Africa (Bodenheimer, 1951), Mozambique (DeFoliart, 2002), CAR (Hoare, 2007)
		Mallodon	molarius Bates		Neo	Mexico (Ramos-Elorduy and Pino Moreno, 2004)

Monochames	maculosus Hald.	Spotted pine sawyer	Nea	N. Am. (DeFoliart, 2002)	
Monochames	scutellatus (Say)	White spotted sawyer	Nea	N. Am. (DeFoliart, 2002)	
Neocerambyx	paris (Wiedemann)		Or	India (DeFoliart, 2002)	
Neoclytus	conjunctus (LeConte)	Western ash borer	Nea	N. Am. (DeFoliart, 2002)	
Neoplocaederus	obesus (Gahan)		Or	Thailand (Hanboonsong, 2010)	
Neoplocaederus	ruficornis (Newman)		Or	Thailand (Hanboonsong, 2010)	
Nothopleurus	maxillosus (Drury)		Neo	Mexico (Ramos-Elorduy and Pino Moreno, 2004)	
Nupserha	fricator (Dalm.)		Or	India (Singh et al., 2007)	
Ornithia	mexicana (sturm)		Neo	Mexico (Ramos-Elorduy and Pino Moreno, 2004)	
Petrognatha	gigas Fabr.	Long-horned beetle	Afr	Gabon (Bergier, 1941), Senegal (Netolitzky, 1919), CAR (Hoare, 2007)	
Plocaedurus	frenatus Fahr.	Long-horned beetle	Afr	Central Afr (Bergier, 1941), South Africa (DeFoliart, 2002)	
Prionacalus	atys White		Neo	Ecuador (Onore, 2005)	
Prionacalus	cacicus White		Neo	Ecuador (Onore, 2005)	
Prionoplus	reticularis White	Huhu beetle	Aus	New Zealand (DeFoliart, 2002)	
Prionus	damicornis F.		Neo	Brazil, Guyana, West Indies (DeFoliart, 2002)	
Prionus	calliforncius Motsch.		Nea	N. Am. (DeFoliart, 2002)	
Prionus	insularis Motsch.		Pal	Japan (Schimitschek, 1968)	
Psacothea	hilaris (Pascoe)		Pal	China (DeFoliart, 2002)	
Psalidognathus	erythrocerus Reiche		Neo	Ecuador (Onore, 2005)	
Psalidognathus	modestus Fries		Neo	Ecuador (Onore, 2005)	
Pseudonemophas	versteegii (Ritsema)		Or	India (Chakravorty et al., 2011)	
Rhagium	lineatum (Ol.)		Nea	N. Am. (DeFoliart, 2002)	
Rosenbergia	mandibularis Ritsema		Aus	Papua (Ramanday and van Mastricht, 2010)	
Sternotomis	itzingeri katangensis Allard	Long-horned beetle	Afr	DR Congo (Malaisse, 1997)	
Stromatium	barbatum F.	Kulsi teak borer	Or	India (Singh et al., 2007)	
Stromatium	longicorne (Newman)		Pal	China (Chen et al., 2009)	
Threnetica	lacrymans (Thoms.)		Or	Laos (Yhoung-Aree and Viwatpanich, 2005)	
Trichoderes	pini Chevr.		Neo	Mexico (DeFoliart, 2002; Ramos-Elorduy and Pino Moreno, 2004)	
Xylotrechus	quadripes Chevr.	Coffee borer	Or	India (Singh et al., 2007)	
Xylotrechus	smei (Lap. and Gory)		Or	India (Singh et al., 2007)	
Xylotrechus	nauticus (Mann.)		Nea	N. Am. (DeFoliart, 2002)	

(*Continued*)

Table A1 Documented Information for 1555 Species of Insects and Spiders (cont.)

Order	Family	Genus	Species	Common Name	Faunal	Distribution and References
		Xystrocera	globosa (Oliv.)	Monkeypod roundheaded borer	Or	India (DeFoliart, 2002)
		Xixuthrus	costatus (Montrouzier)		Aus	New Caledonia (DeFoliart, 2002)
		Zographus	aulicus Bertolini	Long-horned beetle	Afr	DR Congo (Malaisse, 1997), Angola (DeFoliart, 2002)
	Chrysomelidae	Aplosonyx	albicornis Wiedemann		Or	Malaysia, Sabah (Chung et al., 2002)
		Aplosonyx	chalybaeus (Hope)		Or	India (Singh et al., 2007)
		Blepharida	mexicana Jacoby		Neo	Mexico (Ramos-Elorduy and Pino Moreno, 2004)
		Leptinotarsa	decemlineata Say		Neo	Mexico (DeFoliart, 2002; Ramos-Elorduy and Pino Moreno, 2004)
		Caryoborus	serripes (Stum)		Neo	Venezuela (Choo, 2008)
		Caryobrucus	cf scheelaea Bridwell	Palm seed bruchid	Neo	Colombia (DeFoliart, 2002)
		Pachymerus	cardo (Fahr.)		Neo	Brazil (Costa Neto and Ramos-Elorduy, 2006)
		Pachymerus	nucleoum (F.)		Neo	Brazil (Costa Neto and Ramos-Elorduy, 2006)
		Sagra	femorata (Drury)		Or	Laos (Yhoung-Aree and Viwatpanich, 2005)
		Speciomerus	giganteus (Chevr.)		Pal	China (Chen and Feng, 1999)
					Neo	Venezuela (Choo, 2008)
	Curculionidae	Arrhines	hirtus Faust		Or	Thailand (Hanboonsong, 2010)
		Astycus	gestroi Marshall		Or	Thailand (Hanboonsong, 2010)
		Cnaphoscapus check	decoratus Faust		Or	Thailand (Hanboonsong et al., 2000; Hanboonsong, 2010)
		Cosmopolites	sordidus (Germar)		Neo	Ecuador (Onore, 2005)
		Cyrtotrachelus	birmanicus Faust		Or	Thailand (Utsunomiya and Masumoto, 1999)
		Cyrtotrachelus	buqueti (Guér.)		Or	Thailand (Utsunomiya and Masumoto, 1999)
		Cyrtotrachelus	longimanus (F.)	Giant bamboo weevil	Pal	China (Chen et al., 2009)
					Or	Thailand (Utsunomiya and Masumoto, 1999), India (Singh et al., 2007)
		Cyrtotrachelus	rufopectinipes Chevrolat	Bamboo weevil	Pal	China (DeFoliart, 2002)
		Dynamis	borassi (F.)		Or	India (Singh et al., 2007)
		Dynamis	nitidulus (Guérin)		Neo	Ecuador (Onore, 2005)
		Episomus	aurivillusi Faust		Neo	Ecuador (Onore, 2005)
					Or	Thailand (Hanboonsong et al., 2000; Hanboonsong, 2010)
		Eugnoristus	monachus (Oliv.)	Weevil	Afr	Madagascar (Bodenheimer, 1951)
		Hypodisa check	talaca Walk.		Or	Thailand (DeFoliart, 2002)

Genus	species	common name	region	references
Hypomeces	squamosus (Fabr.)		Or	Thailand (DeFoliart, 2002; Hanboonsong, 2010), Laos (Yhoung-Aree and Viwatpanich, 2005)
Larinus	mellificus Jekel		Pal	Iran, Syria (DeFoliart, 2002)
Larinus	onopordi (F.)		Pal	Iran, Iraq (DeFoliart, 2002)
Larinus	rudicollis Petri		Pal	Israel (DeFoliart, 2002)
Larinus	syriacus (Gyll.)		Pal	Iran, Iraq, Syria (DeFoliart, 2002)
Macrochirus	longipes Lacordaire		Pal	China (Chen and Feng, 1999)
Metamasius	cinnamominus Champ.		Neo	Ecuador (Onore, 2005)
Metamasius	dimidiatipennis (Jekel)		Neo	Ecuador (Onore, 2005)
Metamasius	hemipterus (L.)		Neo	Ecuador (Onore, 2005)
Metamasius	spinolae (Gyll.)	Cactus weevil	Neo	Mexico (DeFoliart, 2002)
Otidognathus	davidis Fairm.		Pal	China (Chen and Feng, 1999)
Pachyrrhynchus	moniliferus (Germ.)		Or	Philippines (DeFoliart, 2002)
dera	atomaria Motsch.		Or	Thailand (Hanboonsong, 2010)
Polyclaeis	equestris Boheman	Weevil	Afr	South Afr (Quin, 1959)
Polyclaeis	plumbeus Guérin	Weevil	Afr	South Afr (Quin, 1959)
Rhinostomus	barbirostris (F.)	Bearded weevil	Neo	Brazil (Costa Neto and Ramos-Elorduy, 2006)
Rhynchophorus	phoenicis (Fabr.)	African palm weevil	Afr	Angola (Santos Oliveira et al., 1976), Benin (Tchibozo, 2015), CAR (Hoare, 2007), Cameroun (Bodenheimer, 1951), DR Congo (Hoare, 2007), Congo (Hoare, 2007), Nigeria (Fasoranti and Ajiboye, 1993), Ivory Coast, Niger, Sao Tomé, Guinée, Togo, Liberia, Benin, G. Bissau (Kelemu et al., 2015)
Rhynchophorus	bilineatus (Montr.)	Black palm weevil	Aus	Papua New Guinea (DeFoliart, 2002)
Rhynchophorus	ferrugineus (Oliv.)	Asian palm weevil	Aus	Papua New Guinea (DeFoliart, 2002)
Rhynchophorus		Red palm weevil	Or	Thailand (Utsunomiya and Masumoto, 1999)
Rhynchophorus	ferrugineus schah Fabr.	Red stripe weevil	Pal	China (DeFoliart, 2002)
Rhynchophorus			Or	Thailand, Indonesia, Malaysia, Philippines, Vietnam, Sri Lanka (DeFoliart, 2002; Meer Mohr, 1965; Lukiwati, 2010), Thailand (Hanboonsong, 2010), Laos (Yhoung-Aree and Viwatpanich, 2005)
Rhynchophorus	cruentatus (F.)	Palmetto weevil	Nea	N. Am. (DeFoliart, 2002)
Rhynchophorus			Neo	Mexico (Ramos-Elorduy and Pino Moreno, 2004)
Rhynchophorus	palmarum (L.)	South American palm weevil	Neo	Pan-regional West Indies, Barbados, Trinidad, Mexico (DeFoliart, 2002; Ramos-Elorduy and Pino Moreno, 2004), Brazil (Costa Neto and Ramos-Elorduy, 2006), Ecuador (Onore, 2005), Venezuela (Araújo and Beserra, 2007)

(Continued)

Table A1 Documented Information for 1555 Species of Insects and Spiders (cont.)

Order	Family	Genus	Species	Common Name	Faunal	Distribution and References
		Sipalinus	aloysii-sabaudiae Camerano	Weevil	Afr	Tanzania (Bodenheimer, 1951; Harris, 1940)
		Sitophilus	oryzae (L.)		Or	Whole region (Taylor, 1975; van Huis et al., 2013)
		Scyphophorus	acupunctatus Gyll.	Agave weevil	Neo	Mexico (DeFoliart, 2002)
		Sphaerotrypes	yunnanensis Tsai and Yin		Pal	China (Chen and Feng, 1999)
		Tomicus	piniperda (L.)		Pal	China (Chen et al., 2009)
		Xyleborus	emarginatus (Eichhoff)		Pal	China (Chen and Feng, 1999)
	Dytiscidae	Agabus	fulvipennis Régimbart		Pal	China (Chen et al., 2009)
		Cybister	bengalensis Aubé		Pal	China (DeFoliart, 2002)
		Cybister	brevis Aubé		Pal	Japan (Schimitschek, 1968)
		Cybister	distinctus Reg.	Water beetle	Afr	Senegal, Sierra Leone, DR Congo (Ramos-Elorduy et al., 2009)
		Cybister	ellipticus Le Conte		Nea	United States (Ramos-Elorduy et al., 2009)
		Cybister	explanatus Le Conte		Nea	N. Am. (DeFoliart, 2002)
		Cybister	flavocinctus Aubé		Neo	Mexico (Ramos-Elorduy and Pino Moreno, 2004)
		Cybister	fimbriolatus (Say)		Neo	Mexico (Ramos-Elorduy and Pino Moreno, 2004)
		Cybister	guerini Aubé		Neo	Mexico (Ramos-Elorduy and Pino Moreno, 2004)
		Cybister			Or	Indonesia (Ramos-Elorduy et al., 2009)
		Cybister	japonicus Sharp		Pal	Japan (Ramos-Elorduy et al., 2009), China (DeFoliart, 2002)
		Cybister	lewisianus Sharp		Pal	China, Japan (Schimitschek, 1968; Chen and Feng, 1999; Jäch, 2003; Ramos-Elorduy et al., 2009)
		Cybister	limbatus (F.)	Diving beetle	Pal	China (Ramos-Elorduy et al., 2009)
		Cybister			Or	Thailand (DeFoliart, 2002; Hanboonsong, 2010), Laos (Yhoung-Aree and Viwatpanich, 2005)
		Cybister			Pal	China (DeFoliart, 2002; Chen and Feng, 1999; Ramos-Elorduy et al., 2009)
		Cybister	occidentalis Aubé		Neo	Mexico (Ramos-Elorduy and Pino Moreno, 2004)
		Cybister	owas Laporte	Water beetle	Afr	Madacasgar (Decary, 1937)
		Cybister	rugosus (MacLeay)		Or	Thailand (Hanboonsong, 2010), Laos (Yhoung-Aree and Viwatpanich, 2005)
		Cybister	sugillatus Er.		Pal	Japan (Ramos-Elorduy et al., 2009), China (DeFoliart, 2002)
		Cybister	tripunctatus (Olivier)		Or	Thailand (DeFoliart, 2002), Indonesia (Meer Mohr, 1965), India (Shantibala et al., 2014)

	Cybister	tripunctatus asiaticus Sharp		Pal	Japan (Ramos-Elorduy et al., 2009), China (DeFoliart, 2002)
	Dytiscus	habilis Say		Or	Thailand (Hanboonsong et al., 2000)
	Dytiscus	marginalis L.		Neo	Mexico (Ramos-Elorduy and Pino Moreno, 2004)
	Dytiscus	marginicollis Le Conte		Pal	China, Japan (Ramos-Elorduy et al., 2009)
	Dytiscus	validus Régimbart		Pal	Japan (Ramos-Elorduy et al., 2009), China (DeFoliart, 2002)
	Dytiscus			Neo	Mexico (Ramos-Elorduy and Pino Moreno, 2004)
	Eretes	sticticus (L.)	Water beetle	Pal	Japan (Ramos-Elorduy et al., 2009)
				Afr	Kenya (Ramos-Elorduy et al., 2009)
				Or	Malaysia (Ramos-Elorduy et al., 2009), Myanmar, India (DeFoliart, 2002), Thailand (Hanboonsong, 2010), Laos (Yhoung-Aree and Viwatpanich, 2005)
	Hydaticus	rhantoides Sharp	Diving beetle	Or	Thailand (Hanboonsong et al., 2000; Hanboonsong, 2010)
	Laccophilus	apicalis Sharp		Neo	Mexico (Ramos-Elorduy and Pino Moreno, 2004)
	Laccophilus	pulicarius Sharp		Or	Thailand (Hanboonsong et al., 2000; Hanboonsong, 2010)
	Megadytis	giganteus (Laporte)		Neo	Mexico (Ramos-Elorduy and Pino Moreno, 2004)
	Platambus	guttulus (Régimbart)		Pal	China (Ramos-Elorduy et al., 2009)
	Rhantaticus	congestus (Klug)		Or	Thailand (Hanboonsong et al., 2000; Hanboonsong, 2010)
	Rhantus	atricolor (Aubé)		Neo	Mexico (Ramos-Elorduy and Pino Moreno, 2004)
	Rhantus	consimilis Motsch.		Neo	Mexico (Ramos-Elorduy and Pino Moreno, 2004)
	Rhantus	latus (Faim.)	Water beetle	Afr	Madacasgar (Ramos-Elorduy et al., 2009)
	Rhantus	pulverosus (Stephens)		Pal	Japan (Schimitschek, 1968)
	Thermonectus	basilaris (Harris)		Neo	Mexico (Ramos-Elorduy and Pino Moreno, 2004)
	Thermonectus	marmoratus (Gray)		Neo	Mexico (Ramos-Elorduy and Pino Moreno, 2004)
Elateridae	Tetralobus	flabellicornis	Giant acacia click beetle	Afr	Central Afr (Bodenheimer, 1951)
	Cardiophorus check	aequabilis Candèze	Click beetle	Or	India (Singh et al., 2007)
	Chalcolepidius	lafargi Chevr.		Neo	Mexico (Ramos-Elorduy and Pino Moreno, 2004)
	Chalcolepidius	rugatus Cand.		Neo	Mexico (Ramos-Elorduy and Pino Moreno, 2004)
	Deilelater	mexicanus (Champ.)		Neo	Mexico (Ramos-Elorduy and Pino Moreno, 2004)
	Lacon	mexicana Cand.		Neo	Mexico (Ramos-Elorduy and Pino Moreno, 2004)
	Pyrophorus	pellucens Esch.		Neo	Mexico (Ramos-Elorduy and Pino Moreno, 2004)
Elmidae	Austrelmis	chilensis (Germain)		Neo	Chile (DeFoliart, 2002), Peru (Ramos-Elorduy et al., 2009)

(*Continued*)

Table A1 Documented Information for 1555 Species of Insects and Spiders (cont.)

Order	Family	Genus	Species	Common Name	Faunal	Distribution and References
	Gyrinidae	Austrelmis	condimentarius (Philippi)		Neo	Chile (Ramos-Elorduy et al., 2009), Peru (Ramos-Elorduy et al., 2009)
		Aulonogyrus	strigosus F.		Aus	(DeFoliart, 2002; Jäch, 2003)
		Dineutes	marginatus Sharp		Pal	Japan (Schimitschek, 1968)
		Gyrinus	curtus Motsch.		Pal	Japan (Schimitschek, 1968)
		Gyrinus	japonicus Sharp		Pal	Japan (Schimitschek, 1968)
		Gyrinus	parcus Say		Neo	Mexico (Ramos-Elorduy, 2004)
		Gyrinus	plicatus Régimb.		Neo	Mexico (Ramos-Elorduy and Pino Moreno, 2004)
	Haliplidae	Haliplus	punctatus Aubé		Neo	Mexico (Ramos-Elorduy and Pino Moreno, 2004)
		Peltodytes	mexicanus (Wehncke)		Neo	Mexico (Ramos-Elorduy and Pino Moreno, 2004)
		Peltodytes	ovalis Zimm.		Neo	Mexico (Ramos-Elorduy and Pino Moreno, 2004)
	Histeridae	Hololepta	guidonis Marseul		Neo	Mexico (Ramos-Elorduy, 2009)
	Hydrophilidae	Hydrobiomorpha	spinicollis (Eschscholtz)		Or	Thailand (Hanboonsong et al., 2000; Hanboonsong, 2010)
		Hydrophilus	acuminatus Motsch.		Or	India (Ramos-Elorduy et al., 2009)
		Hydrophilus			Pal	China (DeFoliart, 2002), Japan (Ramos-Elorduy et al., 2009)
		Hydrophilus	bilineatus (MacLeay)		Or	Thailand (Hanboonsong et al., 2000), Vietnam (DeFoliart, 2002)
		Hydrophilus			Pal	China (DeFoliart, 2002), Japan (Ramos-Elorduy et al., 2009)
		Hydrophilus	cavisternum (Bedel)		Or	Thailand (Utsunomiya and Masumoto, 1999), Laos (Yhoung-Aree and Viwatpanich, 2005), Vietnam (Ramos-Elorduy et al., 2009)
		Hydrophilus			Pal	China (DeFoliart, 2002), Japan (Ramos-Elorduy et al., 2009)
		Hydrophilus	hastatus (Herbst)		Or	Vietnam (DeFoliart, 2002), Laos, Thailand, Cambodja, Myanmar (Ramos-Elorduy et al., 2009)
		Hydrophilus			Pal	China (DeFoliart, 2002), Japan (Ramos-Elorduy et al., 2009)
		Hydrophilus	olivaceus F.		Or	India (DeFoliart, 2002), Thailand (Utsusomiya and Masumoto, 1999)
		Hydrophilus	picicornis Chevr.		Or	Philippines (DeFoliart, 2002)
		Hydrophilus	senegalensis (Percheron)	Water scavenger beetle	Afr	Senegal (Ramos-Elorduy et al., 2009)
		Sternolophus	rufipes (F.)		Or	Thailand (Hanboonsong et al., 2000; Hanboonsong, 2010)

				Appendix	321

		Tropisternus	*mexicanus* Lap.		Neo	Mexico (Ramos-Elorduy and Pino Moreno, 2004), Panama (Ramos-Elorduy et al., 2009)
		Tropisternus	*sublaevis* (Le Conte)		Neo	Mexico (Ramos-Elorduy and Pino Moreno, 2004)
		Tropisternus	*tinctus* Sharp		Neo	Mexico (Ramos-Elorduy and Pino Moreno, 2004)
	Lucanidae	*Cyclommatus*	*pahangensis* (Nagel)		Or	India (Singh et al., 2007)
		Cyclommatus	*strigiceps albersi* Kraatz		Or	India (Singh et al., 2007)
		Hexarthrius	*forsteri* (Hope)		Or	India (Singh et al., 2007)
		Lucanus	*cantori* (Hope)		Or	India (Singh et al., 2007)
		Lucanus	*laminifer* Waterhouse		Or	India (Singh et al., 2007)
		Lucanus	*maculifemoratus* Motsch.		Pal	Japan (Schimitschek, 1968)
		Odontolabis	*cuvera* Hope		Or	India (Singh et al., 2007)
		Odontolabis	*gazella* (F.)		Or	India (Chakravorty et al., 2011)
		Odontolabis	*siva* Hope and Wetson		Or	India (Singh et al., 2007)
		Prosopocoilus	*inclinatus* (Motsch.)		Pal	Japan (Schimitschek, 1968)
		Prosopocoilus	*serricornis* (Latr.)	Stag beetle	Afr	Madagascar (Decary, 1937)
		Platycerus	*virescens* (F.)		Neo	Mexico (Ramos-Elorduy and Pino Moreno, 2004)
		Sphaenognathus	*feisthamelii* (Guérin)		Neo	Ecuador (Onore, 2005)
		Sphaenognathus	*lindenii* Murray		Neo	Ecuador (Onore, 2005)
		Sphaenognathus	*metallifer* Bomans and Lacroix		Neo	Ecuador (Onore, 2005)
	Meloidae	*Meloe*	*dugesi* Champ.		Neo	Mexico (Ramos-Elorduy and Pino Moreno, 2004)
		Meloe	*laevis* Leach		Neo	Mexico (Ramos-Elorduy and Pino Moreno, 2004)
		Mylabris	*cichorii* (L.)		Or	India (Singh et al., 2007)
		Mylabris	*himalayaensis* Saha		Or	India (Singh et al., 2007)
		Meloe	*nebulosus* Champ		Neo	Mexico (Ramos-Elorduy and Pino Moreno, 2004)
	Passalidae	*Aceraius*	*helferi* Kuwert		Or	India (Singh et al., 2007)
		Aulacocyclus syn	*bicuspi* Kaup		Or	India (Singh et al., 2007)
		Heliscus	*yucatanus* Bates		Neo	Mexico (Ramos-Elorduy and Pino Moreno, 2004)
		Odontotaenius	*striatopunctatus* Perch.		Neo	Mexico (Ramos-Elorduy and Pino Moreno, 2004)
		Odontotaenius	*yucatanus* Bates		Neo	Mexico (Ramos-Elorduy and Pino Moreno, 2004)
		Odontotaenius	*zodiacus* Truq.		Neo	Mexico (Ramos-Elorduy and Pino Moreno, 2004)
		Oileus	*rimator* Truqui		Neo	Mexico (Ramos-Elorduy and Pino Moreno, 2004)
		Passalus	*interruptus* L.		Neo	Suriname (DeFoliart, 2002)
					Or	India (DeFoliart, 2002)
		Passalus	*interstitialis* Esch.		Neo	Mexico (Ramos-Elorduy and Pino Moreno, 2004)

(*Continued*)

Table A1 Documented Information for 1555 Species of Insects and Spiders (*cont.*)

Order	Family	Genus	Species	Common Name	Faunal	Distribution and References
		Passalus	punctiger Lep. and Serv.		Neo	Mexico (Ramos-Elorduy and Pino Moreno, 2004)
		Passalus	puntatostriatus Perch.		Neo	Mexico (Ramos-Elorduy and Pino Moreno, 2004)
		Paxillus	leachi MacLeay		Neo	Mexico (DeFoliart, 2002; Ramos-Elorduy and Pino Moreno, 2004)
		Verres	corticicola (Truq.)		Neo	Mexico (Ramos-Elorduy and Pino Moreno, 2004)
		Veturius	sinuosus (Drapiez)		Neo	Colombia (Paoletti et al., 2001)
	Scarabaeidae	Adoretus	compressus (Weber)		Or	Thailand (DeFoliart, 2002)
		Adoretus	convexus Burm.		Or	Thailand (DeFoliart, 2002)
		Adoretus	pachysomatus Kobayashi		Or	Thailand (Utsunomiya and Masumoto, 1999)
		Agestrata	orichalca (L.)		Or	Thailand (Hanboonsong et al., 2000; Hanboonsong, 2010)
		Allomyrina	dichotoma (L.)	Japanese rhinoceros beetle	Pal	Japan (Schimitschek, 1968)
		Allomyrina	dichotomus (L.)		Or	India (Singh et al., 2007; Chakravorty et al., 2011)
		Amphimallon	assimile (Herbst)		Pal	China (Chen and Feng, 1999)
		Amphimallon	pini (Ol.)		Pal	Italy, Moldavia, Walachia (DeFoliart, 2002)
		Anoplognathus	viridiaeneus (Donovan)	Christmas beetle	Pal	Italy, Moldavia, Walachia (DeFoliart, 2002)
		Ancognatha	castanea Erichson		Aus	Australia (DeFoliart, 2002)
		Ancognatha	jamesoni Murray		Neo	Ecuador (Onore, 2005)
		Ancognatha	vulgaris Arrow		Neo	Ecuador (Onore, 2005)
		Anomala	anguliceps Arrow		Neo	Ecuador (Onore, 2005)
		Anomala	antiqua Gyll.		Or	Thailand (Utsunomiya and Masumoto, 1999)
		Anomala	bilunulata Fairm. check		Or	Thailand (Utsunomiya and Masumoto, 1999)
		Anomala	blaisei Ohaus		Or	Thailand (Utsunomiya and Masumoto, 1999)
		Anomala	cantori (Hope)		Or	Thailand (Utsunomiya and Masumoto, 1999)
		Anomala	chalcites Sharp		Or	Thailand (Utsunomiya and Masumoto, 1999)
		Anomala	concha Ohaus		Or	Thailand (Utsunomiya and Masumoto, 1999)
		Anomala	corpulenta Motsch.		Or	Malaysia, Sabah (Chung et al., 2002)
		Anomala	coxalis Bates		Pal	China (Chen and Feng, 1999)
		Anomala	cupripes (Hope)		Or	Malaysia, Sabah (Chung et al., 2002)
		Anomala	fusikibia Lin		Or	Thailand (Utsunomiya and Masumoto, 1999), Laos (Yhoung-Aree and Viwatpanich, 2005)
		Anomala	laotica Frey		Or	Thailand (Utsunomiya and Masumoto, 2000)
					Or	Thailand (Utsunomiya and Masumoto, 1999)

Anomala	lasiocnemis Ohaus		Or	Malaysia, Sabah (Chung et al., 2002)
Anomala	latefemorata (Ohaus)		Or	Malaysia, Sabah (Chung et al., 2002)
Anomala	lignea Arrow		Or	Thailand (Utsunomiya and Masumoto, 1999)
Anomala	parallela Benderitter		Or	Thailand (Utsunomiya and Masumoto, 2000)
Anomala	pallidae F.		Or	Thailand (Hanboonsong, 2010)
Anomala	punctulicollis Fairmaire		Or	Thailand (Utsunomiya and Masumoto, 2000)
Anomala	rugosa Arrow		Or	Thailand (Utsunomiya and Masumoto, 2000)
Anomala	shanica Arrow		Or	Thailand (Utsunomiya and Masumoto, 1999)
Anomala	scherei Frey		Or	Thailand (Utsunomiya and Masumoto, 1999)
Anomala	vuilletae Paulian		Or	Thailand (Utsunomiya and Masumoto, 1999)
Aphodius (Pharaphodius)	crenatus Harold		Or	Thailand (Hanboonsong et al., 2000; Hanboonsong, 2010)
Aphodius (Pharaphodius)	marginellus (F.)		Or	Thailand (Hanboonsong et al., 2000; Hanboonsong, 2010)
Aphodius (Pharaphodius)	putearius Reitter		Or	Thailand (Hanboonsong et al., 2000; Hanboonsong, 2010)
Augosoma	centaurus (Fabr.)	Rhinoceros beetle	Afr	Cameroun (Bodenheimer, 1951), Congo (Hoare, 2007), DR Congo (Hoare, 2007)
Brahmina	mikado T. Itoh		Or	Thailand (Utsunomiya and Masumoto, 1999)
Brahmina	parvula Moser		Or	Thailand (Utsunomiya and Masumoto, 1999)
Canthon	humectus hidalgoensis Bates		Neo	Mexico (Ramos-Elorduy and Pino Moreno, 2004)
Cassolus	humeralis Arrow		Or	India (Singh et al., 2007)
Catharsius	birmanensis Lansberge		Or	Thailand (Utsunomiya and Masumoto, 1999)
Catharsius	molossus (L.)		Pal	China (Chen and Feng, 1999; DeFoliart, 2002)
Chaetadoretus	cribratus (White)		Or	Thailand (Utsunomiya and Masumoto, 1999)
Chalcosoma	atlas (L.)		Or	Thailand (Hanboonsong et al., 2000; Hanboonsong, 2010)
Chalcosoma	moellenkampi Kolbe		Or	Malaysia, Sabah (Chung et al., 2002), Indonesia (Meer Mohr, 1965; Lukiwati, 2010)
Chondrorrhina	abbreviata F.	Flower beetle	Afr	Indonesia (Chung, 2010), Malaysia, Sabah (Chung et al., 2002)
Chrysina	macropus (Franc.)		Neo	Benin (Riggi et al., 2013)
Clavipalpus	antisanae Bates		Neo	Mexico (Ramos-Elorduy and Pino Moreno, 2004)
Coelosis	biloba (L.)		Neo	Ecuador (Onore, 2005)
Copris (Microcopris)	reflexus (F.)		Or	Ecuador (Onore, 2005)
Copris (Paracopris)	punctulatus Wiedemann		Or	Thailand (Utsunomiya and Masumoto, 1999)

(*Continued*)

Table A1 Documented Information for 1555 Species of Insects and Spiders (cont.)

Order	Family	Genus	Species	Common Name	Faunal	Distribution and References
		Copris (s.str.)	carinicus Gillet		Or	Thailand (Utsunomiya and Masumoto, 1999)
		Copris	corpulentus Gillet		Or	Thailand (Utsunomiya and Masumoto, 1999), India (Singh et al., 2007)
		Copris	furciceps Felsche		Or	Thailand (Utsunomiya and Masumoto, 1999), India (Singh et al., 2007)
		Copris	sinicus Hope		Or	Thailand (Utsunomiya and Masumoto, 1999)
		Copris (s.str.)	nevinsoni Waterhouse		Or	Thailand (Hanboonsong et al., 2000; Hanboonsong, 2010), Laos (Yhoung-Aree and Viwatpanich, 2005)
		Copris	vitalisi Gillet		Or	India (Singh et al., 2007)
		Cotinis	mutabilis Gory and Perch.		Neo	Mexico (Ramos-Elorduy, 2006)
		Cyclocephala	borealis Arrow	Northern masked chafer	Nea	N. Am. (DeFoliart, 2002)
		Cyclocephala	capitata Hohne		Neo	Mexico (Ramos-Elorduy and Pino Moreno, 2004)
		Cyclocephala	dimidiata Burm.		Nea	N. Am. (DeFoliart, 2002)
		Cyclocephala	fasciolata Bates		Neo	Mexico (Ramos-Elorduy and Pino Moreno, 2004)
		Cyclocephala	guttata Bates		Neo	Mexico (Ramos-Elorduy and Pino Moreno, 2004)
		Democrates	burmeisteri Reiche		Neo	Ecuador (Onore, 2005)
		Dicranocephalus	wallichi bowringi Pascoe		Pal	China (Chen and Feng, 1999)
		Dynastes	hercules (L.)		Neo	Mexico (Ramos-Elorduy and Pino Moreno, 2004), Brazil (Costa Neto and Ramos-Elorduy, 2006), Ecuador (Onore, 2005)
		Dynastes	hyllus Chev.		Neo	Mexico (Ramos-Elorduy and Pino Moreno, 2004)
		Enema	pan (F.)		Neo	Mexico (Ramos-Elorduy and Pino Moreno, 2004)
		Empectida	tonkinensis Moser		Or	Thailand (Utsunomiya and Masumoto, 1999)
		Eupatorus	gracilicornis Arrow		Or	Thailand (Utsunomiya and Masumoto, 1999)
		Exolontha	castanea Zhang		Or	Thailand (Utsunomiya and Masumoto, 1999)
		Exolontha	hypoleuca (Wiedemann)		Or	Indonesia (Meer Mohr, 1965; DeFoliart, 2002; Malaysia, Sabah (Chung et al., 2002)
		Geniatosoma	nigrum (Ohaus)		Neo	Brazil (Costa Neto and Ramos-Elorduy, 2006)
		Golofa	aegeon (Drury)		Neo	Ecuador (Onore, 2005)
		Golofa	eacus Burmeister		Neo	Ecuador (Onore, 2005)
		Golofa	imperialis Thomson		Neo	Mexico (Ramos-Elorduy and Pino Moreno, 2002)
		Golofa	tersander Burmeister		Neo	Mexico (Ramos-Elorduy and Pino Moreno, 2002)
		Golofa	unicolor (Bates)		Neo	Ecuador (Onore, 2005)

Gnathocera	impressa (Oliv.)		Afr	Benin (Riggi et al., 2013)
Gnathocera	varians G. and P.	Flower beetle	Afr	Benin (Riggi et al., 2013)
Goliathus	cacicus Ol.	Goliath beetle	Afr	Central Africa (Bergier, 1941)
Goliathus	regius Klug	Goliath beetle	Afr	Central Africa (Bergier, 1941)
Goliathus	cameronensis check	Goliath beetle	Afr	Central Africa (Bergier, 1941)
Goliathus	goliathus L.	Goliath beetle	Afr	Central Africa (Bergier, 1941)
Gymnopleurus syn	aethiops Sharp		Or	Laos (Yhoung-Aree and Viwatpanich, 2005)
Heliocopris	bucephalus (F.)		Or	Myanmar (DeFoliart, 2002), Thailand (Hanboonsong, 2010), Laos (Yhoung-Aree and Viwatpanich, 2005), India (Singh et al., 2007)
Heliocopris	dominus Bates		Or	Laos (Yhoung-Aree and Viwatpanich, 2005)
Heterogomphus	bourcieri Guérin		Neo	Ecuador (Onore, 2005)
Heteroligus	meles (Billberger)	Yam beetle	Afr	Nigeria (Agbidye et al., 2009)
Heteronychus	lioderes Redtenbac		Or	Thailand (Hanboonsong, 2010)
Holotrichia	cephalotus (Burmeister)		Or	Thailand (Utsunomiya and Masumoto, 1999)
Holotrichia	hainanensis Chang		Or	Thailand (Utsunomiya and Masumoto, 1999)
Holotrichia	nigricollis Brenske		Or	Thailand (Utsunomiya and Masumoto, 1999)
Holotrichia	nigricollis rubricollis Moser		Or	Thailand (Utsunomiya and Masumoto, 1999)
Holotrichia	oblita Faldermann		Pal	China (Chen and Feng, 1999)
Holotrichia	parallela (Motsch.)		Pal	China (Chen and Feng, 1999)
Holotrichia	scrobiculata Brenske		Pal	China (Chen and Feng, 1999)
Hoplosternus	malaccensis Moser		Or	Thailand (Utsunomiya and Masumoto, 1999)
Isonychus	ocellatus Burm.		Neo	Mexico (Ramos-Elorduy et al., 2008)
Lepidiota	anatina Brenske		Afr	Zimbabwe (Chavanduka, 1976)
Lepidiota	bimaculata (Saunders)		Or	Thailand (Utsunomiya and Masumoto, 1999)
Lepidiota	mashona Arrow		Arica	Zimbabwe (Chavanduka, 1976; Gelfand, 1971)
Lepidiota	nitidicollis Kolbe		Afr	Zimbabwe (Chavanduka, 1976)
Lepidiota	punctum Blanch.		Or	Philippines (DeFoliart, 2002)
Lepidiota	stigma (F.)		Or	Thailand (DeFoliart, 2002), Indonesia (Meer Mohr, 1965), India (Singh et al., 2007)
Lepidiota	vogeli Brenske		Aus	Papua New Guinea (DeFoliart, 2002)
Leucopholis	irrorata Chevr.		Or	Philippines (DeFoliart, 2002)
Leucopholis	pulverulenta Burm.		Or	Philippines (DeFoliart, 2002)
Leucopholis	staudingeri Brenske		Or	Malaysia, Sabah (Chung et al., 2002)
Leucopholis	rorida (F.)		Or	Indonesia (DeFoliart, 2002)

(*Continued*)

Table A1 Documented Information for 1555 Species of Insects and Spiders (cont.)

Order	Family	Genus	Species	Common Name	Faunal	Distribution and References
		Liatongus	affinis (Arrow)		Or	Thailand (Utsunomiya and Masumoto, 1999)
		Liatongus syn	rhadamistus (F.)		Or	Thailand (Utsunomiya and Masumoto, 1999; Hanboonsong, 2010)
		Liatongus	tridentatus (Boucomont)		Or	Thailand (Utsunomiya and Masumoto, 1999)
		Liatongus	venator (F.)		Or	Thailand (Utsunomiya and Masumoto, 1999)
		Macrodactylus	dimidiatus Gué.		Neo	Mexico (Ramos-Elorduy et al., 2008)
		Macrodactylus	lineatus Chev.		Neo	Mexico (Ramos-Elorduy et al., 2008)
		Megaceras	crassum Prell		Neo	Colombia (Dufour, 1987; DeFoliart, 2002)
		Megasoma	actaeon (L.)		Neo	Brazil (Costa Neto and Ramos-Elorduy, 2006)
		Megasoma	anubis (Chevrolat)		Neo	Brazil (DeFoliart, 2002)
		Megasoma	elephas (F.)		Neo	Mexico (Ramos-Elorduy and Pino Moreno, 2004)
		Megasoma	elephas occidentalis B. and P.		Neo	Mexico (Ramos-Elorduy and Pino Moreno, 2004)
		Megistophylla	andrewsi Moser		Or	Thailand (Utsunomiya and Masumoto, 1999)
		Mimela	ferreroi Sabatinelli		Or	Thailand (Utsunomiya and Masumoto, 1999)
		Mimela	ignistriata Lin		Or	Thailand (Utsunomiya and Masumoto, 1999)
		Mimela	linpingi Sabatinelli		Or	Thailand (Utsunomiya and Masumoto, 1999)
		Miridiba	tuberculipennis obscura T.		Or	Thailand (Utsunomiya and Masumoto, 1999)
		Oniticellus	cinctus (F.)		Or	Thailand (Utsunomiya and Masumoto, 1999)
		Onitis	castaneus Redtenbacher		Or	India (Singh et al., 2007)
		Onitis	falcatus (Wulfen)		Or	Thailand (Utsunomiya and Masumoto, 1999)
		Onitis	feae Felsche		Or	India (Singh et al., 2007)
		Onitis	niger Lansberge		Or	Thailand (Hanboonsong et al., 2000)
		Onitis	subopacus Arrow		Or	India (Singh et al., 2007), Thailand (Utsunomiya and Masumoto, 1999)
		Onitis	virens Lansberge		Or	Thailand (DeFoliart, 2002)
		Onthophagus	avocetta Arrow		Or	Thailand (Hanboonsong et al., 2000)
		Onthophagus	bonasus (F.)		Or	Thailand (Utsunomiya and Masumoto, 1999)
		Onthophagus	khonmiinitnoi Masumoto		Or	Thailand (Hanboonsong et al., 2000)
		Onthophagus	luridipennis Boheman		Or	Thailand (Utsunomiya and Masumoto, 1999)
		Onthophagus	Oris Harold		Or	Thailand (Utsunomiya and Masumoto, 1999)
		Onthophagus	papulatus Boucomont		Or	Thailand (Hanboonsong, 2010)
		Onthophagus	proletarius Harold		Or	Thailand (Utsunomiya and Masumoto, 1999)
		Onthophagus	rectecornutus Lansberge		Or	Thailand (Utsunomiya and Masumoto, 1999)

Onthophagus	sagittarius (F.)		Or	Thailand (Utsunomiya and Masumoto, 1999)
Onthophagus	seniculus (F.)		Or	Thailand (Utsunomiya and Masumoto, 1999)
Onthophagus	taurinus White		Or	Thailand (Utsunomiya and Masumoto, 1999)
Onthophagus	tragus (F.)		Or	Thailand (Hanboonsong, 2010)
Onthophagus	tricornis (Wiedemann)		Or	Thailand (Hanboonsong, 2010)
Onthophagus	trituber (Wiedemann)		Or	Thailand (Utsunomiya and Masumoto, 1999)
Onthophagus	tragoides Boucomont		Or	Thailand (Hanboonsong, 2010)
Oryctes	boas (Fabr.)	Rhinoceros beetle	Afr	Congo (Bani, 1995; Hoare, 2007), Nigeria (Fasoranti and Ajiboye, 1993), South Afrrica (Bergier, 1941; Netolitzky, 1919), D.R. Congo (DeFoliart, 2002), Ivory Coast, Sierra Leone, Guinée, Liberia, G. Bissau, Botswana, Namibia (Kelemu et al., 2015)
Oryctes	centaurus Sternberg		Aus	Papua New Guinea (DeFoliart, 2002)
Oryctes	rhinoceros (L.)		Aus	Solomon Islands (DeFoliart, 2002)
			Afr	D.R. Congo (Gahukar, 2011)
			Pal	China (Chen et al., 2009)
			Or	Thailand, Myanmar, Philippines, India (DeFoliart, 2002), Malaysia, Sabah (Chung et al., 2002)
Oryctes	monoceros (Oliv.)	Rhinoceros beetle	Afr	South Afr (Bergier, 1941)
Oryctes	nasicornis (Linn.)	European rhinoceros beetle	Afr	Madagascar (Bergier, 1941)
Oryctes	owariensis Beauvois	Rhinoceros beetle	Afr	Congo (Bani, 1995; Hoare, 2007), South Africa (Bergier, 1941; Netolitzky, 1919), D.R. Congo (DeFoliart, 2002), Cameroun (Womeni et al., 2009), Ivory Coast, Sierra Leone, Guinée, Ghana, E. Guinée, Guinée Bissau (Kelemu et al., 2015)
Oxycetonia	jucunda (Falldermann)		Pal	China (Chen and Feng, 1999)
Pachnoda	cordata (Drury)		Afr	Benin (Riggi et al., 2013), Cameroun (Tchibozo, 2015)
Pachnoda	marginata aurantia (Herbst)		Afr	Cameroun (Tchibozo, 2015)
Pachnoda	vossi Kolbe	Flower beetle	Afr	Benin (Riggi et al., 2013)
Pachylomera	femoralis Kirby	Dung beetle	Afr	Zambia (Mbata, 1995)
Paraphytus	hindu Arrow		Or	India (Singh et al., 2007)
Pelidnota syn missp.	nigricauda Bates		Neo	Ecuador (Onore, 2005)
Peltonotus	nasutus Arrow		Or	Thailand (Jameson and Drumont, 2013)
Polyphylla	laticollis Lewis		Pal	China (Chen and Feng, 1999)
Polyphylla	tonkinensis DeWailly		Or	Thailand (Utsunomiya and Masumoto, 1999)
Phyllophaga	fusca (Froelich)	Northern June beetle	Nea	N. Am. (DeFoliart, 2002)

(*Continued*)

Table A1 Documented Information for 1555 Species of Insects and Spiders (cont.)

Order	Family	Genus	Species	Common Name	Faunal	Distribution and References
		Phyllophaga	crinita Le Conte	Ten-lined June beetle	Nea	N. Am. (DeFoliart, 2002)
		Phyllophaga	sp. af lenis (Horn)		Neo	Mexico (Ramos-Elorduy et al., 2008)
		Phyllophaga	mexicana (Blanch.)		Neo	Mexico (Ramos-Elorduy and Pino Moreno, 2004)
		Phyllophaga	rugipennis (Schaufuss)		Neo	Mexico (Ramos-Elorduy and Pino Moreno, 2002)
		Phyllophaga	rubella (Bates)		Neo	Mexico (DeFoliart, 2002)
		Platycoelia	forcipales Ohaus		Neo	Ecuador (Onore, 2005)
		Platycoelia	lutescens Blanchard		Neo	Ecuador (Smith and Paucar, 2000; Onore, 2005)
		Platycoelia	rufosignata Ohaus		Neo	Ecuador (Onore, 2005)
		Platygenia	barbata Afzelius		Afr	D.R. Congo (Adriaens, 1951), CAR (Hoare, 2007)
		Podischnus	agenor (Ol.)		Neo	Colombia, Venezuela (Ruddle, 1973; DeFoliart, 2002)
		Popillia	femoralis Klug		Afr	Cameroun (Bodenheimer, 1951)
		Protaetia	aerata (Erichs.)		Pal	China (Chen and Feng, 1999)
		Rhabdotis	bouchardi Legrand		Afr	Benin (Riggi et al., 2013)
		Scarabaeus	sacer L.		Pal	Egypt (DeFoliart, 2002)
		Sophrops	abscensus Brenske		Or	Thailand (Hanboonsong, 2010)
		Sophrops	bituberculatus (Moser)		Or	Thailand (Hanboonsong, 2010)
		Sophrops	brunneus (Moser)		Or	Thailand (Utsunomiya and Masumoto, 1999)
		Sophrops	excisus T. Itoh		Or	Thailand (Utsunomiya and Masumoto, 1999)
		Sophrops	foveatus (Moser)		Or	Thailand (Utsunomiya and Masumoto, 1999)
		Sophrops	opacidorsalis T. Itoh		Or	Thailand (Utsunomiya and Masumoto, 1999)
		Sophrops	paucisetosa Frey		Or	Thailand (Utsunomiya and Masumoto, 1999)
		Sophrops	planicollis (Burmeister)		Or	Thailand (Masumoto and Utsunomiya, 1998)
		Sophrops	rotundicollis T. Itoh		Or	Thailand (Utsunomiya and Masumoto, 1999)
		Sophrops	simplex Frey		Or	Thailand (Utsunomiya and Masumoto, 1999)
		Sophrops	tonkinensis (Moser)		Or	Thailand (Masumoto and Utsunomiya, 1998)
		Strategus	aloeus aloeus (L.)		Neo	Mexico (Ramos-Elorduy and Pino Moreno, 2004)
		Strategus	aloeus julianus Burm.		Neo	Mexico (Ramos-Elorduy and Pino Moreno, 2004)
		Strategus	fallaciosus Kolbe		Neo	Mexico (Ramos-Elorduy and Pino Moreno, 2004)
		Xyloryctes	corniger Bates		Neo	Mexico (Ramos-Elorduy and Pino Moreno, 2004)
		Xyloryctes	ensifer Bates		Neo	Mexico (Ramos-Elorduy and Pino Moreno, 2004)
		Xyloryctes	furcatus Burm.		Neo	Mexico (Ramos-Elorduy and Pino Moreno, 2004)
		Xyloryctes	teuthras Bates		Neo	Mexico (Ramos-Elorduy and Pino Moreno, 2002)

		Genus	Species	Common name	Region	Location (Reference)
		Xyloryctes	thestalus Bates		Neo	Mexico (Ramos-Elorduy and Pino Moreno, 2002)
		Xylotrupes	gideon (L.)		Aus	Papuna New Guinea (DeFoliart, 2002)
					Or	Thailand, Myanmar, India (DeFoliart, 2002) Malaysia, Sabah (Chung et al., 2002), Thailand (Utsunomiya and Masumoto, 1999), Indonesia (Meer Mohr, 1965), India (Chakravorty et al., 2011), Laos (Yhoung-Aree and Viwatpanich, 2005)
	Staphylinidae	Oxytelus	rugulosus Say		Neo	Mexico (Ramos-Elorduy and Pino Moreno, 2004)
	Tenebrionidae	Eleodes	blapoides Esch.		Neo	Mexico (Ramos-Elorduy and Pino Moreno, 2004)
		Eleodes	spinipes Sol.		Neo	Mexico (Ramos-Elorduy and Pino Moreno, 2004)
		Tenebrio	molitor L.	Mealworm beetle	Neo	Mexico (DeFoliart, 2002)
					Pal	China (Chen and Feng, 1999)
					Or	Laos (Hanboonsong and Durst, 2014)
		Tribolium	castaneum (Herbst)	Red flour beetle	Neo	Mexico (Ramos-Elorduy, 2009)
		Tribolium	confusum Du Val		Neo	Mexico (Ramos-Elorduy, 2009)
		Ulomoides	dermestoides (Fairm.)		Neo	Brazil (Costa Neto and Ramos-Elorduy, 2006)
					Or	Philippines (Adalla and Cleolas, 2010)
		Zophobas	morio F.	Super worm	Neo	Mexico (Ramos-Elorduy and Pino Moreno, 2004)
	Zopheridae	Zopherus	jourdani Sallé		Neo	Mexico (Ramos-Elorduy and Pino Moreno, 2004)
		Zopherus	mexicanus Gray		Neo	Mexico (Ramos-Elorduy and Pino Moreno, 2004)
Blattaria	Blaberidae	Blaberus	craniifer Burm.		Neo	Mexico (Ramos-Elorduy et al., 2008)
	Blattidae	Blatta	oris L.		Pal	Japan (Schimitschek, 1968)
					Or	Thailand (DeFoliart, 2002)
		Neostylopyga	rhombifolia (Stoll)	Harlequin cockroach	Or	Thailand (DeFoliart, 2002)
		Periplaneta	americana L.		Neo	Brazil (Costa Neto and Ramos-Elorduy, 2006), Mexico (Ramos-Elorduy, 2009)
					Pal	China (DeFoliart, 2002), Japan (Schimitschek, 1968)
		Periplaneta	australasiae (F.)		Pal	China (DeFoliart, 2002)
					Neo	Mexico (Ramos-Elorduy et al., 2008)
		Periplaneta	fuliginosa Serville		Pal	Japan (Schimitschek, 1968)
	Ectobiidae	Blatella	germanica (L.)		Neo	Mexico (Ramos-Elorduy and Pino Moreno, 2002)
					Pal	Japan (Schimitschek, 1968)
Mantodea	Empusidae	Gongylus	gongylodes L.		Or	India (Singh et al., 2007)
	Mantidae	Hierodula	coarctata Saussure		Or	India (DeFoliart, 2002)
		Hierodula	patellifera Serville		Pal	Japan (Schimitschek, 1968)
		Hierodula	sternosticta W.-M.		Aus	Papau New Guinea (DeFoliart, 2002)

(*Continued*)

Table A1 Documented Information for 1555 Species of Insects and Spiders (cont.)

Order	Family	Genus	Species	Common Name	Faunal	Distribution and References
		Hierodula	venosa Olivier		Or	Indonesia (Meer Mohr, 1965)
		Hierodula	westwoodi Kirby		Or	India (DeFoliart, 2002)
		Mantis	religiosa (L.)		Pal	Japan (Schimitschek, 1968)
					Or	Thailand (Hanboonsong, 2010), India (Singh et al., 2007)
		Statilia	maculata Thunberg		Pal	Japan (Schimitschek, 1968)
		Tenodera	aridifolia angustipennis Saussure		Pal	Japan (Schimitschek, 1968)
		Tenodera	aridifolia aridifolia Stoll		Pal	Japan (Schimitschek, 1968)
		Tenodera	aridifolia sinensis Saussure		Or	Thailand (Hanboonsong, 2010)
	Tarachodidae	Tarachodes	saussurei Giglio-Tos		Afr	Cameroun (Barreteau, 1999)
	Thespidae	Hoplocorypha	garuana Giglio-Tos		Afr	Cameroun (Barreteau, 1999)
Diptera	Calliphoridae	Chrysomya	megacephala (F.)	Oriental latrine fly	Pal	China (DeFoliart, 2002)
					Or	India (Singh et al., 2007)
	Chaoboridae	Chaoborus	edulis Edwards	Lake flies phantom midges	Afr	East Afrn lakes (Bergier, 1941; Owen, 1973), Tanzania (Bodenheimer, 1951; Harris, 1940), Uganda, Kenya
	Chaoboridae	Chaoborus	pallidipes Theob.	Lake flies Phantom midges	Afr	Uganda (Bergier, 1941)
	Drosophilidae	Drosophila	melanogaster Meigen	Common fruit fly	Neo	Mexico (Ramos-Elorduy, 2009)
	Ephydridae	Ephydra	cinerea Jones	Shore fly	Nea	N. Am. (DeFoliart, 2002)
		Ephydra	hians Say	Alcali fly	Nea	N. Am. (DeFoliart, 2002)
		Ephydra	macellaria er		Neo	Mexico (Ramos-Elorduy, 2009)
		Mosillus	tibialis Cresson	Shore fly	Nea	N. Am. (DeFoliart, 2002)
	Muscidae	Musca	domestica L.		Neo	Mexico (DeFoliart, 2002)
					Neo	Mexico (Ramos-Elorduy, 2009)
		Musca	domestica vicina Macq.	Oriental housefly	Pal	China (DeFoliart, 2002)
					Pal	China (DeFoliart, 2002; Chen et al., 2009)
	Oestridae	Hypoderma	bovis (L.)	Bot fly, warble fly	Nea	N. Am. (DeFoliart, 2002)
		Hypoderma	tarandi (L.)	Reindeer warble fly	Nea	N. Am. (DeFoliart, 2002)
	Simulidae	Simulium	rubrithorax Lutz		Neo	Brazil (DeFoliart, 2002; Costa Neto and Ramos-Elorduy, 2006)
	Stratiomyidae	Hermetia	illucens (L.)		Or	Malaysia, Sabah (Chung et al., 2002)
	Syrphidae	Copestylum	anna (Williston)		Neo	Mexico (Ramos-Elorduy et al., 1998)

		Copestylum	haagii (Jaennicke)		Neo	Mexico (DeFoliart, 2002)
	Tephritidae	Anastrepha	ludens (Loew)	Mexican fruit fly	Neo	Mexico (Ramos-Elorduy, 2009)
	Tipulidae	Holorusia	hespera Arnaud and B.	Giant western crane fly	Nea	N. Am. (DeFoliart, 2002)
		Tipula	derbyi Doane	Crane fly	Nea	N. Am. (DeFoliart, 2002)
		Tipula	paludosa Meig.		Pal	China (Chen and Feng, 1999)
		Tipula	quaylii Doane	Crane fly	Nea	N. Am. (DeFoliart, 2002)
		Tipula	simplex Doane	Crane fly	Nea	N. Am. (DeFoliart, 2002)
Ephemeroptera	Baetidae	Cloeon	dipterum (L.)		Pal	Japan (Schimitschek, 1968)
		Cloeon	kimminsi Hubbard		Or	India (Singh et al., 2007)
	Caenidae	Caenis	kungu Kimm.	Mayfly	Afr	Malawi (Grant, 2001; DeFoliart, 2002)
		Povilla	adusta Navas	Mayfly	Afr	Kenya, Tanzania, Uganda (Grant, 2001)
	Ephemerellidae	Ephemerella	jinghongensis Xu et al.	Mayfly	Pal	China (Chen et al., 2009)
Hemiptera	Aetalionidae	Darthula	hardwickii (Gray)		Pal	China (Chen et al., 2009)
	Alydidae	Leptocorisa	acuta (Thunb.)	Paddy bug	Or	Indonesia (DeFoliart, 2002), India (Singh et al., 2007)
		Leptocorisa	oratorius (F.)	Rice bug	Or	Malaysia, Sabah (Chung et al., 2002), Malaysia (Chung, 2010), Indonesia, Kalimantan (Chung, 2003)
	Aphididae	Hyalopterus	pruni (Geoffr.)	Mealy plum aphid	Nea	N. Am. (DeFoliart, 2002)
	Belostomatidae	Abedus	dilatatus (Say)	Giant water bug	Neo	Mexico (Ramos-Elorduy et al., 1998)
		Abedus	ovatus Stal	Giant water bug	Neo	Mexico (DeFoliart, 2002)
		Belostoma	micantulum (Stal)		Neo	Venezuela (Araújo and Beserra, 2007)
		Limnogeton	fieberi Mayr		Afr	Togo (Tchibozo, 2015)
		Lethocerus	americanus (Leidy)	Eastern toe biter	Nea	N. Am. (DeFoliart, 2002)
		Lethocerus	deyrollei (Vuillefroy)	Giant water bug	Pal	Japan (Schimitschek, 1968)
		Lethocerus	indicus (L. and S.)		Pal	China (DeFoliart, 2002; Chen and Feng, 1999)
					Or	Thailand (Hanboonsong et al., 2000), Myanmar, Malaysia, Vietnam, India (DeFoliart, 2002; Singh et al., 2007; Chakravorty et al., 2011), Laos (Yhoung-Aree and Viwatpanich, 2005)
		Sphaerodema	rusticum (F.)		Or	Thailand (DeFoliart, 2002), Indonesia (Meer Mohr, 1965)
		Sphaerodema	molestum (Duf.)		Pal	China (Chen and Feng, 1999)
	Cercopidae	Aeneolamia	postica (Walker)		Or	Thailand (DeFoliart, 2002)
	Cicadellidae	Goniagnathus	decoratus (Hpt.)		Neo	Mexico (Ramos-Elorduy et al., 1997)
		Opsius	jucundus (Leth.)		Pal	Sinai Desert (DeFoliart, 2002)
	Cicadidae	Afzeliada	afzelii Stal		Pal	Sinai Desert (DeFoliart, 2002)
					Afr	D.R. Congo (Malaisse, 1997)

(Continued)

Appendix

Table A1 Documented Information for 1555 Species of Insects and Spiders (cont.)

Order	Family	Genus	Species	Common Name	Faunal	Distribution and References
		Afzeliada	duplex Diabola		Afr	D.R. Congo (Malaisse, 1997)
		Chremistica	ribhoi Hajong and Yaakob		Or	India (Hajong, 2013)
		Cicada syn	montezuma Distant		Neo	Mexico (Ramos-Elorduy and Pino Moreno, 2002)
		Cicada	flammata Dist.		Pal	China (Chen and Feng, 1999)
		Cosmopsaltria	aurata Duffels		Or	Indonesia (Duffels and van Mastrigt, 1991)
		Cosmopsaltria	papuensis Duffels		Or	Indonesia (Duffels and van Mastrigt, 1991)
		Cosmopsaltria	signata Duffels		Or	Indonesia (Duffels and van Mastrigt, 1991)
		Cosmopsaltria	gigantea occidentalis Duffels		Or	Indonesia (Duffels and van Mastrigt, 1991)
		Cosmopsaltria	waine Duffels		Or	Indonesia (Duffels and van Mastrigt, 1991)
		Cryptotympana	aquila (Walker)		Or	India (Singh et al., 2007)
		Cryptotympana	atrata (F.)		Pal	China (Chen et al., 2009)
		Cryptotympana	facialis (Walker)		Pal	Japan (Schimitschek, 1968)
		Diceroprocta	apache (Davis)	Apache cicada	Nea	N. Am. (DeFoliart, 2002)
		Dundubia	jacoona Distant		Or	Malaysia Sabah (Chung et al., 2002)
		Dundubia	rufivena Walker		Or	Laos Thailand (Sueur, 1988/1999)
		Dundubia	spiculata Noualhier		Or	India (Singh et al., 2007)
		Euterpnosia	crowfooti (Dist.)		Or	India (Chakravorty et al., 2011)
		Fidicinoides	picea Walker		Neo	Mexico (Ramos-Elorduy et al., 2008)
		Graptopsaltria	nigrofuscata (Motsch.)		Pal	Japan (Schimitschek, 1968; DeFoliart, 2002)
		Ioba	horizontalis Karsch		Afr	D.R. Congo (Malaisse, 1997)
		Ioba	leopardina Distant		Afr	D.R. Congo (Malaisse, 1997) Zambia (Mbata, 1995) Zimbabwe (Malaisse, 1997)
		Magicicada	cassini (Fisher)	Periodic cicada	Nea	N. Am. (DeFoliart, 2002)
		Magicicada	septemdecim (L.)	Periodic cicada	Nea	N. Am. (DeFoliart, 2002)
		Magicicada	septendecula (A. and M.)	Periodic cicada	Nea	N. Am. (DeFoliart, 2002)
		Magicicada	tredecassini (A. and M.)	Periodic cicada	Nea	N. Am. (DeFoliart, 2002)
		Magicicada	tredecim (Riley)	Periodic cicada	Nea	N. Am. (DeFoliart, 2002)
		Magicicada	tredecula (A. and M.)	Periodic cicada	Nea	N. Am. (DeFoliart, 2002)
		Meimuna	opalifera (Walker)		Pal	Japan (Schimitschek, 1968)
		Monomatapa	insignis Distant		Afr	Botswana (Roodt, 1993)
		Munza	furva Distant		Afr	D.R. Congo (Malaisse, 1997)
		Okanagana	bella Davis		Nea	N. Am. (DeFoliart, 2002)
		Okanagana	cruentifera (Uhler)		Nea	N. Am. (DeFoliart, 2002)

	Oncotympana	maculaticollis (Motsch.)		Pal	Japan (Schimitschek, 1968)
	Platypedia	areolata (Uhler)		Nea	N. Am. (DeFoliart, 2002)
	Platypleura	adouma Distant		Afr	Congo (Hoare, 2007)
	Platypleura	insignis Distant		Or	Myanmar (DeFoliart, 2002)
	Platypleura	kaempferi (F.)		Pal	China (Chen and Feng, 1999) Japan (Schimitschek, 1968)
	Platypleura	stridula (L.)		Afr	Zambia (Mbata, 1995)
	Pomponia	imperatoria (Westw.)		Or	Malaysia (DeFoliart, 2002)
	Pomponia	linearis (Walker)		Or	India (Singh et al., 2007)
	Pomponia	merula (Distant)		Or	Indonesia (Chung, 2010)
	Pycna	repanda (L.)		Or	India (Chakravorty et al., 2011)
	Quesada	gigas (Olivier)		Neo	Mexico (Ramos-Elorduy and Pino Moreno, 2002)
	Sadaka	radiata (Karsch)		Afr	D.R. Congo (Malaisse, 1997)
	Tanna	japonensis (Distant)		Pal	Japan (Schimitschek, 1968)
	Tibicen	pruinosa (Say)	Annual cicada	Neo	Mexico (DeFoliart, 2002)
	Ugada	giovanninae Boulard		Afr	Congo (Hoare, 2007)
	Ugada	limbalis (Karsch)		Afr	Congo (Nkouka, 1987) D.R. Congo (Malaisse, 1997) Zambia (Mbata, 1995)
	Ugada	limbata (F.)		Afr	Congo (DeFoliart, 2002; Hoare, 2007)
	Ugada	limbimacula (Karsch)		Afr	Congo (Nkouka, 1987; Hoare, 2007) D.R. Congo (Malaisse, 1997)
	Yezoterpnosia	vacua (Olivier)		Pal	Japan (Schimitschek, 1968)
Coccidae	Ceroplastes	sinensis Del Guercio	Chinese wax scale	Pal	China (Donkin, 1977; Chen and Feng, 1999)
Coreidae	Acanthocephala	luctuosa Stal		Neo	Mexico (DeFoliart, 2002)
	Anoplocnemis	phasianus (F.)		Or	Thailand (Hanboonsong et al., 2000)
	Carlisis	wahlbergi Stäl	Tip wilter	Afr	Zimbabwe (Chavanduka, 1976)
	Dalader	acuticosta Amyot and Serv.		Or	India (Chakravorty et al., 2011)
	Mictis	tenebrosa (F.)		Pal	China (Chen et al., 2009)
				Or	India (Chakravorty et al., 2011, 2013)
	Petascelis	remipes Signoret		Afr	Zimbabwe (Chavanduka, 1976)
	Thasus	gigas (Klug)	Giant mesquite bug	Neo	Mexico (DeFoliart, 2002)
Corixidae	Corisella	edulis (Champion)	Water boatman	Neo	Mexico (DeFoliart, 2002)
	Corisella	mercenaria (Say)	Water boatman	Neo	Mexico (DeFoliart, 2002)
	Corisella	texcocana Jacz.	Water boatman	Neo	Mexico (Ramos-Elorduy, 2009)
	Graptocorixa	abdominalis (Say)	Water boatman	Neo	Mexico (Ramos-Elorduy, 2009)

(*Continued*)

Table A1 Documented Information for 1555 Species of Insects and Spiders (cont.)

Order	Family	Genus	Species	Common Name	Faunal	Distribution and References
		Graptocorixa	bimaculata (Guérin)	Water boatman	Neo	Mexico (Ramos-Elorduy, 2009)
		Krizousacorixa	azteca (Jacz.)	Water boatman	Neo	Mexico (Ramos-Elorduy, 2009)
		Krizousacorixa	femorata Guér.	Water boatman	Neo	Mexico (Ramos-Elorduy, 2009)
	Dactylopidae	Dactylopius	confusus (Cockerell)	Cochineal insect	Neo	Mexico (Ramos-Elorduy, 2009)
		Dactylopius	indicus Green	Cochineal insect	Neo	Mexico (Ramos-Elorduy, 2009)
		Dactylopius	coccus Costa	Cochineal insect	Neo	South America, Mexico (Ramos-Elorduy, 2009; Aldama-Aguilera et al., 2005)
		Dactylopius	tomentosus Lamarck	Cochineal insect	Pal	Canary Islands (Aldama-Aguilera et al., 2005)
	Eriococcidae	Apiomorpha syn	pomiformis Froggatt	Cochineal insect	Neo	Mexico (Ramos-Elorduy et al., 1998)
	Flatidae	Flatida	rosea Melichar		Aus	Australia (DeFoliart, 2002)
		Lawana	imitata (Melichar)		Afr	Madagascar (Decary, 1937)
	Fulgoridae	Zanna	tenebrosa Fabr.		Pal	China (Chen et al., 2009)
	Gerridae	Cylindrostethus	scrutator (Kirkaldy)	Water strider	Afr	Madagascar (Decary, 1937)
		Gerris	spinole (Leth.)	Water strider	Or	Thailand (Hanboonsong et al., 2000; Hanboonsong, 2010)
	Kerriidae	Austrotachardia	acaciae (Maskell)		Or	India (DeFoliart, 2002)
		Kerria	lacca (Kerr)	Indian lac insect	Aus	Australia (DeFoliart, 2002)
	Membracidae	Anthiante	expensa Germar		Or	Thailand (DeFoliart, 2002)
		Hoplophorion	monogramma Germar		Neo	Mexico (Ramos-Elorduy et al., 1998)
		Umbonia	reclinata (Germar)		Neo	Mexico (Ramos-Elorduy et al., 1998)
		Umbonia	orizabae Nuckton		Neo	Mexico (Ramos-Elorduy et al., 2007)
		Umbonia	spinosa (F.)		Neo	Mexico (Ramos-Elorduy et al., 2008)
	Naucoridae	Ambrysus	stali La Rivers		Neo	Brazil (DeFoliart, 2002) Ecuador (Onore, 2005) Colombia (Paoletti et al., 2001)
		Ambrysus	usingeri La Rivers		Neo	Venezuela (Araújo and Beserra, 2007)
		Limnocorus	cf minutus De Carlo		Neo	Venezuela (Araújo and Beserra, 2007)
	Nepidae	Laccotrephes	griseus (Guer.)		Neo	Venezuela (Araújo and Beserra, 2007)
		Laccotrephes	japonensis (Scott)	Water scorpion	Or	Thailand (DeFoliart, 2002)
		Laccotrephes	maculatus (F.)		Pal	Japan (Schimitschek, 1968)
		Laccotrephes	robustus Stal		Or	India (Shantibala et al., 2014)
		Laccotrephes	ruber (L.)		Or	Indonesia (Meer Mohr, 1965)
					Or	Thailand (Hanboonsong et al., 2000; Hanboonsong, 2010), Laos (Yhoung-Aree and Viwatpanich, 2005)

	Ranatra	*chinensis* Mayr	Water stick insect	Pal	Japan (Schimitschek, 1968)
	Ranatra	*longipes thai* Lansbury	Water stick insect	Or	Thailand (Hanboonsong et al., 2000; Hanboonsong, 2010)
	Ranatra	*unicolor* Scott	Water stick insect	Pal	Japan (Schimitschek, 1968)
	Ranatra	*varipes* Stal	Water stick insect	Or	Thailand (Hanboonsong et al., 2000; Hanboonsong, 2010)
Notonectidae	*Anisops*	*barbata* Brooks	Backswimmer	Or	Thailand (Hanboonsong et al., 2000; Hanboonsong, 2010)
	Anisops	*bouvieri* Kirkaldy	Backswimmer	Or	Thailand (Hanboonsong et al., 2000; Hanboonsong, 2010)
	Notonecta	*unifasciata* Guér	Backswimmer	Neo	Mexico (DeFoliart, 2002)
Pentatomidae	*Agonoscelis*	*versicolor* (Fabr.)	Sudan millet bug	Afr	Sudan (Van Huis, 2005; Kelemu et al., 2015)
	Aspongopus	*chinensis* Dallas		Or	India (DeFoliart, 2002)
	Aspongopus	*nepalensis* Westw.		Or	India (DeFoliart, 2002; Chakravorty et al., 2011, 2013)
	Bagrada	*picta* (F.)		Or	India (DeFoliart, 2002)
	Brochymena	*tenebrosa* Walker		Neo	Mexico (DeFoliart, 2002)
	Coridius	*viduatus* (Fabr.)	Melon bug	Afr	Sudan, Namibia (van Huis, 2013; Kelemu et al., 2015)
	Cyclopelta	*parva* Distant	Stink bug	Pal	China (Chen and Feng, 1999)
	Cyclopelta	*subhimalayensis* Strickl.		Or	India (DeFoliart, 2002)
	Edessa	*conspersa* Stal		Neo	Mexico (DeFoliart, 2002)
	Edessa	*cordifera* (Walker)		Neo	Mexico (Ramos-Elorduy, 2009)
	Edessa	*mexicana* Stal		Neo	Mexico (DeFoliart, 2002)
	Edessa	*montezuma* Distant		Neo	Mexico (Ramos-Elorduy et al., 1998)
	Edessa	*petersii* Distant		Neo	Mexico (DeFoliart, 2002)
	Edessa	*rufomarginata* (DeGeer)		Neo	Mexico (Ramos-Elorduy and Pino Moreno, 2002)
	Edessa	*taxcoensis* (Ancona)		Neo	Mexico (DeFoliart, 2002)
	Erthesina	*fullo* (Thunb.)		Or	India (DeFoliart, 2002)
	Eurostus	*validus* Dallas		Pal	China (Chen and Feng, 1999)
	Halyomorpha	*picus* (F.)		Or	India (Chakravorty et al., 2011, 2013)
	Nezara	*robusta* Distant		Afr	Malawi (Shaxson et al., 1999; DeFoliart, 2002)
	Nezara	*viridula* (L.)		Afr	Nigeria (Agbidye et al., 2009)
				Aus	Papua New Guinea (DeFoliart, 2002)
				Or	India (Singh et al., 2007; Chakravorty et al., 2011), Malaysia, Sabah Chung et al., 2002), Indonesia, Kalimantan (Chung, 2003)

(*Continued*)

Table A1 Documented Information for 1555 Species of Insects and Spiders (cont.)

Order	Family	Genus	Species	Common Name	Faunal	Distribution and References
		Euschistus	crenator (F.)		Neo	Mexico (DeFoliart, 2002; Yen et al., 2013)
		Euschistus	lineatus Walk		Neo	Mexico (DeFoliart, 2002)
		Euschistus	strenuus Stal		Neo	Mexico (DeFoliart, 2002)
		Mormidea	notulata (H.-S.)		Neo	Mexico (Acuña et al., 2011)
		Pharypia	fasciata (Haglund)		Neo	Mexico (DeFoliart, 2002)
		Proxys	punctulatus (Palisot)		Neo	Mexico (Ramos-Elorduy and Pino Moreno, 2002)
		Rhynchocoris	humeralis Thunberg		Or	India (Chakravorty et al., 2013)
	Pseudococcidae	Phenacoccus	prunicola Borchs		Pal	China (Chen and Feng, 1999)
		Trabutina	mannipara (Ehrenberg)		Pal	Sinai Desert (DeFoliart, 2002)
		Trabutina	serpentinus Green		Pal	Iran Sinai Desert (DeFoliart, 2002)
	Psyllidae	Cyamophila	astragalicola (Gegechkori)		Pal	Iran (Grami, 1998)
		Retroacizzia	mopani (Pettey)		Afr	Botswana (Sekhwela, 1988) Zimbabwe (Weaving, 1973)
		Glycaspis	eucalypti (Dobson)		Aus	Australia (DeFoliart, 2002)
	Pyrrhocoridae	Antilochus	coqueberti (F.)		Or	India (Chakravorty et al., 2011)
	Tessaratomidae	Encosternum	delegorguei Spin.		Afr	South Afr (Faure, 1944; Dzerefos et al., 2009) Zimbabwe (Chavanduka, 1976; Bodenheimer, 1951) Swaziland Mozambique Malawi Botswana Namibia (Kelemu et al., 2015)
		Eusthenes	cupreus (Westw.)		Pal	China (Chen and Feng, 1999)
		Eusthenes	saevus Stal		Pal	China (Chen et al., 2009)
		Lyramorpha	edulis Blöte		Aus	Papua New Guinea (Blöte, 1952)
		Natalicola	pallidus (Westwood)		Afr	Zimbabwe (Weaving, 1973)
		Tessaratoma	javanica (Thunb.)		Or	Thailand (DeFoliart, 2002), Laos (Yhoung-Aree and Viwatpanich, 2005)
		Tessaratoma	papillosa (Drury)	Lichi stink bug	Pal	China (DeFoliart, 2002; Chen et al., 2009)
		Tessaratoma	quadrata Distant		Or	Thailand (DeFoliart, 2002)
		Tessaratoma			Or	India (Chakravorty et al., 2011, 2013), Laos (Hanboonsong and Durst, 2014)
					Or	Laos (Yhoung-Aree and Viwatpanich, 2005), Thailand (Raksakantong et al., 2010)
	Tibicinidae	Carineta	fimbriata Distant		Neo	Ecuador (Onore, 2005)
Hymenoptera	Apidae	Apis	cerana F.	Indian bee	Pal	China (Chen et al., 2009), Japan (Schimitschek, 1968)

Apis	dorsata F.		Or	Thailand Philippines (DeFoliart, 2002; Adalla and Cleolas, 2010) Indonesia (Meer Mohr, 1965) India (Singh et al., 2007) Malaysia Sabah (Chung et al., 2002)
			Pal	China (Chen et al., 2009)
Apis			Or	Thailand, Philippines, India, Nepal (DeFoliart, 2002; Hanboonsong et al., 2000; Singh et al., 2007; Adalla and Cleolas, 2010) Indonesia (Meer Mohr, 1965), Malaysia, Sabah (Chung et al., 2002)
Apis	florae F.		Pal	China (Chen et al., 2009)
			Or	Thailand (DeFoliart, 2002), Laos (Yhoung-Aree and Viwatpanich, 2005)
Apis	mellifera adansoni Latr.	Bee	Afr	D.R. Congo (Takeda, 1990), Tanzania (Harris, 1940), Zambia (Mbata, 1995), CAR (Bahuchet, 1985), Nigeria, Sierra Leone, Ghana, Benin (Kelemu et al., 2015)
Apis	mellifera mellifera L.	Bee	Afr	Senegal (Gessain and Kinzler, 1975), Zambia (Mbata, 1995), Tanzania, D.R. Congo (DeFoliart, 2002), Botswana, Nigeria, Sierra Leone, Ghana, South Sudan, Togo, Lesotho, Benin (Kelemu et al., 2015)
Apis	mellifera L.		Neo	Nearly pan-regional, West Indies (Ramos-Elorduy, 2006; DeFoliart, 2002), Brazil (Costa Neto and Ramos-Elorduy, 2006)
			Pal	Japan (DeFoliart, 2002)
			Or	Thailand, Malaysia, Vietnam, India, Nepal, Sri Lanka (DeFoliart, 2002)
Apotrigona	nebulata (Smith)	Stingless bee	Afr	CAR (Bahuchet, 1985)
Axestotrigona	ferruginea (Lep.)	Stingless bee	Afr	CAR (Bahuchet, 1985)
Axestotrigona	richardsi (Darchen)	Stingless bee	Afr	D.R. Congo (DeFoliart, 2002)
Bombus	atratus Franklin		Neo	Ecuador (Onore, 2005)
Bombus	diligens Smith		Neo	Mexico (DeFoliart, 2002; Ramos-Elorduy, 2009)
Bombus	ecuadorius Meunier		Neo	Ecuador (Onore, 2005)
Bombus	ephippiatus Say		Neo	Mexico (DeFoliart, 2002; Ramos-Elorduy, 2009)
Bombus	funebris Smith		Neo	Ecuador (Onore, 2005)
Bombus	medius Cresson		Neo	Mexico (DeFoliart, 2002; Ramos-Elorduy, 2009)
Bombus	robustus Smith		Neo	Ecuador (Onore, 2005)
Bombus	rufocinctus Cresson		Neo	Mexico (Ramos-Elorduy and Pino Moreno, 2002)
Bombus	appositus Cresson		Nea	N. Am. (DeFoliart, 2002)

(Continued)

Table A1 Documented Information for 1555 Species of Insects and Spiders (cont.)

Order	Family	Genus	Species	Common Name	Faunal	Distribution and References
		Bombus	nevadensis Cresson		Nea	N. Am. (DeFoliart, 2002)
		Bombus	terricola occidentalis Greene		Nea	N. Am. (DeFoliart, 2002)
		Bombus	vosnesenskii Radoszk.		Nea	N. Am. (DeFoliart, 2002)
		Cephalotrigona	capitata (Smith)		Neo	Brazil (Costa Neto and Ramos-Elorduy, 2006), Argentine (Zamudio and Hilgert, 2012)
		Cephalotrigona	femorata (Smith)		Neo	Brazil (Costa Neto and Ramos-Elorduy, 2006)
		Cephalotrigona	zexmeniae (Cockerel)		Neo	Mexico (Ramos-Elorduy and Pino Moreno, 2002), Guatemala (Vit et al., 2004)
		Dactylurina	staudingeri (Gribodo)	Stingless bee	Afr	D.R. Congo (Takeda, 1990)
		Duckeola	ghilianii (Spinola)		Neo	Brazil (Costa Neto and Ramos-Elorduy, 2006)
		Friesella	schrottkyi (Friese)		Neo	Brazil (Costa Neto and Ramos-Elorduy, 2006)
		Frieseomelitta	nigra (Lepeletier)		Neo	Mexico (Reyes-González et al., 2014)
		Frieseomelitta	silvestrii (Friese)		Neo	Brazil (Costa Neto and Ramos-Elorduy, 2006)
		Frieseomelitta	varia (Lep.)	Stingless bee	Neo	Brazil (Ferreira et al., 2010)
		Geotrigona	acapulconis (Strand)		Neo	Mexico (Reyes-González et al., 2014)
		Geotrigona	mombuca (Smith)		Neo	Brazil (Costa Neto and Ramos-Elorduy, 2006)
		Hypotrigona	gribodoi (Magretti)	Stingless bee	Afr	Angola (Crane, 1992), D.R. Congo (Takeda, 1990), CAR (Bahuchet, 1985)
		Hypotrigona	ruspolii (Magretti)	Stingless bee	Afr	Senegal (Gessain and Kinzler, 1975)
		Lestrimelitta	chamelensis Ayala		Neo	Mexico (Reyes-González et al., 2014)
		Lestrimelitta	cubiceps Friese		Afr	Angola (Crane, 1992)
		Lestrimelitta	limao (Smith)		Neo	Mexico (DeFoliart, 2002), Brazil (Costa Neto and Ramos-Elorduy, 2006)
		Lestrimelitta	niitkip Ayala		Neo	Mexico (Ramos-Elorduy and Pino Moreno, 2002)
		Meliplebeia	beccarii (Gribodo)	Stingless bee	Afr	CAR (Bahuchet, 1985)
		Meliponula	bocandei (Spin.)	Stingless bee	Afr	Angola (Crane, 1992), D.R. Congo (DeFoliart, 2002), CAR (Bahuchet, 1985)
		Meliponula	gambiana Moure	Stingless bee	Afr	Senegal (Gessain and Kinzler, 1975)
		Meliponula	lendliana (Friese)	Stingless bee	Afr	D.R. Congo (DeFoliart, 2002)
		Melipona	asilvai Moure		Neo	Brazil (Costa Neto and Ramos-Elorduy, 2006)
		Melipona	atratula Ill.		Neo	Brazil (Costa Neto and Ramos-Elorduy, 2006)
		Melipona	beecheii Bennett		Neo	Guatemala (Vit et al., 2004), Mexico (Ramos-Elorduy and Pino Moreno, 2002)
		Melipona	bicolor Lep.		Neo	Brazil (Costa Neto and Ramos-Elorduy, 2006), Argentine (Zamudio and Hilgert, 2012)

Melipona	bicolor schencki Gribodo		Neo	Brazil (Costa Neto and Ramos-Elorduy, 2006)
Melipona	bilineata Sm		Neo	Brazil (Costa Neto and Ramos-Elorduy, 2006)
Melipona	compressipes (Fabr.)		Neo	Brazil (Costa Neto and Ramos-Elorduy, 2006)
Melipona	compressipes fasciculata Smith		Neo	Brazil (Costa Neto and Ramos-Elorduy, 2006)
Melipona	crinita Moure and Kerr		Neo	Brazil (Costa Neto and Ramos-Elorduy, 2006)
Melipona	eburnea fuscopilosa Friese		Neo	Brazil (Costa Neto and Ramos-Elorduy, 2006)
Melipona	fasciata Latr.		Neo	Guatemala (Vit et al., 2004), Mexico (DeFoliart, 2002; Reyes-González et al., 2014)
Melipona	favosa favosa (Fabr.)		Neo	Venezuela (Vit et al., 2004)
Melipona	grandis Guérin		Neo	Brazil (Costa Neto and Ramos-Elorduy, 2006)
Melipona	interrupta Latr.		Neo	Brazil (Costa Neto and Ramos-Elorduy, 2006), Colombia (Crane, 1992)
Melipona	mandacaia Smith		Neo	Brazil (Costa Neto and Ramos-Elorduy, 2006)
Melipona	marginata Lep.		Neo	Brazil (Costa Neto and Ramos-Elorduy, 2006)
Melipona	melanoventer Schwarz		Neo	Brazil (Costa Neto and Ramos-Elorduy, 2006)
Melipona	nigra Lepeletier		Neo	Brazil (Crane, 1992)
Melipona	oblitescens Cockerell	Stingless bee	Neo	Brazil (Ferreira et al., 2010)
Melipona	obscurior Moure		Neo	Argentine (Zamudio and Hilgert, 2012)
Melipona	paraensis Ducke		Neo	Venezuela (Vit et al., 2004)
Melipona	pseudocentris pseudocentris Cock.		Neo	Brazil (Costa Neto and Ramos-Elorduy, 2006)
Melipona	quadrifasciata Lep.		Neo	Argentine (Zamudio and Hilgert, 2012), Brazil (Costa Neto and Ramos-Elorduy, 2006)
Melipona	rufiventris Lep.		Neo	Brazil (Costa Neto and Ramos-Elorduy, 2006)
Melipona	schwarzi Moure		Neo	Brazil (Costa Neto and Ramos-Elorduy, 2006)
Melipona	scutellaris Latr.		Neo	Brazil (Costa Neto and Ramos-Elorduy, 2006), Mexico (Ramos-Elorduy et al., 2009)
Melipona	seminigra merrillae Cock.		Neo	Brazil (Costa Neto and Ramos-Elorduy, 2006)
Melipona	solani Cock.		Neo	Brazil (Ramos-Elorduy and Pino Moreno, 2002), Guatemala (Vit et al., 2004)
Melipona	trinitatis Cockerell		Neo	Venezuela (Vit et al., 2004)
Melipona	yucatanica Camargo et al.		Neo	Guatemala (Vit et al., 2004)
Nannotrigona	minuta (Lep.)		Or	Indonesia (DeFoliart, 2002)
Nannotrigona	perilampoides (Cresson)		Neo	Brazil (Ramos-Elorduy and Pino Moreno, 2002), Guatemala, Mexico (Vit et al., 2004)

(*Continued*)

Table A1 Documented Information for 1555 Species of Insects and Spiders (cont.)

Order	Family	Genus	Species	Common Name	Faunal	Distribution and References
		Nannotrigona	testaceicornis (Lep.)		Neo	Brazil (Costa Neto and Ramos-Elorduy, 2006), Mexico (Ramos-Elorduy and Pino Moreno, 2002)
		Oxytrigona	mediorufa (Cockerell)		Neo	Guatemala (Vit et al., 2004)
		Oxytrigona	obscura (Friese)		Neo	Brazil (Costa Neto and Ramos-Elorduy, 2006)
		Oxytrigona	tataira (Smith)		Neo	Argentina (Zamudio and Hilgert, 2012), Brazil (Costa Neto and Ramos-Elorduy, 2006)
		Oxytrigona	flaveola	Stingless bee	Neo	Brazil (Ferreira et al., 2010)
		Paratrigona	guatemalensis Schwarz		Neo	Guatemala (Vit et al., 2004)
		Partamona	bilineata (Say)		Neo	Mexico (Ramos-Elorduy and Pino Moreno, 2002)
		Partamona	cf. cupira (Smith)		Neo	Brazil (Costa Neto and Ramos-Elorduy, 2006), Mexico (Crane, 1992)
		Partamona	orizabaensis (Strand)		Neo	Mexico (Ramos-Elorduy and Pino Moreno, 2002)
		Plebeia	emerina (Friese)		Neo	Brazil (Costa Neto and Ramos-Elorduy, 2006)
		Plebeia	frontalis (Friese)		Neo	Mexico (Ramos-Elorduy and Pino Moreno, 2002)
		Plebeia	jatiformis Cockerell		Neo	Guatemala (Vit et al., 2004)
		Plebeia	mexica Ayala		Neo	Mexico (Acuña et al., 2011)
		Plebeia	mosquito (Smith)		Neo	Brazil (Costa Neto and Ramos-Elorduy, 2006)
		Plebeia	remota (Holmb.)		Neo	Brazil (Costa Neto and Ramos-Elorduy, 2006)
		Ptilotrigona	lurida (Smith)		Neo	Brazil (Costa Neto and Ramos-Elorduy, 2006)
		Scaptotrigona	hellwegeri (Friese)		Neo	Mexico (Reyes-González et al., 2014)
		Scaptotrigona	mexicana Guérin		Neo	Guatemala (Vit et al., 2004), Mexico (DeFoliart, 2002)
		Scaptotrigona check	nigrohirta (Moure)		Neo	Brazil (Costa Neto and Ramos-Elorduy, 2006)
		Scaptotrigona	pectoralis (Dalla Torre)		Neo	Guatemala (Vit et al., 2004), Mexico (Ramos-Elorduy and Pino Moreno, 2002)
		Scaptotrigona	polystica (Moure)		Neo	Brazil (Costa Neto and Ramos-Elorduy, 2006)
		Scaptotrigona	postica (Latreille)		Neo	Brazil (Costa Neto and Ramos-Elorduy, 2006; Ferreira et al., 2010)
		Scaptotrigona	tubiba (Smith)		Neo	Brazil (Costa Neto and Ramos-Elorduy, 2006)
		Scaptotrigona	xanthotricha (Moure)		Neo	Brazil (Costa Neto and Ramos-Elorduy, 2006)
		Scaura	latitarsis (Friese)		Neo	Venezuela (Vit et al., 2004)
		Schwarziana	quadripunctata (Lep.)		Neo	Argentine (Zamudio and Hilgert, 2012)
		Tetragonisca	angustula (Latreille)	Stingless bee	Neo	Brazil (DeFoliart, 2002; Ferreira et al., 2010), Ecuador (Onore, 2005), Guatemala, Mexico, Venezuela (Vit et al., 2004)

	Tetragonisca	fiebrigi (Schwarz)		Neo	Argentine (Zamudio and Hilgert, 2012)
	Tetragona	clavipes (Fabr.)	Stingless bee	Neo	Colombia (DeFoliart, 2002) Brazil (Costa Neto and Ramos-Elorduy, 2006; Ferreira et al., 2010), Argentine (Zamudio and Hilgert, 2012), Venezuela (Ruddle, 1973)
	Tetragona	dorsalis (Smith)		Neo	Brazil (Costa Neto and Ramos-Elorduy, 2006), Guatemala (Vit et al., 2004)
	Tetragona	truncata Moure	Stingless bee	Neo	Brazil (Ferreira et al., 2010)
	Tetragona	mombuca (Smith)		Neo	Brazil (Crane, 1992)
	Trigona	biroi Friese		Or	Philippines (Adalla and Cleolas, 2010)
	Trigona	branneri Cockerell		Neo	Brazil (Costa Neto and Ramos-Elorduy, 2006)
	Trigona	chanchamayoensis Schwarz		Neo	Brazil (Costa Neto and Ramos-Elorduy, 2006)
	Trigona	cilipes F.		Neo	Brazil (Costa Neto and Ramos-Elorduy, 2006)
	Trigona	dallatorreeana Friese		Neo	Brazil (Costa Neto and Ramos-Elorduy, 2006)
	Trigona	flaveola Cock.		Neo	Brazil (Costa Neto and Ramos-Elorduy, 2006)
	Trigona	fulviventris Guérin-M.		Neo	Mexico (Costa Neto, 2002), Guatemala (Vit et al., 2004)
	Trigona	fuscipennis Friese		Neo	Mexico (Ramos-Elorduy and Pino Moreno, 2002)
	Trigona	hypogea Silvestri		Neo	Brazil (Costa Neto and Ramos-Elorduy, 2006)
	Trigona	jaty Smith		Neo	Mexico (DeFoliart, 2002)
	Trigona check	leucogaster Cockerell		Neo	Brazil (Costa Neto and Ramos-Elorduy, 2006)
	Trigona	muscaria G. syn ?		Neo	Brazil (Costa Neto and Ramos-Elorduy, 2006)
	Trigona	nigerrima Cresson		Neo	Guatemala (Vit et al., 2004)
	Trigona	nigra nigra Cresson		Neo	Mexico (DeFoliart, 2002)
	Trigona	recursa Smith		Neo	Brazil (Costa Neto and Ramos-Elorduy, 2006)
	Trigona check	senegalensis Darchen	Stingless bee	Afr	Senegal (Gessain and Kinzler, 1975)
	Trigona	silvestriana Vachal		Neo	Guatemala (Vit et al., 2004)
	Trigona	spinipes (F.)		Neo	Brazil (DeFoliart, 2002), Argentine (Zamudio and Hilgert, 2012)
	Trigona	trinidadensis (Prov.)		Neo	Colombia, Venezuela (Ruddle, 1973; DeFoliart, 2002)
	Trigona	vidua (Lep.)		Or	Indonesia (DeFoliart, 2002)
	Trigonisca	pipioli Ayala		Neo	Mexico (Ramos-Elorduy and Pino Moreno, 2002)
	Trigonisca	schultessi (Friese)		Neo	Mexico (Ramos-Elorduy and Pino Moreno, 2002)
	Xylocopa	aestuans (L.)	Drury bee	Or	Thailand (DeFoliart, 2002), Indonesia (Meer Mohr, 1965; Lukiwati, 2010), Malaysia, Sabah (Chung et al., 2002)

(*Continued*)

342 Appendix

Table A1 Documented Information for 1555 Species of Insects and Spiders (*cont.*)

Order	Family	Genus	Species	Common Name	Faunal	Distribution and References
		Xylocopa	*appendiculata circumvolens* Smith	Japanese carpenter bee	Pal	Japan (Schimitschek, 1968)
		Xylocopa	*latipes* (Drury)		Or	Thailand (DeFoliart, 2002) India (Chakravorty et al., 2011), Malaysia, Sabah (Chung et al., 2002, as Platynopoda)
	Cynipidae	*Hedickiana*	*levantina* (Hed.)		Pal	Turkey (DeFoliart, 2002)
		Liposthenus	*glechomae* (L.)		Pal	France (DeFoliart, 2002)
	Diprionidae	*Neodiprion*	*gillettei* (Rohwer)		Neo	Mexico (DeFoliart, 2002)
	Formicidae	*Acromyrmex*	*octospinosus* (Reich)		Neo	Mexico (Ramos-Elorduy and Pino Moreno, 2002)
		Acromyrmex	*rugosus* (Smith)		Neo	Mexico (Ramos-Elorduy and Pino Moreno, 2002)
		Atta	*bisphaerica* Forel syn		Neo	Brazil (Costa Neto and Ramos-Elorduy, 2006)
		Atta	*capiguara* Goncalves		Neo	Brazil (Costa Neto and Ramos-Elorduy, 2006)
		Atta	*cephalotes* (L.)		Neo	Brazil, Colombia, Guyana, Honduras, Nicaragua, Mexico (Dufour, 1987; DeFoliart, 2002; Costa Neto and Ramos-Elorduy, 2006; Ecuador (Onore, 2005), Venezuaela (Araújo and Beserra, 2007)
		Atta	*laevigata* (Smith)		Neo	Brazil, Colombia (Dufour, 1987; DeFoliart, 2002; Costa Neto and Ramos-Elorduy, 2006)
		Atta	*mexicana* (Smith)		Neo	Mexico (Ramos-Elorduy et al., 1998)
		Atta	*sexdens* (L.)		Neo	Brazil, Colombia (Dufour, 1987; DeFoliart, 2002; Costa Neto and Ramos-Elorduy, 2006), Ecuador (Onore, 2005)
		Camponotus	*aurocinctus* (Smith)	Banded sugar ant	Aus	Australia (DeFoliart, 2002)
		Camponotus	*consobrinus* (Erichson)		Aus	Australia (DeFoliart, 2002)
		Camponotus	*gasseri* (Forel)		Aus	Australia (DeFoliart, 2002)
		Camponotus	*gigas* (Latr.)	Giant forest ant	Or	Malaysia (Chung, 2010)
		Camponotus	*inflatus* Lubb.	pot ant	Aus	Australia (DeFoliart, 2002)
		Camponotus	*japonicus* Mayr		Pal	China (Chen and Feng, 1999)
		Camponotus	*maculatus dumetorum* Wheeler		Neo	Mexico (Ramos-Elorduy and Pino Moreno, 2002
		Carebara	*castanea* Smith		Or	Thailand (Hanboonsong et al., 2000; Hanboonsong, 2010)
		Carebara	*lignata* Westw. (species group)		Afr	Southern Africa, Zambia, South Africa, Zimbabwe, Botswana, Sudan, Mozambique, Namibia, South Sudan (Kelemu et al., 2015, 2003)
		Carebara			Pal	China (Chen et al., 2009)

Carebara	vidua Smith	African thief ant	Afr	South Africa (Quin, 1959), Zambia (Mbata, 1995), Zimbabwe (Chavanduka, 1976), Malawi, D.R. Congo (DeFoliart, 2002), Kenya (Kinyuru et al., 2012), Botswana, Sudan, South Sudan (Kelemu et al., 2015)
Crematogaster	dohrni Mayr		Or	India (Meyer-Rochow, 2004)
Crematogaster	vandermeermohri Menozzi		Or	Indonesia (Meer Mohr, 1965)
Dorylus (s.g. Anomma)	nigricans Illiger	Driver ant	Afr	Cameroun (Van Huis, 2003)
Eciton	burchellii (Westwood)		Neo	Venezuela (Araújo and Beserra, 2007)
Formica	aquilonia Yarrow		Pal	China (Chen and Feng, 1999)
Formica	beijingensis Wu		Pal	China (Chen and Feng, 1999)
Formica	fusca L.		Pal	China (Chen and Feng, 1999)
Formica	japonica Motsch.		Pal	China (Chen and Feng, 1999)
Formica	sanguinea Latr.		Pal	China (Chen and Feng, 1999)
Formica	truncicola var. yessensis Wheeler		Pal	China (Chen and Feng, 1999)
Formica	uralensis Ruzsky		Pal	China (Chen and Feng, 1999)
Lasius	flavus (F.)		Pal	China (Chen and Feng, 1999)
Liometopum	apiculatum Mayr	Escamoles	Neo	Mexico (Ramos et al., 1998; DeFoliart, 2002)
Liometopum	apiculatum luctuosum Wheeler		Neo	Mexico (Ramos et al., 1998; DeFoliart, 2002)
Melophorus	bagoti Lubb.	Ant	Aus	Australia (DeFoliart, 2002)
Melophorus	cowlei Frogg		Aus	Australia (DeFoliart, 2002)
Myrmecia	pyriformis Smith		Aus	Australia (DeFoliart, 2002)
Myrmecocystus	melliger Forel		Nea	N. Am. (DeFoliart, 2002)
Myrmecocystus	mexicanus Wesmael		Neo	Mexico (Ramos et al., 1998; DeFoliart, 2002)
			Nea	N. Am. (DeFoliart, 2002)
			Neo	Mexico (DeFoliart, 2002)
Oecophylla	longinoda (Latr.)	Weaver ant	Afr	D. R. Congo (DeFoliart, 2002), Cameroun, Tchad (Kelemu et al., 2015)
Oecophylla	smaragdina (F.)	Green ttree ant	Aus	Australia Papua New Guinea (DeFoliart, 2002)
			Pal	China (Chen et al., 2009)
			Or	Thailand, Myanmar, India (DeFoliart, 2002; Chakravorty et al., 2011, 2013), Thailand (Hanboonsong et al., 2000; Hanboonsong, 2010), Laos (Yhoung-Aree and Viwatpanich, 2005), Malaysia, Sabah (Chung et al., 2002)

(*Continued*)

Table A1 Documented Information for 1555 Species of Insects and Spiders (cont.)

Order	Family	Genus	Species	Common Name	Faunal	Distribution and References
		Pogonomyrmex	barbatus (Smith)		Neo	Mexico (Ramos-Elorduy, 2006)
		Pogonomyrmex	californicus (Buckley)		Nea	N. Am. (DeFoliart, 2002)
		Pogonomyrmex	desertorum Wheeler		Nea	N. Am. (DeFoliart, 2002)
		Pogonomyrmex	occidentalis (Cresson)		Nea	N. Am. (DeFoliart, 2002)
		Pogonomyrmex	owyheei Cole		Nea	N. Am. (DeFoliart, 2002)
		Polyrhachis	dives Smith	Black weever ant	Pal	China (Chen and Feng, 1999; Chen et al., 2009)
		Polyrhachis	illaudata Walker		Pal	China (Chen and Feng, 1999)
		Polyrhachis	lamellidens Smith		Pal	China (Chen and Feng, 1999)
		Polyrhachis	vicina Roger		Pal	China (DeFoliart, 2002)
		Tetramorium	caespitum (L.)		Pal	China (Chen and Feng, 1999)
	Tenthredinidae	Dielocerus	formosus (Ohaus)		Neo	Brazil (Costa Neto and Ramos-Elorduy, 2006)
	Vespidae	Agelaia	angulata (F.)		Neo	Colombia (Dufour, 1987; DeFoliart, 2002)
		Agelaia	areata (Say)		Neo	Mexico (Ramos-Elorduy and Pino Moreno, 2002)
		Agelaia	baezae (Richards)		Neo	Ecuador (Onore, 2005)
		Agelaia	corneliana (Richards)		Neo	Ecuador (Onore, 2005)
		Agelaia	lobipleura (Richards)		Neo	Ecuador (Onore, 2005)
		Agelaia	multipicta Haliday		Neo	Argentina (Zamudio et al., 2012)
		Agelaia	ornata (Ducke)		Neo	Ecuador (Onore, 2005)
		Agelaia	panamensis (Cameron)		Neo	Mexico (Ramos-Elorduy and Pino Moreno, 2002)
		Angiopolybia	paraensis (Spinola)		Neo	Ecuador (DeFoliart, 2002)
		Apoica	pallens (F.)		Neo	Brazil (Costa Neto and Ramos-Elorduy, 2006; Ecuador (Onore, 2005)
		Apoica	pallida (Oliv.)		Neo	Ecuador (Onore, 2005)
		Apoica	strigata Richards		Neo	Ecuador (Onore, 2005)
		Apoica	thoracica Buysson		Neo	Colombia (Dufour, 1987; DeFoliart, 2002), Venezuela (Araújo and Beserra, 2007)
		Brachygastra	azteca (Sauss.)		Neo	Mexico (DeFoliart, 2002)
		Brachygastra	lecheguana (Latr.)		Neo	Mexico (DeFoliart, 2002), Brazil (Costa Neto and Ramos-Elorduy, 2006), Ecuador (Onore, 2005)
		Brachygastra	mellifica (Say)		Neo	Mexico (DeFoliart, 2002)
		Brachymenes	wagnerianus (Saussure)		Neo	Ecuador (Onore, 2005)
		Epipona	media Cooper		Neo	Brazil (Costa Neto and Ramos-Elorduy, 2006)
		Eumenes	petiolata F.	Potter wasp	Or	Thailand (DeFoliart, 2002)

Icaria	*artifex* (De Saus.)	Or	India (Meyer-Rochow, 2004)
Mischocyttarus	*basimacula* (Cameron)	Neo	Mexico (Ramos-Elorduy and Pino Moreno, 2002)
Mischocyttarus	*cubensis* (Saussure)	Neo	Mexico (Ramos-Elorduy and Pino Moreno, 2002)
Mischocyttarus	*pallidipectus* (Smith)	Neo	Mexico (Ramos-Elorduy and Pino Moreno, 2002)
Mischocyttarus	*rotundicollis* (Cameron)	Neo	Ecuador (Onore, 2005)
Mischocyttarus	*tomentosus* Zikan	Neo	Ecuador (Onore, 2005)
Montezumia	*dimidiata* Saussure	Neo	Ecuador (Onore, 2005)
Parachartergus	*apicalis* (F.)	Neo	Mexico (DeFoliart, 2002)
Polistes	*apicalis* Saussure	Neo	Mexico (Ramos-Elorduy and Pino Moreno, 2002)
Polistes	*bicolor* Lep.	Neo	Ecuador (Onore, 2005)
Polistes	*canadensis* (L.)	Neo	Mexico, Colombia, Venezuela (Ruddle, 1973; Ramos-Elorduy et al., 1998, DeFoliart, 2002)
Polistes	*carnifex* (F.)	Neo	Mexico (Ramos-Elorduy and Pino Moreno, 2002)
Polistes	*chinensis antennalis* Perez	Pal	Japan (Schimitschek, 1968)
Polistes	*deceptor* Schulz	Neo	Ecuador (Onore, 2005)
Polistes	*dorsalis* (F.)	Neo	Mexico (Ramos-Elorduy and Pino Moreno, 2002)
Polistes	*erythrocephalus* Latr.	Neo	Colombia (DeFoliart, 2002)
Polistes	*hebraeus* F.	Afr	Mauritius Madagascar (Kelemu et al., 2015)
Polistes	*instabilis* Sauss.	Neo	México (DeFoliart, 2002)
Polistes	*jokahamae* Radoszkowsky	Pal	Japan (Schimitschek, 1968)
Polistes	*kaibabensis* Hayward	Neo	Mexico (Ramos-Elorduy and Pino Moreno, 2002)
Polistes	*mandarinus* Saussure	Pal	Japan (Schimitschek, 1968)
Polistes	*major* Beauv.	Neo	Mexico (DeFoliart, 2002; Ramos-Elorduy et al., 2008)
Polistes	*occipitalis* Ducke	Neo	Ecuador (Onore, 2005)
Polistes	*pacificus* F.	Neo	Colombia Venezuela (Ruddle, 1973; DeFoliart, 2002)
Polistes	*sagittarius* Saussure	Pal	China (Feng et al., 2010; Chen and Feng, 1999)
Polistes	*snelleni* Saussure	Pal	Japan (Schimitschek, 1968)
Polistes	*stigma* (F.)	Hornet	Or	Laos (Yhoung-Aree and Viwatpanich, 2005)
Polistes	*sulcatus* Smith	Pal	China (Feng and Long, 2010)
Polistes	*testaceicolor* Bequart	Neo	Ecuador (Onore, 2005)
Polistes	*versicolor* (Ol.)	Neo	Colombia, Venezuela (Ruddle, 1973; DeFoliart, 2002)
Polybia	*aequatorialis* Zavattari	Neo	Ecuador (Onore, 2005)

(*Continued*)

Table A1 Documented Information for 1555 Species of Insects and Spiders (cont.)

Order	Family	Genus	Species	Common Name	Faunal	Distribution and References
		Polybia	diguetana Buysson		Neo	Brazil (Costa Neto and Ramos-Elorduy, 2006; Ecuador (Onore, 2005)
		Polybia	dimidiata (Oliv.)		Neo	Brazil (Costa Neto and Ramos-Elorduy, 2006; Ecuador (Onore, 2005)
		Polybia	emaciata Lucas		Neo	Ecuador (Onore, 2005)
		Polybia	flavifrons Smith		Neo	Ecuador (Onore, 2005)
		Polybia	ignobilis (Haliday)		Neo	Colombia, Venezuela (Ruddle, 1973; DeFoliart, 2002)
		Polybia	micans Ducke		Neo	Venezuela (Araújo and Beserra, 2007)
		Polybia	occidentalis (Oliv.)		Neo	Brazil (Costa Neto and Ramos-Elorduy, 2006; Venezuela (Araújo and Beserra, 2007)
		Polybia	occidentalis bohemani Holmgren		Neo	Mexico (DeFoliart, 2002)
		Polybia	occidentalis nigratella Buysson		Neo	Mexico (Ramos-Elorduy and Pino Moreno, 2002)
		Polybia	parvulina Richards		Neo	Mexico (Ramos-Elorduy and Pino Moreno, 2002)
		Polybia	rejecta (F.)		Neo	Colombia (Dufour, 1987; DeFoliart, 2002)
		Polybia	striata (F.)		Neo	Mexico (Ramos-Elorduy and Pino Moreno, 2002)
		Polybia check	pygmaea (F.)		Neo	Mexico (Ramos-Elorduy and Pino Moreno, 2002)
		Pseudopolybia	vespiceps (Sauss.)		Neo	Brazil (Costa Neto and Ramos-Elorduy, 2006)
		Provespa	anomala (Saussure)		Or	Indonesia Kalimantan (Chung, 2010)
		Provespa	barthelemyi (Buysson)		Pal	China (Feng and Long, 2010)
		Synoeca	surinama (L.)		Neo	Mexico (Ramos-Elorduy and Pino Moreno, 2002)
		Synoeca	virginea (F.)		Neo	Ecuador (Onore, 2005)
		Vespa	affinis (L.)		Or	Malaysia (Chung et al., 2002)
		Vespa	affinis indosinensis Pérez		Or	Thailand (Hanboonsong, 2010)
		Vespa	analis F.		Pal	China (Chen et al., 2009; Feng and Long, 2010)
		Vespa	basalis Smith		Pal	China (Chen et al., 2009; Feng and Long, 2010)
		Vespa	bicolor F.		Pal	China (Chen et al., 2009; Feng and Long, 2010)
		Vespa	ducalis Smith		Pal	China (Feng and Long, 2010)
		Vespa	mandarinia japonica Radoszk.	Japanese giant hornet	Pal	Japan (Schimitschek, 1968; DeFoliart, 2002)
		Vespa	mandarinia magnifica Smith		Pal	China (Chen et al., 2009)
		Vespa	mandarinia mandarinia Smith		Pal	China (Feng and Long, 2010)
		Vespa	multimaculata Perkins		Or	Indonesia (Meer Mohr, 1965)

		Vespa	oris L.		Or	India (Singh et al., 2007; Chakravorty et al., 2011, 2013)
		Vespa	soror Buysson		Pal	China (Chen et al., 2009; Feng and Long, 2010)
		Vespa	tropica (L.)	Greater banded hornet	Or	Thailand (DeFoliart, 2002) as Vespa cincta, Indonesia (Meer Mohr, 1965), Malaysia, Sabah (Chung et al., 2002)
		Vespa	variabilis Buysson		Pal	China (Chen and Feng, 1999; Feng et al., 2010)
		Vespa	velutina auraria Smith	Asian predatory wasp	Or	Myanmar (DeFoliart, 2002)
		Vespa			Pal	China (Chen and Feng, 1999), Japan (Schimitschek, 1968)
		Vespula	arenaria (F.)		Nea	N. Am. (DeFoliart, 2002)
		Vespula	flaviceps lewisii (Cameron)	Korean yellow jacket	Pal	Japan (DeFoliart, 2002)
		Vespula	pennsylvanica (Saussure)		Nea	N. Am. (DeFoliart, 2002)
		Vespula	squamosa (Drury)		Neo	Mexico (DeFoliart, 2002; Ramos-Elorduy and Pino Moreno, 2002)
		Vespula	vulgaris (L.)		Or	India (Singh et al., 2007)
Isoptera	Hodotermitidae	Microhodotermes	viator (Latr.)		Afr	South Afr (Bodenheimer, 1951)
	Kalotermitidae	Kalotermes	flavicollis (F.) check		Neo	Brazil (Costa Neto and Ramos-Elorduy, 2006)
	Rhinotermitidae	Coptotermes	formosanus Shiraki		Or	Thailand (DeFoliart, 2002)
		Reticulitermes	flavipes (Kollar) check	Subterranean termite	Pal	China (DeFoliart, 2002; Chen et al., 2009)
		Reticulitermes	tibialis Banks		Or	Thailand (DeFoliart, 2002)
	Termitidae	Labiotermes	labralis (Holmgren)		Nea	N. Am. (DeFoliart, 2002)
		Macrotermes	acrocephalus Ping		Neo	Colombia (Paoletti et al., 2001)
		Macrotermes	amplus (Sjöstedt)		Pal	China (Chen et al., 2009)
		Macrotermes	annandalei (Silvestri)		Afr	D.R. Congo (DeFoliart, 2002)
		Macrotermes	barneyi Light		Pal	China (Chen et al., 2009)
		Macrotermes	bellicosus (Smeathman)		Pal	China (DeFoliart, 2002; Chen et al., 2009)
					Afr	CAR (Roulon-Doko, 1998), Congo (Hoare, 2007), D.R. Congo (Bequaert, 1921), Nigeria (Fasoranti and Ajiboye, 1993), Kenya (Ayieko and Nyambuga, 2009), Cameroun, Ivory Coast, Sao Tomé, Guinée, Togo, Liberia, G. Bissau, Burundi (Kelemu et al., 2015)
		Macrotermes	denticulatus Li et Ping		Pal	China (Chen and Feng, 1999)
		Macrotermes	falciger (Gerstäcker)		Afr	Benin (Riggi et al., 2013), Zambia (Mbata, 1995), Zimbabwe (Chavanduka, 1976), South Afr (Bodenheimer, 1951), Burkina Faso, Burundi (Kelemu et al., 2015), Mali, Togo, Cameroun, Congo, Guinée (Tchibozo, 2015)

(*Continued*)

Table A1 Documented Information for 1555 Species of Insects and Spiders (cont.)

Order	Family	Genus	Species	Common Name	Faunal	Distribution and References
		Macrotermes	gilvus (Hagen)		Or	Thailand (Hanboonsong et al., 2000), Malaysia, Sabah (Chung et al., 2002)
		Macrotermes	jinghongensis Ping et Li		Pal	China (Chen and Feng, 1999)
		Macrotermes	menglongensis Han		Pal	China (Chen and Feng, 1999)
		Macrotermes	michaelseni (Sjöstedt)		Afr	East and Southern Afr (Kelemu et al., 2015)
		Macrotermes	natalensis (Haviland)		Afr	Zimbabwe (DeFoliart, 2002), D.R. Congo (DeFoliart, 2002), Nigeria (Banjo et al., 2006), Cameroun, Congo, CAR, Burundi, South Africa, Malawi (Kelemu et al., 2015)
		Macrotermes	subhyalinus (Rambur)		Afr	Angola (Santos Oliveira et al., 1976), Zambia (Mbata, 1995), Kenya (Kinyuru et al., 2012), Togo, Burundi (Kelemu et al., 2015)
		Macrotermes	vitrialatus (Sjöstedt)		Afr	Zambia (Mbata, 1995)
		Macrotermes	yunnanensis Han		Pal	China (Chen and Feng, 1999)
		Microcerotermes	dubius (Haviland)		Or	Malaysia Sabah (Chung et al., 2002)
		Microcerotermes	serratus (Haviland)		Or	Malaysia Sabah (Chung et al., 2002)
		Nasutitermes	corniger (Motschulsky)		Neo	Venezuela (Araújo and Beserra, 2007)
		Nasutitermes	ephratae (Holmgren)		Neo	Venezuela (Araújo and Beserra, 2007)
		Nasutitermes	macrocephalus (Silvestri)		Neo	Venezuela (Araújo and Beserra, 2007)
		Nasutitermes	surinamensis (Holmgren)		Neo	Venezuela (Araújo and Beserra, 2007)
		Odontotermes	angustignathus Tsai et Chen		Pal	China (Chen and Feng, 1999)
		Odontotermes	annulicornis Xia et Fan		Pal	China (Chen and Feng, 1999)
		Odontotermes	badius (Haviland)		Afr	South Afr (Quin, 1959), Zambia (Silow, 1983), Kenya (Ayieko and Nyambuga, 2009)
		Odontotermes	capensis (DeGeer)		Afr	South Afr (DeFoliart, 2002)
		Odontotermes	conignathus Xia et Fan		Pal	China (Chen and Feng, 1999)
		Odontotermes	feae (Wasmann)		Or	India (DeFoliart, 2002)
		Odontotermes	formosanus Shiraki		Or	India (Wilsanand, 2005)
		Odontotermes	foveafrons Xia et Fan		Pal	China (Chen et al., 2009)
		Odontotermes	gravelyi Silvestri		Pal	China (Chen et al., 2009)
		Odontotermes	hainanensis (Light)		Pal	China (Chen and Feng, 1999)
		Odontotermes	obesus (Rambur)		Or	India (Meyer-Rochow, 2004)
		Odontotermes	yunnanensis Tsai and Chen		Pal	China (Chen et al., 2009)

	Pseudacanthotermes	militaris (Hagen)		Afr	Angola (Silow, 1983), Kenya (Kinyuru et al., 2012)
	Pseudacanthotermes	spiniger (Sjöstedt)		Afr	D.R. Congo (Bequaert, 1921), Zambia (Silow, 1983), Kenya (Kinyuru et al., 2012)
	Syntermes	aculeosus Emerson		Neo	Venezuela (Araújo and Beserra, 2007)
	Syntermes	parallelus Silvestri		Neo	Colombia (Dufour, 1987; DeFoliart, 2002)
	Syntermes	snyderi Emerson		Neo	Colombia (Dufour, 1987; DeFoliart, 2002)
	Syntermes	tanygnathus Constantino		Neo	Colombia (Paoletti et al., 2001)
	Termes check (= Cubitermes)	atrox (Smeatman)		Or	Indonesia (DeFoliart, 2002)
	Termes	destructor DeGeer		Neo	Guyana (DeFoliart, 2002)
	Termes	flavicolle Perty		Or	Indonedia (DeFoliart, 2002)
	Termes	mordax L./	Mordax Smeatman	Neo	Brazil (DeFoliart, 2002)
	Termes check	mordax L./	Mordax Smeatman	Or	Indonesia (DeFoliart, 2002)
	Termes check	sumatranum		Or	Indonesia (DeFoliart, 2002)
Lepidoptera	Bombyx	mori (L.)		Neo	Mexico (Ramos-Elorduy, 2009; Ramos-Elorduy et al., 2011)
Bombycidae				Or	Thailand, Myanmar, Vietnam, India (DeFoliart, 2002; Hanboonsong, 2010), Laos (Yhoung-Aree and Viwatpanich, 2005)
				Pal	China, Japan, Korea (DeFoliart, 2002; Chen and Feng, 1999)
Brahmaeidae	Dactyloceras	lucina (Drury)		Afr	D.R. Congo (Malaisse, 2005), Zambia, South Africa, Cameroun, Congo, Angola, Gabon, Sierra Leone, Sao Tomé, E. Guinée (Kelemu et al., 2015)
Castniidae	Castnia	daedalus Cramner		Neo	Ecuador (Onore, 2005)
	Castnia	licoides Boisduval		Neo	Ecuador (Onore, 2005)
	Castnia	licus Drury		Neo	Ecuador (Onore, 2005)
	Castnia	chelone Hopfter		Neo	Mexico (Ramos-Elorduy et al., 2011)
	Catoxophylla	cyanauges Turner		Aus	Australia (DeFoliart, 2002)
Cossidae	Comadia	redtenbacheri (Hammerschmidt)	Red tequila worm	Neo	Mexico (Ramos-Elorduy, 2006)
	Endoxyla	amphiplecta (Turner)		Aus	Australia (DeFoliart, 2002)
	Endoxyla	biarpiti (Tindale)		Aus	Australia (DeFoliart, 2002)
	Endoxyla	cinereus (Tepper)		Aus	Australia (DeFoliart, 2002)
	Endoxyla	encalypti H.-S.		Aus	Australia (DeFoliart, 2002)
	Endoxyla	leucomochla (Turner)		Aus	Australia (DeFoliart, 2002)

(Continued)

Table A1 Documented Information for 1555 Species of Insects and Spiders (cont.)

Order	Family	Genus	Species	Common Name	Faunal	Distribution and References
		Endoxyla	lituratus (Don.)		Aus	Australia (DeFoliart, 2002)
		Holcocerus	vicarius Walker		Pal	Japan (Schimitschek, 1968)
		Zeuzera	pyrina (L.)		Pal	Japan (Schimitschek, 1968)
	Crambidae	Brihaspa	atrostigmella Moore		Or	Vietnam (DeFoliart, 2002)
		Omphisa	fuscidentalis Hampson	Bamboo borer	Or	Thailand (Hanboonsong et al., 2000; Hanboonsong, 2010), Laos (Hanboonsong and Durst, 2014)
	Erebidae	Amastus	ochraceator (Walker)		Neo	Mexico (Ramos-Elorduy et al., 2008, 2011)
		Amerila	madagacariensis (Boisduval)		Afr	Madagascar (Decary, 1937)
		Arctia	caja americana Harris		Nea	N. Am. (DeFoliart, 2002)
		Diacrisia	obliqua (Walker)		Or	India (DeFoliart, 2002)
		Elysius	superba (Druce)		Neo	Mexico (Ramos-Elorduy et al., 2008, 2011)
		Estigmene	acrea (Drury)		Neo	Mexico (Ramos-Elorduy et al., 2008, 2011)
		Pelochyta	cervina (Edwards)		Neo	Mexico (Ramos-Elorduy et al., 2008, 2011)
		Rhypopteryx	poecilanthes Collenette	Slug moth	Afr	D.R. Congo (Malaisse and Parent, 1980; Zimbabwe, Zambia (DeFoliart, 2002)
		Syntomis	phaegea (L.)		Pal	Italy (Zagrobelny et al., 2009)
	Eupterotidae	Hemijana	variegata Rothschild		Afr	South Afr (Egan et al., 2014)
		Striphnopteryx	edulis (Boisduval)		Afr	Southern Afr (Bergier, 1941)
	Gelechiidae	Pectinophora	gossypiella (Saunders)	Pink bollworm	Pal	China (DeFoliart, 2002; Chen and Feng, 1999)
	Geometridae	Achlyodes	pallida (Felder)		Neo	Mexico (Ramos-Elorduy et al., 2011)
		Acronyctodes	mexicaria (Walker)		Neo	Mexico (Ramos-Elorduy et al., 2011)
		Biston	marginata Shiraki		Pal	China (Chen and Feng, 1999)
		Panthera	pardalaria Hübner		Neo	Mexico (Ramos-Elorduy et al., 2011)
	Hepialidae	Abantiades	marcidus Tindale	Bardi grub	Aus	Australia (DeFoliart, 2002)
		Hepialus	cingulatus Yang and Zhang		Pal	China (Chen and Feng, 1999)
		Hepialus	davidi Poujade		Pal	China (Chen and Feng, 1999)
		Hepialus	dongyuensis Liang		Pal	China (Chen and Feng, 1999)
		Hepialus	ferrugineus Li Yang et Shen		Pal	China (Chen and Feng, 1999)
		Hepialus	ganna Hübner		Pal	China (Chen and Feng, 1999)
		Hepialus	jinshaensis Yang		Pal	China (Chen and Feng, 1999)
		Hepialus	litangensis Liang		Pal	China (Chen and Feng, 1999)
		Hepialus	luquensis Yang and Yang		Pal	China (Chen and Feng, 1999)
		Hepialus	macilentus Eversmann		Pal	China (Chen and Feng, 1999)

Hepialus	maikamensis Yang Li and Shen		Pal	China (Chen and Feng, 1999)
Hepialus	nebulosus Alphéraky		Pal	China (Chen and Feng, 1999)
Hepialus	pratensis Yang		Pal	China (Chen and Feng, 1999)
Hepialus	varians Staudinger		Pal	China (Chen and Feng, 1999)
Hepialus	xunhuaensis Yang and Yang		Pal	China (Chen and Feng, 1999)
Hepialus	yeriensis Liang		Pal	China (Chen and Feng, 1999)
Hepialus	yunnanensis Yang and Li		Pal	China (Chen and Feng, 1999)
Hepialus	zhongzhiensis Liang		Pal	China (Chen and Feng, 1999)
Hepialus check	albipictus Yang		Pal	China (Chen and Feng, 1999)
Hepialus check	yuloangensis Liang		Pal	China (Chen and Feng, 1999)
Phassus	triangularis Edwards		Neo	Mexico (DeFoliart, 2002; Ramos-Elorduy et al., 2008)
Phassus	trajesa L.		Neo	Mexico (DeFoliart, 2002)
Thitarodes (Hepialus)	altaicola Wang		Pal	China (Chen and Feng, 1999)
Thitarodes (Hepialus)	armoricanus (Oberthür)		Pal	China (Chen and Feng, 1999; DeFoliart, 2002)
Thitarodes (Hepialus)	baimaensis Liang et al		Pal	China (Chen and Feng, 1999)
Thitarodes (Hepialus)	deqinensis Liang et al		Pal	China (Chen and Feng, 1999)
Thitarodes (Hepialus)	gonggaensis Fu and Huang		Pal	China (Chen and Feng, 1999)
Thitarodes (Hepialus)	kangdingensis Chu and Wang		Pal	China (Chen and Feng, 1999)
Thitarodes (Hepialus)	kangdingroides Chu and Wang		Pal	China (Chen and Feng, 1999)
Thitarodes (Hepialus)	lijiangensis Chu and Wang		Pal	China (Chen and Feng, 1999)
Thitarodes (Hepialus)	meiliensis Liang		Pal	China (Chen and Feng, 1999)
Thitarodes (Hepialus)	menyuanicus Chu and Wang		Pal	China (Chen and Feng, 1999)
Thitarodes (Hepialus)	oblifurcus Chu and Wang		Pal	China (DeFoliart, 2002)
Thitarodes (Hepialus)	renzhiensis Yang		Pal	China (Chen and Feng, 1999)
Thitarodes (Hepialus)	sichuanus Chu and Wang		Pal	China (Chen and Feng, 1999)
Thitarodes (Hepialus)	yunlongensis Chu and Wang		Pal	China (Chen and Feng, 1999)
Thitarodes (Hepialus)	yushuensis Chu and Wang		Pal	China (Chen and Feng, 1999)
Thitarodes (Hepialus)	zhangmoensis Chu and Wang		Pal	China (Chen and Feng, 1999)
Thitarodes (Hepialus)	zhayuensis Chu and Wang		Pal	China (Chen and Feng, 1999)
Trictena	argyrosticha Turner		Aus	Australia (DeFoliart, 2002)
Trictena	atripalpis Walker	Bardi grub	Aus	Australia (DeFoliart, 2002)

(*Continued*)

Table A1 Documented Information for 1555 Species of Insects and Spiders (cont.)

Order	Family	Genus	Species	Common Name	Faunal	Distribution and References
	Hesperiidae	Aegiale	hesperiaris (Walker)	White tequila worm	Neo	Mexico (Ramos-Elorduy et al., 2011)
		Ancistroides	nigrita (Latr.)	Chocolate demon	Or	Malaysia (Chung, 2010), Malaysia, Sabah (Chung et al., 2002)
		Erionota	torus Evans		Pal	China (Chen and Feng, 1999)
		Erionota	thrax (L.)	Banana skipper	Or	Malaysia (Chung, 2010), Malaysia, Sabah (Chung et al., 2002), Thailand (Hanboonsong et al., 2000; Hanboonsong, 2010), Indonesia (Meer Mohr, 1965), Laos (Yhoung-Aree and Viwatpanich, 2005)
		Megathymus	yuccae Boisd. and LeC.	Giant skippers	Nea	N. Am. (DeFoliart, 2002)
		Coeliades	libeon (Druce)		Afr	Congo (Hoare, 2007), D.R. Congo (DeFoliart, 2002)
	Hyblaeidae	Hyblaea	puera (Cramer)	Teak defoliator	Or	Indonesia (DeFoliart, 2002; Lukiwati, 2010)
	Lasiocampidae	Bombycomorpha	pallida Distant	Pepper tree moth	Afr	South Afr (Quin, 1959)
		Borocera	cajani Vinson	Wild silkworm	Afr	Madagascar (Decary, 1937)
		Borocera	madagascariensis Boisduval		Afr	Madagascar (Decary, 1937)
		Catalebeda	jamesoni Bethune-Baker	Jameson's cream spot	Afr	Zambia (Silow, 1976)
		Dendrolimus	houi Lajonquiere		Pal	China (Chen and Feng, 1999)
		Dendrolimus	kikuchii Matsumura		Pal	China (Chen and Feng, 1999)
		Dendrolimus	punctatus (Walker)		Pal	China (Chen and Feng, 1999)
		Dendrolimus	punctatus wenshanensis Ysai and Liu		Pal	China (Chen and Feng, 1999)
		Dendrolimus	spectabilis (Butler)		Pal	Japan (Schimitschek, 1968)
		Eutachyptera	psidii (Sallé)		Neo	Mexico (Ramos-Elorduy et al., 2011)
		Gonometa	postica Walker	Dark chopper	Afr	South Afr (Quin, 1959), Zambia (Silow, 1976), Botswana (Yen, 2015)
		Mimopacha	aff. knoblauchi Dew.		Afr	Central Afr (Silow, 1976)
		Pachymeta	robusta Aurivillius	Msasa moth	Afr	Zimbabwe (Bodenheimer, 1951)
		Pachypasa	bilinea Walker	Twin line lappet	Afr	Zambia (Silow, 1976)
	Limacodidae	Cania	bilinea (Walker)		Pal	China (Chen and Feng, 1999)
		Cnidocampa	flavescens (Walker)		Pal	Japan (Schimitschek, 1968)
		Hadraphe	ethiopica (Bethune-Baker)		Afr	Zambia (Malaisse, 2005)
		Thosea	sinensis (Walker)		Pal	China (Chen and Feng, 1999)
	Lycaenidae	Liphyra	brassolis Westw.		Or	Thailand (Eastwood et al., 2010)
Noctuidae		Agrotis	infusa (Boisd.)	Bogong moth	Aus	Australia (DeFoliart, 2002; Yen, 2015)

	Agrotis	ypsilon (Hufnagel)		Pal	China (Chen and Feng, 1999)
	Ascalapha	odorata (L.)		Neo	Mexico (Ramos-Elorduy et al., 2011)
	Busseola	fusca (Fuller)	African maize stalk borer	Afr	Zambia (DeFoliart, 2002)
	Cerra	sevorsa (Grote)		Neo	Mexico (Ramos-Elorduy et al., 2011)
	Helicoverpa	armigera armigera (Hübner)	Afro-Asian bollworm	Afr	Zambia (DeFoliart, 2002)
	Helicoverpa	zea (Boddie)	American bollworm	Nea	N. Am. (DeFoliart, 2002)
	Hydrillodes	lentalis Guénée		Neo	Mexico (Ramos-Elorduy et al., 2011)
	Homoncocnemis	fortis (Grote)		Pal	China (DeFoliart, 2002)
	Latebraria	amphipyroides Guenée		Nea	N. Am. (DeFoliart, 2002)
	Mocis	punctularis Hübner		Neo	Mexico (Ramos-Elorduy et al., 2011)
	Nyodes	prasinodes Prout		Neo	Colombia (DeFoliart, 2002)
	Sphingomorpha	chlorea (Cramer)	Sundowner moth	Afr	D.R. Congo (Malaisse and Parent, 1980)
	Spodoptera	exempta (Walker)	African army worm	Afr	Zambia (Silow, 1976)
	Spodoptera	exigua (Hübner)	Lesser army worm	Afr	Zambia (Mbata, 1995)
	Spodoptera	frugiperda (Smith)		Afr	Zambia (Mbata, 1995)
				Neo	Mexico (Ramos-Elorduy et al., 2011)
			Black army worm	Nea	N. Am. (DeFoliart, 2002)
	Thysania	agrippina (Cramer)		Neo	Colombia, Mexico (Ramos-Elorduy 2006; DeFoliart, 2002; Ruddle, 1973)
				Neo	Mexico (Ramos-Elorduy et al., 2011)
Notodontidae	Anaphe	infracta Walsingham		Afr	Tanzania (Bodenheimer, 1951; Harris, 1940), Zambia (Silow, 1976)
	Anaphe	reticulata Walker		Afr	Nigeria (Banjo et al., 2006)
	Anaphe	venata Butler	African silkworm	Afr	Nigeria (Ashiru, 1988), Zambia (Silow, 1976), CAR (Hoare, 2007), Ivory Coast, Sierra Leone, Guinée, Liberia, G. Bissau (Kelemu et al., 2015)
	Antheua	insignata Gaede		Afr	D.R. Congo (Hoare, 2007)
	Cerurina	cf marshalii (Hampson)		Afr	CAR (Malaisse, 2005)
	Drapetides	uniformis (Swinhoe)		Afr	D.R. Congo (Hoare, 2007)
	Elaphrodes	lactea (Gaede)		Afr	D.R. Congo (Hoare, 2007)
	Epanaphe	carteri (Walsingham)		Afr	D.R. Congo (Malaisse, 2005), Zambia, Angola, Gabon, Sierra Leone, Sao Tomé, E. Guinée (Kelemu et al., 2015)
	Hypsoides	diego Coquerel		Afr	Madagascar (Decary, 1937)
	Hypsoides	radama Coquerel		Afr	Madagascar (Decary, 1937)
	Leucodonta	bicoloria (D. and S.)		Pal	China (Chen and Feng, 1999)

(Continued)

Appendix 353

Table A1 Documented Information for 1555 Species of Insects and Spiders (cont.)

Order	Family	Genus	Species	Common Name	Faunal	Distribution and References
		Nephele	comma Hopffer		Afr	Zambia (Silow, 1976)
		Notodonta	dembowskii Oberthuer		Pal	China (Chen and Feng, 1999)
		Ochrogaster	lunifer H.-S.		Aus	Australia (Yen, 2005)
		Phalera	assimilis (Bremer and Grey)		Pal	China (Chen and Feng, 1999)
		Phalera	bucephala (L.)		Pal	China (Chen and Feng, 1999)
		Rhenea	mediata Walker		Afr	D.R. Congo (Malaisse and Parent, 1980)
		Semidonta	biloba (Oberthuer)		Pal	China (Chen and Feng, 1999)
	Nymphalidae	Anartia	fatima (F.)		Neo	Ecuador (Onore, 2005)
		Brassolis	astyra Godart		Neo	Ecuador (Onore, 2005)
		Brassolis	sophorae (L.)		Neo	Brazil (Costa Neto and Ramos-Elorduy, 2006; Ecuador (Onore, 2005)
		Caligo	memnon Felder		Neo	Mexico (Ramos-Elorduy and Pino Moreno, 2002)
		Chlosyne	lacinia (Geyer)		Neo	México (Ramos-Elorduy et al., 2011)
		Cymothoe	aramis Hewitson		Afr	CAR (Hoare, 2007)
		Cymothoe	caenis (Drury)		Afr	CAR D. R. Congo (Latham, 2003; Hoare, 2007)
		Danaus	gilippus thersippus (Bates)		Neo	Mexico (Ramos-Elorduy et al., 2011)
		Danaus	plexippus (L.)		Neo	Mexico (Ramos-Elorduy et al., 2011)
		Junonia	lavinia Cramer		Neo	Mexico (Ramos-Elorduy and Pino Moreno, 2002)
		Nymphalis	antiopia (L.)		Neo	Mexico (Ramos-Elorduy and Pino Moreno, 2002)
		Panacea	prola Doubleday		Neo	Ecuador (Onore, 2005)
		Pareuptychia	metaleuca (Boiduval)		Neo	Mexico (Ramos-Elorduy et al., 2011)
		Vanessa	annabella (Field)		Neo	Mexico (Ramos-Elorduy et al., 2011)
		Vanessa	virginiensis (Drury)		Neo	Mexico (Ramos-Elorduy et al., 2011)
	Oecophoridae	Linoclostis	gonatias Meyrick		Pal	China (Chen and Feng, 1999)
	Papilionidae	Papilio	machaon L.		Pal	China (Chen and Feng, 1999)
		Papilio	multicaudata (Kirby)	Two-tailed swallowtail	Neo	Mexico (Ramos-Elorduy et al., 2011)
		Papilio	polyxenes F.	Black swallowtail	Neo	Mexico (Ramos-Elorduy and Pino Moreno, 2002)
		Protographium	philolaus (Boisduval)		Neo	Mexico (Ramos-Elorduy et al., 2011)
	Pieridae	Catasticta	flisa (H.-S.)		Neo	Mexico (Ramos-Elorduy et al., 2011)
		Catasticta	nimbice (Boisduval)		Neo	Mexico (Ramos-Elorduy et al., 2011)
		Catasticta	teutila (Doubleday)		Neo	Mexico (Ramos-Elorduy et al., 2011)
		Catopsilia	(prob. pomona L.)		Or	Indonesia (Meer Mohr, 1965)
		Eucheira	socialis Westw.		Neo	Mexico (DeFoliart, 2002)

	Eurema	(prob. hecabe L.)		Or	Indonesia (Meer Mohr, 1965)
	Eurema	lisa (B. and LeC.)		Neo	Mexico (Ramos-Elorduy and Pino Moreno, 2002)
	Eurema	salome jamapa (Reakirt)		Neo	Mexico (Ramos-Elorduy et al., 2011)
	Leptophobia	aripa (Boisduval)		Neo	Mexico (Ramos-Elorduy et al., 2011)
	Phoebis	agarithe (Boisduval)		Neo	Mexico (Ramos-Elorduy et al., 2011)
	Phoebis	philea (L.)		Neo	Mexico (Ramos-Elorduy et al., 2011)
	Phoebis	sennae marcellina (Cramer)		Neo	Mexico (Ramos-Elorduy et al., 2011)
	Pieris	rapae (L.)		Pal	China (Chen and Feng, 1999)
	Pontia	protodice (B. and LeC.)		Neo	Mexico (Ramos-Elorduy et al., 2011)
Psychidae	Clania	moddermanni Heylaerts		Afr	Equatorial Afr (Bergier, 1941)
	Deborrea	malgassa Druce		Afr	Madagascar (Decary, 1937)
	Eumeta	rougeoti Bourgogne		Afr	D.R. Congo (DeFoliart, 2002)
Pyralidae	Aglossa	dimidiatus (Haworth)		Pal	China (Chen and Feng, 1999)
	Chilo	simplex (Butler)		Pal	Japan (Schimitschek, 1968)
	Laniifera	cyclades (Druce)		Neo	Mexico (DeFoliart, 2002)
	Myelobia (Morpheis)	smerintha Hubner	Bamboo borer	Neo	Brazil (Costa Neto and Ramos-Elorduy, 2006)
	Omphisa	fuscidentalis (Hampson)		Pal	China (Chen and Feng, 1999)
	Ostrinia	furnalis Guence		Pal	China (Chen and Feng, 1999)
	Schoenobius	incertellus (Walker)		Pal	Japan (Schimitschek, 1968)
Saturniidae	Actias	luna (L.)		Neo	Mexico (Ramos-Elorduy et al., 2011)
	Actias	truncatipennis (Sonthonnax)		Neo	Mexico (Ramos-Elorduy et al., 2011)
	Antheraea	assamensis Helfer	Muga silk-moth	Or	India (Chakravorty et al., 2013)
	Antheraea	paphia (L.)	Indian tussah silkworm	Or	India (DeFoliart, 2002)
	Antheraea	pernyi (G.-M.)	Chinese tussah moth	Pal	China (DeFoliart, 2002; Chen and Feng, 1999), Japan (Mitsuhashi, 1997)
	Antheraea	polyphemus (Cramer)		Neo	Mexico (Ramos-Elorduy and Pino Moreno, 2002)
	Antheraea	polyphemus mexicana Hoffmann		Neo	Mexico (Ramos-Elorduy et al., 2011)
	Antheraea	roylei Moore		Or	India (DeFoliart, 2002)
	Antheraea	yamamai (Gérin-M.)	Japanese oak silkworm	Pal	Japan (Mitsuhashi, 1997)
	Antherina	suraka (Boisd.)		Afr	Madagascar (DeFoliart, 2002)
	Argema	mimosae (Boidd.)	African moonmoth	Afr	South Afr (Malaisse, 2005)
	Arsenura	armida (Cramer)		Neo	Mexico (Ramos-Elorduy et al., 2008, 2011)
	Arsenura	polyodonta (Jordan)		Neo	Mexico (Ramos-Elorduy et al., 2011)
	Athletes	gigas (Sonthonnax)		Afr	D.R. Congo (Malaisse and Parent, 1980)

(Continued)

Table A1 Documented Information for 1555 Species of Insects and Spiders (cont.)

Order	Family	Genus	Species	Common Name	Faunal	Distribution and References
		Athletes	semialba (Sonthonnax)		Afr	D.R. Congo (Malaisse and Parent, 1980)
		Bunaea	alcinoë (Stol)	Common emperor	Afr	Cameroun, D.R. Congo, Gabon, Zambia (Silow, 1976; Latham, 2003), South Africa, Zimbabwe, Tanzania (DeFoliart, 2002), CAR (Hoare, 2007), Nigeria (Agbidye et al., 2009), Congo (Kelemu et al., 2015)
		Bunaea	aslauga Kirby		Afr	Tanzania (Malaisse, 2005)
		Bunaeopsis	aurantiaca (Rothschild)		Afr	D.R. Congo (Hoare, 2007), Zambia (Silow, 1976), Congo (DeFoliart, 2002)
		Caio	championi (Druce)		Neo	Mexico (Ramos-Elorduy et al., 2011)
		Caio	richardsoni (Druce)		Neo	Mexico (Ramos-Elorduy and Pino Moreno, 2002)
		Callosamia	promethea Drury		Neo	Mexico (Ramos-Elorduy and Pino Moreno, 2002)
		Cinabra	hyperbius (Westwood)	Banded emperor moth	Afr	D.R. Congo (Hoare, 2007), Zambia (Silow, 1976)
		Cirina	forda (Westwood)	Pallid emperor moth	Afr	Burkina Faso, D. R. Congo, (Latham, 2003; Hoare, 2007), Zambia (Mbata et al., 2002), Congo (DeFoliart, 2002), Nigeria (Banjo et al., 2006), CAR (Hoare, 2007), South Afr, Botswana, Mozambique, Namibia, Ghana, Togo, Tchad (Kelemu et al., 2015), Cameroun (Tchibozo, 2015)
		Coloradia	pandora Blake	Pandora moth	Nea	N. Am. California (Blake and Wagner, 1987; DeFoliart, 2002)
		Cricula	trifenestrata (helfer)		Or	Indonesia (Roepke, 1952)
		Eacles	aff. ormondei yacatanensis	(Lemaire)	Neo	Mexico (Ramos-Elorduy et al., 2011)
		Epiphora	bauhiniae (Guérin)		Afr	D.R. Congo (Malaisse, 2005)
		Eriogyna	pyretorum (Westw.)		Pal	China (DeFoliart, 2002)
		Goodia	kuntzei (Dewitz)		Afr	D.R. Congo (Malaisse and Parent, 1980), Zimbabwe (DeFoliart, 2002)
		Gynanisa	ata Strand		Afr	D.R. Congo (Malaisse and Parent, 1980), Malawi, Zambia (DeFoliart, 2002), South Sudan (Kelemu et al., 2015)
		Gynanisa	maja (Klug)		Afr	Malawi (Munthali and Mughogho, 1992), Namibia (Oberprieler, 1995), South Africa (Quin, 1959), Zambia (Mbata et al., 2002)
		Heniocha	apollonia (Cramer)		Afr	South Africa (Malaisse, 2005)
		Heniocha	dyops (Maassen)		Afr	Southern Africa (Marais, 1996)
		Heniocha	marnois (Rogenhofer)		Afr	Southern Africa (Marais, 1996; Malaisse, 2005)
		Holocerina	agomensis (Karsch)		Afr	Zambia (Silow, 1976)

Hyalophora	euryalus (Boisd.)		Nea	N. Am. (DeFoliart, 2002)
Hylesia	coinopus Dyar		Neo	Mexico (Ramos-Elorduy and Pino Moreno, 2002)
Hylesia	frigida Schaus		Neo	Mexico (Ramos-Elorduy et al., 2011)
Imbrasia	alopia (Westwood)		Afr	D.R Congo (Hoare, 2007), Congo, Guinée (Tchibozo, 2015)
Imbrasia	anthina (Karsch)		Afr	D.R Congo (Hoare, 2007), Congo (Tchibozo, 2015)
Imbrasia	belina (Westwood)		Afr	Malawi (Munthali and Mughogho, 1992), South Africa (Quin, 1959), Southern Africa (Oberprieler, 1995), Zambia (Mbata et al., 2002), Zimbabwe (Chavanduka, 1976), Botswana (DeFoliart, 2002), D.R. Congo (Kelemu et al., 2015)
Imbrasia	butyrospermi Vuillot		Afr	West, Central and Southern Africa (Kelemu et al., 2015), D.R. Congo (Malaisse and Parent, 1980), Mali (Bergier, 1941), Nigeria (Fasoranti and Ajiboye, 1993), South Africa (Quin, 1959), Southern Africa (Oberprieler, 1995), Zambia (Silow, 1976), Zimbabwe (Chavanduka, 1976), Burkina Faso, Ghana (Kelemu et al., 2015)
Imbrasia	cytherea (Fabr.)	Pine tree emperor moth	Afr	Zambia (Silow, 1976)
Imbrasia	dione (Fabr.)		Afr	D.R. Congo (Hoare, 2007)
Imbrasia	epimethea (Drury)		Afr	Congo (Bani, 1995; Tchibozo, 2015), D.R. Congo (Malaisse and Parent, 1980), CAR (Hoare, 2007), Zambia (Mbata et al., 2002), Zimbabwe (Gelfand, 1971; Weaving, 1973), South Africa (DeFoliart, 2002), Cameroun (Kelemu et al., 2015), Guinée (Tchibozo, 2015)
Imbrasia	ertli Rebel		Afr	Angola (Santos Oliveira et al., 1976), D.R. Congo (Latham, 1999; Hoare, 2007), Southern Africa (Oberprieler, 1995), Zimbabwe (DeFoliart, 2002), Zambia, Cameroun, Congo, C.A.R, Botswana (Kelemu et al., 2015)
Imbrasia	hecate Rougeot		Afr	D. R. Congo (Malaisse and Parent, 1980)
Imbrasia	macrothyris (Rothschild)		Afr	D. R. Congo (Malaisse and Parent, 1980)
Imbrasia	nictitans (F.)		Afr	D. R. Congo (Malaisse, 2005)
Imbrasia	obscura Butler		Afr	Congo (Bani, 1995), D.R. Congo (Hoare, 2007), Cameroun (Tchibozo, 2015)
Imbrasia	petiveri Guérin-M.		Afr	D. R. Congo (Hoare, 2007), Congo (Tchibozo, 2015)
Imbrasia	rectilineata (Sonthoanax)		Afr	D. R. Congo (Malaisse and Parent, 1980)
Imbrasia	rhodina Rothsch.		Afr	D.R. Congo (Hoare, 2007)
Imbrasia	richelmanni Weymer		Afr	D.R. Congo (DeFoliart, 2002)

(Continued)

Table A1 Documented Information for 1555 Species of Insects and Spiders (cont.)

Order	Family	Genus	Species	Common Name	Faunal	Distribution and References
		Imbrasia	truncata Aurivilius		Afr	Congo (Bani, 1995), D.R. Congo, CAR (Hoare, 2007)
		Imbrasia	tyrrhea (Cramer)		Afr	Namibia (Oberprieler, 1995) D.R. Congo (Hoare, 2007)
		Imbrasia	wahlbergii Boisd.		Afr	D.R. Congo (Hoare, 2007)
		Imbrasia	zambesina (Walker)		Afr	D. R. Congo (Malaisse and Parent, 1980), Zambia (Mbata et al., 2002), South Africa (DeFoliart, 2002)
		Latebraria	amphipyroides Guenée		Neo	Mexico (DeFoliart, 2002)
		Lobobunaea	angasana (Westwood)		Afr	D. R. Congo (Malaisse and Parent, 1980), Zambia (Silow, 1976)
		Lobobunaea	christyi (Sharpe)		Afr	Zambia (DeFoliart, 2002)
		Lobobunaea	goodi (Holland)		Afr	D.R. Congo (Hoare, 2007)
		Lobobunaea	phaedusa (Drury)		Afr	D.R. Congo (Hoare, 2007)
		Lobobunaea	saturnus (F.)		Afr	D.R. Congo, Zambia (DeFoliart, 2002; Hoare, 2007)
		Melanocera	menippe (Westwood)		Afr	Gabon (Bergier, 1941), South Africa (DeFoliart, 2002)
		Melanocera	nereis (Rothchild)		Afr	D.R. Congo (Hoare, 2007)
		Melanocera	parva (Rothchild)		Afr	D. R. Congo (Malaisse and Parent, 1980), Zambia (Silow, 1976)
		Micragone	ansorgei (Rothchild)		Afr	D. R. Congo (Malaisse, 1997), Zambia (Silow, 1976)
		Micragone	cana (Aurivillius)		Afr	D. R. Congo (Malaisse and Parent, 1980), South Africa (DeFoliart, 2002)
		Micragone	herilla (Westwood)		Afr	Cameroun (Bodenheimer, 1951)
		Nudaurelia	eblis (Strecker)		Afr	D.R. Congo (Latham, 2003)
		Nudaurelia	melanops (Bouvier)		Afr	D.R. Congo, CAR, Congo, Guinée (Tchibozo, 2015)
		Paradirphia	fumosa (Felder)		Neo	Mexico (Ramos-Elorduy et al., 2011)
		Paradirphia	hoegei (Druce)		Neo	Mexico (Ramos-Elorduy et al., 2011)
		Pseudantheraea	discrepans (Butler)		Afr	D.R. Congo (Silow, 1976), Congo (DeFoliart, 2002), CAR (Hoare, 2007)
		Pseudobunaea	irius (F.)		Afr	Namibia (Malaisse, 2005)
		Pseudodirphia	mexicana (Bouvier)		Neo	Mexico (Ramos-Elorduy et al., 2008, 2011)
		Rohaniella	pygmaea (Maassen and Weyding)		Afr	Zimbabwe (DeFoliart, 2002)
		Samia	cynthia (Drury)	Cynthia moth	Or	India (DeFoliart, 2002)

	Samia	ricini (Boisd.)	Eri silk-moth	Or	India (DeFoliart, 2002; Chakravorty et al., 2013)
	Saturnia	marchii check		Afr	Gabon (Bergier, 1941)
	Tagoropsis	flavinata (Walker)		Afr	D.R. Congo (Hoare, 2007)
	Tagoropsis	natalensis Felder		Afr	D.R. Congo (Malaisse, 2005)
	Urota	sinope (Westwood)		Afr	D.R. Congo (Malaisse and Parent, 1980; Hoare, 2007), Gabon (Bergier, 1941), Southern Africa, South Africa (DeFoliart, 2002), Zimbabwe, Botswana, Mozambique, Namibia (Kelemu et al., 2015)
	Usta	terpsichore (Maassen and Weymer)		Afr	Angola (Santos Oliveira et al., 1976), D.R. Congo (Malaisse and Parent, 1980; Hoare, 2007)
	Usta	wallengrennii (Felder and Felder)		Afr	Namibia (Oberprieler, 1995)
Sesiidae	Paranthrene	regalis (Butler)		Pal	Japan (Schimitschek, 1968)
	Synanthedon	cardinalis (Dampf)		Neo	Mexico (Ramos-Elorduy et al., 2011)
Sphingidae	Acherontia	atropos (L.)	Dead's head hawk moth	Afr	D.R. Congo (Hoare, 2007)
	Acherontia	lachesis F.		Or	Indonesia (Meer Mohr, 1965)
	Acherontia	styx Westw.	Tomato hornworm	Pal	Japan (Schimitschek, 1968)
				Or	Indonesia (Meer Mohr, 1965)
	Agrius	convolvuli (L.)	Convolvulus hawk moth	Afr	Botswana (Nonaka, 2009), South Africa (Quin, 1959), Zambia (Silow, 1976), Zimbabwe (DeFoliart, 2002)
				Pal	Japan (Schimitschek, 1968)
				Or	Indonesia (Meer Mohr, 1965)
	Clanis	bilineata (Walker)		Pal	China (DeFoliart, 2002)
	Clanis	bilineata tsingtauica Mell.		Pal	China (Chen and Feng, 1999)
	Clanis	deucalion (Walker)		Pal	China (Chen and Feng, 1999)
	Cocytius	antaeus (Drury)		Neo	Mexico (Ramos-Elorduy et al., 2011)
	Coelonia	fulvinotata (Butler)		Afr	CAR (Malaisse, 2005)
	Coenotes	eremophilae (Lucas)		Aus	Australia (DeFoliart, 2002)
	Deilephila	elpenor lewesi (Butler)	Elephant hawk moth	Pal	Japan (Schimitschek, 1968)
	Erinnyis	ello (L.)		Neo	Venezuela (Araújo and Beserra, 2007)
	Hippotion	celerio (L.)		Or	Malaysia Sabah (Chung et al., 2002)
	Hippotion	eson (Cramer)		Afr	D.R. Congo (Malaisse, 2005)
	Hyles	livornicoides Lucas		Aus	Australia (DeFoliart, 2002)
	Hyles	lineata (F.)	White lined sphinx	Nea	N. Am. (DeFoliart, 2002)
				Neo	Mexico (DeFoliart, 2002)

(*Continued*)

Table A1 Documented Information for 1555 Species of Insects and Spiders (cont.)

Order	Family	Genus	Species	Common Name	Faunal	Distribution and References
		Lophostethus	demolini (Angas)		Afr	CAR (Malaisse, 2005)
		Macroglossum	stellatarum (L.)	Hummingbird hawk moth	Pal	Japan (Schimitschek, 1968)
		Manduca	sexta (L.)	Tabacco hornworm	Nea	N. Am. (DeFoliart, 2002)
		Pachilia	ficus (L.)		Neo	Mexico (Ramos-Elorduy et al., 2008, 2011)
		Platysphinx	stigmatica (Mabille)		Neo	Mexico (Ramos-Elorduy et al., 2011)
					Afr	Tropical Afr (Malaisse, 2005) D.R. Congo, Zambia, Congo, CAR, Sierra Leone, Sao Tomé, E. Guinée, Rwanda, Burundi (Kelemu et al., 2015)
		Psilogramma	increta (Walker)		Pal	Japan (Schimitschek, 1968)
		Smerinthus	planus Walker		Pal	China (Chen and Feng, 1999)
		Theretra	nessus (Drury)		Pal	Japan (Schimitschek, 1968)
		Theretra	oldenlandiae (Fabr.)		Pal	Japan (Schimitschek, 1968)
	Tortricidae	Leguminivora	glycinivorella (Mutsumara)	Soybean pod borer	Pal	China (Chen and Feng, 1999)
	Uranidae	Nyctalemon	patroclus goldiei Druce		Aus	Papua (Ramanday and van Mastricht, 2010)
	Zygaenidae	Zygaena	ephialtes (L.)		Pal	Italy (Zagrobelny et al., 2009)
		Zygaena	transalpina (Esper)		Pal	Italy (Zagrobelny et al., 2009)
Megaloptera	Corydalidae	Acanthacorydalis	oris (McLachlan)		Pal	China (Chen et al., 2009)
		Corydalus	cornutus (L.)		Neo	Mexico (Ramos-Elorduy and Pino Moreno, 2002)
		Protohermes	grandis (Thunberg)	Dobsonfly	Pal	Japan (Schimitschek, 1968; Mitsuhashi, 1997)
Odonata	Aeschnidae	Anax	guttatus (Burm.)	Hairy emperor	Or	Thailand (DeFoliart, 2002), Laos (Césard, 2006), Indonesia (Meer Mohr, 1965)
		Coryphaeschna	adnexa (Hagen)		Neo	Ecuador (Onore, 2005)
		Rhionaeschna	brevifrons (Hagen)		Neo	Ecuador (Onore, 2005)
		Rhionaeschna	marchali (Rambur)		Neo	Ecuador (Onore, 2005)
		Rhionaeschna	multicolor (Hagen)		Nea	N. Am. (DeFoliart, 2002)
					Neo	Mexico (Ramos-Elorduy et al., 1998)
		Rhionaeschna	peralta (Ris)		Neo	Ecuador (Onore, 2005)
	Corduliidae	Epophthalmia	vittigera (Rambur)		Or	Thailand (DeFoliart, 2002)
		Epophthalmia	vittigera bellicose Lieftinck		Or	Thailand (Hanboonsong et al., 2000; Hanboonsong, 2010)
		Lauromacromia	dubitalis (Fraser)		Neo	Venezuela (Araújo and Beserra, 2007)
	Gomphidae	Gomphus	cuneatus Needham		Pal	China (Chen and Feng, 1999)
		Ictinogomphus	rapax (Rambur)		Or	India (Singh et al., 2007; Chakravorty et al., 2011)

	Lestidae	*Lestes*	*praemorsus* Hagen in Selys	Pal	China (Chen et al., 2009)
	Libellulidae	*Acisoma*	*panorpoides* Ram	Or	India (DeFoliart, 2002)
		Brachythemis	*contaminata* (F.)	Or	India (Singh et al., 2007; Chakravorty et al., 2011)
		Cratilla	*lineata assidua* Lieftinck	Or	Indonesia (Césard, 2006)
		Crocothemis	*servilia* (Drury)	Pal	China (Chen et al., 2009)
				Or	Indonesia (Césard, 2006), India (Shantibala et al., 2014)
		Diplacodes	*trivialis* (Rambur)	Or	India (Singh et al., 2007)
		Neurothemis	*ramburii* (Bauer)	Or	Indonesia (Césard, 2006)
		Orthetrum	*glaucum* (Brauer)	Or	Indonesia (Césard, 2006)
		Orthetrum	*sabina* (Drury)	Or	Indonesia (Césard, 2006)
		Pantala	*flaviscens* (L.)	Or	Indonesia (Césard, 2006; Lukiwati, 2010)
		Potamarcha	*obscura* (Ramb.)	Or	Indonesia (Césard, 2006)
		Sympetrum	*darwinianum* Selys	Pal	Japan (Schimitschek, 1968)
		Sympetrum	*eroticum* (Selys)	Pal	Japan (Schimitschek, 1968)
		Sympetrum	*infuscatum* (Selys)	Pal	Japan (Schimitschek, 1968)
		Tramea	*transmarina euryala* Selys	Or	Indonesia (Meer Mohr, 1965)
		Trithemis	*arteriosa* (Burm.)	Afr	D.R. Congo (Malaisse, 1997)
		Trithemis	*aurora* (Burm.)	Or	Indonesia (Césard, 2006)
Orthoptera	Acrididae	*Abracris*	*flavolineata* (DeGeer)	Neo	Colombia Mexico (DeFoliart, 2002)
		Acanthacris	*ruficornis* (Fabr.)	Afr	Sahel, D.R. Congo, CAR (Hoare, 2007), Congo (Bani, 1995; Nkouka, 1987; Niger (Lévy-Luxereau, 1980), Zambia (Mbata, 1995), Zimbabwe (Chavanduka, 1976) Malawi (Shaxson et al., 1999; DeFoliart, 2002), South Africa, Cameroun, Burkina Faso, Mali, Togo, Benin (Kelemu et al., 2015)
		Acanthacris	*ruficornis citrina* (Serville)	Afr	Benin (Riggi et al., 2013), Cameroun (Barreteau, 1999), CAR (Bahuchet, 1985), Niger, Mali, Burkina Faso, Guinée (Tchibozo, 2015)
		Acorypha	*clara* (Walker)	Afr	Cameroun (Barreteau, 1999)
		Acorypha	*glaucopsis* (Walker)	Afr	Cameroun (Barreteau, 1999)
		Acorypha	*nigrovariegata* (Bolivar)	Afr	Zambia (Mbata, 1995)
		Acorypha	*pallidicornis* (Stal)	Afr	South Afr (Van der Waal, 1996)
		Acorypha	*picta* Krauss	Afr	Cameroun (Barreteau, 1999)
		Acrida	*bicolor* (Thunberg)	Afr	Cameroun (Barreteau, 1999), Zimbabwe (Chavanduka, 1976)
		Acrida	*cinerea* (Thunb.)	Pal	Korea (DeFoliart, 2002)

(*Continued*)

Table A1 Documented Information for 1555 Species of Insects and Spiders (cont.)

Order	Family	Genus	Species	Common Name	Faunal	Distribution and References
		Acrida	exaltata (Walker)		Or	Thailand (Hanboonsong et al., 2000; Hanboonsong, 2010)
		Acrida	gigantea (Herbst)		Or	India (Singh and Chakravorty, 2008)
		Acrida	sulphuripennis (Gerstäcker)		Or	India (DeFoliart, 2002)
		Acrida	turrita (L.)	Long-headed grasshopper	Afr	Zambia (Mbata, 1995), South Africa (Van der Waal, 1996)
		Acrida	willemsei Dirsch		Afr	Cameroun (Barreteau, 1999), South Africa (Van der Waal, 1996)
		Acridoderes	strenua (Walker)		Pal	Korea (Meyer-Rochow and Chakravorty, 2013)
		Acrotylus	angulatus Stal		Or	Laos (Yhoung-Aree and Viwatpanich, 2005)
		Acrotylus	blondeli Saussure		Afr	Niger (Lévy-Luxereau, 1980), Sahel (Hoare, 2007)
		Acrotylus	longipes (Charpentier)		Afr	South Afr (Van der Waal, 1996)
		Afroxyrrhepes	procera (Burmeister)		Afr	Niger (Lévy-Luxereau, 1980)
		Aidemona	azteca (Saussure)		Afr	Congo (Nkouka, 1987, Hoare, 2007), South Africa (Van der Waal, 1996)
		Aiolopus	thalassinus tamulus (F.)		Neo	Colombia Mexico (DeFoliart, 2002)
		Aiolopus	thalassinus thalassinus (F.)		Or	Thailand (DeFoliart, 2002)
		Akamasacris	variabilis (Scudder)		Afr	Togo (Tchibozo, 2015), South Africa (Van der Waal, 1996)
		Anacridium	burri Dirsh and Uvarov		Neo	Mexico (Ramos-Elorduy and Mora, 2009)
		Anacridium	wernerellum (Karny)	Sudanese tree locust	Afr	Southern Afr (Malaisse, 1997)
		Anacridium	melanorhodon (Walker)	Sahelian tree locust	Afr	Niger (Lévy-Luxereau, 1980), Sahel (Hoare, 2007)
		Anacridium	moestum Serv.		Afr	Cameroun (Barreteau, 1999), Niger (Lévy-Luxereau, 1980), Sahel, Sudan (Kelemu et al., 2015), South Africa (Van der Waal, 1996)
		Arphia	fallax Saussure		Afr	South Afr (Van der Waal, 1996)
		Arphia	pseudonietana (Thomas)		Neo	Mexico (DeFoliart, 2002)
		Boopedon	flaviventris Bruner		Nea	N. Am. (DeFoliart, 2002)
		Boopedon	sp. af. flaviventris Bruner		Neo	Mexico (DeFoliart, 2002)
		Brachycrotaphus	tryxalicerus (Fischer)		Neo	Mexico (Ramos-Elorduy et al., 1998; DeFoliart, 2002)
		Cacantops	melanostictus Schaum.		Afr	Cameroun (Barreteau, 1999)
					Afr	South Afr (Van der Waal, 1996)

Calliptamus	abbreviatus Ikonn.		Pal	China (Chen and Feng, 1999)
Cardeniopsis	nigropunctatus (Bolivar)		Afr	Zambia (DeFoliart, 2002)
Camnula	pellucidae (Scudder)	Clear grasshopper	Nea	N. Am. (DeFoliart, 2002)
Cataloipus	cognatus (Walker)		Afr	South Afr (Van der Waal, 1996)
Cataloipus	cymbiferus (Krauss)		Afr	Cameroun (Barreteau, 1999)
Cataloipus	fuscocoeruleipus Sjöstedt		Afr	Sahel (Hoare, 2007)
Catantops	annexus Bolivar		Or	India (Singh and Chakravorty, 2008)
Catantops	infuscatus (Haan)		Or	Thailand (DeFoliart, 2002)
Catantops	spissus Walker		Afr	Congo (DeFoliart, 2002; Hoare, 2007), Cameroun (Barreteau, 1999), Sahel (Hoare, 2007)
Ceracris	kiangsu Tsai		Pal	China (Chen and Feng, 1999)
Ceracris	nigricornis nigricornis (Walker)		Or	India (Singh and Chakravorty, 2008)
Chirista	compta (Walker)		Afr	Congo (Nkouka, 1987; Hoare, 2007)
Chondacris	rosea (DeGeer)		Or	India (Chakravorty et al., 2014), Thailand (Hanboonsong et al., 2000; Hanboonsong, 2010), Laos (Yhoung-Aree and Viwatpanich, 2005)
Chortoicetes	terminifera (Walker)		Aus	Australia (DeFoliart, 2002)
Coryphosima	stenoptera (Schaum)		Afr	Congo (Tchibozo, 2015)
Cryptocantops	haemorrhoidalis Krauss		Afr	Niger (Lévy-Luxereau, 1980)
Cyrtacanthacris	aeruginosa (Stoll)		Afr	Nigeria (Fasoranti and Ajiboye, 1993), Zambia (Mbata, 1995), Malawi (Shaxson et al., 1999; DeFoliart, 2002), Guinée (Tchibozo, 2015), South Africa (Van der Waal, 1996)
Cyrtacanthacris	tatarica (L.)	Black spotted grasshopper	Afr	Botswana (Nonaka, 2009), South Africa (Van der Waal, 1996), Zambia (Mbata, 1995)
			Or	Thailand, Indonesia (DeFoliart, 2002; Hanboonsong et al., 2000)
Diabolocatantops	axillaris (Thunberg)		Afr	Cameroun (Barreteau, 1999), Niger (Lévy-Luxereau, 1980), Sahel (Hoare, 2007)
Diabolocatantops	innotabilis (Walker)	Clown grasshopper	Or	India (Chakravorty et al., 2011)
Ducetia	japonica (Thunb.)		Or	Thailand (Hanboonsong, 2010), Laos (Yhoung-Aree and Viwatpanich, 2005)
Duronia	chloronota (Stal)		Afr	South Afr (Van der Waal, 1996)
Encoptolophus	herbaceus Bruner		Neo	Mexico (DeFoliart, 2002)
Eupropacris	cylindricollis (Schaum)		Afr	Zambia (DeFoliart, 2002)
Exopropacris	modica (Karsch)		Afr	Cameroun (Barreteau, 1999)

(*Continued*)

Table A1 Documented Information for 1555 Species of Insects and Spiders (cont.)

Order	Family	Genus	Species	Common Name	Faunal	Distribution and References
		Gastrimargus	africanus (Saussure)		Afr	Cameroun (Barreteau, 1999), Congo (Nkouka, 1987; Hoare, 2007), Niger (Lévy-Luxereau, 1980), Sahel, Lesotho, Liberia (Kelemu et al., 2015), South Africa (Van der Waal, 1996)
		Gastrimargus	determinatus procerus (Gerstäcker)		Afr	Cameroun (Barreteau, 1999), Niger (Lévy-Luxereau, 1980)
		Gastrimargus	marmoratus (Thunbeg)		Or	Indonesia (Meer Mohr, 1965)
		Hadrolecocatantops	quadratus (Walker)		Afr	CAR (Hoare, 2007)
		Harpezocatantops	stylifer (Krauss)		Afr	Cameroun (Barreteau, 1999), Niger (Lévy-Luxereau, 1980)
		Heteracris	coerulescens (Stal)		Afr	CAR (Hoare, 2007)
		Heteracris	guineensis (Krauss)		Afr	Congo (Nkouka, 1987; Hoare, 2007)
		Hieroglyphodes	assamensis Uvarov		Or	India (Singh and Chakravorty, 2008)
		Hieroglyphus	africanus Uvarov		Afr	India (Singh and Chakravorty, 2008)
		Hieroglyphus	concolor (Walker)		Or	India (Singh and Chakravorty, 2008)
		Hieroglyphus	daganensis Krauss	Large rice grasshopper	Afr	Sahel Afr (Van Huis, 2005)
		Hieroglyphus	oryzivorus Carl		Or	India (Singh and Chakravorty, 2008)
		Homoxyrrhepes	punctipennis (Walker)		Afr	Cameroun (Barreteau, 1999), D.R. Congo (DeFoliart, 2002), Togo (Tchibozo, 2015)
		Humbe	tenuicornis (Schaum)		Afr	Niger (Lévy-Luxereau, 1980), South Africa (Van der Waal, 1996)
		Kraussaria	angulifera (Krauss)		Afr	Cameroun (Barreteau, 1999), Niger (Tchibozo, 2015), Sahel (Kelemu et al., 2015)
		Kraussella	amabile (Krauss)		Afr	Cameroun (Barreteau, 1999)
		Lamarckiana	bolivariana (Sauss)		Afr	South Afr (Van der Waal, 1996)
		Lamarckiana	cucullata (Stoll)		Afr	Botswana (Nonaka, 2009)
		Lamarckiana	punctosa (Walker)		Afr	South Afr (Van der Waal, 1996)
		Locusta	migratoria (L.)	Migratory locust	Aus	Papua New Guinea (DeFoliart, 2002)
					Pal	China Morocco (DeFoliart, 2002)
					Or	Thailand (Hanboonsong et al., 2000), Philippines
		Locusta	migratoria capito Sauss.		Afr	Madagascar (Decary, 1937)
		Locusta	migratoria manilensis (Meyen)		Pal	China (DeFoliart, 2002; Chen and Feng, 1999)
		Locusta	migratoria migratorioides (R. and F.)	African migratory locust	Afr	Cameroun (Barreteau, 1999), Congo (Nkouka, 1987; Hoare, 2007), Zimbabwe (DeFoliart, 2002), Sudan, South Sudan (Kelemu et al., 2015), Benin, Burkina Faso (Tchibozo, 2015)

Locustana	*pardalina* (Walker)	Brown locust	Afr	South Afr (Quin, 1959), Southern Africa, Zambia (Mbata, 1995), Zimbabwe, Botswana, Malawi (Kelemu et al., 2015)
			Pal	Libya
Melanoplus	*bivittatus* (Say)		Nea	N. Am. (DeFoliart, 2002)
Melanoplus	*devastator* Scudder		Nea	N. Am. (DeFoliart, 2002)
Melanoplus	*differentialis* (Thomas)		Nea	N. Am. (DeFoliart, 2002)
Melanoplus	*femurrubrum* (DeGeer)	Red-led grasshopper	Nea	N. Am. (DeFoliart, 2002)
Melanoplus	*mexicanus* (Sauss.)		Neo	Mexico (Ramos-Elorduy et al., 1998; DeFoliart, 2002)
Melanoplus	*sanguinipes* (F.)		Neo	Mexico (Ramos-Elorduy et al., 1998; DeFoliart, 2002)
Melanoplus	*spretus* (Walsh)	Rocky mountain grasshopper	Nea	N. Am. (DeFoliart, 2002)
Melanoplus	*sumichrasti* (Saussure)		Nea	USA Canada (Yen, 2015)
Mesopsis	*abbreviatus* (Beauvois)		Neo	Mexico (Ramos-Elorduy and Pino Moreno, 2002)
Metaxymecus	*gracilipes* (Brancsik)		Afr	Cameroun (Barreteau, 1999)
Morphacris	*fasciata* (Thunberg)		Afr	Cameroun (Barreteau, 1999)
Nomadacris	*septemfasciata* (Serville)	Red locust	Afr	South Africa (Van der Waal, 1996), Togo (Tchibozo, 2015)
			Afr	Congo (Nkouka, 1987), South Africa (Quin, 1959), Tanzania (Bodenheimer, 1951; Harris, 1940), Zambia (Mbata, 1995), Zimbabwe (Chavanduka, 1976; Gelfand, 1971), Malawi (Shaxson et al., 1999), Uganda (DeFoliart, 2002), Eastern Africa, Botswana, Nigeria, Mozambique (Kelemu et al., 2015), Cameroun (Tchibozo, 2015)
			Pal	Kuwait Saudi Arabia (DeFoliart, 2002)
Ochrotettix	*cer salinus* Bruner		Neo	Mexico (DeFoliart, 2002)
Oedaleonotus	*enigma* (Scudder)	Valley grasshopper	Nea	N. Am. (DeFoliart, 2002)
Oedaleus	*carvalhoi* I. Bol.		Afr	South Africa (Van der Waal, 1996)
Oedaleus	*flavus* L.		Afr	South Africa (Van der Waal, 1996)
Oedaleus	*interruptus* (Kirby)		Afr	South Africa (Van der Waal, 1996)
Oedaleus	*nigeriensis* Uvarov		Afr	Cameroun (Barreteau, 1999)
Oedaleus	*nigrofasciatus* (De Geer)		Afr	Zambia (Mbata, 1995)
Oedaleus	*senegalensis* (Krauss)		Afr	Niger (Lévy-Luxereau, 1980), Togo (Tchibozo, 2015)
Orphula	*azteca* (Saussure)		Neo	Mexico (Ramos-Elorduy and Pino Moreno, 2002)

(*Continued*)

Table A1 Documented Information for 1555 Species of Insects and Spiders (cont.)

Order	Family	Genus	Species	Common Name	Faunal	Distribution and References
		Ornithacris	cavroisi (Finot)		Afr	Benin (Riggi et al., 2013) Congo (Bani, 1995), Niger (Lévy-Luxereau, 1980), Togo, Mali, Burkina Faso (Tchibozo, 2015), Sahel
		Ornithacris	cyanea (Stoll)		Afr	Zimbabwe (Gelfand, 1971)
		Ornithacris	turbida (Walker)		Afr	South Africa (Van der Waal, 1996)
		Orthacanthacris	humilicrus (Karsch)		Afr	Niger (Lévy-Luxereau, 1980)
		Orthochtha	venosa (Ramme)		Afr	Cameroun (Barreteau, 1999)
		Oxya	chinensis (Thunberg)		Or	Indonesia (Meer Mohr, 1965)
		Oxya	japonica (Thunb.)		Pal	Indonesia (Meer Mohr, 1965)
					Or	Thailand (DeFoliart, 2002), Malaysia, Sabah (Chung et al., 2002)
		Oxya	japonica japonica (Thunb.)		Pal	Japan (DeFoliart, 2002)
		Oxya	velox (F.)		Or	Vietnam (DeFoliart, 2002)
					Pal	Japan, Korea (DeFoliart, 2002), Japan (Schimitschek, 1968)
		Oxya	vicina Brunner von W.		Pal	Japan (Schimitschek, 1968)
		Oxya	yezoensis Shiraki		Pal	Japan (DeFoliart, 2002; Nonaka, 2009)
		Oxycatantops	congoensis (Sjöstedt)		Afr	Congo (Bani, 1995; Nkouka, 1987; Hoare, 2007)
		Paracinema	tricolor (Thunberg)		Afr	Cameroun (Barreteau, 1999), Malawi, Lesotho (Kelemu et al., 2015)
		Parapropacris	notatus (Karsch)		Afr	Congo (Tchibozo, 2015)
		Patanga	avis Rehn and Rehn		Or	Thailand (Hanboonsong, 2010)
		Patanga	japonica (Bolivar)		Or	Thailand (Hanboonsong, 2010)
		Patanga	succincta (Johannson)		Or	Thailand, Philippines (DeFoliart, 2002; Hanboonsong, 2010), Indonesia (Meer Mohr, 1965; Lukiwati, 2010), Laos (Yhoung-Aree and Viwatpanich, 2005)
		Pcynodictya	flavipes Miller		Afr	South Africa (Van der Waal, 1996)
		Phaeocatantops	decoratus (Gers.)		Afr	South Africa (Van der Waal, 1996)
		Phlaeoba	antennata B. v. Wattenwyl		Or	Indonesia (Meer Mohr, 1965), India (Singh and Chakravorty, 2008)
		Pnorisa	squalus Stal		Afr	South Africa (Van der Waal, 1996)
		Rhammatocerus	nobilis Walk.		Neo	Mexico (DeFoliart, 2002)
		Rhammatocerus	schistocercoides (Rhen)		Neo	Brazil (Costa Neto and Ramos-Elorduy, 2006)
		Rhammatocerus	viatorius viatorius (Sauss.)		Neo	Mexico (DeFoliart, 2002)

	Roduniella	inspida (Karsch)		Afr	CAR (Hoare, 2007)
	Schistocerca	cancellata cancellata (Serv.)		Neo	South America (DeFoliart, 2002)
	Schistocerca	cancellata paranensis (Serv.)		Neo	South America Mexico (DeFoliart, 2002)
	Schistocerca	gregaria (Forskål)	Desert locust	Afr	India (DeFoliart, 2002)
				Pal	Pan-regional, S.W. Asia, North Africa (DeFoliart, 2002), Morocco (Kelemu et al., 2015)
				Or	India (DeFoliart, 2002)
	Schistocerca	nitens (Thunberg)		Neo	Mexico (Ramos-Elorduy and Pino Moreno, 2002)
	Schistocerca	shoshone (Thomas)		Nea	N. Am. (DeFoliart, 2002)
	Sherifuria	haningtoni Uvarov		Afr	Cameroun (Barreteau, 1999)
	Shirakiacris	shirakii (Bolivar)		Or	Thailand (Hanboonsong, 2010)
				Pal	China (Chen and Feng, 1999)
	Spharagemon	equale (Say)		Neo	Mexico (DeFoliart, 2002)
	Stenocatantops	splendens (Thunberg)		Or	Indonesia (Meer Mohr, 1965)
	Stenocrobylus	festivus Karsch		Afr	Guinée (Tchibozo, 2015)
	Tmetonota	terrosa Sauss.		Afr	South Africa (Van der Waal, 1996)
	Trilophidia	annulata (Thunberg)		Or	Thailand (Hanboonsong, 2010)
	Trimerotropis	occidentalis Bruner		Neo	Mexico (Ramos-Elorduy and Mora, 2009)
	Tropidacris	cristata (L.)		Neo	Colombia Venezuela (Ruddle, 1973)
	Truxalis	burtti Dirsh		Afr	South Africa (Van der Waal, 1996)
	Truxalis	johnstoni Dirsh		Afr	Cameroun (Barreteau, 1999)
	Truxaloides	constrictus (Schaum)		Afr	South Africa (Van der Waal, 1996), Zimbabwe (Gelfand, 1971)
	Tylotropidius syn	gracilipes Brancsik		Afr	Cameroun (Barreteau, 1999)
	Valanga	nigricornis (Burmeister)		Or	Indonesia (Meer Mohr, 1965; Lukiwati, 2010), Malaysia, Sabah (Chung et al., 2002)
	Valanga	irregularis (Walk.)	Large coast locust	Aus	Papua New Guinea (DeFoliart, 2002)
	Xanthippus	corallipes (Haldeman)		Neo	Mexico (Ramos-Elorduy et al., 1998), West Indies (DeFoliart, 2002)
Anostostomatidae	Borborothis	brunneri Bolivar		Afr	Congo (Bergier, 1941) Southern Afr (Malaisse, 1997)
	Henicus	whellani Chopard		Afr	Zimbabwe (Kelemu et al., 2015)
Gryllacridae	Teleogryllus	commodus (Walker)		Aus	Australia, Papua New Guinea (DeFoliart, 2002)
	Stenopelmatus	fuscus Haldeman	Jerusalem cricket	Nea	N. Am. (DeFoliart, 2002)
Gryllidae	Acheta	domesticus (L.)		Or	Thailand (Hanboonsong et al., 2013; Yen, 2015), Laos (Hanboonsong and Durst, 2014)
				Neo	Mexico (Ramos-Elorduy, 2009)

(*Continued*)

Table A1 Documented Information for 1555 Species of Insects and Spiders (cont.)

Order	Family	Genus	Species	Common Name	Faunal	Distribution and References
		Brachytrupes	membranaceus (Drury)	Giant cricket	Afr	Benin (Tchibozo, 2015; Riggi et al., 2013), Congo (Bani, 1995; Nkouka, 1987; Hoare, 2007), D.R. Congo (Adriaens, 1951), CAR (Hoare, 2007), Nigeria (Fasoranti and Ajiboye, 1993), Tanzania (Bodenheimer, 1951; Harris, 1940), (East, Central and Southern Afr), Zambia (Mbata, 1995), Zimbabwe (Chavanduka, 1976; Gelfand, 1971; Weaving, 1973), Cameroun, Burkina Faso, Angola, Togo (Kelemu et al., 2015)
		Brachytrupes	oris Burmeister		Or	India (Chakravorty et al. 2014)
		Grylloderes	melanocephalus (Serv.)		Or	India (DeFoliart, 2002)
		Gryllus	assimilis (F.)		Nea	N. Am. (DeFoliart, 2002)
		Gryllus	bimaculatus DeGeer	Spotted cricket	Neo	Mexico (Ramos-Elorduy, 2009)
					Afr	Zambia (Mbata, 1995) G. Bissau, Sierra Leone, Guinée, Liberia, Benin, Togo, Nigeria, D.R. Congo, Kenya, South Sudan (Kelemu et al., 2015)
					Or	Thailand, India (DeFoliart, 2002; Hanboonsong, 2010), Laos (Yhoung-Aree and Viwatpanich, 2005)
		Gryllus	peruviensis		Pal	China (Chen et al., 2009)
					Neo	Peru (Koga et al., 1999)
		Gymnogryllus	leucostictus (Burm.)		Or	Indonesia (Meer Mohr, 1965)
		Loxoblemmus	arietulus Saussure		Pal	Japan (Schimitschek, 1968)
		Loxoblemmus	doenitzi Stein		Pal	Japan (Schimitschek, 1968)
		Modicogryllus	confirmatus (Walker)		Or	Thailand (Hanboonsong, 2010), Laos (Yhoung-Aree and Viwatpanich, 2005)
		Nisitrus	vittatus (Haan)		Or	Indonesia (Meer Mohr, 1965)
		Tarbinskiellus	oris (Burmeister)		Or	India (Chakravorty et al., 2011)
		Tarbinskiellus	portentosus (Lichtenstein)		Pal	China (Chen et al., 2009) Japan (Schimitschek, 1968)
					Or	Thailand (Hanboonsong et al., 2000), Myanmar, Indonesia, Vietnam, India (DeFoliart, 2002; Lukiwati, 2010), Laos (Yhoung-Aree and Viwatpanich, 2005)
		Teleogryllus	mitratus (Burm.)		Or	Thailand (Hanboonsong, 2010), Indonesia (Meer Mohr, 1965), Laos (Yhoung-Aree and Viwatpanich, 2005)
					Pal	Japan (Schimitschek, 1968)
		Teleogryllus	occipitalis (Serville)		Or	Thailand (Hanboonsong et al., 2013)

Family	Genus	Species	Common name	Region	Country (References)
	Velarifictorus	aspersus (Walker)		Pal	Japan (Schimitschek, 1968)
Gryllotalpidae	Gryllotalpa	africana Palisot	African mole cricket	Afr	Uganda (Bodenheimer, 1951), Zimbabwe (Chavanduka, 1976; Gelfand, 1971; Weaving, 1973)
				Pal	Japan (Schimitschek, 1968)
				Or	Thailand (Hanboonsong et al., 2000), Philippines, Vietnam, India (DeFoliart, 2002), Indonesia (Meer Mohr, 1965), Laos (Yhoung-Aree and Viwatpanich, 2005)
	Gryllotalpa	oris Burm.		Pal	China (Chen and Feng, 1999)
	Gryllotalpa	unispina Saussure		Pal	China (Chen and Feng, 1999)
	Gryllotalpa	hirsuta Burmeister		Or	Malaysia Sabah (Chung et al., 2002)
Pyrgomorphidae	Atractomorpha	psittacina (de Haan)		Or	Malaysia Sabah (Chung et al., 2002)
	Chrotogonus	hemipterus Schaum.		Afr	South Afr (Van der Waal, 1996)
	Chrotogonus	senegalensis Krauss		Afr	Cameroun (Barreteau, 1999)
	Occidentosphena	uvarovi (Rehn)		Afr	Congo (Tchibozo, 2015)
	Phymateus	viridipes brunneri Bolivar	Green bush locust	Afr	Congo (Bergier, 1941), Southern Africa (Malaisse, 1997), Zambia, South Africa, Zimbabwe, Botswana, Mozambique, Namibia (Kelemu et al., 2015)
	Pyrgomorpha	cognata Krauss		Afr	Cameroun (Barreteau, 1999)
	Pyrgomorpha	vignaudi (Guérin-M.)		Afr	CAR (Hoare, 2007)
	Sphenarium	borrei Bolivar		Neo	Mexico (Ramos-Elorduy et al., 2012)
	Sphenarium	magnum Marquez		Neo	Mexico (DeFoliart, 2002)
	Sphenarium	mexicanum mexicanum Sauss.		Neo	Mexico (Ramos-Elorduy and Pino Moreno, 2002)
	Sphenarium	purpurascens Charp.		Neo	Mexico (DeFoliart, 2002)
	Zonocerus	elegans (Thunberg)		Afr	Mozambique South Afr (Quin, 1959)
	Zonocerus	variegatus (L.)	Stink locust	Afr	CAR (Barreteau, 1999), Nigeria (Fasoranti and Ajiboye, 1993), D.R. Congo, Cameroun, Congo, Côte d'Ivoire, Sao Tomé, Guinée, Ghana, Liberia, G. Bissau (Kelemu et al., 2015), Benin, Togo, Burkina Faso (Tchibozo, 2015)
Romaleidae	Chromacris	colorata (Serville)		Neo	Mexico (DeFoliart, 2002)
	Taeniopoda	auricornis (Walker)		Neo	Mexico (Ramos-Elorduy and Pino Moreno, 2002)
	Taeniopoda	bicristata Burm.		Neo	Mexico (Ramos-Elorduy et al. 2008)
	Titanacris	albipes (De Geer)		Neo	Brazil (Costa Neto and Ramos-Elorduy, 2006)
	Tropidacris	collaris (Stoll)		Neo	Brazil (Costa Neto and Ramos-Elorduy, 2006)
	Tropidacris	cristata (L.)		Neo	Colombia, Venezuela (DeFoliart, 2002; Ruddle, 1973)

(*Continued*)

Table A1 Documented Information for 1555 Species of Insects and Spiders (cont.)

Order	Family	Genus	Species	Common Name	Faunal	Distribution and References
	Schizodactylidae	Tropinotus	mexicanus Brunner		Neo	Mexico (DeFoliart, 2002)
		Schizodactylus	monstrosus (Drury)	Sand crickey	Or	India (Singh et al., 2007; Chakravorty et al., 2011)
		Schizodactylus	tuberculatus Andre		Or	India (Singh and Chakravorty, 2008)
	Tettigoniidae	Anoedopoda	erosa (Karsch)		Afr	CAR (Hoare, 2007)
		Anabrus	simplex Haldeman	Mormon bush cricket	Nea	N. Am. (DeFoliart, 2002)
		Arachnacris	tenuipes Giebel		Or	Malaysia Sabah (Chung et al., 2002)
		Chloracris	brullei Pictet and Saussure		Or	India (Chakravorty et al., 2011)
		Conocephalus	angustifrons (Redt.)		Neo	angustifrons (Redt.)
		Conocephalus	ictus (Scud.)		Neo	Mexico (Ramos-Elorduy et al., 2008)
		Conocephalus	cinereus Thum.		Neo	Mexico (Ramos-Elorduy et al., 2008)
		Conocephalus	maculates (LeGuillou)		Or	Thailand (Hanboonsong, 2010)
		Euconocephalus	incertus (Walker)		Or	Thailand (Hanboonsong, 2010)
		Gampsocleis	buergeri (Haan)		Pal	Korea (Meyer-Rochow and Chakravorty, 2013)
		Gymnoproctus	abortivus (Serville)		Afr	Benin (Riggi et al., 2013)
		Hexacentrus	unicolor Serville		Or	Malaysia Sabah (Chung et al., 2002)
		Holochlora	albida Brunner von Wattenwyl		Or	India (DeFoliart, 2002)
		Holochlora	indica Kirby		Or	India (DeFoliart, 2002)
		Idiarthron	subquadratum S. and P.		Neo	Mexico (Ramos-Elorduy et al., 2012)
		Mecopoda	elongata (L.)		Or	India (DeFoliart, 2002), Indonesia (Meer Mohr, 1965), Thailand, Laos (Hanboonsong, 2010), Malaysia, Sabah (Chung et al., 2002)
		Microcentrum	totonacum (Sauss.)		Neo	Mexico (Ramos-Elorduy et al., 2008)
		Neoconocephalus	triops (L.)		Neo	Mexico (Ramos-Elorduy et al., 2012)
		Petaloptera	zendala Saussure		Neo	Mexico (Ramos-Elorduy et al., 1998)
		Pseudophyllus	titan White		Or	Thailand (Hanboonsong et al., 2000), Laos (Yhoung-Aree and Viwatpanich, 2005)
		Ruspolia	differens (Serville)		Afr	The whole of East (Owen, 1973) and Southern Africa, D.R. Congo (Bergier, 1941; Bequaert, 1921), Cameroun (Womeni et al., 2009), Tanzania (Bodenheimer, 1951; Harris, 1940), Malawi (Shaxson et al., 1999), Zambia (Mbata, 1995), Zimbabwe (Chavanduka, 1976; Gelfand, 1971; Weaving, 1973), Kenya (Kinyuru et al., 2012), South Africa, Uganda, Malawi (Kelemu et al., 2015)

	Ruspolia	*nitidulis vicinus* (Walker)		Afr	The whole of East (Owen, 1973) and Southern Africa, D.R. (Bergier, 1941; Bequaert, 1921)
	Scudderia	*mexicana* (Sauss.)		Neo	Mexico (Ramos-Elorduy et al., 2008)
	Stilpnochlora	*azteca* (Saussure)		Neo	Mexico (Ramos-Elorduy et al., 1998)
	Stilpnochlora	*quadrata* (Scud.)		Neo	Mexico (Ramos-Elorduy et al., 2008)
	Stilpnochlora	*thoracica* (Serville)		Neo	Mexico (Ramos-Elorduy et al., 1998)
Psocoptera					
Psocidae	*Metylophorus*	*barretti* (Banks)		Neo	Mexico (Ramos-Elorduy et al., 2007)
Phasmida					
Heteropterygidae	*Haaniella*	*echinata* (Redtenbacher)		Or	Malaysia Sabah (Chung et al., 2002)
	Haaniella	*grayi* (Westw.)		Or	Malaysia (DeFoliart, 2002; Chung, 2010)
Phasmatidae	*Eurycantha*	*horrida* Boisduval		Aus	Papua New Guinea (DeFoliart, 2002)
	Eurycnema	*versirubra* (Serville)		Or	Malaysia (DeFoliart, 2002)
	Extatosoma	*tiaratum* (MacLeay)	Giant prickly stick insect	Aus	Papua New Guinea (DeFoliart, 2002)
	Platycrana	*viridana* (Oliv.)		Or	Malaysia (DeFoliart, 2002)
Phthiraptera					
Pediculidae	*Pediculus*	*humanus* L.		Neo	Brazil (Costa Neto and Ramos-Elorduy, 2006)
				Or	Indonesia (Roepke, 1952)
				Aus	Papua New Guinea (Meyer-Rochow, 1975)
Plecoptera					
Perlidae	*Kamimuria*	*tibialis* (Pictet)		Pal	Japan (Schimitschek, 1968)
	Paragnetina	*tinctipennis* (McLachnan)		Pal	Japan (Schimitschek, 1968)
Pteronarcyidae	*Pteronarcys*	*dorsata* (Say) check		Or	India (Singh et al., 2007)
Trichoptera					
Hydropsychidae	*Cheumatopsyche*	*brevilineata* (Iwata)	Caddis fly	Pal	Japan (Mitsuhashi, 1997)
Leptoceridae	*Oecetis*	*disjuncta* (Banks)		Neo	Mexico (Ramos-Elorduy and Pino Moreno, 2002)
Stenopsychidae	*Parastenopsyche*	*sauteri* Matsuzaki	Caddis fly	Pal	Japan (Mitsuhashi, 1997)
	Stenopsyche	*griseipennis* McLachlan	Caddis fly	Pal	Japan (DeFoliart, 2002)

Source: Based on data from Jongema (2015).

REFERENCES

Acuña, A.M., Caso, L., Aliphat, M.M., Vergara, C.H., 2011. Edible insects as part of the tradional food system of the Popoloca town of Los Reyes Metzontla, Mexico. J. Ethnobiol. 31 (1), 150–169.

Adalla, C.B.C., Cleolas R., 2010. Philippine Edible Insects: A New Opportunity to Bridge the Protein Gap of Resource-poor Families and to Manage Pests. FAO Regional Officer for Asia and the Pacific, Bangkok, Thailand.

Adriaens, E.L., 1951. Recherches sur l'alimentation des populations au Kwango. Agric. Bull. Belgian Congo 42 (2), 227–270.

Agbidye, F.S., Ofuya, T.I., Akindele, S.O., 2009. Some edible insect species consumed by the people of Benue State, Nigeria. Pak. J. Nutrit. 8 (7), 946–950.

Aldama-Aguilera, C., Lianderal-Cazares, C., Soto-Hernandez, M., Castillo-Marquez, L.E., 2005. Cochineal (*Dactylopius coccus Costa*) production in prickly pear plants in the open and in microtunnel greenhouses. Agrociencia 39 (2), 161–171.

Aldrovandi, U., 1644. De Animalibus Insectis Libri Septem. Apud Clementem Ferronium, Bononiae, Italy.

Araújo, Y., Beserra, P., 2007. Diversidad de invertebrados consumidos por las etnias Yanomami y Yekuana del Alto Orinoco, Venezuela. Interciencia 32 (5), 318–323.

Ashiru, M.O., 1988. The food value of the larvae of *Anaphe venata* Butler (Lepidoptera: Notodontidae). Ecol. Food Nutri. 22, 313–320.

Ayieko, M.A., Nyambuga, I.A., 2009. Termites and lake flies in the livelihood of households within the Lake Victoria region: Methods for harvesting and utilization. Techn. Report for the National Museums of Kenya.

Bahuchet, S., 1985. Les Pygmées Aka et la Forêt Centrafricaine. Selaf, Paris, France.

Bani, G., 1995. Some aspects of entomophagy in the Congo. Foods Insects Newslett. 8, 4–5.

Banjo, A.D., Lawal, O.A., Songonuga, E.A., 2006. The nutritional value of fourteen species of edible insects in Southwestern Nigeria. Afr. J. Biotechnol. 5 (3), 298–301.

Barreteau, D. 1999. Les Mofu-Gudur et leurs criquets. In: Baroin, C., Boutrais, J., (Eds). L'homme et l'Ánnimal dans le Bassin du Lac Tchad. Actes du colloque du rëseua Mega-Tchad, Orléans 15–17 octobre 1997. Collection Colloques et Séminaires, no. 00/354. Université Nanterre, Paris, pp. 133–169.

Bequaert, J., 1921. Insects as food: how they have augmented the food supply of mankind in early and recent times. J. Am. Museum Nat. Hist. 21, 191–200.

Bergier, E., 1941. Peuples Entomophages et Insectes Comestibles: Étude sur les Moeurs de l'Homme et de l'Insecte. Imprimerie Rullière Frères, Avignon, France.

Blake, E.A., Wagner, M.R., 1987. Collection and consumption of Pandora moth, *Coloradia pandora lindseyi* (Lepidoptera: Saturniidae), larvae by Owens Valley and Mono Lake Paiutes. Bull. Entomol. Soc. Am. 33, 23–27.

Blöte, H.L., 1952. On some Oncomerini from New Guinea (Heteroptera, Pentatomidae). Zoolog. Commun. 31, 251–257.

Bodenheimer, F.S., 1951. Insects as Human Food. W. Junk Publishers, The Hague, Netherlands.

Bristowe, W.S., 1932. Insects and other invertebrates for human consumption in Siam. Trans. Entomol. Soc. London 80, 387–404.

Bristowe, W.S., 1945. Spider superstitions and folklore. Trans. CT Acad. Arts Sci. 36, 53–91.

Cerritos, R., 2009. Insects as food: an ecological, social and economical approach. CAB Reviews: Perspectives in Agriculture, Veterinary Science, Nutrition and Natural Resources 4 (27), 1–10.

Césard, N., 2006. Des libellules dans l'assiette; les Insect. Soc. consommés à Bali. Insect. Soc. 1 (140), 3–6.

Chakravorty, J., Ghosh, S., Meyer-Rochow, V.B., 2011. Practices of entomophagy and entomotherapy by members of the Nyishi and Galo tribes, two ethnic groups of the state of Arunachal Pradesh (North-East India). J. Ethnobiol. Ethnomed. 7 (5), 1–14.

Chakravorty, J., Ghosh, S., Meyer-Rochow, V.B., 2013. Comparative survey of entomophagy and entomotherapeutic practices in six tribes of Eastern Arunachal Pradesh (India). J. Ethnobiol. Ethnomed. 9, 50.

Chakravorty, J., Ghosh, S., Jung, C., Meyer-Rochow, V.B., 2014. Nutritional composition of *Chondacris rosea* and *Brachytrupes orientalis*: two common insects used as food by tribes of Arunachal Pradesh, India. J. Asia-Pac. Entomol. 17, 407–415.

Chavanduka, D.M., 1976. Insects as a source of protein to the Africain. Rhodesia Sci. News 9 (7), 217–220.

Chen, X.-M., Feng, Y., 1999. The Edible Insects of China. Science and Technology Publishing House, Shenyang, China.

Chen, X., Feng, Y., Chen, Z., 2009. Common edible insects and their utilization in China. Entomol. Res. 39 (5), 299–303.

Choo, J., 2008. Potential ecological implications of human entomophagy by subsistence groups of the Neotropics. Terr. Arthropod Rev. 1, 81–93.

Chung, A.Y.C., 2010. Edible insects and entomophagy in Borneo, In: Durst, P.B., Johnson, D.V., Leslie, R.N., Shono, K. FAO, Regional Office for Asia and the Pacific, Bangkok, Thailand. pp. 141–150.

Chung, A.Y.C., 2003. A note on the defoliation of lxora coccinea by *Clethrogyna turbata* (Lepidoptera, Lymantriidae). Malaysian Nat. 564, 43–44.

Chung, A.Y.C., Khen, C.V., Unchi, S., Binti, M., 2002. Edible insects and entomophagy in Sabah, Malaysia. Malay. Nature J. 56 (2), 131–144.

Costa Neto, E.M., 2002. Manual de Etnoentomología. Manuales & Tesis SEA, 4. Sociedad Entomológica Aragonesa, Aragón.

Costa Neto, E.M., Ramos-Elorduy, J., 2006. Los insectos comestibles de Brasil: etnicidad, diversidad e importancia en la alimentación. Bull. Entomol. Soc. Aragonesa 38, 423–442.

Crane, E., 1992. The past and present status of beekeeping with stingless bees. Bee World 73 (1), 29–42.

Darwin, E., 1800. Phytologia, or the Philosophy of Agriculture and Gardening with the Theory of Draining Morasses and with an Improved Construction of the Drill Plough. London, UK.

Decary, R., 1937. L'entomophagie chez les indigènes de Madagascar. Bulletin de la Société entomologique de France, p. 168–171.

DeFoliart, G.R., 2002. The human use of insects as food resource: a bibliographic account in progress. Available from: http://www.food-insects.com/book7_31/The%20human%20use%20of%20insects.

Donkin, 1977. Spanish red: an ethnogeographical study of cochineal and the *Opuntia* cactus. Trans. Am. Philosoph. Soc. 67 (Pt. 5), 7–11.

Duffels, J.P., van Mastrigt, J.G., 1991. Recognition of cicadas (Homoptera, Cicadidae) by the Ekagi people of Irian Jaya (Insonesia), with a description of a new species of *Cosmopsaltria*. J. Nat. Hist. 25, 173–182.

Dufour, D.L., 1987. Insects as food: a case study from the Northwest Amazon. Am. Anthropol. 89, 383–397.

Dzerefos, C.M., Witkowski, E.T.F., Toms, R., 2009. Life-history traits of the edible stink bug, *Encosternum delegorguei* (Hem., Tessaratomidae), a traditional food in southern Africa. J. Appl. Entomol. 133, 749–759.

Eastwood, R., Kongnoo, P., Reinkaw, M., 2010. Collecting and eating *Liphyra brassolis* (Lepidoptera: Lycaenidae) in southern Thailand. J. Res. Lepidoptera 43, 19–22.

Egan, B.A., Toms, R., Minter, L.R., Addo-Bediako, A., Masoko, P., Mphosi, M., Olivier, P.A.S., 2014. Nutritional significance of the edible insect, *Hemijana variegata* Rothschild (Lepidoptera: Eupterotidae), of the Blouberg Region, Limpopo, South Africa. Afr. Entomol. 22 (1), 15–23.

Fasoranti, J.O., Ajiboye, D.O., 1993. Some edible insects of Kwara State, Nigeria. Am. Entomol. 39 (2), 113–116.

Faure, J.C., 1944. Pentatomid bugs as human food. J. Entomolog. Soc. South Afr. 7, 110–112.

Feng, Y, Long, S., 2010. Common edible wasps in Yunnan Province, China and their nutritional value. FAO, Regional Office for Asia and the Pacific, Bangkok, Thailand. pp. 93–98.

Ferreira, M.N., Ballester, W.C., Dorval, A., Costa, R.B., 2010. Conhecimento tradicional dos Kaiabi sobre abelhas sem ferrão no Parque Indígena do Xingu, Mato Grosso, Brasil. Tellus 19, 129–144.

FIN, 1998. A fabulous food insect festival for the Entomological Society of America and the Entomological Society of Canada. Food Insects Newslett. 11 (3), 10–11.

Gahukar, R.T., 2011. Entomophagy and human food security. Int. J. Trop. Insect Sci. 31, 129–144.

Gelfand, M., 1971. Insects. In: Gelfand, M. (Ed.), Diet and Tradition in African Culture. E. & S. Livingstone, Edinburgh and London, UK, pp. 163–171.

Gessain, M., Kinzler, T., 1975. Miel et Insect. Soc. à miel chez les Bassari et dáutres populations du Sénégal Oriental. In: Pujol, R. (Ed.), L'Homme et l'Annimal. Premier Colloque d'Ethnozoologie, Paris, France, pp. 247–254.

Ghesquière, J., 1947. Les insectes palmicoles comestibles. In: Lepesme, P., Ghesqère, J. (Eds.), Les Insectes des Palmiers, Lechevalier, Paris, France. Appendice II, pp. 791–793.

Grami, B., 1998. Gaz of Khunsar: the manna of Persia. Econ. Botany 52, 183–191.

Grant, P.M., 2001. Mayflies as food. In: Dominguez, E. (Ed.), Trends in Research in Ephemeroptera and Plecoptera. Springer, pp. 107–134.

Hajong, S.R., 2013. Mass emergence of a cicada (Homoptera: Cicadidae) and its capture methods and consumption by villagers in Ri-bhoi district of Meghalaya. J. Entomol. Res. 37 (4), 341–343.

Hanboonsong, Y. 2010. Edible insects and associated food habits in Thailand. In: Durst, P.B., Johnson, D.V., Leslie, R.N., Shono, K., (Ed.), Edible Forest Insects: Humans Bite Back. Proceedings of a Workshop on Asia-Pacific Resources and their Potential for Development. FAO Regional Office Bangkok, Chiang Mai, Thailand. pp. 173–182.

Hanboonsong, Y., Durst, P.B., 2014. Edible insects in LAO PDR. RAP publication FAO 2014/12.

Hanboonsong, Y., Rattanapan, A., Utsunomiya, Y., Masumoto, K., 2000. Edible insects and insect-eating habits in Northeast Thailand. Elytra 28 (2), 355–364.

Hanboonsong, Y., Jamjanya, T., Durst, P.B., 2013. Six-legged livestock: edible insect farming, collection and marketing in Thailand. RAP Publication 2013/03. FAO regional office for Asia and the Pacific.

Harris, W.V., 1940. Some notes on insects as food. Tanganyika Notes and Records 9, 45–48.

Hoare, A.L., 2007. The Use of Non-Timber Forest Products in the Congo Basin: Constraints and Opportunities. The Rainforest Foundation, London, UK.

Jäch, M.A., 2003. Fried water beetles, Cantonese style. Am. Entomol. 49 (1), 34–38.

Jameson, M.L., Drumont, A., 2013. Aroid scarabs in the genus *Peltonotus* Bürmeister (Coleoptera, Scrarabaeidae, Dynastinae) key to species and new distributional data. Zookeys 320, 63–95.

Jongema, Y., 2015. List of Edible Insects of the World. Wageningen University, Wageningen, The Netherlands. Available from: http://www.ent.wur.nl/UK/Edible+insects/Worldwide+species+list/.

Katz, M.H., 2002. Body of Text: The Emergence of the Sunni Law of Ritual Purity. State University of New York Press, Albany, NY.

Kelemu, S., Niassy, S., Torto, B., Fiaboe, K., Affognon, H., Tonnang, H., Mania, N.K., Ekesi, S., 2015. African edible insects for food and feed: inventory, diversity, commonalities and contribution to food security. J. Insects Food Feed 1 (2), 103–119.

Kinyuru, J.N., et al., 2012. Identification of traditional foods with public health potentional for complementary feeding in Western Kenya. J. Food Res. 1 (2), 148–158.

Kirby, W., Spence, W., 1863. An Introduction to Entomology, seventh ed. Longman, Green, Longman, Roberts and Green, London, UK.

Koga, R., García, F., Carcelén, F., Arbaiza, T., 1999. Valor nutritivo del *Gryllus peruviensis* (Orthoptera: Gryllidae). Peru Vet. Res. Mag. 10 (1), 92–94.

Latham, P., 1999. Edble caterpillars of the Bas Congo Region of the Democratic Region of the Democratic Republic of Congo. Antenna 23 (3), 135–139.

Latham, P., 2003. Edible caterpillars and their food plants in Bas-Congo. Mystole Publications, Canterbury.

Lévy-Luxereau, A., 1980. Note sur quelques criquets de la région de Maradi (Niger) et leurs noms Hausa. Trad. Agric. Log Appl. Bot. 37, 263–272.

Lukiwati, D.R., 2010. Teak caterpillars and other edible insects in Java. In: Durst, P.B., Johnson, D.V., Leslie, R.N., Shono, K. (Eds.), Forest Insects as Food: Humans Bite Back. FAO Regional Office Bangkok, Chiang Mai/Thailand, pp. 99–104.

Malaisse, F., 1997. Se Nourir en Foret Claire Africaine: Approche Ecologique et Nutritionnelle. Les Presses Agronomiques de Gembloux, Gembloux, Belgium.

Malaisse, F., 2005. Human consumption of Lepidoptera, termites, Orthoptera, and ants in Africa. In: Paoletti, M.G. (Ed.), Ecological Implications of Minilivestock. Science Publishers, INC, New Hampshire, pp. 175–230.

Malaisse, F., Parent, G., 1980. Les chenilles comestibles du Shaba meridional. Naturalists Belgians 61, 2–24.

Marais, E., 1996. Omaungu in Namibia: *Imbrasia belina* (Saturniidae: Lepidoptera) as a commercial resource. In: Gashe, B.A. and Mpuchane, S.F. (Eds.) Phane, Proc. 1st Multidisciplinary Symp. On Phane, June 1996 Botswana, Dep. of Biol. Sci. and the Kalahari Conserv. Soc. (org.), 23–31.

Masumoto, K., Utsunomiya, Y., 1998. Edible insects from Northern Thailand. Elytra 26 (2), 443–444.

Mbata, K.J., 1995. Traditional use of arthropods in Zambia: I. The food insects. Food Insects Newslett. 8 (3), 1, 5–7.

Mbata, K.J., Chidumayo, E.N., Lwatula, C.M., 2002. Traditional regulation of edible caterpillar exploitation in the Kopa area of Mpika district in northern Zambia. J. Insect Conserv. 6, 115–130.

Meer Mohr, J.C.v.d., 1965. Insects eaten by the Karo-Batak people (A contribution to entomo-bromatology). Entomol. Posts 6 (1), 101–107.

Meyer-Rochow, V.B., 1975. Local taxonomy and terminology for some terrestrial arthropods in five different ethnic groups of Papua New Guinea and Central Australia. J. Roy. Soc. West. Australia 58, 15–30.

Meyer-Rochow, V.B., 2004. Traditional food insects and spiders in several ethnic groups of northeast India, Papua New Guinea, Australia, and New Zealand. In: Paoletti, M.G. (Ed.), Ecological Implications of Minilivestock: Rodents, Frogs, Snails, and Insects for Sustainable Development. Science Publ., Inc., Boca Raton, FL, pp. 385–409.

Meyer-Rochow, V.B., Chakravorty, J., 2013. Notes on entomophagy and entomotherapy generally and information on the situation in India in particular. Appl. Entomol. Zool. 48, 105–112.

Ministero delle politiche agricole alimentari e forestali, 2015. Disciplinari di produzione prodotti DOP e IGP riconosciuti. 1.3 Formaggi. Politicheagricole.it. Available from: https://www.politicheagricole.it/flex/cm/pages/ServeBLOB.php/L/IT/IDPagina/202.

Mitsuhashi, J., 1997. Insects as traditional food in Japan. Ecol. Food Nutri. 36, 187–199.

Munthali, S.M., Mughogho, D.E.C., 1992. Economic incentives for conservation: bee-keeping and Saturniidae caterpillar utilization by rural communities. Biod. Cons. 1, 143–154.

Netolitzky, F., 1919. Käfer als Nahrung und Heilmittel. Koleopterologische Rundschau 8, 21–26, 47–60.

Nkouka, E., 1987. Les insectes comestibles dans les societes d'Afrique Centrale. Muntu 6 (1), 171–178, Revue Scientifique et Culturelle du CICIBA, ASC LEIDEN.

Nonaka, K., 2009. Feasting on insects. Entomol. Res. 39, 304–312.

Oberprieler, R., 1995. Emperor moths and man. Emperor moths and conservation. Anomalous Emperor. In: Oberprieler, R. (Ed.), The Emperor Moths of Namibia. Ekogilde Ecoguild, South Africa, 91 pp.

Onayade, O.A., 1995. Entomochemical analyses of the larvae of *Anaphe venata* Butler (Lepidoptera, Notodontidae). J. Toxicol. Toxin Rev. 14 (2), 213.

Onore, G., 2005. Edible insects in Ecuador. In: Paoletti, M.G. (Ed.), Ecological Implications of Minilivestock. Science Publishers, INC, New Hampshire, pp. 343–352.

Owen, D.F., 1973. Man's Environmental Predicament. An Introduction to Human Ecology in Tropical Africa. Oxford University Press, London, UK.

Paoletti, M.G., Buscardo, E., Dufour, D.L., 2001. Edible invertebrates among Amazonian Indians: a critical review of disappearing knowledge. Environ. Develop. Sustainability 2000 (2), 195–225.

Quin, P.J., 1959. Food and feeding habits of the Pedi with special reference to identification, classification, preparation and nutritive value of the respective foods. Thesis, Witwatersrand University Press, Johannesburg, South Africa.

Raksakantong, P., Meeso, N., Kubala, J., Sriamornpun, S., 2010. Fatty acids and proximate composition of eight Thai edible terricolous insects. Food Res. Int. 43, 350–355.

Ramanday, E., Mastricht van, H., 2010. Edible insects in Papua, Indonesia: from delicious snack to basic need. In: Durst, P.B., Johnson, D.V., Leslie, R.N., Shono, K. (Eds.), Forest Insects as Food: Humans Bite Back. FAO Regional Office Bangkok, Chiang Mai/Thailand, pp. 105–114.

Ramos-Elorduy, J., 2004. La etnoentomología en la alimentación, la medicina y el reciclaje. In: Llorente, J.B., Morrone, J., Yañez, O.O., Vargas, I.F. (Eds.), Biodiversidad, Taxonomía y Biogeografía de Artrópodos de México: Hacia una Síntesis de su Conocimiento. Vol. 4. UNAM, México D.F., pp. 329–413.

Ramos-Elorduy, J., 2006. Threatened edible insects in Hidalgo, Mexico and some measures to preserve them. J. Ethnobiol. Ethnomed. 2, 51.

Ramos-Elorduy, J., 2009. Anthropo-entomophagy: cultures, evolution and sustainability. Entomol. Res. 39, 271–288.

Ramos-Elorduy, J., Moreno, J.M.P., 2004. Los Coleoptera comestibles de México. Annales del Instituto de Biología de la UNAM. Serie Zoología 75 (1), 149–183.

Ramos-Elorduy, J., Pino Moreno, J.M., 2002. Edible insects of Chiapas, Mexico. Ecol. Food Nutri. 41 (4), 271–299.

Ramos-Elorduy, J., Pino Moreno, J.M., Martinez Camacho, V.H., 2012. Could grasshoppers be a nutritive meal? Food Nutr. Sci. 3, 164–175.

Ramos-Elorduy, J., Moreno, J.M.P., et al., 2011. Edible Lepidoptera in Mexico. J. Ethnobiol. Ethnomed. 7 (2), 1–22.

Ramos-Elorduy, J., Pino Moreno, J.M., Martínez, V.H., 2007. Historia de la antropoentomofagia. In: Navarrete-Heredia, J.L., Quiroz-Rocha, G.A., Fierros-López, H.E. (Coords.), Entomología Cultural: Una Visión Iberoamericana. Universidad de Guadalajara, Guadalajara, pp. 239–284.

Ramos-Elorduy, J., Landero-Torres, I., Murguía-González, J., Pino Moreno, J.M., 2008. Biodiversidad antropoentomofágica de la región de Zongolica, Veracruz, México. J. Trop. Biol. 56 (1), 303–316.

Ramos-Elorduy, J., Pino Moreno, J.M., Martinez Camacho, V.H., 2009. Edible aquatic Coleoptera of the world with an emphasis on Mexico. J. Ethnobiol. Ethnomed. 5 (11).

Ramos-Elorduy, J., Mora, A.V., 2009. Composición vegetal y desarrollo poblacional de algunos acridoideos del municipio de Cuautitlán Izcalli (Estado de México, México), utilizados en la alimentación humana, con énfasis en *Sphenarium purpuracens* Ch. (Insecta: Ortghoptera: Acridoidea) y su conservación. Bull. Entomol. Soc. Aragonesa 44, 587–595.

Ramos-Elorduy, J., Muñoz, J.J., Pino Moreno, J.M., 1998. Determinación de minerales en algunos insectos comestibles de México. J. Chem. Soc. Mexico 42 (1), 18–33.

Ramos-Elorduy, J., Pino, J.M., Prado, E.E., Perez, M.A., Otero, J.L., de Guevara, O.L., 1997. Nutritional value of edible insects from the State of Oaxaca, Mexico. J. Food Compos. Anal. 10, 142–157.

Reyes-González, A., Camou-Guerrero, A., Reyes-Salas, O., Argueta, A., Casas, A., 2014. Diversity, Local Knowledge and Use of Stingless Bees (Apidae: Meliponini) in the Municipality of Nocupétaro. Michoacan, Mexico, DF.

Riggi, L., Veronesi, M., Verspoor, R., MacFarlane, C., 2013. Exploring Entomophagy in Northern Benin. Bugs for life, London, UK.

Roepke, W., 1952. Insecten op Java als menselijk voedsel of als medicijn gebezigd. Entomol. Posts 14, 172–174.

Roodt, V., 1993. The Shell Field Guide to the Common Trees of the Okavanga Delta and Moremi Game Reserve. Shell, Gaborone, Botswana.

Roulon-Doko, P., 1998. Chasse, Cueillette et Cultures Chez les Gbaya de Centrafrique. L'Harmattan, Paris, France.

Ruddle, K., 1973. The human use of insects: examples from the Yukpa. Biotropica 5, 94–110.

Santos Oliveira, J.F., Passos de Carvalho, J., Bruno de Sousa, R.F.X., Madalena Simao, M., 1976. The nutritional value of four species of insects consumed in Angola. Ecol. Food Nutri. 5, 91–97.

Schimitschek, E., 1968. Insekten als Nahrung, in Brauchtum, Kult und, Kultur. In: Helmcke, J-G., Stark, D., Wermuth, H. (Eds.) Handbuch der Zoologie–eine.

Schultz, S.A., Schultz, M.J., 2009. The Tarantula Keeper's Guide: Comprehensive Information on Care, Housing, and Feeding, third ed. Barron's Educational Series, Hauppauge, NY.

Seignobos, C., Deguine, J.P., Aberlenc, H.P., 1996. Les Mofu et leurs insectes. Journal d'Agriculture Traditionnelle et de Botanique Appliquee 38 (2), 125–187.

Sekhwela, M.D.B., 1988. The nutritive value of mophane bread—mophane insect secretion (Maphote or Maboti). Botswana Notes and Records 20, 151–153.

Shantibala, T., Lokeshwari, R.K., Debaraj, H., 2014. Nutritional and antinutritional composition of five species of aquatic edible insects consumed in Manipur, India. J. Insect. Sci. 14, 14.

Shaxson, A., Dickson, P., Walker, J., 1999. The Malawi Cookbook, Insects. Dzuka Publishing Blantyre, Malawi.

Silow, C.A., 1976. Edible and other insects of mid-western Zambia; studies in Ethno-Entomology II.

Silow, C.A., 1983. Notes on Ngangela and Nkoya Ethnozoology. Ants and termites. Etnologiska Studier 36, Goeteborg, Sweden.

Singh, O.T., Chakravorty, J., 2008. Diversity and occurrence of edible Orthopterans in Arunanchal Pradesh, with a comparative note on edible Orthopterans of Manipur and Nagaland. J. Natcon. 20 (1), 113–119.

Singh, O.T., Chakravorty, J., Nabom, S., Kato, D., 2007. Species diversity and occurrence of edible insects with special reference to coleopterans of Arunachal Pradesh. J. Natcon. 19 (1), 159–166.

Smith, A.B.T., Paucar, C.A., 2000. Taxonomic review of *Platycoelia lutescens* (Scarabaeidae: Rutelinae: Anoplognathini) and a description of its use as food by the people of the Ecuadorian Highlands. Ann. Entomol. Soc. Am. 93 (3), 408–414.

Sueur, J., 1988/1999. Les cigales dans la culture vietnamienne: de la biologie aux usages culinaires, médicinaux et symboliques. EPHE, Biol. Evol. Insectes 11/12, 55–63.

Takeda, J., 1990. The dietary repertory of the Ngandu people of the tropical rain forest: an ecological and anthropological study of the subsistence activities and food procurement technology of a slash-and-burn agriculturist in the Zaire river basin. African Study Monographs. Supplementary Issue, 11: 1–75. ASC: (675) 301.185.12Ngandu 330.191.11 392.8.01:58 631.584 639.1.

Taylor, R.L., 1975. Butterflies in My Stomach. Woodbridge Press, Santa Barbara, CA.

Tchibozo, S., LINCAOCNET. 2015. Les insectes comestibles d'Afrique de L'Ouest et Centrale sur Internet. Available from: http://gbif.africamuseum.be/lincaocnet.

Utsunomiya, Y., Masumoto, K., 1999. Edible beetles (Coleoptera) from Northern Thailand. Elytra Tokyo 27 (1), 191–198.

Utsunomiya, Y., Masumoto, K., 2000. Additions to the edible beetles (Coleoptera) from Northern Thailand. Elytra Tokyo 28 (1), 12.

Van der Waal, B.C.W., 1996. Importance of grasshoppers as traditional food in villages in northern Transvaal, South Africa. In: Jain, S.K. (Ed.), Ethnobiology in Human Welfare. Deep Publications, New Delhi, India, pp. 35–41.

Van Huis, A., 2003. Insects as food in Sub-Saharan Africa. Insect Sci. Appl. 23 (3), 163–185.

Van Huis, A., 2005. Insects eaten in Africa. In: Paoletti, M.P. (Ed.), Ecological implications of minilivestock. Science Publishers, Enfield, NH, pp. 231–244.

van Huis, A., 2013. Potential of insects as food and feed in assuring food security. Ann. Rev. Entomol. 58, 563–583.

van Huis, A., Itterbeeck, J.V., Klunder, H., Mertens, E., Halloran, A., Muir, G., Vantomme, P., 2013. Food and Agriculture Organization of the United Nations. Edible Insects: Future Prospects for Food and Feed Security, Food and Agriculture Organization of the United Nations, Rome, Italy.

van Itterbeeck, J., van Huis, A., 2012. Environmental manipulation for edible insect procurement: a historical perspective. J. Ethnobiol. Ethnomed. 8 (3), 1–19.

Vázquez-Dávila, M.A., 2008. *Nephila clavipes* (Nephilidae, Araneae). Primer reporte de aracnofagía en mesoamérica. Etnobiología 6, 91–92.

Vit, P., Medina, M., Enríquez, E., 2004. Quality standards for medicinal uses of Meliponinae honey in Guatemala, Mexico and Venezuela. Bee World 85, 2–5.

Weaving, A., 1973. Edible insects: *Ornithacris* sp. – *Homorocoryphus nitidulus* – *Brachytrypes membranaceus* – *Gryllotalpa africana* – *Petascelis remipes* – *Natalicola pallida* – *Odontotermes* sp. – *Bunaea alcinoe* – *Imbrasia epimethia*. In: A., Weaving, Insects: a Review of Insect Life in Rhodesia, Irwin Press Ltd., Salisbury, plate 6, 7, and 8.

Weiblen, G.D., 2002. How to be a fig wasp. Ann. Rev. Entomol. 47, 299–330.

Wilsanand, V., 2005. Utilization of termite, Odontotermes formosanus by tribes of South India in medicine and food. Explorer 4 (2), 121–125.

Womeni, H.M., Lindner, M., et al., 2009. Oils of insects and larvae consumed in Africa: potential sources of polyunsaturated fatty acids. Oléagineux, Corps Gras, Lipides 16 (4), 230–235.

Yen, A.L., 2005. Insects and other invertebrate foods of the Australian aborigines. In: Paoletti, M.G. (Ed.), Ecological Implications of Minilivestock: Potential of Insects, Rodents, Frogs and Snails. Science Publishers, Inc, Enfield, NH, pp. 367–388.

Yen, A.L., 2015. Insects as food and feed in the Asia Pacific region: current perspectives and future directions. J. Insects Food Feed 1 (1), 33–55.

Yen, A.L., Hanboonsong, Y., van Huis, A., 2013. In: Lemelin, R.H. (Ed.). The management of Insects in Recreation and Tourism. Cambridge Univ. Press, Cambridge, UK.

Yhoung-Aree, J., Viwatpanich, K., 2005. Edible insects in the Laos PDR, Myanmar, Thailand, and Vietnam. In: Paoletti, M.G. (Ed.), Ecological Implications of Minilivestock: Potential of Insects, Rodents, Frogs and Snails. Science Publishers, Enfield, NH, pp. 415–440.

Zagrobelny, M., Dreon, A.L., Gomiero, T., Marcazzan, G.L., Glaring, M.A., Møller, B.L., Paoletti, M.G., 2009. Toxic moths: source of a truly safe delicacy. J. Ethnobiol. Ethnomed. 29 (1), 64–76.

Zamudio, F., Hilgert, N.I., 2012. Descriptive attributes used in the characterization of stingless bees (Apidae: Meliponini) in rural populations of the Atlantic forest (Misiones-Argentina). J. Ethnobiol. Ethnomed. 8, 9.

Zamudio, F., Kujawska, M., Hilgert, N.I., 2012. Honey as medicinal and food resource: comparison between Polish and multiethnic settlements of the Atlantic Forest, Misiones, Argentina. Open Complementary Med. J. 2, 58–73.

Subject Index

A

Absolute ratios, 158
Acanthoplus spiseri, 42
ACE. *See* Angiotensin converting enzyme (ACE)
Acheta domesticus (L.). *See* House crickets (*Acheta domesticus*)
Acheta domesticus densovirus (AdDNV), 123, 179
AdDNV. *See* A. domesticus decimated by infections of the densovirus (AdDNV)
ADF fraction, 77
Administrative regulation of food safety and quality standards, 206, 236
    cleaning and sanitizing procedures, 240
    good manufacturing practices (GMP), 238
    International Administration, 237
    labelling regulation and health claims applicable to insects, 207
    milk sanitation, 237
    pest control, 238–239
    prerequisite programs, 238
    processing and product quality recommendations, 122, 126
    regulatory considerations for insects-as-food ingredient, 204–206
    safety considerations for insects as food, 207–211
    sanitary equipment design, 239
    sanitary facility design, 239
    sanitation, 239
    training and personal hygiene, 238
    United States Department of Agriculture (USDA), 237
    United States Food and Drug Administration (FDA), 237
Advanced glycation end products (AGE), 265
Aeromonas hydrophila Chester, 232
AGE. *See* Advanced glycation end products (AGE)
Agricultural systems, 3
    productivity of, 3
Agricultural technology, 94
Agriprotein, 154
Air circulation network, 183
Air conditioners equipment, 163
Air flow assay and nonair exchange assay, 185
Air movement tendency, 187
Air renovation system, 188
Alcoholic beverages, 249
Alignment, of insect tropomyosins, 261
Allomyrina dichotomus L., 49
Alphitobius diaperinus Panzer, 282
American cockroach *(Penaeus americana)* tropomyosin, 260
American lobster (*Homarus americanus* Mine-Edwards), 262
Amino acids, 63
    composition, 3
    essential, 4
Amnesic shellfish poisoning (ASP), 236
Anabrus simplex, 36
ANF. *See* Antinutrient factors (ANF)
Angiotensin converting enzyme (ACE), 155
Animal and Plant Health Inspection Service (APHIS), 21
Animal based protein, 7
    commodities, 117, 142
Animal-derived food
    large-scale production of, 7
    processing and products, 122, 126
Animal-derived protein
    demand for, 3
Animal feed
    insect farming for, 154
Animal feeding, 273
Animal feed, insects in
    justification, 273–274
    overview of, 274–276
Animal producing farm
    needs for, 122
Animal production, 93
Antheraea mylitta Drury, 280
Antheraea pernyi, 39
Anthropocene, 2
Anthropoentomophagy, 29
Antibiotic resistance, 297
Antibiotics, 212
Antinutrient factors (ANF), 157
Antioxidant enzymes, 155
APHIS. *See* Animal and Plant Health Inspection Service (APHIS)
Aposematism, 41
Aquaculture feeding, 283
Aquatic insects, 36
Arginine kinase, 260
    allergenicity, 260
Artificial light systems, 174
Ascorbic acid, 79
Asiatic rhinoceros beetle, 290
Asiatic Rhinoceros Beetle (*Oryctes rhinoceros* L.), 283
ASP. *See* Amnesic shellfish poisoning (ASP)
Aspergillus flavus, 99
Atlantic Salmon (*Salmo salar* L.), 288
Automatic watering system, 179
Automation control mechanisms, 195–196
    artificial intelligence, 195–196
    derivative effect, 195
    integral effect, 195
    proportional effect, 195
Automation control platforms, 194–195
    programmable automation controllers, 195
    programmable logic controllers, 194–195

B

Bacillus cereus enterotoxin, 234
Backyard food production
    form of, 5
Bacterial foodborne infections, 232
Bacterial hazards, 229
Bacterial population, typical heat inactivation curve, 246
Batch approach
    benefits to, 176
BBC. *See* British Broadcasting Company (BBC)
Bill and Melinda Gates Foundation, 14
Bio-cycle linked observations, 50
Biodiversity
    advantage of, 90
    of plants, 90
Biogeographical realm, 31
Biologically-derived carmine dye, 212
Biomass, 36, 85
Biphenyl-contaminated water, 212
Bivoltine breeds, 99
Black Bullhead (*Ameiurus melas* Rafinesque), 287
Black soldier fly (*Hermetia illucens*), 287, 289, 291
    larvae, 175, 292
        mass production of, 176
    life-history of, 173
    prepupae meal, 283
    production, 173–176
        adult colony, 173
        adult management, 174
        costs, 175
        larval maintenance, 175
        larval production, 175–176
        mating behavior, 174–175
        oviposition, 175
        quality assurance, 176
        storage of prepupae, 176
    production of, 173

Blister beetles, 48
Blue swimmer crab *(Portunus pelagicus)*, 265
Blue Tilapia *(Oreochromis aureus* Steindachner), 291
Bombyx mori L., 39, 49, 52, 63, 67, 72, 74, 90, 120, 153, 155, 258, 274, 276
Boopedon flaviventris, 274
Botulism, 235
Bovine spongiform encephalopathy (BSE), 234
Brachytrupes membranaceus, 43
Breeding animals, 93
Brewer's yeast granules, 177
British Broadcasting Company (BBC), 47
Broiler chickens, 293
Broiler diets
    groundnut cake (GNC), 278
Broilers fed cassava peel-maggot meal mixture, 278
BSE. *See* Bovine spongiform encephalopathy (BSE)
Bug feasts, 47
Bug fests, 47
    at Garfield Park Nature Center, 47
Butterworm *(Chilecomadia moorei* Silva Figueroa), 225

C

CAFO. *See* Confined animal feeding operations (CAFO)
Calcium binding proteins, 263
Camponotus inflatus Lubb., 38
Campylobacter jejuni, 231
    gastroenteritis, 231
Canadian Food Inspection Agency of Health Canada, 205
Cannibalization, 169
Canning, 247
Carbohydrates, 160
Carbon dioxide
    gas, 249
Carbon rich feedstock, 166
Carcass yield, 277
Cellulose, 77
Center for Food Safety and Applied Nutrition (CFSAN), 206
Center of Veterinary Medicine (CVM), 207
Central HVAC system scheme, 189
Certificate of analysis (COA), 209
CFD. *See* Computational fluid dynamics (CFD)
CFSAN. *See* Center for Food Safety and Applied Nutrition (CFSAN)
cGMP. *See* Current good manufacturing practices (cGMP)
Channel Catfish *(Ictalurus punctatus* Rafinesque), 287, 291
Cheap production methods, 105
Chemical hazards, 241
Chemical-intensive methodology, 129
Chemically treated feed, 99
Chemical pesticides, 94
    applications of, 93
Chemical screening, 49
Chicken feed, 100

Chitin, 76, 77, 79, 81
    digestibility of, 299
    increased activity, in innate immune system, 297
Chitin-chitosan, 78
Chitosan, 14, 77–79
    as chelating agent of, 82
    hypolipidemic influence of, 78
Choline, 161
Cirina forda (Emporer moth), 38, 42, 210 Westwood, 280
Clarification, 82
Cleaned rearing boxes, 165
Clostridium sporogenes Metchnikoff PA 3679, 247
COA. *See* Certificate of analysis (COA)
Cobia *(Rachycentron canadum* L.), 294
Cochineal derived coloring
    industrial use of, 43
Cochineal insect *(Dactylopius coccus* L.), 259
Codex Alimentarius, 21, 105
Codex Alimentarius Commission, 237
Coleoptera, 31, 161
    role in nervous system, 161
Collective adaptability, 31
College of Agriculture, 17
College of Health and Human Development, 17
Coloradia pandora lindseyi, 50
Commoditization, 3
Common Green Bottle Fly *(Lucilia sericata* Meigen), 290
Community forestry, 88
Complex coacervation, 157
Complex social system, 2
Computational fluid dynamics (CFD), 190
Confined animal feeding operations (CAFO), 125
Conimbrasia belina, 63, 67
Consumer acceptance panels, 81
Contemporary agricultural systems, 11
Cooking temperatures, 247
Cooling process, 193
Corn-cricket diet, 280
Corn-soybean diet, 278
Corrugated cardboard blocks, 175
Coyote Creek Farm, 142
Crab extracts *(Scylla paramamosain)*, 266
Crayfish *(Procambarus clarkii)*
    thermal treatment of, 266
Creutzfeldt-Jakob disease, 234
Cricket paralysis virus (CrPV), 123
Crickets. *See* House crickets *(Acheta domesticus)*
Cropping systems, 90
CrPV. *See* Cricket paralysis virus (CrPV)
Crude oil
    for transportation, 9
Crustaceans feed, insects, 291
Crustacea species, 264
CSMA housefly formulation, 182
Cultivate feed crops, 96
Cultural ethnocentrism, 50
Current edible-insect food industry, 135
Current global energy production, 7

Current good manufacturing practices (cGMP), 206
Current improved technique, 101
Current insect-based food industry, 130
CVM. *See* Center of Veterinary Medicine (CVM)
Cyanogenic species, 296

D

Dactylopius coccus Costa, 42
DAEC. *See* Diffusely adherent (DAEC)
Danger zone, 130
DANIDA. *See* Danish International Development Agency (DANIDA)
Danish International Development Agency (DANIDA), 204
Database of Origin and Registration, 39
Deccan Mahseer *(Tor khudree* Sykes), 286
Dehumidifiers, 194
Dehydrated silkworm pupae meal, 292
Dermatophagoides pteronyssinus (Trouessart), 257
Diarrhetic shellfish poisoning (DSP), 236
Diatraea grandiosella, 184
Diet, 31
Dietary protein, animal-derived, 62
Dietary restrictions
    rationale for, 34
Dietary Supplement Health and Education Act (DSHEA), 206
Diet containing protein
    Nutrid, 98
Diets containing maggot meal, 279
Diffusely adherent (DAEC), 231
Digestive enzymes, 291
Diphyllobothrium latum L. (raw fish), 233
Diptera, 274, 283
Disease-causing agents, 97
Disgust factor, 13, 19
Disk Abalone *(Haliotis discus* Reeve), 292
DNA
    sequences, from abdominal crayfish muscle, 261
    synthesis of, 161
Docosahexaenoic acid, 44
Drosophila melanogaster, 155
Drum drying, 134
Drying methods, 132–134
    drying pastes, slurries, and liquids to produce powders and meals, 133
    whole insects, 132
DSHEA. *See* Dietary Supplement Health and Education Act (DSHEA)
DSP. *See* Diarrhetic shellfish poisoning (DSP)
Dust allergies, 264
Dytiscus marginalis L., 296

E

Eat-a-Bug Cookbook, 46
ECI. *See* Efficiency of conversion of ingested food (ECI)
Ecological footprint, 90
Economic system, 36, 51
EDI. *See* Expected dietary intake level (EDI)

Edible insect resources, 52
Edible insects, 1–24, 268
   aims, 2
   animal-derived protein, population growth and rising demand for, 3–4
   aquaculture and environment, 7
   call to action, 23
   climate change and agricultural productivity, 6–7
   consumption of, 50
   cultural avoidance of, 2
   current trends in using insects as food, 17–18
   definition of terms, 20
   farms, 163
   funding and legislation, 14–15
   historic relevance of, 2
   increasing recognition in the academic sector, 15–17
   industry, 104, 117
   insects are an important and feasible solution, 11–12
   insects as a living source of protein in space, 10–11
   land use, 4–5
   limits to nonrenewable energy, 7–10
   mass production, 87, 118, 154, 156
   products, 114, 115, 224
      geographical distribution of, 13
   protein worldwide, our sources of reevaluation of, 2–3
   psychological barriers and disgust, 18–20
   research, 15, 116, 118
   role in diet, 2
   and spiders of world, 32
   summary of, 20–23
   treatment of, 1
   urban and vertical agriculture, 5–6, 114
   use of, 37
   water use, 10
   worldwide acceptance of insects as food, 12–14
Edible insects farming, 85–106
   biodiversity and availability of insects, 89–90
   commercial insect farming for mass production, 97–103
     farming in space, 102–103
     indoor farming, 97–101
       house cricket, 100–101
       mulberry silkworm, 98–100
       yellow mealworm, 101
     outdoor farming, 101–102
       bamboo caterpillar, 101–102
       grasshoppers, 102
       other insects, 102
       Palm Weevil/Sago Larva, 101
       weaver ants, 102
   consumption of insects *vs.* livestock, 90–91
   cost of cultivation, 91–93
   current challenges and conclusions, 105–106
   environmental impact, 94–96
   food security/family livelihood, 87–89
   industrial perspective, 96–97, 118, 148
   market potential, 103–105
     export, 104
     new products from farmed insects, 104–105
     retail/local marketing, 103
     replacing livestock with insects as human food, possibility of, 93–94
   safety regulations, 105
Edible species
   farming of, 97
Eeeuwh factor, 19
Efficiency of conversion of ingested food (ECI), 90, 91, 114, 158
EFSA. *See* European Food Safety Authority (EFSA)
Egg yolk cholesterol, 281
ELC. *See* Essential light chains (ELC)
Electronic probes, 186
ELISA. *See* Enzyme-linked immunosorbent assay (ELISA)
Embarrassment factor, 19
Encapsulation technique, 157
Endoxyla leucomochla, 37
Energy
   availability and use, 9
   use for transportation, 9
Energy-intensive (electricity-intensive) equipment, 132
Ensiling, 300
Entamoeba histolytica, 233
Enterobacter asburiae, 224
Enteroinvasive strains, 231
Enteropathogenic strains, 231
Enterotoxin, 234
Enterprise resource planning (ERP), 164
Entomophagous resources, 45
Entomophagy, 29
   advantages of, 46
   educators, 47
   industry, 117, 118
   phenomenon, 47
   role of, 53
   terminology recommendations, 124–125
   by Western cultures, 204
Environmental degradation, 7
Environmental Protection Agency (EPA), 206
Enzyme-linked immunosorbent assay (ELISA), 260, 266, 267
EPA. *See* Environmental Protection Agency (EPA)
Ephestia kuehniella Zell
   rearing room holding, 164
Equipment and automation system, 184
Eri silkworm, 102
ERP. *See* Enterprise resource planning (ERP)
Escherichia coli, 295
Essential amino acid content
   for selected insect species and common high protein commodities, 69
Essential light chains (ELC), 262
Ethanol toxicity, 249
European-American society, 46
European Commission, 15
European Food Safety Authority (EFSA), 21, 156, 204, 225
European Seabass (*Dicentrarchus labrax* L.), 288

European Union (EU), 21
Exorista sorbillans, 99
Expected dietary intake level (EDI), 209
Export-oriented insects, 104
Exposure, to whole-body insects, 213
Extracted insect proteins, 81
Extraction and purification methods, 80

F

FALCPA. *See* Food Allergen Labeling and Consumer Protection Act of 2004 (FALCPA)
FAO. *See* Food and Agriculture Organization (FAO)
Farmed insects
   international trade of, 104
Farming insects, 88
   benefits of, 91
Farm-raised superworm (*Zophobas morio* Fab.), 225
Fatty acids
   composition of insect species, 76
   and common high-protein commodities, 74
   profiles, 293
FCE. *See* Feed conversion efficiency (FCE)
FCR. *See* Feed conversion rate (FCR)
FD&C Act. *See* Federal Food, Drug, and Cosmetic Act (FD&C Act)
Federal Food, Drug, and Cosmetic Act (FD&C Act), 127
Feed conversion efficiency (FCE), 5, 91, 184
Feed conversion rate (FCR), 277
Feed conversion ratio, 90
Feed handling, 164
Feed ingredients, 122
Feeding trials
   on trout diets, 156
Feed manufacturing technologies, 156
Feed-to-product conversion ratios, 9
Fermentation processes, 39
Field crickets (*Gryllus bimaculatus* De Geer), 258
Fig tree, 42
Filters, 191–192
   carbon filter, 192
   electrostatic precipitator, 192
   high-efficiency particle arresting, 192
   mechanical filters, 191
   ultraviolet (UV) filter, 192
Fish feed, insects, 283–290
Fishmeal, 7, 273, 277
   based diets, 156
   replacement
     with maggot meal, 279
   substitution, 278
Flacherie disease, 99
Flow process scheme, 177
Fluidized bed chromatography, 79
Food
   commodity prices, 10
   production of, 4
Food Allergen Labeling and Consumer Protection Act of 2004 (FALCPA), 257

Food allergy, 255–257
  allergen detection methods, 266–267
    enzyme-linked immunosorbent assay (ELISA), 267
    immunological methods, 266
    mass spectrometry methods, 267
    polymerase chain reaction (PCR), 267
  food processing, effects, 131, 264–266
  insects connection, 257–259. *See also* Insect, allergens
Food and agricultural industries, 149
Food and Agriculture Organization (FAO), 2, 4, 30, 203
Food and Drug Administration (FDA), 21, 42, 79, 105, 132
  EPA seafood guidance, 207
  Federal Food, Drug and Cosmetic Act (FFDCA, 1908), 206
Foodborne bacterial intoxications, 234
Foodborne diseases
  hazards
    Bacillus cereus enterotoxin, 234
    bacterial infections, 229–230, 232
    botulism, 235
    campylobacteriosis, 231–232
    chemical intoxications, 235–236
    Clostridium perfringens, 232
    enteroviruses, 233
    Escherichia coli, 231
    foodborne bacterial intoxications, 234
    infectious hepatitis, 233
    insects, 225–229
    listeriosis, 232
    multicellular parasites, 233
    nonbacterial foodborne infections, 232
    physical hazards, 236
    prions, 234
    salmonellosis, 230
    shigellosis, 230
    Staphylococcus aureus enterotoxin, 234
    vibriosis, 230
    yersiniosis, 231
  microbial intoxications, 234
  outbreaks, 229
  transmission, 227
  in United States, 226
Food budgets, 39
Food choice
  umbrella themes, 126
Food conversion
  efficiency of, 167
Food defect action levels, 127
Food-induced anaphylaxis, 256
Food industry, 129
Food ingredient, for poultry, 282
Food Insect Festival in Montreal, 42
Food insects, 14
Food Insects Newsletter, 1, 30
The Food Insects Newsletter, 46
Food plants
  grand diversity of, 6
Food policy makers, 15, 19
Food preservation, 130, 131, 244, 245
  asepsis/removal, 245–246
  drying, 248
  fermentation, 249
  high-temperature processing, 246–247
  irradiation, 249
  low-temperature preservation, 247–248
  modified atmosphere conditions, 246
  preservatives, 248–249
  principles of, 245
Food preservatives, 248
Food processing, 126
  drying methods, 133
  equipment, 13
  methodologies, 127
  regulatory considerations, for insects-as-food ingredient, 216–217
  technology, 157
Food production, 4
  types of, 8
Food products, 79
  "authenticity" of, 17
  availability of, 12
  commercial, 116
Food protein sources, 255
Food reserves, 85
Food resource, 43
Food safety, 127
  issues, 37, 106
Food Safety and regulatory issues, 21
Food Safety Modernization Act (FSMA) of 2011, 206, 243–244
  food safety plan elements, 243
Food Science, 17, 23, 126
Food security, 10, 54, 90, 121
  and sustainability, 4, 90
Food-service operators, 224
Food Standards Agency (FSA), 15
Food system, 2, 6
  global warming potential of, 6
Food taboos, 18
Food waste, 10
Forest-farmed insects, 105
Forest Insects as Food: Humans Bite Back, 116
Fossil fuels, 7
Freeze-drying, 132
Freezing, 130
Freshwater prawns (*Macrobrachium rosenbergii* De Man), 258
Frozen shipping
  for raw insects, 147
FSA. *See* Food Standards Agency (FSA)
Fuel
  world energy consumption by, 9

G

Galleria mellonella (L.), 21, 63
  larval development of, 182
  natural food of, 182
GAP. *See* Good agricultural practices (GAP)
Gastrointestinal tract (GIT), 79
Gender-based inequalities, 5
Generally recognized as safe (GRAS), 206
  determinations, 207
  substances, 235
Genetically modified organisms (GMO), 214
  free blends, 122
Genotoxic assessment, 210
GHGEs. *See* Greenhouse gas emissions (GHGEs)
Giant River Prawn (*Macrobrachium rosenbergii* De Man), 291
GIT. *See* Gastrointestinal tract (GIT)
Global food system, 3
Globalization, 45
Global warming, 94
GLP. *See* Good laboratory practices (GLP)
Gluten sensitivity, 124
GMO. *See* Genetically modified organisms (GMO)
Good agricultural practices (GAP), 224
Good laboratory practices (GLP), 208
GPS-based precision farming, 5
Grain plants, 122
Gran Mitla, 87
GRAS. *See* Generally recognized as safe (GRAS)
Grasshoppers
  farming various species of, 102
Greenhouse gas emissions (GHGEs), 3, 6, 10, 20, 85, 94
  equivalent production, 95
  total global emissions of, 95
Grinding/milling, 134–135
  dry milling, 135
  wet milling, 134
Gryllodes sigillatus (Walker), 123, 154
Gryllus assimilis (Fab.), 154
Guillian-Barré syndrome, 231

H

Habitat management, 88
HACCP. *See* Hazard analysis and critical control points (HACCP) system
Harvested insect, 88
Harvesting, 130
  insect species, industrial method of, 125
  separation apparatus, 183
  techniques, 102
Hazard analysis and critical control points (HACCP) system, 106, 204, 241
  HACCP plan development, 242–243
  principles, 242
HDM. *See* House dust mites (HDM)
Heating, 131
Heat transfer, by conduction, 247
Heat-treated crustacean samples, 264
Heavy metal contamination, 176
Hemocyanin, 263
HEPA. *See* High-efficiency particulate arresting (HEPA) filters
Hepatitis A, 232
  virus, 233
Hermetia illucens L., 274
Hermetia meal, 278
Herodotus, 35
High-density polyethylene (PET)
  films, 82
High-efficiency particulate arresting (HEPA) filters, 187
High-fiber plant sources, 78

High heat exposure, 132
High pressure processing (HPP), 131
High-protein agricultural product, 103
High-protein beverages, 116
High-protein commodities, 67
High-protein feed, 101
High-protein food ingredients, 127
High-quality animal protein, 93
High-quality protein ingredients, 143
High-temperature short time (HTST), 247
Holometabolous insects, 33
   stages of, 91
Hominids
   diet of, 33
Honeybees, 35
Honeypot ants, 38
House crickets *(Acheta domesticus)*, 15, 69, 87, 119, 205, 224
   farming methods, 135
   farming sector, 101
   meal, 280
   nymphs, 79
   production in United States, 179–182
      diets and feeds, 180
      environmental conditions, 180–181
      rearing units, 179–180
      reproduction, 181–182
House dust mites (HDM), 257
House Fly *(Musca domestica)*, 281, 284, 285, 286, 289, 290
   egg collection, 178
   maggots (*M. domestica* larvae), 278
   production, 176–179
      feed and feed ingredients, 177
      flow chart process scheme for, 177
      production process, 177–179
HPP. *See* High pressure processing (HPP)
HTST. *See* High-temperature short time (HTST)
Human alienation, 46
Human consumption
   animal proteins for, 52
Human entomophagy, 85
Human food, 87
   ingredient, 91, 122
   security, 115
   source, 97
   supply, 91
Human-insect relationship, 45
Humectants, 248
Humidifiers, 193–194
   evaporative pads, 193
   nebulization with high and low pressure nozzle, 193
   steam humidifiers, 193–194
   ultrasonic nebulization, 193

I

Imbrasia belina Westwood, 225
Immunoglobulin E (IgE) antibodies, 256
   binding, 265, 266
   binding proteins, 262
   cross-reactivity, with shellfish allergens, 264
   mediated allergies, 256

   sensitization and elicitation steps necessary, 257
Immunological methods, 266
Indoor/semiindoor farming, 88
Industrial farming techniques, 148
Industrial insect mass rearing techniques, 96
Industrial scale insect farms, 118
Industrial system automation, 194
Insect
   allergens, 259–264
      arginine kinase, 260
      hemocyanin, 263
      known aero allergens, 264
      myosin light chain, 262
      novel allergens, 264
      phospholipase, 264
      sarcoplasmic calcium binding protein (SCP), 261–262
      sarcoplasmic endoreticulum calcium ATPase (SERCA), 263
      tropomyosin, 259–260
      troponin C, 263
   based diets, 106
   based food and commodity industry, 141
   beneficial uses of, 37
   consumption by humans, 29
   cultural restrictions, 43–48
      disgust factor, 44–47
      educational campaigns, 47–48
      food taboos and religious/dietary restrictions, 43–44
   edible insects
      as nutraceuticals, 48–50
      of world, 31–33
   ethnoentomology, 50–51
   fiber component of, 14
   final comments and recommendations, 53–54
   as food, 29–54
   functional properties of, 81
   harvesting and cultivation, 51–53
      farming of edible insects, 52–53
      hemimetabolous, 76
   history of insects as human food, 33–36
      age of reason and emergence of scientific era (1700s/18th century), 35–36
      archeological data, 33–34
      early historic times (3600BC-500AD), 34–35
   large-scale processing of, 13, 118, 130
   modern cultural uses, 36–43
      edible insects around world, selected classic examples, 37–42
      indirect or unintentional presence in foods, 42–43
   production, 91
   soil-dwelling feeding, on solid waste, 212
   species
      advantages of, 165
      for feed and food, 155
      proximate analysis data, 63
   used in animal food, 276
   utilization of, 30
   world map showing numbers of insect species, 30
Insect-based food companies, 125
   list of, 136

Insect-based food entrepreneurs, 123
Insect-based food industry, 103, 124, 125, 140, 142, 148, 154
Insect-based food ingredients, 143
   production methods, 127
Insect-based food processors, 142
Insect-based food producers, 148
Insect-based food products, 12, 14, 123, 128, 144
   tasting contest, 47
Insect-based ingredient format, 128
Insect-derived chemicals, 49, 123, 212
Insect-derived protein, 143
Insect-developed powders, 115, 116, 207
Insect-eating cultures, 31, 135
Insect-eating events, 47
Insect-eating populations, 79
Insect farmers, 146
Insect farming, 90, 96
   benefits of, 91
   industry, 118, 123
   systems, 10
   type of, 88
   vs. other livestock, 86
Insect farm machinery, 185
Insect feeds
   manufacturing processes for, 157
Insect fossils, 33
Insect mass production technologies, 153–196
   current methodologies of, 21
   environmental control for efficient production of insects in general, 184–196
      air flow design, 186–190
      automatic control and artificial intelligence
         automation control mechanisms, 195–196
         automation control platforms, 194–195
         automation in insect production industry, 194
      equipment for climate control, 191–194
         dehumidifiers, 194
         filters, 191–192
         humidifiers, 193–194
         temperature conditioners, 192–193
      estimating optimal conditions required for design for insects, 184–186
         continuous air flow assay, 186
         light requirements, 186
         nonair exchange assay, 185–186
         space and location, 186
         thermal requirements for production, 184–185
         vital measurements assay, 185
   insect mass rearing equipment and mechanization, considerations for, 162–166
      cleaning room, 165–166
      compost area, 166
      feeding and watering, 164–165
      production and operation management, 162–163
      rearing area, 163–164
      separation and sorting room and product traceability, 165

Insect mass production technologies (cont.)
  mass-produced insect species and respective applications, 154–155
    insects for food and feed, 154
    insects for medicinal use, 155
    insects for other applications, 155
  nutritional requirements for farmed insects, 158–162
    macronutrients, 158–160
    micronutrients, 160–162
  production techniques by species, 166–184
    black soldier fly, *Hermetia illucens*, production, 173–176
    cricket production in United States, 179–182
    housefly, *Musca domestica*, production, 176–179
    mealworm production technologies, 166–173
    waxworm production, 182–184
  using conventional feedstock for rearing insects, potential of, 156–158
    manufacture of insect feed, 157–158
    principals of feed production
      liquid and semisolid feeds, 157
      solid feed presentations, 157
Insect meals, physical/chemical treatments
  defatting, 300
  disinfection, 300
  drying, 300
  silage, 300
Insect nutritive value, 299
Insect-producing community, 53
Insect production industry, 147, 184
Insect production systems, 190
Insect-rearing facilities, 5
Insects nutrient content
  and health benefits, 61–82
  insect physiology and functionality, 77–79
  insect protein functionality, 81–82
  insects as food ingredient, 61
  nutrient content, 61–76
Integrated pest management (IPM), 238
  campaigns, 52
Integrated rearing system, 163
Interconnected systems
  benefits of, 3
Intercropping systems, 3
Internal filtration system, 188
Internal organ measurements, 277
Intoxications
  foodborne bacterial intoxications, 234
Invertebrates, 31
Iodine containing sanitizers, 240
IPM. *See* Integrated pest management (IPM)
Iron and zinc deficiency, 146
Irrigated crops, 6

J

Japanese Seabass (*Lateolabrax japonicus* Cuvier), 288
JECFA. *See* Joint FAO/WHO Expert Committee on Food Additives (JECFA)
Joint FAO/WHO Expert Committee on Food Additives (JECFA), 209

K

Kuruma prawn (*Marsupenaeus japonicus* Spence Bate), 265

L

Laminar air flow, 187
Laminar flow system, 189
Land sparing, 5
Large-scale insect harvests, 130
Lateral flow devices (LFDs), 266
LCA. *See* Life cycle analysis (LCA)
LED. *See* Light emitting diode (LED) lights
Legal pasteurization, 131
Legislation, 106
  process, 15
Lepidoptera, 274
Lesser meal worms (*Alphitobius diaperinus* Panzer), 225
LFDs. *See* Lateral flow devices (LFDs)
Life cycle analysis (LCA), 96
Light emitting diode (LED) lights, 186
Lipids, 160
Lipoprotein carrier molecules, 160
Liquid feeds, 157
Liquid fuels, 9
Liquid/liquefied feeds, 156
Listeria monocytogenes, 225, 277
Livestock animals, environmental enrichment, 299
Livestock feed ingredients
  animal welfare, 297
  feed security/safety, 295
    antibiotic resistance, 297
    antinutritional compounds, 296
    digestive enzyme inhibitors, 296
    heavy metals/pesticides/contaminants, 295
    insect derived toxins/venoms/allergens, 296
    microorganisms, 295
    preservation/storage, 296
  geographic distribution of papers, 276
  nutritional, 293
    chitin, 294
    lipids/fatty acids, source of, 293–294
    protein/amino acids, source of, 293
    vitamins and minerals, 294–295
  papers per year, 275
  research and technological advancement
    ecological aspects/sustainability, 298
    environmental benefits, 298
    mass rearing, selected species for, 298
    native insects by local farms, 298
    processing and commercial production, 116, 120, 121, 129
Livestock protein, 93
Livestock systems, 3
Locusta migratoria (L.), 39
Locustana pardalina, 52
Lophocampa hickory moth, 213
Low-carbon farming methods, 105
Low-cost rearing houses, 98

M

Macrotermes sp., 286
  Macrotermes nigeriensis, 284
  mounds, 3
Maggot meal, 278
  fish meal replacement, 279
Maggot therapy, 155
Malnutrition, 145
Mammalian gut chitinases, 294
Marine ecosystems
  biodiversity of, 7
Market research firms, 116
Mass-produced insects, 128
Mass production technologies, 20, 154
Mass rearing acclimatization system, 191
Mass spectrometry methods, 267
MDGs. *See* Millennium development goals (MDGs)
Mealworm, 120
  larvae separation system, 170
  in plastic trays, 171
  production technologies, 166–173
    in China, 170–173
    diets and feeds, 167–168
    larvae and adult rearing, 166–167
    rearing density, 169
    separation, 169–170
    watering, 168–169
  rearing in paper containers, 171
  rearing systems, 169
  screened bottom tray system, 168
Megachile nigeriensis, 285
Melanoplus mexicanus, 274
Metabolic processes, 94
Microbes, associated with insects, 224–225
Microbial community, 224
Microbial contamination, 176
Micro-livestock, 298
Microorganisms, 234
Micro tests
  standard panel of, 131
Microwaves, 249
  drying machine, 172
Migratory Locust (*Locusta migratoria* L.), 281, 284
Millennial generation worldwide, 48
Millennium development goals (MDGs), 116, 144
Mini-farm systems, 52
MLC. *See* Myosin light chain (MLC)
Modern insect-based food industry, 113–150
  biodiversity, 115–117
  current insect farming industry, 117–121
    modern industrial mass production of insects, 118–121
      crickets, 119
      current industrial insect farms and predominant farmed species, 118–119
      grasshoppers, 120
      mealworms, superworms, and buffalo worms (lesser mealworms), 120
      other species, 120–121
  efficiency, 114–115
  intriguing larger food industry, uses of insects as industrial food ingredients, 148–149

real pioneers, entrepreneurs in insect-based food space, 135–146
  current companies, farms, and other organizations, 140–146
  selection for aspiring insect-based food producers and insect farmers, recommendations and considerations for, 122–135
    animal welfare, 125
    diseases affecting mass produced/farmed insects, 123
    farming insects, feed formulations and biomass sources for, 122
    "fear factor" important considerations for normalizing insects as mainstream food ingredient beyond the novelty niche, 125–126
    insect processing considerations, 126–134
      drying methods, 132–134
      grinding/milling, 134–135
      harvesting, packaging, and storage, 128–130
      "kill step" mitigating microbiological risk, 131–132
      product decision, 128–130
    NASCAR jacket of food, 123–124
    underutilized biomass amenable as feed ingredients for mass-farmed edible insects, 122–123
  supply chain needs, 146–148
    farms and farmed species, 147
    feed, 146–147
    processing and manufacturing infrastructure, 148
    transportation, storage, and distribution, 147–148
Mollusks feed, insects, 292
MoniQA Food Allergen Reference Material Task Force, 267
Montana State University (MSU), 18
Mopane caterpillars, 87
Mormon Cricket (*Anabrus simplex* Haldeman), 280
Moyiomoyio, 43
Mozambique tilapia (*Oreochromis mossambicus* Peters), 293
MSU. *See* Montana State University (MSU)
Mudfish (*Clarias anguillaris* L.), 285
Muga silkworm, 279
Mulberry silkworm
  constraints and measures, 99–100
  salient points, 98–99
Multiple-product food-insect systems, 53
Musca domestica L., 174–175
Myosin light chain (MLC), 262
Myrmecocystus, 38
Myrmecocystus mexicanus, 38

N

Nano-sized insect powders, 215
NASA. *See* National Aeronautics and Space Administration (NASA)
National Aeronautics and Space Administration (NASA), 241

National policy-making circles, 19
National Symposium on Anthropoentomophagy, 48
Natural host plants, 98
Natural resources, 43
  medicinal compounds from, 49
Natural sciences, 51
NDI. *See* New dietary ingredient (NDI)
Nematodes (roundworms), 233
Neurotoxic shellfish poisoning (NSP), 236
New dietary ingredient (NDI), 207
Niacin, 294
Nile tilapia diets, 283
Nitrous oxide emissions, 95
NLEA. *See* Nutrition Labeling and Education Act (NLEA)
NOAEL. *See* No-observed-adverse-affect-level (NOAEL)
Nonbacterial foodborne infections, 232
Nonfood crop biomass, 147
Nonpathogenic species, 131
Nonprofit/nongovernmental organizations, 144
Nonrenewable energy, 14
  discussion of, 10
Nonrenewable fuels, 9
Nonrenewable sources of energy, 7
No-observed-adverse-affect-level (NOAEL), 207
Nordic Food Lab, 104
Noroviruses, 232, 233
Notable allergen content, taxonomic relationship, 258
Notonecta species, of Ephemeroptera, 293
Noxious substances, 212
NSP. *See* Neurotoxic shellfish poisoning (NSP)
Nuclear polyhedrosis virus, 99
Nucleoproteins, 160
Nutraceutical entomofauna, 48
Nutrients
  content, 115
  isolation, 289
  source of, 115
Nutritional balance, 176
Nutritional biochemists, 31
Nutrition Labeling and Education Act (NLEA), 207
Nutrition security, 126
Nutritive needs, of monogastric species, 273

O

Obesity
  in humans, 77
OECD. *See* Organization of Economic Cooperative Development (OECD)
Oecophylla smaragdina, 37
Oils
  by-product as a biofuel, 23
  from insects, 128
Omega-3 fatty-acid content, 296
Omega-3 fatty acids, 293
Omega-6 fatty acids, 11
Oral tolerance, 256
Organization of Economic Cooperative Development (OECD), 208

Organophosphates/neuroactives, 212
Organophosphorous insecticides, 99
Oriental leafworm moth larvae/caterpillars (*Spodoptera litura* Fab.), 295
Oxidative stress, 297
Oxygen deprivation, 130

P

Pace-setting culinary explorations, 17
Pachilis gigas L., 42
Packaging
  regulatory considerations, for insects-as-food ingredient, 216–217
PACs. *See* Programmable automation controllers (PACs)
Paleo diet, 124
PAP. *See* Processed animal proteins (PAP)
Parallel hypothesis, 44
Paralytic shellfish poisoning (PSP), 236
Parasitoids, 191
PCBs. *See* Polychlorinated biphenyls (PCBs)
PCR. *See* Polymerase chain reaction (PCR)
PDO. *See* Protected Designation of Origin (PDO)
Peer-reviewed articles, 81
Pelleting, 157
Penaeus monodon Fab., 259
Periplaneta americana L., 48, 121
Pesticides, 37
Pest management, documentation of, 239
Phanerotoxic insects, 42
Pharmaceutical compounds
  bioprospecting for, 49
Phassus triangularis Edwads larvae, 293
Phospholipase, 264
Phospholipids, 160
Phytochemicals, 212
PID. *See* Proportional integral derivative controllers (PID) system
Pigs feed, insects, 282–283
Pink shrimp (*Pandalus eous* Makarov), 262
Plant-derived products, 99
Plastic drawers, 100
PLC. *See* Programmable logic controller (PLC); Protein, lipid, and carbohydrate (PLC) ratio
Plerergate, 38
Plesiomonas shigelloides Bader, 232
Poekilocerus pictus (Fab.), 286
Policy makers, 106
Polyacrylamide, 168
Polychlorinated biphenyls (PCBs), 212
Polyculture feed, insects, 290–291
Polymerase chain reaction (PCR), 267
Polypeptide chains, 160
Polyphenoloxidase (PPO), 82
Polystyrene
  biodegradation and mineralization of, 155
Polyunsaturated fatty acids (PUFAs), 33, 44, 160
Polyvinyl chloride (PVC) tubing, 183
Postharvesting processing, 141
Poultry feed, insects, 277–282
PPO. *See* Polyphenoloxidase (PPO)

Predominant human bacterial infections, 229
Pressure water system, 165
Preventative Controls for Human Foods rules, 243
Processed animal proteins (PAP), 143, 204
Processed mealworms, 104, 115
Processing contaminates, 212
Processing machines, 122
Processing methods, 80, 122, 127
Process mechanization, 154
Production and preservation techniques, 145
Product packaging, 217
Programmable automation controllers (PACs), 195
Programmable logic controller (PLC), 191, 194
Propionate salts, 249
Propionic acid, 249
Proportional integral derivative controllers (PID) system, 164, 194
    drivers, 195
Protected Designation of Origin (PDO), 39
Protein, 160
    chemical structure of, 160
    conversion efficiency, 91
    functionality, 81
    low-energy-intensive source of, 7
    production and sustainability, 97
    quality, 77, 115
    solubility, 81
    sources, 3, 114
PROteinINSECT, 204
Protein, lipid, and carbohydrate (PLC) ratio, 158
PROteINSECT, 214
Protozoa, 233
Pseudomonas aeruginosa Migula, 295
Pseudoterranova decipiens Krabbe, 233
PSP. *See* Paralytic shellfish poisoning (PSP)
Psychological barriers
    to insect consumption, 18
PUFAs. *See* Polyunsaturated fatty acids (PUFAs)
Putitor Mahseer (*Tor putitora* Hamilton), 287
PVC. *See* Polyvinyl chloride (PVC) tubing

Q

QACs. *See* Quaternary ammonium compounds (QACs)
QSAR. *See* Quantitative structure-activity relationships (QSAR)
Quantitative structure-activity relationships (QSAR), 209
Quaternary ammonium compounds (QACs), 240

R

Raw agricultural commodities, 245
Raw food materials, 131
Ready-to-eat, 216
Ready-to-use therapeutic food (RUTF), 128, 135
    products, 145
Rearing systems, 53, 166, 184

Rearing techniques, 87, 88
Regulatory considerations, for insects-as-food ingredient, 204
    chemical, toxicological hazards of, 211
        antinutrient source, 213
        exogenous chemical contaminants and residues, 212
        heavy metal contamination, 213
        natural toxicants of, 212
    farming/novel considerations, 214–216
    labelling and health claims applicable, 207
    physical hazards, 213
    present history of use, 204–206
    processing/preparation/packaging/transport, 216–217
    safety considerations, 207
        clinical evaluation, 211
        toxicological assessment, 210–211
        whole body/processed feed, 209–210
Regulatory processes, 12
Relative humidity (RH), 184
Research and development (R&D) programs, 100
    work, 135
RFID tags, 165
Rhantus atricolor, 274
Rhynchophorus phoenicis, 20, 145
Rice-duck farming systems, 3
Rocky Mountain regions, 6
Rohu (*Labeo rohita* Hamilton), 287
Rural inhabitants, 45
Ruspolia differens, 38
Ruspolia nitidula, 87
RUTF. *See* Ready-to-use therapeutic food (RUTF)

S

Safety, regulatory considerations, 207
    clinical evaluation, 211
    toxicological assessment, 210–211
    whole body/processed feed, 209–210
SAGE. *See* Sustainability and Global Environment (SAGE)
Salmonella enterica subsp., 230
Salmonella sp., 277
Sanitation impacts, 239
Sarcoplasmic calcium binding protein (SCP), 261
Sarcoplasmic endoreticulum calcium ATPase (SERCA), 261, 263
SBIR. *See* Small Business and Innovative Research (SBIR) program
SCP. *See* Sarcoplasmic calcium binding protein (SCP)
Screen bottom container systems, 167
Scylla paramamosain Estampador, 260
Scyphophorus acupunctatus, 205
SDS-PAGE separations, 260
Semicultivation, 101
SERCA. *See* Sarcoplasmic endoreticulum calcium ATPase (SERCA)
SHC. *See* Superior Health Council (SHC)

Shelf life, 127, 157
Shellfish
    allergens, 265
    allergy, 257
Shrimp, 263, 266
    abdominal muscle, 262
Silk extraction, 35
Silkworm (*Bombyx mori* L.), 52, 258, 274, 280, 281, 286, 287
    chrysalis of, 120
    hybrids, 99
    pupae, 280, 288
    pupae meal, 280, 281
    rearing, 102
Silver Barb (*Barbonymus gonionotus* Bleeker), 287
SIT. *See* Sterile insect release technique (SIT)
Small Business and Innovative Research (SBIR) program, 14, 130, 135
Small-sized chitin particles, 49
Soil fertilizer, 11
Soil-plant-insect-chicken food chain, 213
Solid feeds, 157
Sorbate salts, 248
Sorbic acid, 248
Soybean meal, 277
Soy cultivation, 273
Sphenarium histrio, 274
Sphenarium purpurascens, 39
Spore-forming bacterium, 234
Spray drying, 133
    efficiency of, 133
Stagnant air, 186
Standard processed insect-containing foods, 146
Staphylococcus aureus
    enterotoxin, 234
    Rosenbach, 295
Sterile insect release technique (SIT), 121, 154
Sterols, 160
Stinging Catfish (*Heteropneustes fossilis* Bloch), 287
Superior Health Council (SHC), 204
Super mealworm (*Zophobas morio*), 116, 205, 283
Sustainability and Global Environment (SAGE), 4
    researcher, 4
Sustainable and secure food supply
    substantial benefits for, 114
Synthetic pesticides, 105
Systema Naturae, 41

T

Tanzania
    chitemene agricultural systems of, 3
Taste barriers, 53
TDT. *See* Thermal death time (TDT)
TEK. *See* Traditional ecological knowledge (TEK)
Teleogryllus mitratus, 281
Temperature conditioners, 192–193
    air conditioners, 192–193
    evaporative panels, 193
    water heating system, 192

Tenebrio molitor L., 3, 63, 116
  biological aspect of, 169
  hosts bacteria, 224
  larvae, water requirements of, 168
  production systems, 169
Termites, 38, 81
Thermal death time (TDT), 246
Thiamin, 76
Three-screen circular separator, 169
Thrombotic thrombocytopenic purpura (TTP), 231
Tobacco hornworm hawk moth (*Manduca sexta* L.), 296
Toxic metals, 99
Toxicological hazards, 211
Toxic Substances Control Act (TSCA), 206
Trabutina mannipara, 34
Traditional agricultural mechanization, 163
Traditional ecological knowledge (TEK), 23
  application of, 24
Traditionally-consumed insects, 37
Transport
  regulatory considerations, for insects-as-food ingredient, 216–217
Trichogramma spp., 21
Tropomyosin, 259, 260
TSCA. *See* Toxic Substances Control Act (TSCA)
Tse-tse fly (*Glossina morsitans* Westwood), 296
TTP. *See* Thrombotic thrombocytopenic purpura (TTP)
Turbulent air flow system scheme, 188

U

Ulomoides dermestoides, 48
Ultraviolet, 249
Underutilized biomass, 180
United Nations (UN), 116
  Food and Agriculture Organization (FAO), 14, 89, 115, 116, 144
    global stakeholder directory, 141
    MDG, 113

United States Department of Agriculture (USDA), 21, 206
  Agricultural Research Service (USDA-ARS), 157
  dietary guidelines 2010, 77
  Food Safety and Inspection Service (USDA-FSIS), 241
  Humane Slaughter Act of 1978, 216
  National Institute of Food and Agriculture, 30
United States Department of Agriculture Small Business Innovation Research (USDA SBIR) funding, 130
Urban agriculture, 5
US cricket industry, 117
USDA. *See* United States Department of Agriculture (USDA)
USDA SBIR. *See* United States Department of Agriculture Small Business Innovation Research (USDA SBIR) funding
US Food and Drug Administration (FDA), 131
US Foodborne Illnesses, 228
US National Academy of Sciences, 7
Uzi powder
  dusting of, 99

V

Validation
  HACCP and preventative controls for human foods, 244
Variegated Grasshopper (*Zonocerus variegatus* L.), 285
Vegetarian philosophies, 18
Venomous insects, 296
Vertebrate livestock-derived foods, 53
Vesicant species, 296
Vibrio anguillarum Bergeman, 297
Vitamin, 161
  content of selected insect species and common high-protein commodities, 72

vitamin A, 294
vitamin B1, 294
vitamin B2, 294
vitamin B12, 294
vitamin C, 294
vitamin E, 294

W

Walking Catfish (*Clarias batrachus* L.), 286
Washington DC-based Oyamel, 144
Waste biomass, 153
Water pulverization, 193
Water resources
  depletion of, 10
Wax moth (*G. mellonella*), 225
Waxworm production, 182–184
  adult rearing and reproduction, 183–184
  larval development and diets, 182–183
Western culture protein sources, 1
Wet milling process, 134
Wet products
  disadvantages of, 134
Whiteleg Shrimp (*Litopenaeus vannamei* Boone), 291
WHO. *See* World Health Organization (WHO)
Whole-animal dried insect-derived foods, 217
World Health Organization (WHO), 144

X

Xenoestrogens, 212

Y

Yeast-derived protein, 101
Yellow Mealworm (*Tenebrio molitor*), 225, 283, 284, 287–289
  proteins, 266
Yersinia enterocolitica Schleifstein, 231

Z

Zero gravity environment, 11
Zophobas morio, 72

Printed in the United States
By Bookmasters